IONOSPHERE AND APPLIED ASPECTS OF RADIO COMMUNICATION AND RADAR

IONOSPHERE AND APPLIED ASPECTS OF RADIO COMMUNICATION AND RADAR

Nathan Blaunstein
Eugeniu Plohotniuc

CRC Press
Taylor & Francis Group
Boca Raton   London   New York

CRC Press is an imprint of the
Taylor & Francis Group, an **informa** business
AN AUERBACH BOOK

CRC Press
Taylor & Francis Group
6000 Broken Sound Parkway NW, Suite 300
Boca Raton, FL 33487-2742

First issued in paperback 2018

ISBN-13: 978-1-4200-5514-6 (hbk)
ISBN-13: 978-1-138-37264-1 (pbk)

Library of Congress Cataloging-in-Publication Data

Blaunstein, Nathan.
　　Ionosphere and applied aspects of radio communication and radar / Nathan Blaunstein and Eugeniu Plohotniuc.
　　　　p. cm.
　　Includes bibliographical references and index.
　　ISBN 978-1-4200-5514-6 (alk. paper)
　　1. Ionospheric radio wave propagation. 2. Radio--Transmitters and transmission--Fading. 3. Radio--Interference. 4. Radar--Interference. 5. Artificial satellites in telecommunication. I. Plohotniuc, Eugeniu. II. Title.

TK6553.B5245 2008
621.384--dc22

2007049103

Visit the Taylor & Francis Web site at
http://www.taylorandfrancis.com

and the CRC Press Web site at
http://www.crcpress.com

Contents

Preface ix

Acknowledgments xiii

Authors xv

Abbreviations xvii

1   The Regular Ionosphere: Main Characteristics and Processes 1
     1.1   Ionosphere Content 2
     1.2   Major Characteristics of the Ionospheric Plasma 7
     1.3   Transport Processes of Plasma in the Ionosphere 13
     1.4   Ionization–Recombination Balance in the Ionosphere 17
     References 20

2   Nonlinear Phenomena and Plasma Instabilities in the Disturbed
     Irregular Ionosphere 23
     2.1   Physical Aspects of Nonlinear Phenomena in the Artificially Heated
           Ionospheric Plasma 27
     2.2   Interaction of Probing Radio Waves with Heating-Induced Plasma
           Instabilities Generated by Strong Radio Waves 34
     2.3   Plasma Instabilities of Natural Origin Generated in the Irregular Ionosphere 38
           2.3.1   Natural Thermal-Induced Instabilities 40
                   2.3.1.1   Recombination Instability 42
                   2.3.1.2   Kinetic Instability 44
                   2.3.1.3   Kinetic Drift-Dissipative Instability 44
           2.3.2   Current-Induced Natural Plasma Instabilities 45
                   2.3.2.1   The Current-Convective Instability 48
                   2.3.2.2   Gradient-Drift Instability 48
                   2.3.2.3   Two-Stream Instability 52
                   2.3.2.4   Convergence Instability 57
                   2.3.2.5   Dissipative Instability 58
                   2.3.2.6   Drift-Dissipative Instability 58
                   2.3.2.7   Rayleigh–Taylor Instability 60
                   2.3.2.8   Hydrodynamic Instabilities 60

2.4   Theoretical Aspects of Generation of Nonlinear Plasma Instabilities 62

    2.4.1   Nonlinear Wave–Wave Interactions in the Ionospheric Plasma 63

    2.4.2   Particle–Wave Interactions in the Ionospheric Plasma............................ 68

        2.4.2.1   Nonlinear Saturation of Electrostatic Instabilities in the
                Ionosphere 73

        2.4.2.2   Numerical Analysis of Nonlinear Effects
                of Instabilities Generation 78

References 101

3   Radio Signal Presentation in the Ionospheric Communication Channel.................. 111

3.1   Bandpath and Baseband Signal Presentation 111

3.2   Narrowband Signal Presentation 114

3.3   Wideband Signal Presentation... 122

References 125

4   Fading Phenomena in Ionospheric Communication Channels................................ 127

4.1   Path Loss: A Mathematical Description 129

4.2   Slow Fading of Radio Signal: A Mathematical Description 130

4.3   Fast Fading: A Mathematical Description 131

4.4   Definition of Fading Phenomena in Static and Dynamic Ionospheric Channel 136

References 138

5   Evolution of Plasma Irregularities in the Ionosphere.. 139

5.1   Diffusion of Plasma Irregularities in the Ionosphere.. 140

    5.1.1   Classical Description of the Diffusion Process in Plasma......................... 141

    5.1.2   Diffusion of Small Plasma Disturbance in Weakly Ionized Plasma 143

    5.1.3   Diffusion of Plasma Disturbances with an Arbitrary Degree
          of Ionization.. 164

    5.1.4   Diffusion Spreading of Plasma Irregularities in the Middle-Latitude
          Ionosphere 172

5.2   Drift of Plasma Irregularities in the Ionosphere... 192

    5.2.1   A Classical Description of Drift in Unbounded Homogeneous Plasma..... 194

    5.2.2   Dynamics and Spreading of Arbitrary Irregularity in the Ionosphere......... 207

5.3   Thermodiffusion of Plasma Irregularities in the Ionosphere 231

    5.3.1   General Problem Statement: Functions and Parameters 232

    5.3.2   One-Dimensional Heating of the E and F Layers of the Ionosphere......... 234

    5.3.3   Evolution of 3-D Heating-Induced Irregularity in the Ionosphere 241

References 248

6   Modern Radiophysical Methods of Investigation of Ionospheric Irregularities........ 253

6.1   Radio and Optical Methods of Studying Artificially Induced Irregularities
     in the Ionosphere 254

    6.1.1   Radio Methods for Diagnostics of Heating-Induced Irregularities............ 254

    6.1.2   Methods of Investigations of Dynamics of Plasma Clouds in the
          Ionosphere by Rocket Injection.. 268

6.2 Natural Ionospheric Irregularities and Methods of Their Parameters' Diagnostics 287
    6.2.1 Radio Methods of Diagnostics of Meteor-Induced and H_E Irregularities in the Middle-Latitude Ionosphere 287
    6.2.2 Radio Sounding of Sporadic Inhomogeneities and Bubbles in the Equatorial Ionosphere 303
    6.2.3 Observations of Irregularities in the High-Latitude Ionosphere 359
References 390

7 Performance of Radio Communication in Ionospheric Channels 403
7.1 Absorption of Radio Waves in the Regular Ionosphere 403
7.2 Radio Scattering Caused by Plasma Irregularities of Various Origins in the Ionosphere 405
    7.2.1 Experimental Investigations of Backscattering and Forward Scattering in Region E of the Ionosphere 406
    7.2.2 Theory of Backscattering of Radio Waves from Anisotropic Plasma Irregularities 432
    7.2.3 Forward Scattering of Radio Waves from Anisotropic Irregularities 436
    7.2.4 Power of H_E-Scatter Radio Signals 442
    7.2.5 Forward and Backscattering from Irregularities of the F Region of the Ionosphere 449
7.3 Long-Distance Radio Propagation through the Ionospheric Channels 455
    7.3.1 Wave Propagation in the Ionosphere Disturbed by Powerful Radio Waves 456
7.4 Capturing of Radio Waves into the Ionospheric Layered Waveguides 469
7.5 Partial Scattering of Radio Waves in the D Region of the Ionosphere 472
7.6 Methods of Design of Ionospheric Radio Propagation Channels 477
References 486

8 Optical and Radio Systems for Investigation of the Ionosphere and Ionospheric Communication Channels 493
8.1 Devices and Systems for Diagnostics of Ionospheric Phenomena 494
    8.1.1 Optical Devices 494
    8.1.2 Incoherent Scatter Radars 499
    8.1.3 SuperDARN 503
    8.1.4 The Global Positioning System in Investigations of the Ionosphere 505
8.2 Diagnostics of the Nonregular Ionosphere by LFM Ionosondes 513
    8.2.1 Developments of LFM Ionosondes in the Historical Perspective 513
    8.2.2 Operating Principles of LFM Ionosondes 515
8.3 Monitoring of Ionospheric Communication Channels 524
References 531

9 Performance of Land–Sattelite Communication Links Passing through the Irregular Ionosphere 539
9.1 Refraction of Radio Waves in Quasi-Regular Ionospheric Plasma 540
9.2 Propagation Effects of Large- and Moderate-Scale Ionospheric Irregularities 541
    9.2.1 Effects of Large-Scale Irregularities on Radio Propagation 541
    9.2.2 Effects of Moderate-Scale Irregularities on Radio Propagation 542

9.3 Fading Effects Caused by Small-Scale Ionospheric Irregularities............................ 545
9.4 Effects of Magnetic Storm on Land–Satellite Communication.............................. 555
9.5 Parameters of Data Signals in the Land–Satellite Communication Links
with Fading.. 562
    9.5.1 Main Parameters of the Information Data Stream.................................... 562
    9.5.2 Data Stream Parameters for Channels with Fading Caused
          by Magnetic Storm 564
References 567

Index 571

Preface

This book is intended for researchers, engineers, and graduate or postgraduate students who are concerned with the practical aspects of ionospheric radio propagation, over-horizon and above-horizon radars, and ionospheric stations used for investigating nonregular phenomena occurring in the ionosphere, as well as with land–satellite and satellite–satellite communications. The phenomena treated include transport processes and photochemistry reactions occurring in the regular homogeneous ionosphere, nonlinear phenomena, and instabilities in the inhomogeneous disturbed ionosphere caused by various ambient natural and man-made (e.g., artificial) sources, and the corresponding plasma irregularities, natural and artificially induced, associated with these plasma instabilities.

In previous books [1–3] written in the former Union of Soviet Socialist Republics and Romania, the reader was introduced to radiophysical and optical methods of observations of artificially induced plasma irregularities, such as ion clouds and beams injected from the geophysical rockets into the background ionospheric plasma [1] or heating-induced irregularities generated in the field of strong pump waves transmitted to the ionosphere from powerful ground-based facilities [2,3]. Unfortunately, all discussions at that time concentrated on techniques used by the Soviet Geophysical Society from the 1960s through the 1980s and only on man-made phenomena, completely ignoring the effects of numerous natural plasma inhomogeneities occurring in the nonregular disturbed ionosphere at various latitudes from the polar cap to the equator. Furthermore, completely absent in previous books were the main aspects of the influence of different plasma irregularities, small-, moderate-, and large-scale, on radio wave propagation through the inhomogeneous ionosphere, and on diagnostics of long-distance communication channels created in the ionosphere at various altitudes. These aspects necessitate additional requirements for the prediction of the main characteristics of ionospheric communication links by using modern radiophysical methods based on different vertical and oblique radar systems, pulse and continue-wave (CW) ionosondes, radiometers, and so on, only several examples of which were presented later in Ref. [3].

Furthermore, in the past two or three decades, a rapidly increasing demand for improving the efficiency of land–satellite and satellite–satellite communication networks, as megacell wireless service systems, was observed. This aspect is strongly related to improvement of the efficiency of wireless networks beyond the third generation (3-G), and, foremost, with the purposes of positioning and localizing of any subscriber located in the area of service of each desired satellite.

To design such systems successfully, it is important to predict strictly propagation characteristics of land–satellite communication links passing the irregular ionospheric plasma perturbed by various natural phenomena such as solar activity and penetration of high-energy particles of solar wind into the ionosphere at high latitudes and the polar cap, and perturbation of the Earth's magnetic field (called "magnetic storms") due to close coupling of the magnetosphere and ionosphere, as well as by strong coupling between high- and middle-ionosphere and the equatorial ionosphere, etc. The precise study of these phenomena, and the corresponding various instabilities (plasma waves) and irregularities caused by these phenomena in the disturbed ionosphere, allows us to obtain a stable prediction of the key parameters of ionospheric radio links and of the information data stream transmitted and received by the two-end terminal antennas of the land–satellite link. An initial effort to describe a few of these main aspects was recently pointed out in a book [4] by one of the authors.

This book presents the reader with pertinent aspects of the diagnostics of irregular plasma structure of the ionosphere by using both well-known standards and modern (e.g., developed recently) radiophysical and optical methods, as well as algorithms of how to predict key parameters of ionospheric communication links, signals, and data streams. The more current aspects have not been described until now in such a complete way, even the best guides to ionospheric communication having suffered from wide information gaps in describing more current ionospheric phenomena, its influence on the key parameters of the channel and signal—such as pass loss, slow and fast fading, and signal-to-noise ratio (SNR)—and, finally, the relationships between key parameters of the channel and the information data stream, such as capacity, spectral efficiency, and bit error rate (BER).

Another problem that also stimulated creation of this book, in the authors' opinion, is that up to the end of the 1970s to the mid-1980s scientific groups investigating ionospheric phenomena in countries of eastern Europe and the former Soviet Union and those in the western countries worked independently and separately without close relations among them. Finally, during *perestroika* in the former Soviet Union the boundaries opened and comparing theoretical and experimental studies carried out by the two sides showed that there was much overlapping between theoretical frameworks and experimental technologies and the methodology of measurements of ionospheric phenomena and diagnostics of ionospheric radio channels using the same or close facilities created over 30–50 years. While being involved in some principal projects carried out in the former Soviet Union during the 1970s–1990s, the authors then investigated theoretical and experimental data obtained by different research groups over the world, mostly in the United States (where investigations of ionospheric communication links have been carried out more intensively compared with other western countries). As a result, they tried to create some "virtual bridges" between existing frameworks, theoretical and experimental, to find correlations between different data through the existing theoretical frameworks and experimental data concerning the corresponding ionospheric phenomena and radio communication through the ionosphere, and, finally, to obtain optimal descriptions of the ionospheric phenomena, diagnostics of ionospheric communication channels, and key parameters of signals and information data, based on a whole spectrum of studies carried out by different scientific groups around the world, including their own theoretical and experimental investigations. Of course, the authors did not review all existing approaches; they have systematically described most adapted theories proved experimentally and the corresponding modern technologies and methods of diagnostics of ionospheric communication links, land–ionosphere–land and land–satellite passing through the disturbed ionosphere, as well as various signals and data passing through such channels.

The material and chapter sequence in the book's text are organized into nine chapters. Chapter 1 introduces a description of a quasi-regular layered ionosphere, based on a brief review of plasma content, the main characteristics of ionospheric plasma, transport processes, diffusion, thermo-diffusion and drift, which predominate at altitudes of the upper ionosphere, as well as on ionization–recombination reactions prevalent at altitudes of the low ionosphere. In Chapter 2, the authors show the full spectrum of instabilities generated in the ionosphere, artificially or naturally, due to the activity of natural ambient phenomena occurring in near-the-earth cosmic space. Chapter 3 introduces the reader to the main parameters of the ionospheric channel and different presentations of signals, pulse, and CW, within such types of channels. Chapter 4 describes the statistical aspects of multipath propagation phenomena by introducing key parameters of channel and signal, such as path loss, slow and fast fading, and their statistical description using different presentations of the probability distribution function (PDF), from Gaussian to Ricean and Rayleigh, as well as their combinations. In Chapter 5, the authors discuss various kinds of ionospheric irregularities, natural and artificially induced, associated with the whole spectrum of instabilities occurring at different latitudes from the polar cap to the equator. Chapter 6 deals with radiophysical and optical methods of investigating ionospheric irregularities of various origins, their dynamics, structure and lifetime, using standard and modern methods and the corresponding equipment. In Chapter 7, the authors describe classical and modern methods of the diagnostics of the ionospheric radio communication channels and radiotraces with complex radio propagation geometry, which differ from those usually passing in the quasi-regular ionosphere along the great circle, using standard and modern radars and ionosondes.

Chapter 8 introduces the reader to different optical and radio devices and apparatus, such as lidars, optical cameras, radiometers, radars, and ionosondes, utilized for diagnostics of the key parameters of radio communication ionospheric channels and the information data inside them. Finally, in Chapter 9, we show how different kinds of plasma irregularities, small-, moderate- and large-scale, occurring at various altitudes of the ionosphere can affect radio signals passing through the corresponding land–satellite channel, depending on the position and elevation of the satellite antenna with respect to the ground-based subscriber antenna. Here, the main parameters of the information data stream, such as capacity, spectral efficiency, and BER, are introduced and, as an example, show how to predict these parameters within the multipath ionospheric communication link with fading caused by the ambient magnetic storm.

References

1. Filipp, N., V. Oraevskii, N. Blaunstein, and Yu. Ruzhin, *Evolution of Artificial Plasma Irregularities in the Earth's Ionosphere*, Kishinev, Moldova: Shtiintza, 1986.
2. Filipp, N., N. Blaunstein, L. Eruhimov, V. Ivanov, and V. Uryadov, *Modern Methods of Investigation of Dynamic Processes in the Ionosphere*, Kishinev, Moldova: Shtiintza, 1991.
3. Ureadov, V., V. Ivanov, E. Plohotniuc, L. Eruhimov, N. Blaunstein, and N. Filipp, *Dynamic Processes in the Ionosphere – Methods of Investigations*, Iasi, Romania: Tehnopress, 2006.
4. Blaunstein, N. and Ch. Christodoulou, *Radio Propagation and Adaptive Antennas for Wireless Communication Links: Terrestrial, Atmospheric and Ionospheric*, NJ: Wiley InterScience, 2006.

Acknowledgments

It is with the kind permission of Eugeniu Plohotniuc, my coauthor, that I write the acknowledgments for this book.

First, this book would never have seen the light of day had Professor Nicolae Filipp not initiated us both into the study of the ionosphere and conditions of radiowave propagation through middle-latitude traces using vertical and oblique sounding of the ionosphere, and their corresponding long-path ionospheric channels. He was not only our inspiration but also the first of our future colleagues and supervisors, who introduced both of us to the study of the ionosphere and the field of ionospheric radio propagation, and for which we express our thanks. His contributions to Chapters 6 and 7 on the description of the effects of meteor trails and H_F irregularities on radio propagation through the ionosphere are invaluable.

I would like to mention that without the immense help of a group of scientists at IZMIRAN (Academy of Sciences, Moscow), especially, Dr. Jacob Dimant, Elena Tsedilina, Victor Vas'kov, and Genadii Bochkarev, led by Professor Alexander Gurevich, and Professors Lev Tsendin and Vladimir Rozhansky at the Leningrad Polytechnic University (USSR). Without their unwavering support, I would not have obtained the knowledge that has enabled me to conduct investigations in plasma physics, especially on nonlinear phenomena occurring in the ionosphere disturbed by different natural and man-made sources. I have had the honor of publishing with these distinguished peers, and also individually, several articles concerning the wide spectrum of the corresponding processes, also described in Chapters 2, 5, and 6, for which I am ever grateful to them.

Many thanks to my colleagues at IZMIRAN, especially Drs. Yurii Ruzhin, Victor Oraevsky, and Eugeny Mishin with whom I investigated the evolution of ion clouds and plasma beams in the ionosphere and published several articles and a book. Some of the results of our collaborative research are described in Chapters 6 and 7.

We also thank our colleagues at NIRFI (Gorky, USSR) and State University (Yoshkar-Ola, USSR), where we were involved in investigations of heat-induced ionospheric irregularities using radar and ionosondes, described in Chapters 6 through 8. Among our colleagues at NIRFI and Mari State University, those deserving special mention include Professors Lev Erukhimov, Valerii Uryadov, Vladimir Ivanov, and Dr. Vladimir Shumaev, with whom we have published several articles and two books during recent decades.

My appreciation for my coauthor, Eugeniu Plohotniuc, professor at the Moldova State University (Beltsy, Moldova), who took over middle-latitude ionosphere research first from Professor Filipp, and then from me, and who is also involved in the further development of the Radiophysical and Geophysical Observatory in Moldova, working closely with several research groups all over the world, including groups from Romania, Russia, Japan, and Israel.

I must also mention that without the friendly and creative atmosphere at the Department of Electrical and Computer Engineering, University of New Mexico, Albuquerque, where I was a visiting professor during my sabbatical in 2007, as well as without the joint discussions I had with Professor Christos Christodoulou, my coauthor of another book and who helped me to finalize Chapters 3 and 4, this book would never have been finished.

We extend our thanks to the reviewers and editors of this book who helped with the presentation of the text with clarity and precision.

Finally, it would have been impossible for both of us to carry out research without the warm support and understanding of our families. This book is dedicated to our families and to the memory of our teachers, colleagues, and supervisors.

Authors

Nathan Blaunstein received his BSc and MSc in radiophysics and electronics from Tomsk University, Tomsk, Russia, in 1972 and 1976, respectively, and his PhD, DSc, and professor in radiophysics and electronics from the Institute of Geomagnetism, Ionosphere, and Radiowave Propagation (IZMIR), Academy of Science USSR, Moscow, Russia, in 1985 and 1991, respectively. From 1979 to 1984, he was an engineer, a lecturer, and then, from 1984 to 1992, a senior scientist, an associate professor, and a professor at Moldavian University, Beltsy, Moldova, in the former USSR. Since 1993 he has been a senior scientist in the Department of Electrical and Computer Engineering and a visiting professor in the Wireless Cellular Communication Program at Ben-Gurion University of the Negev, Beer Sheva, Israel, and then from 2001, professor in its Department of Communication Systems Engineering.

Dr. Blaunstein's research interests include problems of optical and radio wave propagation, diffraction and scattering in various media (subsoil medium, terrestrial environments, troposphere, and ionosphere) for purposes of optical and radiolocation, and aircraft, mobile-satellite and terrestrial communications. The results of his investigations are presented in more than 160 papers in different journals and conference proceedings, 5 books, and 7 patents.

Eugeniu Plohotniuc graduated from State University of Moldova, Chișinău, in 1975 with the physicist diploma. From 1975 to 1978 he worked as a senior laboratory assistant and a researcher in the physics of plasma laboratory of the Department of Optics and Spectroscopy, State University of Moldova, Chișinău. From 1978 he worked as lecturer in the Technical Disciplines Department of the Alecu Russo State University, Moldova, Bălți. During 1982–1986 he continued his studies as a postgraduate student at the Institute of Radiotechnics and Electronics of the Academy of Sciences of the United Soviet Socialists Republic in Moscow, where he investigated problems connected with radiowave propagation. After he defended his thesis for the kandidat in physics and mathematics in 1986, he returned to Alecu Russo State University. In 1989 he was elected associate professor of the Department of Electronics and Informatics. At present he teaches courses in radiophysics, radioelectronics, and computer architecture.

From 1990–1994 Dr. Plohotniuc served as deputy dean of the Faculty of Technics, Physics, Mathematics, and Informatics. In 1994, he was elected dean of the faculty, serving in that post until 2003. In 2005, Professor Plohotniuc was appointed scientific secretary of the University Senate and in 2007, rector of Alecu Russo State University.

Beginning in 1996, Professor Plohotniuc started directing the scientific laboratory of Radiophysics and Electronics, within the framework of which he does research in the field of radio physics (radio wave propagation through natural and artificially modified ionosphere) and radio electronics (designing radio electronic devices, including digital ionosondes). The results of his investigations are presented in more than 90 papers in different journals and conference proceedings.

Abbreviations

1D	one-dimensional, 1-D model
2D	two-dimensional, 2-D model
3D	three-dimensional, 3-D model
ACF	auto-correlation function
AER	artificial electromagnetic radiation
AF	amplitude–frequency
AFC	amplitude–frequency characteristic
AIT	artificial ionospheric turbulence
AIT	artificially induced turbulence
AISRS	Arecibo Incoherent Scatter Radar System
API	artificial periodic irregularity
AWGN	additive white Gaussian noise
BER	bit error rate
BOS	back-oblique sounding
CDF	cumulative distribution function
CCI	current-connective instability
CCDF	complimentary cumulative distribution function
CI	convergent instability
CEEC	Condor Equatorial Electroject Compaign
CUPRI	Cornell University using a portable radar interferometer
CW	continuous wave
DDI	drift-dissipative instability
dB	decibels
DF	Doppler frequency
DHR	disturbed heated region
DKB	decameter wave band
DMSP	Defense Meteorological Satellite Program
DPS	Digisonde portable system
DSTO	Defense Science and Technology Organization
ED	electrodynamic
EGNOS	Euro Geostationary Navigation Overlay Service
EISCAT	European Incoherent SCATter radar
EPB	equatorial plasma bubbles
erfc (*)	error function

FAI	field-aligned irregularities
F-B	Farley–Buneman
FBI	Farley–Buneman instability
FFF	flat fast fading
FIRI	Faraday International Reference Ionosphere
FM	frequency modulated
FPI	Fabry–Perot interferometer
FSF	flat slow fading
FSSF	frequency-selective slow fading
FSFF	frequency-selective fast fading
FUV	far ultraviolet
GDI	gradient-drift instability
GLONASS	global navigation satellite system
GLon	geographic longitude
GoS	grade of service
GNSS	global navigation satellite system
GPS	global positioning system
HF	high frequency
HU	Huancayo
IGY	International Geophysical Year
IEEY	International Equatorial Electrojet Year
ISR	incoherent scatter radar
JULIA	Jicamarca Unattended Long-Term Investigation
K	Ricean parameter (factor)
LEO	low Earth orbit
LFM	linear frequency modulation
LOS	line of sight
LT	local time
MEO	medium Earth orbit
MEOS	moderate Earth-orbit satellite
MHD	magnetohydrodynamic
MIP	moving ionospheric perturbations
MLat	magnetic latitude
MSAS	Multi-Functional Satellite Augmentation System
MLT	magnetic local time
MU	multibeam pulsed radars
MUF	maximum usable frequency
NLOS	non-line-of-sight
NL	nonlinear
O-mode	ordinary mode
OS	oblique sounding
PACE	Polar Anglo-American Coordinate System
PFC	phase-frequency characteristic
PIM	Parameterized Ionospheric Model
PDF	probability density function
PSD	power spectral density
PRI	Post-Rozenbluth instability

RF	radio frequency
RMS	root mean square
ROCSAT-1	Republic of China satellite
ROS	radar operating system
RAI	range–azimuth–intensity (B-scan)
RTI	Rayleigh–Taylor instability
SAPPHIRE	Saskatchewan auroral polametric phased array radar experiment
SESCAT	sporadic E scatter experiment
S4	scintillation index
S/N	signal-to-noise ratio
SNR	signal-to-noise ratio
SPS	striated plasma structures
SD	super dual radar
STARE	Scandinavian Twin Auroral Radar Experiment
SAPS	subauroral polarization streams
SuperDARN	Super Dual Radar Network
TEC	total electronic content
TD	time delay
UHR	upper-hybrid resonance
UHF	ultra high frequency
UT	universal time
VHF	very high frequency
VS	vertical sounding
WAAS	wide area augmentation system
Z-mode	non-ordinary mode

Chapter 1

The Regular Ionosphere: Main Characteristics and Processes

The idea of using the ionosphere (which was called a "conducting ionized layer" by pioneer researchers during the last decades of the nineteenth century and the beginning of the twentieth century [1–11]) for long-distance electromagnetic wave propagation was demonstrated theoretically and practically only in the middle of the twentieth century [12–26]. It is only in recent decades that practical applications were developed, both for long-range land–land radio communications by using the ionosphere as a "reflected screen," as well as for satellite–land and satellite–satellite links. The basic theories of radio wave propagation through the ionosphere, known as the magneto-ionic theories (see Ref. [13]), were established in the middle of the twentieth century, giving a great impetus to investigations of possibilities of ionospheric radio communication by reflection and scattering from ionospheric layers.

The ionosphere had been defined by Plendl [8], and then by Kaiser [12] as a part of the upper atmosphere of Earth, which stretches from 50 km to about 500–600 km. This region is filled with ionized gas, called *plasma*. The upper bound of the ionosphere is determined as the altitude at which the concentration of charged particles of plasma, the electrons and ions, exceeds that of neutral molecules and atoms. At these altitudes the ionosphere transforms continuously into the *magneto-sphere*, which consists of only strongly ionized plasma (i.e., in the absence of neutral molecules and atoms) and strong ambient electric and magnetic fields. The ionosphere is rising as the result of the influence of solar ionized waves and high-energy particles on different gases in Earth's atmosphere [8,13]. The structure and properties of the ionosphere depend essentially on processes occurring in the sun (called *solar activity* [14]), on variations of Earth's magnetic field (called the *geomagnetic field effect* [11]), movements of neutral "wind" in the upper atmosphere due to Earth's rotation, on the effects of electrical current and ambient electrical fields [15–18], on the density and the content of the atmosphere at different altitudes and geographical latitudes [19,20], and so on.

Usually, the ionosphere is separated into five independent regions, sometimes called *layers* [21–26]. The bottom one, from 50 to 80 km, is called the *D layer*; from 80 to 130 km is the *E layer*; and above 150 km is the *F layer*. The latter region is usually separated at the F_1 *layer*, from 130 to 200–250 km, and the F_2 *layer* is above 250 km. Apart from these layers, at 90–120 km, the sporadic layer E_s is observed, whose thickness in the vertical plane along the height is small. Its occurrence is usually explained by the influence of a neutral wind of atmospheric gases on the charged particles of plasma, resulting in exchange of plasma accompanied by stratified wind structure along the height (for more detailed information, see Refs. [21,22]).

Now, taking into account the geomagnetic line structure as well as the irregular distribution of variations of the geomagnetic field above Earth and its influence on the ionosphere, the latter can be categorized as the high-latitude ($|\Phi| > 55°-60°$), middle-latitude ($30° \leq |\Phi| \leq 55°$), and lower-latitude ($|\Phi| < 30°$) ionosphere, with a special zone of geomagnetic equator with $|\Phi| \leq 5°-10°$, where Φ is the magnitude of the geomagnetic latitude [21–26]. Note that in the physics of magnetosphere–ionosphere interactions, the definition "geomagnetic latitudes" relates to the dipole distribution of Earth's magnetic field when the magnetic dipole is located at the center of Earth and the magnitude of $\Phi = \tan^{-1}(\frac{1}{2} \tan I)$, where I is the angle of the magnetic field inclination.

The corpuscular solar radiation penetrates at the altitudes of ionosphere mainly at the high latitudes called the *polar ionosphere*. In the middle-latitude and lower-latitude ionosphere, the properties of plasma are determined by solar wave radiation, mostly for the *D* layer, because the ionization occurs here only in the diurnal periods.

1.1 Ionosphere Content

As is well known [21–26], in the neutral atmosphere in conditions of hydrostatic equilibrium and in the isothermal case ($T_m(z) = const$, where z is the altitude of the atmosphere from Earth's surface), the deviation of concentration of the neutral molecules and atoms in the ionosphere along height z is described by the barometric formula:

$$N_m = N_{m0} \exp\left\{-\frac{(z - z_0)}{H_m}\right\} \tag{1.1}$$

where $N_m = \sum_\beta N_\beta$, where β is the type of neutral particle; N_{m0} is the concentration of neutral particles at the height $z = z_0$ of pressure equilibrium; and $H_m = T/M_m g$ is the height of the homogeneous atmosphere, where $T \equiv T_m$ is the temperature of neutral particle content in energetic units, $M_m = \sum_\beta m_\beta N_\beta / N_m$, and g is the gravitational acceleration ($g = 9.81$ m/s^2, and m_β is the mass of the particle β).

Equation 1.1, as with all macroscopic gas laws, is valid for the ionosphere only under conditions such that the number of interactions between atoms and molecules is sufficient to establish a statically physical balance between them. In the case of Earth's ionosphere, Equation 1.1 is valid up to altitudes of $z_e \approx 1000-1200$ km [23,24]. For $z \geq z_e$, the length of the free path, l_β, of gas particle β between interactions exceeds the height of the regular atmosphere H_m, i.e., $l_\beta \gg H_m$, and neutral particles can move freely without collisions between them, and finally leave Earth's atmosphere. The region with altitude $z \geq z_e$, in which this phenomenon occurs, is called the *exosphere* [25,26].

However, in the real ionosphere the temperature of neutral particles T_m increases with altitude and leads to deviations from the barometric formula (Equation 1.1). The growth and the degree of these deviations with altitude depend on the intensity of the solar radiation, e.g., on the wave radiation in the lower- and middle-latitude ionosphere, and on the corpuscular radiation in the polar zone at higher latitudes. Therefore, the parameter H_m in Equation 1.1 increases with altitude and depends strongly on latitude. The change of atmospheric temperature with height in the middle-latitude ionosphere for moderate solar activity, both for diurnal and nocturnal conditions, is shown in Table 1.1 [27,28], which also presents the molecular content of the ionosphere. It is seen that at altitudes below 100 km the atmosphere contains mainly molecules of nitrogen N_2 and oxygen O_2. At these altitudes the concentration of other neutral components (He, O, H_2, NO) is

Table 1.1 Ionospheric Molecular Components

| z (km) | Concentration, N_m (cm^{-3}) | | | | | | T (K) |
	N_2	O_2	He	O	H	N_m	
	Day (12 h)						
60	$5.5 \cdot 10^{15}$	$1.5 \cdot 10^{15}$	—	—	—	$7.0 \cdot 10^{15}$	270
70	$1.6 \cdot 10^{15}$	$4.2 \cdot 10^{14}$	—	—	—	$2.0 \cdot 10^{15}$	200
80	$2.3 \cdot 10^{14}$	$6.2 \cdot 10^{13}$	—	—	—	$2.9 \cdot 10^{14}$	180
90	$3.1 \cdot 10^{13}$	$8.2 \cdot 10^{12}$	—	—	—	$3.9 \cdot 10^{13}$	190
100	$7.7 \cdot 10^{12}$	$1.9 \cdot 10^{12}$	$5.7 \cdot 10^{7}$	$2.0 \cdot 10^{11}$	$6.2 \cdot 10^{4}$	$9.6 \cdot 10^{12}$	210
110	$1.4 \cdot 10^{12}$	$3.5 \cdot 10^{11}$	$3.8 \cdot 10^{7}$	$1.4 \cdot 10^{11}$	$5.3 \cdot 10^{4}$	$1.9 \cdot 10^{12}$	270
120	$5.8 \cdot 10^{11}$	$1.2 \cdot 10^{11}$	$2.5 \cdot 10^{7}$	$7.6 \cdot 10^{11}$	$4.3 \cdot 10^{4}$	$7.8 \cdot 10^{11}$	360
130	$2.0 \cdot 10^{11}$	$3.8 \cdot 10^{10}$	$1.7 \cdot 10^{7}$	$3.7 \cdot 10^{10}$	$3.2 \cdot 10^{4}$	$2.8 \cdot 10^{11}$	460
150	$4.8 \cdot 10^{10}$	$7.2 \cdot 10^{9}$	$1.0 \cdot 10^{7}$	$1.35 \cdot 10^{10}$	$2.1 \cdot 10^{4}$	$6.9 \cdot 10^{10}$	670
200	$4.8 \cdot 10^{9}$	$5.9 \cdot 10^{8}$	$4.9 \cdot 10^{6}$	$3.0 \cdot 10^{9}$	$1.3 \cdot 10^{4}$	$8.4 \cdot 10^{9}$	1070
250	$1.1 \cdot 10^{9}$	$1.2 \cdot 10^{8}$	$3.4 \cdot 10^{6}$	$1.3 \cdot 10^{9}$	$1.0 \cdot 10^{4}$	$2.5 \cdot 10^{9}$	1250
300	$3.2 \cdot 10^{8}$	$2.7 \cdot 10^{7}$	$2.7 \cdot 10^{6}$	$5.9 \cdot 10^{8}$	$9.3 \cdot 10^{3}$	$9.3 \cdot 10^{8}$	1330
400	$3.5 \cdot 10^{7}$	$2.2 \cdot 10^{6}$	$1.9 \cdot 10^{6}$	$1.6 \cdot 10^{8}$	$8.3 \cdot 10^{3}$	$2.0 \cdot 10^{8}$	1390
500	$4.4 \cdot 10^{6}$	$2.1 \cdot 10^{5}$	$1.4 \cdot 10^{6}$	$5.0 \cdot 10^{7}$	$7.6 \cdot 10^{3}$	$5.6 \cdot 10^{7}$	1400
600	$5.9 \cdot 10^{5}$	$2.1 \cdot 10^{4}$	$1.0 \cdot 10^{6}$	$1.6 \cdot 10^{7}$	$7.1 \cdot 10^{3}$	$1.7 \cdot 10^{7}$	1400
700	$8.5 \cdot 10^{4}$	$2.3 \cdot 10^{3}$	$8.0 \cdot 10^{5}$	$5.2 \cdot 10^{6}$	$6.6 \cdot 10^{3}$	$6.0 \cdot 10^{6}$	1400
800	$1.3 \cdot 10^{4}$	$2.7 \cdot 10^{2}$	$6.1 \cdot 10^{5}$	$1.8 \cdot 10^{6}$	$6.2 \cdot 10^{3}$	$2.4 \cdot 10^{6}$	1400
900	$2.1 \cdot 10^{3}$	$3.3 \cdot 10^{1}$	$4.7 \cdot 10^{5}$	$6.2 \cdot 10^{5}$	$5.8 \cdot 10^{3}$	$1.1 \cdot 10^{6}$	1400
1000	$3.5 \cdot 10^{2}$	4.3	$3.7 \cdot 10^{5}$	$2.2 \cdot 10^{5}$	$5.4 \cdot 10^{3}$	$6.0 \cdot 10^{5}$	1400

(continued)

Table 1.1 (continued) Ionospheric Molecular Components

z (km)	Concentration, N_m (cm^{-3})						T (K)
	N_2	O_2	He	O	H	N_m	
Night (24 h)							
60	$5.5 \cdot 10^{15}$	$1.5 \cdot 10^{15}$	—	—	—	$7.0 \cdot 10^{15}$	270
70	$1.6 \cdot 10^{15}$	$4.2 \cdot 10^{14}$	—	—	—	$2.0 \cdot 10^{15}$	200
80	$2.3 \cdot 10^{14}$	$6.2 \cdot 10^{13}$	—	—	—	$2.9 \cdot 10^{14}$	180
90	$3.1 \cdot 10^{13}$	$8.2 \cdot 10^{12}$	—	—	—	$3.9 \cdot 10^{13}$	190
100	$7.7 \cdot 10^{12}$	$1.9 \cdot 10^{12}$	$5.7 \cdot 10^{7}$	$2.2 \cdot 10^{11}$	$6.2 \cdot 10^{4}$	$9.6 \cdot 10^{12}$	210
110	$1.4 \cdot 10^{12}$	$3.5 \cdot 10^{11}$	$3.8 \cdot 10^{7}$	$1.4 \cdot 10^{11}$	$5.3 \cdot 10^{4}$	$1.9 \cdot 10^{12}$	270
120	$5.8 \cdot 10^{11}$	$1.2 \cdot 10^{11}$	$2.5 \cdot 10^{7}$	$7.6 \cdot 10^{10}$	$4.3 \cdot 10^{4}$	$7.8 \cdot 10^{11}$	360
130	$2.0 \cdot 10^{10}$	$3.7 \cdot 10^{10}$	$1.7 \cdot 10^{7}$	$3.7 \cdot 10^{10}$	$3.2 \cdot 10^{4}$	$2.7 \cdot 10^{11}$	470
150	$4.8 \cdot 10^{10}$	$7.0 \cdot 10^{9}$	$1.0 \cdot 10^{7}$	$1.4 \cdot 10^{10}$	$2.2 \cdot 10^{4}$	$6.9 \cdot 10^{10}$	650
200	$4.7 \cdot 10^{9}$	$5.6 \cdot 10^{8}$	$5.9 \cdot 10^{6}$	$3.2 \cdot 10^{9}$	$1.6 \cdot 10^{4}$	$8.4 \cdot 10^{9}$	850
250	$7.7 \cdot 10^{8}$	$7.1 \cdot 10^{7}$	$4.3 \cdot 10^{6}$	$1.2 \cdot 10^{8}$	$1.4 \cdot 10^{4}$	$2.0 \cdot 10^{9}$	910
300	$1.4 \cdot 10^{8}$	$1.0 \cdot 10^{7}$	$3.3 \cdot 10^{6}$	$4.4 \cdot 10^{8}$	$1.3 \cdot 10^{4}$	$5.9 \cdot 10^{8}$	930
400	$6.0 \cdot 10^{6}$	$2.8 \cdot 10^{5}$	$2.1 \cdot 10^{6}$	$7.0 \cdot 10^{7}$	$1.1 \cdot 10^{4}$	$7.8 \cdot 10^{7}$	940
500	$2.8 \cdot 10^{5}$	$8.6 \cdot 10^{3}$	$1.3 \cdot 10^{6}$	$1.2 \cdot 10^{7}$	$1.0 \cdot 10^{4}$	$1.3 \cdot 10^{7}$	950
600	$1.5 \cdot 10^{4}$	$2.9 \cdot 10^{2}$	$8.7 \cdot 10^{5}$	$2.3 \cdot 10^{6}$	$9.0 \cdot 10^{3}$	$3.2 \cdot 10^{6}$	950
700	$8.4 \cdot 10^{2}$	$1.1 \cdot 10^{1}$	$5.8 \cdot 10^{5}$	$4.4 \cdot 10^{5}$	$8.1 \cdot 10^{3}$	$1.0 \cdot 10^{6}$	950
800	$5.1 \cdot 10$	$4.6 \cdot 10^{-1}$	$3.9 \cdot 10^{5}$	$8.9 \cdot 10^{4}$	$7.3 \cdot 10^{3}$	$4.8 \cdot 10^{5}$	950
900	3.4	$2.1 \cdot 10^{-2}$	$2.6 \cdot 10^{5}$	$1.9 \cdot 10^{4}$	$6.7 \cdot 10^{3}$	$2.8 \cdot 10^{5}$	980
1000	$2.4 \cdot 10^{-1}$	$1.1 \cdot 10^{-3}$	$1.8 \cdot 10^{5}$	$4.2 \cdot 10^{3}$	$6.1 \cdot 10^{3}$	$1.9 \cdot 10^{5}$	950

small and depends on their transport—caused by atmospheric turbulence and wave phenomena—in the atmosphere [24–26].

At altitudes above 100 km turbulences are absent (the level $z_T \approx 100$ km is called a turbo-pause [25,26]). Here the barometric formula (Equation 1.1) is valid up to altitudes z_e of the exosphere, as was mentioned earlier. In the turbo-pause region, as follows from Table 1.1, the main neutral components are atoms of nitrogen and oxygen created by the dissociation of the corresponding molecules: at the altitudes of 100–200 km, due to the dissociation of molecules of the oxygen ($O_2 \rightarrow O + O$), and at altitudes beyond 300 km, due to dissociation of molecules of nitrogen ($N_2 \rightarrow N + N$). At altitudes of 500–600 km the relative concentration of helium, He, sharply

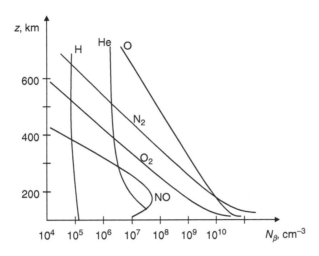

Figure 1.1 Height dependence of the neutral content of the ionospheric plasma.

increases, and above 1000 km the same increase is observed for hydrogen, H. At the altitudes of the exosphere, hydrogen is predominant, and for $z \geq 1500$ km hydrogen is the main component of the neutral gas in Earth's atmosphere (see also Figure 1.1).

As for the concentration of charged particles, the same processes that influence the neutral content of the ionosphere and create its structure are important for creation of charged particle content of ionospheric plasma. Again, they are solar wave radiation, its spectrum and intensity, energy of corpuscular solar radiation, cosmic rays, the content of the atmosphere, chemical processes of ionization and recombination of plasma electrons and ions, and so on [23,24,29–32]. In Table 1.2, the temperature, T_e, and concentration $N_e \equiv N$ of electrons versus the height are presented, as well as the temperature of ions T_i and the relative concentration of ion content in plasma $n = N_i/N$. It is seen that the concentration of electrons and ions increases up to altitudes of 300–400 km and for $z > 400$ km it gradually decreases. The degree of ionization of ionospheric plasma is determined by the ratio of concentration of charged particles in plasma (considering plasma as a quasi-neutral medium, $N_e \approx N_i \equiv N$) and the neutral component of plasma N_m, i.e., by N/N_m. Comparing N from Table 1.2 and N_m from Table 1.1, we can conclude that the degree of ionization of plasma at the lower ionosphere is minimal ($N/N_m \sim 10^{-8}–10^{-6}$), and increases with height z (for $z \geq 300$ km, $N/N_m \sim 10^{-4}$). At altitudes of exosphere ($z \geq 1000$ km) the degree of ionization reaches 10^{-1} and at altitudes of the magnetosphere, plasma becomes fully ionized, with $N/N_m \geq 1$.

In the D layer of the middle-latitude ionosphere the main ions are NO^+ and O_2^+; in the E and F layers, the main ions are O^+ up to 300 km; beyond these altitudes and up to the heights of the exosphere the ions He^+, H^+, N^+, and N_2^+ are also observed, the concentration of which is small with respect to O^+. However, such types of ions may play an important role in the process of generation of charged particles in the ionosphere.

At the same time, the temperature of electrons T_e and ions T_i is not particularly dependent on the time of day and increases smoothly with the height. At altitudes of the lower ionosphere ($z < 150–200$ km) both temperatures are identical, whereas in the higher ionosphere ($z \geq 300–350$ km) T_e can exceed T_i about twofold.

Table 1.2 Ionospheric Structure

z (km)	Day			Relative Concentration of Ions, $n = N_i/N$						
	N (cm^{-3})	T_e (K)	T_i (K)	NO^+	O_2^+	O^+	H_e^+	H^+	N^+	N_2^+
60	80	270	270	0.8	0.2	—	—	—	—	—
70	$2 \cdot 10^2$	200	200	0.8	0.2	—	—	—	—	—
80	10^3	180	180	0.7	0.3	—	—	—	—	—
90	$8 \cdot 10^3$	200	190	0.65	0.35	—	—	—	—	—
100	$8 \cdot 10^4$	240	210	0.62	0.38	—	—	—	—	—
110	$1.2 \cdot 10^5$	320	270	0.55	0.45	—	—	—	—	—
120	$1.3 \cdot 10^5$	400	360	0.39	0.60	0.01	—	—	—	—
130	$1.5 \cdot 10^5$	500	460	0.42	0.55	0.02	—	—	—	0.01
150	$3 \cdot 10^5$	800	670	0.45	0.41	0.13	—	—	—	0.01
200	$5 \cdot 10^5$	1300	1100	0.045	0.045	0.90	—	—	$5 \cdot 10^{-3}$	$5 \cdot 10^{-3}$
250	$1.0 \cdot 10^6$	1700	1300	6.10	5.10	0.98	—	—	$6 \cdot 10^{-3}$	$3 \cdot 10^{-3}$
300	$1.6 \cdot 10^6$	2000	1400	—	—	0.99	—	—	0.01	—
400	$1.5 \cdot 10^6$	2400	1450	—	—	0.97	$5 \cdot 10^{-3}$	0.01	0.02	—
500	$9 \cdot 10^5$	2600	1600	—	—	0.90	0.015	0.03	0.06	—
600	$4 \cdot 10^5$	2700	2100	—	—	0.84	0.02	0.06	0.08	—
700	$2 \cdot 10^5$	2800	2200	—	—	0.75	0.04	0.11	0.10	—
800	10^5	2870	2870	—	—	0.61	0.06	0.21	0.12	—
900	$7 \cdot 10^4$	2940	2940	—	—	0.41	0.09	0.40	0.10	—
1000	$5 \cdot 10^4$	3000	2500	—	—	0.28	0.14	0.51	0.07	—
z (km)	Night			Relative Concentration of Ions, $n = N_i/N$						
	N (cm^{-3})	T_e (K)	T_i (K)	NO^+	O_2^+	O^+	H_e^+	H^+	N^+	N_2^+
60	—	—	—	0.8	0.2	—	—	—	—	—
70	—	—	—	0.8	0.2	—	—	—	—	—
80	10	180	180	0.7	0.3	—	—	—	—	—
90	60	190	190	0.65	0.35	—	—	—	—	—
100	$1.2 \cdot 10^3$	210	210	0.62	0.38	—	—	—	—	—
110	$1.8 \cdot 10^3$	270	270	0.55	0.45	—	—	—	—	—

Table 1.2 (continued)   Ionospheric Structure

z (km)	Night N (cm^{-3})	T_e (K)	T_i (K)	Relative Concentration of Ions, $n = N_i/N$ NO$^+$	O$_2^+$	O$^+$	H$_e^+$	H$^+$	N$^+$	N$_2^+$
120	$2.1 \cdot 10^3$	360	360	0.39	0.60	0.01	—	—	—	—
130	$2.2 \cdot 10^3$	480	470	0.42	0.55	0.02	—	—	—	0.01
150	$2.4 \cdot 10^3$	670	650	0.45	0.41	0.13	—	—	—	0.01
200	$3 \cdot 10^3$	900	850	0.045	0.045	0.90	—	—	$5 \cdot 10^{-3}$	$5 \cdot 10^{-3}$
250	10^4	1000	910	6.10	5.10	0.98	—	—	$6 \cdot 10^{-3}$	$3 \cdot 10^{-3}$
300	10^5	1200	930	—	—	0.99	—	—	0.01	—
400	$3 \cdot 10^5$	1400	950	—	—	0.97	$5 \cdot 10^{-3}$	0.01	0.02	—
500	$2 \cdot 10^5$	1500	1000	—	—	0.90	0.015	0.03	0.06	—
600	$1.3 \cdot 10^5$	1600	1020	—	—	0.84	0.02	0.06	0.08	—
700	$8 \cdot 10^4$	1700	1100	—	—	0.75	0.04	0.11	0.10	—
800	$5 \cdot 10^4$	1800	1200	—	—	0.61	0.06	0.21	0.12	—
900	$3 \cdot 10^4$	1900	1300	—	—	0.41	0.09	0.40	0.10	—
1000	$2 \cdot 10^4$	2000	1400	—	—	0.28	0.14	0.51	0.07	—

The ion content of the ionospheric plasma is not strongly dependent on latitude Φ up to the high-latitude ionosphere (i.e., for $|\Phi| > 55°-60°$). Near the polar cap, the total ion content in the ionosphere decreases sharply with altitude. More details about the structure and properties of the ionosphere, including its neutral and charged particles, can be found in many excellent books [13–15,19,20].

1.2   Major Characteristics of the Ionospheric Plasma

Parameters of the ionospheric plasma such as the concentration and temperature of neutral and charged particles, mentioned earlier, mostly define the content of the ionosphere and its global structure. Apart from these plasma parameters, there are other important characteristics, functions of the main parameters, which are usually introduced to describe physical processes and phenomena occurring in the ionospheric plasma. All these characteristics have their own physical interpretation and can be considered separately.

First, we consider the frequencies of collisions between all neutral and charged particles contained in the ionospheric plasma, the lengths of free path between interactions for each plasma component, the plasma frequency and gyrofrequencies of electrons and ions, the rate of rotation of charged particles around magnetic strength lines, the coefficients of diffusion of neutral and charged particles, the conductivity of plasma components, and so on.

One important parameter of the ionospheric plasma that influences dynamic processes occurring in it and radio wave propagation in the ionosphere is the frequency of electron–neutral collisions, ν_{em}. In the following text, from Refs. [27,28], we present the approximate expressions for ν_{em} for various interactions of plasma electrons with different components of neutral gas in the ionosphere:

$$\nu_{eN_2} = 2.9 \cdot 10^{-7} N_{N_2} T_e \left(1 + T_e^{1/2}\right)^{-1} \tag{1.2a}$$

$$\nu_{eO_2} = 1.6 \cdot 10^{-8} N_{O_2} T_e^{1/2} (1 + 4.5 T_e) \tag{1.2b}$$

$$\nu_{em} = 1.23 \cdot 10^{-7} N_m T_e^{5/6} \quad (N_m = N_{N_2} + N_{O_2}) \tag{1.2c}$$

$$\nu_{eO} = 3.0 \cdot 10^{-8} N_O T_e^{1/2} \tag{1.2d}$$

$$\nu_{eH} = 4.9 \cdot 10^{-8} N_H T_e^{1/2} \tag{1.2e}$$

$$\nu_{eHe} = 5.0 \cdot 10^{-8} N_{He} T_e^{1/2} \tag{1.2f}$$

According to estimations made in Ref. [25], the effective frequency of electron–ion collisions, ν_{ei}, can be presented as a function of the ion concentration and electron temperature:

$$\nu_{ei} = \frac{5.5 N_i}{T_e^{3/2}} \ln \frac{220 T_e}{N_i^{1/3}} \tag{1.3}$$

The effective frequency for ion–neutral collisions (i.e., ions with molecules and atoms) can be presented approximately as [25,27]:

$$\nu_{im} = \beta_{im}^0 (T_i + T_m)^{1/2} N_m \tag{1.4}$$

where the coefficient β_{im}^0 takes into account the mass difference of molecules and the negligible effects of nonelastic ion–neutral collisions, related to polarization of molecules and their over-charged distribution.

Real parameters of the plasma in the middle-latitude ionosphere at altitudes of 90–600 km are shown in Table 1.3 [31–33]; measurements were taken during both the day and night, at moderate solar activity. Here, to deduce formulas and their relationships, as well as the number of parameters presented in the major formulas, we introduce (for future computations and experimental data description) the following notations: $q_H = \omega_H / \nu_{em}$, $Q_H = \Omega_H / \nu_{em}$, $\gamma = m \nu_{em} / M_i \nu_{im}$, where m and M_i, ω_H and Ω_H, ν_{em} and ν_{im} are the mass, gyromagnetic frequency, and the frequency of collisions with neutrals for electrons ($\alpha = e$) and ions ($\alpha = i$) in plasma, respectively (we assume that ions, like electrons, have unit charge and only differ from electrons by the sign of the charge).

As seen from Table 1.3, with increasing ionospheric altitude h, the degree of magnetization of the ionospheric plasma also increases, and is determined by the inequality $\omega_H \Omega_H \gg \nu_{em} \nu_{im}$. For $h < 150$ km, $\omega_H > \nu_{em}$, whereas $\Omega_H < \nu_{im}$; i.e., the electrons are magnetized, but the ions are only weakly magnetized. When $h > 150$ km, $\omega_H \gg \nu_{em}$ and $\Omega_H > \nu_{im}$; i.e., for middle and higher ionosphere altitudes both electrons and ions are strongly magnetized. As was mentioned earlier, the degree of ionization of the ionospheric plasma is usually determined by the ratio of concentration of charged particles, electrons, or ions (plasma is quasi-neutral, $N_e \approx N_i = N$) to that of neutrals, i.e., N/N_m. However, in numerous works concerning the ionosphere [27,28,34–39] the degree of

Table 1.3 Major Characteristics of the Ionospheric Plasma

Altitude (km)	M_i (kg)	T (K)	ν_{em} (s^{-1})	ν_{im} (s^{-1})	q_H	Q_H	γ
Day							
90	$4.82 \cdot 10^{-26}$	200	$2.97 \cdot 10^{5}$	$1.67 \cdot 10^{4}$	29.3	$9.94 \cdot 10^{-3}$	$3.35 \cdot 10^{-4}$
100	$4.72 \cdot 10^{-26}$	219	$5.63 \cdot 10^{4}$	$3.2 \cdot 10^{3}$	153	$5.2 \cdot 10^{-2}$	$3.39 \cdot 10^{-4}$
110	$4.6 \cdot 10^{-26}$	260	$1.34 \cdot 10^{4}$	$7.71 \cdot 10^{2}$	664	0.22	$3.44 \cdot 10^{-4}$
120	$4.5 \cdot 10^{-26}$	420	$5.29 \cdot 10^{3}$	$3 \cdot 10^{2}$	$1.62 \cdot 10^{3}$	0.57	$3.57 \cdot 10^{-4}$
130	$4.4 \cdot 10^{-26}$	600	$2.41 \cdot 10^{3}$	121	$3.55 \cdot 10^{3}$	1.45	$4.22 \cdot 10^{-4}$
140	$4.32 \cdot 10^{-26}$	800	$1.26 \cdot 10^{3}$	61	$6.75 \cdot 10^{3}$	2.94	$4.35 \cdot 10^{-4}$
150	$4.22 \cdot 10^{-26}$	1080	$7.32 \cdot 10^{2}$	35.1	$1.16 \cdot 10^{4}$	5.21	$4.51 \cdot 10^{-4}$
160	$4.12 \cdot 10^{-26}$	1160	$4.57 \cdot 1$	21.5	$1.85 \cdot 10^{4}$	8.65	$4.69 \cdot 10^{-4}$
170	$4.02 \cdot 10^{-26}$	1240	$2.94 \cdot 10^{2}$	14	$2.86 \cdot 10^{4}$	13.6	$4.75 \cdot 10^{-4}$
180	$3.92 \cdot 10^{-26}$	1240	$1.07 \cdot 10^{2}$	9.41	$4.24 \cdot 10^{4}$	20	$4.86 \cdot 10^{-4}$
190	$3.82 \cdot 10^{-26}$	1280	$1.37 \cdot 10^{2}$	6.62	$6.1 \cdot 10^{4}$	20	$4.93 \cdot 10^{-4}$
200	$3.72 \cdot 10^{-26}$	1360	98.6	4.83	$0.4 \cdot 10^{4}$	41.8	$5 \cdot 10^{-4}$
220	$3.54 \cdot 10^{-26}$	1500	53.8	2.66	$1.53 \cdot 10^{5}$	79.3	$5.2 \cdot 10^{-4}$
240	$3.37 \cdot 10^{-26}$	1600	30.5	1.57	$2.7 \cdot 10^{5}$	139.5	$5.25 \cdot 10^{-4}$
260	$3.23 \cdot 10^{-26}$	1700	18.7	0.915	$4.31 \cdot 10^{5}$	248	$4.76 \cdot 10^{-4}$
280	$3.10 \cdot 10^{-26}$	1800	12	0.62	$6.7 \cdot 10^{5}$	376	$5.65 \cdot 10^{-4}$
300	$2.98 \cdot 10^{-26}$	1900	7.81	0.41	10^{6}	595	$5.79 \cdot 10^{-4}$
350	$2.78 \cdot 10^{-26}$	2060	2.72	0.11	$2.8 \cdot 10^{6}$	$236 \cdot 10^{3}$	$6.83 \cdot 10^{-4}$
400	$2.60 \cdot 10^{-26}$	2240	1.03	0.05	$7.35 \cdot 10^{6}$	$5.39 \cdot 10^{3}$	$7.35 \cdot 10^{-4}$
450	$2.42 \cdot 10^{-26}$	2460	0.401	0.02	$1.85 \cdot 10^{7}$	$1.36 \cdot 10^{4}$	$7.36 \cdot 10^{-4}$
500	$2.18 \cdot 10^{-26}$	2520	0.25	$9.6 \cdot 10^{-3}$	$2.9 \cdot 10^{7}$	$3.15 \cdot 10^{4}$	$1.08 \cdot 10^{-3}$
600	$1.78 \cdot 10^{-26}$	2780	0.072	$4.9 \cdot 10^{-3}$	$9.74 \cdot 10^{7}$	$7.1 \cdot 10^{-4}$	$7.6 \cdot 10^{-4}$
Night							
90	$4.82 \cdot 10^{-26}$	200	$2.97 \cdot 10^{5}$	$1.67 \cdot 10^{4}$	29.3	$9.94 \cdot 10^{-3}$	$3.35 \cdot 10^{-4}$
100	$4.72 \cdot 10^{-26}$	219	$5.63 \cdot 10^{4}$	$3.2 \cdot 10^{3}$	153	$5.2 \cdot 10^{-2}$	$3.39 \cdot 10^{-4}$
110	$4.6 \cdot 10^{-26}$	260	$1.34 \cdot 10^{4}$	$7.71 \cdot 10^{2}$	664	0.22	$3.44 \cdot 10^{-4}$
120	$4.5 \cdot 10^{-26}$	340	$5.87 \cdot 10^{3}$	304	$1.46 \cdot 10^{3}$	0.57	$3.8 \cdot 10^{-4}$

(continued)

Table 1.3 (continued) **Major Characteristics of the Ionospheric Plasma**

Altitude (km)	Mi (kg)	T (K)	ν_{em} (s^{-1})	ν_{im} (s^{-1})	q_H	Q_H	γ
130	$4.41 \cdot 10^{-26}$	500	$2.27 \cdot 10^3$	122	$3.77 \cdot 10^3$	1.44	$3.8 \cdot 10^{-4}$
140	$4.3 \cdot 10^{-26}$	560	$1.09 \cdot 10^3$	60.1	$7.8 \cdot 10^3$	2.98	$3.8 \cdot 10^{-4}$
150	$4.2 \cdot 10^{-26}$	600	$5.77 \cdot 10^2$	33.4	$1.47 \cdot 10^3$	5.5	$3.8 \cdot 10^{-4}$
160	$4.1 \cdot 10^{-26}$	650	$3.39 \cdot 10^2$	19.7	$2.49 \cdot 10^4$	5.5	$3.8 \cdot 10^{-4}$
170	$3.96 \cdot 10^{-26}$	700	$2.12 \cdot 10^2$	12.3	$3.93 \cdot 10^4$	15.6	$3.96 \cdot 10^{-4}$
180	$3.85 \cdot 10^{-26}$	750	$1.38 \cdot 10^2$	8.12	$6.06 \cdot 10^4$	24.3	$4 \cdot 10^{-4}$
190	$3.74 \cdot 10^{-26}$	800	94.2	5.54	$8.8 \cdot 10^4$	36.5	$4.14 \cdot 10^{-4}$
200	$3.62 \cdot 10^{-26}$	840	65.3	3.81	$1.27 \cdot 10^5$	54.3	$4.31 \cdot 10^{-4}$
220	$3.44 \cdot 10^{-26}$	940	34	2.02	$2.41 \cdot 10^5$	107.4	$4.45 \cdot 10^{-4}$
240	$3.27 \cdot 10^{-26}$	1000	18.8	1.1	$4.32 \cdot 10^5$	205.5	$4.76 \cdot 10^{-4}$
260	$3.1 \cdot 10^{-26}$	1080	11.1	0.65	$7.26 \cdot 10^5$	361.5	$5 \cdot 10^{-4}$
280	$2.98 \cdot 10^{-26}$	1150	6.72	0.39	$1.2 \cdot 10^6$	620	$5.23 \cdot 10^{-4}$
300	$2.87 \cdot 10^{-26}$	1250	4.15	0.24	$1.9 \cdot 10^6$	$1.04 \cdot 10^3$	$5.46 \cdot 10^{-4}$
350	$2.67 \cdot 10^{-26}$	1345	1.35	0.09	$5.7 \cdot 10^6$	$2.9 \cdot 10^3$	$5.1 \cdot 10^{-4}$
400	$2.47 \cdot 10^{-26}$	1440	0.48	0.28	$1.57 \cdot 10^6$	10^4	$6.42 \cdot 10^{-4}$
450	$2.19 \cdot 10^{-26}$	1520	0.19	$8.6 \cdot 10^{-3}$	$3.8 \cdot 10^7$	$3.6 \cdot 10^4$	$9 \cdot 10^{-4}$
500	$1.78 \cdot 10^{-26}$	1600	0.85	$2.6 \cdot 10^{-3}$	$8.54 \cdot 10^7$	$1.4 \cdot 10^5$	$1.65 \cdot 10^{-3}$
600	$1.3 \cdot 10^{-26}$	1780	0.146	$2.5 \cdot 10^{-3}$	$4.7 \cdot 10^8$	$1.7 \cdot 10^6$	$1.1 \cdot 10^{-3}$

plasma ionization was determined by the parameter of ionization $p = \nu_{ei}/\nu_{em}$, where ν_{ei} is the frequency of electron–ion collisions. Despite the fact that ionospheric plasma is usually at low temperature, the cross section of collisions between electrons and ions significantly exceeds the cross section of collisions of electrons with neutral molecules and atoms. Therefore, electron–ion interactions are essential even when $N/N_m \ll 1$, but $p \geq 1$. As follows from Table 1.4 (see also Figure 1.2), where the parameter of ionization p for $h = 90$–700 km is shown, in the real ionosphere the parameter of ionization can reach high values, though, in the original sense the ionospheric plasma is weakly ionized (i.e., $N \ll N_m$).

At altitudes under consideration ($h = 90$–600 km), $N_m > N_e$ or N_i provides the condition of stability of a full plasma pressure necessary for the transport processes of diffusion, thermodiffusion, and drift in the ionosphere. Under such circumstances different disturbances of neutral molecules and their movement do not influence the transport of electrons and ions in the ionospheric plasma [25,27,33]. That is why, just as in Refs. [25–28,33,39], we will evaluate the degree of ionization of

Table 1.4 The Ionization Parameters of Ionospheric Plasma

Altitude (km)	24 h	Ionization Parameter	Altitude (km)	24 h	Ionization Parameter
90	Day	$7.5 \cdot 10^{-4}$	240	Day	8.4
	Night	$6.9 \cdot 10^{-6}$		Night	0.43
100	Day	$1.75 \cdot 10^{-2}$	260	Day	18.1
	Night	$4.28 \cdot 10^{-4}$		Night	1.01
110	Day	$7.1 \cdot 10^{-2}$	280	Day	33.4
	Night	$1.6 \cdot 10^{-3}$		Night	2.87
120	Day	$9.35 \cdot 10^{-2}$	300	Day	45
	Night	$2.41 \cdot 10^{-3}$		Night	15.5
130	Day	$1.63 \cdot 10^{-1}$	350	Day	91.2
	Night	$3.92 \cdot 10^{-3}$		Night	91.3
140	Day	$2.8 \cdot 10^{-1}$	400	Day	168.5
	Night	$5.75 \cdot 10^{-3}$		Night	300.4
150	Day	$5.3 \cdot 10^{-1}$	450	Day	256.6
	Night	$8.37 \cdot 10^{-3}$		Night	578.3
160	Day	$6.8 \cdot 10^{-1}$	500	Day	333
	Night	$1.05 \cdot 10^{-2}$		Night	999.8
180	Day	1.78	550	Day	384.5
	Night	$3.1 \cdot 10^{-2}$		Night	2112
190	Day	2.19	600	Day	518.5
	Night	$4.44 \cdot 10^{-2}$		Night	2439
200	Day	2.93	650	Day	634.2
	Night	$5.8 \cdot 10^{-2}$		Night	6001
220	Day	4.72	700	Day	704
	Night	$1.8 \cdot 10^{-1}$		Night	8889

the ionospheric plasma by the parameter of ionization p, which is shown in Figure 1.2 for diurnal and nocturnal midlatitude ionosphere during median solar activity. Judging by the figure and the data presented in Tables 1.3 and 1.4, we can see that with increase of altitude the effective values of the collision frequencies, ν_{em} and ν_{im}, fall, whereas the value ν_{ei} increases with height. It is quite clear that with increase of the altitude the number of neutral particles in the ionosphere decreases

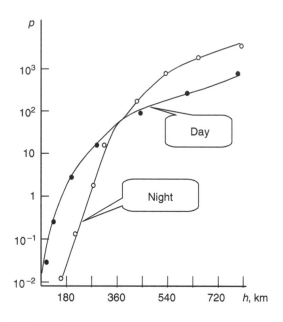

Figure 1.2 Height distribution of the parameter of ionization, *p*, in the diurnal and nocturnal ionosphere observed in Refs. [33,39].

quickly while the total number of ions in the ionosphere increases with height. Hence, the parameter of ionization of plasma also increases in relation to the altitude increase. Thus, the nocturnal ionosphere is weakly ionized ($p \ll 1$) up to $h \sim 250$ km, and the diurnal ionosphere is weakly ionized up to $h \sim 150$ km. At altitudes beyond 350 km, the growth of the degree of ionization of the nocturnal ionosphere exceeds the degree of ionization of the diurnal ionosphere (see Figure 1.2). At the same time, as mentioned in Refs. [31,33], from $h > 500$ km, the growth of the ionization parameter p decelerates with the height, while the degree of plasma magnetization increases monotonically with increase of ionospheric altitudes above 500 km.

Therefore, while analyzing the physical processes and phenomena occurring in the real ionosphere for $h > 150$–160 km, as well as in meteor trails or ion clouds, we have to account for the collisions between the charged particles and degree of magnetization of ionospheric plasma, conditions that are not usually taken into consideration by some existing models when comparing them with real ionospheric experiments. This is one of the reasons for the ambiguous interpretation of experimental data particularly with meteor trails or barium clouds; see Chapters 5 and 6 and also Refs. [38,39] and the references presented there.

Furthermore, according to the empirical models of the ionosphere [31,33], at noon in the ionosphere during periods of low solar activity, the mean free path of charged particles for electrons, λ_e, changes from 30 cm to 12 km for $h = 90$–500 km and that for ions, λ_i, is from 0.5 cm to 2 km for the same altitude range of 90–500 km. Larmor's radius of electrons, ρ_{eH}, at the altitudes under discussion, changes from 1 to 4 cm, and the Larmor's radius of ions, ρ_{iH}, at the same altitudes (of 90–500 km) changes from 0.4 to 7.4 m [30–33].

All these characteristics and parameters will be used when modeling dynamic processes in the ionospheric plasma and choosing methods to solve the corresponding problems.

1.3 Transport Processes of Plasma in the Ionosphere

When investigating physical phenomena occurring in the ionospheric plasma, the two main aspects that usually determine irregular structure of plasma and nonuniform distribution of charge particles in plasma are (a) the transfer processes of plasma components, electrons, and ions, their temperature and energy, and (b) the chemical reactions that determine the ionization–recombination balance of charge particles in plasma with neutral components of ionosphere. In the following text we will consider both these aspects separately.

To separate the phenomena mentioned earlier, we need first to compare initial scales and dimensions of irregular plasma structures and turbulences in the plasma at the altitudes discussed with the characteristic scales and dimensions of plasma presented at the end of the previous section. As mentioned before, such a comparison allows stating the problems and choosing the method for their solution. The qualitative analysis carried out earlier has shown that all characteristic scales of plasma are less than the dimensions of plasma irregularities. Moreover, at altitudes beyond 150–160 km, the transfer processes, diffusion, thermodiffusion, and drift prevail compared with the chemical processes of ionization and recombination. Therefore, for a description of transfer processes beyond altitudes of the upper E layer (i.e., at altitudes of more than 150–160 km), we can use a system of magnetohydrodynamic equations for all components of the ionospheric plasma, electrons, and ions to describe evolution of irregular structures of ionospheric plasma of various types, both artificial and natural.

As had been mentioned in the pioneering works, such processes can be fully described by the continuity equation of charged particles and neutrals, by equation of motion and the equation of heat energy conservation [15–21].

As shown by later investigations [22–39], usually concerning the solution of the problems of dynamic inhomogeneous ionospheric plasma at altitudes of the upper E layer and higher, the absence of processes of ionization, attachment, detachment, and recombination is assumed, as well as the unification between electrons and ions with unit charge and between the ions of plasma irregularities and those of the background ionospheric plasma. In this case a full system of transfer equations of plasma components (denoted by indexes $\alpha = e,i$) and neutral components (denoted by index $\alpha = m$) can be presented in the following form [27,28,33,39]:

$$\frac{\partial N_\alpha}{\partial t} + \nabla \cdot \mathbf{j}_\alpha = 0 \tag{1.5a}$$

$$\frac{\partial N_m}{\partial t} + \nabla \cdot (\mathbf{U}_m N_m) = 0 \tag{1.5b}$$

$$\mathbf{j}_e = -\frac{\hat{\sigma}_e}{e}\left\{\mathbf{E} + \frac{1}{c}[\mathbf{U}_m,\mathbf{B}_0]\right\} - \hat{D}_{ee}\nabla N_e - \hat{D}_{Te}\frac{N_e}{T_e}\nabla T_e + N_e\mathbf{U}_m \tag{1.6a}$$

$$\mathbf{j}_i = \frac{\hat{\sigma}_i}{e}\left\{\mathbf{E} + \frac{1}{c}[\mathbf{U}_m,\mathbf{B}_0]\right\} - \hat{D}_{ii}\nabla N_i - \hat{D}_{Ti}\frac{N_i}{T_i}\nabla T_i + N_i\mathbf{U}_m \tag{1.6b}$$

$$M_m N_m\left[\frac{\partial U_m}{\partial t} + (\mathbf{U}_m,\nabla)\mathbf{U}_m\right]$$
$$= -\nabla(N_m T_m)\frac{\hat{\sigma}_e}{e} - \eta\mathbf{U}_m - \frac{1}{3}\eta(\mathbf{U}_m) - m\hat{\nu}_{em}(N_e\mathbf{U}_m - \mathbf{j}_e) - M_i\hat{\nu}_{im}(N_i\mathbf{U}_m - \mathbf{j}_i) \tag{1.6c}$$

In the ionosphere, when there is any plasma irregularity it always arises in the electrical field and then in the magnetic field closely related to this irregularity. Therefore, Equations 1.5 and 1.6 must be completed by Maxwell's equations:

$$\nabla \cdot \mathbf{E} = 4\pi e(N_i - N_e) \tag{1.7a}$$

$$\nabla \times \mathbf{E} = -\frac{\partial \mathbf{B}}{\partial t} \tag{1.7b}$$

$$\nabla \cdot \mathbf{B} = 0 \tag{1.7c}$$

$$\nabla \times \mathbf{B} = \frac{4\pi e}{c}(\mathbf{j}_i - \mathbf{j}_e) \tag{1.7d}$$

In the last equation (Equation 1.7d) of the system, the displacement current $\frac{1}{4\pi e}\frac{\partial \mathbf{D}}{\partial t}$ is omitted from consideration, because only quasi-stationary processes for the plasma concentration and temperature are taken into account, as well as for the ambient electrical $\mathbf{E}$ and magnetic $\mathbf{B}$ fields. In Equations 1.5 through 1.7 the additional notations are introduced:

$$\hat{D}_{ee} = \hat{D}_e\left(1 + \frac{T_i}{T_e}\right) - \frac{T_e}{e^2 N_e}\hat{\sigma}_e \tag{1.8a}$$

$$\hat{D}_{ii} = \hat{D}_i\left(1 + \frac{T_e}{T_i}\right) - \frac{T_i}{e^2 N_i}\hat{\sigma}_i \tag{1.8b}$$

Here, $\hat{D}_e$, $\hat{D}_i$, $\hat{D}_{Te}$, $\hat{D}_{Ti}$ and $\hat{\sigma}_e$, $\hat{\sigma}_i$ and $\mathbf{j}_i$, $\mathbf{j}_e$ are the tensors of diffusion, thermodiffusion and conductivity, and the current densities for electron and ion components of the plasma, respectively; $\mathbf{U}_m$ is the vector of the hydrodynamic velocity of molecules (atoms) of the atmospheric gas; η is the coefficient of viscosity; c is the speed of light; and $\hat{\nu}_e$, $\hat{\nu}_i$ are the tensors of collisions of electrons and ions, respectively, with molecules and atoms of the neutral gas. For ionospheric altitudes beyond 160–170 km and a high enough degree of plasma ionization ($\nu_{ei} \gg \nu_{em}$), the charged particles, electrons and ions, are magnetized ($\omega_H \gg \nu_{em}$, $\Omega_H \gg \nu_{im}$; see Tables 1.3 and 1.4).

With satisfaction of conditions of quasi-neutrality and ambipolar diffusion of electrons and ions of the ionospheric plasma:

$$N_e \approx N_i = N \tag{1.9a}$$

$$\nabla \mathbf{j}_e = \nabla \mathbf{j}_i = \nabla \mathbf{j} \tag{1.9b}$$

and accounting the temperature distribution of the charged particles during the ambient heating, the distributions of electron and ion concentration become nonuniform:

$$\frac{\partial T_e}{\partial t} = \frac{\nabla(\hat{\kappa}_e \nabla T_e)}{N_e} + \delta_{ei}\nu_{ei}(T_e - T_i) - \delta_{em}\nu_{em}(T_e - T_m) + \frac{2}{3}\frac{Q}{N_e} \tag{1.10a}$$

$$\frac{\partial T_i}{\partial t} = \frac{\nabla(\hat{\kappa}_i \nabla T_i)}{N_e} + \delta_{ei}\nu_{im}(T_i - T_m) + \frac{2}{3}\frac{Q}{N_e} \tag{1.10b}$$

We thus finally obtain the self-matching system of three-dimensional equations of plasma transfer: diffusion, thermodiffusion, and drift.

Under conditions of arbitrary degree of ionization and magnetization of the ionospheric plasma, the tensors of collisions $\hat{v}_{ei} = \hat{K}_{ei}v_{ei}$, $\hat{v}_{em} = \hat{K}_{em}v_{em}$, and $\hat{v}_{im} = \hat{K}_{im}v_{im}$, can be transformed, accounting that $\hat{K}_{ei}, \hat{K}_{em}, \hat{K}_{im} \to \hat{1}$, into the simple scalar values. In this case, called the *elementary theory* [27,28], in the accuracy of magnitudes of order $\gamma = \frac{mv_{em}}{M_i v_{im}} \ll 1$ (working from $h > 120$ km), we can, following Refs. [27,28], from the following equations of macroscopic movement of electrons and ions

$$mN_e\hat{v}_{em}\mathbf{V}_e = -eN_e\mathbf{E} - \frac{e}{c}N_e[\mathbf{V}_e,\mathbf{B}_0] - T_e\nabla N_e + m\hat{v}_{ei}N_e(\mathbf{V}_e - \mathbf{V}_i) \tag{1.11a}$$

$$M_iN_i\hat{v}_{im}\mathbf{V}_e = eN_i\mathbf{E} - \frac{e}{c}N_i[\mathbf{V}_i,\mathbf{B}_0] - T_i\nabla N_i + M_i\hat{v}_{ei}N_i(\mathbf{V}_i - \mathbf{V}_e) \tag{1.11b}$$

obtain the components of tensors of diffusion, thermodiffusion, and conductivity, which in the case of nonisothermal plasma ($T_e \neq T_i$) can be written in the following form:

$$D_{e\|} = \frac{T_e(1 + T_i/T_e)(1 + \gamma p)}{mv_{em}(1 + p)}, \quad D_{i\|} = \frac{T_i(1 + T_e/T_i)(1 + 2p)}{M_iv_{im}(1 + p)}$$

$$D_{e\perp} = \frac{T_e(1 + T_i/T_e)}{mv_{em}A}\left[(1 + p)(1 + 2\gamma p) + Q_{II}^2(1 + 2p)\right]$$

$$D_{i\perp} = \frac{T_i(1 + T_e/T_i)}{M_iv_{im}A}\left[(1 + p)(1 + 2p) + q_H^2(1 + 2\gamma p)\right] \tag{1.12a}$$

$$D_{e\wedge} = \frac{T_e(1 + T_i/T_e)q_H}{mv_{em}A}(1 + Q_H^2 + 3\gamma p)$$

$$D_{i\wedge} = \frac{T_i(1 + T_e/T_i)Q_H}{M_iv_{im}A}(1 + q_H^2 + p/\gamma)$$

$$\sigma_{e\|} = \frac{e^2N_e}{mv_{em}(1 + p)}, \quad \sigma_{i\|} = \frac{e^2N_i}{M_iv_{im}(1 + p)}$$

$$\sigma_{e\perp} = \frac{e^2N_e}{mv_{em}A}(1 + Q_H^2 + p), \quad \sigma_{i\perp} = \frac{e^2N_i}{M_iv_{im}A}(1 + q_H^2 + p/\gamma) \tag{1.12b}$$

$$\sigma_{e\wedge} = -\frac{e^2N_eq_H}{mv_{em}A}(1 + Q_H^2 + \gamma p), \quad \sigma_{i\wedge} = -\frac{e^2N_iQ_H}{M_iv_{im}A}(1 + q_H^2 + p/\gamma)$$

$$A = (1 + p)^2 + q_H^2(1 + 2\gamma p + Q_H^2)$$

$$D_{Te\|} = \frac{T_e(1 + \gamma p)}{mv_{em}(1 + p)}, \quad D_{Ti\|} = \frac{T_i}{M_iv_{im}}$$

$$D_{Te\perp} = \frac{T_e\left[(1 + p)(1 + Q_H^2) + \gamma p^2\right]}{mv_{em}B}$$

$$D_{Ti\perp} = \frac{T_i\left[p^2 + (1 + \gamma p)(q_H^2 + 2p)\right]}{M_iv_{im}B} \tag{1.12c}$$

$$D_{Te\wedge} = -\frac{T_e q_H \left[(1 + Q_H^2) + 2\gamma p\right]}{m\nu_{em}B}, \quad D_{Ti\wedge} = \frac{T_i Q_H \left[\gamma(q_H^2 + 2p)\right]}{M_i \nu_{im}B}$$

$$B = p^2 + (q_H^2 + 2p)(1 + Q_H^2 + 2\gamma p)$$

$$\kappa_{e\|} = \frac{N_e T_e}{m\nu_{em}(1 + p)}, \quad \kappa_{i\|} = \frac{N_i T_i}{M_i \nu_{im}(1 + \nu_{ii}/\nu_{im})}$$

$$\kappa_{e\perp} = \frac{N_e T_e}{m\nu_{em}} \frac{(1 + p)}{\left[q_H^2 + (1 + p)^2\right]}, \quad \kappa_{i\perp} = \frac{N_i T_i}{M_i \nu_{im}} \frac{(1 + \nu_{ii}/\nu_{im})}{\left[Q_H^2 + 1 + (1 + \nu_{ii}/\nu_{im})^2\right]} \quad (1.12d)$$

$$\kappa_{e\wedge} = \frac{N_e T_e}{m\nu_{em}} \frac{q_H}{\left[q_H^2 + (1 + p)^2\right]}, \quad \kappa_{i\wedge} = \frac{N_i T_i}{M_i \nu_{im}} \frac{Q_H}{\left[Q_H^2 + 1 + (1 + \nu_{ii}/\nu_{im})^2\right]}$$

Here in Equation 1.11 $[\mathbf{V}_\alpha, \mathbf{B}_0]$ is the cross product of two vectors $\mathbf{V}_\alpha$ and $\mathbf{B}_0$ for electrons ($\alpha = e$) and ions ($\alpha = i$); in Equation 1.12 parameters $\sigma_{\alpha\|}$, $D_{\alpha\|}$ and $D_{T\alpha\|}$, and $\sigma_{\alpha\perp}$, $D_{\alpha\perp}$ and $D_{T\alpha\perp}$, $\sigma_{\alpha\wedge}$, $D_{\alpha\wedge}$ and $D_{T\alpha\wedge}$ are the longitudinal, perpendicular (Pedersen), and crossing (Hall) components of coefficients of conductivity, diffusion, and thermodiffusion, respectively, relative to the electrical and magnetic fields, accounting that $\mathbf{E} \perp \mathbf{B}_0$ and $\mathbf{E} = \mathbf{E}_0 + \mathbf{E}_p$ is a full electrical field in the ionosphere, where $\mathbf{E}_0$ is the ambient electric field and $\mathbf{E}_p = -\nabla\varphi$ is the intrinsic potential field of polarization (sometimes called the *ambipolar field*), where φ is the potential of this field.

In Equation 1.10a and b, $\delta_{ei} \approx 2m/M_i$, $\delta_{em} \approx 10^{-3}$, and $\delta_{im} \approx 1$ are, respectively, the fractions of energy lost by electrons in collisions with ions and neutrals and also by ions in collisions with neutrals. An external source of heating is associated with high-energy photoelectrons forming in the ionosphere in the process of ionization of the neutral component by solar ultraviolet radiation. As an example, the altitudinal dependence of the coefficients of diffusion is shown in Figure 1.3a and b calculated according to Equation 1.11 for diurnal and nocturnal models of the middle-latitude ionosphere for moderate solar activity, respectively, all parameters of which are presented in Tables 1.3 and 1.4. Here, the results of computation for the virtual case of $p \approx 0$ are also presented. It is seen that the real coefficients of diffusion (i.e., estimated for the case of the real ionosphere for $p > 0$) change with ionosphere altitude nonmonotonically compared with the nonrealistic (virtual) case.

Thus, for $D_{e\|}$ and $D_{i\perp}$ the wide maxima are observed at 200 and 150 km, respectively; for $D_{e\perp}$ a wide minimum is observed at altitudes of 200–250 km and a maximum at altitude of about 300 km. As for the coefficient of ion diffusion along the magnetic field, $D_{i\|}$, it grows monotonically as ionospheric altitude increases.

Moreover, if we use a virtual model of the ionospheric plasma presented by dashed curves in Figure 1.3 without accounting for the real degree of plasma ionization (i.e., setting $p = 0$ in Equation 1.12a), then the difference between $D_{e\|}$ and $D_{i\|}$, $D_{e\perp}$ and $D_{i\perp}$, becomes greater with increase of ionospheric altitude (i.e., with increase of q_H and Q_H). Diffusion in the corresponding directions to $\mathbf{B}_0$ is determined not by the minimal ($D_{i\|}$ and $D_{e\perp}$), but by the maximal ($D_{e\|}$ and $D_{i\perp}$) coefficients of unipolar diffusion of charged particles. Accounting for the real degree of ionization of the ionospheric plasma ($p > 1$), which increases with altitude, the limit of $D_{e\|}$ is $D_{i\|}$ and that of $D_{e\perp}$ is $D_{i\perp}$ (see Figure 1.3), and the character of diffusion fundamentally changes. This aspect will be analyzed in detail in Chapter 5.

We have to point out that the additional evaluations carried out in Refs. [38,39] have shown that components of the tensor of conductivity, $\sigma_{e\perp}$, $\sigma_{i\perp}$, $\sigma_{i\|}$, and $\sigma_{e\|}$, have maxima at altitudes of

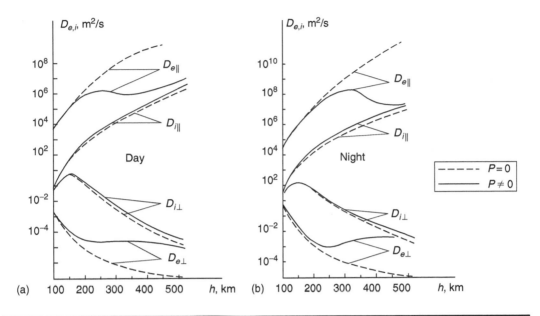

Figure 1.3 Height dependence of the coefficients of diffusion of electron and ion components of the (a) diurnal and (b) nocturnal ionospheric plasma in the case of weakly ionized plasma (dashed curves) and plasma with real degree of ionization (solid line) corresponding to the actual model of the middle-latitude ionosphere.

100, 150, 250, and 300 km, respectively. The same tendency of nonmonotonically height dependence with local maxima was observed for other transport coefficients [38].

Hence, apart from the macro-scale regular distribution of plasma, in the ionosphere there occur naturally and artificially created nonregular distributions of plasma concentration, the so-called macroscopic plasma irregularities, having dimensions ranging from hundreds kilometers (macro-disturbances) to tens and fewer meters (micro-disturbances), the evolution of which can be described by the self-matching system of diffusion, thermodiffusion and drift (Equations 1.5 through 1.10). At the same time, the peculiarities of the processes of relaxation of such plasma irregularities are determined by coefficients of transfer (e.g., diffusion, drift, and thermodiffusion) of electron and ion components of inhomogeneous plasma according to Equation 1.12.

1.4 Ionization–Recombination Balance in the Ionosphere

At the range of altitudes of the E layer of the ionosphere from 90 to 150 km, the effects of ionization and recombination are predominant and must also be taken into account in analysis of transport processes occurring in the background ionospheric plasma. In this case, we need to change equations (Equation 1.5a) written both for electron ($\alpha = e$) and ion ($\alpha = e$) components of plasma, by introducing the term of plasma ionization, I, and the term of recombination of charged particles, R, as follows:

$$\frac{\partial N_\alpha}{\partial t} + \nabla \cdot \mathbf{j}_\alpha = I - R \tag{1.13}$$

We will analyze these processes in the regions of the lower E layer and D layer of the ionosphere, where chemical reactions determining the ionization–recombination balance in the ionosphere prevail.

As is well known [21–23], for the lower ionosphere at altitudes less than 120–130 km (i.e., at the lower E layer and at the D layer), the temporal changes of concentration of electrons N (or ions, because plasma is quasi-neutral, $N_e \approx N_i = N$; see Equation 1.9a) are determined by processes of ionization and recombination, and Equation 1.5a can be reduced to

$$\frac{\partial N}{\partial t} = q_i - \alpha N^2 \tag{1.14}$$

Here, $q_i = q_{iO} + q_{iO_2} + q_{iN_2}$ is the total intensity of ionization of atoms O, O_2, and N_2; α is the effective coefficient of dissociative recombination.

Equations 1.14 and 1.13 describe the balance of number of charged particles in the ionosphere (without and with their transport by diffusion and drift processes, respectively), and determine in the general case (Equation 1.13) ion–molecule reactions between the basic components of the ionospheric plasma and the processes of dissipative recombination [27]. Dissociative recombination is the main chemical reaction in the lower ionosphere and is characterized by attachment of the molecular ion M^+ with electron e^-, which is accompanied by the dissociation of the molecular ion into two excited neutral atoms A_1^* and A_2^* [27,28]:

$$M^+ + e^- \rightarrow A_1^* + A_2^* \tag{1.15}$$

For example, for the molecular ion of oxygen O_2^+ or the oxide of nitrogen NO^+, we have, respectively, reactions of the following type:

$$O_2^+ + e^- \rightarrow O^* + O^*, \quad NO^+ + e^- \rightarrow N^* + O^* \tag{1.16}$$

Therefore, for dissociative recombination, the number of electrons attached to ions in a volume of 1 cm^3 during 1 s can be described in Equation 1.13 by the term $R = \alpha N^2$. Taking into account reactions of (Equation 1.16), we can finally present the effective recombination coefficient as [28,33]:

$$\alpha = \alpha_1 n_{NO^+} + \alpha_2 n_{O_2^+} \tag{1.17}$$

where $n_{NO^+} = N_{NO^+}/N$ and $n_{O_2^+} = N_{O_2^+}/N$ are the relative (to electrons) concentrations of ions of NO^+ and O_2^+, respectively; α_1 and α_2 are reaction coefficients of dissipative recombination of these ions, described by Equation 1.16.

The interactions between main ions of plasma and the neutral molecules such as N_2 and O_2 are determined by the following reactions:

$$N_2 + O_2^+ \rightarrow NO^+ + N, \quad O_2 + O^+ \rightarrow O_2^+ + O \tag{1.18}$$

Such reactions are characterized by the coefficients of ion–molecular reactions, β_1 and β_2, respectively.

In more detail, all processes of ionization and recombination that determine the ionization–recombination balance in the ionospheric plasma are described in Refs. [21–24,27–32]. We will present in the following text only a brief analysis of the main parameters of the recombination and ionization reactions and will estimate their importance for different layers of the ionosphere. Apart from the main reactions of dissociative recombination described by Equations 1.16 and 1.18, there are additional reactions of ions N_2^+ with molecules O_2 and O, such as

$$N_2^+ + O \rightarrow N_2 + O^+, \quad N_2^+ + O_2 \rightarrow N_2 + O_2^+, \quad N_2^+ + O \rightarrow N + NO^+ \tag{1.19}$$

which are characterized by the corresponding coefficients γ_1 to γ_3. These reactions are very fast because the coefficients $\gamma_1 - \gamma_3$ are sufficiently large [27,28], and the reactions (Equation 1.19) lead to dissipation of N_2^+ ions in the ionosphere (see also Table 1.2). At the same time, the reactions lead to the creation of additional ions of O^+, O_2^+, NO^+, which is accounted for by the coefficients of these ions (q'_{iO}, q'_{iO_2}, and q'_{iNO}) creation in the equations of ionization balance. However, the total intensity of ionization remains constant, i.e., the number of ions O^+, O_2^+, and NO^+ created by reactions (Equation 1.19) in a volume of 1 cm^3 during 1 s equals the number of ions of O^+, O_2^+, and NO^+ before the reactions, and thus a total ionization balance occurs:

$$q'_{iO} + q'_{iO_2} + q'_{iNO} = q^0_{iO} + q^0_{iO_2} + q^0_{iNO} = q_i \tag{1.20}$$

It should also be noted that apart from the dissociative recombination in the ionosphere, there also occurs radiative recombination, related to the attachment of the electron by the atomic ion, and radiation of the photon. At the D layer of the ionosphere, the attachment of the electron to the molecule with the creation of the molecular ion and the dissipation of electrons in the D layer of the ionosphere ($e^- + M^0 \rightarrow M^-$) play an important role. The following estimations, carried out in Refs. [27,28], have shown that at the E layer and beyond, these reactions can be ignored. Moreover, in the E layer, the coefficients of dissociative recombination described by Equation 1.16 are $\alpha_1 - 10^{-7}$ cm^3/s, $\alpha_2 = 2 \cdot 10^{-7}$ cm^3/s, whereas those described by Equation 1.18 are $\beta_1 \approx \beta_2 \approx 10^{-13}$ cm^3/s. At the same time, in the F layer of the ionosphere $\alpha_1 = (3 \div 5) \cdot 10^{-8}$ cm^3/s and $\alpha_2 = (2 \div 5) \cdot 10^{-8}$ cm^3/s, whereas for dissociative reactions (Equation 1.18) they are equal: $\beta_1 = 3 \cdot 10^{-12}$ cm^3/s and $\beta_2 = 2 \cdot 10^{-12}$ cm^3/s. The coefficient of *radiative* recombination, $\sim \alpha_R N_e^2$, is $\alpha_R \approx 10^{-12}$ cm^3/s. These estimations show that events in the ionosphere above the D layer are mainly governed by Equation 1.16, which is characterized by the coefficients of *dissociative* recombination, α_1 and α_2. They are also important at altitudes of the E layer comparable to those in the F layer. The radiative recombination is less effective, compared with the dissociative reactions, even at the altitudes of the E layer of the ionosphere. The temperature dependence of coefficients α_1 and α_2 (in Kelvin) can be defined, according to Refs. [27,28], by the following formulas:

$$\alpha_1 \approx 5 \cdot 10^{-7}(300/T_e)^{1.2}, \quad \alpha_2 \approx 2.2 \cdot 10^{-7}(300/T_e)^{0.7} \tag{1.21}$$

From Equation 1.21 it follows that with increase of temperature of plasma (e.g., caused by heating of the ionosphere), the coefficients of recombination decrease, whereas the electron concentration increases in the E layer with increase of temperature of electrons. All formulas obtained for this layer characterize the local balance of ionization because they are correct only when ignoring the transfer processes with respect to chemical processes. However, in the F layer, the transfer processes, such as diffusion, thermodiffusion and drift, prevail and, in accounting these processes, we can find the region of maximal plasma density beyond which the transfer processes prevail and below which the reactions of recombination and ionization prevail.

Further, for the computations of plasma concentration and temperature in the inhomogeneous ionosphere, a more general expression for the coefficient of recombination R was evaluated to cover the altitudes below and above the critical range of 150–160 km, where the process of plasma transfer prevails over chemical processes [39]:

$$R = \frac{N(\beta_1 N_{N_2} + \beta_2 N_{O_2})}{\left(1 + \frac{\beta_1 N_{N_2}}{\alpha_1 N} + \frac{\beta_2 N_{O_2}}{\alpha_2 N}\right)} \tag{1.22}$$

Here, we take again into account the quasi-neutrality of the ionospheric plasma, i.e., $N_e \approx N_{O^+} + N_{NO^+} + N_{O_2^+} \equiv N$. Then, the densities of the molecular ions can be found by the local balance equations of the ions:

$$N_{O^+} = \frac{N}{\left(1 + \frac{\beta_1 N_{N_2}}{\alpha_1 N} + \frac{\beta_2 N_{O_2}}{\alpha_2 N}\right)} \tag{1.23a}$$

$$N_{NO^+} = N_{O^+} \frac{\beta_1 N_{N_2}}{\alpha_1 N} \tag{1.23b}$$

$$N_{O_2^+} = N_{O^+} \frac{\beta_2 N_{O_2}}{\alpha_2 N} \tag{1.23c}$$

For $h > 200$ km Equation 1.22 reduces to the simple formula $R = \alpha N^2$, with α defined by Equation 1.17 obtained in Refs. [27,28], which describes quite closely the actual ionization–recombination balance at altitudes $h = 200$–500 km in the ionosphere. Equation 1.22 is more general and correctly describes chemical reactions in both the lower and higher ionosphere; if this formula is substituted in Equation 1.14 or Equation 1.13, instead of the term $R = \alpha N^2$ presented in the right-hand side of these equations. Hence, Equation 1.13 or Equation 1.14 fully determines the balance of ionization and recombination in the ionosphere at altitudes beyond the D layer, but the latter equation only up to altitudes of the upper E layer.

Finally, we can conclude that during analysis of the process of formation and further evolution of ionospheric plasma irregularities of either natural origin—such as meteor trails and sporadic H_E plasma structures—or artificial origin—such as heating of the ionosphere by powerful radio wave or injection of chemical active substances (we will discuss these aspects in Chapters 6 and 7)—it is necessary to account for the whole complex of physical–chemical and transport processes that form large-scale, average-scale, and small-scale regular and irregular plasma structures in the nonuniform ionospheric background.

References

1. Gauss, C.F., *General Theory of Terrestrial Magnetism*, 1839; English translation in *Scientific Memoirs* (Ed. by R. Taylor), London, Vol. 2, 1841, pp. 184.
2. Stewart, B., Hypothetical views regarding the connection between the state of the sun and terrestrial magnetism, in *Encyclopedia Britannica*, Edinburgh, 9th ed., Vol. 16, 1882, pp. 181–184.
3. Lodge, O., Mr. Marconi's results in day and night wireless telegraphy, *Nature*, Vol. 66, 1902, p. 222.

4. Watson-Watt, R.A., Weather and wireless, *Q. J. Royal Meteor Soc.*, Vol. 55, 1929, pp. 273–301.
5. Appelton, E.V. and Barnett, M.A.F., Local reflection of wireless waves from the upper ionosphere, *Nature*, Vol. 115, 1925, pp. 333–334.
6. Hulburt, E.O., Ionization in the upper atmosphere of the Earth, *Phys. Rev.*, Vol. 31, 1928, pp. 1018–1037.
7. Appelton, E.V. and Naismith, R., Some further measurements of upper atmospheric ionization, *Proc. Roy. Soc.*, London, Vol. A150, 1935, pp. 685–708.
8. Plendl, H., Concerning the influence of eleven-year solar activity period upon the propagation of waves in wireless technology, *Proc. Inst. Radio Eng.*, Vol. 20, 1932, pp. 520–539.
9. Dellinger, J., Sudden disturbances of the ionosphere, *Proc. Inst. Radio Eng.*, Vol. 25, 1937, pp. 1253–1290.
10. Chapman, S., The absorption and dissociative or ionizing effect of monochromatic radiation in an atmosphere on a rotating Earth, *Proc. Phys Soc.*, London, Vol. 43, 1931, pp. 26–45.
11. Chapman, S. and Ferraro, V.R., A new theory of magnetic storms, *Terrestrial Magnetism and Atmospheric Electricity*, Vol. 38, 1933, pp. 79–96.
12. Kaiser, T.R., The first suggestion of an ionosphere, *J. Atmos. Terr. Phys.*, Vol. 24, 1962, pp. 865–872.
13. *Modern Ionospheric Science*, Ed. by Kohl, H., Ruster, R., and Schlogel, K., Berlin: Luderritz and Bauer, Germany, 1996.
14. Tinsley, B.A., Brown, G.M., and Scherrer, P.H., Solar variability influences on weather and climate: possible connections thorough cosmic rays fluxes and storm intensification, *J. Geophys. Res.*, Vol. 94, 1989, pp. 14783–14792.
15. Martyn, D.F., Electric currents in the ionosphere, III. Ionization drift due to winds ans electric fields, *Phil. Trans. Roy. Soc.*, Vol. A246, 1953, pp. 306–320.
16. Maeda, K., Dynamo-theoretical conductivity and current in the ionosphere, *J. Geomagn. Geoelect.*, Vol. 4, 1952, pp. 63–82.
17. Baker, S.G. and Martyn, D.F., Electric currents in the ionosphere, I. Conductivity, *Phil. Trans. Roy. Soc.*, Vol. A246, 1953, pp. 281–294.
18. King, J.W. and Kohl, H., Upper atmosphere winds and ionospheric drifts caused by neutral air pressure gradients, *Nature*, Vol. 206, 1965, pp. 899–701.
19. Ratclife, J.A., The formation of the ionospheric layers *F*-1 and *F*-2, *J. Atmos. Terr. Phys.*, Vol. 8, 1956, pp. 260–269.
20. Yonezawa, T., A new theory of formation of the *F2* layer, *J. Radio Research Labs*, Vol. 3, 1956, pp. 1–16.
21. Ratcliffe, J.A., *Physics of the Upper Atmosphere*, New York-London: Academic Press, 1960.
22. Gershman, B.N., *Dynamics of the Ionospheric Plasma*, Moscow: Nauka, 1974.
23. Volland, H., *Atmospheric Electrodynamics*, Heidelberg: Springer-Verlag, 1984.
24. Al'pert, Ya.L., *Propagation of Electromagnetic Waves in the Ionosphere*, Moscow: Nauka, 1973.
25. Gershman, B.N., Erukhimov, L.M., and Yashin, Yu.Ya., *Wave Phenomena in the Ionosphere and Cosmic Plasma*, Moscow: Nauka, 1984.
26. Richmond, A.D., *Ionospheric Electrodynamics*, pp. 249–290, In: Volland, H. (Ed.), *Handbook of Atmospheric Electrodynamics*, Vol. II, Boca Raton (FL): CRC Press, 1995.
27. Gurevich, A.V. and Shwarzburg, A.B., *Nonlinear Theory of Radiowave Propagation in the Ionosphere*, Moscow: Nauka, 1973.
28. Gurevich, A.V., *Nonlinear Phenomena in the Ionosphere*, Berlin: Springer-Verlag, 1978.
29. Ivanov-Kholodnii, G.S. and Nikol'skii, *The Sun and the Ionosphere*, Moscow: Nauka, 1969.
30. Whitten, R.C. and Poppoff, I.G., *Physics of the Lower Ionosphere*, New York: Prentice Hall, 1965.
31. Fatkullin, M.N., Zelenova, G.L., Kozlov, V.K., Legen'ka, A.D., and Soboleva, T.N. *Empirical Models of the Middle-Latitude Ionosphere*, Moscow: Nauka, 1981.
32. Rees, H., *Physics and Chemistry of the Upper Atmosphere*, Cambridge: Cambridge University Press, 1989.
33. Filipp, N.D., Blaunshtein, N.Sh., Erukhimov, L.M., Ivanov, V.A., and Uryadov, V.P., *Modern Methods of Investigation of Dynamic Processes in the Ionosphere*, Kishinev (Moldova): Shtiintza, 1991.
34. Ginzburg, V.L., *Propagation of Electromagnetic Waves in Plasmas*, New York: Pergamon Press, 1964.

35. Belikovich, V.V., Benediktov, E.A., Tolmacheva, A.V., and Bakhmet'eva, N.V., *Ionospheric Research by Means of Artificial Periodic Irregularities*, Berlin: Copernicus GmbH, 2002.
36. Likhter, Ya.L., Gul'el'mi, A.V., Erukhimov, L.M., and Mikha'lova, G.M., *Wave Diagnostics of the Near-Earth's Plasma*, Moscow: Nauka, 1989.
37. Mishin, E.V., Ruzhin, Yu.Ya., and Telegin, V.A. *Interaction of Electron Fluxes with the Ionospheric Plasma*, Moscow: Edition of Hydrometeorology, 1989.
38. Filipp, N.D., Oraevskii, V.N., Blaunshtein, N.Sh., and Ruzhin, Yu.Ya., *Evolution of Artificial Plasma Irregularities in the Earth's Ionosphere*, Kishinev (Moldova): Shtiintza, 1986.
39. Ureadov, V., Ivanov, V., Plohotniuc, E., Eruhimov, L., Blaunstein, N., and Filip, N. *Dynamic Processes in the Ionosphere—Methods of Investigations*, TehnoPress, Iasi (Romania), 2006.

Chapter 2

Nonlinear Phenomena and Plasma Instabilities in the Disturbed Irregular Ionosphere

In the regular ionosphere, a plasma homogeneous continuum consisting of gas molecules, electrons, and ions, nonlinear effects are weak, because plasma is in an equilibrium conditions, between its components. In the real ionosphere, plasma is not in an equilibrium state, and strong deviations from its equilibrium (also called the *steady state* regime in the literature) lead to formations of plasma wave perturbations, which create different kinds of instabilities in the ionosphere.

The physical processes that accompany nonlinear phenomena in the disturbed nonregular ionosphere, and the corresponding plasma instabilities, may be caused by natural environmental sources, such as magnetic storms or solar and seismic activity [1–5], as well as by human activities resulting in the interaction of powerful radio waves (called *pump waves*) sent from the Earth's surface into the ionosphere, using ground-based facilities [6–8], and those related to active space experiments involving injection of electron beams and fluxes [9] and ion clouds [10–11] into the ionospheric plasma and excitation of fields, acoustic waves and currents in the ionosphere [2,4,12], and so on.

Nonlinear phenomena in the real ionospheric plasma can be classified as weak and strong, depending on how strong ambient electromagnetic effects are and on how the spatial and temporal characteristics of interactions between plasma particles, electrons, and ions, with electromagnetic fields passing through the ionospheric plasma, are coupled. Nowadays, weak and strong nonlinearity, described correspondingly by linear, quasilinear, and nonlinear theories, cannot be differentiated [1–4]. The existence of even weak ambient electric and magnetic fields can generate instabilities in plasma, and the related strong intrinsic electric fields and plasma density disturbances, which essentially change the dispersion properties of plasma.

Therefore, the conditions of *weak nonlinearity* are defined in the literature as those in which *dispersion properties of plasma are constant* in the time domain. In this case, the equations of movement (Equation 1.6), thermoconductivity (Equation 1.10), and continuity (Equation 1.5) of plasma electrons and ions take into account linear and quadratic dependences of currents induced in plasma on ambient electric field $\mathbf{E}_0$ of various origins. Correspondingly, the Ohm law, $\mathbf{j} = \sigma \mathbf{E}_0$, consists of terms proportional to E_0, disregarding terms containing E_0^2, E_0^3, and higher degrees of E_0 [1–4,13–15].

For example, if the field of the probing alternate radio wave passing through the ionosphere, $\mathbf{E}_\omega$, is weak enough and fluctuations of plasma density (electrons and ions, as plasma is quasineutral) are weak or have other origins, the induced current, $\mathbf{j} \sim \mathbf{E}_\omega$, and effects of attenuation, scattering, and transformation of wave energy into plasma waves are weak, and do not depend on the field strength $E_\omega \equiv |\mathbf{E}_\omega|$. Thus, probing radio waves usually used for radio communication weakly change properties of ionospheric plasma, and the wave characteristics within the ionospheric channel can be described by using linear theory.

Where the electromagnetic wave is strong enough to essentially change the parameters of plasma, its density, frequencies of collisions between plasma particles, their movement velocities, and so on, the currents induced in such a plasma become proportionate to terms with higher degrees of E_ω. The existence of terms with quadratic, cubic, and higher degrees of E_ω in the Ohm law leads to nonlinear effects related to nonlinear terms existing in the equations mentioned previously. That is, the existence of nonlinear terms in the equation of thermoconductivity leads to a change in the temperature of plasma electrons and ions. This occurs if the powerful high-frequency electromagnetic wave heats the plasma electrons in the ionosphere. The corresponding "heater" is proportional to E_ω^2. This type of nonlinearity is called in the literature the *thermal* or *heated* nonlinearity [1–5,13–15]. As shown in the following text, this nonlinearity is caused by dissipative processes taking place when the characteristic time of nonlinear interaction of strong waves with plasma exeeds the characteristic time of interactions between charged particles $\left(\nu_e^{-1} = (\nu_{em} + \nu_{ei})^{-1}\right)$, and the characteristic spatial scales of wave–plasma interactions, L, significantly larger than the free path of plasma electrons (in isotropic plasma, when ambient magnetic field is weak, $\mathbf{B}_0 \to 0$) or than the longitudinal scale, $\ell_{||}$ (along $\mathbf{B}_0$) of plasma disturbances (in anisotropic plasma, $\mathbf{B}_0 > 0$). As also shown subsequently, heating of plasma electrons leads to a change in plasma pressure in the heated area of the ionosphere. Moreover, high-energy electrons after heating leave the heated area and, interacting with ions due to the creation of polarization intrinsic electric fields in plasma, called *ambipolar fields* [6,7], generate plasma disturbances in the ionosphere. These plasma disturbances are responsible for nonlinear effects occurring during propagation of probing radio waves through the heated ionospheric regions, such as self-focusing and self-defocusing effects, nonlinear refraction, scattering and transformation of plasma modes, generation of high-order plasma modes, and so on, the effects of which will be described in Chapters 6 and 7.

We also consider in the following paragraphs the so-called the *striction nonlinearity*, which occur only in the collisionless plasma, even if the ionospheric plasma is the collisional plasma. This is discussed because it is usually generated by artificial heating of ionospheric plasma by electromagnetic pump waves from powerful ground-based facilities. This nonlinearity appears because of the modulation of movements of plasma particles in the oscillatory field of pump waves. The appearance of striction nonlinear forces displaces the circular trajectory of electrons and ions with the radius, $\mathbf{r}_\omega$, proportional to the oscillatory high-frequency component of the alternate wave field $\mathbf{E}_\omega$; i.e., $\mathbf{r}_\omega \sim \mathbf{E}_\omega$. As shown further, this striction nonlinearity is described by new nonlinear terms in Equation 1.6, such as $(\mathbf{u}_{e\omega} \nabla)\mathbf{u}_{e\omega}$ and $\mathbf{u}_{e\omega} \times \mathbf{B}_\omega$, where $\mathbf{u}_{e\omega}$ is the oscillation velocity of plasma electrons in the field of powerful high-frequency waves and $\mathbf{B}_\omega$ is the strength of the magnetic field in the high-frequency wave.

The corresponding nonlinearities are closely related with instabilities created in plasma under ambient forces of various origins. There is a wide spectrum of plasma instabilities which can be generated in the ionosphere by natural geophysical and artificial manmade phenomena. In our description of ionospheric instabilities as a source of plasma irregularities of various scales and degrees of perturbations, which affect the propagation of probing radio waves through the real ionosphere, we will limit this spectrum of instabilities to only those that are important to researchers in understanding the nature of the creation of plasma irregularities responsible for multiray propagation caused by multiscattering, multidiffraction, and multireflection occurring in the ionospheric radio channels (see Chapters 6 and 7). Therefore, we will not enter into the subject of beam instability generation and the corresponding Cherenkov radiation (or Landau damping) occurring in the collisionless laboratory plasma.

In a further description of instabilities under consideration, we will focus mostly on linear and quasilinear theoretical approaches and thereby establish only linear conditions for the generation of plasma instabilities and the initial stages of evolution of the corresponding plasma density perturbations associated with these instabilities. Despite this fact, the results presented here allow for the finding of initial growth rates of instabilities in the space and time domains, defined as increments (or decrements) of plasma turbulences [1–5].

As for nonlinear theory, until now the basic concepts of this theory have not yet been formulated. Some analytical nonlinear solutions have been obtained only for the description of weak turbulences and for collisionless plasma. However, using such approximate theories, some positive information can be obtained about further evolution of plasma instabilities in the partly ionized ionosphere (i.e., taking into account interactions between charged and neutral particles). This can be done using some specific mathematical methods with elements of statistical and quantum mechanics. In other words, based on some idealized analytical approaches in nonlinear theory, we will briefly explain experimentally observed interactions of plasma waves in the collisional ionospheric plasma. A full analysis of the evolution of plasma irregularities in the nonhomogeneous ionosphere, taking into account its real properties and electrical coupling between the main regions, *E* and *F*, of the ionosphere, will be presented in Chapter 5 based on 2-D and 3-D numerical analysis.

Now, let us classify the main instabilities associated with those plasma disturbances affecting radio propagation in the ionospheric communication channels and, therefore, sufficiently interesting to expand on.

As mentioned previously, during propagation of strong electromagnetic or electrostatic waves in the ionospheric plasma (called *pump waves*), the nature of creation of plasma waves (instabilities) can be understood by the oscillatoring movements of the plasma particles (electrons and ions) within the strong electromagnetic field of pump waves. The corresponding instabilities are called the *artificial parametric*. In specific conditions, some mentioned previously and others described subsequently, *artificial striction* instabilities can be created by strong electromagnetic waves entering into the ionospheric layer.

Natural plasma instabilities, which have a macroscopic character and can be simply described, by movements of plasma, as a liquid, are usually called *hydrodynamic* instabilities. For a description of these instabilities, magnetohydrodynamic equations are usually used. Because in the ionosphere there are dissipative processes due to the real conductivity of plasma, the processes of diffusion and recombination, the so-called *dissipative hydrodynamic* instabilities, occur.

Then, from currents in the ionosphere, mostly in the polar and equatorial regions, E and F, the *two-stream current* instabilities and *drift current* instabilities, respectively, can be generated. Taking into account the altitudinal redistribution of plasma density in the ionosphere and the height dependence of transport coefficients in plasma (see Chapter 1), we find that *gradient-drift* instabilities are usually generated.

Another group of instabilities occurs in the nonequilibrium ionospheric plasma with non-Maxwellian distribution of charge particles (mostly of plasma electrons). In this case, plasma waves can be amplified due to interactions with highly energetic electrons or ions, or both. These interactions lead to redistribution of energy in plasma to the so-called Landau damping and, finally, to an increase of attenuation of probing radio waves passing through the disturbed ionospheric region. Such instabilities are usually called *kinetic* instabilities because they can be described only by using the kinetic theory and the corresponding kinetic equations with collisional integrals described by using elements of statistical and quantum mechanics.

In the theoretical analysis of the processes involved in the formation of plasma instabilities, a general problem is whether to present plasma waves as (1) monochromatic plane waves or (2) wave packets (signals). In both cases of plasma wave field presentation, the main mathematical difficulty is to obtain a dispersion equation as a strict relation between the plasma parameters and the frequency and phase velocity of plasma waves.

As mentioned in the Refs. [1–4], there are two ways to take into account the collisional absorption and analyze the corresponding instabilities in plasma. In the first approach, the wave vector $\mathbf{k}$ is assumed to be real but the wave frequency ω is assumed to be a complex value, $\omega = \mathrm{Re}(\omega) + j\,\mathrm{Im}(\omega) = \varpi + j\gamma$. Here, for $\gamma > 0$, where γ is defined as the *increment* of growth (excitation) of plasma waves, the corresponding exponential growth of plasma wave with increment $|\gamma|$ occurs in the time domain according to the law $\sim e^{|\gamma| t}$. In the inverse case, $\gamma < 0$, an exponential damping of plasma waves in the time domain ($\sim e^{-|\gamma| t}$) occurs. Here, γ is defined as a *decrement* of damping of plasma waves. We have to note here that some authors define the excitation of plasma wave (instability) by $\gamma < 0$, and its damping by $\gamma > 0$, taking into account the law $e^{-\gamma t}$. To unify different definitions in our further description of the problem, we introduced the absolute value of γ in the exponential law and inserted the corresponding sign which finally determines the character of the process, either excitation ($\sim e^{|\gamma| t}$) or damping ($\sim e^{-|\gamma| t}$).

In the other limited approach, the wave frequency ω is assumed to be a real value, but the wave number k is assumed to be a complex value, $k = \mathrm{Re}(k) + j\,\mathrm{Im}(k) = k_0 n_r + j k_0 n_i$, where $k_0 = \frac{2\pi}{\lambda} = \frac{\omega}{c}$ is the wave number in free space, c is the light speed, λ is the wavelength, and n_r and n_i are the real and imaginary components, respectively, of the refractive index in plasma. In this approach, plasma is not stable even if it exists in one part of the interval of the wave numbers, $\Delta k = k_2 - k_1$, and the increment will be negative, i.e., $\gamma(k) < 0$. The main question is how to describe the character of evolution of those unstable waves in real time, t. Could one present plasma instability as a plane monochromatic wave or is there another mathematically stricter description of the problem? As mentioned in the Refs. [1–4], a more precise mathematical approach dictates "dealing" with wave packets (i.e., signals). In such a mathematical description, the wave field will be a function of the coordinate vector $\mathbf{r}$ and time t; i.e., $\mathbf{E}_\omega = \mathbf{E}_\omega(\mathbf{r},t)$. Expanding this field using Fourier's integral over the real $\mathbf{k}$ and complex $\omega(k)$, where $\omega(k)$ can be found from the corresponding dispersion equation [1–4], and assuming that the Fourier transformance inside this expansion, $\tilde{\mathbf{E}}(\omega) = g(\mathbf{r},t)e^{i\omega(k)t}$, is a limited function for all values of the wave number k, one can find that each component of the field strength $\mathbf{E}(\mathbf{r},t)$ approaches zero when $\mathbf{r} \to \infty$. This means that the wave packet is *spatial-localized*. If, in this case, localization in time does not exist, the instability is called *absolute*. Because of absolute instability the field amplitude, $E_\omega = \tilde{\mathbf{E}}(\omega)$ will grow in time t at every point $\mathbf{r}$ according to the law $\sim e^{|\gamma| t}$, and self-excitation of plasma waves will be observed. The question of the statement of plasma, containing such self-excited waves (e.g., instabilities), can only be resolved by using nonlinear approaches [1–4,11–15].

In contrast, if the instability generated, i.e., a wave packet in the previous discussion, is localized both in the space and time domains, such instability is called *convective*. Mathematically, using linear approximation, this means that each component of the field strength $\mathbf{E}_\omega(\mathbf{r},t)$ tends to zero when $\mathbf{r} \to \infty$ and $t \to \infty$. At the same time, because the processes grow at each point of the space domain for every point in time, such a plasma system has amplifier properties for these kinds of instabilities. We have to note here that a strict differentiation of instabilities between *absolute* and *convective* needs a very complicated mathematical tool based on corresponding theoretical derivations carried out in the literature [1–4]. Therefore, we do not enter into a detailed analysis of the problem, but only recomend the reader to these exellent books.

Because the most complicated physical mechanisms are generated in the presence of powerful alternate high-frequency electromagnetic waves, creating plasma instabilities there due to the heating of regular ionospheric plasma, we will start our description of plasma instabilities with those that are accompanied by heating phenomena in the ionosphere.

2.1 Physical Aspects of Nonlinear Phenomena in the Artificially Heated Ionospheric Plasma

In ionospheric plasma under interaction with powerful alternate radio waves, an oscillatory speed of plasma electrons (ions—plasma is quasineutral; see Section 1.2), $\mathbf{v}_\omega$, is generated in the field of the high-frequency radio wave (we consider here the HF/VHF frequency band), where the angular frequency of the wave $\omega = 2\pi f$. The radiated wave frequency f essentially exceeds the characteristic plasma frequencies such as the gyrofrequency of electrons, ω_H, or the plasma frequency, ω_p, defined in Chapter 1. Then, for the incident field presented in the form of the plane wave [6],

$$\mathbf{E}_t \propto \mathbf{E}_\omega \exp\{i(\omega t - \mathbf{kr})\} \tag{2.1}$$

and, by following its relations with the amplitude of the oscillatory field and the frequency, we can obtain for the magnitude of the oscillatory speed as

$$|\mathbf{v}_\omega| = \frac{e|\mathbf{E}_\omega|}{m\omega} \tag{2.2}$$

Here, as earlier, e and m are the charge and mass, respectively, of the plasma electron. Consequently, the probability distribution function (pdf) of electrons $f(\mathbf{v}')$ becomes dependent on $\mathbf{v}_\omega$; i.e., $f(\mathbf{v}') \Rightarrow f(\mathbf{v}' + \mathbf{v}_\omega)$, and its main momentums, $N = \int f(\mathbf{v}')d\mathbf{v}'$, $N\mathbf{v} = \int f(\mathbf{v}')\mathbf{v}'d\mathbf{v}'$, and $NT \approx \int f(\mathbf{v}')(\mathbf{v}')^2 d\mathbf{v}'$, are also changed. The changes of the first three momentums, for concentration, impulse, and temperature (energy), lead to hydrodynamic effects in plasma. At the same time, the deformation of the corresponding pdf is responsible for the occurrence of kinetic effects in plasma. All these effects can be illustrated by new hydrodynamic and kinetic equations, differing from those presented in Sections 1.2 and 1.3 for the description of processes occurring in regular, nondisturbed ionospheric plasma. Following Ref. [6], we will present them with illustrations of the specific role of each term in the description of different nonlinear phenomena occurring in the field of powerful alternate radio waves in the ionosphere.

Thus, to describe the hydrodynamic effects, we follow the same method [6], accounting for plasma as a two-species quasineutral fluid (i.e., $N_e \approx N_i = N$, $\nabla \mathbf{j}_e \approx \nabla \mathbf{j}_e = \nabla \mathbf{j}$; see Section 1.2):

For total electron velocity $\mathbf{v} = \mathbf{v}' + \mathbf{v}_\omega$,

$$\frac{\partial \mathbf{v}}{\partial t} = \underbrace{\nu_{e\alpha}(T_e)(\mathbf{v} - \mathbf{v}_\alpha)}_{\text{Friction force, } \mathbf{F}_c} - \underbrace{(\mathbf{v}_\omega \nabla)\mathbf{v}_\omega - \frac{e}{mc}[\mathbf{v}_\omega, \mathbf{B}_\omega]}_{\text{Striction force, } \mathbf{F}_{st}} - \underbrace{N \nabla T_e}_{\text{Heating force, } \mathbf{F}_T} + \cdots \tag{2.3}$$

For the electron temperature,

$$\frac{\partial T_e}{\partial t} = \underbrace{\frac{2}{3}\frac{\hat{\sigma}_e}{N} E_\omega E_\omega^*}_{\text{Nonlinear heating}} - \underbrace{\delta_{e\alpha}(T_e - T_\alpha)}_{\text{Cooling}} + \underbrace{\hat{\kappa}_{e\alpha}\Delta T_e}_{\text{Thermoconductivity}} + \cdots \tag{2.4}$$

For quasineutral plasma concentration,

$$\frac{\partial N}{\partial t} + \underbrace{\nabla \cdot (N_\omega \mathbf{v}_\omega)}_{\substack{\text{Nonlinear flux} \\ \mathbf{j}_n = -eN_\omega \mathbf{v}_\omega}} + \underbrace{\cdots\cdots\cdots}_{\text{other effects}} = \underbrace{q'(T_e)}_{\substack{\text{additional} \\ \text{ionization}}} - \underbrace{\alpha'(T_e)N^2}_{\substack{\text{additional} \\ \text{recombination}}} \tag{2.5}$$

Kinetic effects are now described by the following equation:

$$\frac{\partial f}{\partial t} = \frac{e}{m}\nabla\varphi\frac{\partial f}{\partial v} - \frac{e}{m}E_\omega\frac{\partial f}{\partial v} \tag{2.6}$$

Let us examine the corresponding nonlinear effects. In hydrodynamic approximation, the first term in Equation 2.3 describes heating effects caused by collisions between electrons and ions and neutral molecules and atoms. It is an additional source of a new type of nonlinearity occurring in the ionospheric plasma, caused by the ambient alternate electric field of radio waves perturbing the regular ionospheric plasma. The nature of this kind of nonlinearity, defined earlier as *thermal nonlinearity*, is easy to understand if one takes into account that the mass of electrons is less than that of ions and atoms (see Section 1.1). Thus, electrons cannot transfer the energy of heating to ions and atoms during their mutual collisions. Finally, the electron mean free path becomes equal to the mean path between collisions, during which it can be heated, even in a weak electrical field of the wave. In this case, the permittivity and conductivity of plasma become dependent on the field strength magnitude; i.e., $\varepsilon = \varepsilon(E_\omega)$ and $\sigma = \sigma(E_\omega)$. In this situation, the Ohm law does not satisfy the well-known relation between the current density $\mathbf{j}$ and the vector of electric field of the wave, $\mathbf{j} = \sigma E_\omega$, and becomes dependent on the field strength magnitude. As a result of such dependence, thermal nonlinearity occurs in plasma [6,13,14], changing electromagnetic processes in plasma and, therefore, influencing the propagation of probing radio waves through the thermally perturbed ionospheric regions.

Consequently, all this leads to generation of nonlinear disturbances of plasma density and changes of plasma permittivity, which can be estimated as [13,14]:

$$\Delta \varepsilon_T \propto \frac{|E_\omega|^2}{E_p^2} \tag{2.7}$$

where

$$E_p^2 = 3m\kappa T_e \delta_0 \frac{(\omega^2 + \nu_{e\alpha}^2)}{e^2} \approx 17.6 \cdot 10^{-10} \cdot T_e \delta_0 (\omega^2 + \nu_{e\alpha}^2) \tag{2.8}$$

is the intensity of the plasma intrinsic field caused by penetrating alternate electric fields of the powerful radio wave with the radiated frequency $f = \omega/2\pi$ and with the plasma temperature determined by electrons, T_e (in degrees); δ_0 is the average fraction of energy lost by each electron during one collision; $\nu_{e\alpha} = \nu_{em} + \nu_{ei}$ is the effective frequency of electron collisions with ions ($\alpha = i$) and neutrals ($\alpha = m$).

Estimations presented in Refs. [13,14] have shown that in the ionosphere at altitudes from 100 to 400 km, with an increase in the frequency of carrier signals from 5 to 100 MHz and in the effective radiated power of radio waves from 10^2 to 10^6 kW, the parameter of plasma permittivity variations, $\Delta\varepsilon_T$, can change its magnitude from 10^1 to 10^3 at the lower ionosphere ($h < 150$–160 km) and from 10^{-5} to 10^1 at the upper ionosphere ($h > 250$–300 km), respectively.

At the same time, nonlinearity in hydrodynamic approximation occurs due to the pressure of light, called the *striction force* and described by the second term in Equation 2.3. Usually, this force plays an important role in the upper regular, nondisturbed ionosphere in the rare collisionless plasma. In the disturbed ionosphere, under the powerful field-heating conditions, this force prevails in conditions in which the characteristic scales of field-perturbed plasma irregularities are less than the electron mean free path. This force is called a *light pressure* because the inhomogeneous ambient, alternating electric field of the wave compresses the plasma due to pressure on its electron component [15,16]. Consequently, the plasma density and, therefore, the dielectric permittivity of plasma become dependent on the amplitude E_ω of this wave field—i.e., $N = N_0 + \delta N(E_\omega)$ and $\varepsilon = \varepsilon_0 + \Delta\varepsilon(E_\omega)$—finally, causing the nonlinearity of the electrodynamic processes occurring in the ionospheric plasma.

This nonlinear mechanism is frequently called *striction nonlinearity*, which occurs in plasma in the field of powerful radio waves. The corresponding striction force can be found after averaging the intensity of the radio wave, $E_\omega E_\omega^*$, over the high frequency [6]:

$$|\mathbf{F}_{st}| = \frac{\varepsilon - 1}{8\pi} \nabla |E_\omega|^2 \tag{2.9}$$

The corresponding striction nonlinearity leads to local disturbances of plasma permittivity [16]:

$$\Delta\varepsilon_{st} \propto \frac{e^2 |E_\omega|^2}{8m T_e \omega^2} \tag{2.10}$$

In a couple of studies [13,14], such plasma permittivity perturbations were estimated for different magnitudes of radiowave power and radiation frequencies. Thus, at ionosphere heights beyond 150–160 km for radiated frequencies of 5 to 100 MHz, $\Delta\varepsilon_{st}$ varies in the range of 10^{-2} to 10^{-7}, respectively. Hence, with an increase in radiated frequencies of powerful radio waves penetrating the ionospheric plasma, deviations of plasma permittivity become indispensable.

A comparison between these two kinds of plasma nonlinearity showed that in the ionospheric plasma at altitudes under consideration, $h = 50$–400 km, thermal nonlinearity significantly exceeds striction nonlinearity and plays a major role in the description of most nonlinear processes accompanying plasma heating by a powerful radio wave. Therefore, in our subsequent discussions, we will pay special attention to the thermal collision processes causing thermal nonlinearities in the perturbed ionosphere. However, we have to point out that striction nonlinearity is adequate not only in the rare upper ionospheric plasma, at altitudes beyond 400 km, but also in nonstationary processes of short pulses propagation in plasma [17], the duration of which is less than the time of the electron-free run between collisions with ions and atoms.

The last term in Equation 2.3, containing the gradient of temperature, ∇T_e, describes a pressure force, $F_T = -N\nabla T_e$, caused by heating of plasma in the field of a powerful alternate radio wave penetrating into the depth of the ionosphere. As for Equation 2.4, it contains the additional term $\sim \hat{\sigma}_e E_\omega E_\omega^*$, which determines plasma nonlinear heating in the field of powerful alternate radio waves, due to electron collisions with ions and atoms of plasma, described by the second term of Equation 2.4, where $\delta_{ea} = (\delta_{ei}\nu_{ei} + \delta_{em}\nu_{em})/(\nu_{ei} + \nu_{em})$. Associated with the gradient of temperature, ΔT_e, thermodiffusion occurs in plasma, which is described by the third term of Equation 2.4, where κ is the thermoconductivity. Together with thermodiffusion and cooling, due to the relaxation of electron temperature during collision, the effect of further heating by high-power waves changes the balance of charged particles in the ionospheric plasma, finally, causing the generation of nonlinear current through plasma ($\sim \mathbf{j}_N = -eN_\omega \mathbf{v}_\omega$), as described by the terms in the left-hand side of Equation 2.5 for conservation of charge particles in plasma. In this equation, apart from the electron temperature dependence of thermo-ionization, $q(T_e)$, and of the recombination rate, $\alpha(T_e)$, a nonlinear current, $\mathbf{j}_N$, may introduce a sufficient contribution to the development of artificial ionospheric turbulence (AIT) at lower ionospheric altitudes due to the occurrence there of longitudinal plasma waves ($N_\omega = \mathbf{k}_\omega \cdot \mathbf{v}_\omega N/\omega$) [18,19].

The heating process, following from Equation 2.5, decreases the rate of dissociative recombination, $\alpha_d(T_e) \propto T^{-n}$, where $n = 0.7–1.2$, and, consequently, increases plasma concentration in regions perturbed by powerful radio waves, mostly at heights of 130 to 180 km, where this recombination process is predominant.

The first conclusion that follows from the qualitative analysis of only those hydrodynamic effects in which the changes of pdf described by the kinetic Equation 2.6 do not play any special role is that the existence of thermal nonlinearity leads to further heating of the ionospheric plasma and, in turn, to the redistribution of plasma density caused by the heating. This effect is more evident in the lower- and middle-latitude ionosphere. The same effects, but for the high-latitude ionosphere, can be observed due to the gradient of pressure of light $\sim |E_\omega|^2$, where the striction force causes striction nonlinearity [18,19]. At lower ionospheric altitudes ($h < 100$ km), changes of the recombination coefficient $\alpha(T_e)$ in Equation 2.5 relate to temperature dependence of the coefficient of attachment of electrons to molecules and atoms, $\beta(T_e)$, and the creation of negative ions [7]:

$$\beta(T_e) = \beta_0(T_{e0}) \cdot \frac{T_{e0}}{T_e} \exp\left\{\frac{700}{T_{e0}} - \frac{700}{T_e}\right\} \tag{2.11}$$

Here, T_{e0} is the nondisturbed temperature of electrons in the regular ionospheric plasma, and $T_e = T_{e0} + \delta T$, the perturbed electron temperature caused by the heating process of plasma in the field of alternate high-power electromagnetic wave.

Furthermore, heating of electrons by powerful alternate radio waves leads to change in frequency of collisions between electrons and ions, $\nu_{ei}(T_e) \propto T_e^{-3/2}$, and between electrons and neutrals, $\nu_{em}(T_e) \approx \nu_{em}(T_{e0})(T_e/T_{e0})^{5/6}$ [6]. It is clear that the first frequency decreases with an increase in electron temperature at altitudes above 150 km, where this type of collision prevails, whereas the second frequency increases as the electron temperature rises at altitudes below 120 to 130 km, where this type of collisions is predominant. Such temperature dependences of collision frequencies change the absorption of probing radio waves propagating through the heat-perturbed regions of the ionosphere. This effect depends on the relationship between the frequencies of probing waves, $f_{pr} = \omega_{pr}/2\pi$, and the collision frequency, $\nu_e = \nu_{ei} + \nu_{em}$. Thus, for $\omega_{pr} > \nu_e$, absorption increases sharply with an increase in the energy of heating (i.e., temperature T_e). When a strong wave penetrates into the ionospheric plasma (e.g., its strength E_0 exceeds the plasma field E_p defined by Equation 2.8), the wave cannot do so deeply due to strong absorption of its energy. As shown in the Refs. [13,14], a saturation of the field intensity is produced in the interior of the plasma layer, where the wave amplitude ceases to depend on the heating power of the wave incident at the boundary of this plasma layer. In the other limiting case of $\omega_{pr} > \nu_e$, the absorption decreases with an increase of power in the wave field. In this limiting case the ionospheric plasma may become transparent to the high-power radio waves.

Moreover, the changes in times of power of strong radio wave, as a heater of plasma, modulate the amplitude of probing radio wave passing through the perturbed heated plasma layers, finally changing its own modulation. This means that if a high-power wave with frequency ω_1 affects the ionospheric plasma and then some probing wave with frequency ω_2 passes through the perturbed ionospheric area, the law of its absorption and propagation is changed according to the modulation effect of the powerful wave. This effect can be considered as an interaction of two waves in the same area of plasma. For $E_p \leq E_0$, this change of modulation, which induces a large absorption of the probing wave and leads to a *cross-modulation* effect between waves, can be significant and is defined in the literature as the Luxemburg-Gorky effect [13,14,19]. This nonlinear interaction also leads to the generation of new radio waves with combining frequencies of $\omega_2 \pm \omega_1$, $\omega_2 \pm 2\,\omega_1$, and so on, and is often used for investigation of parameters of the lower ionosphere by analyzing waveforms of radio pulses reflected from the D and E layers of the ionosphere. At the same time, for $E_p \leq E_0$, a strong radio wave can suppress other radio waves propagating through the perturbed heated ionospheric regions.

The dependence of ν_{em} on T_e can be an additional source of temperature dependence of the friction force $F_c \propto \nu_{e\alpha}(T_e)(\mathbf{v} - \mathbf{v}_\alpha)$, which causes changes in the electron-speed component oriented perpendicular to the lines of the geomagnetic field, $\mathbf{B}_0$, with unit vector $\mathbf{b} = \mathbf{B}_0/|\mathbf{B}_0|$. The source of such electron movements may be either the ambient electric field $\mathbf{E}_0$ or interactions between plasma electron (ions; plasma is quasineutral) and neutral molecules moving with the speed $\mathbf{U}_m$ of the neutral gas; i.e., $\mathbf{v}_\alpha \equiv \mathbf{U}_m$, $\alpha = e,i$ [20,21]. So we get [6]:

$$\mathbf{v}_{\alpha\perp} = \frac{\nu_{em}}{m_\alpha\left(\nu_{\alpha m}^2 + \omega_{\alpha H}^2\right)}\left\{\mathbf{F}_\alpha + \frac{\omega_{\alpha H}}{\nu_{\alpha m}}[\mathbf{F}_\alpha,\mathbf{b}]\right\} \tag{2.12}$$

and

$$\mathbf{F}_\alpha = e_\alpha\mathbf{E}_0 + m_\alpha\nu_{\alpha m}\mathbf{U}_m \tag{2.13a}$$

is the electromotive force in plasma. Further,

$$\omega_{\alpha H} = \frac{e_\alpha|\mathbf{B}_0|}{m_\alpha c} \tag{2.13b}$$

gives the hydromantic frequencies of electrons ($\alpha = e$) and ions ($\alpha = i$), where the corresponding sign ("$-$" for electrons and "$+$" for ions) must be taken into account. The relative movement of ions and electrons in plasma, i.e., the current, affected by the modulated radiation of the alternate powerful wave with frequency Ω, can also be amplitude-modulated with the same frequency. As a result, an artificial low-frequency radiation of the ionosphere is recorded by the ground-based facilities during experimental heating of the ionosphere by strong radio waves [22] and is successfully used for diagnostics of currents occurring in the modified ionosphere [23]. In the literature, this is also called the detection effect of the ionosphere [13,14]. Generally speaking, nonlinear interactions between powerful and probing waves can explain the generation of other types of electromagnetic (namely, whistlers), magnetohydrodynamic, electro-acoustic, and ion-acoustic waves in the perturbed ionosphere by powerful radiowaves, called *nonlinear wave transformation* in the literature [7,8,13,14].

If the heating region of the ionosphere is so extensive that one can eliminate processes of thermoconductivity, thermodiffusion, and runaway of plasma electrons, the local heating effects described previously are predominant. Conversely, in the limiting case of small heated areas, the effects of nonlocal heating become predominant, causing spatial distribution of plasma density.

The effect of nonlocal heating stems from analysis of Equation 2.4 for the electron temperature T_e, the simple solution of which can be obtained in the form of its spatial Fourier component [6]:

$$T_{e\kappa} = \frac{2e^2}{3m_e v^2 \delta_T} \left(1 - e^{-\delta_T \nu_e t}\right) \cdot |E_\omega|^2 \tag{2.14}$$

where

$$\delta_T = \delta + \kappa_\perp^2 \rho_{He}^2 + \kappa_\parallel^2 \Lambda_e^2 \tag{2.15}$$

and $\rho_{He} = v_{Te}/\omega_{He}$ is the Larmor radius of electrons in the ambient magnetic field, v_{Te}, the thermal velocity of electron, Λ_e, the length of the free path of the electron between collisions with ions and atoms, ω_{He}, the hydromagnetic frequency of the electron, δ, the fraction of energy transmitted by electrons during their nonelastic interactions with other components of plasma, $\kappa_\parallel = 2\pi/\ell_\parallel$ and $\kappa_\perp = 2\pi/\ell_\perp$, $\ell_\parallel$ and $\ell_\perp$ are the characteristic scales of temperature plasma inhomogeneities in the direction of the magnetic vector $\mathbf{B}_0$, normal to the plane containing vector $\mathbf{B}_0$. From Equations 2.14 and 2.15 it can be seen that the nonlocal effects of the electron temperature T_e are significant when $\kappa_\perp^2 \rho_{He}^2 + \kappa_\parallel^2 \Lambda_e^2 > \delta$ and $\delta_T \nu_e \tau \gg 1$, where τ is the characteristic time of relaxation of the temperature plasma disturbances caused by the field of strong waves.

Because redistribution of plasma density under the thermal source $F_T = -N\nabla T_e$ in Equation 2.3 is determined by the process of diffusion, the characteristic time of diffusion process, $\tau_D \propto \min\left\{\left(\kappa_\parallel^2 D_{e\parallel}\right)^{-1}, \left(\kappa_\perp^2 D_{e\perp}\right)^{-1}\right\}$, is determined by the coefficients of diffusion of electrons along, and normal to, the magnetic field lines, $D_{e\parallel}$ and $D_{e\perp}$. The role of the chemical recombination process is predominant if the time of this processes, τ_{ch}, is less than the time of diffusion, i.e., $\tau_{ch} < \tau_D$. The role of photo-chemistry is predominant if τ_{ch} is less than $\tau_D \propto l_\parallel^2 \nu_{im}/v_{Ti}^2 (D_{i\parallel} \approx 2T_i/m_i\nu_{im})$, particularly for the lower ionosphere, where $\nu_{im} \gg \omega_{Hi}$ and now $\tau_D \propto \min\left\{\left(\kappa_\parallel^2 D_{i\parallel}\right)^{-1}, \left(\kappa_\perp^2 D_{i\perp}\right)^{-1}\right\}$ is determined by coefficients of diffusion of ions. Taking into account the striction force F_{st} defined by Equation 2.9, we can compare the relative contribution of F_{st} and $F_T = -N\nabla T_e$ in plasma dynamics. Thus, using the above expressions we get [6]

$$\frac{F_T}{F_{st}} = \begin{cases} 1/\delta_T, & t > (\delta_T \nu_e)^{-1} \\ \nu_e t, & t < (\delta_T \nu_e)^{-1} \end{cases} \tag{2.16}$$

It is clear that the condition $F_T \gg F_{st}$ for $t > (\delta_T \nu_e)^{-1}$ can be satisfied for sufficiently small-scale disturbances of plasma with the wave vector oriented perpendicular to $\mathbf{B}_0$ (e.g., with $\kappa \approx \kappa_{\parallel}$).

We will now consider another interesting phenomenon which is determined by the wavelength λ of the powerful radio wave, i.e., by the minimum scale of the field of this wave, l. This phenomenon is related to the occurrence of the standing electromagnetic wave created by interference of the direct radio wave and the wave reflected from the ionosphere (see Figure 2.1).

For example, if a probing radio wave with a frequency of 6 MHz is reflected from the ionosphere, the characteristic scale equals $l = \lambda/2$, or 25 m, at the altitude where the refractive index equals unity, i.e., $n(f) \approx 1$. In the proximity of the reflection level due to the decrease of refractive index n caused by the heating process, the scale l can be increased. Simultaneously, the amplitude of the wave field is also increased because of temporal compression of the wave packet, i.e., according to the decrease of the group velocity of the wave. Due to a run of heated plasma from regions of maximum of the wave field in the ionosphere, a new kind of plasma perturbation, the periodical structures of plasma density, are generated, which strongly scatter probing radio waves sent from a diagnostic ground-based transmitter.

Using this scattering effect, it is easy to measure the major parameters of such artificial periodical irregularities (API) [24–30]. The nonlinear effects occurring during interference of the direct and the reflected radio waves from the ionosphere can be a good example and a good illustration of the propagation phenomena discussed previously and presented schematically in Figure 2.1.

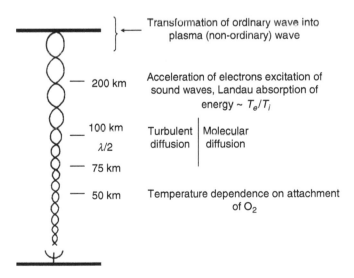

Figure 2.1 Processes and phenomena accompanying propagation of radio waves through the ionosphere.

As mentioned in Section 1.1, the parameters of the ionosphere, such as its content, frequencies of interactions, and the lengths of free paths of plasma species, are changed with an increase of ionospheric altitude. That is, at heights of $h \approx 50$–70 km, the main contribution to generation and disintegration of API introduces the attachment of electrons to, and the detachment of electrons from, molecules/atoms [7]. The characteristic time of such processes at a height of 65 km does not exceed 0.5 s [7]. At altitudes of $h \approx 70$–95 km the main contribution to the disintegration of API is determined by turbulent diffusion of the neutrals, and at altitudes $h \approx 95$–130 km the process is controlled by ion laminar diffusion along magnetic field lines (mostly for $h > 100$ km, i.e., at the heights of *turbopause*) [7].

Finally, at altitudes of the higher ionosphere where the length of free path of plasma species significantly exceeds the dimensions of API, the main process of generation and disintegration of periodical plasma structures is the runaway of plasma along magnetic field lines under the influence of striction forces described by Equation 2.3. Because the noncollision regime describes the existence of ion-sound waves, the nonstationary API can excite these waves [24]. The efficiency of excitation of ion-sound waves by API depends on collision or noncollision damping (called *Landau attenuation* [13]) of these unstable waves (for details see Ref. [24]).

2.2 Interaction of Probing Radio Waves with Heating-Induced Plasma Instabilities Generated by Strong Radio Waves

It is important to note that near the level of reflection of the powerful radio wave from the ionosphere some new effects occur under its interaction with ionospheric plasma [31–56]. Here, a wide spectrum of waves can be generated, such as slow plasma waves with the frequency close to the frequency of radio wave. Therefore, it is possible to observe experimentally a mutual transformation of these waves and the electromagnetic radiation from the ionosphere. Let us briefly describe such effects and the corresponding instabilities occurring in the perturbed ionosphere.

In plasma consisting of different kinds of disturbances, the absorption of additional energy from powerful radio waves at the frequencies which are in resonance with the plasma frequencies, the energy from the waves can be transferred to plasma disturbances to increase their density. Finally, the so-called *resonance instability* in regions where such resonances take place is generated in the field of strong radio waves. As shown in Ref. [40], such kinds of instabilities stimulate the generation of small-scale plasma irregularities elongated along the geomagnetic field lines in the region of reflection of the ordinary wave. Such inhomogeneities can be detected by observation of the scattering of probing waves at VHF and UHF frequency bands from the perturbed regions of the ionosphere (see Chapter 7). To understand this phenomena, we depicted schematically in Figure 2.2 at the plane $\{z, n_z^2\}$ the dispersive curves for high-frequency ordinary (O) and non-ordinary (z-mode) electromagnetic waves in the ionosphere, as well as the quasilongitudinal component (with respect to geomagnetic lines) of the O-mode (dashed curve). Here, $n_z^2 = (\omega_{e0}/\omega)^2$ is the square of the refraction coefficient, $\omega_{e0} = \sqrt{\frac{4\pi e^2}{m_e} N(z)}$ the plasma frequency, which is increased with the height at the lower part of the F layer, and z the height above Earth's level. The curves are presented for $(\omega_{He}/\omega)^2 \ll 1$. The curve for z-mode corresponds to different angles of its wave vector with respect to the lines of the geomagnetic field defined by the unit vector $\mathbf{b} = \mathbf{B}_0/|\mathbf{B}_0|$. Thus, $n_z \to \infty$ for the wave with the angle $\alpha = \cos^{-1}(\mathbf{k} \cdot \mathbf{b})$ close to $\pi/2$ and at the frequencies close to upper-hybrid resonance of $\omega_{uh} = \sqrt{\omega_{He}^2 + \omega_{e0}^2}$. In the region $n_z \gg 1$ these

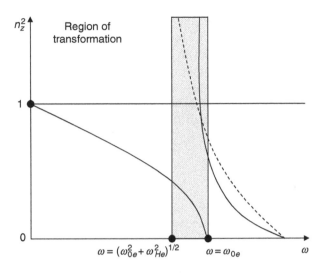

Figure 2.2 Frequency dependence of the refractive index for different wave modes in the ionosphere-dispersive curves.

waves become purely longitudinal, and for waves with wave vector $\mathbf{k}_l = \mathbf{k}_{\|}$, i.e., parallel to $\mathbf{b}$, $n_z \rightarrow \infty$ for $\omega = \omega_{e0}$, whereas for a plasma wave with $\omega \cong \omega_{uh}$ the wave vector, $\mathbf{k}_e \perp \mathbf{b}$, i.e., perpendicular to $\mathbf{b}$ [14].

From Figure 2.2 it can be seen that in the height region, where $\omega_{uh} \leq \omega \leq \omega_{e0}$, it is possible to transform the probing wave to other types of waves without adequate changes in frequency (jumping from one dispersive curve to another along the vertical line). For such a transformation it is necessary that in the medium of the first wave (powerful one) the alternate current is excited with the frequency $\omega \approx \omega_{e0}$ and the wave number $\kappa \approx k_{\|} \equiv k_l$.

From laws of conservation of energy and impulse in the approximation of interaction of three waves, it follows that for this case $\omega_e = \omega + \Omega$ and $\kappa = \mathbf{k}_l \pm \mathbf{k}$, where $\mathbf{k}$ is the wave vector, Ω and κ are the frequency and the wave spectrum, respectively, of the Fourier component of the plasma wave (instability) polarized in the field of the powerful wave. Transforming a radio wave to a plasma wave with $n_l \equiv n_z \gg 1$, we have $k_l \gg k$ and $\kappa \approx k_l$. The latter means that for such transformation in plasma the instabilities, having scales $\ell = 2\pi/k$ significantly less than λ, can be easily generated. Therefore, in plasma nonlinear phenomena with the characteristic scales of the field essentially less than λ can be generated, because in the case of a strong plasma wave the spatial structure of interaction will be determined by $\lambda_l \approx f/cn_l, f = \omega/2\pi$ [36,37]. The plasma waves that generate strong plasma inhomogeneities can be sufficiently intensive even when the comparably small transformation of radio waves into plasma waves occurs. Thus, in the homogeneous ambient plasma the field strength of plasma waves in the region of their interaction equals $E_l \approx E_\omega \delta N \omega / \nu_e$, where δN is the relative amplitude of plasma inhomogeneities responsible for such wave transformation. The characteristic length of induced plasma (nonordinary) waves can be changed from $\lambda_0 = f/c$ up to units of a few centimeters, i.e., close to the Debay radius of electrons $r_{De} = v_{Te}/\omega_{eo}$, which, in the ionosphere, changes from 1 cm to few centimeters. In the inverse case of $k_l r_{De} > 1$, a strong collisionless absorption of plasma waves at the thermal electrons is observed [56]. The longitudinal waves with $\mathbf{k}_l \| \mathbf{b}$ have scales much less than the length of the free path; therefore, their effects on plasma are caused exclusively by the striction force F_{st}. In the region of upper-hybrid resonance (UHR), where $\mathbf{k}_l \perp \mathbf{b}$, the main contribution in the nonlinear dynamics of plasma for $|\mathbf{k}_l| \rho_{He} > 1$ introduces the thermal force F_T, but when $|\mathbf{k}_l| \rho_{He} < 1$, again $F_{st} > F_T$, and striction nonlinearity is predominant compared with thermal nonlinearity. What is important is that the plasma wave created by the transformation of a radio wave in natural plasma fluctuations or in the sharp regular gradients of plasma concentration can, due to interference with the "mother" radio wave, create the quasistanding structures, which can be considered as a generator of the forces F_T and F_{st}, by averaging them at the high frequency. Finally, the intensity of these plasma perturbations is

$$\nabla|E_\omega|^2 \propto \nabla\langle(E_l E_t^* + E_t^* E_l)\rangle \propto k_e\langle(E_l E_t^* + E_t^* E_l)\rangle \qquad (2.17)$$

As a result, due to runaway of plasma under the action of F_{st} or due to plasma thermodiffusion under the action of F_T, new disturbances of plasma density are created, which amplify the wave transformation. This is called *parametric instability*, if the striction mechanism is predominant, or *thermal parametric instability*, if thermodynamic processes are predominant. Both instabilities are upper limited. However, their upper limit does not depend on contrast of perturbations, because F_{st} and F_F are proportional to E_l, i.e., proportional to δN linearly and increases linearly with increase of ionospheric height. At the same time, the process of relaxation of such plasma structures determined by the gradient of pressure is also proportional to δN. In this case, the upper limits of instability excitation are defined by other dissipative processes.

First, we will consider the nondissipative parametric instability, which occurred in the plasma-resonance regions due to the striction effect. The time of generation of such instabilities is less than the electron's free-path time and leads to the absorption and attenuation of plasma waves. Taking into account Equation 2.2 with $\omega = \omega_{e0}$, the constraint $F_{st} > F_T = (T_e + T_i)\nabla N$, and the fact that the longitudinal electric field of plasma waves equals $E_l \propto \frac{N\omega}{N_0\nu_e}\eta^{1/2}$, we finally get:

$$\frac{E_0^2}{8\pi N_0(T_e + T_i)} > \frac{\nu_e}{\omega\eta} \qquad (2.18a)$$

Here, the parameter η is introduced to take into account the inhomogeneous plasma effects; N_0 is the concentration of background nonperturbed ionospheric plasma. Strict kinetic calculations made in Ref. [18] lead, for $\eta = 1$, to the existence in the right-hand side of (Equation 2.18a) of a product ~ 4 in the dominator. Note that one of the solutions for nondissipative parametric instability in plasma is the ion-sound waves generation due to transformation of radio waves to plasma waves. The ion-sound waves lead to nonlinear generation of noise and additional absorption of wave energy in plasma [7,8]. However, in plasma, with an increase of T_e/T_i, the attenuation of ion-sound waves, due to collisions of electrons with ions, is decreased, and the level of instability is decreased compared with Equation 2.18a. For $T_e/T_i > 7$, in the right-hand side of this inequality, exists the small term $\approx \sqrt{8\pi m_e/M_i}$ [6]. For thermal parametric instability, following from relation between F_T and F_{st} (2.16), $F_T \propto N\nabla \ T_e \propto F_{st}/\delta$, and from the condition $F_T \gg F_{st} \propto (T_e + T_i)\nabla N$, we get [6]:

$$\frac{E_0^2}{8\pi N_0(T_e + T_i)} > \frac{\delta\nu_e}{\omega\eta} \qquad (2.18b)$$

The increment of *parametric instability* is easy to estimate from the following assumptions. Combining the equations of movements of ions and electrons, it is easy to find the velocity of plasma under constraint $F_{st} > (T_e + T_i) \nabla N$ (see Refs. [13,14]). It should be noted that in the case of absence of collisions the time of force action is determined by the ion free-path time. Thus,

$$v_i \approx \frac{[F_{st} - (T_e + T_i)\nabla N]}{M_i v_{Ti} \eta N_0} \tag{2.19}$$

Taking into account this expression, we find from Equation 2.5 [with $\frac{\partial N}{\partial t} \to \gamma N$] that the increment of parametric instability is

$$\gamma = \eta v_{Ti} \frac{\omega' - \omega}{\omega_{eo}} \frac{(T_e - T_i)}{T_i} \tag{2.20}$$

where $\omega' = E_\omega^2 / 8\pi N_0 (T_e - T_i)$. For $\eta = 1$, Equation 2.20 for γ is the same, with an accuracy of the coefficient $\sim\sqrt{2/\pi}$, as that obtained strictly in Refs. [14,52].

In the same manner we can obtain the expression for the increment of *thermal parametric instability*. The difference is that in this case the ion velocity in plasma

$$v_i \approx \frac{[F_T - (T_e + T_i)\nabla N]}{M_i N_0} \tag{2.21}$$

is determined by the corresponding regime of diffusion (defined either by the longitudinal-to-magnetic-field coefficient of ion diffusion in the lower ionosphere or by the perpendicular-to-magnetic-field coefficient of ion diffusion in the upper ionosphere [10]). It should be noted that the increment of thermal parametric instability is essentially smaller compared with the increment of parametric instability; the latter can achieve $10^2 - 10^3$ s^{-1}. The plasma waves generated in such a manner are scattered at the inhomogeneities of plasma density in the directions $\kappa = \mathbf{k}_e - \mathbf{k}_e'$, where $\mathbf{k}_e'$ is the wave vector of the scattered plasma waves. The scattered waves must satisfy the corresponding dispersion equation, which, ignoring electromagnetic effects ($k_l \gg k_0$), is the following:

$$\omega_e^2 = \omega_{e0}^2 + \omega_{He}^2 \sin^2 \alpha + 3v_{Te}^2 k_l^2, \quad \omega_{He} \ll \omega_{e0} \tag{2.22}$$

In general, the dispersion equation for plasma waves has the following form [57]:

$$3uwk_l^4 - (1 - u - w + uw\cos^2 \alpha)k_l^2 + [2(1 - w)^2 + uw\cos^2 \alpha - u(2 - w)]k_0^2 = 0 \tag{2.23}$$

where

$$u = \omega_{He}^2 / \omega^2, \quad w = \omega_{e0}^2 / \omega^2, \quad k_0 = \omega/c.$$

Therefore, during scattering from the ionospheric plasma region with $n_l > 1$ containing stationary inhomogeneities (i.e., without changes of plasma frequency ω_{e0}), a decrease in the absolute value of the wave vector is accompanied by an increase in the angle of scattering α. In the region of $\omega \approx \omega_{e0}$, where $\mathbf{E}_\omega \| \mathbf{b}$ and plasma waves with small α are excited intensively, a further scattering process with an increase of α leads to the occurrence of plasma waves with small wave numbers k_l. Consequently, the plasma waves generated first must be with "short-lengths," i.e., $k_l \rho_{He} > 1/\sqrt{3}$ [38], to orient the vector of scattering wave k_l to be perpendicular to $\mathbf{B}_0$.

Simultaneously, the change of plasma concentration and frequency in time caused by runaway of plasma particles from the perturbed regions can lead, in the region $\omega \approx \omega_{e0}$, to an increase in the scattering angle α up to $\alpha \approx \pi/2$. This change is related to the inhomogeneous character of the field and its amplification near the level of reflection. The existence of waves with $\alpha \approx \pi/2$ near the level of reflection can lead to the existence of small-scale thermal scattering from ionospheric plasma. Such inhomogeneities of heat propagate with the velocity of longitudinal electron diffusion to magnetic field lines in the region of upper-hybrid resonance (UHR), accelerating in this region the transformation of the probing radio wave in plasma waves. In the region of UHR scattering of plasma waves occurs with decrease of angle α and with increase of k_l. Here, scattering can lead to a tendency for k_l to be isotropic in the plane orthogonal to $\mathbf{B}_0$. When the amplitudes of plasma waves achieve magnitudes for which $E_l \approx \eta^{1/2} E_\omega \delta N_\omega / \nu_e > E_\omega$, the forces F_T and F_{st} will be fully determined by the interference of longitudinal plasma waves. Here, new effects occur, such as the explored fast increase of the amplitude of small-scale instabilities of plasma, where $\frac{\partial N}{\partial t} \propto N_\omega^2 E_\omega^2$ [36], and the disintegration of plasma wave on the same wave but with a lower frequency ($\omega_e = \omega'_e + \Omega$), i.e., on the ion-sound waves with $\Omega = \kappa v_s$, where v_s is the sound speed, the attachment of waves, and so on [39–55]. As mentioned previously, waves due to scattering at the plasma inhomogeneities are transformed into radio waves, which leads to the existence of artificial electromagnetic radiation (AER) of ionosphere, the frequency of which differs from that of the exciting wave [19,42–44]. In some ionospheric heating experiments AER was decoded at frequencies $f = 100-300$ kHz [43,44].

High-frequency plasma instability with $\mathbf{k}_l \| \mathbf{b}$, having phase velocities close to velocities of photo-electrons, can lead to their acceleration up to energies corresponding to the energy of ionization of atmospheric neutral atoms (molecules) at the altitudes of F layer of the ionosphere and the lightness of the ionosphere [45]. However, such effects cannot be described by the hydrodynamic Equations 2.3 through 2.5 and we need to use the strict kinetic approach based on Equation 2.6.

We should also mention other interesting effects accompanied by active modification of the ionosphere by powerful radio waves, such as the oblique sounding of waves [46–49], effects of high-frequency waves, effects related to sporadic E layer [50,51], and so on (see Chapter 7), each of which is based on physical phenomena described previously at the qualitative level to understand the major nonlinear effects occurring in the ionospheric plasma perturbed by heating processes. The preceding explanation is enough to illustrate to the reader the fact that under the influence of high-power alternate electromagnetic waves a complex of new phenomena and corresponding instabilities can be generated in the ionosphere. Investigations into these allow us not only to obtain information about the parameters of plasma but also to perform ionospheric experiments that can excite such phenomena in the ionospheric plasma, which either do not exist or are not observed in the pristine natural conditions of the ionosphere [39–56].

2.3 Plasma Instabilities of Natural Origin Generated in the Irregular Ionosphere

We will now put the question, can the same heating-induced or current-induced plasma instabilities having a natural origin be generated by natural phenomena occurring in the ionosphere, and what types of instabilities are responsible for the creation of plasma irregularities observed from the equator to the polar cap in the real ionosphere? The problem here is that there is a wide spectrum of small-, moderate-, and large-scale natural plasma structures (turbulences) observed experimentally in the real ionosphere at different altitudes of regions E and F [20,59–79], covering all regions from the equator to the polar cap [80–96]. This fact excludes the possibility of describing their creation and further temporal and spatial evolution by any unified physical mechanism or universal theoretical approach. Following from previous sections, heating-induced plasma irregularities can be generated by local heating of the ionosphere with powerful radio waves, which finally generates *parametric* and *striction* instabilities in plasma.

The same effects can be observed in the real ionosphere in other natural conditions, so that nonisothermal conditions occur in the ionospheric plasma [58] during heating of plasma particles by electric fields (due to fluctuations of the ambient electric field as a local heater of ionospheric plasma [59,60]). Plasma instabilities can also be generated by the wind of neutral molecules (atoms) [61], by drift in crossing ambient electrical and magnetic fields (usually denoted as $\mathbf{E} \times \mathbf{B}$ drift [70–76]), and by other current-induced two-stream instabilities [97–101]. At the same time, most natural small-, moderate-, and large-scale plasma structures observed experimentally in the auroral and equatorial ionosphere can be explained by the existence of plasma waves caused by convergent-current and gradient-current instabilities [20,65–79,98,102–118]. All these phenomena can be unified as the natural sources of plasma instabilities in the regular ionosphere.

We now briefly discuss heating-induced (or thermal) and current-induced instabilities as the main sources of the small-, moderate-, and large-scale plasma density perturbations, δN, which can be either positive or negative compared to non-disturbed background ionospheric plasma density N_0. Usually, a case in which $\delta N > 0$ is called *plasma density enhancements*, and one in which $\delta N < 0$ is called *plasma density depletions* [70–77].

Before describing different kinds of instabilities of the ionospheric plasma caused by a wide spectrum of morphological and dynamical processes occurring in the ionosphere, we will first give some physical examples. As suggested in Refs. [71–74], in the equatorial ionosphere, the electrojet region is unstable for gradient-drift processes, when $\mathbf{E} \cdot \nabla N_0 > 0$. The gradient-drift instability generates long-length perturbations (e.g., waves) with wavelengths of 100 m and more, up to 10 km, when the diffusion relaxation process (which usually decreases growth of such instabilities) is sufficiently weak. Such large-scale plasma perturbations as primary waves observed in the equatorial electroject can generate small-scale secondary gradient-drift waves propagating at large flow angles to the ambient $\mathbf{E}_0 \times \mathbf{B}_0$ drift. Thus, the nonlinear cascade wave interactions can be assumed to be the main mechanism of generation of small-scale plasma irregularities usually observed in the equatorial ionosphere [71–74].

As for the high-latitude ionosphere, because of a strong electric field and neutral winds that are sharply height-structured, plasma perturbations with scales from thousands of kilometers to tens of centimeters are observed experimentally, and can be explain also by the interaction of large-scale and small-scale plasma waves, i.e., by the cascade mechanisms of plasma instability generation [70,75–77].

Plasma perturbations (i.e., waves) of a large range of scales, from tens to hundreds of kilometers creating diffuse spread-F traces observed by ionosondes in the middle-latitude ionosphere [78,79], are also evidence of a complicated nonlinear mechanism of plasma waves interaction. The unstable processes mentioned previously cannot be understood using linear theory alone [80–101]. Based on nonlinear phenomena and the corresponding physical models, the effects of the generation of small-scale plasma perturbations in the background of large-scale plasma disturbances may finally be explained by strict physical phenomena [98,100–118]. We will later discuss both linear and nonlinear approaches to these phenomena and deal with the analysis of plasma instabilities growth, interactions, and dumping.

We shall start with a brief description of *linear regimes* of plasma instability generation in the ionospheric plasma due to natural geophysical and morphological processes occurring in the real E and F regions of the ionosphere, from the equatorial ionosphere to the polar (auroral) ionosphere.

2.3.1 Natural Thermal-Induced Instabilities

We will start with an analysis of those thermal instabilities that can be described by nonlinear equations of hydrodynamics and kinetic theory (Equations 2.3 through 2.6). Here, as in Section 2.1, we have to differentiate between the phenomena and effects that can be described by a purely hydrodynamic approach based on Equations 2.3 through 2.5, and by an approach based on kinetic equations, such as Equation 2.6 for the distribution functions of electrons and ions. At the same time, as shown in Refs. [20,65,66], the methodology of hydrodynamic and kinetic description of waves in the ionospheric plasma produces approximately the same physical results if the spatial dimensions of plasma instabilities along and across the geomagnetic field $\mathbf{B}_0$ exceed the free-path length, $\ell_\alpha = v_{T\alpha}/\nu_\alpha$, and the Larmor radii, $\rho_{H\alpha} = v_{T\alpha}/\omega_{H\alpha}$, of plasma electrons ($\alpha = e$) and ions ($\alpha = i$), respectively. All these parameters are introduced in Section 1.1.

Furthermore, some limits for temporal scales of plasma parameter deviations must be satisfied. Thus, for plasma waves with frequency ω_p and wave vector $\mathbf{k}_p$, a condition to use only the hydrodynamic approach based on the systems (Equations 2.3 through 2.5) can be written as [20,65]:

$$\frac{\left|\omega_p - \mathbf{k}_p\mathbf{V}_{\alpha 0}\right|}{\nu_\alpha} \ll 1 \tag{2.24a}$$

Here, for $\alpha = e$, $\nu_e = \nu_{em} + \nu_{ei}$, and, for $\alpha = i$, $\nu_i = \nu_{im} + \nu_{ei}$, $\mathbf{V}_{\alpha 0}$ are the nondisturbed velocities of plasma electrons ($\alpha = e$) and ions ($\alpha = i$) in the quasi-uniform nondisturbed ionosphere. Comparison between results of purely kinetic and purely hydrodynamic analysis of instabilities caused by natural sources in the ionosphere showed that near the threshold of instability of plasma, the limitation (Equation 2.24a) corresponds to weaker [67] and even still weaker equations [68], respectively:

$$\frac{\left|\omega_p - \mathbf{k}_p\mathbf{V}_{\alpha 0}\right|}{\nu_\alpha} \leq 0.1 \quad \text{or} \quad \frac{\left|\omega_p - \mathbf{k}_p\mathbf{V}_{\alpha 0}\right|}{\nu_\alpha} \leq 1 \tag{2.24b}$$

We will now start to investigate thermal instabilities for which Equation 2.24 are valid, and we can describe the processes of their generation based on a purely hydrodynamic approach.

It is obvious that during local heating of the ionosphere we first observe fluctuations of temperature and concentration of the ionospheric plasma, as the first-order parameters of plasma (see definitions introduced in Chapter 1). These deviations create changes of the second-order parameters of plasma (see Chapter 1) such as collisional frequencies of the plasma-charged particles, $\nu_\alpha(\alpha = e,i)$, and the coefficient of recombination $\beta = \alpha_{eff}N^2$ (where α_{eff} denotes the rate of recombination of charged particles). These changes of frequencies, $\delta\nu_{\alpha m}$ and $\delta\nu_{ei}$, and the coefficient of recombination, $\delta\beta$, can be presented as [58–64]:

$$\delta\nu_{\alpha m} = \frac{1}{2}\nu_{\alpha m}^{(0)}\frac{\delta T_\alpha}{T_{\alpha 0}}, \quad \alpha = e,i \tag{2.25a}$$

$$\delta v_{\alpha i} = v_{ei}^{(0)} \left(\frac{\delta N}{N_0} - \frac{3}{2} \frac{\delta T_e}{T_{e0}} \right) \tag{2.25b}$$

$$\delta \beta = \frac{d\beta'}{dT} \frac{\delta T_e}{T_{e0}} \tag{2.25c}$$

Now, following Refs. [58,64], we can present the linearization of Equations 2.3 through 2.5, obtained for weak disturbances of electron and ion components of plasma concentration, δN_α, and their velocities, $\delta \mathbf{V}_\alpha$ ($\alpha = e, i$). Thus, the linear equations of continuity of plasma-charged particles, the equations of their movements, and the equations of their thermal balance in the ionosphere have the following form:

$$\frac{\partial N_\alpha}{\partial t} + \nabla \cdot (\delta N_\alpha \mathbf{V}_{\alpha 0} + N_0 \delta \mathbf{V}_\alpha) = -\beta' \delta N_\alpha \tag{2.26a}$$

$$\frac{\partial \delta \mathbf{V}_\alpha}{\partial t} + \mathbf{V}_{\alpha 0} \nabla \cdot \delta \mathbf{V}_\alpha = -\frac{\nabla \delta p_\alpha}{m_\alpha N_0} + \frac{q_\alpha e}{m_\alpha} \left(\delta \mathbf{E} + \left[\delta \mathbf{V}_\alpha \times \frac{\mathbf{B}_0}{|\mathbf{B}_0|} \right] \right) - v_\alpha \delta \mathbf{V}_\alpha \tag{2.26b}$$

$$\frac{\partial \delta T_e}{\partial t} + \frac{2}{3} T_{e0} \nabla \cdot \delta \mathbf{V}_e = -\frac{\nabla \hat{\chi}_e \nabla \delta T_e}{N_0} + \frac{2}{3} \delta Q_e - \frac{2}{3} \delta_{ei} v_{ei}^{(0)} \left\{ (T_{eo} - T_{io}) \left(\frac{\delta N}{N_0} - \frac{3}{2} \frac{\delta T_e}{T_{e0}} \right) + \delta T_e \right\}$$
$$- \frac{2}{3} \delta_{em} v_{em}^{(0)} \left\{ \frac{1}{2} (T_{eo} - T_{io}) \frac{\delta T_e}{T_{e0}} + \delta T_e \right\}$$

$$\frac{\partial \delta T_i}{\partial t} + \frac{2}{3} T_{i0} \nabla \cdot \delta \mathbf{V}_i = -\frac{\nabla \hat{\chi}_i \nabla \delta T_i}{N_0} + \frac{2}{3} \delta Q_i - \frac{2}{3} v_{em}^{(0)} \left\{ \frac{1}{2} (T_{io} - T_{mo}) \frac{\delta T_i}{T_{i0}} + \delta T_i \right\} \tag{2.26c}$$

Here, δp_α is the disturbance of the pressure of plasma particles, $\alpha = e, i$, $\delta_{ei} = 2m_e / M_i$, $\delta_{em} = m_e / M_m$, $\hat{\chi}_\alpha$ is the tensor of thermoconductivity of charged particles, $\delta Q_e = \delta(\mathbf{j}_e \cdot \mathbf{E}/N)$ and $\delta Q_i = \delta(\mathbf{j}_i \cdot \mathbf{E}/N)$ are the disturbances of power of the thermal sources during local heating of electrons and ions, $\mathbf{j}_\alpha = q_\alpha e N \left(\mu_{\alpha P}^2 E_\perp + \mu_{\alpha \parallel}^2 E_\parallel \right)$ is the current created in plasma in heating-induced conditions, and $\mathbf{E} = \mathbf{E}_0 + \delta \mathbf{E}$, $N = N_0 + \delta N$ are the total electrical field and the plasma concentration, including both ambient nonperturbed and perturbed parts; all other parameters are as defined in Section 1.1.

Here also the parameter $\beta' = 2\alpha_{eff} N_0$ is the coefficient associated with the recombination process at the ionospheric altitudes where the interactions of molecular ions are predominant with respect to those of atomic ions. Conversely, the parameter $\beta' = \beta$ is observed at those altitudes where the concentration of the atoms and atomic ions is predominant compared with that for molecular ions. As mentioned in Chapter 1, the second situation with the parameter β' occurs at higher ionospheric altitudes, where $h > 150$–200 km, and the first situation is more significant for those altitudes that are less than 100–120 km (see Chapter 1).

Changes of velocities of charged particles due to temperature fluctuations can be presented as [59–64]:

$$\delta \mathbf{V}_e = e\hat{\mu}_e \mathbf{E}_e - \frac{\hat{D}_e \mathbf{B}_0 \nabla \hat{\mu}_e}{q_H} \frac{\nabla \delta N}{N_0} + e\delta v_e \frac{d\hat{\mu}_e \mathbf{E}_0}{dv_e} \tag{2.27a}$$

$$\delta \mathbf{V}_i = Ze\hat{\mu}_i \mathbf{E}_i - \frac{\hat{D}_e \mathbf{B}_0 \nabla \hat{\mu}_i}{Q_H} \frac{\nabla \delta N}{N_0} + Ze\delta v_e \frac{d\hat{\mu}_i \mathbf{E}_0}{dv_i} \tag{2.27b}$$

where $\delta\nu_\alpha$ ($\alpha = e,i$) is defined by Equation 2.25b. In the preceding equations, we also assumed that the electron charge is e, whereas the ion charge is Ze, where Z is the charged order number of the ion compared to an equal unit for electrons. We assume that the disturbance of the electric field in the ionosphere, $\delta\mathbf{E}$, is electrostatic, i.e., $\delta\mathbf{E} = -\nabla\delta\varphi$. Here also $\hat{\mu}_e$ and $\hat{\mu}_i$ are tensors of mobility of electrons and ions, respectively. Their absolute values in the isotropic plasma are $\mu_e = e/m_e(\nu_{em} + \nu_{ei})$ and $\mu_i = Ze/M_i$ ($\nu_{im} + \nu_{ei}$), where all parameters are described in Section 1.1. They are simply related with usually used tensors of conductivity of charged particles (see Chapter 1):

$$\hat{\sigma}_e = eN\hat{\mu}_e \quad \text{and} \quad \hat{\sigma}_i = ZeN\hat{\mu}_i \tag{2.28}$$

Presented before the system of Equations 2.26 and 2.27 is the basic system in our further analysis of heating-induced plasma instabilities. Mathematical derivations used for this analysis, are very complicated (see Refs. [57–67]). Therefore, we will deal with the final results of various researches [57–67] and discuss only the physical mechanisms of generation of nonisothermal plasma oscillations.

2.3.1.1 Recombination Instability

We start with the recombination instability, which creates the small-scale plasma disturbances in the ionosphere [63,64]. As shown in Refs. [63,64], such instability develops due to heating of plasma electrons by intrinsic (polarized) electric fields and, therefore, due to a decrease in the rate of recombination of plasma-charged particles within the areas of heating of the low (usually E region) ionosphere. We will now present the plasma wave frequency in the complex form, by introducing the increment of instability growth, γ, i.e., $\omega = \omega_r + i\gamma$. Then, following the solutions of the dispersion equation of plasma wave obtained in Refs. [63,64], we obtain the corresponding expressions for the nondisturbed frequency ω_r and for the increment of recombination instability γ_R of the disturbed ionospheric plasma:

$$\omega_r = \frac{\mathbf{k}\mathbf{V}_{i0}\mathbf{k}\hat{\mu}_e\mathbf{k} + \mathbf{k}\mathbf{V}_{e0}\mathbf{k}\hat{\mu}_i\mathbf{k}}{\mathbf{k}\hat{\mu}\mathbf{k}} \tag{2.29a}$$

$$\gamma_R = -\frac{3}{2}\frac{d\beta'}{dt}\mu_{Pe}\frac{\mathbf{k}\mathbf{E}_0\mathbf{k}\mathbf{U}_0}{\delta_{em}\nu_{em}\mathbf{k}\hat{\mu}\mathbf{k}} - \beta' - \frac{D_0 B}{Q_H}\frac{\mathbf{k}\hat{\mu}_i\mathbf{k}\mathbf{k}\hat{\mu}_e\mathbf{k}}{\mathbf{k}\hat{\mu}\mathbf{k}} \tag{2.29b}$$

Here, $\mathbf{U}_0 = \mathbf{V}_{i0} - \mathbf{V}_{e0}$ is the difference between the nondisturbed velocity of plasma ions and electrons, $D_0 = C_s^2/\nu_i$ is the coefficient of ion diffusion, $\nu_i = \nu_{im} + \nu_{ei}$, the total frequency of ion–neutral and ion–electron collisions, $C_s^2 = (T_e + T_i)/M_i$, the velocity of ion sound waves, $\mu_{Pe} = e[p + (1 + Q_H^2)]/m_e\nu_{em}[(1 + p^2) + q_H^2(1 + Q_H^2)]$, the Pedersen component of the mobility of electrons (other components can be obtained from relation 2.28 and the corresponding expressions of components of the tensor of conductivity of electrons and ions presented in Section 1.1).

Recombination instability can be developed in the region of high-latitude ionosphere caused by ambient electric fields, as heater, of $E_0 > 40$ mV/m [61]. Thermal instabilities occur due to increase of the polarized field at the sites of plasma depletion and its heating by polarization electric fields [59,60,62,64].

We now consider the source of the free energy of plasma as a heater equaling the difference between the full energy and the internal plasma energy, as its source nonisothermally (e.g., $T_{e0} > T_{i0} \approx T_{m0}$). Then the condition of plasma instability in the absence of ambient electric fields can be simplified as [59]

$$\frac{3}{2} + \frac{2T_{e0}}{T_{e0} + T_{i0}} - \frac{T_{e0}}{T_{e0} - T_{i0}} > 0 \tag{2.30}$$

From the solution of this inequality (Equation 2.30), it follows that thermal instability develops for $T_{e0}/T_{i0} > 1.58$ [59].

When longitudinal electric fields exist in the ionosphere, they heat electrons and can generate instability of isothermal plasma if the strength of the longitudinal field exceeds the threshold that equals [60]

$$E_{0kr\parallel} = \left[\frac{T_{e0}^2}{e^2}\left(1 + \frac{k_\perp^2}{k_\parallel^2}\frac{1}{q_H Q_H}\right)\left(k_\parallel^2 + \frac{2}{3}\frac{\delta_{em}m_e\nu_{em}(1+p)^2}{q_H Q_H}\right)\right]^{1/2} \tag{2.31}$$

For the small-scale inhomogeneities strongly stretched along the geomagnetic field $\mathbf{B}_0$ (for which $k_\perp/k_\parallel > \sqrt{Q_H q_H}$), the magnitude of this critical longitudinal field was estimated as $E_{kr\parallel} \approx 3 \cdot 10^{-4}$ mV/m [58–60]. Now analyzing the Pedersen component of ions' mobility that is easily obtain from Equation 2.28, and formulas presented in Section 1.1, we get $\mu_{Pi} = Ze[1 + p + q_H^2]/M_i\nu_{im}[(1 + p)^2 + q_H^2(1 + Q_H^2)]$. Comparing this with that obtained from the electron component of plasma introduced earlier, one can find that μ_{Pi} exceeds μ_{Pe} at altitudes above 100 km (see also Section 1.1). This effect leads to the heating of ions by the perpendicular ambient electric fields occurring in the ionosphere. In this case, if the vector $\mathbf{k}_\perp$ is oriented along the ambient electric field $\mathbf{E}_0$, the condition of development of the thermal instability has the following form [60]:

$$E_{0kr\perp} \geq \left[\frac{T_{i0}^2}{Z^2e^2}\left(1 + \frac{q_H Q_H k_\parallel^2 + k_\perp^2}{k_\perp^2}\right)\left(k_\perp^2 + \frac{2}{3\rho_{Hi}^2}\right)\right]^{1/2} \tag{2.32}$$

where, as in Section 1.1, ρ_{Hi} is the gyroradius of ions.

Moreover, at altitudes of the E layer of the ionosphere, the wind of neutral molecules and atoms orients them along the geomagnetic field $\mathbf{B}_0$ because of interactions with plasma ions, and causes additional polarization of plasma disturbances. In the sites of depletions (i.e., with the negative deviations of concentration, $\delta N < 0$), the polarization field increases the temperature of plasma. Finally, by action of thermodiffusion forces, plasma is pushed out from the depletion areas [64]. The threshold of values of the neutral wind, starting from where the increase of plane plasma waves (e.g., instabilities) is observed orthogonal to $\mathbf{B}_0$ with $k_\perp \leq 2$ m^{-1} at altitudes $h > 110$ km equaling 80–140 m/s [58–64].

In the foregoing discussion, thermal instabilities were described by using a pure hydrodynamic approach. Unfortunately, in the nonequilibrium, nonstationary, high-latitude ionosphere small-scale plasma disturbances (turbulences) can be generated, with perpendicular (to $\mathbf{B}_0$) dimensions less than, or at the same order of, the gyroradius of plasma ions. The hydrodynamic description is not valid for analyzing such plasma disturbances, and we need to use kinetic equations such as Equation 2.6 for distribution functions of plasma electrons and ions. The kinetic approach was used by many authors regarding the analysis of current instabilities [65,66], of ion-cyclotron instability in the E layer of the ionosphere for waves with both $k > \rho_{Hi}^{-1}$ and $k \le \rho_{Hi}^{-1}$ [67,68]. In the following text we discuss briefly such kinds of instabilities for which the hydrodynamic approach is not valid.

2.3.1.2 Kinetic Instability

As mentioned previously, when the drift velocities of plasma exceed the thermal velocities of plasma ions, i.e., $V_d > v_{Ti}$, reaction of interactions between the ions with neutrals leads to anisotropy of velocity of ions, and their distribution function becomes anisotropy [69]

$$f_i(v) = \frac{1}{(2\pi v_{Tm}^2)^{3/2}} I_0\left(\frac{v_\perp V_d}{v_{Tm}^2}\right) \exp\left\{-\frac{v_\parallel^2 + v_\perp^2 + V_d^2}{v_{Tm}^2}\right\} \tag{2.33}$$

We took into account here the following approximate relations [69,80]: $T_\perp / T_\parallel \approx 1 + V_d^2/2v_{Tm}^2$. In Equation 2.33 $I_0(\bullet)$ is the Bessel function of zero order. As shown in Refs. [69,80], the temperature anisotropy leads to the generation of so-called the *Post–Rozenbluth* plasma instability (defined by names of the corresponding researchers of [80]). The condition of creation of such a thermal instability is [80]

$$\int \frac{\partial f_i(v)}{\partial v_\perp} \frac{dv_\perp}{v_\perp} > 0 \tag{2.34}$$

Introducing Equation 2.33 into this inequality (Equation 2.34) gives

$$\left[\frac{V_d^2}{v_{Tm}^2} - 2\right] I_0\left(\frac{V_d^2}{4v_{Tm}^2}\right) > \frac{V_d^2}{v_{Tm}^2} I_2\left(\frac{V_d^2}{v_{Tm}^2}\right) \tag{2.35}$$

where $I_2(\bullet)$ is the Bessel function of second order. From the constraint (Equation 2.35) it follows that $V_d/v_{Tm} > 1.8$ [69]; i.e., the instability can be generated for drift velocities of $V_d \ge 1000$ m/s in the ionospheric plasma. This electrostatic instability has frequency ω_0 and increment γ of the order of low-hybrid frequency of ions Ω_{Hi} and a wave number of $k \propto \omega_0/V_d$; i.e., a wave length of $\lambda \sim 0.1$ m. Here, the ratio of $k_\perp/k_\parallel \approx 1/100$; i.e., the high-anisotropic plasma inhomogeneities can be created from the generation of Post–Rozenbluth plasma thermal instabilities. Despite the fact that two-stream instability was fully investigated by using simple [98–100] and rigorously consistent [101,102] kinetic theory, physical conditions of its generation are not the same as mentioned earlier. Therefore, we will consider the two-stream instability growth and damping separately. In the following paragraphs, we will analyze a kind of kinetic instability differing from two-stream instability.

2.3.1.3 Kinetic Drift-Dissipative Instability

If in the ionosphere the sharp small-scale gradients of plasma density are presented with the scale $l < 100$ m, then the drift-dissipative instability can be developed [81–83]. In this case the frequency of excited plasma waves ω_0 is in the order of low-hybrid frequency of ions, and the increment of such plasma instability γ_{CD} can be presented in a more complicated form [81–83]:

$$\gamma_{CD} = -\left(\frac{\omega_0^2}{kV_{di}}\right)\left[\frac{\sqrt{\pi}(\omega_0 - kV_{di})}{Y_0 kv_{Ti}} + \frac{T_i(\nu_{em} + \nu_{ei})(1 - Y_0)}{T_e\omega_0}\left(1 + \frac{T_e kV_{di}}{T_i\omega_0}\right)\right] \tag{2.36}$$

where $V_{di} = v_{Ti}^2/\omega_{Hi}l$ is the drift velocity of ions, $Y_0 = I_0\left(\frac{k^2\rho_{He}^2}{2}\right)\exp\left[-\frac{k^2\rho_{He}^2}{2}\right]$. This instability can be developed only if the following condition takes place:

$$\left|\frac{V_{di}}{v_{Ti}}\right| > \left[\frac{(\nu_{em} + \nu_{ei})}{\sqrt{\pi}\omega_{Hi}}\left(\frac{m_e}{M_i}\right)^{1/2}\frac{(1 - Y_0)}{k\rho_{He}}\frac{(2 + k^2\lambda_{Di}^2 - Y_0)(2 + k^2\lambda_{Di}^2)}{2(1 - Y_0) + k^2\lambda_{Di}^2}\right] \qquad (2.37)$$

where $\lambda_{Di}^2 = v_{Ti}^2/2\omega_{Pi}^2$. Such instability generates waves with a wavelength less than that of gyroradius of plasma ions. The threshold of such instability development defined by inequality (Equation 2.37) depends essentially on the plasma concentration because $l \sim N_0^{-1/3}$ and, for $N_0 = 10^4$ cm^{-3}, $l \approx 100$ m. Plasma irregularities of such scales with $|\delta N/N_0| \approx 1$ have often been observed in the F region of the auroral ionosphere [84].

2.3.2 Current-Induced Natural Plasma Instabilities

In the analysis of thermal instabilities presented in the previous section, we assumed that plasma oscillations are nonisothermal. Now we will put the question, what kinds of isothermal instabilities can be created in the real ionosphere, and what are the natural sources of such instabilities. As mentioned elsewhere [20,65,66], such low-frequency electrostatic instabilities, which can then be associated with small- and moderate-scale plasma perturbations, can be created by different types of currents and their convergence, as well as by gradient-drift mechanisms. To analyze these phenomena, we will use both the hydrodynamic and kinetic approaches. Let us start with a general presentation of the problem. We use the full system of linear Equations 2.25 through 2.27 obtained for weak plasma disturbances and small perturbations of velocities of plasma particles and ambient electric fields, which have a form of plane waves; i.e.,

$$\delta N, \delta\mathbf{E}, \delta\mathbf{V}_\alpha \sim \exp\{i(\mathbf{k r} - \omega t)\} \qquad (2.38)$$

These conditions are valid for ionospheric altitudes up to 250–300 km (see Chapter 1). Moreover, the components of tensor of mobility of electrons $\hat{\mu}_e$ and ions $\hat{\mu}_i$ do not depend on spatial coordinates. For linear analysis of current-based instabilities occurring in the ionospheric plasma, we use the coordinate system (x,y,z) with z-axis oriented along the geomagnetic field $\mathbf{B}_0$. We also suggest that the ambient electric field $\mathbf{E}_0$ and the gradient of plasma density $\kappa = \nabla \ln N_0$ lie at the plane (x,y). There is also the longitudinal current $j_{0\parallel}$ along z-axis (along $\mathbf{B}_0$), created by thermal electrons. From Equations 2.26b we can find that

$$\delta\mathbf{V}_e = e\tilde{\hat{\mu}}_e\delta\mathbf{E} - i\frac{D_e B}{q_H}\tilde{\hat{\mu}}_e\mathbf{k}\frac{\delta N}{N_0} \qquad (2.39a)$$

$$\delta\mathbf{V}_i = Ze\tilde{\hat{\mu}}_i\delta\mathbf{E} - i\frac{D_i B}{Q_H}\tilde{\hat{\mu}}_i\mathbf{k}\frac{\delta N}{N_0} \qquad (2.39b)$$

where the tensors of mobility $\tilde{\hat{\mu}}_\alpha$ of charged particles, electrons ($\alpha = e$), and ions ($\alpha = i$) can obtain from the corresponding tensors $\hat{\mu}_\alpha$, defined earlier, by changes of real frequencies of collision ν_α on complex frequencies; i.e.,

$$\tilde{\nu}_\alpha = \nu_\alpha - i(\omega - \mathbf{k}\cdot\mathbf{V}_{\alpha 0}) \qquad (2.40)$$

As we showed previously,

$$\left| \frac{\omega - \mathbf{k} \cdot \mathbf{V}_{e0}}{\nu_{em} + \nu_{ei}} \right| \ll \left| \frac{\omega - \mathbf{k} \cdot \mathbf{V}_{i0}}{\nu_{im} + \nu_{ei}} \right| \tag{2.41}$$

and in most cases we must assume that $\tilde{\hat{\mu}}_e \approx \hat{\mu}_e$, i.e., $\tilde{\nu}_e \approx \nu_e$. Elimination of the imaginary term in $\tilde{\nu}_i$, according to Equation 2.40, is equivalent to the rejection of the ion current (two-stream) instability in the general dispersion equation [20,58]. Taking into account that $\frac{\omega - \mathbf{k} \cdot \mathbf{V}_{i0}}{\nu_{im} + \nu_{ei}} \ll 1$, and from Refs. [67,68], we obtain:

$$\mathbf{k}\tilde{\hat{\mu}}_e\mathbf{k} = \mathbf{k}\hat{\mu}_e\mathbf{k} + i\frac{\omega - \mathbf{k} \cdot \mathbf{V}_{i0}}{\nu_{im} + \nu_{ei}}\left[\mu_{Pi}k_\perp^2 \frac{1 - Q_H}{1 + q_H} + \mu_{\|i}k_\|^2 \right] \tag{2.42}$$

where $l_\| \sim k_\|^{-1}$ and $l_\perp \sim k_\perp^{-1}$ are the longitudinal and transverse dimensions of plasma perturbation (with respect to geomagnetic field $\mathbf{B}_0$ lines). Introducing Equation 2.39 in Equation 2.26, rewritten for ions taking into account Equations 2.38 and 2.42, we can after some straightforward derivations obtain the dispersion equation (not presented here due to its complexity), the frequency of plasma density oscillations, ω_0, and the increment of plasma current-generated instability, γ. Thus,

$$\omega_0 = \frac{\mathbf{k}\mathbf{V}_{i0}\mathbf{k}\hat{\mu}_e\mathbf{k} + \mathbf{k}\mathbf{V}_{e0}\mathbf{k}\hat{\mu}_i\mathbf{k}}{\mathbf{k}(\hat{\mu}_e + \hat{\mu}_i)\mathbf{k}} \tag{2.43}$$

describes the dispersive properties of plasma waves created by current-generated instabilities in the real ionosphere. For plasma waves propagating close to orthogonal direction (to $\mathbf{B}_0$ lines), we get the following conditions:

$$\frac{\mathbf{k}\hat{\mu}_e\mathbf{k}}{\mathbf{k}(\hat{\mu}_e + \hat{\mu}_i)\mathbf{k}} \approx \frac{Q_H}{q_H} \ll 1 \quad \text{and} \quad \frac{\mathbf{k}\hat{\mu}_i\mathbf{k}}{\mathbf{k}(\hat{\mu}_e + \hat{\mu}_i)\mathbf{k}} \approx 1 \tag{2.44}$$

Such waves as will propagate with phase velocity equals approximately the projection of electron velocity at the direction of the vector $\mathbf{k}$ [20,65,66]. From Equation 2.43 it follows that the frequencies of electron and ion plasma component oscillations, $\tilde{\omega}_\alpha = \omega - kV_{\alpha0}$, $\alpha = e, i$, equal

$$\tilde{\omega}_e = kU_0 \frac{\mathbf{k}\hat{\mu}_e\mathbf{k}}{\mathbf{k}(\hat{\mu}_e + \hat{\mu}_i)\mathbf{k}} \tag{2.45a}$$

$$\tilde{\omega}_i = kU_0 \frac{\mathbf{k}\hat{\mu}_i\mathbf{k}}{\mathbf{k}(\hat{\mu}_e + \hat{\mu}_i)\mathbf{k}} \tag{2.45b}$$

where $U_0 = V_{i0} - V_{e0}$. From Equations 2.44 and 2.45, the validity of inequality 2.24 follows, and this finally differentiates between the two main approaches, the hydrodynamic and the kinetic.

If the plasma wave now propagates under some angle ϑ to the magnetic field $\mathbf{B}_0$, which is less than some critical angle, $\vartheta_{kr} = \tan^{-1}(\mu_\| / \mu_P)$, we get

$$\frac{\mathbf{k}\hat{\mu}_e\mathbf{k}}{\mathbf{k}(\hat{\mu}_e + \hat{\mu}_i)\mathbf{k}} \approx 1 \quad \text{and} \quad \frac{\mathbf{k}\hat{\mu}_i\mathbf{k}}{\mathbf{k}(\hat{\mu}_e + \hat{\mu}_i)\mathbf{k}} \approx \frac{Q_H}{q_H} \ll 1 \tag{2.46}$$

and the phase velocity of the wave propagation in the background plasma equals $V_{ph} \approx kV_{i0}/k$. In the E layer of the ionosphere, inequality 2.24 will be satisfied with good accuracy. As for F layer of the ionosphere, this inequality is not so correct. However, it can be shown that the contribution of the term, proportional to ω_e/ν_e, in the increment of the current-generated instability is smaller than the term, proportional to ω_i/ν_i.

As for the increment of the total current-generated instability, and following Refs. [20,58,65,66], we will differentiate all instabilities existing in its expression that can occur in the ionosphere:

$$\gamma = \frac{k_{\parallel} \mathbf{V}_{e0\parallel} \kappa \hat{\mu}_i \mathbf{k}}{\mathbf{k}(\hat{\mu}_e + \hat{\mu}_i)\mathbf{k}} + \frac{\hat{\mu}_e \mathbf{k} \kappa \hat{\mu}_i \mathbf{k} - \hat{\mu}_i \mathbf{k} \kappa \hat{\mu}_e \mathbf{k}}{\mathbf{k}(\hat{\mu}_e + \hat{\mu}_i)\mathbf{k}} \mathbf{k}(\hat{\mu}_e + \hat{\mu}_i)\mathbf{E}_0$$
$$- \frac{\tilde{\omega}_e \tilde{\omega}_i}{(\nu_{im} + \nu_{ei})\mathbf{k}(\hat{\mu}_e + \hat{\mu}_i)\mathbf{k}} \times \left[\mu_{Pi} k_{\perp}^2 \frac{1 - Q_H}{1 + q_H} + \mu_{\parallel i} k_{\parallel}^2 \right]$$
$$- \frac{\nabla \cdot \mathbf{V}_{i0} \mathbf{k} \hat{\mu}_e \mathbf{k} + \nabla \cdot \mathbf{V}_{e0} \mathbf{k} \hat{\mu}_i \mathbf{k}}{\mathbf{k}(\hat{\mu}_e + \hat{\mu}_i)\mathbf{k}} + \frac{D_0(\kappa \times \mathbf{k})^2}{Q_H \mathbf{k}(\hat{\mu}_e + \hat{\mu}_i)\mathbf{k}}$$
$$+ \frac{M_i}{Ze} \frac{\kappa \hat{\mu}_e \mathbf{k} \mathbf{k} \hat{\mu}_i \mathbf{k} \mathbf{k} \hat{\mu}_i \mathbf{g}}{\mathbf{k}(\hat{\mu}_e + \hat{\mu}_i)\mathbf{k}} - \frac{D_0 B_0 \mathbf{k} \hat{\mu}_e \mathbf{k} \mathbf{k} \hat{\mu}_i \mathbf{k}}{Q_H \mathbf{k}(\hat{\mu}_e + \hat{\mu}_i)\mathbf{k}} - \beta' \qquad (2.47)$$

Here, $\kappa = \nabla(\ln N_0)$ is the gradient of plasma density in the logarithm scale, $D_0 = C_s^2/(\nu_{im} + \nu_{ei})$, the ion coefficient of diffusion, and $C_s = \sqrt{(T_e + T_i)/M_i}$, the velocity of ion sound.

Equation 2.47 is more general for an increment (decrement) that describes generation (dissipations or damping) of the electrostatic low-frequency plasma waves occurring in the ionosphere. The *first term* in Equation 2.47 describes the contribution of convective-current instability, which can be created by longitudinal currents when gradients of plasma density are an essential feature in the ionosphere [88,107]. Such instabilities generate collisional ion-cyclotron waves in the ionospheric plasma [89,90]. The *second term* in Equation 2.47 describes effects of gradient-drift instability, which is usually generated when drift of ions and electrons in the ambient electric field occur in the inhomogeneous ionospheric plasma, parameters of which, as well as the transport coefficients, have complicated height dependence [91–97]. The *third term* in Equation 2.47 describes the contribution of two-stream instability, usually called Farley–Buneman instability [67,68,98–102]. During the generation of this kind of instability, both hydrodynamic and kinetic effects must be taken into account [98–102]. The *fourth term* takes into account redistribution of plasma density due to convergence of velocities of charged particles, which leads to negative divergence of electron and ion current [20,103,104]. The *fifth term* describes the existence of drift-dissipative instability in plasma, which also depends on the existence of an ambient electric field of sufficient magnitudes in the ionosphere [95,108–113]. The *sixth term* describes the development of the so-called Rayleigh–Taylor instability, occuring in the ionosphere in the presence of sharp and strong plasma density gradients [104,105,114–118]. The *last two terms* describe dissipation of plasma waves due to diffusion and recombination [58,61,64].

To all these electrostatic instabilities we can also add those having an electrodynamic nature, when drift in crossing electrical and magnetic fields is developed and, instead of electrostatic polarized electrical fields in the plasma, the rotated electrical field is observed [98,106]. Such instabilities are significant for plasma perturbations with scales more than 20–30 m. Results presented by numerous researchers of ionospheric plasma have shown that nonequilibrium processes occur in the ionospheric plasma leading to generation of low-frequency electrostatic and electrodynamic plasma waves. Their excitation is determined by the ambient electrical and magnetic field, and the velocity of their movement is determined by the existence of cumulative effects of several kinds of instabilities, which gives the same contribution to cumulative increment γ. Therefore, we will briefly discuss each of the previously mentioned instabilities in the following paragraphs.

2.3.2.1 The Current-Convective Instability

This plasma instability and the corresponding plasma waves occur in the ionospheric regions where the "run-in" and the "run-out" longitudinal currents (along $\mathbf{B}_0$) are observed with the existence of sharp gradients of the plasma density [107]. The increment of such kind of instability equals (here, as in the general case, the z-axis is elongated along $\mathbf{B}_0$):

$$\gamma_{CC} = \frac{k_{\parallel} \mathbf{V}_{e0\parallel} [\kappa \times \mathbf{k}]_z}{\mathbf{k}(\hat{\mu}_e + \hat{\mu}_i)\mathbf{k}} \approx \frac{\mathbf{V}_{e0\parallel} \kappa}{q_H} \frac{k_{\perp}}{k_{\parallel}} \equiv \frac{\mathbf{V}_{e0\parallel} \kappa}{q_H} \frac{l_{\parallel}}{l_{\perp}} \qquad (2.48)$$

This increment depends strongly on the intensity of the longitudinal current and on plasma density. Thus, for $j_{0\parallel} \approx 3 \cdot 10^{-6}$ A/m^2, $N_0 = 10^{11}$ m^{-3}, and, for the parameter of anisotropy $l_{\parallel}/l_{\perp} = \sqrt{q_H Q_H} > 10$, the increment $\gamma_{cc} = 10^{-3}$ s^{-1}. As mentioned elsewhere [104], in the turbulent ionospheric plasma at high (polar) latitudes, it is possible to generate plasma disturbances with smaller anisotropy than at the middle latitudes.

At the same time, in Ref. [88] it was shown that the longitudinal current with density $j_{0\parallel} > j_{0\parallel kr} \approx 3 \cdot 10^{-6}$ A/m^2 generates the ion-cyclotron waves (ICW) in the upper ionosphere. The possibility of generation of ICW at maximum altitudes of the F layer and below was investigated in a couple of studies [89,90]. Following one of these [90] and, based on linear Equations 2.26a through b, we obtain the dispersion equation for ICW:

$$\omega_{Hi}^2 - \omega^2 + C_s^2 k^2 + i \frac{m_e(\nu_{em} + \nu_{ei})}{M_i} \frac{k^2}{k_{\parallel}^2} (\omega - k_{\parallel} V_{e0\parallel}) = 0 \qquad (2.49)$$

which yields

$$\omega_0^2 = \omega_{Hi}^2 + C_s^2 k^2 \qquad (2.50a)$$

$$\gamma_{CC} = \frac{m_e(\nu_{em} + \nu_{ei})}{2M_i} \frac{k^2}{k_{\parallel}^2} \left(\frac{k_{\parallel} V_{e0\parallel}}{\omega_0} - 1 \right) \qquad (2.50b)$$

From Equation 2.50a, it follows that the wave frequency ω_0 is bigger than the gyrofrequency of ions; i.e., $\omega_0^2 \gg \omega_{Hi}^2$. As also clearly seen from Equation 2.50b, the increment of instability of ionospheric plasma is positive; i.e., $\gamma > 0$, and the phase velocity of waves, $V_{ph} = \omega_0/k_{\parallel}$, along the ambient magnetic field $\mathbf{B}_0$, is smaller than the electron velocity $V_{e0\parallel} = -j_{0\parallel}/eN_0$.

2.3.2.2 Gradient-Drift Instability

In the real ionosphere, for strongly elongated plasma structures where the longitudinal (along $\mathbf{B}_0$) scale of plasma structure exceeds the scale of height of neutral atmosphere H_m (see definitions in Chapter 1), the inhomogeneity of background plasma density $N_0(z)$ and its main parameters along the height of the ionosphere significantly influence the creation and further evolution of plasma structures of various origins. Finally, in the ionosphere there exist excited plasma waves, depending on the altitude of their generation, which in the literature are called *gradient-drift instabilities* (GDI) [91–97]. We also need to differentiate the effects of such instabilities generation at altitudes of the E region from those occurring at altitudes of the F region of the ionosphere. It was shown both theoretically and experimentally [91–97] that in the E layer of the ionosphere gradient-drift instability occurs when plasma density disturbances, δN, are converted to the direction of ∇N_0 by the perturbation drift with velocity $\delta V = \delta E \times B_0 / |B_0|^2$.

As shown in Chapter 1, the first-order (density and temperature), the second-order (collision frequencies between particles) and the third-order (transport and recombination coefficients) parameters of ionospheric plasma are functions of the ionospheric altitude h (usually denoted by the coordinate z). From general equations of continuity (Equation 2.3 or 2.26b) and taking into account (Equation 2.39), it follows that, for plasma disturbances with dimensions that exceeded the standard height H_m along B_0, i.e., for $l_\| > H_m \approx 10{-}50$ km ($k_\| > H_m^{-1} \approx 10^{-1}{-}2 \cdot 10^{-2}$ km^{-1}), and for $l_\perp \geq 10$ km ($k_\perp < 10^{-1}$ km^{-1}) across B_0, we get

$$l_\| / l_\perp \ll \sqrt{q_H Q_H} = 10{-}100 \tag{2.51}$$

The parameters of GDI can be obtained as [91–97]:

$$\omega_0 = \frac{\mathbf{k}\mathbf{V}_{i0}\mathbf{k}\hat{\mu}_e\mathbf{k} - \mathbf{k}\mathbf{V}_{e0}\mathbf{k}\hat{\mu}_i\mathbf{k}}{\mathbf{k}(\hat{\mu}_e + \hat{\mu}_i)\mathbf{k}} - \frac{D_i}{H_m}\frac{k_\|}{\mathbf{k}(\hat{\mu}_e + \hat{\mu}_i)\mathbf{k}} \tag{2.52a}$$

$$\gamma_{GD} = \frac{[\kappa \times \mathbf{k}]\left(\mathbf{k}\cdot\mathbf{U}_0 - \frac{D_\perp}{H_m}k_\|\right)}{|\mathbf{B}_0|\cdot\mathbf{k}(\hat{\mu}_e + \hat{\mu}_i)\mathbf{k}} - \frac{(\omega - \mathbf{k}\mathbf{V}_{e0})\mu_{i\|}k_\|}{H_m \cdot \mathbf{k}(\hat{\mu}_e + \hat{\mu}_i)\mathbf{k}}$$
$$- \frac{D_e Q_H}{H_m q_H}\frac{k_\|\mathbf{k}\hat{\mu}_e\mathbf{k}}{\mathbf{k}(\hat{\mu}_e + \hat{\mu}_i)\mathbf{k}} - \frac{D_0|\mathbf{B}_0|}{Q_H}\frac{\mathbf{k}\hat{\mu}_e\mathbf{k}\mathbf{k}\hat{\mu}_i\mathbf{k}}{\mathbf{k}(\hat{\mu}_e + \hat{\mu}_i)\mathbf{k}} \tag{2.52b}$$

For the waves, which Equation 2.51 satisfies, we have

$$\mathbf{k}(\hat{\mu}_e + \hat{\mu}_i)\mathbf{k} \approx q_H k_\|^2 / |\mathbf{B}_0| \tag{2.53}$$

Then, from Equation 2.52, we get

$$\omega' \equiv (\omega - \mathbf{k}\mathbf{V}_{\alpha 0}) \approx \mathbf{k}\mathbf{U}_0 - \frac{D_i}{H_m}k_\| \tag{2.54}$$

where, again, $\mathbf{U}_0 = \mathbf{V}_{e0} - \mathbf{V}_{i0}$ and D_0 is the coefficient of ion diffusion as defined earlier; all other parameters have also been introduced previously.

Taking into account relationships 2.53 and 2.54, we can rewrite Equation 2.52b as

$$\gamma_{GD} = \frac{Q_H}{q_H H_m}\frac{\mathbf{k}\mathbf{U}_0}{k_\|} + \frac{D_i\kappa}{q_H H_m}\frac{k_\perp}{k_\|} + \frac{\kappa|\mathbf{U}_0|}{q_H}\frac{k_\perp^2}{k_\|^2}$$
$$- D_0 k_\|^2 \left(1 - \frac{T_e}{T_e + T_i}\frac{1}{H_m k_\|} - \frac{T_e}{T_e + T_i}\frac{Q_H}{q_H}\frac{1}{H_m^2 k_\|^2}\right) \tag{2.55}$$

The solution for the increment of instability creation, described by Equation 2.55, was obtained in a study [91] using of the so-called *local* approximation. The last term in Equation 2.55 describes the relaxation of diffusion spreading of plasma inhomogeneities for waves with arbitrary direction of longitudinal wave vector component $k_{||}$. At the same time, from expression 2.55 it follows that, for plasma waves propagating downward ($k_{||} < 0$) in the ionosphere, inhomogeneity of the background ionospheric plasma increases the velocity of growth of plasma waves and decreases their attenuation due to the diffusion process ($\sim D_i$). For plasma waves propagating upward ($k_{||} > 0$), the effect of altitudinal dependence of parameters of plasma is the opposite: their dependence on the height decreases the velocity of growth of plasma waves and increases their attenuation due to diffusion process. This occurs because of changing of waves' phase velocity, $V_{ph} \approx V_{i0} = V_d + U_0$, with altitude z. At the lower altitudes of the ionosphere, the phase velocity is greater and plasma waves grow faster, and their diffusion relaxation is slower. From Equation 2.55 it can also be seen that dissipation of waves due to the diffusion process is minimum when $k_{||} \rightarrow H_m^{-1}$ ($l_{||} \rightarrow H_m$). However, for plasma disturbances with $l_{||} \approx H_m$, the application of local approximation becomes incorrect.

In a study [92], for a description of GDI-induced plasma perturbations the *quasilocal* approach was used. It has been suggested that plasma density perturbation, δN, has a form of plane wave with the wave vector $\mathbf{k}_\perp$ orthogonal to $\mathbf{B}_0$ and with amplitude $N_\mathbf{k}$ depending on the height coordinate z, which again was directed along $\mathbf{B}_0$; i.e.,

$$\delta N = N_\mathbf{k}(z) \exp(i\mathbf{k}_\perp \mathbf{r} - i\omega t) \tag{2.56}$$

In the same study [92] it was shown that along $\mathbf{B}_0$ the diffusion is unipolar ($\mathbf{j}_e = \mathbf{j}_i$), whereas across $\mathbf{B}_0$ diffusion is ambipolar ($\nabla \mathbf{j}_e = \nabla \mathbf{j}_i = \nabla \mathbf{j}$). We use here the unified definitions introduced in Chapter 1, in which diffusion spreading of plasma disturbances was analyzed. Following the analysis [92], we suggest that $\kappa = \nabla \ln N_0(z) \sim 10^{-1}$ km^{-1} and the large-scale gradients of plasma density become weaker with height at distances of order of $H_m \approx 30-50$ km. Based on assumptions made in the study [92], we can present the equation for the amplitude of plasma waves generated due to GDI, in the following form:

$$\frac{\partial N_\mathbf{k}}{\partial t} + (\beta - \gamma_k)N_\mathbf{k} - \frac{\partial}{\partial z}\left(D_0 \frac{\partial N_\mathbf{k}}{\partial z} + \frac{|\mathbf{g}|}{\nu_{im}} N_\mathbf{k}\right) \tag{2.57}$$

where, as earlier, $|\mathbf{g}| = 9.81$ m/s^2, $D_0 = C_s^2/(\nu_{im} + \nu_{ei})$ is the coefficient of ion diffusion, the increment of plasma waves can be presented as [92]

$$\gamma_\mathbf{k} = \frac{\mathbf{k} \cdot \mathbf{E}_0 [\kappa \times \mathbf{k}_\perp]_z}{k_\perp^2} - \frac{C_s^2 k_\perp^2}{\Omega_H} \tag{2.58}$$

Equation 2.57 was obtained with the assumption that in the real high-latitude ionosphere $k_\perp < 10^{-1}$m^{-1} and the second (negative) term in Equation 2.58 is small enough. Then, in the first-order approximation it can be assumed that γ_k does not depend on the coordinate z. In Refs. [93,94] the same analysis, but for a simpler case of $\gamma_k = 0$, was carried out. We will not enter into mathematical solutions of Equation 2.57 obtained in Ref. [92], but will only point out that since in the high-latitude ionosphere $\gamma_k^{-1} \approx 10^2-10^3$ s, the height profile of growth plasma waves differs from that of electron concentration of regular ionospheric plasma.

The preceding approximations were carried out for local stability analysis, and the corresponding eigenmodes of GDI with the complex eigenfrequency $\omega = \omega_r + j\gamma$ were derived at the local position $\mathbf{r}_0$, using local approximation, which requires that $|kL| \gg 1$. Here, $L = N_0(\partial N_0/\partial z)^{-1}$ is the scale length of the plasma disturbance in the steady-state (equilibrium) regime, and k is the wave number of the perturbed plasma wave (e.g., GDI). In Ref. [93], the GDI was analyzed using nonlocal linear approximation for the description of type II irregularities associated with such instability, which were observed experimentally in the E and F equatorial ionospheric regions (details in Chapter 6). The authors of Ref. [93] in their study took into account the altitudinal dependence of ion–neutral (ν_{im}) and electron–neutral (ν_{em}) frequencies, as well as the altitudinal dependence of mobilities of plasma electrons (μ_e) and ions (μ_i). Then, the global eigenmodes of the GDI were investigated using realistic profiles of the Pedersen (σ_P) and Hall (σ_H) conductivities of plasma with the framework of a physical model of the daytime equatorial electrojet.

Considering the complexity of the expressions for oscillation frequency ω_r and growth rate γ, they are not presented here; we only present a simple qualitative analysis of the phenomena under consideration. The corresponding computations of these two main parameters of the GDI have shown that the growth rate parameter has sharp diffusional damping at large wave numbers (where the horizontal component of wave vector $k_{\parallel} > 0.2$ m^{-1}), i.e., for the wave modes with wavelengths less than about 30 m. Its preferential range of waves excitation varies from 60 to 600 m, which corresponds to $0.01 < k_{\parallel} < 0.1$ m^{-1}. Its negative values described the damping process of unstable waves lying at the range of wavelengths larger than 3 km (i.e., for modes with $k_{\parallel} < 2 \cdot 10^{-3}$ m^{-1}). The kilometer wave structures, similar to most unstable modes, were often observed as wave packets composed of a few, or several, oscillations. Presenting such GDI as wave packets, it was found in a study [93] that the localization and main characteristics of plasma modes are not sensitive to details of the steady-state plasma density profile $N_0(z)$. What is also important to note is that the nonlocal growth rates obtained for plasma disturbances with a wavelengths of 1 km (or a few kilometers) are still smaller than those for wavelengths of the order of tens or hundreds of meters. This effect can explain the predominance of long wavelengths in the electrojet spectrum observed in the study mentioned [93].

In another study [94], the analysis of nonlocal approximation mentioned previously was continued, and it was shown that many features observed experimentally can find their physical explanation only in the framework of full nonlocal nonlinear equations. As the task of evaluating nonlinear equations nowadays is a complicated issue, it was proposed in the study [94] that the following approach gives a good explanation for the main features of the large number of experimental observations. It is based on introduction into the formulation of the long-wavelength problem of the additional electron mobility and diffusion caused by short-wavelength turbulence occurring in the plasma background. Introducing such a procedure, the authors retained all the relevant nonlocal effects and analyzed analytically and numerically the linearized version of the renormalized large-scale equations. The corresponding oscillatory frequency and increment of GTI excitation and growth were derived for wavelengths typically of the order of 1 km, i.e., for wave numbers of the order of $k_0 = (\sigma_H/\sigma_P) \cdot L^{-1}$, through the Hall $\sigma_H = eN_0(\mu_{iH} - \mu_{eH})$ and Pedersen $\sigma_P = eN_0(\mu_{iP} - \mu_{eP})$ conductivities of plasma, the vertical component (i.e., along z-axis) of the total field, E_z, and the background (equilibrium) plasma density $N_0(z)$:

$$\omega_r = -\mu_{eH} E_z k_{\parallel} \left[1 + \frac{\mu_{eP}}{\mu_{iP} - \mu_{eP}} \frac{1 + k_0 \mu_{eH}/\mu_{eP} L k_{\parallel}^2}{1 + (k_0/k_{\parallel})^2} \right] \tag{2.59a}$$

$$\gamma = -\frac{\mu_{iH} - \mu_{eH}}{(\mu_{iP} - \mu_{eP})(1 + \varsigma)} \frac{\mu_{eH} E_z}{L\left[1 + (k_0/k_\parallel)^2\right]} - 2\alpha N_0 \tag{2.59b}$$

Here, as earlier, $\varsigma = 1/q_H Q_H$, and α is the coefficient of the radiative recombination (see Chapter 1, in which all additional parameters presented in Equation 2.59 are also defined). The long-wavelength unstable modes described by Equation 2.59 appear to be dispersive. As the quadratic term is presented in the denominator of Equation 2.59b, the increment of plasma modes growth decreases as the magnitude of $k_\parallel$ becomes comparable to or smaller than k_0. Finally, at even longer wavelengths, all the modes will eventually be stabilized due to the effects of recombination. Additional analysis carried out in the study [94] has shown that the dispersive nature of the kilometer-wavelength plasma modes depends critically on the value of k_0, i.e., on the scale length of the equilibrium plasma density gradient, L, and the ratio of the Hall and Pedersen conductivities. The authors of Ref. [94] have finally obtained the same principal result as was obtained in Ref. [93] that the localization and qualitative properties of the unstable modes do not depend on the type of height profile of the background ionospheric plasma $N_0(z)$ or on its derivative $\partial N_0(z)/\partial z$. Finally, we have to point out that because the long-wavelength nonlocal modes in the framework of the linear approach demonstrate growth rates monotonically decreasing with $k_\parallel$ and are smaller in magnitude than those of the moderate- or small-wavelength modes (which is in contradiction with long-wavelength experimental observations), the nonlinear approach has to be used to explain all features observed experimentally. This will be done in the next section by introducing into the framework of nonlinear theory the anomalous (turbulent) mobility and diffusiuon coefficients in the nonlinear transport equations of plasma species, electrons, and ions.

2.3.2.3 Two-Stream Instability

The physical concept of how such instabilities are generated in the ionosphere is based on the existence of high relative velocities of electrons and ions, i.e., strong currents (even in the homogeneous plasma) caused by strong ambient electric field or winds of neutral molecules (atoms). In the ionospheric plasma, under a strong ambient electric field of $E_0 \geq 30-50$ mV/m, a two-stream instability is mostly created in the high-latitude ionosphere, in the perturbed middle-latitude ionosphere, and in the equatorial electrojet [67,68,98–102]. Such an instability is the main source of plasma disturbance generation in the ionosphere (associated with irregularities of type I; see Chapter 6), and we will analyze it more consistently compared with other kinds of ionospheric instabilities. The two-stream instabilities are usually called the Farley–Buneman (FB) instabilities because Buneman first analyzed the conditions of current instability formation in the collisionless adiabatic plasma in the absence of an ambient magnetic field [99] and, then, Farley found the criteria for formation of such current instabilities in region E of the equatorial and polar ionosphere [71,98], assuming the existence of magnetized (where electrons are magnetized and the ions are not) and isothermal plasma ($T_e \approx T_i = T$). In another study [100], the two-stream instability was analyzed based on the fluid theory extending Buneman's model by considering both electron and ion waves in the dispersion equation, i.e., taking into account different temperatures for electrons and ions. According to all these criteria, in the isothermal plasma ($T_e \approx T_i = T$) the appearance of two-stream instability (defined also as *ion-acoustic wave*) occurs if the relative electron stream (compared with the ion stream) exceeds the ion-sound velocity, $C_s = \sqrt{T/M_i}$, and, in the nonisothermal plasma ($T_e \neq T_i$), it exceeds the thermal velocity of ions v_{T_i} [71,98–100]. Both authors, Buneman and Farley, analyzed stability and instability of low-frequency electromagnetic wave perturbations ($\varpi \ll \nu_{im} < \nu_{em}$) for the case of magnetized electrons ($q_H \gg 1$) and weakly magnetized ions ($Q_H \ll 1$), i.e., for conditions in the E layer of the ionosphere (see notations in Section 1.1).

Before entering into the problem using Equations 2.43 and 2.47, we will outline briefly the main aspects of pioneer theories described in Refs. [98–100]. Thus, Farley, using Boltzman's kinetic equation and Maxwell's electromagnetic equation, has analyzed weakly ionized (when $\nu_{ei} \ll \nu_{im} < \nu_{em}$) and magnetized ($\mathbf{B}_0 \neq 0$) plasma instability excitation, where plasma perturbations are strongly stretched along $\mathbf{B}_0$, taking into account the conditions of ionospheric plasma at altitudes in region E of the ionosphere and assuming isothermality of plasma ($T_e \approx T_i = T$). Despite the fact that Farley used isothermal approximation, his theory treats the dynamics of both the electron and ions. As for Buneman, he used in his investigation the hydrodynamic approach based on the Navier–Stokes (fluid) equations for electrons and ions and analyzed the nonizothermic case, where $T_e > T_i$. Despite all the differences in the two approaches, the results obtained were roughly the same. Some additional coefficients obtained by Buneman in dispersive equations (in the order of ~5/3), compared with coefficients derived by Farley, are significant in relation to the Buneman's statement that the plasma electron perturbations are adiabatic in contrast to the isothermal perturbations of plasma ions considered by Farley. It should be noted that Buneman's adiabatic approximation for plasma electrons is usually true when high-collision frequency approximation is valid, because for the plasma isothermality effect about ~10^3 collisions between particles are required to obtain a significant exchange between plasma electrons and neutral molecules (atoms) in region E of the ionosphere. Furthermore, all computations made by Farley [98] were based on the simplified kinetic approach, in which the Boltzmann collisional integral in the kinetic equations was derived using the Maxwellian distribution function for plasma electrons. In Buneman's simplified hydrodynamic approach, the behavior of the ion component of plasma was described in terms of a simple fluid model for the same ambient conditions as in Farley's model. Therefore, both approaches gave close results in evaluating the threshold value of an ambient electric field for two-stream instability excitation, and both authors obtained roughly the same results in the theoretical analysis of two-stream (i.e., for the relative electron-ion current) instability, now called in the literature Farley–Buneman (FB) instability.

In the analysis of FB instability, we follow Equation 2.43 for the frequency, $\omega_0 \equiv \omega_{FB}$, of such plasma instabilities and Equation 2.47 for their increment of growth, $\gamma \equiv \gamma_{FB}$, assuming, as did Farley and Buneman, that plasma disturbances strongly stretched along $\mathbf{B}_0$, that is, $l_{\parallel}/l_{\perp} > \sqrt{q_H Q_H} \approx 10{-}100$. In this case Equation 2.43 yields

$$\omega_{FB} = \frac{q_H Q_H}{1 + q_H Q_H} \mathbf{k} \cdot \mathbf{V}_d \tag{2.60a}$$

and Equation 2.47 is reduced to

$$\gamma_{FB} = \frac{1}{\nu_{im}(1 + q_H Q_H)} \left[\omega_0^2 - k^2 C_s^2 \right] \tag{2.60b}$$

where $\mathbf{V}_d$ is the drift velocity of plasma electrons; all other parameters are as defined and introduced in Chapter 1.

From Equation 2.60b it follows that with an increase in the wave number k of plasma waves the increment of FB instability increases proportionally to k^2. This occurs only for small values of k, because for large enough k values the hydrodynamic approach is not valid for the description of plasma instabilities, and we need to use the kinetic approach. According to Refs. [65,100], Equation 2.60b for the increment of two-stream instability can be modified as

$$\gamma_{FB} = \frac{1}{\nu_{im}(1 + q_H Q_H)} \left[\omega_0^2 \left(1 - \frac{6k^2 v_{Ti}^2}{\nu_{im}^2} \right) - k^2 C_s^2 \right] \tag{2.61}$$

where v_{Ti} is the magnitude of thermal velocity of plasma ions. For wave vectors **k**, oriented along the drift velocity $\mathbf{V}_d$, Equation 2.61 is maximized for

$$k_{max} = \frac{\nu_{im}}{v_{Ti}|\mathbf{V}_d|} \left[\frac{1}{12} \left(V_d^2 - C_d^2 (1 + 1/q_H Q_H)^2 \right) \right]^{1/2} \tag{2.62}$$

and tends to zero as $k \to 0$ and $k \approx \nu_{im}/2.45 v_{Ti}$. From Equation 2.62 it also follows that the maximum of the increment shifts to the range of large wave numbers with an increase of drift velocity, $\mathbf{V}_d$, of plasma electrons. It is important to point out that as shown by many researchers dealing with FB instability [67,68,98–100], the hydrodynamic and kinetic descriptions of this kind of instability give the same result only for $0 < k < k_{max}$. With an increase in the ambient electric field E_0, the drift velocity increases and becomes predominant (compared with v_{Ti}), and k_{max} also increases, broadening the applicability of the hydrodynamic approach.

In Ref. [12] it was shown that, observed experimentally, small-anisotropic meteor-induced plasma inhomogeneities in the middle-latitude ionosphere can be explained by the generation of FB instabilities. For such meteor-induced inhomogeneities with a relatively small degree of anisotropy ($l_{\parallel}/l_{\perp} \ll \sqrt{q_H Q_H} \approx 10$) we can show that

$$\omega_{FB} = \frac{\mathbf{k} \cdot \mathbf{V}_d}{\frac{(1 + q_H Q_H)}{q_H Q_H} + \frac{q_H k_{\parallel}^2}{Q_H k_{\perp}^2}} \tag{2.63}$$

That is, with a decrease in anisotropy of plasma disturbances, the frequency, ω_0, and the phase velocity, $V_{ph} = \omega_0/k$, of plasma waves decrease correspondingly, according to standard models [98,99]. For conditions $V_d < C_s$, two-stream instability is not generated in the homogeneous plasma.

Furthermore, with the existence of altitudinal dependence of the parameters of plasma (i.e., in the inhomogeneous ionosphere) and of gradients of plasma density, $\kappa = \nabla \ln N_0$, orthogonal to ambient magnetic field $\mathbf{B}_0$ and parallel to ambient large-scale electric field $\mathbf{E}_0$, plasma waves can grow due to the joint effect of gradient-drift and two-stream instability. The frequency of these waves will be described by the same equations, either Equation 2.60a or 2.63. The increment of such complex GDI-FB instability can be presented as [58,65]:

$$\gamma_{\Sigma} = \frac{1}{\nu_{im}(1 + q_H Q_H)} \left[\omega_0^2 + \frac{\omega_H \nu_{im}}{\nu_{em}(1 + p)} \frac{|\kappa \cdot \mathbf{E}_0|}{|\mathbf{B}_0|} - C_s^2 k^2 \right] \tag{2.64}$$

where, $p = \nu_{ei}/\nu_{em}$ is the parameter of ionization introduced in Chapter 1. For the value of drift velocity $V_d \approx E_0/B_0 = 200$ m/s and, in the case of nonisothermal plasma, $C_s = [(T_e + T_i)/M_i]^{1/2} \approx 350$ m/s, and for the wave number $k \approx 1$ m^{-1} [20,58], taking into account the parameters of plasma at the considered altitudes of $\nu_{im} = 2 \cdot 10^3$ s^{-1}, $\nu_{em}(1 + p) \approx 1.6 \cdot 10^5$ s^{-1}, $\Omega_H = 2 \cdot 10^2$ s^{-1}, $\omega_H = 10^7$ s^{-1} (see Chapter 1), plasma instability can be developed if $\kappa > 3.5 \cdot 10^{-3}$ m^{-1}. The horizontal gradients of such plasma structures were observed in the polar ionosphere [108].

However, as shown by Dimant and Sudan [101,102], the standard approach mentioned previously, called the Farley–Buneman standard model, cannot explain many features observed experimentally in the E layer of the polar and equatorial ionosphere. As pointed out by the authors [101,102], this can be done only by using consistent kinetic theory for the description of low-frequency ($\omega \approx kV_d \ll \nu_{em}$) cross-field ($\mathbf{E}_0 \times \mathbf{B}_0$) two-stream instability in collisional weakly ionized plasmas, dominated in the E layer of the ionosphere. In Ref. [101] the threshold of FB instability excitation was obtained more precisely using rigorous kinetic theory only concerning the *linear stage* of growth of two-stream instabilities based on a short-wave asymptotic limit, i.e., analyzing plasma waves with wavelengths less than approximately 10 m. At the same time, it was shown [101,102] that for the long-wave limit (i.e., for unstable plasma waves with wavelengths varying from tens to hundreds of meters) the perturbed electron distribution function is close to the Maxwellian, so that rigorous kinetic theory fully covers, in this long-wave range the simplified standard Farley kinetic model. Generally speaking, some essential corrections were effected in the theory of temporal evolution of FB instabilities in Refs. [101,102] based on a self-consistent kinetic theory developed [13,14], where it was shown that the parameters of plasma can change dramatically because the electron distribution functions are essentially subject to non-Maxwellian perturbations, which can only be described on the basis of rigorous theory. Thus, using the consistent kinetic theory developed in Refs. [101,102], a significant increase (up to 100%) of the threshold value of the ambient electric field $\mathbf{E}_0$, needed for excitation of plasma instabilities, was obtained with respect to the results of the standard FB theory. This is only the first correction effected from the consistent theory, which is valid when plasma waves propagate in directions close to those of electron drift velocity $\mathbf{V}_d$. Another correction relates to excitation of waves (instabilities) propagating in directions near the bisector between the direction of drift velocity $\mathbf{V}_d$ and the ambient electric field $\mathbf{E}_0$. Furthermore, according to rigorous kinetic theory, the excitation of combined FB–GD long-wave instabilities occurs in the direction near the bisector between the directions of the ambient electric field $\mathbf{E}_0$ and the $\mathbf{E}_0 \times \mathbf{B}_0$ drift, and this new, exciting mechanism is related to the perturbed electron current caused by the Pedersen conductivity discussed previously. All these features cannot be explained on the basis of the standard FB theory framework.

We are not entering into complicated mathematical descriptions of the problem and the results obtained using the consistent kinetic theory because this subject is beyond the scope of our book. It is more important to give the reader a qualitative explanation of the physical mechanism of the generation of FB instability and show under what ambient conditions of the ionospheric plasma such instability can be excited or damped. First, it was shown in the studies mentioned [101,102] that the growth of FB instability occurs when the following restriction is valid:

$$|\mathbf{V}_d| \cdot \cos \vartheta > C_s \left[1 + \frac{1}{1 + \cos \theta} \Psi(\theta) \right] \tag{2.65}$$

where

$$\cos \theta = \frac{k_\parallel}{k} = \frac{k_\parallel}{\sqrt{k_\parallel^2 + k_\perp^2}}, \quad \cos \vartheta = \frac{\mathbf{k}_\perp \cdot \mathbf{V}_d}{k_\perp V_d} \tag{2.66a}$$

and

$$\Psi(\theta) = \frac{1}{q_H Q_H} \left(1 + q_H^2 \tan^{-1}(\theta) \right) \tag{2.66b}$$

Here, $k_{\parallel}$ and $k_{\perp}$ are the wave numbers parallel and perpendicular to $\mathbf{B}_0$; other parameters are defined in Chapter 1. Note that, under conditions $\omega_r \approx kV_d \ll \nu_{em}$, the condition $|\gamma| \ll \omega_r$ in equation $\omega = \omega_r + j\gamma$, which is related to the standard FB theory, is automatically satisfied with $\omega_r \equiv \omega_{FB}$, as defined by Equation 2.60a.

When analyzing the total electron and ion flux densities it was found in a study [102] that the intrinsic polarization field occurring between moving electrons and ions consists of two different parts, i.e.,

$$\delta \mathbf{E} = \delta \mathbf{E}_1 + \delta \mathbf{E}_2 \tag{2.67}$$

The greater part, $\delta \mathbf{E}_1$ ($\delta \mathbf{E}_1 > \delta \mathbf{E}_2$), which determines the phase velocity of the plasma wave (instability) is responsible for acceleration and deceleration of electron and ion streams necessary for the constancy of flux densities associated with large velocities of the two fluids in the electron-ion comoving coordinate system (see corresponding Equations 2.13 and 2.14 in Ref. [102]). The smaller part, $\delta \mathbf{E}_2$, provides the equality of small diffusive fluxes (defined as the ambipolar diffusion) caused by the pressure gradients. Rejecting this term in the original equations, the authors [101,102] have found that $\delta \mathbf{E}_1$ is related to the electron and ion Pedersen current associated with the corresponding Pedersen conductivity of charged particles. Then, the corresponding phase velocity can simply be associated with the drift velocity which we present in our unified notation as

$$\mathbf{V}_{ph} \approx \frac{q_H Q_H}{1 + q_H Q_H} \mathbf{V}_d \tag{2.68}$$

and the corresponding frequency of plasma modes oscillations is

$$\omega_r \approx \frac{1}{1 + \Psi(\theta)} \mathbf{k} \cdot \mathbf{V}_d \tag{2.69}$$

It can be seen that in this case the plasma waves propagate strictly along the electron drift velocity $\mathbf{V}_d$, because the components perpendicular to the wave vector $\mathbf{k}$ do not affect the density perturbations. It was shown in Refs. [101,102] that if the vector $\mathbf{k}$ has even a small component along the magnetic field lines (along $\mathbf{B}_0$) the situation becomes more complicated and the term $\delta \mathbf{E}_2$, which is associated with ambipolar diffusion, leads to additional "pressure" associated with the ion's kinetic energy $\sim M_i \nu_{iT}^2 / 2$. Because the ion flux velocity is maximum at the sites of minimum plasma density (and, conversely, minimum where the perturbations of plasma density are maximum), the gradient of the additional pressure is directed oppositely to the gradient of pressure of plasma associated with the thermal (chaotic) motion of plasma ions with velocity ν_{iT}. This fact, shown in Refs. [101,102], stimulates excitation of two-stream instability with a growth rate of

$$\gamma \approx \frac{\Psi(\theta)}{1 + \Psi(\theta)} \frac{\omega_r^2 - k^2 C_s^2}{\nu_{im}} \tag{2.70}$$

At the same time, and as shown in Ref. [102], the situation is not so simple even in the framework of the consistent kinetic theory developed in Ref. [101]. Thus, with regard to Equation 2.70, the term in the numerator $\sim k^2 C_s^2$, which is responsible for chaotic thermal motion, undergoes a change. It was shown that because the electron velocity distribution function becomes non-Maxwellian, the electron Predersen conductivity in the ambient (nonperturbed) electric field $\mathbf{E}_0$

becomes sufficient for the generation of moderate and long (more than 10 m) plasma waves (instabilities) excited along the bisector between the perpendicular directions of the electric field $\mathbf{E}_0$ and an unperturbed electron drift velocity $\mathbf{V}_d \sim \mathbf{E}_0 \times \mathbf{B}_0$. These are the principal results observed experimentally, which do not have any physical explanation when the standard FB model is used. Moreover, the authors [101,102] investigated the regions in the frequency domain in which such instabilities can be generated and, conversely, in which they have a tendency to damp. Briefly speaking, when the frequency shift $\Gamma_{\omega,\mathbf{k}} \equiv |\Delta\omega_{\omega,\mathbf{k}}|$ and the frequencies of electron–electron collisions are smaller than those for electron–neutral collisions, i.e., for

$$\Gamma_{\omega,\mathbf{k}}, \, \nu_{ee} < \delta_{em}\nu_{em} \tag{2.71}$$

the situation is entirely Maxwellian and leads to isothermality of plasma, and thus we return to the simple Farley–Buneman model of growth of instabilities in the isothermal plasma. As shown in Ref. [102], this situation is valid only for the altitudes of the nocturnal ionospheric region E. However, the actual velocity dependence of the electron–neutral collisions shown in Refs. [13,14] can dramatically change the conditions of FB instability growth and excitation. Following Refs. [13,14], it was shown [101,102] that when the constraint

$$\Gamma_{\omega,\mathbf{k}} < \delta_{em}\nu_{em} < \nu_{ee} \tag{2.72}$$

or, especially,

$$\delta_{em}\nu_{em} < \Gamma_{\omega,\mathbf{k}} < \nu_{ee} \tag{2.73}$$

is valid, the electron temperature T_e undergoes noticeable plasma wave perturbations, similar to electron density waves. In such conditions, distributions of plasma species are far from the classical Maxwellian, and all additional features found in the strict consistent analysis made in the study mentioned [101,102] now have a clear physical and distinct explanation.

Thus, in the consistent linear kinetic theory framework, Dimant and Sudan [101,102] analyzed precisely the conditions of excitation and damping of two-stream instabilities in the ionosphere responsible for plasma irregularity generation in the E layer of auroral and equatorial zones where ambient electromagnetic fields are sufficiently strong to create plasma perturbations (turbulences).

2.3.2.4 Convergence Instability

The convergence or negative divergence of velocities of charged particles in the ionospheric plasma leads to the redistribution of plasma density and to the growth of amplitude of plasma waves [20,103]. The increment of convergent instability (CI) can be found as follows [103]:

$$\gamma_{CI} = \frac{\nabla \cdot \mathbf{V}_{i0}\mathbf{k}\hat{\mu}_e\mathbf{k} + \nabla \cdot \mathbf{V}_{e0}\mathbf{k}\hat{\mu}_i\mathbf{k}}{\mathbf{k}(\hat{\mu}_e + \hat{\mu}_i)\mathbf{k}} \tag{2.74}$$

For strongly anisotropic plasma perturbations, when $l_\parallel/l_\perp \geq \sqrt{Q_H q_H}$,

$$\gamma_{CI} = -\left(\nabla \cdot \mathbf{V}_{i0}\frac{Q_H}{q_H} + \nabla \cdot \mathbf{V}_{e0}\right) \tag{2.75a}$$

and for weakly anisotropic plasma perturbations, when $l_{\parallel}/l_{\perp} \ll Q_H/q_H$

$$\gamma_{CI} = -\left(\nabla \cdot \mathbf{V}_{e0}\frac{Q_H}{q_H} + \nabla \cdot \mathbf{V}_{i0}\right) \tag{2.75b}$$

Such an increment depends weakly on the anisotropy of plasma disturbances. It was shown in Ref. [103] that the CI can be related to changes in the frequency of ion–neutral collisions and, for $\nu_{im} \approx 0.3$ s^{-1} and $H_m \approx 30$ km, $\nabla \cdot \mathbf{V}_{i0} \approx g/\nu_{im}H_m \approx 10^{-3}$ s^{-1}. Convergence of $\mathbf{V}_{e0}$ can also be related to the height dependence of velocity of thermal electrons transferring the longitudinal current. It follows from the relationship $\mathbf{V}_{e0\parallel} = -j_{0\parallel}/eN_0$ that for $j_{0\parallel} = const$ [58]:

$$\nabla \cdot \mathbf{V}_{e0\parallel} = -\mathbf{V}_{e0\parallel}\kappa_{\parallel} \tag{2.76}$$

It can be seen from Equation 2.76 that in regions of run-in longitudinal currents, the divergence $\nabla \cdot \mathbf{V}_{e0\parallel}$ will be negative above the maximum of F layer but in the region of run-out longitudinal currents it will be negative below this maximum. In Refs. [58,103] an estimation is presented of CI for $j_{0\parallel} = 10^{-6}$ A/m^2, $N_0 = 10^{10}$ m^{-3}, and $H_p = 10^2$ km, which equals $\nabla \cdot \mathbf{V}_{e0\parallel} \approx 6 \cdot 10^{-3}$ s^{-1}. Accordingly, a more convenient scenario for development of convergence instability is the existence of longitudinal currents in regions with lower plasma density.

2.3.2.5 Dissipative Instability

The last two negative terms in Equation 2.47 for the total increment of instabilities occurring in the ionosphere represent the attenuation of plasma waves due to the recombination process and diffusion. As numerous researchers [20,58,65] have shown, at altitudes in the range of 200–450 km, where $Q_H \geq 10^3$ and $q_H \geq 10^4$ (see Chapter 1), as a rule, the coefficient of recombination is less than $\beta_{kr} \approx 10^{-9}$ m^3/s, and the recombination process in computations of the increment of instability generation described by Equation 2.47 is not accounted for. As for the negative term of diffusion in Equation 2.47, we can present it as follows [20,58]:

$$\gamma_D = \frac{D_0 B_0 \mathbf{k}\hat{\mu}_e \mathbf{k}\mathbf{k}\hat{\mu}_i \mathbf{k}}{Q_H \mathbf{k}(\hat{\mu}_e + \hat{\mu}_i)\mathbf{k}} = \begin{cases} \frac{2D_0}{Q_H q_H}k_{\perp}^2, & k_{\perp}^2 \geq Q_H q_H k_{\parallel}^2 \\ D_0 k_{\parallel}^2, & k_{\perp}^2 < Q_H q_H k_{\parallel}^2 \end{cases} \tag{2.77}$$

For strongly anisotropic plasma structures with $k_{\perp} = \pi \cdot 10^{-2}$ m^{-1} and $D_0 = 10^6$ m^2/s, we find that $\gamma_D \approx 2 \cdot 10^{-4}$ s^{-1}, which is sufficiently strong. In such scenarios, the redistribution of plasma by the electrostatic (e.g., potential) field is not sufficient to compensate for diffusion of plasma and, therefore, relaxation of plasma perturbations is observed at an increasing rate.

2.3.2.6 Drift-Dissipative Instability

GDI leads to the growth of plasma waves close to the orthogonal direction of the geomagnetic field $\mathbf{B}_0$ [95,98,104–109]. At the same time, as shown in pioneer reseach [95,98], in low-density weakly collisional ionospheric plasma when dissipative processes such as diffusion and recombination are predominant (in conditions of region E of the ionosphere, see Chapter 1), instabilities called *drift-dissipative*, should still occur. Therefore, the development of GDI has to be considered together with drift-dissipative instability (DDI). The corresponding plasma waves have a phase velocity in the same order as a drift speed perpendicular to the geomagnetic field $\mathbf{B}_0$. Such instability can be generated by even small electric currents flowing perpendicular to $\mathbf{B}_0$. In such a situation, the growth rate of DDI exceeds the loss rate caused by diffusion and recombination processes. In Ref. [98] also it was considered that in weakly ionized plasma (with $\nu_{em} \gg \nu_{ei}$ and $\nu_{im} > \nu_{ei}$) DDI occurs at low frequencies in a plasma containing a density gradient, linearly changed with altitude. Then, in Ref. [95], this assumption of linear density gradient was extended to exponential variations of plasma density with altitude, and the conditions necessary for DDI generation in E and F regions of the ionosphere were found.

To explain the generation mechanism of plasma DDI responsible for ionospheric F layer, Cunnold [95] generalized the results obtained for weakly magnetized ionosphere of region E to those that deal with the magnetized ionosphere of the F region by introducing magnetohydro-dynamic (MHD) effects. It was found that MHD effects determine the electric-field component in the direction of $\mathbf{B}_0$. However, this component's contribution to the redistribution of plasma ionization is weak. A more important mechanism of DDI generation occurs in a case in which the gradient of plasma density possesses a component in the $\mathbf{B}_0$ direction. In this case a different instability mechanism appears, in which a component of the ambient electric field in the direction of $\mathbf{B}_0$ (denoted by $\mathbf{E}_{0\|}$) causes plasma electron transport. This effect will be considered later when the so-called MHD instabilities are analyzed.

We do not examine in depth the results obtained in Refs. [58,65,95,104], but only point out that the results allow the description of the amplification mechanism for the production of the plasma perturbations responsible for the sporadic E layer at equatorial latitudes observed experimentally in the 120–130 km region, as well as the F spread sporadic structures observed in the nocturnal ionosphere (see Chapters 6 and 7).

The joint increment of DDI equals [95]

$$\gamma_{DDI} = \frac{[\kappa \times \mathbf{k}]_z (\mathbf{k} \cdot \mathbf{U}_0)}{\mathbf{k}(\hat{\mu}_e + \hat{\mu}_i)\mathbf{k}} + \frac{D_0}{Q_H |\mathbf{B}_0|} \frac{[\kappa \times \mathbf{k}]^2}{\mathbf{k}(\hat{\mu}_e + \hat{\mu}_i)\mathbf{k}} \tag{2.78}$$

In Ref. [104], the development of the DDI was connected to phase velocity of waves generated in the inhomogeneous ionospheric plasma in the absence of ambient currents. Therefore, the estimations of phase velocity, $V_{ph} < 0.5$ m/s [104], give the impression that with such low phase velocities DDI cannot be generated. However, as shown by other researchers [58,65], in the ionospheric plasma, the phase velocity of plasma waves is determined not by the gradient of plasma density (i.e., plasma inhomogeneity), but mostly by an ambient electric field (see Equation 2.43). Drift waves are transferred by moving plasma, and the development of modified instability is a real event. In Equation 2.78, the first term is positive for gradients κ directed toward the drift velocity of plasma $\mathbf{V}_d = [\mathbf{E}_0 \times \mathbf{B}_0]/B_0^2$ and is negative in the opposite direction. The second term in Equation 2.78 is always positive. So, toward $\mathbf{V}_d$ the GDI and DDI together cause growth of plasma waves, whereas in the direction opposite to $\mathbf{V}_d$ DDI causes growth of plasma waves and GDI causes their attenuation. The increment (given in Equation 2.78) is maximum when the ambient electric field $\mathbf{E}_0$ and the wave vector of plasma waves $\mathbf{k}$ are orthogonal and equal:

$$\gamma_{DD\,\text{max}} = \kappa \cdot \mathbf{V}_d + \frac{C_s^2 \kappa^2}{\nu_{im} + \nu_{ei}} \tag{2.79}$$

Estimations show [58,65] that, for $V_d = 10^3$ m/s, $C_s \approx 800$ m/s, $\nu_{im} = 0.3$ s^{-1}, and $\kappa = 5 \cdot 10^{-5}$ m^{-1}, this maximum equals $\gamma_{max} \approx 5 \cdot 10^{-3} + 5 \cdot 10^{-3}$ (s^{-1}) $= 10^{-2}$ (s^{-1}). For plasma disturbances with a smaller degree of anisotropy, i.e., $l_{\parallel}/l_{\perp} < \sqrt{Q_H q_H}$, we obtain

$$\gamma_{DD} = \frac{q_H}{Q_H} \frac{k_{\perp}^2}{k_{\parallel}^2} \left[\kappa V_d + \frac{C_s^2 \kappa^2}{\nu_{im} + \nu_{ei}} \right] \qquad (2.80)$$

2.3.2.7 Rayleigh–Taylor Instability (RTI)

This kind of instability can only be generated in sharp concentration gradients of ionospheric plasma, of the order of $\kappa = 10^{-3}$ m^{-1} together with an ambient electric field or gravity (or both) [105,106,114–118]. Such plasma structures were observed both in the polar and the equatorial ionosphere at region F, called *spread F layer* [58,105,114–118]. When gravity is dominant, such an instability is defined as the collisional Rayleigh–Taylor (RT) instability [116–118]; otherwise, it is defined as the $\mathbf{E} \times \mathbf{B}$ instability or gradient-drift instability described earlier. Taking into account the angle of inclination of the geomagnetic field denoted by I, we can present the increment of RTI by the following equation [58,105]:

$$\gamma_{RT} = \frac{|\mathbf{g}| \kappa}{\nu_{im}} \cos I \qquad (2.81)$$

Estimations presented in Ref. [105] state that $\gamma_{RT} \approx (7 \div 8) \cdot 10^{-3}$ s^{-1}. At altitudes higher than the maximum of F layer, RT instability causes growth of plasma waves at altitudes of 500–700 km where $\nu_{im} \approx 10^{-2}$ s^{-1} [105]. Later, several authors [114,115], on the basis of the linear theory of instability generation in the equatorial ionospheric plasma of the Rayleigh–Taylor type, investigated the creation of large- and moderate-scale plasma depletions, called *bubbles*, in the F region of the equatorial ionosphere which form the equatorial spread F observed experimentally. The evidence of such plasma structures follows from numerous radar observations, for example, those presented in Ref. [118] using an Altair radar scan (this aspect will be discussed in more detail in Chapters 6 and 7 regarding equatorial F spread and bubbles). Then, by a generalization of this theory to the nonlinear case [116–118], the existence of experimentally observed small-scale plasma structures during continuous ionosonde and radar observations [119–122] was proved. According to Ref. [117], such small-scale plasma waves can be explained by the generation of the so-called *interchange instability* that includes both RT and $\mathbf{E}_0 \times \mathbf{B}_0$ or GD instabilities.

2.3.2.8 Hydrodynamic Instabilities

Instabilities considered previously are low-frequency electrostatic instabilities. In the real ionosphere, plasma waves can be considered as electrostatic only as a "first approximation," because during the development of plasma instability, the fluctuations of current, $\delta\mathbf{j}$, lead to oscillations of the magnetic field, $\delta\mathbf{B}$, and to the generation of rotating electric field, $\mathbf{E}_c$; i.e., pure electrodynamic effects are caused. Therefore, the total electric field can be presented as the sum of the electrostatic ($\mathbf{E}_p$) (i.e., potential) and rotating ($\mathbf{E}_c$) electric fields; i.e., $\mathbf{E}_{\Sigma} = \mathbf{E}_p + \mathbf{E}_c$, where $\mathbf{E}_p = \mathbf{k} (\mathbf{k}\delta\mathbf{E})/k^2$ and

$$k^2 \delta \mathbf{E} - \mathbf{k}(\mathbf{k}\delta \mathbf{E}) = \frac{\omega^2}{c^2}\delta \mathbf{E} + i\mu_0\omega\delta\mathbf{j} \tag{2.82}$$

Thus, in the F region of the ionosphere, for waves with $k = 10^{-3}$ m^{-1}, the longitudinal component of the curled field $\mathbf{E}_{c\|}$ exceeds the corresponding component of the potential field $\mathbf{E}_{p\|}$ by two orders of magnitude. The electrodynamic effects connected with the curled field lead to an increase in the velocity of instability growth, causing an increase of plasma waves with $l_\|/l_\perp \ll \sqrt{q_H Q_H}$.

We will now describe some important aspects of the generation of nonpotential or electro-dynamic waves in the homogeneous and inhomogeneous ionosphere, following results obtained elsewhere [95–97]. As in Ref. [95], we direct the z-axis along wave vector $\mathbf{k}$ and the x-axis orthogonal to $\mathbf{B}_0$. The angle between $\mathbf{k}$ and $\mathbf{B}_0$ will be denoted by ϑ. Following some complicated derivations [95–97], which is beyond the scope of our book, we can present the increment of instability of electrodynamic (*ED*) plasma waves as follows:

$$\begin{aligned}
\gamma_{ED} = \frac{\omega_{Pi}^2}{c^2 k^2 (\nu_{im} + \nu_{ei})} \Bigg\{ & \left((\omega - \mathbf{k}\mathbf{V}_{e0})\sin\vartheta + kV_{e0y}\cos\vartheta\right)^2 \\
& + \frac{kV_{e0y}^2 v_{Te}^2}{(\nu_{im} + \nu_{ei})}\sin\vartheta\cos\vartheta\left(\frac{\kappa_x}{q_H}\sin\vartheta + \kappa_x\sin\vartheta\cos\vartheta + \kappa_z\cos^2\vartheta\right)\Bigg\} \\
& - D_0 k^2 \cos^2\vartheta
\end{aligned} \tag{2.83}$$

All parameters in Equation 2.83 have been defined in this chapter. For $\kappa = 0$, this expression is limited to that for nonpotential waves obtained in Ref. [96]. From Equation 2.83 it follows that the nonpotential waves corresponding to inequality $\vartheta < \tan^{-1}(q_H Q_H)$ in the homogeneous ionosphere can be developed from orthogonal (to $\mathbf{B}_0$) currents, when $(\omega - \mathbf{k}\mathbf{V}_{e0})\sin\vartheta \gg kV_{e0y}\cos\vartheta$, and from longitudinal currents, when $(\omega - \mathbf{k}\mathbf{V}_{e0})\sin\vartheta \leq kV_{e0y}\cos\vartheta$. In the study cited [96] it was shown that the orthogonal current can generate weakly ionized plasma perturbations only in the lower part of F region, where the condition $(C_s^2/V_d^2)(c^2k^2/\omega_{Pi}^2)Q_H^2\cos^2\vartheta < 1$ is satisfied. As for longitudinal currents, the generation of nonpotential waves by these currents is possible only when

$$\frac{V_{e0\|}^2}{C_s^2}\frac{\omega_{Pi}^2}{c^2 k^2}\sin^2\vartheta > 1 \tag{2.84}$$

In cases where the preceding condition (Equation 2.84) is satisfied, the creation of plasma perturbations with $l_\|/l_\perp \leq 10$ is possible for the ionosphere with lower plasma density (i.e., for higher altitudes). Thus, estimations made in Ref. [96] showed that inequality 2.84 is satisfied for $j_{0\|} = 4\cdot 10^{-6}$ A/m^2 and $N_0 = 10^{10}$ m^{-3}.

In the inhomogeneous ionospheric plasma with large gradients of plasma density $\kappa = \kappa_x$, the condition of development of nonpotential plasma waves will change to the inequality

$$\frac{\mathbf{k}V_{e0\|}^2\mathbf{k}}{k^2 C_s^2}\frac{\omega_{Pi}^2}{c^2 k^2}\frac{\kappa_x v_{Te}^2}{\omega_H}\sin^2\vartheta > \cos\vartheta \tag{2.85}$$

As shown in Ref. [96], for $j_{0\|} = 4\cdot 10^{-5}$ A/m^2, $N_0 = 10^{10}$ m^{-3}, $\kappa = 10^{-4}$ m^{-1}, and $k = 10^{-3}$ m^{-1}, instability can be generated for $\cos\vartheta \approx 0.01$, i.e., for $l_\|/l_\perp \approx 100$. In another case where $\kappa = \kappa_z$

(i.e., $\boldsymbol{\kappa} \| \mathbf{k}$), plasma disturbances with a degree of anisotropy of $l_\| / l_\perp \approx 10$ can be developed for longitudinal currents $j_{0\|} \approx 10^{-6}$ A/m^2. Hence, in streams of intense longitudinal currents of $j_{0\|} \approx 10^{-5}$ A/m^2 it is possible to generate nonpotential plasma waves with $l_\perp / l_\| \approx 10^{-1}$. With plasma gradients (i.e., for the inhomogeneous ionosphere) and, for $j_{0\|} \approx 10^{-6}$ A/m^2, electrodynamic nonpotential instabilities lead to plasma disturbances with a degree of anisotropy of $l_\| / l_\perp \approx 10^2$. Electrodynamic effects are essential for plasma waves with wave numbers of $k < \pi \cdot 10^{-3} \text{m}^{-1}$. For nonpotential waves with larger k values, the ratio $\frac{\omega_{Pi}^2}{c^2 k^2}$, from inequality 2.85, is not sufficient to stimulate the curled electric field to compensate for the longitudinal diffusion in plasma.

2.4 Theoretical Aspects of Generation of Nonlinear Plasma Instabilities

The theoretical analysis presented previously concerning generation of different kinds of instabilities in the ionospheric plasma was based on linear approaches and models, more adequately describing the evolution of weak plasma structures (turbulencies) in the real ionospheric plasma. At the same time, the authors of these approaches have pointed out that a wide spectrum of features and phenomena accompanying experimentally observed plasma irregularities in the real ionosphere, from the polar to the equatorial, cannot be explained by using only linear physical models and the corresponding linearized mathematical equations.

In particular, in pioneer research [95–107], the authors found that the linear theory of plasma wave growth cannot predict the intensity of plasma perturbations, their scales, and the shape of their spectra observed experimentally by using linearized equations describing evolution of the two-stream, $\mathbf{E}_0 \times \mathbf{B}_0$, and gradient-drift instabilities, as the main instabilities associated with type I and type II irregularities observed in the ionospheric plasma (see details in Chapter 6). It was shown that such features can be understood only by using a nonlinear description of plasma species dynamics. Thus, in their analysis of two-stream instability, Farley [98] and Buneman [99], in the course of explaining the occurrence of *two-stream* instabilities in a weakly ionized plasma with strongly magnetized elecrtrons and weakly magnetized ions (i.e., for E layer conditions of the ionosphere), have found that the threshold for the generation of an instability defined by $|\mathbf{V}_d| = C_s$ (where $|\mathbf{V}_d| = |[\mathbf{E}_0 \times \mathbf{B}_0]| / B_0^2$ is the absolute value of drift velocity of electrons relative to the ions and $C_s = \sqrt{k_B(T_e + T_i)/M_i} = \sqrt{(m_e/M_i)u_e^2 + u_i^2} \approx u_i$) is the acoustic velocity, which approximately equals the velocity of ion-acoustic waves because $m_e/M_i \ll 1$.

The existence of a physically explained threshold of unstable wave excitation in plasma was shown independently by Simon [106] and Ried [108], who analyzed instabilities generated by plasma density and electric potential gradients called gradient-drift instabilities (see Section 2.2). Later [73], using linear low-pressure two-fluid (electron and ion) collisional models for the plasma, linearized expressions for the eigenfrequency $\omega_{\mathbf{k}r} = \text{Re}(\omega_\mathbf{k})$ and the increment of growth (or decrement of damping) $\gamma_\mathbf{k} = \text{Im}(\omega_\mathbf{k})$ of the mode with wave vector $\mathbf{k}$ were obtained:

$$\omega_\mathbf{k} = \omega_{\mathbf{k}r} + i\gamma_\mathbf{k} \tag{2.86}$$

and

$$\omega_{\mathbf{k}r} = \frac{\mathbf{k} \cdot \mathbf{V}_d}{(1 + \zeta)} = \frac{k V_d \cos\theta}{(1 + \zeta)} \tag{2.87a}$$

$$\gamma_{\mathbf{k}} = \frac{\zeta}{(1+\zeta)} \left[\frac{Q_H \cos^2 \theta}{L(1+\zeta)} - \frac{k^2 C_s^2}{(\nu_{im} + \nu_{ie})} \right] \qquad (2.87b)$$

Here, $\zeta = (q_H Q_H)^{-1}$, $\zeta \leq 1$ for E layer, and $\zeta \ll 1$ for the F layer of the ionosphere; $L^{-1} = (1/N_0)$ (dN_0/dh) is the vertical plasma density scale height (plasma is quasineutral; $N_e \approx N_i = N_0$); θ is the angle between the wave vector $\mathbf{k}$ and the plasma drift velocity $\mathbf{V}_d$; all other parameters and unified notations have been introduced earlier in this chapter.

Farley [98] and then Sudan et al. [125,127,128] found that there are threshold conditions for unstable wave growth, depending on oscillation frequencies, and on the orientation of wave vectors with respect to ambient electric and magnetic fields and currents in plasma beyond which linearized kinetic equations for electrons and ions become incorrect. In this case, the nonlinear effects have to be taken into account—such as interactions between waves, the particle–wave interactions, etc., which can significantly change the Maxwellian distribution functions of plasma electrons and ions and disturb their orbits and trajectories and, as a result, perturb ambient fields in plasma [123–137]. Therefore, to determine the course of plasma wave growth in real time, it is necessary to include nonlinear processes in a complete picture of the evolution of plasma instabilities. For example, in an analysis of drift waves, ion-acoustic waves or ion-cyclotron waves in plasma, i.e., waves having a narrow spectrum both in the phase velocity and in the wave frequency domains, the theory of "weak turbulence" based on linear [123] and quasilinear [124] approaches becomes inapplicable.

However, a strict nonlinear theory, applicable to the description of temporal and spatial evolution of "strong turbulence," has as yet not been established. There exist some models based on results obtained for collisionless plasma, which have been modified to describe the nonlinear processes occurring in the strongly disturbed collisional ionospheric plasma. Among these are models that describe the effects of nonlinear transformation of the energy of linear increasing waves into that of the linear damping waves, the wave–wave interactions, the cascade Kolmogorov effects observed in the turbulent atmosphere, the perturbation of trajectories of charged plasma particles, and so on. In the following text we will discuss the effects of wave–wave interactions, and the effects of particle–wave interactions based on results obtained in various studies [123–137]. We will then analyze natural phenomena that lead to nonlinear stabilization of plasma instabilities in the ionosphere on the basis of previous research [138–145]. Finally, we will analyze some general results concerning growth of instabilities in the ionospheric plasma [146–165] based on results of numerical simulation of the nonlinear phenomenon of instability excitation in the ionospheric plasma.

2.4.1 Nonlinear Wave–Wave Interactions in the Ionospheric Plasma

As shown in Refs. [123–131], nonlinear interaction of plasma waves leads to transfer of energy from the region of generation of plasma waves (defined by positive increment of growth, i.e., by $\gamma_k > 0$) to the region of linear damping of plasma waves (defined by $\gamma_k < 0$). So, at the nonlinear stage of instability evolution, the amplitude of linear damping waves can grow in the time domain. Sudan et al. [125,127–129] have extended the Kolmogorov's theory of "weak turbulence" [123], which is valid only for collisionless plasma [124], to the case of partially ionized plasma (e.g., taking into account collisions between neutral and charged particles) and have analyzed the possibility of nonlinear amplification of growth of damping waves at the linear stage of their evolution due to gradients of plasma concentration and polarization fields of large-scale plasma waves, growing linearly in the time and space domains. Following a renormalized mode-coupling theory developed by Kadomtsev [124] and based on Kraichnan's direct interaction approximation [126], it was shown [125,127–130] that the system of waves with densities $n_{\mathbf{k}} = (2\pi)^2 \int \delta n \cdot \exp\{-i\mathbf{k} \cdot \mathbf{r}\} \mathrm{d}^2 r$ has a strong nonlinearity which cannot be explained in terms of Kolmogorov's "weak turbulence" theory [123] ($\delta n = \delta N/N_0$ is the relative perturbation of the background plasma density). This nonlinearity (i.e., "strong turbulence") now deals with strong disturbances of wave amplitudes $n_{\mathbf{k}}$, at the same time conserving a linearity of the real part, $\omega_{\mathbf{k}r}$, of eigenfrequencies ($\omega_{\mathbf{k}} = \omega_{\mathbf{k}r} + i\gamma_{\mathbf{k}}$) and of wave modes spectrum width, $\Delta\omega_{\mathbf{k}}$, which is proportional to the linear growth (damping) rate $\gamma_{\mathbf{k}}$. In other words, even in the presence of strong turbulence, there are no changes in $\omega_{\mathbf{k}r} = \mathbf{k} \cdot \mathbf{V}_d$ (because $\zeta = 1/q_H Q_H$ from Equation 2.87a is usually smaller than a unit). At the same time, the wave spectrum width spread, $\Delta\omega_{\mathbf{k}} \equiv \Gamma_{\mathbf{k}}$, caused by nonlinear wave interactions, exceeds significantly the increment (decrement) $\gamma_{\mathbf{k}}$. Thus, it was obtained that in region E of the ionosphere secondary waves can grow if the relative amplitude of plasma concentration in the primary waves, $n_{\mathbf{k}}$, achieves 5%–10% and $\Gamma_{\mathbf{k}}$ exceeds $\gamma_{\mathbf{k}}$ [125,127–130]. In this case, the wave vectors of secondary waves, $\mathbf{k}'$, are oriented orthogonally (or close to that direction) to the wave vectors of primary linear growing waves, $\mathbf{k}$. Generally speaking, nonlinear interactions of waves occur if the conditions of synchronism, which are satisfied between two waves, three waves, and so on, are satisfied [124]:

$$\mathbf{k} = \mathbf{k}' + \mathbf{k}'', \quad \mathbf{k} = \mathbf{k}' + \mathbf{k}' + \mathbf{k}''', \dots \tag{2.88a}$$

$$\omega_k = \omega_{k'} + \omega_{k''}, \quad \omega_k = \omega_{k'} + \omega_{k''} + \omega_{k'''}, \dots \tag{2.88b}$$

Such low-frequency (or long-wavelength) waves for which Equation 2.43 is valid cause excitation of plasma waves in the ionosphere in the definite interval of wave numbers k and frequencies defined by Equation 2.88b, and for definite orientation of vectors k, defined by Equation 2.88a relative to vectors of ambient electric $\mathbf{E}_0$ and magnetic $\mathbf{B}_0$ fields and drift velocities for electrons $\mathbf{V}_{e0}$ and ions $\mathbf{V}_{i0}$ (or to their relative velocity $\mathbf{U}_0 = \mathbf{V}_{e0} - \mathbf{V}_{i0}$). On the other hand, as shown in Refs. [125,127–129], for low-frequency (long-wavelength) waves in the ionospheric plasma, where the related plasma density disturbances, $\delta n = \delta N/N_0$, do not exceed 10%, the four- (and more) wave interactions render a smaller effect in the total process of instability temporal evolution compared with that for triad-wave interactions. In the real ionosphere, a triad-wave interaction is the main process of nonlinear interaction of plasma waves, and, for description of such interactions, a theory of "not strong turbulence" [124–126] with some modifications [125,127–130] is valid. Therefore, we will review those works in which a triad-wave interaction was fully analyzed. In these studies [125,127–130] a two-step mechanism of wave interactions was proposed, according to which primary long-wavelength instability generates additional secondary short-wavelength instabilities caused by the gradients of the primary wave orthogonal to the direction of propagation of the primary wave. It was found that plasma becomes strongly turbulent through the nonlinear interaction of a triad of waves $\mathbf{k}$, $\mathbf{k}'$, and $\mathbf{k}''$ allowing validation of the constraint in Equation 2.88a with $k \approx k' \approx k''$. Such a process, during its evolution in the time domain, leads to isotropization of waves and generates additional waves in the linearly stable region of the wave number domain. In the approximation of triad-wave interactions, the system of equations for the relative amplitude $n_{\mathbf{k}} = N_{\mathbf{k}}/N_0$ (N_0 is the concentration of the background ionospheric nondisturbed plasma) can be described as follows [125,127–130]:

$$\frac{\partial n_{\mathbf{k}}}{\partial t} - \gamma_{\mathbf{k}} n_{\mathbf{k}} + \sum_{\mathbf{k}'} V_{\mathbf{k},\mathbf{k}'} n_{\mathbf{k}'} n_{\mathbf{k}-\mathbf{k}'} = 0 \tag{2.89}$$

where $\gamma_{\mathbf{k}}$ is the linear increment of the wave with the wave vector $\mathbf{k}$, and $V_{\mathbf{k},\mathbf{k}'}$, the coefficient of a nonlinear triad-wave interaction, which can be complex [131]. In the case of the ionospheric collisional plasma, when $|\omega - \mathbf{k} \cdot \mathbf{v}_d| \ll \nu_\alpha$ ($\alpha = e, i, m$), the imaginary part of $V_{\mathbf{k},\mathbf{k}'}$ exceeds its real part [131]. Moreover, the authors [125,127–130] found that a triad-wave interaction leads to nonlinear growth (or damping) of waves, and in this case the increment (or decrement) $\Gamma_{\mathbf{k},\omega}$ has to be added to (or subtracted from) the term $(\omega - \omega_{\mathbf{k}} - \Gamma_{\mathbf{k},\omega})$ for the intensity of plasma waves, $I_{\mathbf{k},\omega} = \langle |N_{\mathbf{k}}/N_0|^2 \rangle$, described by the system of equations; i.e.,

$$|\omega - \omega_{\mathbf{k}} - \Gamma_{\mathbf{k},\omega}|^2 I_{\mathbf{k},\omega} = \frac{1}{2} \sum_{\mathbf{k}'} \int |\widetilde{V}_{\mathbf{k},\mathbf{k}'}|^2 I_{\mathbf{k}',\omega'} I_{\mathbf{k}-\mathbf{k}',\omega-\omega'} \, d\omega' \qquad (2.90)$$

where

$$\Gamma_{\mathbf{k},\omega} = -\sum_{\mathbf{k}'} \int \frac{\widetilde{V}_{\mathbf{k},\mathbf{k}-\mathbf{k}'} \widetilde{V}_{\mathbf{k}-\mathbf{k}',\mathbf{k}} I_{\mathbf{k}',\omega'}}{\omega - \omega' - \omega_{\mathbf{k}-\mathbf{k}'} - \Gamma_{\mathbf{k}-\mathbf{k}',\omega-\omega'}} \, d\omega' \qquad (2.91)$$

and $\widetilde{V}_{\mathbf{k},\mathbf{k}'} = V_{\mathbf{k},\mathbf{k}'} + V_{\mathbf{k},\mathbf{k}-\mathbf{k}'}$. The difference between notations introduced in works referred to [125,127–130] and those presented here is that, instead of integration over vector $\mathbf{k}'$, we introduced summation over all wave vectors $\mathbf{k}$ varying from $-\infty$ to ∞, and also took into account that $\mathbf{k}'' = \mathbf{k} - \mathbf{k}'$. We do not introduce here complicated formulas describing coefficients $V_{\mathbf{k},\mathbf{k}'}$; we only give their estimations separately for regions E and F of the ionosphere.

Thus, for the weakly ionized and weakly nonregular plasma in the E layer of the ionosphere

$$V_{\mathbf{k},\mathbf{k}'} = i \frac{[\mathbf{k} \times \mathbf{k}'] \cdot \omega_{\mathbf{k}'}}{Q_H k'^2} + O([\mathbf{k} \times \mathbf{k}'],(\mathbf{k} \cdot \mathbf{k}'),(\mathbf{k}' \cdot \mathbf{k}'')) \qquad (2.92)$$

Here, the second term takes into account small effects of nonlinearities in movements of electrons and ions in plasma. Relations between these effects depend on orientation of wave vectors in the ambient geomagnetic field $\mathbf{B}_0$. Here also it has taken into account that at the region E electrons are strongly magnetized ($q_H \gg 1$) but ions are weakly magnetized ($Q_H \leq 1$).

In region F where both electrons and ions are magnetized (see Chapter 1), we find, correspondingly, that

$$V_{\mathbf{k},\mathbf{k}'} = -i \frac{[\mathbf{k} \times \mathbf{k}'] \cdot k' \mathbf{u}_0}{|\mathbf{B}_0| k' \hat{\mu} \mathbf{k}'} + O([\mathbf{k} \times \mathbf{k}'],(\mathbf{k}' \hat{\mu}_i \mathbf{k}'),(\mathbf{k}' \hat{\mu}_e \mathbf{k}')) \qquad (2.93)$$

Here, as earlier, $\hat{\mu}_\alpha$ is the tensor of mobilities for electrons ($\alpha = e$) and for ions ($\alpha = i$), $\hat{\mu} = \hat{\mu}_e - \hat{\mu}_i$. The second term describes small effects of nonlinearities caused by nonregular plasma in the F layer, by oscillations of plasma ions in the wave field, and by nonpotentiality of plasma waves, i.e., the existence of ambipolar (potential) electric fields and of whirlwind electric fields in plasma. The system of Equations 2.90 through 2.91 determines a spectrum of plasma turbulence caused by nonlinear instabilities following Kraichnan's approximation of direct interaction of waves [126].

On the assumption that $I_{k,\omega}$ does not depend on the direction of vector **k** relative to plasma drift velocity $\mathbf{V}_d$ (i.e., on angle θ), the nonlinear frequency spreading has a Gaussian form [127,128], and the wave intensity can be expressed as

$$I_{k,\omega} = \frac{I_k}{\left(2\pi\Gamma_k^2\right)^{1/2}} \exp\left\{-\frac{(\omega - \omega_k)^2}{2\Gamma_k^2}\right\} \tag{2.94}$$

with

$$\Gamma_k = -i \int_0^{2\pi} \Gamma_{k,\omega_k} \frac{d\theta}{2\pi} \tag{2.95}$$

With this assumption, the simple expression for nonlinear increment of growth (or decrement of damping) was derived, which yields

$$\Gamma_k = \frac{C}{Q_H(1-\zeta)} V_0 k^2 I_k^{1/2} \tag{2.96}$$

where C is the constant of order unity and V_0 the mean electron drift velocity normal to $\mathbf{B}_0$ [127,128]; all other parameters are as defined earlier. Formulas 2.94 through 2.96 were obtained on the assumption that $\Gamma_k \gg |\gamma_k|$, where γ_k is the linear growth (damping) of the wave modes. Presented in a simple form (Equation 2.96), the nonlinear frequency spread parameter, Γ_k, can now be interpreted in terms of the mixing length theory of Kolmogorov [123] as the inverse of the eddy turnover time $\tau \sim l/\delta V$, where l is the eddy scale size, and δV is the eddy velocity. As, according to Refs. [128,129], the electron velocity can be presented in a simple form as

$$|\mathbf{V}_k| = \frac{V_d}{Q_H(1+\zeta)} \frac{N_k}{N_0} \frac{|\mathbf{h} \times \mathbf{k}|}{k^2} \tag{2.97}$$

where **h** is the vertical axis defining the height of the ionosphere, we can now estimate Γ_k from the other side, using Ref. [123] and Equation 2.97. Thus, for $V_d \equiv V_0$ [128,129],

$$\Gamma_k = \tau^{-1} = \delta V/l \approx k\delta V \approx k\left[k^2\langle|V_k|^2\rangle\right]^{1/2} = \frac{V_0}{Q_H(1+\zeta)} k^2 I_k^{1/2} \tag{2.98}$$

which is the same result as obtained in Refs. [128,129]. In other words, the theory developed [123,124] based on "weak coupling" equations, which describe the interaction between eddies, is equivalent to Kaichnan's approximation [126] of "direct interactions of plasma waves" generated in the ionosphere under natural ambient phenomena. Additionally, it was shown [128,129] that there exists the possibility of transferring the energy of linear growing modes to nonlinear damping modes due to triad-wave interactions. Thus, for two-stream instability occurring in the E layer of the ionosphere, for $V_d/C_s \geq (1+\zeta) \approx 1.4 - 1.5$, i.e., roughly above the critical level $V_d = C_s$, obtained by Farley [98], the direct linear mechanism of wave excitation and the quasilinear mechanism described in linear theory are not sufficient to bring plasma to a steady-state regime. Therefore, as shown in Refs. [128–130], there is a single mechanism, which has clear and simple physical explanation. This is the nonlinear transfer of energy from waves with shorter k (or longer wavelengths) to waves with higher k (or shorter wavelengths), as mentioned previously. What is important, the steady-state regime can now be established in the region $\Gamma_k \gg |\gamma_k|$. In this case, with a wave number interval of size k, $\Delta k' \approx k$; it was found [128] that $k^2 I_k \approx const$ and

$$I_k \sim k^{-2} \tag{2.99}$$

Further estimations [128–130] give the steady-state level of wave intensity in the order of $I_k \approx 10^{-5} - 10^{-4}$.

The inverse limited case where $V_d/C_s < (1 + \zeta) \approx 1$ was analyzed [128] using the quasilinear approach for evaluation of instability growth rate and wave intensity. It was found that in a range of k numbers in which damping and growth are absent, the energy flux from waves with longer wavelengths to those with shorter wavelengths is conserved as the energy cascades to shorter wavelengths. In this regime, a new condition $k^4 I_k \approx const$ is valid, which finally yields

$$I_k \sim k^{-8/3} \tag{2.100a}$$

$$\Gamma_k \sim k^{2/3} \tag{2.100b}$$

In the region $k > k_c$ of the k-domain in which wave modes with short wavelengths are damped, where k_c can be written as

$$k_c = \left(\frac{q_H(\nu_{im} + \nu_{ie})}{(1 + \zeta)L} \frac{V_d}{C_s^2} \right)^{1/2} \tag{2.101}$$

it can be shown that $\gamma_k = -(1 + q_H Q_H)^{-1} k^2 C_s^2 / (\nu_{im} + \nu_{ie})$ and, according to Refs. [128–130],

$$I_k \sim k^{-8/3} \left[k_d^{2/3} - k^{2/3} \right]^2 \tag{2.102a}$$

$$\Gamma_k \sim k^{2/3} \left[k_d^{4/3} - k^{4/3} \right] \tag{2.102b}$$

Here, k_d is the limiting wave number where I_k vanishes (called the cutoff value [128,129]).

Now, following Refs. [128–130], we can present a breakpoint $k = k^*$, at which the polynomial dependence of wave intensity I_k is changed from $\sim k^{-8/3}$ for $k < k^*$ to $I_k \sim k^{-2}$ for $k > k^*$, in the following manner:

$$k^* = k_0 \left[\frac{(1 + \zeta)}{2k_0^2 L} \frac{Q_H(\nu_{im} + \nu_{ie})}{V_d} \right]^{3/4} \left(1 - \frac{2C_s^2}{V_d^2(1 + \zeta)^2} \right)^{-3/4} \tag{2.103}$$

In Ref. [131], following the results obtained in the investigations mentioned [128–130], the possibility of transferring energy in triad–wave interaction was investigated based on the general relation between the nonlinear increment (decrement) of wave–wave interaction, the wave intensity, and the coefficients of triad-mode interaction:

$$\Gamma_{k'} \approx \left| \widetilde{V}_{k',k-k'} \widetilde{V}_{k-k',k} \right|^{1/2} I_k^{1/2} \tag{2.104}$$

Here, the energy of a linearly growing wave with wave vector **k**, propagating in a direction approximately orthogonal to $\mathbf{B}_0$ (with $k_\perp^2/k_\parallel^2 > q_H Q_H$), is transferred, as a result of triad–wave interactions, into energy of two linear damping wave modes with vectors **k′** and $\mathbf{k''} = \mathbf{k} - \mathbf{k'}$ (with $k_\perp'^2/k_\parallel'^2 \ll q_H Q_H$). In this case, the decrement of linear damping, $\gamma_{\mathbf{k'}}$, of these two wave modes is controlled by the longitudinal (to $\mathbf{B}_0$) diffusion, i.e.,

$$\gamma_{\mathbf{k'}} \approx -D_\parallel k_\parallel'^2 \tag{2.105}$$

At the steady-state level, linear damping is fully compensated for by nonlinear wave interactions; i.e.,

$$\Gamma_{\mathbf{k'}} = |\gamma_{\mathbf{k'}}| \tag{2.106}$$

Now substituting Equation 2.93 into Equation 2.104, we can, for region F of the ionosphere, estimate the nonlinear increment of wave excitation:

$$\Gamma_{\mathbf{k'}} \approx \frac{[\mathbf{k} - \mathbf{k'}]_\parallel (\mathbf{k} \cdot \mathbf{E}_0)/B_0}{k_\perp^2} I_{\mathbf{k}}^{1/2} \tag{2.107}$$

We can also estimate the degree of anisotropy of plasma disturbances generated by such instabilities in the F layer of the ionosphere. Thus, substituting Equations 2.104 and 2.105 in Equation 2.106, we get

$$\frac{k_\perp'^2}{k_\parallel'^2} \approx \frac{k' k_\perp^2 D_\parallel B_0}{[\mathbf{k} - \mathbf{k'}]_\parallel (\mathbf{k} \cdot \mathbf{E}_0) I_{\mathbf{k}}^{1/2}} \tag{2.108}$$

For $k' \approx k = \pi \cdot 10^{-3}$ m^{-1}, $V_d \approx 10$ m/s, $D_\parallel \approx 10^6$ m^2/s, and $I_{\mathbf{k}} \approx 10^{-4}$, the relation in Equation 2.108 gives $k_\perp'^2/k_\parallel'^2 \approx \pi \cdot 10^2$ [131]. These estimations explain the existence of weakly anisotropic plasma disturbances observed experimentally in the high-latitude F layer ionosphere [131]. We should stress here that these estimations were obtained on the assumption of a random phase of complex coefficients $V_{\mathbf{k},\mathbf{k'}}$ and a steady-state level of plasma turbulence.

Finally, we should note that a full analysis of equations for amplitudes of three interacting waves have shown that the phases of complex coefficients of wave interactions, $V_{\mathbf{k},\mathbf{k'}}$, determine the character of nonlinear instability growth (damping), whereas the steady-state level of wave intensity (amplitude) is defined by modules of these coefficients, i.e., by $|V_{\mathbf{k},\mathbf{k'}}|$. It was also shown that in the ionospheric plasma the imaginary parts of $V_{\mathbf{k},\mathbf{k'}}$, defined by Equations 2.92 and 2.93, are predominant compared with their real parts. Therefore, the steady state level (if it exists) is determined by the imaginary parts of $V_{\mathbf{k},\mathbf{k'}}$, but the character of the nonlinear process of instability growth and the time of instability saturation (i.e., time of steady-state level achievement) is defined by the real parts of $V_{\mathbf{k},\mathbf{k'}}$.

2.4.2 Particle–Wave Interactions in the Ionospheric Plasma

As mentioned previously, for $V_d > C_s$, $k > k^*$, yielding some other saturation mechanism leading from steady-state conditions. This effect relates to the so-called *particle–wave* interactions in which strong short-wavelength waves (instabilities) in plasma perturb electron orbits and thus cause an anomalous (turbulent) diffusion across ambient field lines. Such a wave-generated anomalous diffusion of charged particles in plasma is equivalent to an increase in the effective collision frequency which can essentially change the dispersion relations of waves in the perturbed plasma. The nonlinear particle–wave interaction leads to modification of the electron collisional frequency, ν_e, to a new value, $\nu_e + \nu_e^*$. As shown in the following text, the additional term ν_e^*, depending on the amplitude of the wave spectrum, finally leads to faster vanishing of increment $\gamma_{\mathbf{k}}$ in the saturated state. This process has been investigated in detail by Dupree [132], who estimated the influence of turbulent diffusion on the evolution of plasma instabilities. His approach was then applied [133,134] in an analysis of the nonlinear stage of ionospheric plasma instability and in describing the saturation process of plasma instabilities in the ionosphere. Sudan [128] has modified the results obtained [133,134] by a stricter investigation of the expressions for the dispersion relations. We will present an outline of the methods of perturbed orbits and discussions on the final results. We recommend that readers interested in further details of this method address themselves to the original research [128,132–134].

If fluctuations of the electric fields of plasma turbulence have a random and isotropic character, the interaction of waves and plasma species has the character of turbulent diffusion [132], which was used [133,134] for the analysis of the nonlinear stage of FB and GD instabilities of the ionospheric plasma based on the method of perturbed orbits of plasma species, electrons, and ions, called in Ref. [132] *particle trapping*. The later nonlinear effect for drift waves (e.g., for GDI) was found [132] predominant for waves of the ion-acoustic type, particularly for two-stream instabilities. Based on the theory developed in Ref. [132], it was shown [133,134] that both the wave–wave interactions and perturbations of plasma particle trajectories due to these waves are the dominant nonlinear effects exciting plasma waves, and govern wave growth. Thus, it was found [133,134] that the trajectory $\mathbf{r}(t)$ of the plasma species α ($\alpha = e, i$), after interacting with waves generated by a perturbed electrostatic electric field ($\delta\mathbf{E} = -\nabla\varphi$, where φ is the potential of this field) at moment t, can be defined as

$$\mathbf{r}(t) = \mathbf{r}_0 + \mathbf{V}_0 t \pm \int_0^t dt_1 \int_0^{t_1} dt_2 \hat{\mu}_\alpha \delta E(\mathbf{r}(t_2), t_2) \tag{2.109}$$

where $\mathbf{r}_0$ is the radius of the initial trajectory of the plasma species, δE, the perturbed electric field, $\hat{\mu}_\alpha$, the mobility of species $\alpha = e, i$, and $\mathbf{V}_0$, the minimum drift velocity at which plasma is not damped. It should be noted that Equation 2.109 was analyzed [132–134] in the weak coupling approximation where the nonlinear terms proportional to $(\delta E)^2$ and higher-order have been neglected. The last term in Equation 2.109 takes into account the influence of plasma turbulence on charged-particle movements. Accounting for random fluctuations in the field and its influence on the movement of plasma particles, we can present wave oscillations of charged particles by the following exponent:

$$\exp[i\mathbf{k}\mathbf{r}(t) - \omega t] = \exp[i(\mathbf{k}\mathbf{r}_0 - \omega t)] \exp[i\mathbf{k}\Delta\mathbf{r}(t)] \tag{2.110}$$

Here, all notations are as introduced earlier; $\Delta\mathbf{r}(t)$ equals the last term on the right-hand side of Equation 2.110, and describes random deviations of the particles' orbits.

Derivations made [132–134] have shown that, for the normal distribution of vector $\Delta\mathbf{r}$, we get

$$\langle \exp i\mathbf{k}\Delta\mathbf{r} \rangle = \exp\left(-\mathbf{k}\hat{D}_\alpha^T t\right) \tag{2.111}$$

The corresponding components of the tensors of turbulent diffusion, $\hat{D}_\alpha^T$, and the plasma species can be determined in recurrent forms [132–134], which we simplify by presenting them in one unified expression:

$$
\begin{Vmatrix} \hat{D}_{\alpha\wedge}^T \\ \hat{D}_{\alpha\perp}^T \\ \hat{D}_{\alpha\|}^T \end{Vmatrix} = \int d\mathbf{k} \int d\tau \langle \varphi_\mathbf{k}(t)\varphi_{-\mathbf{k}}(t-\tau)\rangle \begin{Vmatrix} \frac{\mathbf{k}\times\mathbf{B}_0}{B_0} : \frac{\mathbf{k}\times\mathbf{B}_0}{B_0} \\ \frac{q_\alpha e}{m_\alpha} k_\| \frac{[\mathbf{k}\times\mathbf{B}_0]_\|}{B_0} \tau \\ \frac{1}{3}\frac{e^2}{m_\alpha^2} k_\|^2 \tau^2 \end{Vmatrix}
$$

$$
\times \exp\{-i(\omega - \mathbf{k}\mathbf{V}_{\alpha 0})\tau - \mathbf{k}\hat{D}_\alpha^T\mathbf{k}\tau\} \tag{2.112}
$$

Here, $\tau = t_2 - t_1$ is the period of wave–particle interaction; $q_\alpha = -1$ for $\alpha = e$ and $q_\alpha = +1$ for $\alpha = i$ (i.e., one-charged ions with the charge number $Z = 1$); the sign ":" corresponds to the direct product of vectors leading to the corresponding tensor; subscripts in the components of turbulent diffusion correspond to the direction along parallel ($\|$), perpendicular ($\perp$), and across ($\wedge$) the ambient magnetic field $\mathbf{B}_0$. As shown in Refs. [133,134], the term $\mathbf{k}_\perp\mathbf{k}_\perp : \hat{D}_\perp$ is nonlinear in the wave amplitude $|\mathbf{E_k}|$ and determines the nonlinear particle-damping term in general dispersion equations and, therefore, modifies expressions for oscillatory frequency and increment (or decrement) of nonlinear growth (or damping) of wave eigenmode of vector $\mathbf{k}$ (which we denote as nonlinear γ_{NL} instead of the linear one, γ_L):

$$
\gamma_{NL} = \gamma_L - \mathbf{k}(\hat{D}_i^T + \hat{D}_e^T)\mathbf{k} \tag{2.113}
$$

where γ_L is the linear increment (or decrement) of the wave with wave number k defined either by Equation 2.55 for gradient-drift instability (GDI) or by Equation 2.60 for two-stream (FB) instability. The corresponding plasma disturbances (irregularities) associated with such nonlinear instabilities will be analyzed in Chapter 5. Here, we will only stress that estimation of plasma disturbances, $\delta N/N_0$, which can be obtained from the saturation condition of GDI, $\gamma_{NL} = 0$, carried out in Ref. [135], gives

$$
\left\langle \left(\frac{\delta N}{N_0}\right)^2 \right\rangle \approx \frac{\chi^2}{\langle k_\|^2 \rangle} \tag{2.114}
$$

where, as earlier, $\chi = (1/N_0)(dN_0/dz)$ is the longitudinal (to the vertical axis) plasma density gradient. Based on experimental data [135] it was found that χ lies between 10^{-3} m^{-1} and 10^{-2} m^{-1}. Introducing in the preceding relation (Equation 2.114) $\langle k_\|^2 \rangle \approx 9 \cdot 10^{-4}$ m^{-2}, obtained experimentally, it was found that plasma density fluctuations change in a range of about 0.02–0.22; i.e., the generated nonlinear plasma perturbances can exceed the background ionospheric plasma by 2%–22%.

In studies [136–138] based on the quasilinear theory of plasma instability excitation, the limits of accuracy of method of the perturbed orbits were analyzed, and it was shown that the method is useful only for the description of weak turbulences with a wide spectrum in the wave-number domain when the spatial and temporal correlation of electric field perturbations is small enough. The authors [136–138] obtained considerably lower amplitude of plasma perturbations associated with such instabilities—about 1%–2%. As was also shown [139], in a wide spectrum of plasma turbulence ($\Delta k > k$) plasma particles, moving in the wave field with the wave vector $\mathbf{k}$, are scattered in the waves with wave numbers $k' > k$, and the wave field with the wave numbers $k'' < k$ affects them as an ambient field. In the repeated cascade method developed in Ref. [139], the plasma concentration N and the velocities of charged particles $\mathbf{V}_\alpha$ are presented in the form of sum of sets (cascades) of order β (defined as ranks of randomization of the process). For each definite rank β, the parameters $N^{(\beta')}$ and $\mathbf{V}_\alpha^{(\beta')}$ with ranks $\beta' > \beta$ can be considered as fluctuations leading to randomization (chaotization) of the process, and those with ranks $\beta' < \beta$ can be considered as deterministic parameters changing in a regular manner.

Based on the cascade theory for analysis of turbulence of weakly inhomogeneous plasma (particularly, for GDI), it was shown [140] that cascades with higher ranks $(\beta + 1)$ limit growth of perturbations in the definite cascade of rank β due to turbulent (anomalous) diffusion, whereas cascades with lower ranks $(\beta - 1)$ increase the rate of wave growth. For GDI occurring in the ionospheric plasma, the influence of the cascade $(\beta - 1)$ on growth of perturbations in cascade β can be of the same order as in the case of the initially nonequilibrium plasma with a one-dimensional spectrum of turbulence of $U_N \approx k^{-p}$, with $p < 2$ [140].

As we noted previously, in Ref. [128] following Ref. [132], results obtained in earlier studies [133–135] were improved by introducing more exact coefficients of anomalous (turbulent) diffusion for low-frequency (long-wavelength) wave interaction with highly magnetized electrons:

$$D^T = \frac{c^2}{B_0^2} \sum_{\mathbf{k}'} |\mathbf{E_k}|^2 \frac{|\mathbf{k}' \times \mathbf{k}|^2 (k'/k)^2 \Delta\omega_{\mathbf{k}'}}{(\omega_{\mathbf{k}'}^{NL} - \mathbf{k}'\mathbf{V}_0)^2 + (\Delta\omega_{\mathbf{k}'})^2} \tag{2.115}$$

where $\Delta\omega_{\mathbf{k}} = k^2(D + D^T)$, $D = T_e \nu_e / m_e \Omega_e^2$ is the classical expression for the coefficient of electrons' unipolar diffusion, and $D^T = T_e \nu_e^T / m_e \Omega_e^2$ is that for anomalous (turbulent) diffusion, so $\nu_e^T / \nu_e = D_e^T / D_e$. The corresponding eigenmode oscillatory frequency and the increment of mode growth in such a nonlinear (NL) case can be written as

$$\omega_k^{NL} = \frac{kV_0 \cos\theta}{1 + s + s^T} \tag{2.116a}$$

$$\gamma_k^{NL} = \frac{(s + s^T)}{\nu_i (1 + s + s^T)} \left[(\omega_{\mathbf{k}}^{NL})^2 - k^2 C_s^2 \right] \tag{2.116b}$$

Here again, $s = 1/q_H Q_H$, and $s^T = 1/q_H^T Q_H$, $q_H^T = \omega_H / \nu_e^T$; all other parameters are well known and have been introduced previously.

In the saturated steady (equilibrium) state, $\gamma_k^{NL} = 0$ and, consequently,

$$\omega_k^{NL} = kC_s = \frac{kV_0 \cos\theta}{1 + s + s^T} \tag{2.117}$$

In other words, it was shown [128] that s^T increases the probability of shutting the instability in plasma and, simultaneously, causes a shift in phase (due to $\omega_{\mathbf{k}}^{NL}$) to keep the nonlinear phase velocity at ion-acoustic velocity C_s. These features, the isotropy of the process and the limiting value of phase velocity, are explained elsewhere [128] using a more precise mathematical tool approach. At the same time, in that study [128] a result close to that shown earlier [136–138] was obtained, concerning the low saturated level of density perturbation, of about a few percent, in region E of the ionosphere. In this region of the ionosphere, for the case of $V_0 \geq C_s$, as shown earlier, following

Ref. [128], the spectrum I_k of the corresponding instabilities with wave-number spread $\Delta k' \sim k$ is fully described by Equation 2.99.

As mentioned in Refs. [141,142], because of different theories, sometimes with close, and sometimes different, results, and based on kinetic and hydrodynamic (e.g., two-fluid, electron and ion) approaches, it is difficult to present any clear physical picture of the anomalous transport processes accompanying excitation of either two-stream (FB) or $\mathbf{E}_0 \times \mathbf{B}_0$ (GD) instabilities (or both of them) in the equatorial and polar E region of the ionosphere. Moreover, some authors have excluded the anomalous diffusion and the corresponding effective frequencies of collisions between plasma species and neutral molecules (atoms), despite the fact that wave–wave and wave–particle interactions can play an important role, as shown earlier, in the saturation process of growth of unstable waves (i.e., instabilities) in the ionospheric plasma. Therefore, the authors [141,142] have tried to unify the theory of anomalous resistivity of plasma caused by the anomalous (turbulent) diffusion of plasma species and corresponding changes of collisional frequencies due to interaction of unstable waves with particles, based on fluid theory and comparing their approach with those obtained [127–129,143–145] on the basis of kinetic theory.

As shown by Sudan and coauthors [101,102,127–129] and based on simplified and also on consistently strict kinetic theory, wave–wave and wave–particle interactions can modify significantly the characteristic properties of unstable waves in the ionospheric plasma of the auroral or equatorial region E, both FB and GD instabilities, first in their growth rate, phase velocity and oscillatory frequency of eigenmodes (see previous discussions) and, finally, the change of transport processes in plasma, in which additional (to ordinary) anomalous diffusion, called turbulence, always occurs. The close effects were obtained by Robinson and coauthors [143–145], analyzing the effect of the interaction between unstable waves and particles and also using the framework of kinetic theory. It was shown that through wave–particle interactions, plasma waves can essentially change the diffusive properties of ionospheric plasma, the coefficients of diffusion (from ordinary to turbulent) and the frequencies of interaction of species, as well as the plasma resistivity, described as anomalous.

In some studies [141,142] based on fluid theory and a fluid isotropic two-dimensional turbulent model and built upon the Sudan's algorithm, the anomalous (turbulent) collisional electron–neutral frequency was found in the same way as in a previous work [128]. The authors then modified the method, as well as the corresponding coefficient of turbulent diffusion for the strongly anisotropic turbulent model. We will not go into all the mathematical statements in the studies referred to [141,142], but will only point out the common physical aspects leading from these different theories. Considering the FB and GD unstable wave excitation, the same broadband shifting nonlinear effects through diffusion due to the presence of small-scale waves were analyzed [141,142] as done in an earlier investigation [128]. Thus, for the fluid isotropic turbulent model, it was shown that the anomalous frequency of electron–neutral collisions

$$\nu_e^T = \frac{\varsigma}{1 + \varsigma} \frac{Q_H}{2} \omega_H \left| \frac{\delta N}{N_0} \right|^2 \tag{2.118}$$

can be modified by introducing into the denominator the parameter $\varsigma + \varsigma^T$ instead of $\varsigma = 1/q_H Q_H$, where, as earlier, $\varsigma^T = 1/q_H^T Q_H$ ($q_H^T = \omega_H/\nu_e^T$), i.e., the same as obtained in the previous work [128]. But as was shown in Refs. [141,142], in the case of strongly anisotropic turbulence when anomalous diffusion is enhanced only in the direction of the original electric field, such a replacement in Equation 2.118 is impossible. It was also shown [142] that the diffusion process in the $\mathbf{E}_0 \times \mathbf{B}_0$

drift direction reduces the rate of growth in the eigenfrequency, which is executed through the factor ς in Equation 2.118 with only the linear value of $\varsigma = 1/q_H Q_H$ in the one-dimensional case (when eigenmodes $\delta n_{\mathbf{k}}$ or $\delta E_{\mathbf{k}}$ of plasma density and electric field perturbations are delta functions with the peak at $\theta = 0^0$) and for a nonlinear regime. On the assumption of isotropic plasma density distribution, and when $\varsigma + \varsigma^T$ is also isotropic, the results in Refs. [127–129] and those in Refs. [141,142] give, for relations between perturbations of electric field $\delta \mathbf{E}$ and of plasma density δN (called standard isotropic turbulence model), close results; i.e.,

$$\frac{\delta E}{E_0} = \frac{\cos \theta}{Q_H (1 + \varsigma + \varsigma^T)} \frac{\delta N}{N_0} \tag{2.119}$$

Instead of using the nonlinear term $\varsigma + \varsigma^T$, for a description of the coefficient of anomalous diffusion, the factor ς' was chosen as nonlinear [141,142], expressed by $\varsigma' = [\nu_e + \nu_e^T (\theta' + \pi/2)]/Q_H \omega_H$. So, the authors [141,142] instead of using ν_e^T in Equation 2.116 used the more complicated term $\nu_e^T (\theta' + \pi/2)$, accounting for the main anomalous diffusion effects in the direction $\theta = \pi/2$, i.e., in the $\mathbf{E}_0 \times \mathbf{B}_0$ drift direction. In this case, the only difference between the theories will be in the frequency broadening, $\Delta \omega_{\mathbf{k}'}$, which was chosen to be $\Delta \omega_{\mathbf{k}'} = k'^2 (D + D^T)$ [127–129]. According to Refs. [141,142], we get

$$\Delta \omega_{\mathbf{k}'} = k'^2 \frac{D\varsigma}{(\varsigma + \varsigma^T)} \approx k'^2 D \tag{2.120}$$

So, using fluid theory and assuming that $\varsigma' \geq 1$ (instead of $\varsigma < 1$ [127–129]), the authors [141,142] included the nonlinear effects by choosing an anomalous frequency $\nu_e + \nu_e^T$ instead of ν_e used in Refs. [127,128], and therefore obtained the same effects in broadening of frequency due to turbulence in Equation 2.120, replacing D by $D + D^T$. Again, both kinetic and fluid theories give close results even though the physical concepts diverge.

2.4.2.1 Nonlinear Saturation of Electrostatic Instabilities in the Ionosphere

In the foregoing we discussed the saturation of instability growth due to turbulent diffusion caused by wave–wave interaction and transformation of energy from long-wavelength to short-wavelength modes, as well as by wave–particle interactions, both based on kinetic and fluid theories. It is interesting to point out that a decrease in increment of instability, i.e., stabilization of plasma perturbations in the ionosphere, can be observed in several other interesting cases, when fluctuations of electric fields and plasma densities lead to (1) decrease of eigenfrequencies $\omega_{\mathbf{k}}$ (which corresponds to respective decrease of drift velocity V_d), (2) decrease of plasma density gradient $\chi = \nabla \ln N_0$ leading to a steady-state (equilibrium) configuration of the average plasma concentrations, (3) increase of temperature of plasma species, electrons, and ions, and (4) generation of drift velocity shear produced by the electric field component in the direction of the density gradient.

In Refs. [146,147] estimations were obtained of the quasi-equilibrium level of plasma turbulence due to stabilization of GDI of plasma by decreasing the gradient of plasma density, $\chi = \nabla \ln N_0$, and by decreasing the drift velocity of plasma species, V_d. It was shown that for low-frequency (long-wavelength) electrostatic waves with relative amplitudes not exceeding 10% (i.e., $\delta N/N_0 < 0.1$), the amplitude of disturbances of plasma δN and the potential of electric field $\delta \varphi$ (because $\delta \mathbf{E} = -i \mathbf{k} \delta \varphi$) are related through the following expression:

$$\delta N = \delta a \delta \varphi \exp\{-i\Delta\Phi\} \tag{2.121}$$

Here, $\Delta\Phi$ is the phase shift between oscillations of $\delta\varphi$ and δN; the amplitude parameter δa was estimated [146,147] for the lower and upper ionosphere. Thus, for the E region of the ionosphere, where plasma is weakly magnetized,

$$\delta a = \frac{\mu_{iP}k^2}{(i(\omega - \mathbf{k}\mathbf{V}_{0i}) - D_i k^2)} \tag{2.122a}$$

whereas for the strongly magnetized pasma of the F region,

$$\delta a = \frac{\mathbf{k}\hat{\mu}\mathbf{k}}{\left(-i\mathbf{k}\mathbf{U}_0 + \frac{D_i B_0}{Q_H}\mathbf{k}\hat{\mu}_i\mathbf{k}\right)} \tag{2.122b}$$

where $\hat{\mu} = \hat{\mu}_i - \hat{\mu}_e$ is the relative tensor of plasma species mobilities and μ_{iP}, the Pedersen mobility of ions; all other parameters have been introduced previously in this chapter.

The total electric field $\mathbf{E}$ in the turbulized plasma of the total density N can now be presented through the disturbed plasma parameters $\delta\mathbf{E}$ and δN, the nondisturbed parameters $\mathbf{E}_0$ and N_0, and the additional parameters $\tilde{\mathbf{E}}(t)$ and $\tilde{N}(t)$. The last describe time variations of average values of the field and concentration during the growth of wave amplitude in plasma. Finally, we get

$$\mathbf{E} = \mathbf{E}_0 + \delta\mathbf{E} + \tilde{\mathbf{E}}(t) \tag{2.123a}$$

and

$$N = N_0 + \delta N + \tilde{N}(t) \tag{2.123b}$$

The latter term in Equation 2.123 can be obtained by averaging the main equations of conservation of plasma particles in plasma over the spatial ($\sim |k|^{-1}$) and temporal ($\sim |\omega_\mathbf{k}|^{-1}$) scales and by subtraction from the main equations the one written for the nonturbulent equilibrium plasma. Taking into account Equation 2.121, after straightforward computations [146,147] the decrement of damping of eigenmodes in the turbulized plasma was obtained. This describes the effect of stabilization of plasma GDI growth with decrease of the plasma density gradient and of the drift velocity of plasma species:

$$\gamma_\mathbf{k} = -4\mu_{eH}\frac{k_\perp \cdot k_\parallel^2}{\tilde{\chi} \cdot \delta a}\left|\frac{\delta N}{N_0}\right|^2 \sin\Delta\Phi \tag{2.124}$$

Here, as earlier, $\tilde{\chi} = N_0^{-1}(\partial\tilde{N}(t)/\partial y)$ is the gradient of average density of turbulent plasma, the y-axis is the vertical axis, which, for simplicity of the problem, was directed along the magnetic field $\mathbf{B}_0$, and subscripts $\parallel$ and $\perp$ determine components of wave vectors parallel and perpendicular to $\mathbf{B}_0$.

The estimation of the quasi-equilibrium level of waves in plasma can be found from the saturation condition of GDI due to full stabilization of disturbed plasma and its transformation to the steady-state regime (the first option), which yields

$$\chi + \tilde{\chi} = 0 \tag{1.125}$$

For conditions of turbulized plasma analyzed in these studies [146,147], i.e., for $\Delta\Phi = \pi/2$, $k = 2\pi/\Lambda \approx 1$ m^{-1} ($\Lambda \approx 6$ m), $\chi = 10^{-2}$ m^{-1}, $k_{\parallel} \approx k_{\perp}$, and $\mu_{iP}/\mu_{iH} \approx 10^{-1}$, for the condition $\chi = \tilde{\chi}$ (i.e., the gradients of plasma density in a turbulized state and in equilibrium states are equal), we find from Equation 2.124 that $|\delta N/N_0| \approx 10^{-2}$, i.e., about 1%. The saturation of GDI by the turbulent field $\tilde{E}(t)$ yields the following estimation:

$$\left|\frac{\delta N}{N_0}\right| \approx \frac{Q_H^2}{2}\frac{V_d - V_0}{V_0} \tag{2.126a}$$

where V_0 is the minimum drift velocity under which the wave is not damped.

The same estimations made for two-stream (FB) instability yield

$$\left|\frac{\delta N}{N_0}\right| \approx \frac{Q_H^2}{2}\frac{V_d - C_s}{C_s} \tag{2.126b}$$

for small values exceeding the threshold, i.e., either $V_d/V_0 \approx 1.2$ (for GDI) or $V_d/C_s \approx 1.2$ (for FBI), and for $Q_H = \Omega_H/\nu_{im} \approx 10^{-1}$, i.e., for conditions of the E layer, i.e., about 3%. Plasma irregularities of such amplitudes were observed both in the equatorial and auroral ionosphere (details in Chapters 6 and 7).

The process of thermostabilization of current instabilities (FB or GD) in region E of the ionosphere was analyzed [148] using the quasilinear approach for low-frequency electrostatic instability of ionospheric plasma. The corresponding change in the equilibrium temperature, ΔT_{α}, of plasma electrons ($\alpha = e$) and plasma ions ($\alpha = i$) with respect to those occurring in the equilibrium regime, $T_{0\alpha}$, as well as the level of turbulence enough to establish a steady-state regime were estimated [148] based on the hydrodynamic (fluid) theory. The estimations were made separately for E and F layers. Without entering into mathematical details we will present in the following text estimations of the preceding parameters to finally give the amplitude of plasma disturbances (with respect to background steady-state plasma density) usually observed in region E of the ionosphere (see Chapter 6 and results of observations presented there). We will only note that in this investigation [148] a model of homogeneous turbulence and the hydrodynamic equations for the temperature of plasma species, electrons, and ions was used. After averaging these equations over the spatial scale, larger than the wavelength in plasma, and after subtracting from these same equations, but written for the non-turbulent equilibrium plasma (i.e., using the same method as in Refs. [146,147]), the author has obtained expressions for the disturbances of electron and ion temperatures, which we estimate separately for lower and upper ionospheric plasma.

Thus, in region E of the ionosphere, where plasma is weakly magnetized and ionized ($q_H Q_H \geq 1$, $\eta = m_e \nu_{em}/M_i \nu_{im} \sim 10^{-2}$ and $\nu_{ei} < \nu_{im} < \nu_{em}$; see Chapter 1), and following Ref. [148] we obtain

$$\Delta T_e \approx \varepsilon_0 \frac{\omega_{iP}^2}{\omega_{iH}^2 N_0}\langle|\delta E|^2\rangle \leq T_{eo} \tag{2.127a}$$

$$\Delta T_i \approx \varepsilon_0 \frac{\omega_{iP}^2}{\nu_{im}^2 N_0}\langle|\delta E|^2\rangle \ll T_{io} \tag{2.127b}$$

Here ε_0, T_{e0}, and T_{i0} are the plasma permittivity and electron and ion temperatures, respectively, in the steady-state regime (i.e., in equilibrium), ω_{iP} and ω_{iH}, the Pedersen and Hall frequencies of ion oscillations in plasma, and $\langle|\delta E|^2\rangle$, the mean square of the electric field perturbation in plasma; other parameters are as given previously in this chapter.

It can be seen from Equation 2.127 that the plasma waves (instabilities) heat mainly the electron component of plasma.

The corresponding quasi-equilibrium level of plasma turbulence can be found according to the well-known constraint for the quasilinear increment of plasma instability $\gamma_{QL} = 0$ presented in the following form [148]:

$$\gamma_{QL} \equiv \gamma_L - D_0 k^2 - \alpha_{eff} N_0 - \frac{\varepsilon_0 \omega_{iP}^2 k^2}{M_i \nu_{im} \omega_{iH}^2 N_0} \langle|\delta E|^2\rangle \left(1 + Q_H^2\right) \qquad (2.128)$$

The corresponding plasma disturbances associated with such instabilities were estimated as $|\delta N/N_0| \approx 3 \cdot 10^{-2}$, i.e., of about 3%.

As for region F of the ionosphere, the plasma of which is strongly magnetized and ionized, i.e., $q_H Q_H \gg 1$ and $\nu_{ei} \gg \nu_{em} \gg \nu_{im}$, estimations show significant, and equal, heating of both components of plasma, electrons and ions:

$$\Delta T_e \approx \Delta T_i \approx \frac{M_i}{B_0^2} \langle|\delta E|^2\rangle \qquad (2.129)$$

The same estimation of the quasi-equilibrium level of plasma turbulence determined by Equation 2.128 shows that plasma disturbances associated with this kind of instability will not exceed $|\delta N/N_0| \approx 1.5 \cdot 10^{-2}$, i.e., 1.5%. These estimations made for the lower and upper ionosphere are in good agreement with those made in Refs. [146,147], presented previously.

Concluding this paragraph, we note that the stabilizing effects of the electric field components along the plasma density gradient χ play an important role in "saturating" the $\mathbf{E}_0 \times \mathbf{B}_0$ instability [149–151] generated in region F during artificial experiments involving injection of plasma clouds in the ionosphere, observed experimentally [152–156] and obtained analytically and numerically [157–169]. The authors [157–159] have investigated the generation of $\mathbf{E}_0 \times \mathbf{B}_0$ instability based on the model of one- and two-layered ionosphere, where in the upper layer the barium cloud was artificially injected as background equilibrium plasma disturbance. They therefore dealt with the one- and two-dimensional models of plasma structure generated in the homogeneous background ionospheric plasma. In such a statement of the problem, the equation of continuity (i.e., conservation) of the ion component of plasma and the equation of the nondivergence of current have been integrated along the ionospheric altitude on the assumption that the electrostatic potential Φ does not depend on the height z, which extended along the geomagnetic field lines. In the initial conditions of the problem for plasma density during numerical simulations, artificially induced initial small perturbations were introduced and their spatial and temporal evolution was analyzed. It was shown that if the integral Pedersen conductivity of the E layer (the bottom layer) is higher than that of the F layer (where the barium cloud was injected), the stratified small-scale structure cannot be developed because the polarization fields of each strata can be neutralized by short-circuit currents passing through the background ionosphere between two layers. The result obtained leads from the assumption made in these investigations [157–159] that the potential Φ does not depend on the height of the ionosphere for instabilities of various scales. As shown in Refs. [160–162], based on the three-dimensional model, and again in Refs. [163–166], where the effects of short-circuit currents and their role in the evolution of strong plasma disturbances have been investigated both analytically and numerically, for small-scale plasma perturbations this statement is not correct. For small-scale perturbations of the electric field for the transverse scales (to $\mathbf{B}_0$) of plasma perturbations less than 1 km, the polarization fields cannot penetrate from the F layer to the E layer. This example shows the limitation of the two-dimensional model of plasma instability evolution in the ionosphere. We will discuss the barium cloud evolution and its accompanying features in Chapter 6.

We now present the effect of stabilization of growth of plasma GDI, following results obtained in Refs. [151,169] by analyzing drift processes of plasma clouds in the ionosphere. It was shown that the stabilization effect can be observed at the initial stage of plasma cloud evolution due to the creation of the shear of the azimuthal drift velocity created in the background plasma by azimuthal currents. Their stabilizing effect is observed if the azimuthal drift velocity exceeds its threshold value [169]:

$$V_{th} \sim V_A[A/(A + 2)] \tag{2.130}$$

Here, $A = \delta N\,(0,0,0)/N_0$ is the relative initial plasma density perturbation within the barium cloud, $V_A \sim c^{-1}\,\mathbf{E}_{0\perp} \times \mathbf{B}_0$ is the drift velocity in the azimuthal direction, c, the speed of light, and $\mathbf{E}_{0\perp}$, the ambient transverse (to $\mathbf{B}_0$) electric field component. It was also shown that the shear of the azimuthal drift velocity decreases with an increase of the plasma structure dimensions along $\mathbf{B}_0$. According to Drake et al. [151], the criterion of plasma drift stabilization can be presented (using notations introduced in this chapter) as follows:

$$\frac{T_e(1 + T_i/T_e)}{eE_{0\perp}} > \left(\frac{\ell_{||}}{\ell_{\perp}}\right)^2 \frac{(1 - V_{th})}{q_H Q_H} \tag{2.131}$$

From the preceding inequality (Equation 2.131), with an increase in $E_{0\perp}$, i.e., with an increase in azimuthal drift velocity V_A, the threshold of stabilization is also increased. Hence, the absence of stabilizing effects can occur, accompanied by drift instability growth, in a strong ambient electric field. At the same time, with increasing parameters of plasma density perturbations A within the initial plasma structure ($A > 10$), the factor $A/(A+2) \approx 1$ and, as seen from the previous inequality (Equation 2.131) for $E_{0\perp} = const$, the threshold of plasma drift stabilization becomes smaller than in the inverse case of $A < 10$, where $A/(A+2) < 1$. So, with increasing plasma perturbations inside the plasma structure (cloud, beam, etc.) the effect of plasma drift stabilization is strongly manifested and a smooth spreading of plasma structure in the process of its evolution is observed (see Refs. [167–169] and Chapter 6).

It was also found in several studies [151,160,167–169] that the effect of stabilization of $\mathbf{E}_0 \times \mathbf{B}_0$ instability strongly depends on the criterion of "stretching" of GDI, the amplitude of initial plasma instability, and its transversal dimension to $\mathbf{B}_0$. It was found that the maximal GDI growth rate that is needed for the development of the stretching mechanism is of the order [151,160]

$$\gamma_{max} \sim cE_{0\perp}/B_0\ell_{\perp} \tag{2.132}$$

At the same time, as was shown by Blaunstein et al. [167–169], GDI will develop if the growing perturbation of ambient electric field δE is not cancelled by longitudinal electron currents due to the short-circuit effect in the background plasma. In other words, when the time of the "mean" electron's run along $\mathbf{B}_0, \tau_{||} = \ell_{||}/V_{||}$, exceeds the inverse increment of GTD instability, $\tilde{\tau}_{||} = X\gamma_{max}^{-1}$, i.e., $\tau_{||} > \tilde{\tau}_{||}$. Here, $V_{||} = e\delta E_{||}/m_e(\nu_{em} + \nu_{ei})$, the electron's speed along $\mathbf{B}_0$, and $1 < X < 10$ is a numerical factor characterizing the perturbation's growth from the initial (noise) level; all other parameters have been introduced earlier in this chapter. Because $\delta E_{||}/\delta E_{\perp} \sim \ell_{\perp}/\ell_{||}$, we have $\tau_{\parallel} \approx m_e(\nu_{em} + \nu_{ei})\ell_{\parallel}^2/e\ell_{\perp}\delta E_{\perp}$ [169], and the stretching criterion can be formulated as follows:

$$\left(\frac{\ell_{\parallel}}{\ell_{\perp}}\right) \geq \left(\frac{\omega_H}{\nu_{em} + \nu_{ei}} X \frac{\delta E_{\perp}}{E_{0\perp}}\right)^{-1/2} \quad (2.133)$$

Using estimations made in Refs. [167–169] that $X\delta E_{\perp}/E_{0\perp} \approx 0.1$, we finally obtain the stretching criterion for any plasma structure, which completely depends on plasma conditions at the altitude of its generation:

$$\left(\frac{\ell_{\parallel}}{\ell_{\perp}}\right) > \left(0.1 \frac{\omega_H}{\nu_{em} + \nu_{ei}}\right)^{-1/2} \quad (2.134)$$

Thus, for ionospheric altitudes of about 200 km, close to experimental barium cloud rocket injection heights [152–156], $\omega_H \sim 8 \cdot 10^6$ s^{-1} and $\nu_e = \nu_{em} + \nu_{ei} \sim 10^2$ s^{-1} (see Chapter 1). Substituting these parameters into the previous inequality (Equation 2.134) finally allows the stretching criterion for the development of the GDI and corresponding stratification of the plasma structure on strata to be achieved: $\ell_{\parallel}/\ell_{\perp} \geq 3 \cdot 10^1$. This is why some numerical results obtained in Refs. [160–162] for a plasma cloud strongly stretched along $\mathbf{B}_0$ with the parameter of anisotropy $\ell_{\parallel}/\ell_{\perp} \approx 4 \cdot 10^2$ and for $E_{0\perp} \approx 5$ mV/m do not satisfy the preceding criteria (Equations 2.133 and 2.134). At the same time the striation mechanism observed in Ref. [160] for very anisotropic plasma structures is fully confirmed by the analysis presented here and by estimations obtained from the expansion given in Equations 2.132 through 2.134). At the same time, following from these estimations and with a decreasing parameter of initial plasma structure anizotropy, i.e., for $\ell_{\parallel}/\ell_{\perp} < 3 \cdot 10^1$, the threshold of drift stabilization is smaller, and its saturation will affect weakly stretched plasma structures at an ealier stage of their evolution.

So, using the formulas numbered 2.130–2.134 and results obtained in various studies [160–162,167–169], we can predict both the process of GDI stabilization in strongly ionized $(A > 10)$ weakly anisotropic $(\ell_{\parallel}/\ell_{\perp} < 3 \cdot 10^1)$ plasma structures' evolution in a weak ambient electric field $(E_{0\perp} \leq 1-5$ mV/m), and the process of GDI generation during evolution of a weakly ionized $(A < 10)$, strongly anisotropic $(\ell_{\parallel}/\ell_{\perp} > 3 \cdot 10^1)$ plasma structure drifting in a strong ambient electric field $(E_{0\perp} > 5$ mV/m).

2.4.2.2 Numerical Analysis of Nonlinear Effects of Instabilities Generation

As mentioned previously, due to the complicity of analytical investigation of the whole complex of physical processes accompanying the generation and further evolution of plasma waves (instabilities) and associated plasma disturbances in the ionosphere, several groups of researchers, including those mentioned here, separately and idependently investigated numerically two- and four-order systems of equations for eigenmodes based on both kinetic and hydrodynamic approaches.

Thus, some authors [170,171] investigated GDI and the associated plasma irregularities in the equatorial region of the ionosphere based on a two-dimensional numerical simulation of the equations governing the dynamics of electron and ion plasma components in the quasineutral, quasihomogeneous, and isothermal background ionosphere, i.e., the equations of conservation of charged particles and of nondivergence current, presented here using our unified notations employed throughout this chapter:

$$\frac{\partial N}{\partial t} + \mu_{eH} \frac{\mathbf{B}_0 \times \nabla\Phi}{B_0^2} \nabla N = \frac{D_i}{Q_H^2} \nabla^2 N \quad (2.135a)$$

$$\mu_{eH}^2 \frac{\mathbf{B}_0 \times \nabla\Phi}{B_0^2} \nabla N + \mu_{iP}\left(\nabla\Phi \cdot \nabla N + N\nabla^2\Phi\right) = \left(\frac{D_i}{Q_H^2} - \frac{D_e}{q_H^2}\right)\nabla^2 N \qquad (2.135b)$$

We also assumed that the plasma concentration $N = N(x,y,t)$ and the potential of electric field $\Phi = \Phi(x,y,t)$ do not depend on the coordinate z situated along the geomagnetic field $\mathbf{B}_0$. In region E of the ionosphere, where $\omega_H \gg \nu_{em}$, $\Omega_H \ll \nu_{im}$, and $\nu_{ei} \ll \nu_{im} \ll \nu_{em}$ (see also Chapter 1) and taking into account that for the scale length of the density perturbation, L, the following constraints are valid, $L \gg (\lambda/2\pi)/Q_H$, $0 < z < L$, and $L \gg V_e^2/V_d\omega_H$, the simplified expressions for the oscillatory frequency and for the increment of GDI growth were obtained in [170]:

$$\omega_k = \frac{kV_d}{1+s} \qquad (2.136a)$$

$$\gamma_k = \frac{s}{1+s}\left[\frac{q_H V_d \cos\theta}{L(1+s)} - \frac{k^2 C_s^2}{\nu_{im}}\right] \qquad (2.136b)$$

Here, as earlier, $s = 1/q_H Q_H$, $\cos\theta = k_y/k$ the angle which determines the direction of unstable wave propagation with respect to electron drift velocity, $\mathbf{V}_d \propto \mathbf{E}_0 \times \mathbf{B}_0$, and C_s, the speed of the acoustic wave, which we present now in another way through the gyrofrequencies and velocities of electrons and ions, i.e., $C_s = \sqrt{V_i^2 + (\Omega_H/\omega_H)V_e^2}$. Equation 2.136b, taken from Refs. [127,128] with our unified notations, is more general than presented in Refs. [170,171] as it takes into account the direction of unstable wave propagation with respect to drift velocity associated with GDI dynamics. At the same time, Equation 2.136b for the increment is different compared with that obtained in Refs. [127,128], where ion inertia was taken into account and the term $-k^2 C_s^2 \nu_{im}$ replaced by $[\omega^2(k) - k^2 C_s^2]/\nu_{im}$. According to the statements of the problem, for the diurnal equatorial region E, where the constraint $V_d^2 \ll (1+s)^2 C_s^2$ is usually satisfied, $\omega^2(k) \ll k^2 C_s^2$, and the ion inertia terms in the equation of ion fluid can safely be neglected. In this case Equation 2.136b is sufficiently correct for numerical analysis of GDI temporal and spatial evolution. The parameters chosen for the simulations were such that only the two or three modes with the longest wavelengths possible in the system had positive $\gamma(k)$; other modes have a tendency to damp fast. The problem was numerically simulated at the limited square plate with scale $L_0 = 75 - 256$ m, the boundaries of which were chosen so that $N(\mathbf{r},t) = N_0$ and $\nabla\Phi(\mathbf{r},t) = -\mathbf{E}_0$. In the initial conditions for the concentration N, small 2-D plasma disturbances of scale $L_0/2\pi$ were introduced. The initial amplitude of these perturbations was changed from 1% to 3% compared with that for the background ionospheric plasma N_0. The nine models were analyzed numerically with different numbers of mesh points for the size L_0 and the drift velocity V_d. One of the models called model (3) of such computations is shown in Figure 2.3 for $V_d = 125$ m/s [170]. Contour lines represent equal increments in δN between maximum (plus signs) and minimum (minus signs). The horizontal axis is y-directed along the electron drift velocity and the vertical axis is z-directed along the altitude of the ionosphere. Density contour plots at $t = 0, 2, 8,$ and 16 s appear in Figure 2.3a through d. It was found that up to $t = 6$ s the density perturbation reaches a maximum of about 12.5% compared with N_0. Short-wavelength growth was strongest around $t = 8$ s, then damping due to turbulences with velocities v_t, reaching the original electron drift velocity and finally approaching a steady state after $t = 12$ s.

Defined solutions of the Equation 2.135, $N(x,y,t) = \delta N(x,y,t) + N_0$ and $\Phi(x,y,t)$, were then expanded in Ref. [171] in the Fourier series and 2-D *spatial* spectra of fluctuations of plasma density (see also the definition introduced previously according to Refs. [127,128]) were derived as follows:

$$I_\mathbf{k}(\mathbf{k},t) = (L_0/2\pi)^2\langle|\delta N(\mathbf{k},t)/N_0|^2\rangle = (L_0/2\pi)^2\langle|n_\mathbf{k}(t)|^2\rangle \qquad (2.137)$$

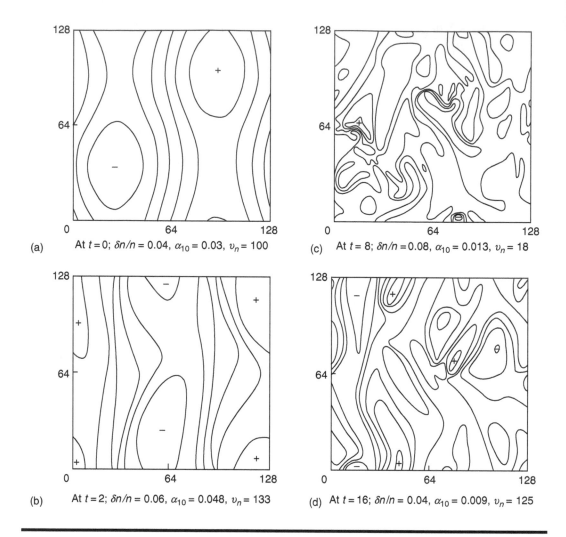

(a) At $t = 0$; $\delta n/n = 0.04$, $\alpha_{10} = 0.03$, $v_n = 100$

(c) At $t = 8$; $\delta n/n = 0.08$, $\alpha_{10} = 0.013$, $v_n = 18$

(b) At $t = 2$; $\delta n/n = 0.06$, $\alpha_{10} = 0.048$, $v_n = 133$

(d) At $t = 16$; $\delta n/n = 0.04$, $\alpha_{10} = 0.009$, $v_n = 125$

Figure 2.3 Two-dimensional plot of disturbances of ionospheric plasma density versus normalized spatial coordinates.

where L_0^2 is the area of the mesh of numerical computations. In the same manner, in Ref. [171] the 2-D spectra of the potential electric field

$$|\Phi(\mathbf{k},t)|^2 = (L_0/2\pi)^2 |\Phi_{\mathbf{k}}(t)|^2 \qquad (2.138)$$

and the angle-average temporal power spectra

$$I_k \equiv I(k) = (2\pi)^{-1} \int_0^{2\pi} I(k,\theta) \, d\theta \qquad (2.139)$$

were derived. The spectrum I_k was computed for values of wave numbers k, varying from $k_0 = 2\pi/L_0 \equiv 2\pi/128$ (m^{-1}), the minimum value, to its maximum, $k = 2\pi/6.4$ (m^{-1}). Figure 2.4a through d illustrates the time evolution of spectra of plasma turbulence computed at

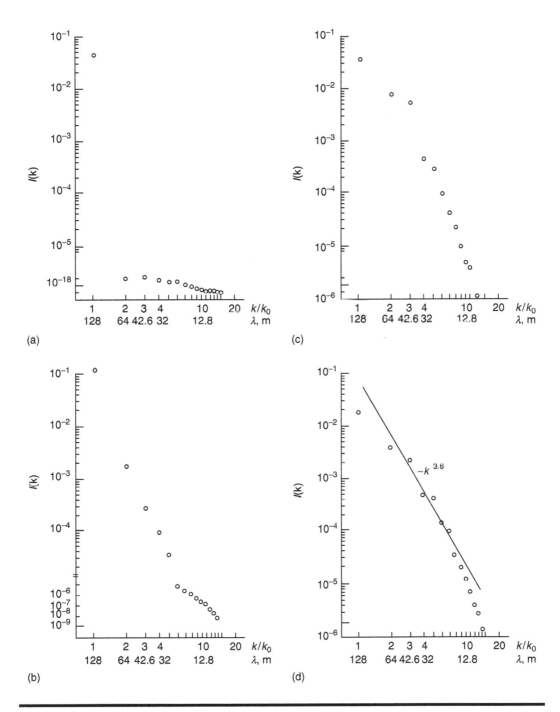

Figure 2.4 **Dependence of intensity of heating-induced plasma instabilities versus the normalized wave number k/k_0 or wavelength λ.**

$t = 0, 2, 13, 25$ s for the drift velocity $V_d = 100$ m/s (i.e., for the model 2 [170,171]). It can be seen that after 13 s from the turbulence generation, its spectrum is not changed significantly, i.e., the time of achievement of a steady-state level (i.e., quasi-equilibrium level) is less than 20 s.

For the analysis of nonlinear broading, the frequency spectra $I_k(\mathbf{k}, \omega)$ were derived in Ref. [171] by using Fourier transformance of the spectra $I_k(\mathbf{k}, t)$ at the time interval of 20.5 s after achievement of the quasi-equilibrium state. In Figure 2.5a through c the frequency spectra $I_k(\mathbf{k}, \omega)$ are shown for the mode of wavelength $\Lambda = 9.2$ m and for the angles $\theta = \tan^{-1}(k_y/k_x) = \pi/6, \pi/4, \pi/2$. Here k_y is the wave number for the mode propagating along the direction of the electron drift velocity, V_d, and k_x is the wave number of the mode propagating in the vertical direction (along altitudes of the ionosphere). The illustration presented shows that the nonlinear interaction of plasma modes (associated with GDI) breaks down the direct dependence between ω and $\mathbf{k}$. Each wave vector $\mathbf{k}$ in such interactions corresponds in the frequency spectra not only to one frequency $\omega_\mathbf{k}$, described by the dispersive Equation 2.43, but also to a wide band of frequencies of width $\Delta\omega \propto \Gamma_\mathbf{k}$. As shown earlier, following previous investigations [127,128], the value of $\Delta\omega$ generally depends on the orientation of the vector $\mathbf{k}$ and achieves maximum magnitude for $\mathbf{k}$ orthogonal to drift velocity V_d (i.e., for $\theta = \pi/2$), as seen from the illustration presented in Figure 2.5c. The average width of the frequency spectra can be found as follows [127,128,171]:

$$\frac{\langle \Delta\omega \rangle}{2\pi} = \frac{1}{(2\pi)^2} \int_0^{2\pi} \Delta\omega(\theta) \, d\theta \tag{2.140}$$

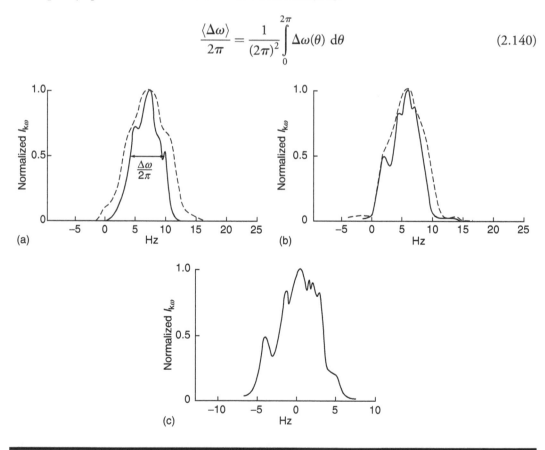

Figure 2.5 Gaussian-shaped dependence of intensity spectrum characteristics of instabilities in frequency domain; dashed curve corresponds to experiments and solid curve corresponds to theoretical prediction.

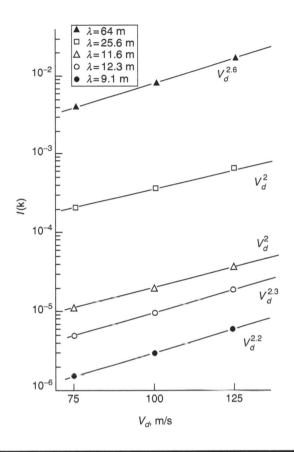

Figure 2.6 **Height dependence of the intensity spectrum of plasma instabilities obtained for the lower ionosphere.**

The dependence of intensity I_k and the average bandwidth spreading $\langle \Delta \omega \rangle$ on the drift velocity of electrons is presented in Figures 2.6 and 2.7, respectively. As can be seen, the specral intensity I_k is increased nonlinearly with drift velocity approximately as $\sim V_d^m$, where $2 < m < 2.5$, whereas the average bandwidth $\langle \Delta \omega \rangle$ is increased roughly $\sim V_d^2$ with an increase of the above-threshold level. Numerically, in Refs. [170,171] the possibility of nonlinear transformation of energy from linearly excited modes to the linear damping modes was predicted. However, the effects of saturation of linear damping modes were not obtained in Ref. [171] even when using the quasilinear approach, but were obtained in Refs. [127,128], where, from a numerical modeling of the evolution of strong GDI in region E of the ionosphere, the nonlinear effects were stressed. In a framework of a strongly turbulent two-dimensional convection of two-fluid plasma, electrons, and ions, based on the same initial conditions of the problem, as in Refs. [170,171], giving the same basic system of Equation 2.135, in Refs. [127,128] the nonlinear wave damping, the steady-state spectra intensity, and the the nonlinear frequency broading were analyzed more correctly. Previously, we described results of quantitative analytical analysis carried out by Sudan and Keskinen [127] based on a two-fluid theory. The calculations were based on the "direct interaction" approximation pioneered by Kraichnan [126] and on the "renormalized nonlinear wave interaction" theory developed by Kadomtsev [124]. Because the main aspects of the theory of strong turbulence [127] had been introduced and discussed earlier, here we will only repeat some principal results of these

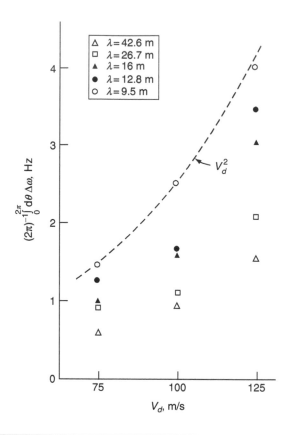

Figure 2.7 Average bandwidth of plasma instability distribution versus the drift velocity of electrons in plasma.

investigations and will then compare results leading from the analytical solutions found in Refs. [127,128] with those obtained numerically by solving the Equation 2.135 [127,171].

As found theoretically [127,128], because of the difference in collisionality of two plasma species, electrons, and ions, in the E layer, the ambient electric field $\mathbf{E_0}$ gives rise to a current $\mathbf{J_0} = J_0\mathbf{y}$ resulting from the drift of the electrons with velocity $\mathbf{V}_d \propto \mathbf{E_0} \times \mathbf{B_0}$. As shown earlier by Simon [106], such a plasma configuration is unstable to electrostatic fluctuations (waves), $\sim \Phi(x)\exp\left[i(k_y y - \omega t)\right]$, if $\nabla N_0 \cdot \nabla(-e\,\Phi_0) > 0$ (here we rearranged coordinates of the problem to confirm results obtained in Refs. [127,171]). Such waves, associated with GDI, are analogous to the Rayleigh–Taylor instability described earlier by replacing the gravitational potential Ψ on the electrostatic potential Φ_0. The nonlinear evolution of these unstable waves generated by the current $\mathbf{J_0} = J_0\mathbf{y}$ leads to a 2-D equilibrium turbulent state (usually called steady-state regime). Based on these assumptions in Refs. [127,171] it was found that:

1. The steady-state 2-D plasma density fluctuation spectrum $I_{\mathbf{k}}(\mathbf{k},t)$ defined by Equation 2.137 or Equation 2.139 is isotropic and proportional to $|\mathbf{k}|^{-p}$, where p lies between 3 and 4, and closer to $p = 3.2$ obtained from experiments.
2. The spectral power of plasma density fluctuations is proportional to the electron drift velocity as $I_{\mathbf{k}} \propto V_d^n$, where $2 \leq n < 2.3$ at the small wavelengths Λ, ranging from 3 to 12 m (see Figure 2.6).

3. The nonlinear damping rate or the frequency width of the spectrum from Equation 2.140 is given by $\left(\gamma_{\mathbf{k}}^2 + \Gamma_{\mathbf{k}}^2\right)^{1/2}$, where $\Gamma_{\mathbf{k}}$ defines the resonance broading of the power spectrum in the frequency domain, and is proportional to $|\mathbf{k}|^{2-p/2}$, $3 \leq p \leq 4$, and $\gamma_{\mathbf{k}}$, the linear increment of wave growth (damping), is defined by Equation 2.136b.

Based on the self-consistent theory [128], we analyzed previously the behavior of $\Gamma_{\mathbf{k}}$ and $I_{\mathbf{k}}$ in the frequency domain, in relation to the wavelength. We will now present the approximate formula which can be used successfully to describe the broadening of the plasma power spectra in the nonlinear regime, corresponding to the theory of strong turbulence. Without going into details of the description of all parameters and notations used during complicated derivations of the non-linear dispersive equation and the corresponding wave–wave interaction coefficients, the predicted approximate formula for $\Gamma_{\mathbf{k}}$ is [127,171]

$$\Gamma_{\mathbf{k}} = const \cdot \frac{q_H}{1 + q_H Q_H} V_d k^2 I_{\mathbf{k}}^{1/2} \tag{2.141}$$

In Figure 2.8 the numerically obtained angle-averaged spectral width, $\langle \Delta\omega \rangle / 2\pi$, defined by Equation 2.140 is shown versus k/k_0 and the corresponding wavelengths Λ for the three electron drift velocities $V_d = 75$, 100, 125 m/s (i.e., for models 1–3, respectively, introduced in Refs. [170,171]). Here, dots, circles, and triangles show $\langle \Delta\omega \rangle / 2\pi$ as a result of simulation; the corresponding solid lines show behavior of nonlinear damping rate $\left(\gamma_{\mathbf{k}}^2 + \Gamma_{\mathbf{k}}^2\right)^{1/2}$, and the dotted curve corresponds to linear growth rate $\gamma_{\mathbf{k}}$. It can be seen that the nonlinear theory, by introducing the broadening parameter of the spectrum $\Gamma_{\mathbf{k}}$ in the nonlinear damping rate, gives the best fit to direct results of simulation of the spectral width, of the order $\langle \Delta\omega \rangle / 2\pi = (1 \div 4)$ Hz (solid lines) with respect to the linear damping rate $\gamma_{\mathbf{k}}$, following from the linear theory (dotted curve). This tendency becomes weaker with an increase in wavelengths of plasma disturbances. Furthermore, in Refs. [127,170] a comparison was made of the nonlinear damping rate (i.e., nonlinear frequency broadening due to turbulence): (1) derived from the full computed spectral width $\Delta\omega$ at half-maximum of

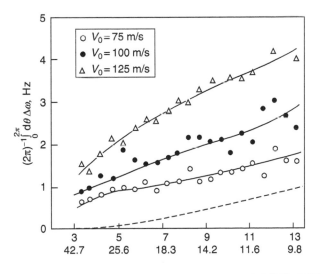

Figure 2.8 The angle-averaged spectral width, $\langle \Delta\omega \rangle / 2\pi$, versus k/k_0 and the corresponding wavelengths Λ for the three electron drift velocities $V_d = 75$, 100, 125 m/s.

the Gaussian power spectrum I_k, taken at $\theta = \pi/4$ to $\mathbf{J}_0 = J_0\mathbf{y}$, defined as $\Gamma_{sw} = \Delta\omega/(8\ln 2)^{1/2}$, (2) due to frequency broadening, from decay of test waves, damping according to the law $\sim \exp(-t/\tau)$ and launching at $\theta = \pi/4$, to $\mathbf{J}_0 = J_0\mathbf{y}$, defined as $\Gamma_{tw} = \tau^{-1/2}$, and (3) from the nonlinear frequency broadening parameter Γ_k computed using Equation 2.141, with all parameters from the nonlinear theory obtained in Refs. [127,171]. The results are as follows. For waves with wavelengths $\Lambda = 9 \div 10$ m for model 2 ($V_d = 100$ m/s), it was found that $\Gamma_{sw} = 1.8 \div 1.7$ Hz, $\Gamma_{tw} = 4.4 \div 1.5$ Hz, and $\Gamma_k = 1.1 \div 0.9$ Hz. Again, there is a good agreement between simulated nonlinear damped rate, Γ_{sw}, and the same parameter, Γ_k, obtained theoretically from Equation 2.141 by using the quasilinear approach, is evident.

To solve the Equation 2.135 numerically, in Refs. [127,128], a spatial Fourier transformation to perturbed plasma concentration, $N(x,y,t) = \delta N(x,y,t) + N_0$, and to the potential of the electric field $\Phi(x,y,t)$ was used over the coordinate x orthogonal both to $\mathbf{B}_0$ and to the gradient of plasma density $\chi = \nabla \ln N_0$, directed parallel to $\mathbf{E}_0$. As the result of such a transformation, a system of differential equations in partial derivatives for the modes N_k and Φ_k was obtained. The functions $N_k = N_k(y,t)$ and $\Phi_k = \Phi_k(y,t)$ are now functions of time and the coordinate y directed parallel to $\mathbf{V}_d$ (see the statement of the problem introduced according to Refs. [170,171]). The system obtained was solved numerically for the discrete set of wave vectors $k_j = k_0 \cdot j$, where $j = 0, 1, 2, \ldots$; k_0 is the wave vector with the maximum linear increment γ_{kmax}; i.e., the wave mode with $j = 0$ described the quasilinear changes of plasma and the electric field (e.g., the potential). The degree of the above-the-threshold value of GDI excitation was estimated by the ratio of V_d/V_0, where V_0 is the minimum velocity for which the most unstable wave has zero-order linear increament. In Figure 2.9

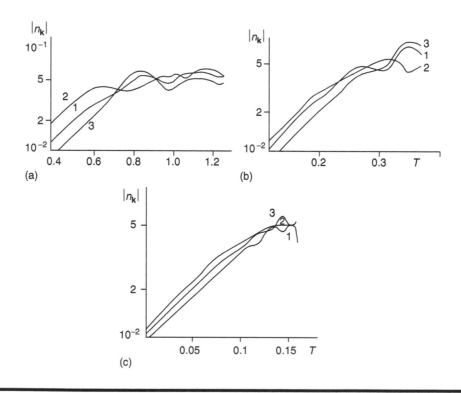

Figure 2.9 The temporal dependence of the modules of the relative plasma wave amplitudes, $|n_k|$, for the three normalized drift velocities, $V_d/V_0 = 5, 10^2, 10^4$.

the temporal changes of the modules of the relative plasma wave amplitudes, $|n_k|$, were shown for the three above-the-threshold levels, $V_d/V_0 = 5, 10^2, 10^4$. Here, the normalized time T along the horizontal axis is presented in the units of $t_0 = l_0/V_d$, where l_0 is the length of the mesh along y-axis. It can be seen that with an increase of the level above the threshold of instability generation, the time of the steady-state regime when the wave system limits its quasi-equilibrium is decreased. Numerical analysis of the Equation 2.135 was carried out in Ref. [172], where the geometry of the problem was assumed to be the following: The gradient of plasma density ∇N_0 as well as the ambient electric field E_0 were directed along the z-axis, and the vector B_0 along the x-axis. The obtained numerical solution of the system (Equation 2.135) for the zero-mode wave (with $j = 0$) showed that turbulization of background ionospheric plasma decreases the large-scale gradient of plasma density and the large-scale electric field E at the place of localization of the initial maximum gradient $\nabla N_0(t = 0)$. In addition to this effect, the moving pulses of this type of solutions usually occur along the y-axis, i.e., in the direction orthogonal to both the ambient electric and magnetic fields. In the system mentioned (Equation 2.135), the nonlinear effects, producing real values of coefficients of nonlinear wave–wave interactions, $V_{k,k'}$, were not taken into account. In Ref. [173] it was shown that even for three-mode interactions with the pure imaging coefficients $V_{k,k'}$, and with $\gamma_k = const$, the linear growing waves do not achieve their quasi-equilibrium level if the nonlinear effects are omitted from the problem under consideration.

So, in the numerical models presented in Refs. [127,170–172], the limitation of the growth of linear unstable modes take place due to quasilinear and nonlinear processes. The same (as for GDI) quasilinear limitation of turbulence growth was obtained during numerical simulation of two-stream (FB) instability in the homogeneous ionosphere [174]. Here, the 2-D system of equations of conservation of plasma species and their movement was numerically solved, where the ion inertia and the viscosity forces were taken into account. The 2-D functions computed, $N(x,y,t)$ and $\Phi(x,y,t)$, allowed for finding the average current $\langle J \rangle = -e\langle N\hat{\mu}\nabla\Phi \rangle$, where $\hat{\mu} = \hat{\mu}_e - \hat{\mu}_i$, the corresponding average electric field denoted by $\langle E \rangle$. It was shown that the component of $\langle E \rangle$ along y-axis, i.e., in the direction of the ambient field $E_0 \equiv E_{0y}y$ coming close to the threshold level, E_{th}, usually occurs for two-stream instability (see previous text) when the turbulence achieves its quasy-equilibrium level. This result shows the efficiency of the quasilinear and nonlinear effects for stabilization of linearly growing wave modes.

This is an important result, because in Ref. [174] the level above the threshold was taken to equal $E_{0y}/E_{th} = 1.06$, sufficiently small. Therefore, in these computations the maximim of wave amplitude has achieved more than 20% (compared to background plasma), and it is very important to point out that, in Refs. [127,173,174], taking into consideration the effects of strong turbulences and nonlinearity in the statement of the problem, it was shown that for higher levels, above the threshold, $V_{dth} = V_0$ (for GDI) and $V_{dth} = C_s$ or $V_{dth} = v_{Ti}$ (for FBI); the more effective quasi- and nonlinear mechanisms of turbulence stabilization should be taken into account to limit the growth of unstable waves and for their saturation.

All results obtained in the studies mentioned [127,128,170–174] correspond to short-scale plasma turbulences. Numerical simulation of large-scale plasma turbulences existing in the daytime equatorial zone of the ionosphere was carried out in an investigation [175], taking into account for the close relation with the nonlinear effects of short-scale turbulence on large-scale plasma disturbances analyzed in Ref. [94], following the strong turbulence theory developed earlier [127]. As shown previously for short-scale turbulences, a nonlinear saturation of GDI and the isotropic distribution of the associated waves in the plane perpendicular to the magnetic field occur, and the linear eigenfrequency ω_{rk} remains simultaneously unchanged even in the presence of strong turbulence with large spectral frequency width, $\Delta\omega(k)$. For large-scale turbulences, however, this theory was modified in Ref. [175] because, as found during the theoretical analysis of the problem, wavelengths of plasma density perturbations become of the same order of the equilibrium electron density scale height. The principal modification was the inclusion of the effects of shear in the electron drift velocity generating small-scale vertical structures. In another study [93], the nonlocal effects modifying the linear growth rate and affecting the long-wavelength perturbations were analyzed, and their influence on the large-scale dynamics predicted in Ref. [94]. The same system of equations, as in Refs. [93,94], describing evolution of isothermal plasma and neglecting inertial terms in them, was used in Ref. [175] to investigate numerically nonlinear nonlocal evolution of large-scale plasma structures, developing and growing out of initial random perturbations (instabilities) in the ionospheric plasma.

To take into account effects of strong small-scale turbulence on large-scale dynamics, the renormalized (turbulent) values for the electron Pedersen mobility, μ_P^*, and diffusion coefficients, D_P^*, were introduced despite those usually used in the linear theory based on weak turbulence effects; i.e.,

$$\mu_P^* = \frac{1}{2} \frac{\psi}{1+\psi} \left(\frac{\mu_{eH}}{\mu_{eP}}\right)^2 \left\langle \left|\frac{\delta N(z)}{N_L}\right|^2 \right\rangle_s \mu_{eP} \qquad (2.142a)$$

$$D_P^* = -\frac{1}{2} \frac{\psi - \psi^2}{(1+\psi)^2} \left(\frac{\mu_{eH}}{\mu_{eP}}\right)^2 \left\langle \left|\frac{\delta N(z)}{N_L}\right|^2 \right\rangle_s D_{eP} \qquad (2.142b)$$

where $\psi = |\mu_{eP}/\mu_{iP}|$, $\langle|\delta N(z)/N_L|^2\rangle_s$ is the turbulent spectrum of small-scale fluctuations (defined by subscript s), and N_L, the large-scale total plasma density; other parameters have been introduced earlier. Then, we can write that $\mu_P \equiv \mu_{eP} - \mu_P^*$ and $D_P \equiv D_e - D_P^*$. Following an earlier study [94], in Ref. [175] a 2-D two-fluid model was introduced to find the large-scale density, $N_L = N_0(z) + \delta N(x,z,t)$, and the gradients of the electrostatic potential $\nabla\Phi_L = -E_0(z) + \nabla\Phi(x,z,t)$, where $N_0(z)$ and $\mathbf{E}_0(z) = E_{0x}\mathbf{x} + E_{0z}(z)\mathbf{z}$ are the steady-state solutions of the background plasma when the vertical currents are usually equal to zero. The ambient magnetic field $\mathbf{B}_0$ is directed along the y-axis; i.e., $\mathbf{B}_0 = B_{0y}\mathbf{y}$.

Introducing Fourier transform to the system, as in Equation 2.135, over the x-coordinate, the corresponding eigenmodes of the concentration and electric field perturbations were found and used to obtain the expression for the perturbed current density and the corresponding plasma density fluctuation spectra. The main result obtained in these studies [94,175] is that there is a strong coupling between interactions of short- and large-scale waves; by introducing nonlinear modifications in the linearized steady-state equations, the authors described the evolution of equilibrium density and electric field gradients. In the strongly turbulent regime, a nonsteady saturated state is reached faster due to damping of large-scale plasma disturbances strongly affected by shear of drift velocity which tends to distort and stretch the plasma perturbations in the horizontal direction (to the vertical axis). These mechanisms coupled with the diffusion process leads to fast damping of all perturbations with wavelengths comparable to the equilibrium gradient length of around 1 km. At the same time, during daytime conditions, recombination prevents growth of instability at longer wavelengths. This result, obtained in the studies mentioned [94,175], is in contradiction to those obtained using the local linear theory of GDI generation, according to which all modes with wavelengths greater than a few tens of meters are unstable. As shown in Refs. [94,175], even in the linear but nonlocal theory, the complete spectrum of global eigenfunctions of the GDI is stable because the coupled effect of all modes, small- and large-scale, leads to a system of steady-state waves. At the same time, in the strongly nonlinear regime the wave system tends to the nonsteady saturated state.

Now large-scale perturbations can survive the action of velocity shear due to modification of the linear shear term. This effect is evident from Figure 2.9, in which the quasiperiodic pattern of the electron horizontal velocity is clearly seen in the time domain (initially with strong velocity shear for $t = 0$ to steep layers with opposite shear in each layer at $t = 60$ s). This quasi periodic pattern can decrease the stabilization effect of the horizontal shear. With time, when processes of collisional diffusion and the recombination mechanism are involved in plasma instability evolution, the striation effect of velocity shear becomes weaker (see bottom of figure for $t = 90$ s), so that the nonlocal features involved in the nonlinear equations describing evolution of plasma perturbations in the ionosphere can slow down the effects of stabilization and saturation of instability growth caused by the horizontal velocity shear. Such an effect of neutralization of velocity shear effect was found in numerical simulations of evolution of plasma clouds in the ionosphere [160–162] as a large-scale plasma structure, and was mentioned previously based on a theoretical analysis [169].

Finally, we have to point out that the dependence of the average power spectrum versus k of $I_k \propto k^{-(1.76 \pm 0.3)}$ numerically obtained in Refs. [94,175] is fully in agreement with that obtained by Sudan using the analytical approach [128] (this dependence from $k^{-8/3}$ to k^{-2} was discussed earlier in this chapter).

Simultaneously, using nonlinear theory of GDI generation, Keskinen Ref. [176] has shown numerically that the existence of short-scale turbulences observed experimentally in the ionospheric plasma containing plasma shears [130,160] as a stabilization mechanism can be explained by the nonlinear cascade of wave energy from the large-scale (e.g., with long wavelengths) to the small-scale (e.g., with short wavelengths) turbulences. In Ref. [177], based on the quasilocal theory of GDI generation, it was shown that not only according to linear theory but also based on quasilocal mode approximation, the existence of short-wavelength unstable waves can be explained even in the presence of velocity shear, which exponentially damps the short-wavelength modes. We do not go into the mathematical description of this approximation and the corresponding direct numerical analysis of eigenmode equations, but will only mention that a new quasilocal wave mode not attenuating exponentially, as do local and nonlocal modes [93,94,175], was found in Ref. [177] and is responsible for the linearly exciting short GDI (modes) despite the damping effects of velocity shear, the aspect outlined earlier as one of the mechanisms of instability stabilization. These quasilocal unstable waves (modes) are evolved from the local wave modes (i.e., from those fully explained by the linear theory). At the same time, due to nonlinear increase of wave numbers with time, these modes grow and saturate even in the linear regime of GDI evolution, whose effect cannot be explained by using only the concept of generation of local modes in the ionospheric plasma, i.e., based only on the linear theory.

If we now compare results obtained in Refs. [160,177] with those obtained in Refs. [93,94,175] using a nonlocal approach, as well as by different numerical and physical models, the former based on the quasilocal, nonexponential solution of two-order linear equations describing temporal and spatial evolution eigenmodes, and the latter models employing the numerical initial-value method for the fourth-order differential equations describing nonlocal growth of unstable waves, we will find some principally differing results obtained by these groups of researchers. Following other studies [178,179], in which the numerical results otained earlier [93,94,175] were compared with those from Refs. [160,175], and again without going into a mathematical description of the problem but only trying to understand the physical aspects of the problem, we can explain the results obtained through the prism of the experimentally observed data. Thus, most experiments carried out in the equatorial regions of the ionosphere have found the predominance of large-scale plasma perturbations of the order of about 1 km associated with long-wavelength unstable waves with the horizontal wave numbers of $k_x \approx (6.0 - 6.3) \cdot 10^{-3}$ m. These results were in contradiction

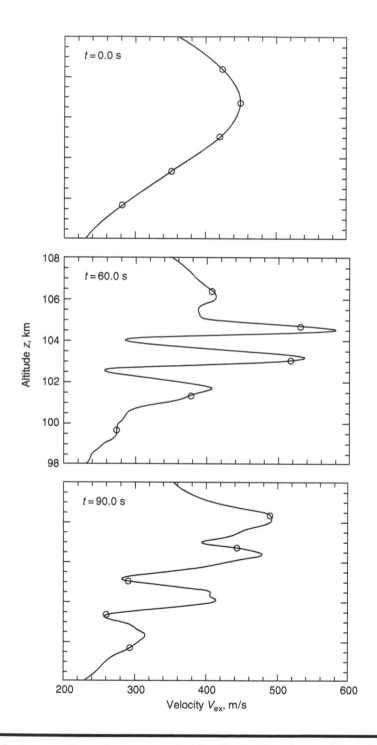

Figure 2.10 The large-scale mode structures height distribution observed in Refs. [178,179] in the lower region E of the equatorial electrojet.

to those obtained earlier [93], where it was found that the nonlocal increment of growth of the large-scale (more precisely, the kilometer-scale) unstable waves is smaller than that for the small-scale waves (with scales of a few hundreds of meters). To confirm this contradiction with experiments in further study [94,175], the effects of short-wavelength turbulence on long-wavelengths was modeled (the so-called the subgrid model), because the linear theory does not explain the evolution of small-scale unstable modes (see previous discussion in this section); in equations for long-wavelength modes were artificially included additional turbulent mobility and diffusion coefficients and the height dependence of the laminar (regular) mobilities, and collisional frequencies of the plasma species were involved in the two-fluid theory. As a result, the essential increase of the increment rate and the corresponding transiently growing large-scale plasma structures were found. At the same time, in Refs. [160,177], using the quasilocal-mode concept and based on a numerical solution of two-order equations for eigenfunctions, it was found that the growth rate of large-scale unstable waves prevailed above the short-scale waves, and the later unstable modes damp faster due to velocity shear and recombination processes occurring in the lower E region of the ionosphere. In Refs. [178,179] two approaches were more carefully compared using the same numerical initial-value approach as in Ref. [93] and dealing with eigenmodes as done in previous studies [160,177], but taking into account a more realistic profile of the parameters of the ionosphere, such as plasma density gradient, mobilities of plasma species, and the ambient electric field. It found more general features than those concluded in previous studies indicating that the large-scale mode structures observed in the E region of the equatorial electrojet are spaced over a wide height range of several kilometers from 100 to 106 km (see Figure 2.10), whereas the short-scale mode structures fill the narrow height range of about 100–300 m around the height of 107 km (called the "upper electrojet"; see Figure 2.11). At these altitudes, the effects of the velocity shear and recombination process are weaker than those of the exitation of rapidly growing short-wavelength instabilities due to a strong plasma density gradient (see Figure 2.12) and the maximum of electric field distribution (at an altitude of 106–107 km (see Figure 2.13). So, taking into account more realistic profiles of the parameters of the ambient field and plasma species parameters, the predominance of large-scale plasma perturbations at the E layer equatorial

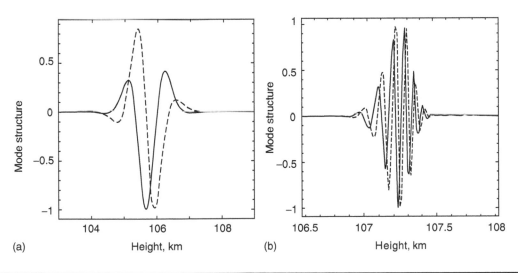

(a)   (b)   Height, km

Figure 2.11 The same, as in Figure 2.10, but for higher E and F₁ layer of the ionosphere. (Extracted from Refs. [178,179].)

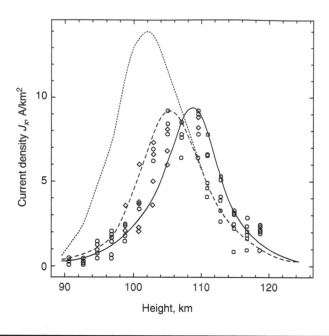

Figure 2.12 Electric current distribution obtained experimentally in Refs. [178,179] at the lower E layer of the equatorial ionosphere.

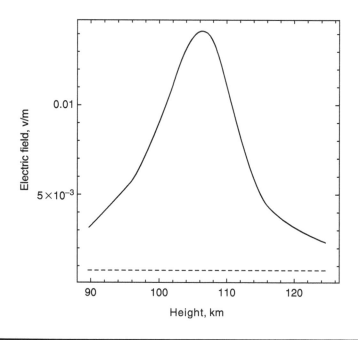

Figure 2.13 Electric field strength height distribution obtained in Refs. [178,179] at the lower E layer of the equatorial ionosphere.

electrojet, observed experimentally, can be predicted correctly based on nonlocal linear theory, even using the two different approaches mentioned earlier, which are based on different mathematical tools.

It is interesting to mention that the saturation mechanism of other kinds of long- and short-wavelength plasma instabilities, such as Rayleigh–Taylor instability, described earlier can explain their growth and saturation by including in the process of instability evolution the presence of time-dependent eigenmodes and density gradient enhances. The importance of these effects was proved theoretically by Basu [180] based on a numerical analysis of the system of quasilinear equations of plasma species transport.

All the researchers mentioned previously used the two-fluid theory based on the hydrodynamic approach, whereas in Refs. [101,102] a consistent kinetic theory was proposed for the description of FBI evolution in the E layer of the lower ionosphere. The consistent kinetic theory presented there was clarified in a study [181] using a simple fluid model associated with a new type of instability generation by Joule heating of electrons in the perturbed electric field occurring in the E region of the high-latitude ionosphere. Such a perturbed field can modulate temperature of plasma in the same manner as artificially done by a powerful alternate field described in Section 2.1. Such temperature perturbations finally lead to the creation of a new kind of instability caused by two independent mechanisms. The first mechanism related to the changes of the sign of plasma pressure to be opposed to the plasma density perturbation and is very sensitive to signs of flow angle ϑ introduced in Refs. [101,102]: $\cos \vartheta \sim \mathbf{k} \cdot \mathbf{V}_d$, ϑ is the angle between the $\mathbf{k}$-vector and the drift velocity $\mathbf{V}_d$ [181]. The corresponding instability growth (i.e. unstable modes generation) is found at the negative angles $\theta = \pi/2 - \vartheta$ ($\sim \mathbf{k} \cdot \mathbf{E}_0$) and its growth stabilization at positive angles θ ($\sim \mathbf{k} \cdot \mathbf{E}_0$). The second mechanism is associated with thermal perturbations of the electron Pedersen conductivity due to the monotonically increasing dependence of the electron–neutral collision frequency on temperature, the effect that plays an essential role in the FB instability evolution obtained from the strict and consistent kinetic theory [101,102]. This thermal mechanism of plasma heating and the corresponding instability generation was earlier predicted and proved theoretically by Gurevich with coauthors [182,183], respectively, for high and moderate ambient electric fields occurring at higher and lower altitudes of the E layer.

The zone of instability growth in the flow angle domain (e.g. θ-domain) covers two wide ranges of θ around the bisectors between the direction of $\pm \mathbf{E}_0$ and the drift velocity $\mathbf{V}_d \propto \mathbf{E}_0 \times \mathbf{B}_0$. The same features of instability evolution were found for FB turbulence in some studies [101,102], as briefly described in Section 2.3. The new instability, found theoretically in Ref. [181], occurs in the broad height range between 75 and 105 km, i.e., covering the D layer and the lower E layer, and the associated unstable waves have wavelengths ranging from several to tens of meters (called longwaves in Ref. [181]). As this instability is generated by the cumulative effect of the Pedersen electron flux directed opposite the electric field, $\mathbf{J}_P \propto \sigma_P(\mathbf{E}_0 + \delta \mathbf{E})$, and the Hall electron flux, directed along $\mathbf{V}_d \propto \mathbf{E}_0 \times \mathbf{B}_0$ because $J_H \propto (N_0 + \delta N)\mathbf{V}_d$, i.e., by currents driven across the ambient magnetic field, this new instability was called by the authors [181] the "cross-field current driven instability." As was noticed in Ref. [181], this new instability can be a preferable candidate of the source generating the meter- and longer-scale turbulences in the lower ionosphere; as observing by VHF radars is problematic, only rocket measurements can be used. In fact, in Refs. [184,185] possible evidence of the existence of "cross-field current-driven instability" was found by rocket observations of an anomalously high level of the long-wavelength plasma structures occurring at altitudes above 78 km, associated with anomalously high values of the measured ambient electric field.

Because the gradient-drift (GD) and two-stream (FB) are the main sources of plasma irregularities in the ionosphere, we return to the recent theories that were developed from the mid-nineties up to the present time. These theories were developed by Hamza and St-Maurice [186–192], who made an overview of existing theories and their drawbacks and advantages, and proposed a self-consistent analysis of plasma instabilities, GDI and FBI, occurring both in equatorial and auroral (polar) ionospheric zones. We will not enter into a detailed analysis of the mathematical aspects of the solutions of fourth-order coupling nonlinear equations of two-fluid dynamics, which sometimes (in convenient conditions) have been reduced to the quasilinear and linear equations for eigenfunctions. Most of the foregoing authors used closed mathematical approaches. For us it is very interesting to show the reader additional elegant ideas of how to transfer results obtained in studies [127–129,170–175] for the equatorial E and F regions of the ionosphere to the high-latitude ionosphere, where an artificially introduced concept of anomalous diffusion was proposed for explaining nonlinear fast growth and stabilization of plasma waves observed experimentally. Moreover, the direct application of the models for the equatorial to the high-latitude ionosphere, introduced previously, has exposed the principal difficulties in explaining the new features of plasma observed experimentally in auroral and polar zones, such as wide broadening of plasma density spectra, nonlinear Doppler shift, and changing thresholds of GDI and FBI depending on ambient electromagnetic conditions, which significantly change densities, temperature, mobilities and diffusion coefficients of plasma species, electrons, and ions.

In Refs. [186–192], after analyzing existing theories, the strict concept of wave–wave and wave–particle interactions proved by Sudan and his coauthors [94,101,102,127–129,170–175,181], was taken as a basic concept using both self-consistent theory of two-fluid dynamic and consistent kinetic theory. Thus, in one study [186], a fully nonlinear theory based on a two-fluid model for the electrons and ions was presented and analyzed following the wave-coupling theory [127,128] based on the "direct approximation principle" of Kraichnan and the corresponding Depree's resonance broading theory (we have described these approaches in Sections 2.3 and 2.4). Drift mechanism both along $\mathbf{B}_0$ (when $k_{||} \neq 0$) and across $\mathbf{B}_0$ (when $k_{||} = 0$) was analyzed. Nonlinear effects occur due to the introduction of additional anomalous collisional frequencies, mobilities, and conductivities of plasma species which were introduced to understand the correction of the proposed mathematical tool to explain the anomalous Doppler shift and frequency broading caused by nonlinear effects in unstable perturbed plasma, as well as the equilibrium spectrum observed in the steady-state regime of saturation of unstable waves due to strong turbulization of plasma. It was shown that the nonlinear mode-coupling mechanism fully predicts nonlinear effects in the perturbed plasma. Therefore, in further research [187–192], the effects of the anomalous intercharge collision frequencies and the anomalous diffusion of plasma species caused by them were disregarded. We will now show additional elegant ideas proposed in Refs. [187–192] of how, based only on wave-coupling and wave–particle interaction mechanisms, we could explain redistribution of energy between primary and secondary unstable waves generated in the high-latitude ionosphere from various ambient sources, which finally does not change overall energy balance of the ionosphere as a steady-state global structure. Thus, wave–wave and wave–particle interactions leading to the steady-state regime (system equilibrium) were analyzed [187] by using the analogy of a sectional bucket into which water flows (see Figure 2.14). This simple sketch illustrates the flow of energy in a steady-state turbulent regime in which interaction between the linearly unstable primary waves and the linearly stable secondary waves occurs. A hole in the middle of the bucket demonstrates coupling between these modes. The energy of the primary and secondary waves fill the left and the right reservoirs. During nonresonant wave–particle interactions, shown at the bottom of the bucket, the energy of waves transfer to the plasma species causing Joule heating of the plasma as shown in Refs. [141,142] (this subject is beyond the scope of this book, so we refer the reader to these exellent works). From the beginning, where perturbations of the ambient electric field are strong and strong currents occur in the high-latitude ionosphere flowing both along and

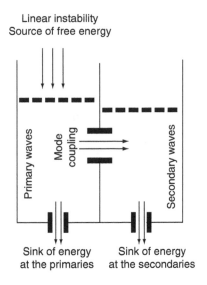

Figure 2.14 Schematic presentation of the wave–wave and wave–particle interactions extracted from Refs. [187,188] using the analogy of a sectional bucket into which water flows.

across the ambient magnetic field (which is close to the vertical axis at these latitudes), the free energy from currents is transferred to the linearly unstable primary waves, as shown at the top left-side of the bucket. Due to coupling between primary and secondary waves in the steady-state regime, the energy from the former modes is transferred to the latter ones. The energy of both primary and secondary modes is extracted to the particles during mode–particle interactions, which explain the two bottom holes of the bucket. In Ref. [187] the steady-state turbulent regime by zero-order increment of unstable wave growth was defined, and it was postulated that the overall rates of free energy from currents to the primary waves and of its loss during interaction with particles are the same. As seen from the sketched figure, in equilibrium the energy of secondary waves is less than that of primary waves. This effect can be understood if we use the results obtained in Ref. [193], in which the interaction between modes in the anisotropic magnetized plasma was investigated using Feunman's diagrams and Bouret's approximation for a mean double propagator. It was found that slowly growing primary unstable waves due to coupling can transfer their energy to the fast secondary stable modes, the energy of which has the tendency to increase after interaction (they becomes unstable). Conversely, due to coupling, the fast unstable primary modes can transfer their energy to the slow secondary modes with the tendency of their damping. Finally, the redistribution of the energy of the secondary waves is not the same in the steady-state configuration as that for the primary waves. It was also pointed out in Ref. [193] that the overall time required to reach the equilibrium energy distribution is of the order of the magnitude $(\varepsilon\sigma)^{-1}$ both for short- and long-wavelength modes, and for a weak turbulent regime (e.g., for a medium with small random fluctuations) the transfer of energy from primary to secondary waves occurs over a long period. As shown in Ref. [187], this happens because the time scales related to such equilibrium configurations have to cover a large number of mode-coupling events to obtain a steady-state system with further stabilization of the process. The preceding qualitative analysis fully predicts the energy conservation in the steady-state plasma configuration during FBI and GDI generation, which, as was shown elsewhere [180–182], can be converted to other instabilities in the perturbed ionospheric plasma at low and high latitudes.

The same qualitative analysis can be made on the basis of previous research [187–191] to explain nonlinear growth rate, spreading of frequency width of the power spectra of unstable modes and the nonlinear Doppler shift, as well as evolution of the plasma-perturbed structures in the space and time domains, observed experimentally [194–203]. What is interesting to note is that a more general self-consistent two-fluid theory performed in Refs. [187–191] is adopted to explain non-linear evolution of both kinds of main plasma instabilities, two-stream (FB) and gradient-drift (GD), and accounts for the vector of plasma gradient κ, which in the general case occurs along and across the vertical axis (i.e., the magnetic field $\mathbf{B}_0$ for the moderate- and high-latitude ionosphere). We present the increment of nonlinear growth $\gamma_{\mathbf{k}}^{NL}$ (or decrement of nonlinear damping) obtained in the studies mentioned [187–191] through the linear increment $\gamma_{\mathbf{k}}^{L}$ (decrement) using our unified notation introduced previously in this chapter:

$$\gamma_{\mathbf{k}}^{NL} = \frac{\Delta\omega_{\mathbf{k}}^2}{\nu_{im}(1 + q_H Q_H)} + \frac{q_H \mathbf{k} \cdot \kappa\overline{\omega}_{\mathbf{k}}}{k^2(1 + q_H Q_H)} + \frac{(\overline{\omega}_{\mathbf{k}}^2 - k_\perp^2 C_s^2)}{\nu_{im}(1 + q_H Q_H)}$$
$$= \frac{(\overline{\omega}_{\mathbf{k}}^2 + \Delta\omega_{\mathbf{k}}^2 - (\omega_{\mathbf{k}}^L)^2)}{\nu_{im}(1 + q_H Q_H)} + \frac{q_H \mathbf{k} \cdot \kappa(\overline{\omega}_{\mathbf{k}} - \omega_{\mathbf{k}}^L)}{k^2(1 + q_H Q_H)} + \gamma_{\mathbf{k}}^{L} \tag{2.143}$$

Here, we rewrite all parameters and functions, following Refs. [187–191] but also using our notations:

The linear oscillation frequencies:

$$\omega_{\mathbf{k}}^L = \frac{q_H Q_H}{(1 + q_H Q_H)} \mathbf{k} \cdot \tilde{\mathbf{V}}_d \tag{2.144a}$$

The linear increment (decrement) of linear mode growth (damping):

$$\gamma_{\mathbf{k}}^L = \frac{q_H Q_H}{(1 + q_H Q_H)} \left\{ \frac{((\omega_{\mathbf{k}}^L)^2 - k_\perp^2 C_s^2)}{\nu_{im} q_H Q_H} + \frac{\mathbf{k} \cdot \kappa\omega_{\mathbf{k}}^L}{Q_H k^2} \right\} \tag{2.144b}$$

where all parameters of plasma are the same as earlier. The drift velocity and gradient density vector are presented in a more general form as in Refs. [127,128]; i.e.,

$$\tilde{\mathbf{V}}_d = \frac{c}{B_0^2} \mathbf{E}_0 \times \mathbf{B}_0 - \frac{c q_H}{B_0^2} (\mathbf{E}_0 \cdot \mathbf{B}_0)\mathbf{B}_0 \tag{2.145a}$$

$$\kappa = \frac{c}{B_0^2} \nabla N_0 \times \mathbf{B}_0 - \frac{c q_H}{B_0^2} (\nabla N_0 \cdot \mathbf{B}_0)\mathbf{B}_0 \tag{2.145b}$$

The frequency spectrum of plasma density perturbations was characterized in the previously mentioned works [187–191] by an average frequency and a frequency broadening (spread) defined as follows:

$$\overline{\omega}_{\mathbf{k}} = \frac{1}{N_{\mathbf{k}}} \sum_{\omega} \omega |n_{\mathbf{k}}|^2 \tag{2.146a}$$

$$\Delta\omega_{\mathbf{k}}^2 = \frac{1}{N_{\mathbf{k}}} \sum_{\omega} (\omega - \overline{\omega}_{\mathbf{k}})^2 |n_{\mathbf{k}}|^2 \tag{2.146b}$$

where the wave energy "reservoirs" can be defined through the normalized amplitude of eigenmodes, $n_{\mathbf{k}} = N_{\mathbf{k}}/N_0$, as follows:

$$N_{\mathbf{k}} = \sum_{\omega} |n_{\mathbf{k}\omega}|^2 \quad \text{and} \quad N = \sum_{\mathbf{k}\omega} |n_{\mathbf{k}\omega}|^2 \tag{2.147}$$

With these definitions, the conservation equation of energy density becomes

$$\frac{dN}{d\tau} = 2 \sum_{\mathbf{k}} \gamma_{\mathbf{k}}^{NL} N_{\mathbf{k}} \tag{2.148}$$

Equation 2.148 describes energy balance in the perturbed ionosphere, where the left-hand side gives the total rate of change of energy transferred to the electrostatic modes of the turbulent plasma. In a steady-state situation (i.e., in equilibrium) this rate of change tends to zero, which occurs only when $\gamma_{\mathbf{k}}^{NL} = 0$. Now using this constraint and Equation 2.143, we will present in the following paragraphs the qualitative analysis of the nonlinear Doppler frequency shift and power spectral spread for various scenarios occurring in the ionospheric plasma, weakly, moderately, and strongly turbulized by ambient factors. Without loss of generality, we will assume in our qualitative analysis that the term in Equation 2.143 responsible for plasma density gradient effects is zero; i.e., the growing process due to plasma gradients can vanish. If so, let us start with the quasilinear case of the instability growing process. From Equation 2.143, in the case of weakly turbulent plasma, when the growth of unstable modes can be described using linear theory, i.e., dealing with the linearly unstable eigenmodes, described by Equations 2.144a and b, we get $\Delta\omega_{\mathbf{k}} \to 0$; i.e., a very narrow spread of plasma spectrum occurs, and the average oscillating frequency caused by random fluctuations of plasma density power is limited to the linear eigenfrequency, described by Equation 2.144a; i.e., $\overline{\omega}_{\mathbf{k}} \approx \omega_{\mathbf{k}}^L$. This situation is illustrated in the simpler manner at the right side of Figure 2.15. At this quasilinear regime the primary waves, extracting free energy from currents or electric field perturbations, lead to the steady-state regime changing in their instantaneous amplitude with a slow-to-moderate linear growth rate, as illustrated at the left side of Figure 2.15. The signal time

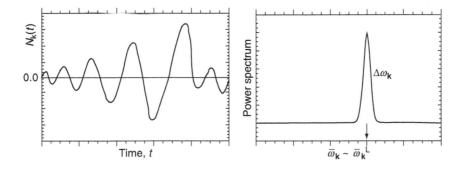

Figure 2.15 Instability power spectrum distribution in the time (left side) and the frequency (right side) domains in the unbounded plasma with absence of ambient sources.

oscillations occur with the linear eigenfrequency obtained from Equation 2.144a. So, in this extremal regime,

$$\overline{\omega}_{\mathbf{k}} \approx \omega_{\mathbf{k}}^L \approx k_\perp C_s \qquad (2.149)$$

Therefore, in the case of weak turbulence, the width of the plasma density spectrum is narrower than the mean Doppler shift, the result illustrated at the right side of Figure 2.15. As is seen from the left side of Figure 2.15, after several oscillations with $\omega_{\mathbf{k}}^L$, the unstable plasma wave attains its maximum and becomes sufficient for transferring energy of the primary wave to the secondary wave due to coupling between modes (according to the scheme sketched in Figure 2.14). After completing the energy transfer, the amplitude of the primary wave decreases sharply, as can be seen from the left side of Figure 2.15, and the coupling process disappears. If, again, the primary unstable wave extracts free energy from the ambient sources, increasing to its instantaneous amplitude, the process of coupling between primary and secondary waves will be repeated in the same manner. What happens in the case of weak turbulence embedded in the medium with a lot of free energy, i.e., in a medium consisting of high-energy sources, as occurs in the auroral and polar ionospheric regions? In this case, due to coupling, the secondary damped waves will quickly extract energy from primary linearly growing waves, and the frequent wave-coupling occurrence leads to faster linear growth rate and faster repletion of the process, as illustrated in the left side of the sketch in Figure 2.16. According to results obtained in Refs. [187–190] for a steady turbulent state regime, and substituting in Equation 2.143 $\gamma_{\mathbf{k}}^{NL} = 0$ by ignoring effects of plasma gradients, we finally get

$$\overline{\omega}_{\mathbf{k}}^2 + \Delta \omega_{\mathbf{k}}^2 = k_\perp^2 C_s^2 \qquad (2.150)$$

As evident from the right side of Figure 2.16, in this case, a nonzero spread of the power spectrum occurs. Equation 2.150 helps obtain a nonlinear phase velocity expressed as

$$V_{ph}^{NL} = \frac{\overline{\omega}_{\mathbf{k}}}{k_\perp} = \frac{C_s}{\sqrt{1 + (\Delta \omega_{\mathbf{k}} / \overline{\omega}_{\mathbf{k}})^2}} \qquad (2.151)$$

That is, it becomes smaller than the ion-acoustic velocity C_s in the plasma. The same effects will result with the mean Doppler shift, which becomes slower than C_s. It was found in Refs. [187–189], both theoretically and experimentally, that to obtain the saturation regime the Doppler width $\Delta \omega_{\mathbf{k}}$ has to

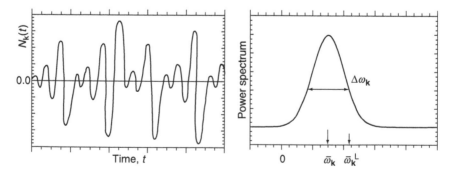

Figure 2.16 Same as Figure 2.15, but for unbounded plasma consisting of high-energy sources.

be equal to, or even larger than, the average Doppler frequency $\overline{\omega}_{\mathbf{k}}$. From this it is easy to obtain the limit at which the phase velocity of linearly damping secondary modes, which extract energy from the linearly growing primary modes due to nonlinear coupling between them, can be smaller than C_s but not more than by a factor given approximately by

$$V_{ph(thr)}^{NL} \approx \frac{C_s}{\sqrt{2}} = 0.7 \cdot C_s \tag{2.152}$$

Now, when a strong turbulent condition occurs, which was associated, in Refs. [187,188], with a situation in which the Doppler spectral width $\Delta\omega_{\mathbf{k}}$ is broad enough but the mean Doppler shift frequency tends to zero; i.e., $\overline{\omega}_{\mathbf{k}} \to 0$. This strong turbulent regime is sketched on the right side of Figure 2.17. In this case, due to the wave-coupling process, the wave modes with eigenfrequencies close to zero periodically receive energy from the primary unstable modes, obtaining instantaneous amplitudes in a very short time. It can easily be shown that in this nonlinear regime we can obtain a negative increment of linear modes growth; i.e., $\gamma^L < 0$. Indeed, when in the steady state configuration $\gamma_{\mathbf{k}}^{NL} = 0$, we get from Equation 2.143, assuming the absence of the plasma gradient term,

$$\gamma_{\mathbf{k}}^{NL} = \frac{\Delta\omega_{\mathbf{k}}^2}{\nu_{im}(1 + q_H Q_H)} + \gamma_{\mathbf{k}}^L \equiv 0 \tag{2.153}$$

from which we get the threshold of the Doppler spectral width for a zero nonlinear increment of wave growth

$$\Delta\omega_{\mathbf{k}}^2 = -\nu_{im}(1 + q_H Q_H)\gamma_{\mathbf{k}}^L \tag{2.154}$$

So, we can see that to satisfy Equation 2.154, the increment of linear modes growth must be negative; i.e., $\gamma^L < 0$. This is the corresponding threshold between growth and damping of the secondary modes due to nonlinear coupling process. So, as was found in Ref. [188], the linearly stable secondary modes with Doppler frequency spread $0 < \Delta\omega_{\mathbf{k}}^2 < -\nu_{im}(1 + q_H Q_H)\gamma_{\mathbf{k}}^L$ will still be nonlinearly damping.

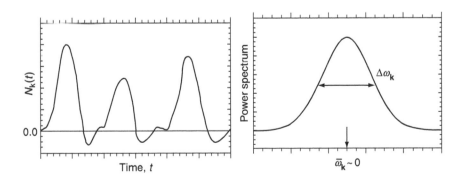

Figure 2.17 Same as Figure 2.15, but for a strong turbulent regime occurring in the unbounded plasma.

Generally speaking, Equation 2.143 covers different scenarios occurring in the high- and low-latitude ionosphere depending on the effects of ambient sources of plasma perturbation. It gives the relation between nonlinear Doppler shift frequencies, the Doppler spectral spread width and the linear eigenfrequencies, as $\overline{\omega}_{\mathbf{k}} \approx \omega_{\mathbf{k}}^{L}$:

$$\left(\omega_{\mathbf{k}}^{NL}\right)^2 = \overline{\omega}_{\mathbf{k}}^2 + \Delta\omega_{\mathbf{k}}^2 \tag{2.155}$$

From this result it follows that for the pure drift process, ignoring the gradient of plasma density, a drift velocity threshold required for instability generation in linear regime will be smaller than that required for generation of nonlinear instability. Moreover, the corresponding constraint of unstable waves generation along the flow angle θ predicted by linear and quasilinear theories [101,102, 127–129,169–175] (we repeat it again using our notations)

$$V_d \cos\theta > (1 + \varsigma)C_s \tag{2.156}$$

was now generalized in Refs. [187–190] for the nonlinear case, following qualitative analysis presented earlier:

$$\widetilde{V}_d \cos\theta > (1 + \varsigma)C_s\sqrt{1 - \frac{\Delta\omega_{\mathbf{k}}^2}{k^2 C_s^2}} \tag{2.157}$$

which, for the special case of a strong turbulent steady-state regime, when $\overline{\omega}_{\mathbf{k}} \to 0$ and, according to Equation 2.150, $\Delta\omega_{\mathbf{k}} \approx k_\perp C_s$, reduces to $\widetilde{V}_d \cos\theta \geq 0$. This result is clearly seen from the illustration sketched in Figure 2.17. The results obtained are consistent with spectral width observations of plasma perturbations with velocities from 350 to 600 m/s obtained by separate groups of investigators using various high-latitude radars in the corresponding wave frequencies [189,194,196,200,203].

At the same time, it was shown experimentally that plasma instabilities, such as FB and GD instabilities, can be observed at the wide spectrum of angles around the flow angle θ. Therefore, following the nonlinear self-consistent fluid theory performed in Refs. [186–188] and proved experimentally in Ref. [189], the extension of this theory on the large aspect angles θ (larger than 10°) was made and analyzed with regard to growth and stabilization mechanisms of FB instabilities generated in the high-latitude ionosphere [190–192]. The effects of the parallel (to $\mathbf{B}_0$) currents on the perpendicular drifts and the influence of the flow angle spread was analyzed based on the three-wave coupling model. We do not enter into complicated mathematical analysis, but will only summarize that the general results obtained in Refs. [186–192] using the self-consistent nonlinear two-fluid theory cover results achieved by linear and quasilinear models performed in other studies [157–163,172–179] and are in good agreement with those obtained in Refs. [101,102,171,181] based on the consistent quasilinear fluid and kinetic theories. The theories and corresponding models discussed in this chapter predict more accurately various scenarios of plasma instabilities and the associated plasma perturbations evolution in the space, time, and frequency domains in the different ranges of the ionosphere, from the equator to the polar cap.

Results obtained during the theoretical analysis carried out in the preceding text allow us to conclude that the instability of the ionospheric plasma is the main source generating small- to large-scale plasma structures in the E and F regions of the ionosphere, among which are H_E irregularities occurring in the middle-latitude ionosphere and plasma field-aligned irregularities observed in the high-latitude ionosphere. The existence of weakly and strongly anisotropic plasma structures at the altitudes of 200–400 km (i.e., in the F region and above) can be explained by development of current-gradient, current, and convergence instabilities. Strictly speaking, all instabilities contribute more or less to the cumulative effect of plasma perturbations. Therefore, it is difficult to find some universal mechanism for the creation of plasma irregularities in the ionosphere, as the reader can well understand from the discussion of a wide spectrum of different theories, models, and physical concepts presented in this chapter.

We notice that the main aspects of the creation and further evolution of weakly anisotropic (with dimensions along and across $\mathbf{B}_0$ exceeding 1–10 km) and strongly anisotropic plasma structures (with degree of anisotropy exceeding 100), generated by various natural and artificial sources of their creation, will be considered in Chapter 6.

References

1. Mikhailovskii, A.V., *Theory of Plasma Instabilities*, Vol. 1: *Instabilities in a Homogeneous Plasma*; English Edition by Barbour, J.B., New York-London: Consultants Bureau, 1975.
2. Gershman, B.N., Erukhimov, L.M., and Yashin, *Wave Phenomena in the Ionosphere and Cosmic Plasma*, Moscow: Science, 1984.
3. Fedorchenko, A.M. and Kotzarenko, N.Ya., *Absolute and Convective Instabilities in Plasma and in Solid Materials*, Moscow: Science, 1981.
4. Mikhailovskii, A.V., *Theory of Plasma Instabilities*, Vol. 2: *Instabilities in a Inhomogeneous Plasma*, English Edition by Barbour, J.B., New York-London: Consultants Bureau, 1975.
5. Whitten, R.C. and Poppoff, I.G., *Physics of the Lower Ionosphere*, New York: Prentice Hall, 1965.
6. Filipp, N.D., Blaunstein, N.Sh., Erukhimov, L.M., Ivanov, V.A., and Uryadov, V.P., *Modern Methods of Investigation of Dynamic Processes in the Ionosphere*, Kishinev (Moldova): Shtiintza, 1991.
7. Belikovich, V.V., Benediktov, E.A., Tolmacheva, A.V., and Bakhmet'eva, N.V., *Ionospheric Research by Means of Artificial Periodic Irregularities*, Berlin: Copernicus GmbH, 2002.
8. Likhter, Ya.L., Gul'el'mi, A.V., Erukhimov, L.M., and Mikha'lova, G.M., *Wave Diagnostics of the Near-Earth's Plasma*, Moscow: Science, 1989.
9. Mishin, E.V., Ruzhin, Yu.Ya., and Telegin, V.A., *Interaction of Electron Fluxes with the Ionospheric Plasma*, Moscow: Edition of Hydrometeorology, 1989.
10. Filipp, N.D., Oraevskii, V.N., Blaunstein, N.Sh., and Ruzhin, Yu.Ya., *Evolution of Artificial Plasma Irregularities in the Earth's Ionosphere*, Kishinev (Moldova): Shtiintza, 1986.
11. Blaunstein, N., The character of drift spreading of artificial plasma clouds in the middle-latitude ionosphere, *J. Geophys. Res.*, Vol. 101, No. A2, 1996, pp. 2321–2331.
12. Filipp, N.D., *Mirror Scattering of Ultra Short Waves by Middle Latitude Ionosphere*, Kishinev (Moldova): Shtiintsa, 1978.
13. Gurevich, A.V. and Shwarzburg, A.B., *Nonlinear Theory of Radiowave Propagation in the Ionosphere*, Moscow: Science, 1973.
14. Gurevich, A.V., *Nonlinear Phenomena in the Ionosphere*, Berlin: Springer-Verlag, 1978.
15. Richmond, A.D., *Ionospheric Electrodynamics*, pp. 249–290, in: Volland, H. (ed), *Handbook of Atmospheric Electrodynamics*, Vol. II, Boca Raton (FL): CRC Press, 1995.
16. Pitaevskii, L.P., Electric forces in the dispersive medium, *Sov. Phys. JETF*, Vol. 12, 1960, pp. 1008–1022.
17. Georges, T.M., Interaction of pulsed radio waves in ionosphere, *Proc. Int. Conf. Phys. Low Ionosphere*, Ottawa, Canada, 1966, pp. 289–293.
18. Grach, S.M., Karashtin, A.N., Mityakov, N.A., Rapoport, V.O., and Trakhtengerts, V.Yu., Parametric interaction of electromagnetic waves with ionospheric plasma, *Izv. Vuz., Radiophyzika*, Vol. 20, 1977, pp. 1827–1842.

19. Erukhimov, L.M., Metelev, S.A., Myasnikov, E.N. et al., Artificial ionospheric turbulence, *Izv. Vuz., Radiophyzika*, Vol. 31, No. 2, 1987, pp. 208–225.

20. Gershman, B.N., *Dynamics of the Ionospheric Plasma*, Moscow: Science, 1974.

21. Al'pert, Ya.L., *Propagation of Electromagnetic Waves in the Ionosphere*, Moscow: Science, 1973.

22. Belyaev, P.P., Kotik, D.S., Mytiakov, N.A. et al., Generation of electromagnetic signals of combination frequencies, *Izv. Vuz., Radiophyzika*, Vol. 31, No. 2, 1987, pp. 248–267.

23. Getmantsev, G.G., Zuikov, A.A., and Kotik, D.S. et al., Combination frequencies in the interaction between high-power short-wave radiation and ionospheric plasma, *JETF Lett.*, English Translation, Vol. 20, 1974, pp. 229–230.

24. Vilenskii, I.M., Izraileva, N.P., Plotkin, V.V., and Freiman, M.E., *Artificial Quasiperiodic Inhomogeneities in the Lower Ionosphere*, Novosibirsk: Science, 1987.

25. Belikovich, V.V., Benediktov, E.A., and Itkina, M.A. et al., Scattering of radio waves by periodic artificial ionospheric irregulatities, *Radiophys. Quantum Electron.*, English Translation, Vol. 20, 1978, pp. 1250–1253.

26. Belikovich, V.V., Benediktov, E.A., Getmantsev, G.G., Ignat'ev, Y.A., and Komrakov, G.P. Scattering of radio waves from the artificially perturbed F region of the ionosphere, *JETF Lett.*, English Translation, Vol. 22, 1975, pp. 243–244.

27. Rietveld, M.T. and Goncharov, N.P., Artificial periodic irregularities from the Tromso heating facility, *Advances in Space Research*, Vol. 21, 1998, pp. 693–696.

28. Rietveld, M.T., Turunen, E., Matveinen, H., Goncharov, N.P., and Pollari, P., Artificial periodic irregularities in the auroral ionosphere, *Annales Geophysicae*, Vol. 14, 1996, pp. 1437–1453.

29. Djuth, F.T., Groves, K.M., and Elder, J.H. et al., Measurements of artificial periodic inhomogeneities at HIPAS observatory, *J. Geophys. Research*, Vol. 102, 1997, pp. 24.023–24.025.

30. Belikovich, V.V., Benediktov, E.A., Goncharov, N.P., and Tolmacheva, A.V., Diagnostics of the ionosphere and neutral atmosphere at E-region heights using artificial periodic inhomogeneities, *J. Atmos. Solar-Terr. Phys.*, Vol. 59, 1997, pp. 2447–2460.

31. Budilin, A.V., Getmantsev, G.G., and Kapustin, P.A. et al., Localization of heights of nonlinear currents responsible for the low-frequency radiation in the ionosphere, *Izv. Vuz., Radiophyzika*, Vol. 20, No. 1, 1977, pp. 83–89.

32. Rietveld, M.T. and Stubbe, P., Ionospheric demodulation of powerful pulsed radio waves: A potential new diagnostic for radar suggested by Tromso heater result, *Radio Sci.*, Vol. 22, No. 6, 1987, pp. 1084–1090.

33. Belikovich, V.V., Benediktov, E.A., and Terina, G.I., Diagnostics of the lower ionosphere by method of resonance scattering of radio waves, *Sov. Phys. JETP*, Vol. 48, No. 11, 1986, pp. 1247–1253.

34. Belikovich, V.V., Benediktov, E.A., and Getmantsev, G.G. et al., Ionospheric electron density measurement using radio wave scattering from artificial plasma inhomogeneities, *Radiophys. Quantum Electron.*, English Translation, Vol. 21, 1979, pp. 853–854.

35. Belikovich, V.V., Benediktov, E.A., Gol'tsova, Y.K., Komrakov, G.P., and Tolmacheva, A.V., Determination of ionospheric parameters in the F-region by means of resonance scattering, *Izv. Vuz., Radiophyzika*, Vol. 29, 1986, pp. 131–138.

36. Mityakov, N.A., Editor, *About Possible Intensification Mechanisms of Periodic Plasma Grid Meshes in the Ionosphere*, Moscow: Science, 1978.

37. Artsimovich, L.A. and Sagdeev, R.Z., *Physics of Plasma for Physicians*, Moscow: Atomizdat, 1979.

38. Mityakov, N.A., Excited temperature scattering of electromagnetic waves in ionospheric plasma, *Izv. Vuz., Radiophyzika*, Vol. 24, No. 6, 1981, pp. 671–674.

39. Vas'kov, V.V., Gurevich, A.V., and Karashtin, A.N., Self-focusing instability under the oblique incidence of radio waves on the ionosphere, *Geomagn. Aeronom.*, Vol. 16, 1976, pp. 322–325.

40. Vas'kov, V.V. and Gurevich, A.V., Resonance instability of small-scale plasma perturbations, *Sov. Phys. JETF*, Vol. 46, 1977, pp. 903–912.

41. Vas'kov, V.V. and Gurevich, A.V., Nonlinear resonance instability of plasma in the reflection region of ordinary electromagnetic wave, *Sov. Phys. JETF*, Vol. 42, 1975, pp. 91–103.

42. Vas'kov, V.V. and Gurevich, A.V., Saturation of self-focusing instability for radio wave beams in plasma, *Sov. J. Plasma Phys.*, Vol. 3, 1977, pp. 185–194.

43. Bochkarev, G.S., Eremenko, V.A., and Lobachevskii, L.A. et al., Nonlinear interaction of decametre radio waves at close frequencies in oblique propagation, *J. Atmos. Terr. Phys.*, Vol. 44, No. 12, 1982, pp. 1137–1141.

44. Bochkarev, G.S., Eremenko, V.A., and Lobachevskii, L.A. et al., Nonlinear interaction of radiowaves in oblique propagation, in: *Effects of Artificial Modification of the Earth's Ionosphere, Moscow*: Science, 1983, pp. 53–71.

45. Erukhimov, L.M. and Frolov, V.L., *Dynamical and Spectral Characteristics of Artificial Radiation of Radio Waves by Ionospheric Plasma*, Gorky: NIRFI, Preprint No. 185, 1984.

46. *Radio Science*, Special issue, Vol. 9, No. 11, 1974, pp. 881–1090.

47. Wong, A.Y. and Brandt, R.G., Ionospheric modification—An outdoor laboratory for plasma and atmospheric science, *Radio Science*, Vol. 25, 1990, pp. 1251–1267.

48. Riggin, D., Swartz, W.E., Providakes, J., and Farley, D.T., Radar stadies of long-wavelength waves associated with mid-latitude sporadic E layers, *J. Geophys. Res.*, Vol. 91, 1986, pp. 8011–8024.

49. Fejer, J.A., Gonzales, C.A., and Ierkic, H.M. et al., Ionospheric modification experiments with the Arecibo heating facility, *J. Atmos. Terr. Phys.*, Vol. 47, 1985, pp. 1165–1179.

50. Gershman, B.N. and Grigor'ev, G.I., Travelling ionospheric disturbances and their relation to ionospheric waves, *J. Ionospheric Research*, Vol. 25, 1978, pp. 5–15.

51. Ignat'ev, Y.A., Calculation of uinization distribution in a mid-latitude sporadic E-layer, *Izv. Vuz., Radiophyzika*, Vol. 14, 1971, pp. 554–561.

52. Vas'kov, V.V. and Gurevich, A.V., Excitation in the ionosphere of parametric instability of hydrodynamic type, *Geomagn. Aeronom.*, Vol. 15, 1975, pp. 194–198, 339–344, 448–453.

53. Fejer, J. A. and Leer, E., Excitation of parametric instabilities by radio waves in the ionosphere, *Radio Sci.*, Vol. 7, 1972, pp. 481–493.

54. Mityakov, N.A., Rapoport, V.O., and Trakhtengerts, V.Yu., Heating of the ionosphere by an electromagnetic field under developed parametric instability, *Izv. Vuz., Radiophyzika*, Vol. 18, 1975, pp. 27–35.

55. Perkins, F.W., Oberman, C., and Valeo, E.J., Parametric instabilities and ionospheric modification, *J. Geophys. Res.*, Vol. 79, 1974, pp. 1478–1487.

56. Getmantsev, G.G., Komrakov, N.P., and Korobkov, P.P. et al., Some results of investigations of nonlinear phenomena in the F-layer of the ionosphere, *JETF Lett.*, Vol. 18, 1973, pp. 364–365.

57. Ginzburg, V.L., *Propagation of Electromagnetic Waves in Plasmas*, New York: Pergamon Press, 1964.

58. Gelberg, M.G., *Irregularities of the High-latitude Ionosphere*, Novosibirsk: Science, 1986.

59. Poliakov, S.V. and Yakhno, V.G., About thermo-diffusion mechanism of generation of inhomogeneities of electron concentration in the F-layer of the ionosphere, *Physics of Plasma*, Vol. 6, 1980, pp. 383–387.

60. Erukhimov, L.M., Kagan, L.M., and Miasnikov, E.H., About heating mechanism of origin of inhomogeneities of F-layer of the ionosphere, *Geomagn. Aeronom.*, Vol. 22, 1982, pp. 721–724.

61. Rumiantsev, S.A. and Smirnov, V.S., About origin of recombination instability in the high-latitude ionosphere, *Geomagn. Aeronom.*, Vol. 20, 1980, pp. 1107–1109.

62. Fel'dshtein, A.Ya., About influence of heating of particles at the Farley-Buneman instability in the auroral ionosphere, *Geomagn. Aeronom.*, Vol. 20, 1980, pp. 333–334.

63. D'Angelo, N., Recombination instability, *Phys. Fluids*, Vol. 10, 1967, pp. 719–723.

64. Erukhimov, L.M., Kagan, L.M., and Savina, O.N., About heating mechanism of formation of small-scale plasma inhomogeneities at the heights of E-layer, *Izv. Vuz., Radiophysika*, Vol. 24, 1983, pp. 1032–1034.

65. Gershman, B.N., Ignat'ev, Yu.A., and Kamenetskaya, G.H., *Mechanisms of Formation of Ionospheric Sporadic E_s-layer at Various Latitudes*, Moscow: Science, 1976.

66. Gershman, B.N., Kazimirovsky, E.S., Kokourov, V.D., and Chernobrovkina, N.A., *Phenomena of F-scattering in the Ionosphere*, Moscow: Science, 1984.

67. Schmidt, M.J. and S.P. Gary, Density gradients and Farley–Buneman instability, *J. Geophys. Res.*, Vol. 78, 1974, pp. 8261–8265.

68. Zarnitsky, Yu.F. and Harkova, T.N., About hydrodynamic description of Farley's turbulence in the polar ionosphere, in book: *Mathematical Modeling of Complex Processes*, Apatity: KFAR USSR, 1982, pp. 35–47.

69. Ott, E. and Farley, D.T., Microstabilities and production of short wavelength irregularities in the auroral *F*-region, *J. Geophys. Res.*, Vol. 80, 1975, pp. 4599–4602.

70. Fejer, B.G., *Natural Ionospheric Plasma Waves*, in book: *Modern Ionospheric Science*, Ed. by Kohl, H., Ruster, R., and Schlegel, K., Berlin: Luderitz and Bauer, 1996, pp. 216–273.

71. Farley, D.T., Theory of equatorial electrojet plasma waves: New developments and current status, *J. Atmos. Terr. Phys*, Vol. 47, 1985, pp. 729–744.

72. Swartz, W.E. and Farley, D.T., High resolution radar measurements of turbulent structure in the equatorial electrojet, *J. Geophys. Res.*, Vol. 99, 1994, pp. 309–317.

73. Rogister, A. and D'Angelo, N., Type 2 irregularities in the equatorial electrojet, *J. Geophys. Res.*, Vol. 75, 1970, pp. 3879–3887.

74. Kelley, M.C. and Fukao, S., Turbulent upwelling of the mid latitude ionosphere, 2. Theoretical framework, *J. Geophys. Res.*, Vol. 96, 1991, pp. 3747–3753.

75. Fejer, B.G. and Providakes, J.F., High latitude E-region irregularities: New results, *Physica Scripta*, Vol. T18, 1987, pp. 167–178.

76. Tsunoda, R.T., High-latitude F region irregularities: A review and synthesis, *Rev. Geophys.*, Vol. 26, 1988, pp. 719–760.

77. Haldopis, C., A review of radio studies of auroral E region ionospheric irregularities, *Annales Geophys.*, Vol. 7, 1989, pp. 239–258.

78. Bowman, G.G. and Hajkowicz, L.A., Small-scale ionospheric structures associated with mid-latitude spread-*F*, *J. Atmos. Terr. Phys.*, Vol. 53, 1991, pp. 4447–4457.

79. Yamamoto, M.S., Fukao, S., Woodman, R.F., Tsuda, T., and Kato, S., Mid latitude E-region field-aligned irregularities observed with the MU radar, *J. Geophys Res.*, Vol. 96, 1991, pp. 15943–15949.

80. Post, R.F. and Rosenbluth, M.N. Electrostatis instability in finite mirror confined plasmas, *Phys. Fluids*, Vol. 9, 1966, pp. 730–749.

81. Sperling, J.I. and Goldman, S.R., Electron collisional effects on lower-hybrid-drift instabilities in the ionosphere, *J. Geophys. Res.*, Vol. 85, 1980, pp. 3494–3498.

82. Huba, J.D. and Ossakow, S.L., On the generation of 3 m irregularities during equatorial spread F by low frequency drift waves, *J. Geophys. Res.*, Vol. 84, 1979, pp. 6697–6709.

83. Keskinen, M.J. and Huba, J.D., Generation of lower hybrid waves by inhomogeneous electron streams, *J. Geophys. Res.*, Vol. 88, 1983, pp. 3109–3115.

84. Frihagen, J. and Jacobson, T., In-situ observations of high-latitude F-region irregularities, *J. Atmos. Terr. Phys.*, Vol. 33, 1971, pp. 519–522.

85. Erukhimov, L.M., Maximenko, O.I., and Miasnikov, E.N., in book: *Ionospheric Reseach*, Moscow: *Sov. Radio*, No. 30, 1980, pp. 27–48.

86. Woodman, R.F. and LaHoz, C., Radar observations of *F* region equatorial irregularities, *J. Geophys. Res.*, Vol. 81, 1976, pp. 5447–5466.

87. Mellory, D.K., Note on the theory of equatorial sporadic E, *J. Atmos. Terr. Phys.*, Vol. 27, 1985, pp. 641–643.

88. Kindel, S.M. and Kennel, C.F., Topside current instabilities, *J. Geophys. Res.*, Vol. 76, 1971, pp. 3055–3078.

89. Chaturvedi, P.K. and Kaw, P.K., Current driven ion-cyclotron waves on collisional plasma, *Plasma Phys.*, Vol. 17, 1975, pp. 447–452.

90. Chaturvedi, P.K., Collisional ion-cyclotron waves in the auroral ionosphere, *J. Geophys. Res.*, Vol. 81, 1976, pp. 6169–6171.

91. Gershman, B.N. and Gmur, L.E., About effects of gravitation force on conditions of generation of some instabilities in F region of the ionosphere, *Izv. Vuz., Radiophysika*, Vol. 21, 1978, pp. 1572–1581.

92. Gel'berg, M.G., Formation of small-scale inhomogeneities in *F*-region of the high-latitude ionosphere, in book: *Ionospheric Research*, Moscow: Science, 1985.

93. Ronchi, C., Similon, P.L., and Sudan, R.N., A nonlocal linear theory of the gradient drift instability in the equatorial electrojet, *J. Geophys. Res.*, Vol. 94, 1989, pp. 1317–1326.

94. Ronchi, C., Sudan, R.N., and Similon P.L., Effect of short-scale turbulence on kilometer wavelength irregularities in the equatorial electrojet, *J. Geophys. Res.*, Vol. 95, 1990, pp. 189–200.

95. Cunnold, D.M. Drift-dissipative plasma instability and equatorial spread *F*, *J. Geophys. Res.*, Vol. 74, 1969, pp. 5709–5720.

96. Gershman, B.N. and Kamenetskaya, G.H., About current mechanism of formation of inhomogeneities leading diffusion of the region *F* at high latitudes, *Izv. Vuz., Radiophysika*, Vol. 16, 1973, pp. 988–996.

97. Gel'berg, M.G., Gradient-drift instability of ionospheric plasma, in book: *Propagation of Radio Waves in the Polar Ionosphere*, Apatiti: KFAN USSR, 1977, pp. 3–37.

98. Farley, D.T., The plasma instability resulting in field-aligned irregularities in the ionosphere, *J. Geophys. Res.*, Vol. 68, 1963, pp. 6083–6097.

99. Buneman, O., Excitation of field-aligned sound waves by electron streams, *Phys Rev. Lett.*, Vol. 10, 1963, pp. 285–287.

100. Wang, T.N.C. and Tsunoda, R.T., On a crossed field two-stream plasma instability in the auroral plasma, *J. Geophys. Res.,* Vol. 80, 1975, pp. 2172–2182.

101. Dimant, Y.S. and Sudan, R.N., Kinetic theory of low-frequency cross-field instability in a weakly ionized plasma, I, *Phys. Plasmas*, Vol. 2, 1995, pp. 1157–1168; *ibid* II, pp. 1169–1181.

102. Dimant, Y.S. and Sudan, R.N., Kinetic theory of the Farley–Buneman instability in the E region of the ionosphere, *J. Geophys. Res.*, Vol. 100, No. A8, 1995, pp. 14,605–14,623.

103. Gershman, B.N., Mechanisms of occurrence of ionospheric irregularities in the *F* region, in book: *Ionospheric Research*, Moscow: Sov. Radio, 1980, pp. 17–26.

104. Gel'berg, M.G., Formation of small-scale inhomogeneities above the maximum of F-layer of the auroral ionosphere, *Geomagn. Aeronom.*, Vol. 19, 1979, pp. 629–632.

105. Gershman, B.N., About conditions of formation of Rayleigh–Taylor instability in the region F of the ionosphere, in the book: *Ionospheric Irregularities*, Yakutsk, Academy of Scienece of USSR, 1981, pp. 3–15.

106. Simon, A., Instability of partially ionized plasma in crossing electric and magnetic fields, *Phys. Fluids*, Vol. 6, 1963, pp. 382–388.

107. Ossakow, S.L. and Chaturvedi, P.K., Current convective instability in the diffuse aurora, *Geophys. Res. Lett.*, Vol. 6, 1973, pp. 332–334.

108. Ried, G.C., The formation of small-scale irregularities in the ionosphere, *J. Geophys. Res.*, Vol. 73, 1968, pp. 1627–1640.

109. Unwin, R.S., The evening diffuse radio aurora, field-aligned currents and particle precipitation, *Planet Space Sci.,* Vol. 28, 1980, pp. 547–557.

110. Woodman, R.F., Yamamoto, M., and Fukao, S., Gravity waves modulations of gradient drift instabilities in mid latitude sporadic-E irregularities, *Geophys. Res. Lett.*, Vol. 18, 1991, pp. 1197–1200.

111. Labelle, J., Kelley, M.C., and Seyler, C.E., An analysis of the role of drift waves in equatorial spread F, *J. Geophys. Res.*, Vol. 91, 1986, pp. 5513–5525.

112. Fejer, B.G., Low latitude electro-dynamic plasma drifts: A review, *J. Atmos. Terrest. Phys.*, vol. 53, 1991, pp. 677–693.

113. Huang, C.-S. and Kelley, M.C., Nonlinear evolution of equatorial spread F, 1. On the role of plasma instabilities and spatial resonance associated with gravity wave seeding, *J. Geophys. Res.*, Vol. 101, 1996, pp. 283–292.

114. Kelley, M.C., Haerendel, G., and Kappler, H. et al., Evidence for a Rayleigh–Taylor type instability and upwelling of depleted density regions during equatorial spread F, *Geophys. Res. Lett.*, Vol. 3, 1976, pp. 448–451.

115. Ossakow, S.L., Spread-*F* theories—A review, *J. Atmos. Terrest. Phys.*, Vol. 43, 1981, pp. 437–452.

116. Migliuolo, S., Nonlocal dynamics in the collisional Rayleigh–Taylor instability: Applications to the equatorial spread F, *J. Geophys. Res.*, Vol. 101, 1996, pp. 10975–10984.

117. Zargham, S. and Seyler, C.E., Collisional and inertial dynamics of the ionospheric interchange instability, *J. Geophys. Res.*, Vol. 94, 1989, pp. 9009–9027.

118. Huang, C.-S. and Kelley, M.C., Nonlinear evolution of equatorial spread F, 2. Gravity waves seeding of the Rayleigh–Taylor instability, *J. Geophys. Res.*, Vol. 101, 1996, pp. 293–302.

119. Farley, D.T., Balsley, B.B., Woodman, R.F., and McClure, J.P., Equatorial spread F: Implications of VHF radar observations, *J. Geophys. Res.*, Vol. 75, 1970, pp. 7199–7216.

120. Jayachandran, B., Balan, N., Rao, P.B., Sastri, J.H., and Bailey, G.J., HF Doppler and ionosonde observations on the onset conditions of equatorial spread F, *J. Geophys. Res.*, Vol. 98, 1993, pp. 13741–13750.

121. Farley, D.T., Swartz, W.E., Hysell, D.L., and Ronchi, C., High-resolution radar observations of daytime kilometer-scale wave structure in the equatorial electrojet, *J. Geophys. Res.*, Vol. 99, 1994, pp. 299–307.

122. Aggson, T.L., Laasko, H., Maynard, N.C., and Pfaff, R.F., In situ observations of bifurcation of equatorial ionospheric plasma depletions, *J. Geophys. Res.*, Vol. 101, 1996, pp. 5125–5132.

123. Kolmogorov, A.N., The local structure of turbulence in incompressible viscous fluid for very large Reynolds numbers, *C.R. (Dokl.) Acad. Sci. URSS*, Vol. 30, 1941, pp. 301–314.

124. Kadomtsev, B.B., *Plasma Turbulence*, New York: Academic Press, 1969.

125. Sudan, R.N., Akinrimisi, J., and Farley, D.T., Generation of small-scale irregularities in the equatorial electroject, *J. Geophys. Res.*, Vol. 78, 1973, pp. 240–248.

126. Kraichnan, R.H., The structure of isotropic turbulence at very high Reynolds numbers, *J. Fluid Mech.*, Vol. 5, 1959, pp. 497–510.

127. Sudan, R.N. and Keskinen, M.J., Theory of strong turbulent two-dimensional convection of low-preasure plasma, *Phys. Fluids*, Vol. 22, 1979, pp. 2305–2314.

128. Sudan, R.N., Unified theory of type I and II irregularities in the equatorial electrojet, *J. Geophys. Res.*, Vol. 88, 1983, pp. 4853–4860.

129. Kulsrud, R.M. and Sudan, R.N., On Kraichnan's 'direct interaction approximation' and Kilmogoroff's theory in two dimensional plasma turbulence, *Comments Plasma Phys. Contr. Fusion*, Vol. 7, 1982, pp. 42–51.

130. Keskinen, M.J. and Ossakow, S.L., Nonlinear evolution of plasma enhancement in the auroral ionosphere, *J. Geophys. Res.*, Vol. 87, 1982, pp. 144–153.

131. Gel'berg, M.G. and Fedorov, V.P., Generation of weakly-anizotropic inhomogeneities in the high-latitude ionosphere, *Geomagn. Aeronom.*, Vol. 23, 1983, pp. 230–233.

132. Dupree, T.H., Nonlinear theory of low-frequency instabilities, *Phys. Fluids*, Vol. 11, 1968, pp. 2680–2688.

133. Skadron, G. and Weinstock, J., Nonlinear stabilization of a two-stream plasma instability in the ionosphere, *J. Geophys. Res.*, Vol. 74, 1969, pp. 5113–5121.

134. Weinstock, J. and Sleeper, A., Nonlinear saturation of 'type I' irregularities in the equatorial electrojet, *J. Geophys. Res.*, Vol. 77, 1972, pp. 3621–3628.

135. Rogulien, T.D. and Weinstock, J., Nonlinear saturation of the gradient drift instability in the equatorial electrojet, *J. Geophys. Res.*, Vol. 78, 1973, pp. 6808–6810.

136. Benford, G. and Thomson, J.J., Probabilistic model of plasma turbulence, *Phys. Fluids*, Vol. 15, 1972, pp. 1496–1502.

137. Rogister, A. and Jamin, E., Two-dimensional nonlinear processes associated with 'type I' irregularities in the equatorial electrojet, *J.Geophys. Res.*, Vol. 80, 1975, pp. 1820–1827.

138. Rogister, A., Nonlinear theory of 'type I' irregularities in the equatorial electrojet, *J. Geophys. Res.*, Vol. 76, 1971, pp. 7754–7759.

139. Then, C.M., Repeated cascade theory of strong turbulence in a magnetic plasma, *Plasma Physics*, Vol. 18, 1976, pp. 609–626.

140. Gel'berg, M.G., Nonlinear effects of gradient drift instability, *Izv. Vuz., Radiophyzika*, Vol. 22, 1979, pp. 295–304.

141. St-Maurice, J.-P., Schlegel, K., and Banks, P.M., Anomalous heating of the polar *E* region by unstable plasma wave, 2, Theory, *J. Geophys. Res.*, Vol. 86, 1981, pp. 1453–1462.

142. St-Maurice, J.-P., A unified theory of anomalous resistivity and Joule heating effects in the presence of uinispheric *E* region irregularities, *J. Geophys. Res.*, Vol. 92, 1987, pp. 4533–4542.

143. Robinson, T.R., Towards a self-consistent nonlinear theory of radar auroral backscatter, *J. Atmos. Terr. Phys.*, Vol. 48, 1986, pp. 417–423.

144. Robinson, T.R. and Honorary, F., A resonance broading kinetic theory of the modified-two-stream instability: Implications for radar auroral backscatter experiments, *J. Geophys. Res.*, Vol. 95, 1990, pp. 1073–1085.

145. Robinson, T.R., Effects of the resonance broadening of Farley–Buneman waves on electron dynamics and heating in the auroral *E*-region, *J. Atmos. Terr. Phys.*, Vol. 54, 1992, pp. 749–757.

146. Sato, T., Nonlinear theory of the cross-field instability explosive mode coupling, *Phys. Fluids*, Vol. 14, 1971, pp. 2426–2435.

147. Sato, T., Nonlinear stabilization and nonrandom behaviour of macroinstabilities in plasmas. II. Numerical verification, *Phys. Fluids*, Vol. 17, 1974, pp. 3162–3169.

148. Kamenetskaya, G.H., About excitation of longitudinal waves by current of the equatorial current stream, *Geomagn. Aeronom.*, Vol. 9, 1969, pp. 351–353.

149. Perkins, F.W. and Doles, J.H., Velocity shear and the **E** × **B** instability, *J. Geophys. Res.*, Vol. 80, 1975, pp. 211–214.

150. Simon, A., Growth and stability of artificial ion clouds in the ionosphere, *J. Geophys. Res.*, Vol. 75, 1970, pp. 6287–6294.

151. Drake, J.F., Mubbrandon, M., and Huba, J.D., Three-dimentional equilibrium and stability of ionospheric plasma clouds, *Phys. Fluids*, Vol. 31, 1988, pp. 3412–3424.

152. Volk, H.J. and Haerendel, G., Striations in ionospheric ion clouds, *J. Geophys. Res.*, Vol. 76, 1971, pp. 4541–4559.

153. Rosenberg, N.W., Observation of striation formation in a barium ion clouds, *J. Geophys. Res.*, Vol. 76, 1971, pp. 6856–6864.

154. Perkins, F.W., Zabusky, N.J., and Doles, J.H., Deformation and striation of plasma cloud in the ionosphere. *J. Geophys. Res.*, Vol. 78, 1973, pp. 697–710.

155. Dzubenko, N.I. et al., Dynamics of artificial plasma clouds in Spolokh experiments: Movement pattern, *Planet Space Sci.*, Vol. 31, 1983, pp. 849–858.

156. Andreeva, L.A. et al., Dynamics of artificial plasma clouds in Spolokh experiments: Cloud deformation, *Planet Space Sci.*, Vol. 32, 1984, pp. 1045–1052.

157. Perkins, F.W., Zabusky, N.J., and Doles, J.H., Deformation and striation of plasma cloud in the ionosphere. 2. Numerical simulation of a nonlinear two-dimensional model, *J. Geophys. Res.*, Vol. 78, 1973, pp. 711–724.

158. Doles, J.H., Zabusky, N.J., and Perkins, F.W., Deformation and striation of plasma clouds in the ionosphere. 3. Numerical simulation of a multilevel model recombination chemistry, *J. Geophys. Res.*, Vol. 81, 1976, pp. 5987–6004.

159. Lloyd, K.H. and Haerendel, G., Numerical modeling of the drift and deformation of ionospheric plasma clouds and their interaction with other layers of the ionosphere, *J. Geophys. Res.*, Vol. 78, 1973, pp. 7389–7415.

160. Huba, J.D., Ossakow, S.L., Satyanarayana, P., and Guzdar, P.N., Linear theory of the instability with an inhomogeneous electric field, *J. Geophys. Res.*, Vol. 88, 1983, pp. 425–434.

161. Zalezak, S.T., Drake, J. F., and Huba, J.D., Dynamics of three-dimensional ionospheric plasma clouds, *Radio Sci.*, Vol. 23, 1988, pp. 591–598.

162. Zalezak, S.T., Drake, J.F., and Huba, J.D., Three-dimensional simulation study of ionospheric plasma clouds, *Geophys. Res. Lett.*, Vol. 17, 1990, pp. 1597–1600.

163. Kudeki, E., Farley, D.T., and Fejer, B.G., Long wavelength irregularities in the equatorial electrojet, *Geophys. Res. Lett.*, Vol. 9, 1982, pp. 684–687.

164. Rozhansky, V.A. and Tsendin, L.D., Evolution of strong ionospheric inhomogeneities of ionospheric plasma. I., *Geomagn. Aeronom.*, Vol. 24, 1984, pp. 414–419.

165. Rozhansky, V.A. and Tsendin, L.D., Evolution of strong ionospheric inhomogeneities of ionospheric plasma. II., *Geomagn. Aeronom.*, Vol. 24, 1984, pp. 598–602.

166. Rozhansky, V.A., Veselova, I.Y., and Voskoboynikov, S.P., Three-dimensional computer simulation of plasma cloud evolution in the ionosphere, *Planet Space Sci.*, Vol. 38, 1990, pp. 1375–1386.

167. Blaunstein, N., Tsedilina, E.E., Mishin, E.V., and Mirzoeva, L.I., Drift spreading and stratification of inhomogeneities in the ionosphere in presence of an electric field, *Geomagn. Aeronom.*, Vol. 30, 1990, pp. 656–661.

168. Blaunstein, N., Milinevsky, G.P., Savchenko, V.A., and Mishin, E.V., Formation and development of striated structure during plasma cloud evolution, *Planet Space Sci.*, Vol. 41, 1993, pp. 453–460.

169. Uruadov, V., Ivanov, V., Plokhotnuk, E., Erukhimov, L., Blaunstein, N., Iasi (Romania), and Filipp, N., *Dynamic Processes in Ionosphere – Methods of Investigations*, Iasi (Romania): Technopress, 2006.

170. Ferch, R.L. and R.N. Sudan, Numerical simulation of type II gradient drift irregularities in the equatorial electrojet, *J. Geophys. Res.*, Vol. 82, 1977.

171. Keskinen, M.J., Sudan, R.N., and Ferch, R.L., Temporal and spatial power spectrum studies of numerical simulations of type II gradient drift irregularities in the equatorial electrojet, *J. Geophys. Res.*, Vol. 84, 1979, pp. 1419–1430.

172. Sato, T. and Ogawa, T., Self-consistent studies of two-dimensional large-scale (∼100 m) electrojet irregularities. *J. Geophys. Res.*, Vol. 81, 1976, pp. 3248–3253.

173. Fedorov, V.P., Numerical modling of three-wave interaction in the ionospheric plasma, in book: *Structure of Magneto-Ionospheric and Auroral Perturbations*, Leningrad: Science, 1977, pp. 58–66.

174. Newman, A.L. and Ott, E., Nonlinear simulations of type I irregularities in the equatorial electrojet, *J. Geophys. Res.*, Vol. 86, 1981, pp. 6879–6891.

175. Ronchi, C., Sudan, R.N., and Farley, D.T., Numerical simulations of large-scale plasma turbulence in the daytime equatorial electrojet, *J. Geophys. Res.*, Vol. 96, 1991, pp. 21,263–21,279.

176. Keskinen, M.J., Nonlinear theory of the $\mathbf{E} \times \mathbf{B}$ instability with an inhomogeneous electric field, *J. Geophys. Res.*, Vol. 89, 1984, pp. 3913–3921.

177. Fu, Z.F., Lee, L.C., and Huba, J.D., A quasi-local theory of the $\mathbf{E} \times \mathbf{B}$ instability in the ionosphere, *J. Geophys. Res.*, Vol. 91, 1986, pp. 3263–3269.

178. Wang, X.H. and Bhattacharjee, A., Gradient drift eigenmodes in the equatorial electrojet, *J. Geophys. Res.*, Vol. 99, 1994, pp. 13,219–13,226.

179. Hu, S., Bhattacharjee, A., and Harrold, B.G., Linear gradient drift instabilities in the daytime equatorial electrojet, *J. Geophys. Res.*, Vol. 102, 1997, pp. 337–346.

180. Basu, B., Nonlinear saturation of Rayleigh–Taylor instability in the presence of time-dependent equilibrium, *J. Geophys. Res.*, Vol. 104, 1999, pp. 6859–6866.

181. Dimant, Y.S. and Sudan, R.N., Physical nature of a new cross-field current-driven instability in the lower ionosphere, *J. Geophys. Res.*, Vol. 102, 1997, pp. 2551–2563.

182. Gurevich, A.V. and Karashtin, A.N., Small-scale thermal diffusion instability in the lower ionosphere, *Geomagn. Aeronom.*, Vol. 24, 1984, pp. 733–741.

183. Gurevich, A.V., Borisov, N.D., and Zybin, K.P., Ionospheric turbulence induced in the lower part of the E region by the turbulence of the neutral atmosphere, *J. Geophys. Res.*, Vol. 102, 1997, pp. 379–388.

184. Blix, T.A., Thrane, E.V., Kirkwood, S., and Schlegel, K., Plasma instabilities in the low E-region observed during the DYANA campaign, *J. Atmos. Terr. Phys.*, Vol. 56, 1994, pp. 1853–1858.

185. Blix, T.A., Thrane, E.V., Kirkwood, S., Dimant, Y.S., and Sudan, R.N., Experimental evidence for unstable waves in the lower *E*/upper *D*-region excited near the bisector between the electric field and the drift velocity, *Geophys. Res. Lett.*, Vol. 23, 1996, pp. 2137–2139.

186. Hamza, A.M., A nonlinear theory for large aspect angle echoes in the auroral *E* region, *J. Geophys. Res.*, Vol. 97, 1992, pp. 16,981–16,993.

187. Hamza, A.M. and St-Maurice, J.-P., A turbulent theoretical framework for the study of current-driven E region irregularities at high latitudes: Basic derivation and application to gradient free situations, *J. Geophys. Res.*, Vol. 98, 1993, pp. 11,587–11,599.

188. Hamza, A.M. and St-Maurice, J.-P., Self-consistent fully turbulent theory of auroral *E* region irregularities, *J. Geophys. Res.*, Vol. 98, 1993, pp. 11,601–11,613.

189. St-Maurice, J.-P., Prikryl, P., Danskin, D.W. et al., On origin of narrow non-acoustic coherent radar spectra in the high-latitude *E*-region, *J. Geophys. Res.*, Vol. 99, 1994, pp. 6447–6474.

190. Hamza, A.M. and St-Maurice, J.-P., Large aspect angles in auroral E region echoes: A self-consistent turbulent fluid theory, *J. Geophys. Res.*, Vol. 100, 1995, pp. 5723–5732.

191. St-Maurice, J.-P. and Hamza, A.M., A new nonlinear approach to the theory of *E* region irregularities, *J. Geophys. Res.*, Vol. 106, 2001, pp. 1751–1759.

192. Hamza, A.M. and Imamura, H., On the excitation of large aspect angle Farley–Buneman echoes via three-wave coupling: A dynamical system model, *J. Geophys. Res.*, Vol. 106, 2001, pp. 24,745–24,754.

193. Blaunstein, N., Theoretical aspects of wave propagation in random media based on quanty and statistical field theory, *J. Progress in Electroman. Research, PIER*, Vol. 47, 2004, pp. 135–191.

194. Haldopis, C.A., Nielsen, E., and Ierkis, H.M., STARE Doppler spectral studies of westward electrojet radar aurora, *Planet Space Sci.*, Vol. 32, 1984, pp. 1291–1300.

195. Koehler, J.A., Sofko, G.J., and Mehta, V., A statistical study of magnetic aspect effects associated with VHF auroral backscatter, *Radio Sci.*, Vol. 20, 1985, pp. 689–698.

196. Hall, G. and Moorcroft, D.R., Doppler spectra of the UHF diffuse radio aurora, *J. Geophys. Res.*, Vol. 93, 1988, pp. 7425–7440.

197. Moorcroft, D.R. and Schlegel, K., E-region coherent backscatter at short wavelength and large aspect angles, *J. Geophys. Res.*, Vol. 93, 1988, pp. 2005–2014.

198. Moorcroft, D.R. and Schlegel, K., Height and aspect sensitivity of large aspect angle coherent backscatter at 933 MHz, *J. Geophys. Res.*, Vol. 95, 1990, pp. 19,011–19,021.

199. Schlegel, K., Turunen, T., and Moorcroft, D.R., Auroral radar measurements at 16-cm wavelength with high range and time resolution, *J. Geophys. Res.*, Vol. 95, 1990, pp. 19,001–19,009.

200. Haniuse, C., Villan, J.P., Cerisier, J.C. et al., Statistical study of high-latitude F region Doppler spectra obtained with the SHERPA HF radar, *Annalous Geophys.*, Vol. 9, 1991, pp. 273–285.

201. Rietveld, M.T., Collins, P.N., and St-Maurice, J.-P., Naturally enhaced ion acoustic waves in the auroral ionosphere observed with the EISCAT 933 MHz radar, *J. Geophys. Res.*, Vol. 96, 1991, pp. 19,291–19,301.

202. Foster, J.C., Tetenbaum, D., and Del Pozo, C.F. et al., Aspect angle variations in intensity, phase velocity, and altitude for high-latitude 34-cm E-region irregularities, *J. Geophys. Res.*, Vol. 97, 1992, pp. 8601–8617.

203. Del Pozo, C.F., Foster, J.C., and St-Maurice, J.-P., Dual-mode E region plasma wave observations from Millstone Hill, *J. Geophys. Res.*, Vol. 98, 1993, pp. 6013–6032.

Chapter 3

Radio Signal Presentation in the Ionospheric Communication Channel

In this chapter, we will discuss the signal within any multipath ionospheric link with fading. First, we will consider both the band-pass or radio frequency (RF) signal and the complex baseband. Then, in Section 3.2, we will define the continuous wave (CW), or narrowband signal, and describe the main parameters of such signals. Finally, in Section 3.3, we will describe the wideband signal by introducing its mathematical description and its main parameters, which, compared with the same parameters of the radio channel, can determine the type of fading occuring in the ionospheric wireless communication link. We will then move on to representations for the *narrowband* baseband signal and the *wideband* (or *pulse*) signal.

3.1   Bandpath and Baseband Signal Presentation

Each RF signal, defined in the literature as a band-pass signal [1–8], can be expressed in the following form:

$$x(t) = A(t) \cos[2\pi f_c t + \phi(t)] \tag{3.1}$$

where
   $A(t)$ is defined as the envelope
   $\phi(t)$ denotes the phase of a real signal $x(t)$

Using the geometric expansion of Equation 3.1 yelds

$$x(t) = a(t) \cos 2\pi f_c t - b(t) \sin 2\pi f_c t \tag{3.2}$$

where

$$a(t) = A(t) \cos \phi(t)$$
$$b(t) = A(t) \sin \phi(t)$$

(3.3)

are the *envelopes* of the two quadratic components, *I* and *Q*, defined as the *in-phase* and *quadrature-phase* components, respectively [2–7]. Usually, the real band-pass signal can be presented in an alternative, complex form containing both parameters, the envelope and the phase of the real signal $x(t)$, i.e.,

$$y(t) = A(t)\, e^{j\phi(t)}$$

(3.4)

The preceding equation expresses the complex baseband representation of the real signal $x(t)$, called the *baseband signal* in Refs. [2–4,7–10], which we denote as $y(t)$. Using complex notation, it is clear that the real signal can always be recovered from the complex baseband signal by simply multiplying it by the exponent with the time factor involving the carrier frequency f_c and finding the real part of the product, i.e.,

$$x(t) = \mathrm{Re}\big[y(t)e^{j2\pi f_c t}\big] = \mathrm{Re}\Big[A(t)\, e^{i(2\pi f_c t + \phi(t))}\Big]$$

(3.5)

where now

$$y(t) = a(t) + jb(t)$$

(3.6)

In the literature [6–10] the baseband waveform $y(t)$ is usually referred to as a complex envelope or the complex low-pass equivalent of the band-pass signal $x(t)$. In other words, $y(t)$ is a *phasor* of $x(t)$.

The corresponding frequency spectra for the baseband and band-pass signals are determined as [2–4,6–10]

$$Y(f) = \int_{-\infty}^{\infty} y(t)e^{-j2\pi ft}\, dt = \mathrm{Re}[Y(f)] + j\,\mathrm{Im}[Y(f)]$$

(3.7a)

and

$$X(f) = \int_{-\infty}^{\infty} x(t)e^{-j2\pi ft}\, dt = \mathrm{Re}[X(f)] + j\,\mathrm{Im}[X(f)]$$

(3.7b)

Substituting for $x(t)$ in integral (Equation 3.7b) from Equation 3.5 gives

$$X(f) = \int_{-\infty}^{\infty} \mathrm{Re}\big[y(t)e^{j2\pi f_c t}\big]e^{-j2\pi ft}\, dt$$

(3.8)

Taking into account that the real part of any arbitrary complex variable w can be written as

$$\text{Re}[w] = \frac{1}{2}[w + w^*]$$

where w^* is the complex conjugate, we can rewrite Equation 3.8 in the following form:

$$X(f) = \frac{1}{2} \int_{-\infty}^{\infty} \left[y(t)e^{j2\pi f_c t} + y^*(t)e^{-j2\pi f_c t} \right] e^{-j2\pi f t} \, dt \tag{3.9}$$

After comparing Equations 3.9 and 3.7, we get

$$X(f) = \frac{1}{2}[Y(f - f_c) + Y^*(-f - f_c)] \tag{3.10}$$

Figure 3.1 shows how the complex baseband signal spectrum can be schematically viewed as a frequency-shifted version of the real band-pass (RF) signal, with the same spectral shape in the frequency domain.

In other words, the spectrum of the real band-pass signal $x(t)$ can be represented by the real part of the complex baseband signal $y(t)$ with a shift of $\pm f_c$ along the frequency axis. It is clear that the baseband signal has its frequency content centered around "zero" frequency. Because initially we deal with narrowband signals, the bandwidth of which, Δf, is always less than the carrier frequency, f_c, the two components of $X(f)$ presented in Figure 3.1 do not overlap in frequency.

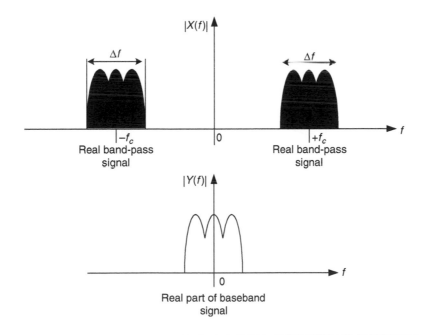

Figure 3.1 Schematic reppresentation of band-pass and baseband signals in the frequency domain.

3.2 Narrowband Signal Presentation

If now the band-pass (or RF) signal is transmitted within the satellite-stationary or satellite-mobile communication link, containing a set of frequencies within a bandwidth that is narrow compared to the carrier frequency, f_c, such a band-pass (RF) signal is called a *narrowband* or *continuous-wave* (CW) signal [1–4,6–14].

We will now give a mathematical description of CW signals. Several mathematical statistical models have been proposed to describe such signals for various kinds of channels: terrestrial, atmospheric, or ionospheric. To describe such narrowband signals mathematically, the 2-D (flat) Clarke model was used in Ref. [11] for a land mobile communication link. It cannot be successfully used for description of the interference of direct (incident) waves and waves reflected from different kinds of plasma structures occurring in the ionosphere, because it does not take into account the elevation of the satellite or any other vehicle embedded in the ionosphere (such as a rocket or missile). For this purpose we present here a more general 3-D model that is based on scattering from obstructions both in the azimutal and elevation planes. This stastistical model was worked out by Aulin and is also described in Ref. [12] for land communication links with high-elevated base station antennas. As was shown in Ref. [10], we can use it for the description of multiple scattering of radio waves within the ionospheric communication channel. Let us briefly describe this model.

The Aulin model assumes a fixed ground-based transmitter with a vertically polarized isotropic (called omnidirectional) antenna and a moving vehicle (satellite or rocket) as a receiver, which also has an omnidirectional antenna (see Figure 3.2). The signal at the receiver is assumed to consist of N horizontally traveling plane waves, each wave with a number i having amplitude A_i, and with statistically independent angle-of-arrival, azimuth (φ_i) and elevation (ϑ_i), and phase angle (ϕ_i) distributions. Aulin has generalised 2-D Clarke model by the inclusion of the angle ϑ_i (see Figure 3.2), which in the Clarke model was always zero.

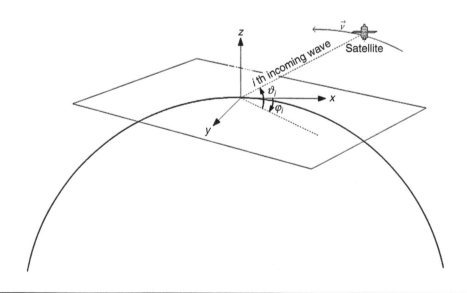

Figure 3.2 Geometry of the land–satellite communication link.

The vertically polarized plane electromagnetic waves arriving at the moving terminal antenna usually have one E-field component (according to the problem presented in Figure 3.2, it is E_z) and two H-field components, H_x and H_y, elongated along the x-axis and y-axis, respectively (see Figure 3.2). In the 3-D case, the angle ϑ_i describes the spatial orientation of the ith incoming wave and φ_i is its projection on the XY-plane (i.e., simply the azimuthal angle). According to Ref. [12], we assume that both ϑ_i and φ_i are homogeneously and independently distributed over the ranges $[0,\pi/2]$ and $[0,2\pi]$, respectively. The same homogeneous distribution is assumed for the ith wave phase ϕ_i over the range $[0,2\pi]$; i.e., the probability density function (PDF) of such a distribution is $P(\phi_i) = 1/(2\pi)$.

Without any loss of generality of the problem, because the same technique is used for each field component, let us consider only the E-field component. In this case, for each electrical field component wave with number i, we can obtain a general expression at the arbitrary point (x_0,y_0,z_0) according to Ref. [12], if an unmodulated RF signal was transmitted:

$$E_i(t) = A_i \cos[\omega_c t - k(x_0 \cos \varphi_i \cos \vartheta_i + y_0 \sin \varphi_i \cos \vartheta_i + z_0 \sin \vartheta_i) + \phi_i] \qquad (3.11)$$

and the resulting E-field component of the received electromagnetic wave is given by

$$E(t) = \sum_{i=1}^{N} E_i(t) \qquad (3.12)$$

where
$k = 2\pi/\lambda$, λ is the wavelength
$\omega_c = 2\pi f_c$, f_c the carrier frequency
ϕ_i the initial phase of ith component of the total signal

For the case when the electric field component is aligned along the z-axis, Equation 3.11 reduces to the case presented in Ref. [11]:

$$E_{zi}(t) = A_i \cos[\omega_c t - k(x_0 \cos \varphi_i + y_0 \sin \varphi_i) + \phi_i] \qquad (3.13)$$

The same expressions can be obtained for the magnetic field components of each multipath wave with number i. We present these components in the 2-D case, according to Ref. [11], when all of them lie in the XY-plane:

$$H_{xi}(t) = -\frac{A_i}{120\pi} \cos[\omega_c t - k \sin \varphi_i(x_0 \cos \varphi_i + y_0 \sin \varphi_i) + \phi_i]$$

$$H_{yi}(t) = \frac{A_i}{120\pi} \cos[\omega_c t - k \cos \varphi_i(x_0 \cos \varphi_i + y_0 \sin \varphi_i) + \phi_i] \qquad (3.14)$$

The resulting H-field components of the received electromagnetic wave are given in the same manner as in Equation 3.12, i.e.,

$$H_x(t) = \sum_{i=1}^{N} H_{xi}(t), \quad H_y(t) = \sum_{i=1}^{N} H_{yi}(t) \qquad (3.15)$$

Equations 3.11 through 3.15 determine the total electromagnetic field at the receiver and cover both the cases of verical and horizontal wave polarization. Following from these equations, and as shown in Refs. [11,12], all three electromagnetic field components are randomly but independently oriented along the corresponding axis in the space domain, and due to random summation of each component with random phase ϕ_i and spatial angles, φ_i and ϑ_i, they are mutually uncorrelated. The mean square value of the amplitude A_i of such uniformly distributed individual waves is constant

$$E\left(A_i^2\right) \equiv \langle A_i^2 \rangle = \frac{E_0^2}{N} \tag{3.16}$$

where the operator in brackets, $\langle \bullet \rangle$, describes the procedure of averaging over the set of signal strengths ($\sim A_i$) or intensities ($\sim A_i^2$). Because $N =$ constant, the real amplitude of the local average field E_0 is also assumed to be a constant.

We now assume that the vehicle (satellite or rocket) moves with velocity v in the XY-plane, making an angle θ to the x-axis (as shown in Figure 3.3 for the vector of the vehicle velocity directed along the x-axis). Then, after unit time, the coordinates of the the vehicle are $\{v \cos\theta, v \sin\theta, z_0\}$ and the Doppler shift, $\omega_i = 2\pi f_{Di}$, of each ith component of E- and H-field can be presented as

$$f_{Di} = \frac{\omega_i}{2\pi} = \frac{v}{\lambda} \cos(\theta - \varphi_i) \cos\vartheta_i = f_{D\max} \cos(\theta - \varphi_i) \cos\vartheta_i \tag{3.17}$$

Because the Doppler shift is comparatively small even for satellites (about hundreds of kHz) with respect to the carrier frequency of the transmitter (about hundreds of MHz to tens of GHz), all field components may be modeled as narrowband random processes and approximated as Gaussian random variables, if $N \rightarrow \infty$, with a uniform phase distribution in the interval $[0, 2\pi]$. If so, the E-field component can be expressed in the following form:

$$E_z = C(t) \cos(\omega_c t) - S(t) \sin(\omega_c t) \tag{3.18}$$

where $C(t)$ and $S(t)$ are the in-phase and quadrature components that would be detected by a suitable receiver:

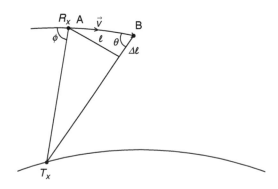

Figure 3.3 Geometrical representation of the Doppler effect in the land–satellite communication link.

$$C(t) = \sum_{i=1}^{N} A_i \cos(\omega_i t + \beta_i)$$

$$S(t) = \sum_{i=1}^{N} A_i \sin(\omega_i t + \beta_i) \qquad (3.19)$$

According to the preceding assumptions, both components $C(t)$ and $S(t)$ are independent Gaussian random processes that are completely characterized by their mean value and auto-correlation function [12]. Moreover, they are also uncorrelated zero-mean Gaussian random variables; i.e.,

$$\langle S \rangle = \langle C \rangle = \langle E_z \rangle = 0 \qquad (3.20)$$

with an equal variance σ^2 (the mean signal power) given by

$$\sigma^2 \equiv \langle |E_z|^2 \rangle = \langle S^2 \rangle = \langle C^2 \rangle = E_0^2/2 \qquad (3.21)$$

The envelope of each received component of the total signal, $E_z(t)$, $H_x(t)$, $H_y(t)$, can be presented by using these in-phase and quadrature components as was done in Refs. [11,12]; i.e., for the electric field component we get:

$$|E_z| = \sqrt{C^2(t) + S^2(t)} = r(t) \qquad (3.22)$$

As was shown in Refs. [11,12], as components $C(t)$ and $S(t)$ are independent Gaussian random variables that satisfy Equations 3.20 through 3.21, the random received signal envelope $r(t)$ has Rayleigh (for NLOS case) and Ricean distributions for (quasi-LOS case). We will discuss the different kinds of signal distribution in the next chapter, considering different kinds of fading occuring in the multipath channel.

In the earlier expressions, new parameters ω_i and ϑ_i were introduced for the 3-D case:

$$\omega_i = \frac{2\pi v}{\lambda} \cos(\theta - \varphi_i) \cos\vartheta_i \qquad (3.23a)$$

$$\beta_i = \frac{2\pi z_0}{\lambda} \sin\vartheta_i + \phi_i \qquad (3.23b)$$

If we now assume that waves arrive from all angles in the azimuth XY-plane with equal probability, we obtain, according to Ref. [12], the *PDFs* of angle-of-arrival φ and:

$$PDF(\varphi) = \frac{1}{2\pi} \qquad (3.24)$$

$$PDF(\vartheta) = \frac{\cos\vartheta}{2\sin\vartheta_m}, \quad |\vartheta| \leq |\vartheta_m| \leq \frac{\pi}{2}$$

$$PDF(\vartheta) = 0, \quad \text{elsewhere} \qquad (3.25)$$

Here, each ϑ_m relates to the corresponding maximum Doppler shift $f_{Dm} = v/\lambda$. Dependence of $PDF(\vartheta)$ for different ϑ_m according to Equation 3.25 is depicted in Figure 3.4.

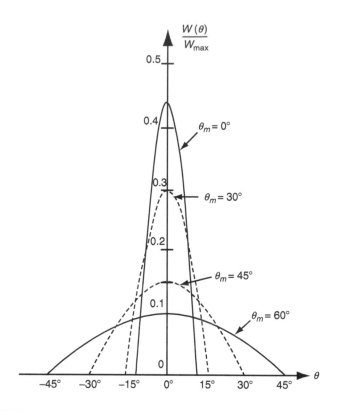

Figure 3.4 Normalized signal power distribution in the elevation domain for various directions of the base station antenna main loop.

In narrowband links, dealing with CW (narrowband) signals, the receivers do not usually detect the previously introduced components $C(t)$ and $S(t)$. They record the envelope $r(t)$ of the total signal $E(t)$ described by Equations 3.11 through 3.12; Equation 3.22 is also used. The complex envelope of the received CW signal can be presented within the multipath channel as the phasor sum of N baseband individual multiray components arriving at the receiver with the corresponding time delay, τ_i, $i = 0, 1, 2, \ldots$ [2–4,6–10]:

$$r(t) = \sum_{i=0}^{N-1} r_i(t) = \sum_{i=0}^{N-1} A_i \exp(j\phi_i(t,\tau_i)) \tag{3.26}$$

If we assume that during the vehicle movements over a local area the variations of amplitude A_i are sufficiently small, whereas phases φ_i vary greatly due to changes in propagation distance over the space, then as a result, we obtain large random oscillations of the total signal $r(t)$ during the vehicle's movement over small distances. Finally, as $r(t)$ is the phasor sum (Equation 3.26) of the individual multipath components, the instantaneous phases of the multipath components cause the large fluctuations that typify small-scale fast fading for CW signal $r(t)$. The average received power over a local area is then given by Refs. [7–12]:

$$\langle P_{CW} \rangle \approx \sum_{i=0}^{N-1} \langle A_i^2 \rangle + 2 \sum_{i=0}^{N-1} \sum_{i,j \neq i} \langle A_i A_j \rangle \langle \cos(\phi_i - \phi_j) \rangle \tag{3.27}$$

Here, the time averaging was carried out for CW measurements made by a mobile vehicle (the receiver) over the local measured area.

Furthermore, in narrowband communication links, most of the energy of the received signal is concentrated near the carrier frequency, f_c, and the power spectra is defined as the RF spectra of such a narrowband signal. At the same time, as was shown in Section 3.1, the RF narrowband signal can be converted into the complex baseband signal that has its frequency content centered around zero frequency (see Figure 3.1). Below, we will present the power baseband spectra distribution in the Doppler shift domain. For this purpose we introduce the autocorrelation function $K(t, t + \tau) = \langle E(t) \cdot E(t + \tau) \rangle$ of signal fading in the time domain [11,12]. This can be done by using, for example, the E-field components' presentation according to Equation 3.18 in terms of a time delay τ, as

$$\begin{aligned} K(\tau) &= \langle C(t) \cdot C(t + \tau) \rangle \cos \omega_c \tau - \langle S(t) \cdot S(t + \tau) \rangle \sin \omega_c \tau \\ &= c(\tau) \cos \omega_c \tau - s(\tau) \sin \omega_c \tau \end{aligned} \tag{3.28}$$

According to Equation 3.28 the correlation properties of signal fading are fully described by functions $c(\tau)$ and $s(\tau)$ [12]:

$$c(\tau) = \frac{E_0}{2} \langle \cos \omega_c \tau \rangle, \quad s(\tau) = \frac{E_0}{2} \langle \sin \omega_c \tau \rangle \tag{3.29}$$

Following the assumption of a homogeneous distribution of incoming waves over the angles of arrival and phase, mentioned earlier, we immediately obtain for the 3-D statistical model proposed by Aulin [12] that $s(\tau) = 0$, and

$$c(\tau) = \frac{E_0}{2} \int_{-\pi}^{+\pi} J_0(2\pi f_{Dm} \tau \cos \vartheta) PDF(\vartheta) \, d\vartheta \tag{3.30}$$

where $J_0(w)$ is the zero-order Bessel function, $PDF(\vartheta)$ is described by Equation 3.25 and f_{Dm} is the maximum Doppler shift, $f_{Dm} = v/\lambda$.

As is well known [8–13], the power spectrum of the resulting received signal can be obtained as the Fourier transform of the temporal autocorrelation function (Equation 3.28). Taking the Fourier transform only for the component (Equation 3.30) of the autocorrelation function of the total signal (Equation 3.28), i.e., converting a 3-D model to a 2-D model, we obtain the corresponding complex baseband power spectrum in the Doppler frequency shift domain:

$$W_0(f) = \frac{E_0}{4\pi f_{Dm}} \frac{1}{\sqrt{1 - \left(\frac{f}{f_{Dm}}\right)^2}}, \quad |f| \leq f_{Dm}$$

$$W_0(f) = 0, \quad \text{elsewhere} \tag{3.31}$$

As follows from Equation 3.31, the 2-D spectrum (defined in the literature as a classical "U-shaped" spectrum) is strictly band-limited within the range $|f| \leq f_{Dm}$, and the power spectral density becomes infinite at $f = \pm f_{Dm}$. This spectrum is a classical Doppler spectrum that is usually used in many situations in various radio communication channels. As for the Aulin model which deals with both components of Equation 3.28, because of the angle of arrival spread in the vertical direction, most of the energy shifts from the peaks at $f = \pm f_{Dm}$ and toward lower Doppler frequencies, therefore reducing the fading rate. In fact, from the Aulin 3-D model, more complicated baseband power spectral density distributions in the frequency domain were obtained in Ref. [12] using Equations 3.28 and 3.29 with Equation 3.25. This spectrum can only be evaluated numerically [12] and analytically [14] for the quasi-3-D case, denoted in Ref. [14] as $W_1(f)$:

$$W_1(f) = 0, \qquad\qquad\qquad\qquad |f| > f_{Dm}$$

$$W_1(f) = \frac{E_0}{4 \sin \vartheta_m f_{Dm}}, \qquad\qquad f_{Dm} \cos \vartheta_m \leq |f| \leq f_{Dm} \qquad (3.32)$$

$$W_1(f) = \frac{E_0}{f_{Dm}} \left[\frac{\pi}{2} - \sin^{-1} \frac{2 \cos^2 \vartheta_m - 1 - (f/f_{Dm})^2}{1 - (f/f_{Dm})^2} \right], \quad |f| < f_{Dm} \cos \vartheta_m$$

Comparison of the 2-D spectrum (Equation 3.31), shown in Figure 3.5 by a continuous curve, with that described by Equation 3.32, shows that two spectra are strictly band-limited to $|f| \leq f_{Dm}$, but in the quasi-3-D case the power spectrum density is always finite. The spectrum $W_1(f)$ from Equation 3.32 is actually limited by $f_{Dm} \cos \vartheta_m \leq |f| \leq f_{Dm}$. In contrast, the 2-D spectrum $W_0(f)$ is infinite at $|f| = f_m$. As was mentioned in Refs. [12,14], the low-frequency content is greatly increased even when ϑ_m is small.

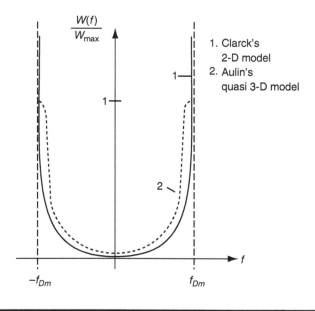

Figure 3.5 Normalized signal power spectrum distribution in the Doppler frequency domain. 1 is according to the Clarke 2-D model (solid curve), and 2 is according to the Aulin 3-D model (dashed curve).

Earlier we talked only about the E_z component. Now we will present the magnetic field components. As for the 3-D model it is very complicated to obtain analytical expressions for the baseband signal power spectrum, we will follow the Clarke statistical 2-D model [11] to show differences between various signal components. A similar analysis as was made for the E_z component, this time using Equations 3.28 through 3.32 with Equation 3.14, yields the expressions for the baseband spectra $W_0(f)$ distribution in the frequency domain for the magnetic field components H_x and H_y:

$$W_0(f) = \frac{E_0}{4\pi f_{Dm}} \sqrt{1 - \left(\frac{f}{f_{Dm}}\right)^2} \quad \text{for} \quad H_x$$

$$W_0(f) = \frac{E_0}{4\pi f_{Dm}} \frac{\left(\frac{f}{f_{Dm}}\right)^2}{\sqrt{1 - \left(\frac{f}{f_{Dm}}\right)^2}} \quad \text{for} \quad H_y \tag{3.33}$$

The signal power spectra (Equation 3.33) versus the radiated frequency is shown in Figure 3.6 with the spectrum of the E_z component included for comparison.

As is shown in the illustrations, the E_z component, like the H_y component, has a minimum at the radiated carrier frequency and both are strictly limited to infinity at the range $\pm f_{Dm}$ around the carrier frequency, whereas the H_x component has a maximum at the carrier frequency and is strictly limited to zero at the range $\pm f_{Dm}$.

We conclude therefore that the baseband signal spectrum of the real passband signal is strictly band-limited to the range $\pm f_{Dm}$ around the "zero" frequency, and within those limits the power spectral density depends on the PDFs associated with the spatial angles of arrival φ and ϑ.

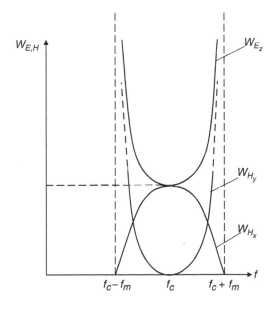

Figure 3.6 Normalized signal power spectrum distribution in the Doppler frequency domain for all three components of the electric and magnetic fields.

The limits of the Doppler spectrum can be quite high. For example, if a satellite moves with velocity $\sim$9 km/s and receives a signal at a carrier frequency of 1 GHz, the maximum Doppler shift is $f_m = v/\lambda = \frac{9 \cdot 10^3 \text{ m/s}}{0.3 \text{ m}} = 30 \text{ kHz}$. Frequency shifts of such a magnitude can cause interference with any message information.

3.3 Wideband Signal Presentation

An example of a power profile of an actual *wideband* (or *pulse*) signal at the receiver is presented in Figure 3.7a. When plotted in the time domain, the signal envelope power is defined as the *power delay profile*. If we now apportion the time delay axis into equal segments, or *bins*, as shown in Figure 3.7b, then there will be a number of signals in each bin corresponding to the different paths whose times of arrival are within the bin duration. If so, these signals can be represented vectorially using delta-function presentation [7–13] centered around each bin, as shown in Figure 3.8. This time-varying discrete-time impulse response, which completely characterizes the communication channel, can be expressed as a function of both the current time t and time delay τ:

$$h_b(t,\tau) = \left\{ \sum_{i=0}^{N-1} A_i(t,\tau)\exp(-j2\pi f_c\tau_i(t) - j\phi_i(t,\tau))\delta(\tau - \tau_i(t)) \right\} \quad (3.34)$$

The variable t represents the time variations due to the motion of one of the terminal antenna, whereas τ represents the channel multipath delay for a fixed value of t. If the channel impulse response is assumed to be time invariant, or is at least stationary over a small-scale time or displacement interval, then the impulse response (Equation 3.34) can be simplified as

$$h_b(\tau) = \sum_{i=0}^{N-1} A_i(\tau)\exp(-j\theta_i)\delta(\tau - \tau_i) \quad (3.35)$$

where
$\theta_i = 2\pi f_c\tau_i + \phi_i(\tau)$
$A_i(\tau)$ is the amplitude of each ith ray

If so, for a wideband probing signal, the total received power is simply related to the sum of the powers of the individual multipath components, where each component has a random amplitude

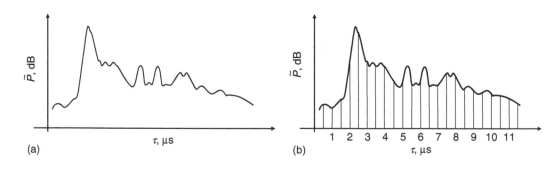

(a)

(b)

Figure 3.7 Signal power presentation of the pulse signal in the time delay domain (a), and its presentation after division into short time bins (b).

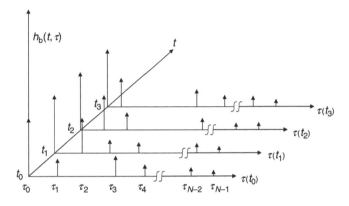

Figure 3.8 Sliced signal power presentation of the pulse signal in the time delay domain where each slice corresponds to real time t_i.

and phase at any time t, Then, the received power delay profile averaged over a small area (i.e., a small-scale received power) can be written as

$$\langle P_{pulse} \rangle = \left\langle \left| \sum_{i=0}^{N-1} A_i^2 | \exp(-j\theta_i)|^2 \right\rangle \approx \sum_{i=0}^{N-1} \langle A_i^2 \rangle \right. \tag{3.36}$$

Hence, in the multipath wideband propagation channel the small-scale received power is simply the sum of the powers received in each multipath component. We may add that if the bandwidth is sufficiently large to resolve all multiray components, then there is no need for spatial averaging.

A comparison between the CW and pulse small-scale power representations, Equations 3.36 and 3.27, shows that when $\langle A_i A_j \rangle = 0$ and/or $\langle \cos(\phi_i - \phi_j) \rangle = 0$, the average power for a CW signal is equivalent to the average received power for a pulse signal in the small-scale region. This can occur when either the path amplitudes are uncorrelated, i.e., when each multipath component is independent after reflection or scattering, or when multipath phases are independently and uniformly distributed over $[0, 2\pi]$. Thus, we can conclude that in the UHF/X-wave band, when the multipath components traverse differential path lengths of hundreds of wavelengths, *the received local ensemble average power of wideband and narrowband signals are equivalent*. This statement will be used for adaptation of the proposed propagation models for further description of fading characteristics both for narrowband and wideband signals.

The more important parameters of wideband radio communication channels that can be determined from a signal power delay profile are *mean excess delay*, *rms delay spread*, and *excess delay spread* for the concrete threshold level X (in decibels, dB) of the channel. Other parameters of the pulses, related to these main parameters, are the *coherent bandwidth*, B_c, the time of coherency, T_c, and the *Doppler spread*, B_D.

The *mean excess delay* is the first moment of the power delay profile of the pulse signal and is defined as

$$\langle \tau \rangle = \frac{\displaystyle\sum_{i=0}^{N-1} A_i^2 \tau_i}{\displaystyle\sum_{i=0}^{N-1} A_i^2} = \frac{\displaystyle\sum_{i=0}^{N-1} P(\tau_i)\tau_i}{\displaystyle\sum_{i=0}^{N-1} P(\tau_i)} \tag{3.37}$$

The *rms delay spread* is the square root of the second central moment of the power delay profile and is defined as

$$\sigma_\tau = \sqrt{\langle \tau^2 \rangle - \langle \tau \rangle^2} \qquad (3.38)$$

where

$$\langle \tau^2 \rangle = \frac{\sum_{i=0}^{N-1} A_i^2 \tau_i^2}{\sum_{i=0}^{N-1} A_i^2} = \frac{\sum_{i=0}^{N-1} P(\tau_i)\tau_i^2}{\sum_{i=0}^{N-1} P(\tau_i)} \qquad (3.39)$$

These delays are measured relative to the first detectable signal arriving at the receiver at $\tau_0 = 0$. Note that these parameters are defined from an average delay profile, which was obtained after temporal or local (small-scale) spatial averaging of the measured impulse response of the channel.

As is well known, the power delay profile in the time domain and the power spectral response in the frequency domain are related through the Fourier transform [5]. Therefore, for a full description of the signal inside the multipath channel with fading, the *time delay* parameters in the time domain and the *coherence bandwidth* in the frequency domain are used simultaneously. The *coherence bandwidth* is the statistical measure of the frequency range over which two frequency signals are strongly amplitude-correlated. Depending of the degree of amplitude correlation of two frequency-separated signals, there are different definitions of this parameter.

According to the *first definition*, the *coherence bandwidth*, B_c, is a bandwidth over which the frequency correlation function is above 0.9% or 90%; then it equals

$$B_c \approx 0.02\sigma_\tau^{-1} \qquad (3.40a)$$

According to the *second definition*, the *coherence bandwidth*, B_c, is a bandwidth over which the frequency correlation function is above 0.5% or 50%; then it equals

$$B_c \approx 0.2\sigma_\tau^{-1} \qquad (3.40b)$$

Earlier we considered two parameters, *delay spread* and *coherence bandwidth*, which describe the time-dispersive nature of the multipath communication channel in a small-scale area. To obtain information about the time-varying nature of the channel caused by movements of the transmitter or of the receiver, or scatterers located along radio path, new parameters such as *Doppler spread* and *coherence time* are usually introduced to describe time- and frequency-dispersive properties of the multipath channel with fading.

Doppler spread, B_D, is a measure that is defined as a range of frequencies over which the received Doppler spectrum is essentially nonzero. It shows the spectral spreading caused by the time rate of change of the dynamic radio channel due to relative motions of vehicles (or scatters) with respect to the receiver. According to the Doppler effect, the Doppler spread B_D depends on the maximal Doppler shift, $f_{Dm} = v/\lambda$, and on the angle θ between the direction of motion of moving vehicle and direction of arrival of the reflected and scattered waves, i.e.,

$$B_D = 2f_{Dm} \cos\theta = 2(v/\lambda)\cos\theta \qquad (3.41)$$

Coherence time T_c is the time-domain dual of *Doppler spread* and is used to characterize the time-varying nature of the frequency-dispersive properties of the channel in time coordinates. There is a simple relationship between these two channel characteristics; i.e.,

$$T_c \approx \frac{1}{f_{Dmax}} \approx \frac{\lambda}{v} \tag{3.42}$$

We can also define the *coherence time* more strictly as the time duration over which two multipath components of receiving signal have a strong potential for amplitude correlation.

If so, one can, as was done earlier for coherence bandwidth, define the coherence time as the time over which the correlation function of two different signals in the time domain is above 0.5% (or 50%). Then, we get [7]:

$$T_c \approx \frac{9}{16\pi f_m} = \frac{9\lambda}{16\pi v} = 0.18\frac{\lambda}{v} \tag{3.43a}$$

This definition can be improved for modern digital communication channels by combining Equations 3.42 and 3.43a as their geometric mean; i.e.,

$$T_c \approx \frac{0.423}{f_m} = 0.423\frac{\lambda}{v} \tag{3.43b}$$

The definition of coherence time implies that two signals, arriving at the receiver with a time separation greater than T_c, are affected differently by the channel.

References

1. Denveport, W.B. and Root, W.L., *An Introduction to the Theory of Random Signals and Noise*, McGraw-Hill, New York, 1958.
2. Steele, R., *Mobile Radio Communications*, Pentech Press, New York, 1993.
3. Jakes, W.C., *Microwave Mobile Communications*, New York: John Wiley and Son, 1974.
4. Lee, W.Y.C., *Mobile Communication Engineering*, New York: McGraw Hill Publications, 1985.
5. Proakis, J.G., *Digital Communications*, New York: McGraw-Hill, 1983.
6. Stuber, G.L., *Principles of Mobile Communication*, Kluwer Academic Publishers, Boston-London, 1996.
7. Rappaport, T.S., *Wireless Communications*, New York: Prentice Hall, 1996.
8. Saunders, S.R., *Antennas and Propagation for Wireless Communication Systems*, John Wiley and Sons, New York, 1999.
9. Cox, D.C. Universal digital portable radio communication, *Proc. IEEE*, Vol. 75, 1987, pp. 436–477.
10. Blaunstein, N. and Christodoulou, Ch., *Radio Propagation and Adaptive Antennas for Wireless Communication Links: Terrestrial, Atmospheric and Ionospheric*, Wiley InterScience, New Jersey, 2006.
11. Clarke, R.H., A statistical theory of mobile-radio reception, *Bell Syst. Tech. J.*, Vol. 47, 1968, pp. 957–1000.
12. Aulin, T., A modified model for the fading signal at a mobile radio channel, *IEEE Trans. Veh. Technol.*, Vol. 28, No. 3, 1979, pp. 182–203.
13. Aulin, T., Characteristics of of a digital mobile radio channel, *IEEE Trans. Veh. Technol.*, Vol. 30, No. 3, 1981, pp. 45–53.
14. Parsons, L.D., *The Mobile Radio Propagation Channels*, Pentech Press, New York-Toronto, 1992.

Chapter 4

Fading Phenomena in Ionospheric Communication Channels

Although direct line-of-sight (LOS) radio propagation between the terminals, transmitter, and receiver occurs in the ionosphere—mostly for satellite–land and satellite–satellite communications—multidiffraction, multireflection, and multiscattering effects are usually observed owing to obstructions in the form of natural and artificial plasma structures. These plasma density irregularities cause additional losses compared to those observed when the ionospheric plasma is a uniform, homogeneous, regular-layered medium, and also cause multipath fading of the signal strength/intensity observed at the receiver [1–8].

In real ionospheric communication links, the field forming the complicated interference picture of received radio waves arrives via several paths, thus the various waves arrive with different time delays (Figure 4.1).

In Figure 4.1 the normalized signal strength (to *rms*; see definitions in the following text) is plotted against decoder real time. At the receiver such waves combine vectorially at any given frequency to give an oscillating resultant signal, the character of whose variations depends on the distribution of phases among the incoming component waves. The signal amplitude random variations are known as the *fading effect* [1–16]. Although the fading is basically a spatial phenomenon (Figure 4.2), such variations in the space domain are experienced as temporal variations by a receiver moving through the multipath field. Therefore, *large-scale* and *small-scale* fading in the space domain correspond to *slow* and *fast* fading, respectively, in the time domain. In the literature, both these definitions are usually used (see, for example, Refs. [8–16]). Thus, we can discuss random space-domain and time-domain variations of an EM-field in ionospheric communication links.

Numerous theoretical and experimental investigations of the spatial and temporal variations of radio waves in ionospheric communication links have shown that the spatial variations of signal level have three attributes (see Figure 4.2a through c).

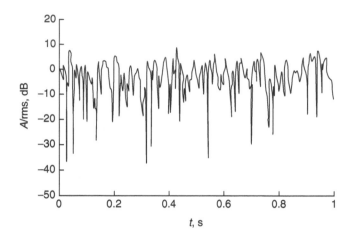

Figure 4.1   Temporal deviations of the real signal passing through a multipath channel with fading.

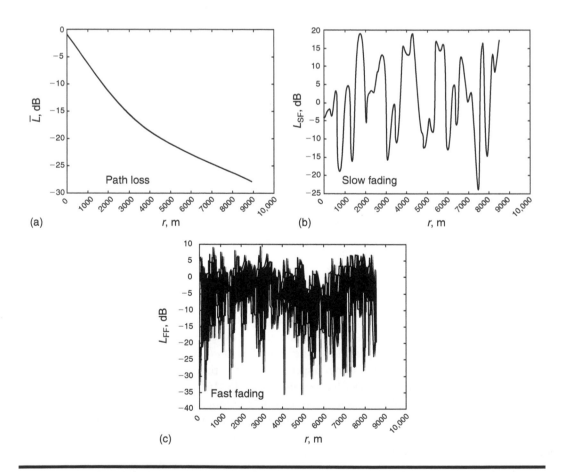

Figure 4.2   Spatial deviation of the main parameters of the real signal: (a) path loss, (b) slow fading or shadowing, and (c) fast fading.

The first is the *path loss* (Figure 4.2a), which can be defined as an overall decrease in signal strength as the distance between the two terminals, the transmitter and the receiver, increases. The cause of path loss is the spreading of the electromagnetic wave radiating outward in space. The spatial and temporal variations of the signal path loss are large and slow.

The second phenomenon, superimposed on the path loss, is *slow fading* (Figure 4.2b), which changes more rapidly, with significant random variations over distances between the terminal antennas. Slow fading is caused by diffractions from small-scale plasma structures (these effects will be discussed in the Chapter 9), all showing slow random variations in signal amplitude with a tendency to a lognormal or Gaussian distribution if the signal fading is expressed in decibels (dB). The spatial scale of the slow variations is up to several tens of meters, which is lognormally distributed [7,8,15]. Such slow random variations of the total field describe its structure within zones of diffraction and are also called *large-scale fading*.

The *third attribute* is fast random variations of the signal (Figure 4.2c), which are caused by the mutual interference of the wave components of the multiray field. The characteristic scale of such waves in the space domain is changed from a fraction of a wavelength to several wavelengths [10–16]. They are therefore usually defined as *small-scale* or *fast-fading* signals in the literature.

4.1 Path Loss: A Mathematical Description

The path loss is the phenomenon most investigated in literature, both theoretically and experimentally [1–16]. It is a principal characteristic determining the effectiveness of the propagation channel in various environments. It defines variations of the signal amplitude or field intensity along the propagation trajectory (*path*) from point to point within the communication channel. The main classical aspects of the path loss phenomenon in the ionospheric communication channel are described in detail in Ref. [2], and we will consider its characteristics and implications for different situations when the ionosphere is considered a homogeneous uniform-layered environment (see Chapter 6).

For its quantitative evaluation we will assume that the signal/wave amplitude at the point $\mathbf{r}_1$ along the propagation path is $A_1(\mathbf{r}_1)$ or the signal/wave intensity $J(\mathbf{r}_1) = A_1^2(\mathbf{r}_1)$. In the process of propagation along the path at any farther point $\mathbf{r}_2$ the signal/wave amplitude is $A_2(\mathbf{r}_2)$ or intensity $J(\mathbf{r}_2) = A_2^2(\mathbf{r}_2)$. In the literature the path loss is defined as the logarithmic difference between the amplitude/power at any two different points $\mathbf{r}_1$ and $\mathbf{r}_2$ along the propagation path in the medium [15,16]. Actually, these points determine the position of both terminals at the ends of the propagation channel, namely, the transmitter and the receiver.

In other words, path loss, which is denoted by L and is measured in decibels (dB), can be evaluated as follows [15].

For signal/wave amplitude $A(\mathbf{r}_j)$ at two points $\mathbf{r}_1$ and $\mathbf{r}_2$ along the propagation path

$$
\begin{aligned}
L &= 10 \cdot \log \frac{A^2(\mathbf{r}_2)}{A^2(\mathbf{r}_1)} = 10 \cdot \log A^2(\mathbf{r}_2) - 10 \cdot \log A^2(\mathbf{r}_1) \\
&= 20 \cdot \log A(\mathbf{r}_2) - 20 \cdot \log A(\mathbf{r}_1) \ [\text{dB}]
\end{aligned}
\tag{4.1}
$$

For signal/wave intensity $J(\mathbf{r}_j)$ at two points $\mathbf{r}_1$ and $\mathbf{r}_2$ along the propagation path

$$
L = 10 \cdot \log \frac{J(\mathbf{r}_2)}{J(\mathbf{r}_1)} = 10 \cdot \log J(\mathbf{r}_2) - 10 \cdot \log J(\mathbf{r}_1) \ [\text{dB}]
\tag{4.2}
$$

If we take point $\mathbf{r}_1$ as the origin of the radio path (the transmitter location) and assume $A(\mathbf{r}_1) = 1$, then

$$L = 20 \cdot \log A(\mathbf{r}) \ [\text{dB}] \tag{4.3a}$$

and

$$L = 10 \cdot \log J(\mathbf{r}) \ [\text{dB}] \tag{4.3b}$$

The next question is what the units are by which the losses are measured at the receiver. Let us assume that the signal/wave amplitude is measured in millivolts (mV) per meter and the signal/wave power in milliwatts (mW). In this case the resulting output value has to be presented in decibels above 1 mV/m for measured amplitude and in decibels above 1 mW for measured power, i.e.,

$$[L]_A = \text{dB}/(\text{mV})$$
$$[L]_J = \text{dB}/(\text{mW}) = \text{dBm} \tag{4.4}$$

Finally, the resulting output value is denoted in dB/(V/m), dB/(mV/m), and dB/(μV/m) if the signal/wave amplitude is measured in decibels corresponding to measured values V/m, mV/m, and μV/m, respectively. In the same way, the resulting output value is denoted in dB, dBm, and dBμ, if the signal/wave power is measured in decibels and W (watt), mW and μW, respectively.

4.2   Slow Fading of Radio Signal: A Mathematical Description

The *large-scale* or *slow fading* in mean signal level, shown in Figure 4.2b, occurs over much larger distances with a spatial "period" of several to tens of meters. As has been shown [2–9,15,16], due to the *slow* random spatial signal variations, as a result of their averaging within some individual small area, the average signal amplitude or power variation (expressed in volts/meter or watts) tends to be lognormally distributed (expressed in decibels, dB) with a standard deviation depending on the degree of perturbation of the ionospheric region along the radio path (see Chapters 6 and 7).

Following Refs. [8–16], we define a Gaussian or normal distribution of a random variable x (representing, let us say, the signal strength or voltage) by introducing the following probability density function (*PDF*) of the received signal random level x:

$$PDF(x) = \frac{1}{\sigma\sqrt{2\pi}} \exp\left\{ -\frac{(x - \bar{x})^2}{2\sigma^2} \right\} \tag{4.5}$$

Here, $\bar{x} \equiv \langle x \rangle$ is the mean value of the random signal level, and $\sigma^2 = \langle x^2 - \bar{x}^2 \rangle$ is the variance or time-averaged power ($\langle x \rangle$ is a sign of averaging of variable x) of the received signal envelope in dB. This *PDF* can only be obtained as a result of the random interference of a large number of signals with randomly distributed amplitudes (voltage) and phases. If the phase of the interfering signals is uniformly distributed over the range $[0,2\pi]$, then one can talk about a zero-mean Gaussian distribution of the random variable x. In this case we define the *PDF* of such a process by Equation 4.5 with $\bar{x} = 0$ and $\sigma^2 = \langle x^2 \rangle$ in it. The definition of the mean value and the variance of random signal envelope is also used for the description of the so-called cumulative distribution function (*CDF*). This function describes the probability of the event that the envelope of received signal strength (in dB) does not exceed a specified value *X*, i.e.,

$$CDF(X) = \Pr(x \le X) = \int_{-\infty}^{X} PDF(x) \, dx$$

$$= \frac{1}{\sigma\sqrt{2\pi}} \int_{-\infty}^{X} \exp\left\{-\frac{(x-\bar{x})^2}{2\sigma^2}\right\} dx = \frac{1}{2} + \frac{1}{2}\operatorname{erf}\left\{\frac{X-\bar{x}}{\sqrt{2}\sigma}\right\} \tag{4.6}$$

where the error function is defined by

$$\operatorname{erf}(w) = \frac{2}{\sqrt{\pi}} \int_{0}^{w} \exp\left(-y^2\right) dy \tag{4.7}$$

As follows from Equation 4.5, the value $\bar{x} = 0$ corresponds to the maximum of the *PDF*, which equals $PDF(0) = 1/\sigma\sqrt{2\pi}$. For $x = \sigma$, it follows from Equation 4.5 that $PDF(\sigma) = 1/\sigma\sqrt{2\pi e}$, where $e \approx 2.71 \ldots$. As also seen from Equations 4.6 and 4.7, the value $\bar{x} = 0$ corresponds to $CDF = 1/2$, which is then limited to unity for $x \gg 1$.

4.3 Fast Fading: A Mathematical Description

As follows from the signal envelope presented in Figure 4.2, small-scale fading is observed over distances of about half or one wavelength, i.e., of a few meters to tens of centimeters for the HF/UHF-frequency band. The ionospheric channels are usually dynamic channels. In the case of a *dynamic multipath* situation in which either the terminal antennas are in movement or the plasma structures move because of wind and ambient electrical and magnetic fields (see Chapters 1 and 2), the spatial variations of the resultant signal at each receiver will be seen as temporal variations by the receiver as it moves through the multipath field. The signal received by the mobile at any spatial point may consist of a large number of signals having randomly distributed amplitudes, phases, and angles of arrival, as well as different time delays. All these features change the relative phase shifts as a function of spatial location of the moving antennas and finally cause the signal to fade simultaneously in the space and time domains. The temporal fading relates to a shift of frequency radiated by the moving transmitter or receiver. In fact, the time variations or dynamic changes of the propagation path lengths are related to the Doppler effect, which is due to a relative movement between a stationary transmitter and a moving receiver, or vice versa. To describe mathematically these kinds of fading, the 3-D statistical model was introduced in Chapter 3, which together with some specific statistical distributions fully describe fast fading occurring in the ionospheric communication link. In the following text we will present the most useful distributions usually used for modeling ionospheric communication channels [2–8,15,16].

Rayleigh PDF and CDF. In wireless communication channels, stationary or mobile, the Rayleigh distribution is commonly used to describe the signal's spatial and temporal fast (small-scale) fading. As is well known [2,8–16], a Rayleigh distribution can be obtained mathematically as the limit envelope of the sum of two quadrature Gaussian noise signals. In practice, this distribution can be predicted in the wireless communication channel if each multipath component of the received signal is independent and if the phases of multipath components are uniformly distributed over the range $[0,2\pi]$. Then, we may deal with a Rayleigh distribution of random variable x, the *PDF* of which can be presented in the following form [1–8]:

$$PDF(x) = \frac{x}{\sigma^2} \exp\left\{-\frac{x^2}{2\sigma^2}\right\} \quad \text{for} \quad x \geq 0 \tag{4.8}$$

All parameters in Equation 4.8 are the same as for the Gaussian *PDF* according to Equation 4.5. According to Equation 4.8, the maximum value of $PDF(x) = \exp(-0.5)/\sigma = 0.6065/\sigma$ corresponds to the random variable $x = \sigma$. One can also operate with the so-called the *expected* or *mean value*, and the *rms value* of random variable x. The definition of both parameters follows from the Rayleigh *CDF* presentation:

$$CDF(X) = \Pr(x \leq X) = \int_0^X PDF(x) \, dx = 1 - \exp\left\{-\frac{X^2}{2\sigma^2}\right\} \tag{4.9}$$

All parameters in Equation 4.9 are the same as for the Gaussian *CDF* according to Equation 4.6. The expected or mean value of the Rayleigh distributed signal strength (voltage), x_{mean}, which in the literature is also denoted as $E[x]$, can be obtained from the following conditions:

$$x_{mean} \equiv E[x] = \int_0^\infty x \cdot PDF(x) \, dx = \sigma \cdot \sqrt{\frac{\pi}{2}} \approx 1.253\sigma \tag{4.10}$$

At the same time, the variance or average power of the received signal envelope for the Rayleigh distribution can be determined as [15,16]

$$\sigma_x^2 \equiv E[x^2] - E^2[x] = \int_0^\infty x^2 \cdot PDF(x) \, dx - \frac{\pi\sigma^2}{2}$$

$$= \sigma^2\left(2 - \frac{\pi}{2}\right) \approx 0.429\sigma^2 \tag{4.11}$$

If so, the *rms value* of the signal envelope is defined as the square root of $2\sigma^2$, i.e.,

$$rms = \sqrt{2} \cdot \sigma \approx 1.414\sigma \tag{4.12}$$

We should note here that in practice in satellite wireless systems' servicing, the *median value* is often used [2].

Rician PDF and CDF. As was mentioned earlier, usually in an ionospheric radio wireless link, it is not only multipath components that arrive at the receiver due to multiple reflection, diffraction, and scattering from various plasma obstructions along the radio path. A line-of-sight (LOS) component describing signal loss along the path of direct visibility (called the *dominant path* [2]) between both antennas is often found at the receiver. This dominant component of the received signal may significantly decrease the depth of the interference picture of the signal envelope variations. The *PDF* of such a received signal is usually said to be Rician.

The Rician *PDF* distribution of the signal strength or voltage, denoted by arbitrary random variable *x*, can be defined as [8,15,16]

$$PDF(x) = \frac{x}{\sigma^2} \exp\left\{-\frac{x^2 + A^2}{2\sigma^2}\right\} \cdot I_0\left(\frac{Ax}{\sigma^2}\right) \quad \text{for} \quad A > 0, x \geq 0 \tag{4.13}$$

where *A* denotes the peak strength or voltage of the dominant component envelope and $I_0(\bullet)$ is the modified Bessel function of the first kind, and is zero-order.

To estimate the contribution of each component, dominant (or LOS) and multipath, for the resulting signal at the receiver, the Rician parameter *K* is usually introduced as a ratio between these components, i.e.,

$$K = \frac{\text{LOS component power}}{\text{Multipath component power}} \tag{4.14}$$

According to Equation 4.14, one can now rewrite the parameter *K*, which was defined earlier as the ratio between the dominant component power and the multipath component power. It is given by [2–8,15,16]

$$K = \frac{A^2}{2\sigma^2} \tag{4.15}$$

Let us rewrite the Rician *PDF* by introducing the Rician *K*-factor from Equation 4.15 into Equation 4.14 as a function of only *K*:

$$PDF(x) = \frac{x}{\sigma^2} \exp\left\{-\frac{x^2}{2\sigma^2}\right\} \cdot \exp(-K) \cdot I_0\left(\frac{x}{\sigma}\sqrt{2K}\right) \tag{4.16}$$

It must be noted that the Rician distribution consists of all the other distributions mentioned earlier. Thus, when $K = 0$, i.e., there is no dominant component but only scattered and reflected components reach the receiver, the communication channel is Rayleigh with the *PDF* described by Equation 4.8. Conversely, under conditions of good line-of-sight between the two terminals with no multipath components, i.e., $K \to \infty$, the Rician fading approaches Gaussian fading, yielding a "Dirac-delta-shaped" *PDF*. This is illustrated by Figure 4.3, depicted according to Refs. [15,16], where a set of *PDFs* that describe the envelope fading profiles for different values of *K* versus the signal level are presented, normalized to the corresponding *rms*.

Thus, we can define the Rician distribution as the general case of signal fading along the radio path between two terminal antennas that cover both Rayleigh and Gaussian distributions of signal strength envelope. As the dominant component at the receiver becomes weaker, the resulting signal becomes closer to that obtained after multiple reflections and scattering, which has an envelope described by the Rayleigh law. In this case the fades with high probability are very deep, whereas if *K* is limited to infinity the fades are very shallow.

The Rician *CDF* can be defined in the same manner as the Gaussian *CDF* (Equation 4.6) and the Rayleigh *CDF* (Equation 4.9), using the same quadrature formula given by Refs. [15,16].

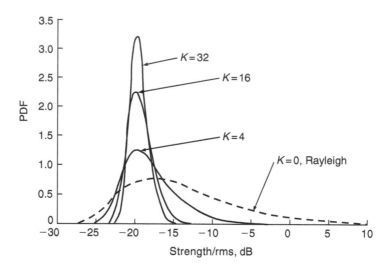

Figure 4.3 Rician probability density function (*PDF*) distribution versus the normalized signal strength, covering the Rayleigh *PDF* for *K* = 0 (dashed curve).

Finally, the formula for Rayleigh *CDF* is more complicated because of the summation of an infinite number of terms such as [15,16]:

$$CDF(x) = 1 - \exp\left\{-\left(K + \frac{x^2}{2\sigma^2}\right)\right\} \cdot \sum_{m=0}^{\infty} \left(\frac{\sigma\sqrt{2K}}{x}\right)^m \cdot I_m\left(\frac{x}{\sigma}\sqrt{2K}\right) \qquad (4.17)$$

Here, $I_m(\bullet)$ is the modified Bessel function of the first kind and m-th order. Once more, the Rician *CDF* depends on one parameter, K, and limits the Rayleigh *CDF* and Gaussian *CDF* for $K = 0$ and for $K \to \infty$, respectively. Hence, if the K factor is known, the signal fading envelope's distributions, *PDF* and *CDF*, are perfectly described.

Nakagami-m PDF. We now present another statistical distribution that is, as are the Rayleigh and Rician distributions, frequently used to characterize the signal strength or power envelope fading in multipath communication links: the Nakagami m-distribution. The PDF for such statistics is given as [12–14]

$$PDF(x) = \frac{2}{\Gamma(m)} \left(\frac{m}{\Sigma}\right)^m x^{2m-1} \exp\left\{-\frac{mx^2}{\Sigma}\right\} \qquad (4.18)$$

where Σ is defined as mean variance $\Sigma = \langle X^2 \rangle$ where X is a specified value and the parameter m is defined as a ratio of moments, called the *fading figure*:

$$m = \frac{\Sigma^2}{\langle (X^2 - \Sigma)^2 \rangle} \qquad (4.19)$$

$\Gamma(m)$ is the gamma function, defined as

$$\Gamma(m) = \int_0^\infty t^{m-1} e^{-t} \, dt, \quad m > 0$$

$$\Gamma(m) = (m-1)! \quad m \text{ an integer}, \ m > 0$$

$$\Gamma\left(\tfrac{1}{2}\right) = \sqrt{\pi}, \quad \Gamma\left(\tfrac{3}{2}\right) = \tfrac{1}{2}\sqrt{\pi}$$

(4.20)

By setting $m = 1$, Equation 4.18 reduces to a Rayleigh *PDF*. Introducing $m = 1/2$ in Equation 4.19 gives the one-sided Gaussian distribution. For $\tfrac{1}{2} < m < 1$ the *PDF*s have larger tails than the Rayleigh distributed random variable, whereas for $m > 1$ the tails for the *PDF* decay faster than for the Rayleigh.

Suzuki PDF. Suzuki [13] suggested that the statistics of the signals within mobile communication links can be described by a mixture of the Rayleigh and lognormal distributions in the form of a Rayleigh distribution with a lognormally varying mean, i.e.,

$$PDF(x) = \int_0^\infty \frac{x}{\sigma^2} \exp\left\{-\frac{x^2}{2\sigma^2}\right\} \frac{M}{\sigma \alpha \sqrt{2\pi}} \exp\left\{-\frac{\log^2(\sigma/\sigma_0)}{2\alpha^2}\right\} d\sigma$$

(4.21)

where
  σ is the standard deviation of the Rayleigh distribution (see earlier text)
  σ_0 is the shape parameter of the lognormal distribution
  $M = \log e = 0.434$

The mean square and mean values are, respectively,

$$\langle x^2 \rangle = 2\sigma_0^2 \exp\left\{\frac{2\lambda^2}{M^2}\right\}$$

$$\langle x \rangle = \sqrt{\frac{\pi}{2}} \sigma_0 \exp\left\{\frac{\lambda^2}{2M^2}\right\}$$

(4.22)

Equation 4.21 is the integral of the Rayleigh distribution over all possible values of σ, weighted by the *PDF* of σ, and this attempts to provide a transition from local to global statistics.

Lorenz PDF. To evaluate the Suzuki *PDF* (Equation 4.21), which is presented in a very complicated integral form, Lorenz introduced the following variables [14]:

$$F = 20 \log x$$

$$F_{OR} = 20 \log\left(\sqrt{2} \cdot \sigma\right)$$

$$F_{OS} = 20 \log\left[\sqrt{2} \cdot \sigma_0 \exp\left(\frac{2\lambda^2}{M^2}\right)\right]$$

(4.23)

which represent the signal strength in dB, the *rms* value of the Rayleigh distribution (see previous text) in dB, and the *rms* of the Suzuki distribution in dB, respectively. In terms of the parameters, $s = 20\lambda$ and $M_1 = 20M$, the Suzuki PDF as formulated by Lorenz is [14]

$$PDF(x) = \int_0^\infty \frac{2}{M_1} \exp\left\{ \frac{2}{M_1}(F - F_{OR}) - \exp\left[\frac{2}{M_1}(F - F_{OR}) \right] \right\}$$

$$\times \frac{1}{s\sqrt{2\pi}} \exp\left\{ -\frac{[F_{OR} - F_{OS} + s^2/M^2]^2}{2s^2} \right\} dF_{OR} \qquad (4.24)$$

To decide which statistical model, the Rice, Nakagami, Rayleigh, or lognormal, best fits measurement data, Suzuki and Lorenz suggested the use of a moment method that equates the theoretical means and variances of the *PDF* to the sample mean and variance of the experimental data. We will not discuss this subject further; the interested reader can consult Refs. [2–8,15,16].

4.4 Definition of Fading Phenomena in Static and Dynamic Ionospheric Channel

It is clear from the foregoing definitions that the type of signal fading within the dynamic radio channel depends on the relation between the signal parameters, the *bandwidth B_s*, and the *symbol period T_S*, and the corresponding parameters of the channel, such as *coherence bandwidth B_c* and *rms delay spread σ_τ* (or *Doppler spread B_D* and *coherence time T_c*). So, different kinds of transmitted signals (defined by B_S and T_S) will undergo different types of fading. There are *four possible effects* due to the time- and frequency-dispersion mechanisms in a mobile radio channel, which are manifested depending on the balance of the aforementioned parameters of the signal and of the channel. The multipath time delay spread leads to *time-dispersive* and *frequency-selective fading*, whereas Doppler frequency spread leads to *frequency-dispersive* and *time-selective fading*. Separation between these four types of small-scale fading for the impulse response of a multipath radio channel is explained in Table 4.1 [8,15]:

  A. *Fading due to multipath time delay spread.* Time dispersion due to multipath phenomena causes small-scale fading, either *flat* or *frequency selective*.
  A.1. Small-scale fading is characterized as a flat, if the mobile channel has a constant-gain and linear-phase impulse response over a bandwidth that is *greater* than the bandwidth of the transmitted signal. Here, $T_S \gg \sigma_\tau$. Obviously, a flat-fading channel can be defined as a narrowband channel, because the bandwidth of the applied signal is narrow with respect to the channel flat-fading bandwidth in the frequency domain; i.e., $B_S \ll B_c$. At the same time, the flat-fading channel is an *amplitude-varying* channel because there is a deep fading of the transmitted signal occurring within the channel.
  A.2. Small-scale fading is characterized as frequency selective if the mobile channel has a constant-gain and linear-phase impulse response over a bandwidth that is *smaller* than the bandwidth of the transmitted signal in the frequency domain, as well as if its impulse response has a multiple delay spread greater than the bandwidth of the transmitted signal waveform, i.e., $T_S \ll \sigma_\tau$. Obviously, a frequency-selective fading channel can be defined as a wideband channel, because the bandwidth of the spectrum

Table 4.1   Types of Fading Depending on Relations between the Signal and the Channel Parameters

Type of Fading within the Channel	Relations between Signal and Channel Parameters
(A1): Flat fast fading (FFF)	(A.1.1): $B_S \ll B_c$, $B_S < B_D$
	(A.1.2): $T_S \gg \sigma_\tau$, $T_S > T_c$
(A.2): Frequency-selective fast fading (FSFF)	(A.2.1): $B_S > B_c$, $B_S < B_D$
	(A.2.2): $T_S < \sigma_\tau$, $T_S > T_c$
(B1): Flat slow fading (FSF)	(B.1.1): $B_S \ll B_c$, $B_S \gg B_D$
	(B.1.2): $T_S \gg \sigma_\tau$, $T_S \ll T_c$
(B.2): Frequency-selective slow fading (FSSF)	(B.2.1): $B_S > B_c$, $B_S \gg B_D$
	(B.2.2): $T_S < \sigma_\tau$, $T_S \ll T_c$

$S(f)$ of the transmitted signal is *greater* than the channel frequency-selective fading bandwidth in the frequency domain (the coherence bandwidth); i.e., $B_S > B_c$.

B. *Fading due to Doppler spread.* Depending on how rapidly the transmitted baseband signal changes with respect to the rate of change of the channel, a channel may be classified as a fast-fading or slow-fading channel.

B.1. The channel in which the channel impulse response changes rapidly within the pulse (symbol) duration is called a fast-fading channel. In other words, the coherence time in such a channel is smaller than the symbol period of the transmitted signal. At the same time, the Doppler spread bandwidth of the channel in the frequency domain is greater than the bandwidth of the transmitted signal; i.e., $B_S < B_D$ and $T_c < T_S$. These effects cause frequency dispersion (also called time-selective fading) due to Doppler spreading, leading to signal distortion.

B.2. The channel in which the channel impulse response changes at a rate *slower* than the transmitted baseband signal $u(t)$, is called a slow-fading channel. In this case, the channel may be assumed to be static over one or several bandwidth intervals. In the time domain, this implies that the reciprocal bandwidth of the signal is smaller than the coherence time of the channel, and in the frequency domain, the Doppler spread of the channel is less than the bandwidth of the baseband signal; i.e., $B_S \gg B_D$ and $T_S \ll T_c$. It is important to note that the velocity of the moving vehicle or moving obstructions within the channel, as well as the baseband signal, determines whether a signal undergoes fast or slow fading.

Usually, for real-world situations in wireless communication links, we deal with time- and frequency-dispersive channels, and therefore cannot separate fading into flat or frequency-selective and fast or slow. There could simultaneously occur *flat slow fading* (FSF) or *frequency-selective slow fading* (FSSF), when the channel is time dispersive, and *flat fast fading* (FFF) or *frequency-selective fast fading* (FSFF). These situations occur within any wireless channel, land, atmospheric, or ionospheric, and are shown in Table 4.1. Following this table and using the corresponding relations between parameters of the signal, period T_s and bandwidth B_s, and parameters of the channel, delay spread σ_τ, time of coherency T_c, coherent bandwidth and Doppler spread B_D, we can strictly define the type of fading occurring within the specific communication channel, and therefore precisely predict the corresponding *PDF* and *CDF*, which, together with evaluated average path loss, allows one to calculate the total link budget within each propagation channel.

References

1. Lee, W.Y.C., *Mobile Cellular Telecommunications Systems*, McGraw Hill, New York, 1989.
2. Saunders, S.R., *Antennas and Propagation for Wireless Communication Systems*, John Wiley and Sons, New York, 1999.
3. Bertoni, H.L., *Radio Propagation for Modern Wireless Systems*, Prentice Hall PTR, New Jersey, 2000.
4. Feuerstein, M.L. and Rappaport, T.S. *Wireless Personal Communication*, Boston-London: Artech House, 1992.
5. Parsons, L.D., *The Mobile Radio Propagation Channels*, Pentech Press, New York-Toronto, 1992.
6. Steele, R., *Mobile Radio Communication*, IEEE Press, 1992.
7. Yacoub, M.D., *Foundations of Mobile Radio Engineering*, CRC Press, 1993.
8. Rappaport, T.S., *Wireless Communications*, New York: Prentice Hall PTR, 1996.
9. Clarke, R.H., A statistical theory of mobile-radio reception, *Bell Syst. Tech. J.*, Vol. 47, 1968, pp. 957–1000.
10. Aulin, T., A modified model for the fading signal at a mobile radio channel, *IEEE Trans. Veh. Technol.*, Vol. 28, No. 3, 1979, pp. 182–203.
11. Papaulis, A., *Probability, Random Variables, and Stochastic* Processes, McGraw Hill, New York, 1991.
12. Proakis, J.G., *Digital Communications*, McGraw Hill, New York, 1995.
13. Suzuki, H., A statistical model for urban propagation, *IEEE Trans. Communication*, Vol. 25, 1977, pp. 673–680.
14. Lorenz, R.W., Theoretical distribution functions of multipath fading processes in mobile radio and determination of their parameters by measurements, *Technischer Bericht 455*, TBr 66, Forschungsinstitut der Deutschen Bundespost (in German).
15. Blaunstein, N., *Wireless Communication Systems*, Ch. 2 In book: *Engineering Electromagnetics Applications*, Ed. by Bansal, R., New York: CRC Taylor and Frances, 2004, pp. 37–97.
16. *Handbook of Engineering Electromagnetics*, Ed. by Bansal, R., New York: Marcel Dekker, 2004.

Chapter 5

Evolution of Plasma Irregularities in the Ionosphere

Diffusion plays an important role in the evolution of macroplasma density structures (usually described in the literature as plasma irregularities, inhomogeneities, or turbulences), regardless of how they were created, as it determines the longest initial stage of relaxation of the structures (compared to drift and thermodiffusion) [1–22]. When we call them *macrostructures*, we take into account the fact that the dimensions of such plasma irregularities are large compared to the Debye radius, $r_D = (T/4\pi e^2 N_e)^{1/2}$. Here, all parameters were defined in Chapter 1. Numerous experimental investigations of natural and artificial plasma inhomogeneities in the high-latitude ionosphere [23–26], weakly ionized H_E plasma structures, and meteor trails in the middle-latitude ionosphere [27–29], and plumes and bubbles in the equatorial regions [30–32] establish the presence of anisotropic irregularities of plasma stretched along the geomagnetic field $\mathbf{B}_0$ with a wide spectrum of scales transverse to $\mathbf{B}_0$. Such plasma irregularities are also formed when ion clouds or beams are artificially injected from rockets at the ionospheric altitudes [33–49], as well as in the field of powerful radio waves radiated from ground-based facilities [50–57] (see also the bibliographies in Refs. [20–22]).

In Refs. [19–22,39–46], the complicated deformation and stratification of barium clouds caused by inhomogeneous properties of the ionosphere in the ambient electric and magnetic fields were explained on the basis of classical theories of plasma dynamics [1–15]. At the same time, while studying artificial heating-induced irregularities in the ionosphere, a satisfactory explanation was obtained for the large degree of anisotropy of small-scale turbulences and quasi-periodic plasma structures formed in the strong field of powerful radio waves, as well as of the complicated relaxation and lifetime dependence on their initial dimensions and the degree of ionization [50–57,75–78]. Satisfactory physical models were created to explain the weak degree of anisotropy of large-scale ionospheric H_E inhomogeneities and meteor trails in the middle-latitude ionosphere

[29,79–82], and plasma bubbles in the equatorial ionosphere [30–32]. All these effects will be discussed in the following text, and the discussion will be continued in Chapters 6 and 7, where the radio and optical methods of investigation of natural and artificial plasma irregularities in the real ionosphere are presented. In these chapters, the effects of various types of plasma irregularities on radio wave propagation are described systematically, following the historical perspective of iono-spheric radio channels performance.

We will now introduce the main equations and the corresponding solutions analyzing the processes of diffusion, drift and thermodiffusion of plasma disturbances, large-, moderate-, and small-scale, in the homogeneous and inhomogeneous ionospheric plasma and will describe the peculiarities of each process separately.

5.1 Diffusion of Plasma Irregularities in the Ionosphere

The spreading of an arbitrary plasma disturbance in the ionospheric plasma in the presence of an ambient electric or magnetic field has not been satisfactorily explained by existing theory despite the significant development of the theory of diffusion and the number of experiments [21,50–57] performed to investigate it. Only in recent decades have the corresponding theoretical models been developed that allow the prediction of many specific features of the evolution of experimen-tally observed plasma disturbances (see bibliographies in Refs. [20–22]). All modifications in the theory of plasma transport are based on the classical description of processes of quasi-neutral and ambipolar diffusion and drift in the magnetic field obtained in Refs. [2–14], mostly for small-scale plasma structures in the presence of an infinite homogeneous background plasma, as well as based on presentation of diffusion as unipolar (*electronic* along $\mathbf{B}_0$ and *ionic* across $\mathbf{B}_0$), the rate and character of which are determined by short-circuit currents through the background iono-spheric plasma [7–14]. In Ref. [15], where generalized results obtained in Refs. [7–14] for a description of diffusion spreading for both strong ($\delta N > N_0$, N_0 being the density of background plasma) and weak ($\delta N \leq N_0$) plasma disturbances (denoted by δN) in the presence of an infinite homogeneous plasma.

On the basis of the real model of the middle-latitude ionosphere, diffusion of weak and strong plasma inhomogeneities in the presence of homogeneous and inhomogeneous background plasmas was investigated theoretically and numerically in Refs. [16–19]. It was found that the character of diffusion along and across the magnetic field $\mathbf{B}_0$ depends significantly on the degree of anisotropy of the initial irregularity, and finally changes occurrence from electron-controlled for plasma structures greatly stretched out along $\mathbf{B}_0$ to ion-controlled for those elongated across $\mathbf{B}_0$. At the same time, analyzing a nonuniform problem of plasma diffusion and taking into account the altitude depend-ence of the transport coefficients of charged plasma particles, some new effects in irregularity diffusion spreading in inhomogeneous middle-latitude background plasma compared with homo-geneous plasma were obtained theoretically. There is the deformation of the initial form of the irregularity, the splitting into two or three plasmoids (depending on the spatial orientation of crossing electrical and magnetic fields [16–19]) in the course of spreading and the elongation of the irregularity upward along the geomagnetic field lines in the upper ionosphere, and so on. Before going deeper into the subject of diffusion spreading of plasma disturbances in the real ionospheric plasma, we will analyze briefly some classical aspects of the diffusion process in the unbounded homogeneous plasma.

5.1.1 Classical Description of the Diffusion Process in Plasma

In the absence of the ambient electrical ($\mathbf{E}_0 = 0$) and magnetic ($\mathbf{B}_0 = 0$) fields, the diffusion process is defined by the free paths of plasma electrons λ_e and ions λ_i, as steps of random (chaotic) movements of the charged particles, which finally determine the coefficients of diffusion

$$D_e \approx \lambda_e^2 \nu_{em}, \quad D_i \approx \lambda_i^2 \nu_{im} \tag{5.1a}$$

and the coefficients of mobility

$$b_e \approx e/m\nu_{em}, \quad b_i \approx Ze/M_i\nu_{im} \tag{5.1b}$$

As in Chapter 2, we will consider simple monoatomic ions and therefore assume $Z = 1$; all other parameters are defined in Chapter 1.

In the case where dimensions of plasma perturbation are sufficiently *small* compared to the Debye radius introduced earlier—i.e., when $(\nabla N/N_0)^{-1} < r_D$, plasma quasi-neutrality is infringed significantly and the self-consistent intrinsic electric field, caused by the displacement of charge particles in plasma, is insufficient and cannot affect the character of movements of charge particles—the potential energy of the intrinsic electric field is less than the kinetic energy of thermal (chaotic) movements of plasma charged particles; i.e., $e\varphi/T \ll 1$, where $T = T_e \approx T_e$ is the equilibrium temperature of electrons and ions expressed in the energetic units.

In the *inverse case*, where $(\nabla N/N_0)^{-1} \gg r_D$, diffusion changes its character significantly. Thus, when the initial dimensions of plasma disturbance are larger than the Debye radius, the charged particles strive for displacement due to the higher mobility of electrons compared to that for ions. Even a small decompensation of charged particles leads to the creation of a self-consistent electric field in plasma, thus preventing further displacements of charged particles. In this case $eE_a L/T \gg 1$, where the characteristic size of plasma perturbation is defined as $L = (\nabla N/N_0)^{-1}$, and E_a is the intrinsic electrostatic field in plasma. As a result, plasma perturbation spreads in such a manner that concentration of the electrons N_e and the ions N_i are similar; i.e., $N_e \approx N_i = N$.

Effects of the Magnetic Field on Plasma Diffusion. The solution of the problem of plasma diffusion in the *absence of the magnetic* field ($\mathbf{B}_0 = 0$) was obtained by Schottky in 1924 [1]. He showed that the condition of plasma quasi-neutrality leads to the condition of ambipolarity of diffusion, which is defined by the equality of flows of electrons and ions at each point of the space filled by plasma; i.e., $\mathbf{\Gamma}_e = \mathbf{\Gamma}_i$. In this case, as shown in Refs. [2–4], we can exclude in the equations of diffusion the inner electrostatic ($\mathbf{E}_a = -\nabla\varphi$) electric field $\mathbf{E}_a$, i.e.,

$$\frac{\partial N_\alpha}{\partial t} + \nabla\mathbf{\Gamma}_\alpha = 0 \tag{5.2a}$$

$$\mathbf{\Gamma}_\alpha \equiv N_\alpha\mathbf{V}_\alpha = -D_\alpha\nabla N_\alpha \pm N_\alpha b_\alpha\mathbf{E}_a \tag{5.2b}$$

In Equation 5.2, $\mathbf{V}_\alpha$ is the directed velocity of electrons ($\alpha = e$) and ions ($\alpha = i$). It should be noted that in the case of $\mathbf{B}_0 = 0$ the ordinary linear equation follows from the constraint $\mathbf{\Gamma}_e = \mathbf{\Gamma}_i$, and the self-consistent electric (ambipolar) field does not exist:

$$\frac{\partial N}{\partial t} = D_a \Delta N \tag{5.3}$$

Additionally, the profile of plasma concentration does not depend on the current flowing through the plasma. Thus, because diffusion is slower than other transport processes, the electric field is electrostatic (potential) and the equations (Equation 5.2) for the plasma concentration N and the potential φ can be separated, i.e.,

$$\nabla(b_e - b_i)N\nabla\varphi = (D_e - D_i)\Delta N \qquad (5.4)$$

that is, the profile of the plasma density can be found independently of the profile of the potential. In the other words, the plasma profile $N(\mathbf{r},t)$ does not depend on the currents flowing through the plasma. In Equation 5.3 the effective coefficient of the ambipolar diffusion

$$D_a = \frac{D_e b_i + D_i b_e}{b_e + b_i} \approx (1 + T_e/T_i)D_i \qquad (5.5)$$

equals the coefficient of ion diffusion increasing in $(1 + T_e/T_i) \approx 2$ times (for $T_e \approx T_e$) caused by the influence of the electron component of the plasma.

Thus, when $\mathbf{B}_0 = 0$, ambipolar diffusion can be described by the linear equations (Equations 5.3 and 5.4) for the plasma density profile and for the inner electric field, respectively, and can be determined by ions as particles with lower mobility. The influence of plasma electrons, as particles with higher mobility, and of the self-consistent (ambipolar) electric field, is described by the term $(1 + T_e/T_i)$.

In the case of a sufficiently *strong magnetic field*, when $\mathbf{B}_0 \neq 0$, even in the case of ordinary plasma consisting of electrons, monoatomic ions, and neutrals of one kind, the situation changes sharply.

a. In the case introduced earlier, where dimensions of plasma perturbation are small compared to the Debye radius, i.e., where $(\nabla N/N_0)^{-1} < r_D$, the step of chaotic movement (i.e., diffusion) along $\mathbf{B}_0$ is defined by the free paths of electrons λ_e and ions λ_i, whereas transverse to $\mathbf{B}_0$ is determined by the Larmor radii of electrons ρ_{He} and ions ρ_{Hi}. In a sufficiently strong magnetic field where $\rho_{He} \ll \lambda_e$ and $\rho_{Hi} \ll \lambda_i$, the transverse coefficient of diffusion of electrons and ions can be presented through these radii as

$$D_{e\perp} = T_e b_{e\perp}/e \approx \rho_{He}^2 \nu_{em}, \quad D_{i\perp} = T_i b_{i\perp}/e \approx \rho_{Hi}^2 \nu_{im} \qquad (5.6)$$

Because $\rho_{He} \sim B_0^{-1}$ and $\rho_{Hi} \sim B_0^{-1}$, the corresponding transverse coefficients of diffusion are sharply decreased $(\sim B_0^{-2})$ with growth in the strength of the magnetic field. At the same time, $\rho_{He} \sim m_e$ and $\rho_{Hi} \sim M_i$, where m_e and M_i are the mass of electrons and ions, respectively. Because $M_i/m_e \geq 10^3 \gg 1$, the Larmor radius of ions exceeds the Larmor radius of electrons significantly. Therefore, in a strong magnetic field (where $\rho_{He}\rho_{Hi} \ll \lambda_e\lambda_i$) the diffusion of charged particles is anisotropic, and $D_{e\perp} \ll D_{i\perp}$, $b_{e\perp} \ll b_{i\perp}$.

b. When the scale of plasma density disturbance exceeds the Debye radius, i.e., $(\nabla N/N_0)^{-1} \gg r_D$, diffusion spreading significantly changes its character. Electrons of plasma disturbance freely move along magnetic field lines, but their movement across $\mathbf{B}_0$ is hard to achieve. They fill an ellipsoid with semi-axes $L_e = \sqrt{4(1 + T_e/T_i)D_{e\parallel}t}$ (along $\mathbf{B}_0$) and $d_e = \sqrt{4(1 + T_e/T_i)D_{e\perp}t}$ (across $\mathbf{B}_0$), where $D_{e\parallel} \gg D_{e\perp}$. The same ellipsoid, but now with semi-axes $L_i = \sqrt{4(1 + T_e/T_i)D_{i\parallel}t}$ and $d_i = \sqrt{4(1 + T_e/T_i)D_{i\perp}t}$, where $D_{i\parallel} \gg D_{i\perp}$, is filled by ions of initial plasma disturbance. Because the ion mobility across $\mathbf{B}_0$ exceeds the electron mobility ($\sim \rho_{Hi}/\rho_{He} \gg 1$), and electrons along $\mathbf{B}_0$ move faster than ions ($\sim \lambda_e/\lambda_i \gg 1$), both components of plasma disturbance strive for separation along and across $\mathbf{B}_0$. However, as shown in Refs. [7–15], it is impossible to perform the potential inner (ambipolar) electric field that will simultaneously hamper faster electrons along $\mathbf{B}_0$ and faster ions across $\mathbf{B}_0$, and will equalize the electron and ion flows at each point in the area filled by plasma. As shown by the authors of Refs. [5,6], the condition of plasma quasi-neutrality, $N_e \approx N_i = N$, leads in the presence of the magnetic field to the equality of divergence of these flows, i.e., to $\nabla \cdot \mathbf{\Gamma}_e = \nabla \cdot \mathbf{\Gamma}_i$. The latter constraint does not prohibit the flow of the vortex current through the background plasma, defined in Refs. [9–15] as *short-circuit current*. This current is created by the background electrons along $\mathbf{B}_0$ and by the background ions across $\mathbf{B}_0$. These particles secure the condition of plasma quasi-neutrality in the region occupied by the plasma disturbance, i.e., in the electronic ellipsoid due to the transverse flow of background ions and in the ion ellipsoid by the longitudinal flow of background electrons. As a result, the depletion regions occur in the background plasma [9–15], and the total plasma concentration is disturbed significantly both in the volume of electronic and ionic ellipsoids. The condition of quasi-neutrality does not limit the volume occupied by the disturbance of plasma density. In a strong magnetic field it can exceed the analogous volume, just as when the magnetic field is absent. The potential of the ambipolar electric field is also disturbed within this volume (compared with the case $\mathbf{B}_0 = 0$). This volume corresponds to propagation of more mobile particles in corresponding directions to $\mathbf{B}_0$: electrons along and ions across $\mathbf{B}_0$. Depending on the degree of plasma perturbation, nonlinear diffusion can be divided at the intermediate and the strong diffusion process, depending on the degree of depletion of the background plasma, weak or strong.

5.1.2 Diffusion of Small Plasma Disturbance in Weakly Ionized Plasma

In the previous text, we sketched a physical picture of the plasma disturbance diffusion spreading in the infinite background plasma both in the presence and absence of the magnetic field. We will now prove this picture theoretically, following the classical theory of diffusion spreading of plasma perturbations of different origins in the infinite background plasma with various degrees of ionization and magnetization [4–15].

Diffusion Spreading of Weak Plasma Disturbances. First, we consider diffusion of a weak plasma density perturbation, $\delta N(\mathbf{r},t) = N(\mathbf{r},t) - N_0$, having in an initial time $t=0$ the following *point* profile, $\delta N(\mathbf{r},0) = n_0 \delta(\mathbf{r})$, where N_0 is the density of the background plasma, n_0 the number of charged particles in the initial plasma perturbation, and $\delta(\mathbf{r})$ is the delta function. The perturbation of plasma density in any time is described by the following expression [5,6,9,15]:

$$\delta N(\mathbf{r},t) = n_0 G(\mathbf{r},t) = n_0 \int \exp\left\{ i\mathbf{k} \cdot \mathbf{r} - D_a(\mu^2)k^2 t \right\} \, \mathrm{d}^3 k \tag{5.7}$$

where $G(\mathbf{r},t)$ is Green's function of the following diffusion equations of quasi-neutral plasma, and which can be described in units of T/e [9,11,15]:

$$\begin{aligned} \frac{\partial N}{\partial t} &= \nabla \hat{\mathbf{D}}_e (\nabla N - N \nabla \Phi) \\ \frac{\partial N}{\partial t} &= \nabla \hat{\mathbf{D}}_i (\nabla N - N \nabla \Phi) \end{aligned} \tag{5.8}$$

Here Φ is the dimensionless potential in units of T/e, i.e., $\Phi = (e/T)\phi$; $\hat{D}_e$ and $\hat{D}_i$ are the tensors of electron and ion components of plasma, respectively. Their components can be obtained, as was done in Chapter 1, based on the framework of elementary theory and on the simple geometry of the problem analyzed in Refs. [7,8,15], where the vector $\mathbf{B}_0$ has been directed vertically upward along the z-axis

$$D_{e\|} = \frac{T}{m_e \nu_{em}}, \quad D_{e\perp} = \frac{T}{m_e \nu_{em}(1 + q_H^2)}, \quad D_{e\Lambda} = \frac{T \cdot q_H}{m_e \nu_{em}(1 + q_H^2)} \tag{5.9a}$$

$$D_{i\|} = \frac{T}{M_i \nu_{im}}, \quad D_{i\perp} = \frac{T}{M_i \nu_{im}(1 + Q_H^2)}, \quad D_{i\Lambda} = \frac{T \cdot Q_H}{M_i \nu_{im}(1 + Q_H^2)} \tag{5.9b}$$

All notations and parameters in Equation 5.9 are the same as those that were introduced in Chapter 1.

As for the coefficient of ambipolar diffusion, $D_a(\mu^2)$, where $\mu^2 = \cos^2\beta$ and $\cos\beta = \mathbf{k} \cdot \mathbf{B}_0/|\mathbf{k}||\mathbf{B}_0|$ is the angle between vectors $\mathbf{k}$ and $\mathbf{B}_0$, it can be analyzed directly only for a simple case, where spreading of a small perturbation with the sinusoidal initial profile occurs, spreading in plasma with the wave-vector $\mathbf{k}$ [9,10]

$$\delta N(\mathbf{r},t) \sim \exp\{i\mathbf{k} \cdot \mathbf{r} - D_a(\mu^2)k^2 t\} \tag{5.10}$$

Then, the expression (Equation 5.7) is simply a superposition of the sinusoidal plasma modes with the profile (Equation 5.10). For such plasma perturbations, in Refs. [9,10] an expression was obtained describing the tensor of ambipolar diffusion $D_a(\mu^2)$ in the infinite weakly ionized isothermal plasma, which, using our unified notations of plasma parameters introduced in Chapter 1, will be presented as

$$D_a(\mu^2) = \frac{2T(1 + q_H^2\mu^2)(1 + Q_H^2\mu^2)}{M_i \nu_{im}(1 + Q_H^2)(1 + q_H^2\mu^2) + m_e \nu_{em}(1 + q_H^2)(1 + Q_H^2\mu^2)} \tag{5.11}$$

In the sufficiently strong magnetic field, corresponding to the constraints $\rho_{He}\rho_{Hi} \ll \lambda_e\lambda_i$ and $q_H Q_H \gg 1$, and taking into account for Equation 5.9, the tensor of the ambipolar diffusion can be presented as

$$D_a(\mu^2) = \begin{cases} 2[D_{e\|}\mu^2 + D_{e\perp}(1 - \mu^2)], & \mu \ll \mu_0 \\ 2[D_{i\|}\mu^2 + D_{i\perp}(1 - \mu^2)], & \mu \geq \mu_0 \end{cases} \tag{5.12}$$

where $\mu_0 = \sqrt{b_{i\perp}/b_{e\|}} \ll 1$. From Equation 5.12 it follows that the tensor of the ambipolar diffusion is determined by the larger unipolar coefficients of diffusion (Equation 5.9) in the direction defined by $\mu = \cos\beta$. Thus, for $\mu > \mu_0$ the character of diffusion is determined by ions with $D_a(\mu^2) \approx 2D_{i\|}$ and with sinusoidal plasma perturbation spreading under the angle $\gamma = \tan^{-1}[(D_{i\perp}/D_{i\|})\tan\beta]$ to $\mathbf{B}_0$. The diffusion flow of plasma electrons along $\mathbf{B}_0$, $\Gamma_{e\|}$, essentially exceeds the longitudinal ion flow, $\Gamma_{i\|}$, and for $\mu \ll \mu_0$ we get $D_a(\mu^2) \approx 2D_{e\|}$. That means that the regions with maximum concentration have a positive charge, creating the inner electric field that conserves the condition of quasi-neutrality along $\mathbf{B}_0$. As a result, the current occurs in directions perpendicular to the vector $\mathbf{k}$ of spreading of the sinusoidal plasma perturbation along the lines with $\delta N = const$. At the same time, as was shown in Refs. [9,10], the inner electric field is absent for $\mu = \mu_0$. Thus, in the simple case of sinusoidal perturbation spreading, the character of diffusion differs essentially from ambipolar diffusion in the absence of the magnetic field (according to Schottky [1]). Even using the usual constraints of ambipolar diffusion, $\mathbf{\Gamma}_e = \mathbf{\Gamma}_i$, for the case under consideration, i.e., $\mathbf{\Gamma}_{e\parallel} = \mathbf{\Gamma}_{i\parallel}$ and $\mathbf{\Gamma}_{e\perp} = \mathbf{\Gamma}_{i\perp}$, we will then get, instead of Equation 5.12, the expression of the coefficient of ordinary diffusion in the anisotropic plasma [4–6]

$$\widetilde{D}(\mu^2) = D(\mu^2 = 1) \cdot \mu^2 + D(\mu^2 = 0) \cdot (1 - \mu^2) \tag{5.13}$$

where from Equation 5.11 it follows that $D(\mu^2 = 1) \equiv D_{a\parallel} = 2T/(M_i \nu_{im} + m_e \nu_{em})$ is the coefficient of the longitudinal diffusion ($\mu = 1$, $\beta = 0°$) equaling the analogous coefficient of diffusion for isotropic plasma defined by Equation 5.9b and taking it into account in $M_i \nu_{im}/m_e \nu_{em} \gg 1$.

As for the coefficient of the ambipolar diffusion transverse to $\mathbf{B}_0 (\mu = 0, \beta = 90°)$, defined by $D(\mu^2 = 0) \equiv D_{a\perp} = 2T/(M_i \nu_{im} Q_H^2 + m_e \nu_{em} q_H^2)$, it is similar to the coefficient of diffusion of electrons in the sufficiently strong magnetic field, increasing by $(1 + T_e/T_i) \approx 2$ times due to the influence of ions [7–9]. It should be noted that $\widetilde{D}(\mu^2)$ from Equation 5.13 differs significantly from $D_a(\mu^2)$ (Equation 5.11), and this difference is stronger at angles corresponding to $\mu \approx \mu_0$. Results of numerical analysis of both coefficients carried out in Ref. [9] showed that for $0 < \mu < 1$ the value of $\widetilde{D}(\mu^2)$ is smaller than that for $D_a(\mu^2)$, and this difference increases with the growth of the degree of magnetization of plasma (i.e., with increase of Q_H). Thus, for $q_H Q_H \gg 1$, a strong anisotropy occurs in the spreading of the sinusoidal plasma perturbation (Equation 5.10), and diffusion along $\mathbf{B}_0 (\mu \to 1, \beta \to 0°)$ occurs with the velocity of diffusion of ions increasing $(1 + T_e/T_i) \approx 2$ times due to the influence of electrons. In the directions transversal to $\mathbf{B}_0 (\mu \to 0, \beta \to 90°)$ the coefficient of ambipolar diffusion $D_a(\mu^2)$ decreases sharply to a value of $2D_{i\perp}\sqrt{\gamma}$. This peculiarity in the dependence of $D_a(\mu^2)$ on angles of β close to $90°$ significantly affects the character of diffusion spreading of plasma disturbance in a strong magnetic field. Even for realization of the equation (Equation 5.13) describing ambipolar diffusion in anisotropic plasma, the nonpotential (i.e., vortex) inner electric field is needed to secure the short-circuit current through background plasma. This new effect relates to a non-single-dimension problem that we will consider briefly on the basis of results obtained in Refs. [7–10].

Thus, for *small distances*, when $r \ll \sqrt{D_a t}$, Equation 5.7 yields

$$G(0,t) = \frac{1}{8\pi^{3/2} t^{3/2}} \int_0^1 \frac{d\mu}{D_a^{3/2}(\mu^2)} \tag{5.14}$$

Following Equation 5.14, the small weak plasma disturbance in its center decreases in time with $\sim t^{-3/2}$, i.e., in the same manner as when the magnetic field was absent [1–3]. The value of $G(0,t)$ that characterizes speed of damping of initial plasma perturbation depends essentially on the diffusion rate of electrons and ions. Thus, as mentioned earlier, in the strong magnetic field plasma electrons fill the ellipsoid with dimensions L_e and d_e, and plasma ions fill the ellipsoid with dimensions L_i and d_i. To oppose the particles with higher mobilities, electrons along $\mathbf{B}_0$ and ions across $\mathbf{B}_0$, the inner self-consistent electric field has to be quadruple-shaped with two symmetrical maxima of electric potential at points $z = \pm z_0$ along the vertical axis z and vector $\mathbf{B}_0$, and with one maximum at the point $\rho = \rho_0$ (i.e., in the plane $z = 0$ transverse to $\mathbf{B}_0$) [8–10]. In this case the flow of background electrons along $\mathbf{B}_0$ under the influence of the inner electric field for $z > z_0$ is directed toward the diffusion flow of electrons inside the disturbance, increasing the diffusion flow of electrons along $\mathbf{B}_0$. In the same manner, for $\rho > \rho_0$ this electric field increases the diffusion flow of ions as particles with higher mobility across $\mathbf{B}_0$. Finally, the plasma disturbance occupies the area consisting of two overlapping ellipsoids, electron and ion, as shown in Figure 5.1 according to Refs. [8–10]. This figure illustrates the relative profile of point perturbation of plasma density, $\delta N(\mathbf{r},t)/\delta N(0,0)$, where the corresponding coordinates along and across $\mathbf{B}_0$ are denoted by the relative units $\tilde{z} = z/L_i$ and $\tilde{\rho} = \rho/L_i$, computed for $q_H = 30$ and $Q_H = 0.3$. The dark areas with negative values of plasma concentration correspond to depletion regions of background plasma. Numerical results obtained in Refs. [8–10] as well as results described earlier indicate that the electron and ion components of a weak small plasma perturbation spread independently in a strong magnetic field, so that the requirement of quasi-neutrality is absent. This quasi-neutrality is secured in the electronic ellipsoid by background ions flowing into it across $\mathbf{B}_0$ and in the ionic ellipsoid by background electrons flowing into it along $\mathbf{B}_0$. These flows of background particles are caused by an inner electric field occurring inside the initial perturbation. At an angle of about $\beta_0 = \cos^{-1}(\mu_0)$ the perturbation of electric potential of this field is limited to zero, i.e., from this direction background particles leave their initial sites under the influence of the electric field creating the depletion regions, as shown in Figure 5.1. The number of particles leaving depletion areas corresponds to the number of particles inside the initial plasma perturbation. We will now estimate the volume filled by the perturbed plasma and the concentration of plasma particles within this volume.

In the absence of the magnetic field ($\mathbf{B}_0 = 0$), the effective volume occupied by the initial perturbation δN of plasma is of the same order as the ambipolar volume, i.e.,

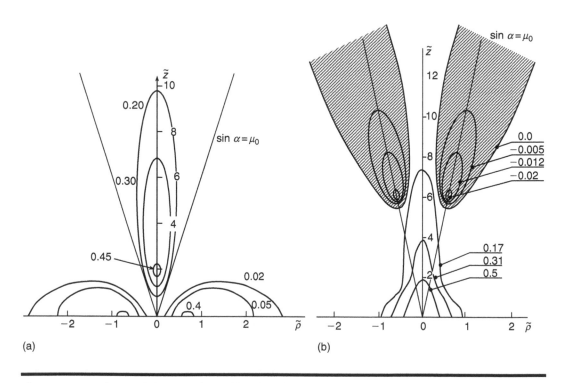

(a) (b)

Figure 5.1 Equipotential (a) and equidensity (b) contours of the point plasma density in the unbounded weakly ionized magnetized plasma.

$$V_a \propto \{(1 + T_e/T_i)D_{i\|}t\}^{3/2} \tag{5.15}$$

In the presence of the magnetic field, we get overlapping of two ellipsoids where the volume of the electronic ellipsoid is $V_e \approx L_e d_e^2 \approx D_{e\|}^{1/2} D_{e\perp} \{(1 + T_e/T_i)t\}^{3/2}$, and the same value for the ion ellipsoid is $V_i \approx L_i d_i^2 \approx D_{i\|}^{1/2} D_{i\perp} \{(1 + T_e/T_i)t\}^{3/2}$. The ratio of these volumes is

$$\tilde{V} = \frac{V_e + V_i}{V_a} \propto \left[\frac{q_H^{1/2}}{Q_H^{3/2}} + \frac{1}{(1 + Q_H)} \right] \tag{5.16}$$

This ratio characterizes changes in the effective volume occupied by plasma perturbation in the magnetic field. It can be seen that the volume occupied by plasma disturbance may essentially increase compared with the case when the magnetic field is absent.

The plasma concentration in the near zone ($r \ll \sqrt{D_a t}$) following Equation 5.14 can be presented as

$$\frac{\delta N(0,t)}{N_0} - \frac{n_0}{8\pi^{3/2} t^{3/2}} \int_0^1 \frac{d\mu}{D_a^{3/2}(\mu^2)}$$

$$\approx \frac{(1 + T_e/T_i)^{3/2}}{(8\pi D_{i\|}t)^{3/2}} \left(1 + Q_H^2 + q_H^{1/2} Q_H^{3/2} \right) \tag{5.17}$$

This ratio is of the order of $\left(V_e^{-1} + V_i^{-1} \right) n_0/N_0$; i.e., the concentration of plasma in the zone of the overlapping of the two ellipsoids is determined by the volume of the smaller ellipsoid. From Equation 5.17 it follows that as the magnetic field increases, the value of the integral in Equation 5.17, which we denoted by I, increases proportionally to B_0^2. In the sufficiently strong magnetic field (when $q_H \gg 1$ and $Q_H \geq 1$) for the case of isothermal plasma ($T_e \approx T_i = T$), the expression for I is

$$I = \int_0^1 \frac{d\mu}{D_a^{3/2}(\mu^2)} \approx \frac{1}{\sqrt{(2D_{i\|})(2D_{i\perp})(2D_{e\perp})}} \tag{5.18}$$

Comparing this expression with that for ordinary diffusion in the anisotropic plasma, $I = (D_\| D_\perp^2)^{-1}$, we get the effective coefficient of diffusion across the magnetic field $\mathbf{B}_0$ as the mean square of doubled coefficients of transverse diffusion of electrons and ions, i.e., $D_\perp = \sqrt{(2D_{i\perp})(2D_{e\perp})}$.

In the presence of the ambient magnetic field, the volume occupied by plasma perturbation increases significantly compared to when the magnetic field is absent, despite the fact that at the same time the corresponding velocities of plasma diffusion across the magnetic field slow down significantly.

In the close zone corresponding to the overlapping of two ellipsoids, as shown in Refs. [8–10], the plasma concentration and the potential of the inner electric field can be calculated by using two separated components of plasma, electrons and ions; i.e.,

$$\delta N(\mathbf{r},t) = N_e + N_i \tag{5.19a}$$

$$\Phi(\mathbf{r},t) = \frac{e}{T}\varphi = \ln\left(\frac{N_e}{N_0} + 1\right) - \ln\left(\frac{N_i}{N_0} + 1\right) \tag{5.19b}$$

where

$$\frac{N_e}{N_0} = \frac{\exp\left\{-\frac{z^2}{8D_{e\|}t} - \frac{\rho^2}{8D_{e\perp}t}\right\}}{(8\pi t)^{3/2} D_{e\|}^{1/2} D_{e\perp}} \tag{5.20a}$$

$$\frac{N_i}{N_0} = \frac{\exp\left\{-\frac{z^2}{8D_{i\|}t} - \frac{\rho^2}{8D_{i\perp}t}\right\}}{(8\pi t)^{3/2} D_{i\|}^{1/2} D_{i\perp}} \tag{5.20b}$$

Thus, in the area of the overlapping of electron and ion ellipsoids, the profile of concentration inside the plasma disturbance attenuates according to the exponential law with the coefficient of diffusion $\sim 2D_{e\perp}$, i.e., in the same manner as during ordinary ambipolar diffusion.

In the *far zones*, where $r \gg \sqrt{D_a t}$, the asymptotic solution according to Refs. [8–10] is as follows:

$$\delta N(\mathbf{r},t) = n_0(G_1(\mathbf{r},t) + G_2(\mathbf{r},t)) \tag{5.21}$$

where

$$G_1(\mathbf{r},t) = \frac{D_{i\perp}t\cos^4\alpha}{2\mu_0^4 r^5} \frac{\left(\frac{9\sin^4\alpha}{2\mu_0^4} - \frac{36\sin^2\alpha}{\mu_0^2} + 12\right)}{\left(1 + \frac{\sin^2\alpha}{\mu_0^2}\right)^{9/2}} \tag{5.22a}$$

$$G_2(\mathbf{r},t) = \frac{9D_{i\perp}t\cos^4\alpha}{2\pi r^5}\left(\frac{\mu_0\cos^2\alpha}{\sin^3\alpha} - \frac{\mu_0}{\sin\alpha}\right) \tag{5.22b}$$

where α is the angle between radius-vector r and the magnetic field $\mathbf{B}_0$, i.e., $\cos\alpha = \mathbf{r}\cdot\mathbf{B}_0/|\mathbf{r}||\mathbf{B}_0|$. The expression (Equation 5.22b) for $G_2(\mathbf{r},t)$ was obtained under the assumption that $\sin\alpha > \mu_0$ and that for $\sin\alpha \leq \mu_0$ we have $G_2 \ll G_1$. It was also noted that asymptotic expressions (Equation 5.22) give the dependence of plasma disturbance on distance as $\sim r^{-5}$ in the far zones for any direction α between $\mathbf{r}$ and $\mathbf{B}_0$, i.e., a damping according to polynomial law. The asymptotic solution in the far zones for the potential of the self-consistent electric field has the following form [8–10]:

$$\Phi = \frac{e}{T}\varphi = n_0(\Phi_1 + \Phi_2) \tag{5.23}$$

where

$$\Phi_1 = \frac{1}{4\pi\mu_0^2 r^3 N_0} \frac{\left(\frac{\sin^2\alpha}{\mu_0^2} - 2\right)}{\left(1 + \frac{\sin^2\alpha}{\mu_0^2}\right)^{5/2}}, \quad \sin\alpha \leq \mu_0 \tag{5.24a}$$

$$\Phi_2 = \frac{\mu_0}{4\pi r^3 N_0}, \qquad \sin\alpha > \mu_0 \qquad (5.24b)$$

It can be seen that in the far zone the distribution of the potential has a quadruple character. In fact, according to Equations 5.23 and 5.24, the potential $\varphi \sim r^{-3}$ and the flows of background plasma caused by the electric field (called the *field flows*) are $\Gamma_b \sim \varphi \cdot r^{-1} = r^{-4}$. Then, the perturbation of plasma concentration δN caused by these flows according to $\partial \delta N/\partial t \approx \nabla \Gamma_b$ will be proportional to $t \cdot r^{-5}$, a result that fully corresponds to that described by Equations 5.21 and 5.22. We can again point out that in the far zones the character of diffusion spreading of point weak plasma disturbances is determined by the short-circuit currents through plasma.

In the intervals $\sin\alpha \approx \mu_0$, from which the background particles come to the volume containing the initial plasma disturbance, are created the areas with the negative plasma concentration; these are the *depletion zones*. Thus, in the zones far from the initial plasma disturbance, its profile attenuates not according to the exponential law, but according to the polynomial law, $\sim r^{-5}$. In Refs. [8–10], the physical description of this phenomenon was analyzed in detail. It was shown that the reason for this behavior—of plasma disturbance spreading at large distances—is that the process of its relaxation is determined not only by diffusion of its electrons and ions along and across $\mathbf{B}_0$, but also by the influence of spatial electrical charge of background plasma, i.e., by the flow of vortex (short-circuit) currents through the background plasma. The relative depth h of the depletion zones, from which background electrons and ions come to the initial plasma disturbance to secure conditions of plasma quasi-neutrality, is of the order of the ratio of the amount of particles inside the disturbance to the amount of background particles in the depletion zones with the volume, $V_b \sim L_e d_i^2$, which greatly exceeds the volume of ellipsoids traced by electrons and ions of the initial disturbance, i.e.,

$$h = \frac{\delta N}{N_0} \sim \frac{n_0}{16\sqrt{2}\pi^{3/2} N_0 D_{e\parallel}^{1/2} D_{i\perp} t^{3/2}} \qquad (5.25)$$

The results of numerical computations of the strict formula (Equation 5.7), of the approximate formulas for the close zone (Equations 5.19a through 5.20) and for the far zone (Equations 5.21 and 5.22), and of the ambipolar exponential profile obtained in the case of $\mathbf{B}_0 = 0$ are shown together in Figure 5.2 by circles, solid, and dashed curves, respectively. The profile of spreading of a weak initially small plasma disturbance versus the dimensionless distance $x = r/L_i$ is shown for the two cases: (a) $\cos\alpha = 0$ (curves 1), and (b) $\cos\alpha = 0.9$ (curves 2). The first case corresponds to diffusion in the direction transverse to $\mathbf{B}_0$ and the second one to diffusion along $\mathbf{B}_0$. Computations were made for sufficiently magnetized plasma ($Q_H = 1$ and $q_H \approx 333$, i.e., $q_H Q_H \gg 1$). Here, the boundary of overlapping of ellipsoids corresponds to the constraint $0 < \cos\alpha < 1$, and the boundary of the larger ion ellipsoid (across $\mathbf{B}_0$) corresponds to the constraint $D_{e\perp}/D_{i\parallel} \approx 0.99$.

In this area, a sharp gradient (with the scale $\sim d_e$) of plasma concentration occurs in the inner electric field, which hinders faster particles (ions) in this direction. Thus, in the areas of the overlapping of the ellipsoids, the profile of plasma concentration is ambipolar and attenuates according to the exponential law, as in the case of $\mathbf{B}_0 = 0$, which is seen from Figure 5.2 for $x < 0.1$. Apart from areas of electron and ion ellipses, the profile of plasma concentration falls according to the polynomial law $\sim r^{-5}$, which is shown in Figure 5.2 for $x > 0.1$. In the far zones, the quick exponential damping of Green's function, $G(r,t) = \delta N(r,t)/n_0$, becomes weaker and decreases according to Equations 5.21 and 5.22 as x^{-5}. It can also be seen that for small x, Green's function falls faster in directions orthogonal to $\mathbf{B}_0$ ($\cos\alpha = 0$). This fact can be understood taking

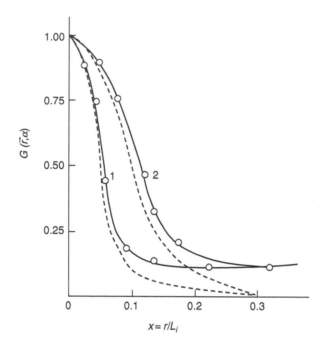

Figure 5.2 Spatial normalized 2-D plasma density distribution for strongly (solid curves) and weakly (dashed curves) ionized plasma.

into account that the magnetic field sharply hinders coefficients of diffusion across $\mathbf{B}_0$. The anisotropy of spreading for $x < 0.1$ is significant, whereas for $x \gg 0.1$ it becomes insufficient. This fact is clearly seen from illustrations of the shape of plasma disturbance shown as lines of equal density, $G(x,t) = const$, versus normalized coordinate x both in horizontal and vertical directions. The form of plasma disturbance in the process of its spreading is shown in Figure 5.3a in the logarithmic scale, and in Figure 5.3b the curve $G(x,t) = 0.03$ is shown in the ordinary scale. The rotation of the angle, $\cos\alpha$, commences from the vertical axis. Thus, we again see that the shape of the initial plasma disturbance during its spreading takes on the anisotropic form, covering the area of overlapping of both electron and ion ellipsoids; i.e., its maximal semi-axis along $\mathbf{B}_0$ corresponds to $L_e = \sqrt{8D_{e\|}t}$, and its maximal semi-axis across $\mathbf{B}_0$ corresponds to $d_i = \sqrt{8D_{i\perp}t}$. Hence, the direct computations presented in Figure 5.3 are in full agreement with the qualitative picture of diffusion spreading of a weak small plasma disturbance in the infinite weakly ionized plasma depicted in Figure 5.1.

Diffusion of Strong Small Plasma Disturbance in the Magnetic Field. In the nonlinear case of strong plasma disturbances, when the concentration of the background plasma is not sufficient to guarantee quasi-neutrality, the spreading of plasma disturbance occurs according to other scenarios. The depletion zones become sufficiently broad (compared to the linear case) and deep. The mechanism of short-circuit current becomes insufficient and the ambipolar regime is predominant; i.e., ambipolar diffusion in the anisotropic plasma occurs [1–3]. The fully consistent analysis of diffusion spreading of small strong ($\delta N \gg N_0$) or intermediary ($\delta N \leq N_0$ and $\delta N \geq N_0$) plasma disturbance in infinite weakly ionized plasma was analyzed in Refs. [10–14]. Without going into the mathematical details of this analysis, we will describe briefly peculiarities occurring during spreading of such plasma disturbances in strong magnetic fields.

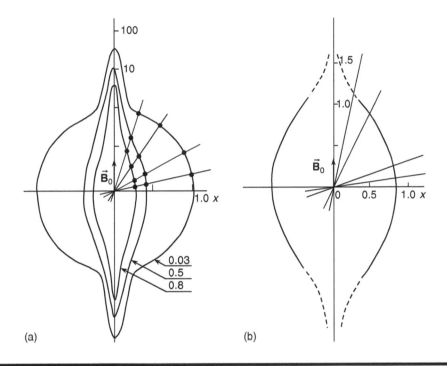

(a)　　　　　　　　　　　　　(b)

Figure 5.3 Shape of weakly ionized point plasma disturbance in the process of diffusion for different normalized time moments of spreading.

In the case of *intermediary* plasma disturbance, the quasi-linear description of the process is valid because, as in the linear case ($\delta N \ll N_0$), the plasma disturbance in the volume of electron and ion ellipsoids exceeds disturbances in the surrounding areas. Thus, the nonlinear effects occur in the overlapping of these ellipsoids. In this region the solution can be presented as follows:

$$\delta N(\mathbf{r},t) = N - N_0 = N_e + N_i + N_e \cdot N_i \cdot N_0^{-1} \tag{5.26a}$$

$$\Phi(\mathbf{r},t) = \frac{e}{T}\varphi(\mathbf{r},t) = \ln\left(\frac{N_e}{N_0}+1\right) - \ln\left(\frac{N_i}{N_0}+1\right) \tag{5.26b}$$

where N_e and N_i are defined by Equation 5.20a and b, respectively. As can be seen, the difference between the linear solution (Equation 5.19) and the quasi-linear solution (Equation 5.26) is in the existence of the last term $\sim N_e N_i N_0^{-1}$, which equals the product of the ambipolar profile, N_a, and the relative depth of depletion region, h, and corresponds to the region of overlapping of electron and ion ellipsoids with their minimum semi-axes along and across $\mathbf{B}_0$, respectively, where N_a is given by

$$N_a = n_0 \frac{\exp\left\{-\frac{z^2}{L_i^2} - \frac{\rho^2}{d_e^2}\right\}}{\pi^{3/2} L_i d_e^2} = N_0 h^2 \frac{D_{i\perp}}{D_{e\perp}}\sqrt{\frac{D_{e\parallel}}{D_{i\parallel}}}\exp\left\{-\frac{z^2}{L_i^2} - \frac{\rho^2}{d_e^2}\right\} \tag{5.27}$$

Outside these ellipsoids, the plasma disturbance is sufficiently small and the linear solution, $\sim t \cdot r^{-5}$, takes place. In other words, diffusion in the overlapping region is defined by the ambipolar mechanism according to which $\mathbf{\Gamma}_e = \mathbf{\Gamma}_i$ at each point of the ambipolar area. Equation 5.26 is valid when $h < 1$, and in the case of intermediate disturbances. In the ambipolar area there is a small share of total plasma particles n_0 of disturbance, proportional to h, with concentration defined by the term $N_0 h^2 D_{i\perp} / D_{e\perp} \sqrt{D_{e\|}/D_{i\|}}$ from Equation 5.27. Outside the overlapping zone, the concentration and the flows in the electron and ion ellipsoids are similar to those for the linear case of a weak disturbance; i.e., they can be described by Equation 5.19, because in Equation 5.26 the term $\sim N_d h$ is sufficiently small compared to other terms.

We will now summarize the physical aspects of the quasi-linear process occurring when an intermediate plasma disturbance is injected into infinite weakly ionized plasma. In this case, because of a weak effect of nonlinearity, until the relative depth of the depletion zone h is sufficiently small, the effects of short-circuit currents through the background plasma caused by electrons and ions from depletion areas are essential. The maximum value of plasma concentration in the electron and ion ellipsoids is limited by the number of particles of background plasma that can come from the depletion areas to secure the statement of quasi-neutrality. Ions of background plasma enter the electron ellipsoid across $\mathbf{B}_0$ from distances $\sim d_i$, and the number of ions filling the volume of electron ellipsoid is of the order $\sim N_0 L_e d_i^2$. Thus, the concentration in this ellipsoid is

$$\delta N_e \sim N_0 h D_{i\perp} / D_{e\perp} \tag{5.28}$$

At the same time, background electrons enter the ion ellipsoid along $\mathbf{B}_0$ from distances $\sim L_e$, and the number of ions filling the volume of the electron ellipsoid is of the order $\sim N_0 L_i d_e^2$. Thus, the concentration in this ellipsoid is

$$\delta N_i \sim N_0 h \sqrt{D_{e\|}/D_{i\|}} \tag{5.29}$$

The background ions collected in the electron ellipsoid are conserved there by a strong transverse (to $\mathbf{B}_0$) electric field that decreases with the scale $\sim d_e$. This leads to strong depletion of the background plasma close to the electron ellipsoid. At the same time, the transverse diffusion of background ions increases the transverse dimension of the depletion area to d_i (for $\mathbf{B}_0 = 0$ was d_e; see earlier text). Simultaneously, the background electrons collected in the ion ellipsoid are conserved there by a strong longitudinal (to $\mathbf{B}_0$) electric field that decreases with the scale $\sim L_i$. This leads to strong depletion of background plasma close to the ion ellipsoid, and also to an increase of the longitudinal dimension of the depletion area to L_e (for the case of $\mathbf{B}_0 = 0$ was L_i; see earlier text). Hence, because of short-circuit flows through the background plasma, the total concentration of plasma is disturbed significantly both in the volume of electron and ion ellipsoids and the condition of quasi-neutrality does not necessarily limit the volume occupied by a weak or an intermediary plasma disturbance because, as shown earlier analytically, the processes are similar for these kinds of plasma disturbances.

If now a *strong* disturbance is injected into the plasma, the background plasma becomes depleted significantly and the constraint $h \gg 1$ usually occurs in this scenario. The character of diffusion changes completely, because the number of particles in the central (i.e., in a zone of overlapping of ellipsoids) becomes equal to the total number of injected charged particles in a plasma disturbance. In this case, the plasma concentration is described by the ambipolar profile (Equation 5.27), and the disturbance spreads, as when the magnetic field is absent ($\mathbf{B}_0 = 0$). Only a negligible percentage of charged particles spread in the electron and ion ellipsoids caused by the short-circuit mechanism. Apart from ellipsoids (in the far zone), the asymptotic $\delta N \sim r^{-5}$ is also valid. In the case of strong nonlinearity ($h \gg 1$) the plasma disturbance at the boundary of this region becomes of the order of N_0, and the asymptotical nonlinear solution δN^{NL} simply equals the product of the linear asymptotic solution δN^L and the inverse depth of the depletion zone, i.e.,

$$\delta N^{NL}(\mathbf{r},t) \propto \delta N^L h^{-1} \tag{5.30}$$

where δN^L is described by Equations 5.21 and 5.22. Hence, for the extreme case when $\delta N \gg N_0$ and $h \gg 1$, the influence of the background plasma is insufficient and plasma disturbance spreads as in the case of $\mathbf{B}_0 = 0$ (according to Schottky [1–3]).

Diffusion of Plasma Disturbances of Finite Dimensions in the Weakly Ionized Unbounded Plasma. Based on the theory of diffusion spreading of weak and strong disturbances of a weakly ionized unbounded plasma in the absence or presence of the ambient magnetic field, a set of simple models of spreading of plasma perturbations with arbitrary initial configuration, infinite [14] and cylindrical [7,15] with limit dimensions L and d, oriented under the angle θ to $\mathbf{B}_0$, was performed. These models will be described briefly in the following text.

We consider now a two-dimensional (2-D) plasma disturbance of infinite length along $\mathbf{B}_0$ oriented at the angle θ to $\mathbf{B}_0$, as shown in Figure 5.4a. In the case of a weak disturbance of the background plasma ($\delta N \ll N_0$), the solution is a 2-D analogue of the problem discussed earlier for a point weak disturbance. The form of such a disturbance is shown schematically in Figure 5.4a, which can simply be presented as a superposition of two ellipsoids, the electronic with semi-axes $\tilde{L}_e - \sqrt{8\tilde{D}_{e\|}t}$ and $d_e = \sqrt{8D_{e\perp}t}$, and the ionic with semi-axes $\tilde{L}_i = \sqrt{8\tilde{D}_{i\|}t}$ and $d_i = \sqrt{8D_{i\perp}t}$, where

$$\begin{aligned}
\tilde{D}_{e\|} &= D_{e\|}\sin^2\theta + D_{e\perp}\cos^2\theta \\
\tilde{D}_{i\|} &= D_{i\|}\sin^2\theta + D_{i\perp}\cos^2\theta
\end{aligned} \tag{5.31}$$

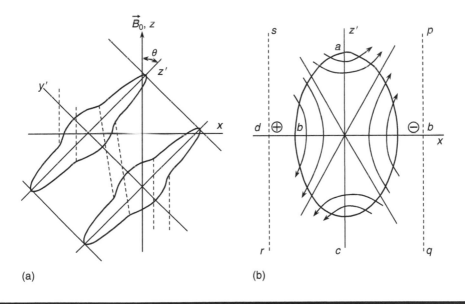

(a) (b)

Figure 5.4 The shape of 2-D plasma disturbance (a) strongly elongated under angle to the ambient magnetic field, and the corresponding electric field spatial distribution (b).

The solution in the close zone (in the zone of overlapping of both ellipsoids) is described by the same expressions as Equations 5.19 and 5.20. The asymptotes in the far zone (outside ellipsoids) can be obtained by using Equations 5.22 and 5.23 to change the longitudinal coefficients of diffusion in them according to Equation 5.31. These asymptotic solutions are valid for angles θ for which $\widetilde{D}_{e\parallel} > \widetilde{D}_{i\parallel}$, i.e., for $\theta > \mu_0 \equiv \sqrt{D_{i\perp}/D_{e\parallel}}$. If the inclination angle θ is sufficiently small, i.e., $\theta < \mu_0$, then $\widetilde{L}_e < \widetilde{L}_i$ and the depletion regions vanish. Then diffusion is limited to the 1-D ambipolar diffusion across $\mathbf{B}_0$ both in the case of weak and strong plasma density disturbance. The dimensions of the disturbed region are quickly decreased, forming the ambipolar volume $\sim L_i d_e^2$. However, in the case of strong plasma disturbance ($\delta N > N_0$), for the initial geometry of the problem, the additional Hall flows occur and become significant. Because these flows are proportional to B_0^{-1}, they can significantly accelerate the transport process across $\mathbf{B}_0$. This is correct only for a small plasma disturbance. In the case presented by Figure 5.4, the situation is more complicated. As shown in Ref. [14], for $\theta > \theta_0 \equiv \min\left[Q_H^{-1}, \sqrt{\mu_0}\right]$ the Hall flows can be short-circuited by the background electronic flow along $\mathbf{B}_0$ and the background ionic flow across $\mathbf{B}_0$. The difference between the point disturbance described previously and the disturbance shown in Figure 5.4a is that here, diffusion occurs in the plane $z'0x$ (see Figure 5.4b). If the disturbance of plasma concentration does not greatly exceed the concentration of the background plasma ($\delta N \geq N_0$), the diffusion in the plane $z'0x$ is determined by short-circuit currents through the background plasma. In the case of strong plasma disturbance ($\delta N \gg N_0$), an ambipolar diffusion (for $\mathbf{B}_0 = 0$) takes place.

Finally, for $\theta > \theta_0$ (corresponding illustrations in Figure 5.4a and b), two cases can be realized: (1) for the relative depth of the depletion regions $h = A/\left(8\pi N_0 t \sqrt{D_{i\perp}/D_{e\parallel}}\right) < 1$, the mechanism of short-circuit currents through plasma occurs, and (2) for $h > 1$ the ambipolar diffusion for $\mathbf{B}_0 = 0$ (according to Schottky [1–3]) takes place. Here, A is the number of particles at the unit length of the plasma disturbance. Thus, in Figure 5.4b the electron flows along $\mathbf{B}_0$ are shown by curved arrows, which, together with Hall flows, create the short-circuit current through the ambipolar ellipsoid, with the positive volume charge in direction sr and with the negative volume charge along line pq (see Figure 5.4b). Divergence of electronic flow, associated with perturbations of the inner field potential, $\delta\varphi$, is proportional to $b_{e\parallel}\delta\varphi N\theta^2/\widetilde{L}_i^2$. Therefore, for $\theta > \sqrt{\mu_0}$, $Q_H < 1$ (ions are weakly magnetized) it is possible even for a small potential disturbance, $\delta\varphi \propto \varphi_0\mu_0/\theta^2$, to compensate for the divergence of electronic Hall flow. The ionic Hall flow is circuited across $\mathbf{B}_0$ in the same manner as the electronic Hall flow.

Only for small angles $\mu_0 < \theta < \theta_0$ does diffusion in the crossing $\mathbf{E}_0 \times \mathbf{B}_0$ fields (Hall current) lead to a fundamental change of the character of diffusion of strong plasma disturbance (with $\delta N > N_0$). The polarization ($pqrs$) (see Figure 5.4b) cannot now be circuited along $\mathbf{B}_0$, and the Hall flows are generated in this field. A divergence of these flows is proportional to $eD_{e\perp}\delta\varphi(T_e d_e\widetilde{L}_i)^{-1}$ at the scale of the ambipolar ellipsoid, and is small compared to $\nabla\cdot\mathbf{\Gamma}_{e\parallel}$ but exceeds $\nabla\cdot\mathbf{\Gamma}_{e\perp}$. Therefore, the electric field changes sign across $\mathbf{B}_0$, and the transverse scale of strong plasma disturbance increases from d_e to L_i. Without going deeply into the theoretical analysis of the process in this case, which is presented in detail in Ref. [14], we point out that for angles of inclination of the initial strong ($\delta N > N_0$) corresponding to the constraint $\mu_0 < \theta < \theta_0$, the Hall flows lead to acceleration of the diffusion process compared to the previous case of $\theta > \theta_0$.

In the case of small inclination angles, where $\theta < \mu_0$, as mentioned earlier, the disturbance, weak or strong, spreads according to 1-D (across $\mathbf{B}_0$) ambipolar diffusion.

In Ref. [14], some numerical computations of spreading of the initial 2-D plasma disturbance with the finite dimensions $L = d$ in the (x,y)-plane and infinite length along z-axis, with the inclination angle θ to $\mathbf{B}_0$, were carried out for wide range of angles θ. The initial profile of plasma disturbance was Gaussian:

$$\delta N(\mathbf{r},0) = 2n_0 \exp\left\{-0.6\frac{x^2+y^2}{L^2}\right\} \tag{5.32}$$

where $\delta N = 3N_0$, which corresponds to the moderate disturbance of the background plasma. As found by numerical analysis, the potential of the internal field has a quadruple character. For large angles θ, its extrema lie practically at the z'- and x-axes, and the profile of concentration is symmetrical relative to x- and z'-axes (see Figure 5.4a). Figure 5.5 shows the distribution of the plasma concentration within the disturbed region (solid curves in the upper plane) and the potential of the inner field (dashed curves at the bottom plane) for the case of small $\theta = 1°50'$. The coordinates z' and x are normalized on L. It is evident that the essential asymmetry of the profile of concentration and potential is observed for small angles θ of the initial disturbance (relative to $\mathbf{B}_0$) infinitely stretched along the z-axis. Because the asymmetry is essential only in the regions where $\delta N > N_0$, it is negligible in the far zones. Estimations made in Ref. [14] give the critical values of $\mu_0 \approx 1°$ and $\theta = 7°$, beyond which the effects of magnetic field are insufficient. The speed of decay of δN along the z-axis increases sharply during the transition from $\theta = 0°$ to $\theta = 1°$, becoming sufficiently small for $1° < \theta < 7°$. Further changes of the angle θ from $7°$ to $90°$ do not affect the speed of concentration decay, which coincides with the qualitative picture presented previously.

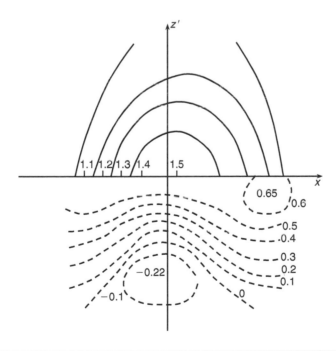

Figure 5.5 Equidensity (solid curves) and equipotential (dashed curves) contours of 3-D plasma density disturbance of finite dimensions in the unbounded weakly ionized magnetized plasma.

Such critical dependence of orientation of the initial plasma density relative to $\mathbf{B}_0$ was found separately in Ref. [7]. We describe briefly the results obtained there. In Ref. [7] diffusion and drift of plate and cylindrical plasma disturbances in the weakly ionized unbounded background plasma were studied numerically. While solving the problem, the 3-D equations of anisotropic diffusion for electronic and ionic components of plasma disturbance were analyzed numerically. These equations can be presented in the following form:

$$\frac{\partial N}{\partial t} = D_\perp \left(\frac{\partial^2 N}{\partial x^2} + \frac{\partial^2 N}{\partial y^2} \right) + D_\| \frac{\partial^2 N}{\partial z^2} \tag{5.33}$$

During the computations, the critical angle $\theta_c = \tan^{-1}[(D_{i\perp} - D_{e\perp})/(D_{e\|} - D_{i\|})]$ of inclination of the initial disturbance relative to $\mathbf{B}_0$ was found, leading the change of the electronic character of diffusion (for $\theta < \theta_c$) to an ionic one (for $\theta > \theta_c$) across $\mathbf{B}_0$. For plasma with a sufficient degree of magnetization ($D_{e\|} \gg D_{i\|}$, $D_{i\perp} \gg D_{e\perp}$) we get $\theta_c \approx \sqrt{D_{i\perp}/D_{e\|}}$, i.e., the same expression as for the critical angle μ_0 introduced in the previous case in accordance with Ref. [14]. Estimations made in Ref. [7] give the same value of $\theta_c \approx \mu_0 = 1°30'$. For $\theta < \theta_c$ the magnetic field essentially hampers the speed of diffusion across $\mathbf{B}_0$, whereas for $\theta > \theta_c$ the character of the diffusion spreading becomes similar to the case $\mathbf{B}_0 = 0$ and diffusion is controlled across $\mathbf{B}_0$ by ions since $\rho_{\mathrm{Hi}} \gg \rho_{\mathrm{He}}$ and $\mu_{i\perp} \gg \mu_{e\perp}$. As is known, in the latter case a magnetic field does not affect the diffusion spreading of the plasma disturbance. A consistent numerical analysis was also carried out in Ref. [7] for the case of a cylindrical plasma disturbance with the limit dimensions $x = \pm L$ and $y = \pm L$ (i.e., with transverse scales $2L$) and with the central axis oriented along the z-axis at an angle θ to the vector $\mathbf{B}_0$. The distribution of plasma concentration was assumed to be Gaussian for the dimensionless density $\rho = N/n_0$:

$$\rho(x,y,0) = \exp\{-\lambda(\tilde{x}^2 + \tilde{y}^2)\} + \varepsilon \tag{5.34}$$

where $\lambda = 0.6$ and $\varepsilon = 0.5$ were taken from numerical simulations of the problem; n_0 is the density of the charged particles along the axis of the cylinder. The coordinates and time were also taken to be dimensionless, i.e., $x = R\tilde{x}$, $y = R\tilde{y}$, and $t = (R^2/D_{i\|})\tilde{t}$, where R is the initial transverse dimension of the disturbance. The dimensionless potential was determined by a similar expression, $\Phi = e\varphi/T$. In computations the initial disturbance was chosen to be moderate, i.e., $\delta N = (3–10)$ N_0. The boundary conditions were taken as follows:

$$\Phi = 0, \quad \frac{\partial \rho}{\partial \tilde{x}} = 0, \quad \tilde{x} = \pm \frac{L}{R}$$

$$\Phi = 0, \quad \frac{\partial \rho}{\partial \tilde{y}} = 0, \quad \tilde{y} = \pm \frac{L}{R} \tag{5.35}$$

In Figures 5.6 through 5.9 the lines of equal normalized concentration (solid curves) and the normalized potential (dashed curves) are presented for $\tilde{t} = 0$, (a) 0.5, (b) and 1 (c), and for $\theta = 0°$, $1°10'$, $1°50'$, and $90°$, respectively. The values of the dimensionless potential Φ shown in the figures were increased 10^4 times for a clearance. For $\theta = 0°$ (Figure 5.6a through c) the field of the spatial charge was directed opposite the z-axis, being negative at the axis of the cylindrical disturbance ($\Phi < 0$). This means that the diffusion across $\mathbf{B}_0$ is defined by electrons of a plasma disturbance as particles with smaller mobility across $\mathbf{B}_0$ compared to ion mobility in this direction. The case $\theta = 1°10' < \theta_c = 1°30'$ is shown in Figure 5.7a through c. Here, we can observe the

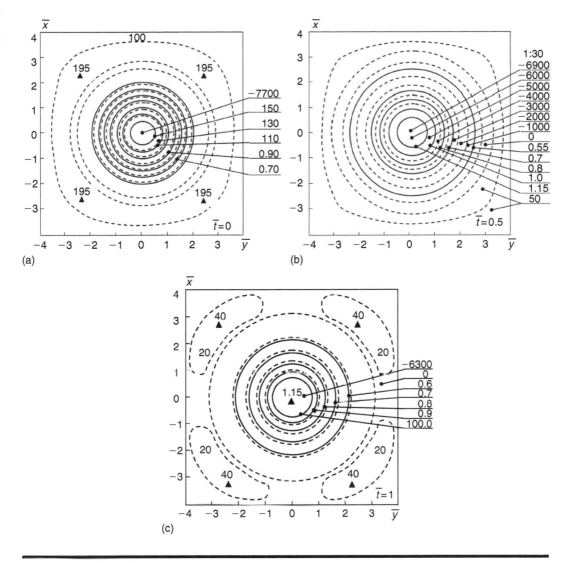

Figure 5.6 The normalized concentration (solid curves) and the normalized potential (dashed curves) for $\tilde{t} = 0$, (a) 0.5, (b) and 1 (c), for $\theta = 0°$.

essential asymmetry and distortion of the lines of equal concentration and potential. The maximal anisotropy of the spherical initial shape of plasma disturbance occurs in a direction parallel to the magnetic field lines. In this case the diffusion across $\mathbf{B}_0$ is again determined by electrons, because the spatial charge and the potential of the inner field at the axis of the plasma disturbance are negative. For $\theta = 1°50' > \theta_c = 1°30'$ (Figure 5.8a through c) the electric field along the cylinder is directed from inside to the outside of the disturbed plasma region; i.e., the spatial charge and the corresponding potential of this electric field are positive. This means that the character of diffusion across $\mathbf{B}_0$ is determined by ions as particles with higher mobility across $\mathbf{B}_0$ compared with electron mobility in this direction. A further increase of the angle of inclination $\theta(\theta \gg \theta_c)$ does not change the character of diffusion spreading of the cylindrical plasma disturbance. Even where the magnetic

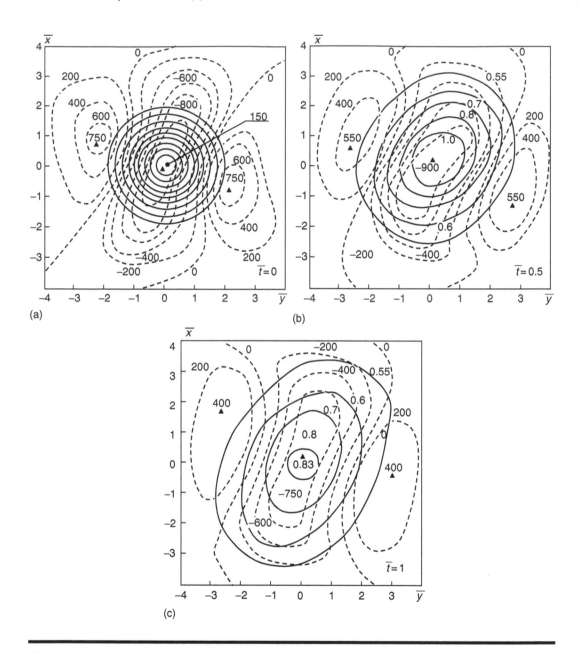

Figure 5.7 The same as in Figure 5.6, but for $\theta = 1°10'$.

field $\mathbf{B}_0$ is perpendicular to the z-axis of the cylindrical disturbance and directed along the $\tilde{x}$-axis, the velocity of spreading along the $\tilde{x}$-axis differs insignificantly compared to that along the $\tilde{y}$-axis (see Figure 5.9a through c for $\theta = 90°$). Following a qualitative description of the process presented previously, we can point out that in the case of $\theta \gg \theta_c$ the Hall flows in the crossing $\mathbf{E}_0 \times \mathbf{B}_0$ fields become insufficient. In this case the equipotential lines become symmetrical relative to the plane ($\tilde{x}0\tilde{y}$). The small degree of anisotropy in direction $\mathbf{B}_0$ can be explained by

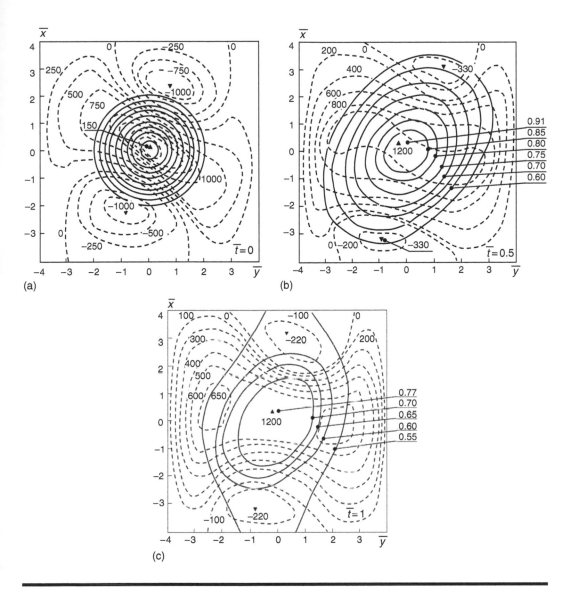

Figure 5.8 The same as in Figure 5.6, but for $\theta = 1°50'$.

the following. According to latest geometry ($\theta = 90°$) the Hall current, even though weak, may flow in the direction of the $\tilde{x}$-axis and therefore cannot distort and displace the spatial charge inside the initial plasma disturbance.

Now we will consider, following Ref. [94], the diffusion of a strong cylindrical disturbance of weakly ionized unbounded background plasma, the initial profile of which is symmetrical relative to z-axis, i.e.,

$$N(z, \rho, 0) = \delta N(z, \rho, 0) + N_0 = An_0 \exp\left\{-\frac{z^2 + \rho^2}{L^2}\right\} \qquad (5.36)$$

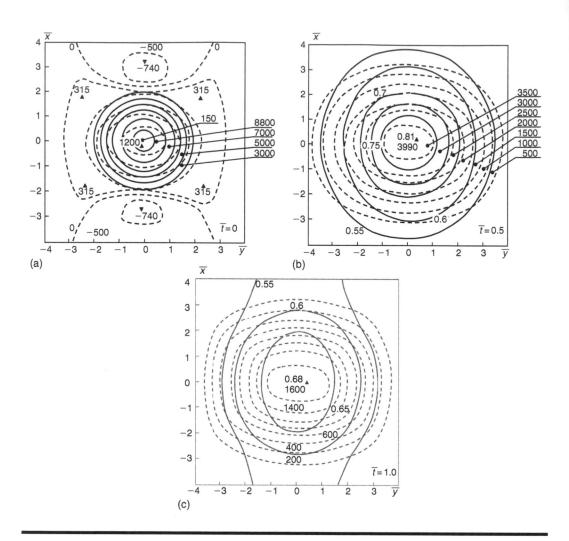

Figure 5.9 The same as in Figure 5.6, but for $\theta = 90°$.

Here, L is the characteristic scale of the initial plasma disturbance, A is the degree of perturbation of plasma density defined earlier, and ρ is the cylindrical coordinate. The boundary conditions of the problem were considered to be the following [94]:

$$N(\infty,t) = N_0, \quad \Phi(\infty,t) = 0 \tag{5.37}$$

For the numerical computations, the transport coefficients were chosen to be $D_{i\parallel} = D_{i\perp} = D_{i0}$ (ions are not magnetized), $D_{i\parallel} = 10 D_{i0}$, and $D_{e\perp} = 0.1 D_{i0}$.

Apart from the numerical analysis, in Ref. [94] simple analytical expressions describing non-linear spreading of the strong plasma disturbances of finite dimensions were obtained and analyzed. Thus, in the process of diffusion spreading of arbitrary plasma disturbance until its initial scale $L \leq \sqrt{D_{i\perp} t}$, depletion regions are not so important, and the diffusion is close to the one-dimensional ambipolar along $\mathbf{B}_0$:

$$N(z,\rho,t) = \frac{An_0 \exp\left\{-\frac{z^2}{L^2 + L_i^2} + \frac{\rho^2}{L^2 + d_e^2}\right\}}{\sqrt{L^2 + L_i^2}\left(L^2 + d_e^2\right)} + N_0 \qquad (5.38a)$$

$$\Phi(z, \rho, t) = \Phi(t) + \frac{\rho^2 - z^2}{L^2 + L_i^2} \qquad (5.38b)$$

Here, the transverse diffusion is insufficient [11,12]. The function $\Phi(t)$ relates to screening effects of the electrostatic field made by the disturbance δN. For large time t, depletion regions can be created, and the character of diffusion spreading becomes directly dependent on the degree of perturbation A of the background plasma. As mentioned earlier, the convenient characteristics of plasma disturbance such as the maximal depth of the depletion zones h (Equation 5.25) can be taken into account and rewritten according to Ref. [94] as

$$h = \frac{AL^3}{8\sqrt{2}D_{e\|}^{1/2}D_{i\perp}\, t^{3/2}} \qquad (5.39)$$

If $h \gg 1$, the depth of the depletion zones is significant, and the character of spreading is determined by the ambipolar diffusion along $\mathbf{B}_0$ described by Equation 5.38. In the case of moderate disturbance, where $h \geq 1$ or $h \leq 1$, the process of diffusion is controlled by short-circuit currents through the depletion regions and the profile of plasma concentration in the electron and ion ellipsoids is close to that described in Ref. [94]:

$$N(z,\rho,t) = \delta N_e + \delta N_i + \delta N_e \delta N_i N_0^{-1} + N_0 \qquad (5.40)$$

where the components δN_e and δN_i are described by Equation 5.20. In Figure 5.10a the dependence of concentration at the origin of the coordinate system ($z = 0$, $\rho = 0$) is shown versus the dimensionless time $\tau = D_{i\|}t/L^2$ according to numerical calculations made in Ref. [94]. Here, curve 1 corresponds to direct calculations of the system (Equation 5.8) with initial (Equation 5.36) and boundary (Equation 5.37) conditions, curve 2 indicates the approximation solution (Equation 5.38), and curve 3 relates to the approximation (Equation 5.40) with Equation 5.20. Parameter A in computations varied from 1 to 100. In Figure 5.10a, the parameter A was taken to equal 10. It can be seen that curve 1 is approximated closely by formula Equation 5.38 up to time $\tau = \tau_0 = 0.46$ and by Equation 5.40 beyond τ_0. The point τ_0 is the moment of switching on the mechanism of short-circuit current generation through the background plasma. For $\tau < \tau_0$, the character of diffusion is close to the ambipolar one described by Equation 5.38. In the case of strong disturbance, $A = 100$, where the character of spreading is presented in Figure 5.11a and b, the direct numerical solution of the system (Equation 5.8) (solid curve) differs insignificantly compared to the ambipolar solution (Equation 5.38) (dashed curves) during the time of computation. The profiles of the concentration along $\mathbf{B}_0$ (curve 1) and across $\mathbf{B}_0$ (curve 2) for $\tau = 0.1$ and $A = 100$ are shown in Figure 5.11a. The continuous curves correspond to the direct numerical computations of Equation 5.8 and the dashed curves to the ambipolar solution Equation 5.38. It is evident that the character of diffusion is fully described by the ambipolar diffusion along $\mathbf{B}_0$ (or for the case of $\mathbf{B}_0 = 0$ [1–3]). Apart from the ambipolar ellipsoid, the solution is changed, and only a dozen parts of the total particles of the initial strong disturbance are spreading unipolarly because of flows from

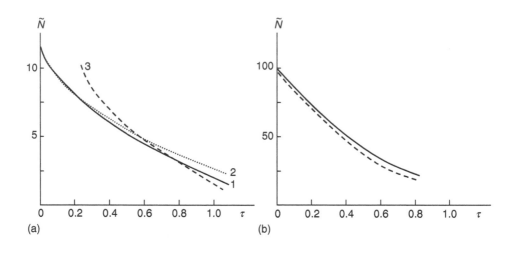

Figure 5.10 The dependence of concentration at the origin of the coordinate system ($z = 0$, $\rho = 0$) versus the dimensionless time $\tau = D_{i\parallel}t/L^2$; direct calculations of system (Equation 5.8) are denoted by curve 1, approximation solution (Equation 5.38) are by curve 2, and the approximation (Equation 5.40) is by curve (3). Parameter A was taken to equal 10.

the depletion regions. The influence of the background plasma in these regions on the character of spreading is essential, mostly for the potential Φ (see Figure 5.12). In Figure 5.12, the profiles of concentration (upper part *a*) and the potential (bottom part *b*) for $\tau = 0.1$ and $A = 100$ are shown. Polarization of the background plasma screens the ambipolar potential (Equation 5.38), which has cut-off points at infinity and therefore guarantees the condition $\Phi(\infty, t) = 0$ for arbitrary degree of plasma disturbance (for any parameter *A*). Similar curves are presented in Figure 5.13 for the case of moderate disturbance ($A = 10$) and for $\tau = 0.7$ for the concentration (upper part *a*) and the potential (bottom part *b*). Comparison between Figure 5.12 (for the strong disturbance) and Figure 5.13 (for the moderate disturbance) indicates the essential change of the profile of the electric field

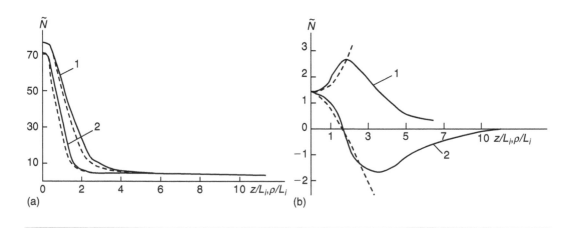

Figure 5.11 Same as Figure 5.10, but for $A = 100$.

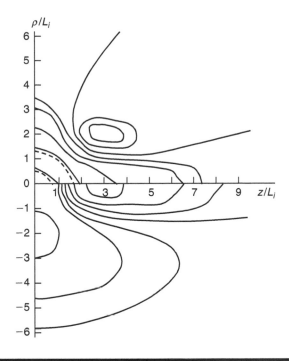

Figure 5.12 2-D contours for the concentration (upper part a) and the potential (bottom part b) for $\tau = 0.1$ and $A = 100$ (strong disturbance).

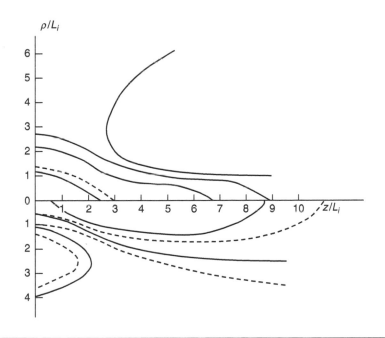

Figure 5.13 Same as Figure 5.12, but for $A = 10$ (moderate disturbance) and $\tau = 0.7$.

in the latter case. Thus, for $A = 100$ (Figure 5.12) the profile of the potential Φ changes (compared to the case $A = 10$, Figure 5.13) in such a manner as to guarantee maximal flows from the depletion regions. In all the figures, solid curves correspond to the direct numerical solution and dashed curves to the approximate solution (Equation 5.38). The maximum depth of the depletion for $A = 10$ is approximately 4%, whereas for $A = 100$ it achieves about 20% and for $0.1 < \tau < 0.7$ is practically unchanged.

5.1.3 Diffusion of Plasma Disturbances with an Arbitrary Degree of Ionization

Despite the fact that in Refs. [4–15] the theory of the arbitrary disturbance of plasma density in the presence of the ambient magnetic field was consistently presented, it cannot explain all the peculiarities occurring in the process of the spreading of irregularities in plasma with an arbitrary degree of ionization. In other words, the previous discussion does not consider the effects of electron–ion collisions occurring in the real ionospheric plasma at altitudes more than 140–150 km. Therefore, based on results obtained in Refs. [16–22,47], we will analyze the main peculiarities of diffusion spreading of arbitrary plasma irregularities in the ionosphere, taking into account the empirical distribution of the main parameters of the real model of the middle-latitude ionosphere described in Chapter 1. We use the data for the middle-latitude ionosphere because they are much more systematically presented in the literature [20–22,27,83].

First, we will consider a simple case of diffusion spreading of a small weak plasma disturbance in the infinite homogeneous ionosphere, now taking into account the real parameter of ionization $p = \nu_{ei}/\nu_{em}$ introduced and described in Chapter 1 for the real model of the middle-latitude ionosphere. In this case the coefficient of ambipolar diffusion differs from that described by Equation 5.11. Thus, for the case of the isothermic ($T_e \approx T_i = T$) background plasma with an arbitrary degree of ionization, and according to Refs. [16,17,20], the coefficient of ambipolar diffusion equals

$$D_a(\mu^2) = \frac{2T\left[(1+p)^2 + q_H^2(1+2\eta p)\mu^2 + Q_H^2 q_H^2 \mu^4\right]}{M_i \nu_{im}\left[(1+p)^2 + Q_H q_H(1+p) + q_H^2(1+\eta p + Q_H^2 \mu^4)\right]} \tag{5.41}$$

where all parameters are defined in Chapter 1. As in the case of weakly ionized plasma ($p = 0$) discussed earlier, a tendency of plasma diffusion in the strong magnetic field is found [17]. Thus, the dependence of the coefficient of ambipolar diffusion (Equation 5.41) in the magnetized plasma ($Q_H q_H \gg 1$) in the direction of the vector **k** of spreading in the ambient magnetic field is sufficiently complicated and fundamentally differs from the similar coefficient of diffusion in the anisotropic medium

$$D_0(\mu^2) = D_{\parallel}\mu^2 + D_{\perp}(1 - \mu^2) \tag{5.42}$$

from which for $\mu = 1$ ($\beta = 0°$) and $\mu = 0$ ($\beta = 90°$), we get, respectively,

$$D_{\parallel} \equiv D(\mu = 1) = \frac{2T}{M_i \nu_{im}(1 + \eta)} \tag{5.43a}$$

$$D_\perp \equiv D(\mu = 0) = \frac{2T}{M_i \nu_{im}(1 + p + \eta q_H^2)} \tag{5.43b}$$

From Equations 5.42 and 5.43, it can be seen that the velocity of longitudinal diffusion in the magnetic field is the same as that of ions in the isotropic plasma, $D_{i\parallel} = \frac{2T}{M_i \nu_{im}}$, because $\eta = m_e \nu_{em}/M_i \nu_{im} \ll 1$, and the collisions of the charged particles do not influence diffusion along $\mathbf{B}_0$. Also following from Figure 5.14, in the case of weakly ionized plasma ($p = 0$), the coefficient of ambipolar diffusion $D_a(\mu^2)$ defined by Equation 5.31 differs significantly from that of anisotropic diffusion $D_0(\mu^2)$ defined by Equation 5.42. Thus, with an increasing degree of ionization of plasma ($p > 1$), the difference between $D_a(\mu^2)$ and $D_0(\mu^2)$ becomes weaker and insufficient for $p \gg 1$ (more strictly for $\eta p > 1$). The corresponding dependences, $D_a(\mu^2)/D_a(0)$ and $D_0(\mu^2)/D_a(0)$, determined by Equations 5.41 and 5.43b and by Equations 5.42 and 5.43b, are shown by a family of solid and dashed curves, respectively. Hence, in the ionospheric plasma with an arbitrary degree of ionization, the ambipolar diffusion simply transforms into an ordinary diffusion process in the anisotropic medium. This peculiarity is evident from analysis of the coefficients of unipolar diffusion described in Equation 1.12a, which was made taking into account the real model of the middle-latitude ionosphere (see Figure 1.3). From Figure 1.3 it can also be seen that for a weakly ionized plasma the short-circuit currents through the background plasma cause higher speed of diffusion (dashed curves) comparing to that for plasma with a real degree of

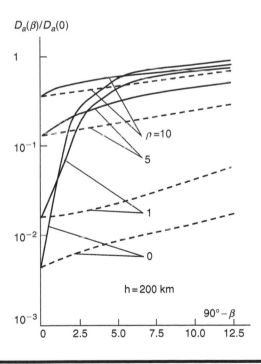

Figure 5.14 Normalized coefficient of ambipolar diffusion versus ionospheric altitude for the real model of the middle-latitude ionosphere (solid curves) and virtual weakly ionized ionospheric plasma (dashed curves).

ionization (solid curves), because it is controlled by faster electrons along $\mathbf{B}_0(D_{e\parallel}/D_{i\parallel} \approx 10^3 - 10^4)$ and faster ions across $\mathbf{B}_0(D_{i\perp}/D_{e\perp} \approx 10^2 - 10^3)$. Accounting for the real degree of ionization of the ionospheric plasma (solid curves in Figure 1.3), where the difference between coefficients of unipolar diffusion of electrons and ions along and across $\mathbf{B}_0$ becomes negligible, we observe the inverse process. Here, the character of diffusion changes from the ambipolar in the presence of the magnetic field (with $D_{e\parallel}$ along $\mathbf{B}_0$ and $D_{i\perp}$ across $\mathbf{B}_0$) to the ordinary diffusion in the anisotropic medium (with $D_{i\parallel}$ along $\mathbf{B}_0$ and $D_{e\perp}$ across $\mathbf{B}_0$).

The foregoing allows for generalizing a tensor of ambipolar diffusion described by Equation 5.12 to the case of real ionospheric plasma with an arbitrary degree of ionization

$$D_a(\mu^2) = \begin{cases} 2\left[D_{e\parallel}\mu^2 + \frac{1+p}{1+2p}D_{e\perp}(1-\mu^2)\right], & \mu^2 \geq 0 \\ 2\left[D_{i\parallel}\mu^2 + \frac{1+p}{1+2p}D_{i\perp}(1-\mu^2)\right], & \mu^2 \leq 1 \end{cases} \tag{5.44}$$

by introducing the additional term $(1+p)/(1+2p)$ before the transverse components of tensor $\hat{D}_{e,i}$ [16,17]. This presentation of the tensor of the ambipolar diffusion allows for transferring continuously from weakly ($p=0$) to strongly ($p \gg 1$) ionized plasma.

Thus, as shown in Refs. [16,17] and clearly seen from Equation 5.44, for the case $p=0$ for angles $\beta = \cos^{-1}(\mathbf{k} \cdot \mathbf{B}_0/kB_0)$ close to the transverse to $\mathbf{B}_0$ direction ($\beta \to 90°$, $\mu \to 0$) the tensor $D_a(\mu^2)$ is the doubled tensor of electron diffusion (for isothermal plasma)

$$D_a(\mu^2) = 2[D_{e\parallel}\mu^2 + D_{e\perp}(1-\mu^2)] \tag{5.45a}$$

which for $\beta = 90°$, $\mu = 0$ equals $2D_{e\perp}$. For $\beta \to 0°$, $\mu \to 1$ the tensor $D_a(\mu^2)$ is the doubled tensor of ion diffusion (for isothermal plasma)

$$D_a(\mu^2) = 2[D_{i\parallel}\mu^2 + D_{i\perp}(1-\mu^2)] \tag{5.45b}$$

which for $\beta = 0°$, $\mu = 1$ equals $2D_{i\parallel}$. These results fully describe diffusion of the point or sinusoidal disturbances in the isothermal weakly ionized infinite plasma reported in Refs. [4,5,9,11].

At the same time, for the case of ionospheric plasma with $p \gg 1$ (more strictly for $\eta p > 1$) for the transverse diffusion ($\beta = 90°$, $\mu = 0$) $D_a(\mu^2) = D_{e\perp} \approx D_{i\perp}$, and for the longitudinal diffusion ($\beta = 0°$, $\mu = 1$) $D_a(\mu^2) = 2D_{i\parallel}$. These results are in agreement with those discussed earlier and also obtained in Refs. [16,17].

We can now estimate the critical parameter of ionization of the ionospheric plasma for which the transformation of diffusion process, from ambipolar in weakly ionized plasma described by Equation 5.12 to that in the case of the ionospheric plasma with arbitrary degree of ionization described by Equation 5.44. Thus, from conditions $D_{e\perp} \approx D_{i\perp}$ and $D_{e\parallel} \approx 2D_{i\parallel}$ with an accuracy of $\eta \approx 10^{-3} \ll 1$, we get according to Ref. [16]:

$$p_{cr} = \frac{1 + Q_H^2 - \eta q_H^2}{2(\eta^2 q_H^2 - Q_H^2)} \tag{5.46}$$

We have to note that the growth of the parameter of ionization cannot be considered up to infinity, because in the case of fully ionized plasma the definition of the longitudinal diffusion does not make sense [4,5]. Thus, following from Equation 1.12, for the components of unipolar diffusion of charged particles, for $p \rightarrow \infty$ (i.e., $\nu_{em} \rightarrow 0$ and $\nu_{im} \rightarrow 0$), the coefficients $D_{e\parallel} \rightarrow \infty$ and $D_{i\parallel} \rightarrow \infty$, i.e., the longitudinal diffusion in the fully ionized plasma, are impossible. The main reason for this condition is that for diffusion in plasma it is necessary that the total pressure, described by a sum of the gas $(P_e + P_i)$ and the magnetic $(P_B = B^2/8\pi)$ pressure, should be constant, i.e., $\nabla(P_e + P) + \nabla P_B = 0$. However, in the fully ionized plasma $\nabla P_B = 0$, and, therefore, $\nabla(P_e + P) = 0$ or $(T_e \nabla N_e + T_i \nabla N_i) = 0$. In the case of isothermal $(T_e \approx T_i = T)$ and quasi-neutral $(N_e \approx N_i = N)$ plasma, these conditions are satisfied if $\nabla N = 0$; i.e., the definition of the longitudinal diffusion loses its physical meaning.

However, the coefficient of the ambipolar diffusion $D_a(\mu^2)$ does not give any information about the shape and form of the spreading plasma disturbance of arbitrary configuration and only describes damping of disturbance with the initial profile of Fourier harmonic type $\sim\exp\{i\,\mathbf{kr} - D_a (\mu^2)k^2 t\}$. In the spreading of the 3-D disturbance, δN, as mentioned previously, new effects occur related to the 1-D problem. Thus, for the point disturbance and following Equation 5.7, Green's function, which characterizes the velocity and the time of spreading of arbitrary disturbance, essentially depends on the velocity of diffusion of electron and ion components of plasma disturbance. In fact, as shown earlier, in strong magnetized plasma $(q_H \gg 1, Q_H > 1)$ the plasma disturbance occupies an area having the shape of the crossing electron and ion ellipsoids with characteristic semi-axes, L_e, d_e and L_i, d_i, respectively. The dimensions of the initial plasma disturbance along $\mathbf{B}_0$ are determined by the length $\sim 2L_e$ and the width $\sim 2d_i$ across $\mathbf{B}_0$. These dimensions and the shape of the initial disturbance depend significantly on the degree of ionization of background plasma. Thus, in Figure 5.15 the shape of the disturbance is presented in the form of the overlapping of electron and ion ellipsoids. The corresponding scales are also shown for each

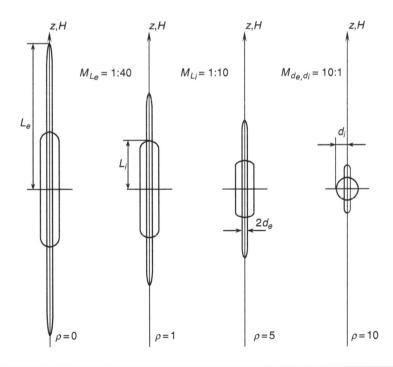

Figure 5.15 The shape of overlapping electron and ion ellipsoids versus the parameter of ionization, $p = 0$, 1, 5, and 10.

dimension of these ellipsoids for different parameters of ionization p and the parameters of the diurnal ionosphere corresponding to the altitude $h = 200$ km (see Tables 1.1 through 1.3) for a moment $t = 0.01$ s from the beginning of the spreading process. It can be seen that with the growth of the degree of ionization, p, the width of the electron ellipsoid is limited to the width of an ion. Along $\mathbf{B}_0$, the dimension of the disturbed area is defined by electrons. Finally, the shape of the disturbance is limited to the that of the ellipsoid, characterized by ordinary diffusion in the anisotropic medium (we should not forget that for $p = 0$ it has the spindle-shaped form; see Figure 5.3). Moreover, the dimensions and shape of the disturbance in the real ionosphere with an arbitrary degree of ionization differ significantly from those in the weakly ionized plasma (for $p = 0$). Thus, in Figure 5.16, the dashed lines correspond to the weakly ionized plasma and the continuous curves to the real degree of ionization and plasma parameters corresponding to the altitudes $h = 200$ km and $h = 300$ km (see Tables 1.1 through 1.3). It can be seen that for $h = 200$ km and $t = 0.01$ s and $p = 0$, $L_e = 400$ m, $d_e = 0.01$ m, $L_i = 10$ m, $d_i = 0.3$ m, whereas for $p = 3$ (corresponding to the altitude of 200 km) $L_e = 200$ m, $d_e = 0.02$ m, $L_i = 11$ m, $d_i = 0.4$ m. For $h = 300$ km and $p = 0$, $L_e = 1700$ m, $d_e = 0.002$ m, $L_i = 40$ m, $d_i = 0.07$ m and for $p = 45$ (corresponding to $h = 300$ km) $L_e = 250$ m, $d_e = 0.02$ m, $L_i = 60$ m, $d_i = 0.08$ m. From these estimations it follows that if the real degree of plasma ionization is not taken into account, this can lead to fundamental mistakes in prediction of the shape and dimensions of the plasma disturbance of various origins created in the ionospheric plasma.

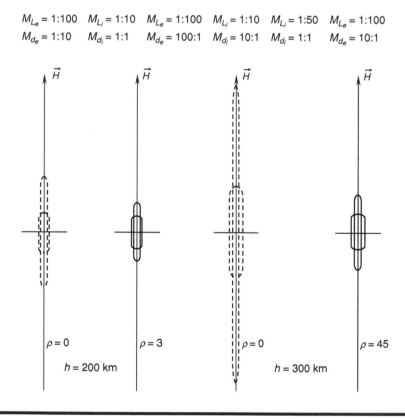

Figure 5.16 Comparison of shape of plasma disturbance generated in the real middle-latitude ionosphere (solid-line contours) and in the virtual weakly ionized ionosphere (dashed-line contours).

Consequently, the volume occupied by plasma disturbance is also fundamentally changed. As mentioned earlier for the case of weakly ionized plasma spreading in a strong magnetic field, the effective volume occupied by plasma perturbation essentially grows and significantly exceeds the volume occupied by the same plasma perturbation when a magnetic field is absent. In the case of plasma with an arbitrary degree of ionization, the volume occupied by the initial plasma disturbance can be found on the basis of the results obtained in Refs. [16,17] and taking into account Equation 1.12a. Thus, the related (to the case of the absence of the magnetic field, $\mathbf{B}_0 = 0$) volume of the overlapping electron and ion ellipsoids equals:

$$\widetilde{V} = \frac{V_e + V_e}{V_{\mathbf{B}=0}} = \frac{(1+p)^{3/2}(1+2\eta p)^{1/2}[(1+p)(1+2\eta p) + Q_H^2(1+p)]}{\eta^{3/2}(1+2p)^{3/2}[(1+p)^2 + q_H^2(1+2\eta p + Q_H^2)]} +$$
$$+ \frac{(1+p)(1+2\eta p)^{1/2}[(1+p)(1+2p) + q_H^2(1+2\eta p)]}{(1+2p)[(1+p)^2 + q_H^2(1+2\eta p + Q_H^2)]} \tag{5.47}$$

For $p \leq 1$ the first term in Equation 5.47 exceeds unity, i.e., $\widetilde{V} > 1$, i.e., the same as for $p = 0$ considered previously. For $p > 1$, the value of $\widetilde{V}$ is of the order of unity, and for $p \gg 1$ we get $\widetilde{V} = 1$. Thus, with growth in the degree of plasma ionization, the volume occupied by the plasma disturbance is limited to the analogous volume in the case of $\mathbf{B}_0 = 0$ (according to Schottky [1–3]).

The character of diffusion spreading of arbitrary plasma disturbance essentially depends on the degree of ionization of the background plasma. Without loss of generality, let us consider a weak $(\delta N \ll N_0)$ small $(\delta N(\mathbf{r}, 0) = n_0 \delta(\mathbf{r}))$ plasma disturbance injected into the background plasma with an arbitrary degree of ionization. In this case, a system of hydrodynamic equations (Equation 1.6) can be linearized, taking into account the quasi-neutrality of plasma, ambipolarity of diffusion, and the constraint $(\nabla \hat{\sigma}_e \nabla)(\nabla \hat{\sigma}_i \nabla) = (\nabla \hat{\sigma}_i \nabla)(\nabla \hat{\sigma}_e \nabla)$ according to Refs. [5,9,11], i.e.,

$$(\nabla \hat{\sigma}_e \nabla + \nabla \hat{\sigma}_i \nabla) \frac{\partial \delta N}{\partial t} = \left[(\nabla \hat{\sigma}_e \nabla)(\nabla \hat{D}_i \nabla) + (\nabla \hat{\sigma}_i \nabla)(\nabla \hat{D}_e \nabla) \right] \delta N \tag{5.48}$$

The solution of Equation 5.48 has the same form as Equation 5.7 for weakly ionized plasma but with the coefficient of ambipolar diffusion $D_a(\mu^2)$ now described by Equation 5.41 or 5.44. We present this solution in the spherical coordinate system $\{k, \mu, \psi\}$, as a frame in the wave-vector space, which after integration over k yields

$$G(r,\alpha,t) \equiv \frac{\delta N(r,\alpha,t)}{n_0} = \frac{1}{8\pi^{5/2}t^{3/2}} \int_0^1 \frac{d\mu}{D_a^{3/2}(\mu^2)} \int_0^\pi e^{-\frac{r^2 B^2}{4D_{ai}t}} \left\{ 1 - \frac{r^2 B^2}{2D_a t} \right\} d\psi \tag{5.49}$$

Here $B = \mu \cos \alpha + \sqrt{1 - \mu^2} \sin \alpha \cos \psi$; ψ is the angle between planes $(\mathbf{r}, \mathbf{B}_0)$ and $(\mathbf{k}, \mathbf{B}_0)$; α the angle between the radius vector $\mathbf{r}$ and vector $\mathbf{B}_0$, i.e., $\cos \alpha = \mathbf{r} \mathbf{B}_0 / |\mathbf{r}||\mathbf{B}_0|$, $\mu = \cos \beta$. For small distances from the origin of the frame $(r \ll \sqrt{D_a t})$ from Equation 5.49, we immediately obtain Equation 5.14. It was shown in Refs. [9,11] that for the case of weakly ionized plasma, the linear solution (Equation 5.49) up to distances $r \leq 10L_i$ can be presented by a simple superposition of solutions obtained for the electron and ion ellipsoids:

$$G(x,\alpha) = G_e(x,\alpha) + G_i(x,\alpha) \tag{5.50}$$

where

$$G_e(x,\alpha) \equiv \frac{\delta N_e(x,\alpha)}{n_0} = \frac{\exp\left\{-x^2\left[\frac{D_{i\parallel}}{D_{e\parallel}}\cos^2\alpha - \frac{D_{i\parallel}}{D_{e\perp}}\sin^2\alpha\right]\right\}}{\pi^{3/2}L_i^2(D_{e\parallel}/D_{i\parallel})^{1/2}(D_{e\perp}/D_{i\parallel})} \tag{5.51a}$$

$$G_i(x,\alpha) \equiv \frac{\delta N_i(x,\alpha)}{n_0} = \frac{\exp\left\{-x^2\left[\cos^2\alpha - \frac{D_{i\parallel}}{D_{i\perp}}\sin^2\alpha\right]\right\}}{\pi^{3/2}L_i^2(D_{i\perp}/D_{i\parallel})} \tag{5.51b}$$

Here, as before, $x = r/L_i$ and $L_i = \sqrt{8D_{i\parallel}t}$ (plasma is isothermal; $T_e \approx T_i = T$). In Figure 5.17 the dependence of $G(x,\alpha)$ on parameter x is shown for different angles α and for two times: $t = 0.01$ s and $t = 0.1$ s. All parameters characterizing the ionospheric plasma at the height $h = 200$ km are taken from Tables 1.1 through 1.3 (see Chapter 1). At the right side of the picture the shape and the characteristic scales of dimensions of plasma disturbance are shown for two different times, 0.01 s (solid curves) and 0.1 s (dashed curves). As can be seen, in the zones of electron and ion ellipsoids, i.e., in the close zone ($x \leq 1$), the Green function $G(x,\alpha)$ falls sharply according to the exponential law, mostly in the direction orthogonal to $\mathbf{B}_0$ (or close to this direction, particularly, for $\alpha \geq 60°$). In this zone, strong anisotropy in the initial disturbance spreading is observed; the speed of diffusion depends on the angle α, and for $\alpha = 0°$ (along $\mathbf{B}_0$) the damping of $G(x,\alpha)$ occurs more slowly than for $\alpha = 90°$ (across $\mathbf{B}_0$). The disturbance has the spindle-shaped form with dimensions $L_e \approx 19L_i$, $d_e \approx 10^{-3}L_i$, and $d_i \approx 2 \cdot 10^{-2}L_i$.

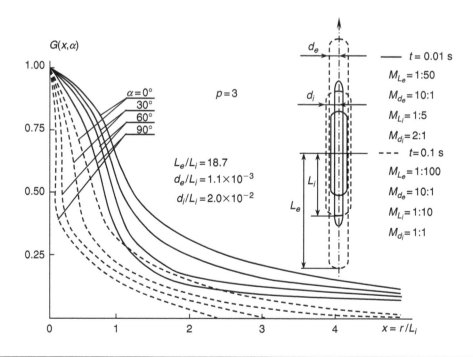

Figure 5.17 Spatial distribution of plasma density within a small disturbance in the case of the real model of middle-latitude ionosphere for $t = 0.01$ s (solid curves) and for $t = 0.1$ s (dashed curves).

Beyond electron and ion ellipsoids (for $x > 1$), the Green function $G(x,\alpha)$ decreases, as is expected in the far zone [5,9,11], proportionally to x^{-5}. In this zone the anisotropy is not as vivid and becomes weaker with an increase of dimensionless distance x. The form of the inhomogeneity in the far zone is limited to ellipsoidal, with the scale $2d_e$ across $\mathbf{B}_0$ and $2L_i$ along $\mathbf{B}_0$, which corresponds to that in the case of diffusion in an anisotropic medium. For a strong degree of plasma ionization, the shape of disturbance in a far zone can even achieve a spherical form analogous to that in isotropic plasma occurring in the absence of a magnetic field. So, for strong plasma ionization, the profile of plasma disturbance concentration falls exponentially both in close and far zones, as in the case of ordinary diffusion in the absence of a magnetic field.

With an increase in the time of the spreading process, $G(x,\alpha)$ decreases quickly. Thus, following from the given illustration, for the same values of the function $G(x, \alpha)$ for $t = 0.1$ s are smaller than the analogous values for $t = 0.01$ s. The concentration of plasma inside the disturbance falls proportionally to $t^{-3/2}$.

From the analysis presented above it follows that in the far zone for $r > \sqrt{D_a t}$ the value of δN decreases as r^{-5} in longitudinal and transverse directions to $\mathbf{B}_0$. With an increase of the magnetic field (with altitude of the ionosphere) the asymptotic of δN increases proportionately with B_0^2. However, for $B_0 \to 0$ this asymptotic has a tendency to change the law of damping versus r from polynomial to exponential. This law for the nonmagnetized plasma [5,9] can be described as

$$G(r,t) = \frac{1}{8(\pi t)^{3/2} D_{\parallel}^{1/2} D_{\perp}} \exp\left\{ -\frac{r^2}{4D_{\parallel} t} \right\}$$ (5.52)

as in the case of the usual diffusion for $\mathbf{B}_0 = 0$ described by the ambipolar equations obtained by Schottky [1–3]. We have mentioned earlier and now want to reiterate that the changes in asymptotic behavior of δN in the far zone follow from the analytical properties of the Fourier components (harmonics) of δN for $\mathbf{k} = 0$. The components of δN in a singularity $\mathbf{k} = 0$ correspond to the changes of the total number of particles in plasma. The physics of such phenomena can be described by the following mechanism. The character of plasma disturbance δN in the far zone (beyond ellipsoids) is determined not by the ambipolar field but by distribution of the spatial charge of the background plasma, i.e., by short-circuit currents through the background plasma. This is the same result that was found in Refs. [5–11] and mentioned earlier when considering weakly ionized plasma in a strong magnetic field.

The lifetime of the plasma disturbance τ was defined in Refs. [5–11] as the time during which the initial concentration of plasma disturbance in its maximum, $\delta N(0,0)$, decreases in the n times ($n = e, 5, 10, \dots$), i.e.,

$$\delta N(0,0) = n \delta N(0,\tau)$$ (5.53)

Then the lifetime of plasma disturbance can be found by using Equation 5.14 according to the following formula:

$$\frac{\tau}{\tau_0} = \left\{ \int_0^1 \left[\frac{D_{i\parallel}(1 + T_e/T_i)}{D_a(\mu^2)} \right]^{3/2} d\mu \right\}^{2/3}$$ (5.54a)

Here, $\tau_0 = \left[\frac{n_0 n}{\delta N(0,0)}\right]^{3/2}/4D_{i\parallel}(1 + T_e/T_i)$ is the lifetime of plasma disturbance in the absence of a magnetic field; the coefficient of ambipolar diffusion $D_a(\mu^2)$ is defined by Equation 5.41 or 5.44, and n_0 is the number of particles in the initial irregularity.

This formula for plasma with an arbitrary degree of ionization can be rewritten as [16,29]:

$$\frac{\tau}{\tau_0} = \left\{\int\limits_0^1 \left[\frac{(1 + 2p)[(1 + p)^2 + q_H Q_H(1 + p) + q_H^2(1 + \eta p + Q_H^2)\mu^2]}{(1 + 2p)[(1 + p)^2 + q_H^2(1 + 2\eta p)\mu^2 + q_H^2 Q_H^2 \mu^4]}\right]^{3/2} d\mu\right\}^{2/3} \quad (5.54b)$$

This formula becomes a simple formula for weakly ionized plasma (when $p = 0$) as

$$\frac{\tau}{\tau_0} = \left\{\int\limits_0^1 \left[\frac{1 + Q_H^2}{1 + Q_H^2\mu^2} + \frac{1 + q_H^2}{1 + q_H^2\mu^2}\right]^{3/2} d\mu\right\}^{2/3} \quad (5.54c)$$

Analysis of Equation 5.54b carried out in Refs. [16,29] for ionospheric plasma according to the ionospheric model described in Chapter 1 for altitudes $h = 90$–500 km has shown that the increase of the lifetime of plasma disturbance of more than one order, compared to when a magnetic field is absent, is observed at altitudes $h > 120$–130 km. For such altitudes the conditions of plasma magnetization ($q_H Q_H \gg 1$) are satisfied. With the increase of the degree of plasma ionizations, the time of relaxation of initial plasma disturbance increases, mostly in directions transverse (or close) to the magnetic field.

5.1.4 Diffusion Spreading of Plasma Irregularities in the Middle-Latitude Ionosphere

We will now introduce the reader to a physical framework concerning diffusion spreading of various irregularities in the real ionospheric parameters that were used for the middle-latitude ionosphere in the description of moderate solar activity presented in Chapter 1, in accordance with Refs. [16,20], and which the authors usually used in their theoretical and experimental research on the middle-latitude ionosphere. First, we will show the new effects and features of diffusion spreading of plasma irregularities caused by the inhomogeneous ionosphere, the transport parameters of which, as was shown in Chapter 1, are complicated functions of the ionospheric altitudes. We will then use quasi-homogeneous background ionospheric plasma to show peculiarities in diffusion spreading of strongly elongated irregularities in the middle-latitude ionosphere and will analyze the effects of their initial dimensions on the nature of diffusion spreading.

Inhomogeneous Ionosphere. We will deal with an inhomogeneous ionosphere and plasma irregularities with various degree of anisotropy, using rigorous inhomogeneous nonlinear equations of plasma diffusion, which cover both strong ($\delta N > N_0$) and weak ($\delta N < N_0$) plasma inhomogeneities and can be obtained from the full system (Equation 1.6) described in Chapter 1, taking into account Equation 1.7. We then examine diffusion spreading of an arbitrary quasi-neutral plasma irregularity with initial concentration $\delta N(\mathbf{r}, t = 0)$ created in the background ionospheric plasma with density $N_0(z)$, i.e., $N(\mathbf{r}, t) = \delta N(\mathbf{r}, t) + N_0(z)$. Here, z is the altitude of the real ionosphere, along which parameters of diffusion vary considerably on a scale up to the order of tens of kilometers (see Figure 5.18). Here, in Figure 5.18, we again present the altitudinal distribution

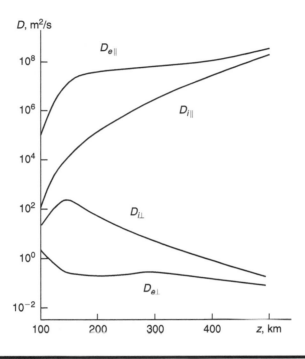

Figure 5.18 Height distribution of the diffusion coefficients according to the model of the middle-latitude ionosphere.

of unipolar coefficients of electron and ion components of plasma in the daytime middle-latitude ionosphere taken from Figure 1.3 as being more convenient for the reader to follow our analysis of effects of the nonregular inhomogeneous ionosphere compared to the regular homogeneous ionosphere. It is evident from Figure 5.18 that, due to complicated altitudinal dependence of diffusion coefficients, we should wait a determining influence of the inhomogeneity of the background ionospheric plasma as a function of altitude z, on the spreading of irregularities stretched along the geomagnetic field $\mathbf{B}_0$.

As mentioned previously, for ionospheric irregularities with scales exceeding the characteristic scales of the plasma (the electron and ion mean free paths, λ_e and λ_i, along $\mathbf{B}_0$ or their Larmor radii, ρ_{He} and ρ_{Hi}, transverse to $\mathbf{B}_0$), a system of hydrodynamic equations (Equation 1.6) can be simplified, taking into account conditions of the plasma's quasi-neutrality (Equation 1.9a) and quasi-stationarity ($T_e \approx T_i = T$), as well as of the diffusion ambipolarity (Equation 1.9b):

$$
\begin{aligned}
\frac{\partial N}{\partial t} &= \frac{1}{2}(\nabla \hat{\mathbf{D}}_e(z)\nabla + \nabla \hat{\mathbf{D}}_i(z)\nabla)N + \frac{1}{2e}(\nabla \hat{\sigma}_i(z)\nabla - \nabla \hat{\sigma}_e(z)\nabla)\phi \\
\frac{1}{e}(\nabla \hat{\sigma}_e(z)\nabla &+ \nabla \hat{\sigma}_i(z)\nabla)\phi = (\nabla \hat{\mathbf{D}}_e(z)\nabla - \nabla \hat{\mathbf{D}}_i(z)\nabla)N
\end{aligned}
\tag{5.55}
$$

Here ϕ is the potential of the internal self-consistent (ambipolar [2]) electric field $\delta \mathbf{E} = -\nabla \phi$; the components of tensors diffusion $\hat{\mathbf{D}}_\alpha(z)$ and conductivity $\hat{\sigma}_\alpha(z)(\alpha = e, i)$ are now functions of the altitude z. Their dependence is presented in Figure 5.18 for the real model of the middle-latitude ionosphere with parameters introduced in Chapter 1. The system (Equation 5.55) is general because it describes diffusion spreading of arbitrary plasma inhomogeneity, weak ($\delta N \ll N_0$), moderate ($\delta N < N_0$), ($\delta N > N_0$), and strong ($\delta N \gg N_0$) in the inhomogeneous ionospheric plasma with density N_0.

The initial density profile of the irregularity $\delta N(\mathbf{r}, t=0)$ was set to be the Gaussian in the cylindrical original frame (ρ, z, φ) with the z-axis aligned along $\mathbf{B}_0$, and the potential $\Phi = (e/T)$ $\phi = const$, i.e.,

$$\delta N(\rho, z, t = 0) = AN_0 \exp\left\{ -\frac{\rho^2}{d^2} - \frac{(z - z_0)^2}{L^2} \right\}$$

$$\Phi(\rho, z, t = 0) = \Phi_0$$

(5.56)

Here, we assume the cylindrical symmetry of the problem, i.e., $\frac{\partial \delta N}{\partial \varphi} = 0$. Then the total plasma density at the initial moment does not depend on φ and is equal to

$$N(\rho, z, 0) = \delta N(\rho, z, 0) + N_0(z)$$

(5.57)

where L and d are the characteristic scales of plasma irregularity, T is the temperature of an isothermic plasma in energy units, $A = \delta N(0,0,0)/N_0$ the degree of the initial irregularity power compared with background ionospheric plasma, and z_0 the height of the initial plasma irregularity.

The boundary conditions to the system (Equation 5.55) are chosen according to commonly accepted physical suggestions, as follows. At the lower boundary, along altitude z, $z = z_1$, the plasma density perturbation δN vanishes because of a strong recombination process, discussed in Chapter 1, i.e.,

$$\delta N(\rho, z = z_1, t) = 0, \quad N(\rho, z = z_1, t) = N_0(z_1)$$

(5.58)

At the upper boundary along z, $z = z_2$, the charged particle flux is supposed to be constant, since the mean free path of electrons and ions along $\mathbf{B}_0$ increases with altitude, i.e.,

$$D_{\alpha\|} \left. \frac{\partial \delta N(\rho, z, t)}{\partial z} \right|_{z=z_2} = const, \quad \alpha = e, i$$

(5.59)

In the directions transverse to the field $\mathbf{B}_0$, when $\rho_1 > (D_{e\perp} t)^{1/2}$, the plasma density perturbation δN vanishes, i.e.,

$$\delta N(\rho = \rho_1, z, t) = 0, \quad N(\rho = \rho_1, z, t) = N_0(z)$$

(5.60)

The introduced boundary conditions with respect to the altitude z correspond to the computations of distributions of the charged particle density in the ionosphere usually used in the modeling of the upper region E and region F of the ionosphere [84–86].

The potential of the internal electric field $\delta\Phi$ created by the perturbation δN equals zero at both lower and upper boundaries of the ionosphere, i.e.,

$$\Phi = \delta\Phi + \Phi_0 = \Phi_0$$

(5.61)

In Ref. [17], the approximate numerical model was also developed instead of Equation 5.55 for the case of diffusion of weak irregularity ($\delta N \ll N_0$, N_0 is the background plasma density) in the unbounded inhomogeneous background ionospheric plasma. According to this model, the density disturbance $\delta N(\mathbf{r},t)$ in the process of spreading of a quasi-neutral irregularity with initial electron density $\delta N(e)$ or ion density $\delta N(i)$ (i.e., $\delta N(\mathbf{r},t) = \delta N(e) = \delta N(i)$) can be represented in a weakly disturbed plasma (the linear case; see earlier text) as a sum of two terms $\delta N(\mathbf{r},t) = \delta N_e + \delta N_i$. Each term is found from the usual diffusion equation in an anisotropic medium

$$\partial \delta N_e / \partial t = \nabla \left(2 \hat{D}_e(z) \nabla \right) \delta N_e - \alpha(z) \delta N_e$$
$$\partial \delta N_i / \partial t = \nabla \left(2 \hat{D}_i(z) \nabla \right) \delta N_i - \alpha(z) \delta N_i \tag{5.62}$$

Here, the components of tensors of diffusion for electrons $\hat{D}_e$ and ions $\hat{D}_i$ are functions of the vertical coordinate z as the result of the inhomogeneity of the ionospheric plasma along altitudes. The effective coefficient of recombination $\alpha(z)$ describing recombination processes usually significant in the low ionosphere at altitudes $h < 160$–180 km, can be presented according to Refs. [17,20,27] as

$$\alpha(z) = \alpha_0 \exp\left\{ -\frac{z - z_0'}{a} \right\} \tag{5.63}$$

where $a = 35$, $\alpha_0 = 10^{-4}$ (daytime) and $a = 25$, $\alpha_0 = 2 \cdot 10^{-4}$ (nighttime), and $z_0' = 300$ km.

In the system Equation 5.55, as well as in Equation 5.62, we have ignored the effect of gravity, which is usually significant in problems concerning the dynamics of the F layer (see, for example, Ref. [8]). This is related, first of all, to the fact that in contrast to problems concerning the formation of the F layer (the one-dimensional case), here, we examine a three-dimensional model following Refs. [17,20]. In the case under consideration, the spreading of the irregularity is determined by vortex (i.e., short-circuit) currents of charged particles through the background plasma (see previous text), for whose formation the transverse diffusion, which largely determines the spreading of the plasma disturbance, is important. Second, in contrast to the problem of the F layer, the spreading of the irregularity is determined by two components, the electronic and ionic. For the electronic component the influence of gravity is insignificant, because for electrons the height of the uniform gas $H_g \approx 10^4$ km greatly exceeds the height $H_i \approx 50$ km for ions, and, therefore, also the effective longitudinal diffusion size of the electronic component in the irregularity $\Lambda_{D\parallel} \approx 50$–$100$ km. We also note that from Equation 5.62, if gravity is included, it follows that the ratio of the magnitude of the diffusion term $F_{D\parallel}$ to the gravity term F_g is around $H_g / \Lambda_{D\parallel} \approx 100$–$200$; i.e., $F_{D\parallel}$ greatly exceeds F_g. For the ionic component in Equation 5.62, δN_i, the principal term is that consisting of the transversal diffusion coefficient. For this reason, the influence of gravity on the ionic component could be significant only in a one-dimensional (1-D) problem. However, as an analysis made in Refs. [17–19] shows, the effect of this term is of little consequence even in this case because the effective longitudinal size of the irregularity along $\mathbf{B}_0$ in the one-dimensional case does not actually increase with time and remains smaller than the characteristic size of the F layer. Hence, the effects of gravity can be omitted in the analysis of diffusion spreading of irregularities strongly stretched along $\mathbf{B}_0$ (also called *field-aligned* irregularities in the literature).

Now taking into account the cylindrical symmetry of the problem and the Gaussian profile of the initial irregularity along and across $\mathbf{B}_0$ according to Equation 5.56 and the initial conditions introduced earlier, we can rewrite the system (Equation 5.62) in the form:

$$
\frac{\partial \delta N_e}{\partial t} = 2\gamma(z)D_{e\perp}(z)\frac{1}{\rho}\frac{\partial}{\partial \rho}\left(\rho\frac{\partial \delta N_e}{\partial \rho}\right)
$$

$$
+\frac{\partial}{\partial z}\left(2D_{e\|}(z)\frac{\partial \delta N_e}{\partial z}\right) - \alpha(z)\delta N_e
$$

$$
\frac{\partial \delta N_i}{\partial t} = 2\gamma(z)D_{i\perp}(z)\frac{1}{\rho}\frac{\partial}{\partial \rho}\left(\rho\frac{\partial \delta N_i}{\partial \rho}\right)
$$

$$
+\frac{\partial}{\partial z}\left(2D_{i\|}(z)\frac{\partial \delta N_i}{\partial z}\right) - \alpha(z)\delta N_i
$$

(5.64)

Here, the factor $\gamma(z) = (1 + p(z))/(1 + 2p(z))$ was introduced, as before, to account for continuous transition from weakly to strongly ionized background plasma. In other words, this coefficient was introduced in Refs. [17,19] to obtain the correct limiting transition in the transverse direction in completely ionized plasma.

It was shown in Refs. [17,19] that the approximate solution for weak plasma disturbances of various origin quite accurately describes the diffusion spreading of plasma irregularities over a large period ($0 < t < 100$ s) and coordinate ($x = r/L_i = 10-30$) intervals. In the following text we will compare the results obtained from an exact solution of the system (Equation 5.55) and from the approximate ones described by the systems (Equation 5.62 or 5.64).

As shown in Refs. [17,19], the inhomogeneous nonlinear system (Equation 5.55) as well as the inhomogeneous linear system (Equation 5.64) with initial and boundary conditions (Equations 5.56 through 5.61) cannot be solved by rigorously using well-known analytical methods, because parameters of the nonlinear (or linear) equations are inhomogeneous, when only numerical methods can be used to solve a problem. This was actually done in Refs. [17,19,20] by using the trial-run and variable direction methods, in which the steps in time and space were chosen to justify the stability conditions of the trial-run and variable direction methods (see also Refs. [17,19,20]). Before entering into numerical computations of the problem concerning spreading of both weak (linear case) and strong (nonlinear case) plasma disturbances, because many factors should be taken into account in the inhomogeneous ionospheric plasma, we present simple analytical formulas for the case of spreading of weak disturbances in the quasi-homogeneous background plasma, and compare the results obtained for the inhomogeneous plasma with those obtained analytically for quasi-homogeneous background plasma.

Quasi-Homogeneous Ionosphere. At the same time, as also shown in Refs. [18,20], the analytical solution can only be obtained for quasi-homogeneous background ionospheric plasma where changes of transport coefficients of plasma particles (described by Equation 1.12a through c) with altitude are weak enough and can be taken as constant along the longitudinal scales of the initial plasma perturbation. In this case, the approximate 3-D solution can be obtained analytically [18,19]. We now briefly describe this solution, because we will use it in our further description of the processes of relaxation of different natural and artificially induced plasma inhomogeneities (see Chapter 6) and their effects on radio propagation through the ionosphere (see Chapter 7).

According to this approximate model, the density distribution $\delta N(\mathbf{r},t)$ in the process of diffusion spreading of a quasi-neutral plasma irregularity with initial electron density $\delta N(e)$ or ion density $\delta N(i)$ can be presented for weak plasma disturbances ($\delta N < N_0$) as the sum of two terms, i.e., $\delta N = \delta N_e + \delta N_i$. Each term is found from the usual diffusion equation in infinite homogeneous anisotropic plasma

$$\frac{\partial \delta N_e}{\partial t} = \nabla(2\hat{\mathbf{D}}_e \nabla)\delta N_e$$

$$\frac{\partial \delta N_i}{\partial t} = \nabla(2\hat{\mathbf{D}}_i \nabla)\delta N_i \tag{5.65}$$

where the components of the tensors of electron and ion diffusion are assumed to be constant and corresponding to an altitude in the ionosphere where the maximum of concentration of the irregularity is located. We also take into account here that plasma is isothermal and in all expressions for the unipolar diffusion coefficients the conditions of $T_e = T_i = T$ should be taken into account. Finally, we consider the process of diffusion spreading of weak disturbances ($\delta N \ll N_0$), stretched along $\mathbf{B}_0$ 3-D irregularities (with a degree of anisotropy of $\chi = L/D \geq 10^2$) for which, as emphasized earlier, the influence of gravity on the spreading process can be ignored.

As in the general case considered earlier, we consider here the approximate model (Equation 5.64) and will study the process of relaxation of a weak cylindrical irregularity with a Gaussian profile (Equation 5.56), which is symmetrical relative to the vector $\mathbf{B}_0$ and oriented along the z-axis at the moment $t > t_0$. Here, $z = z_0$ corresponds to the height h of the maximum $\delta N_{\max}$ ($\rho = 0$, $z - z_0$) of plasma density disturbance δN. The solution of a uniform homogeneous system (Equation 5.65), where all unipolar coefficients of diffusion have constant values, can be presented in the form of the Fourier transformation:

$$\delta N(\mathbf{r},t) = \frac{1}{(2\pi)^3} \int \delta N_{\mathbf{k}}(0) \exp\left\{ i\mathbf{k}\mathbf{r} - \left(D_{\parallel}k_{\parallel}^2 + D_{\perp}k_{\perp}^2 \right)t \right\} \, \mathrm{d}^3k \tag{5.66}$$

where $\delta N_{\mathbf{k}}(0) = \int \delta N(\mathbf{r},0) \exp\{-i\mathbf{k}\mathbf{r}\} \, \mathrm{d}\mathbf{r}$. Accounting for the Gaussian form of the initial profile of concentration $\delta N(\mathbf{r},0)$ of the irregularity (Equation 5.56), we get, following Refs. [18–20], the expression of the total profile of plasma disturbance

$$\delta N(\rho,z,t) = \delta N_0 L d^2 \{\delta N_e(\rho,z,t) + \delta N_i(\rho,z,t)\}$$

$$\delta N_e(\rho,z,t) = \frac{\exp\left[-\frac{z^2}{(L^2+L_e^2)} - \frac{\rho^2}{(d^2+d_e^2)} \right]}{(L^2 + L_e^2)^{1/2}(d^2 + d_e^2)} \tag{5.67}$$

$$\delta N_i(\rho,z,t) = \frac{\exp\left[-\frac{z^2}{(L^2+L_i^2)} - \frac{\rho^2}{(d^2+d_i^2)} \right]}{(L^2 + L_i^2)^{1/2}(d^2 + d_i^2)}$$

Here $L_e^2 = 8D_{e\parallel}t$, $L_i^2 = 8D_{i\parallel}t$, $d_e^2 = 8D_{e\perp}t$, and $d_i^2 = 8D_{i\perp}t$ are the characteristic longitudinal and transverse scales of diffusion at time $t > t_0$ of the initial point irregularity of electrons and ions, respectively. Following on Equation 5.67, we should note that the character of diffusion spreading of initial plasma disturbance δN at time t is very sensitive to the ratio of the initial scales of the irregularity L and d and the characteristic scales $L_{e,i}(t)$ and $d_{e,i}(t)$. Because in the ionosphere for $h > 120$ km (see Figure 1.3), $D_{e\parallel} > D_{i\parallel}$, $D_{i\parallel} > D_{i\perp}$, $D_{e\parallel} > D_{i\perp}$, and $D_{i\perp} > D_{e\perp}$, we finally get

$$L_e \gg L_i, \quad L_i \gg d_i, \quad L_e \gg d_e, \quad d_i \gg d_e \qquad (5.68)$$

Taking into account relations (Equation 5.68) for the relative maximum $\delta N_m = \delta N_{max}/AN_0$ at the centre of the irregularity ($z = z_0$, $\rho = 0$), accounting for Equation 5.67, we obtain

I. For $L_i^2 \ll L_e^2 \ll L^2$, and $d_e^2 \ll d_i^2 \ll d^2$ (1),

$$\delta N_m \approx 2$$

II. For $L_i^2 \ll L_e^2 \ll L^2$, and $d_e^2 \ll d^2 \ll d_i^2$ (2),

$$\delta N_m \approx 1$$

III. For $L_i^2 \ll L_e^2 \ll L^2$, and $d^2 \ll d_e^2 \ll d_i^2$ (3),

$$\delta N_m \approx d^2/d_e^2$$

IV. For $L_i^2 \ll L^2 \ll L_e^2$, and (1):

$$\delta N_m \approx 1 + L/L_e$$

V. For $L_i^2 \ll L^2 \ll L_e^2$, and (2):

$$\delta N_m \approx \left(L/L_e + d^2/d_i^2 \right) \qquad (5.69)$$

VI. For $L_i^2 \ll L^2 \ll L_e^2$, and (3):

$$\delta N_m \approx \left(Ld^2/L_e d_e^2 + d^2/d_i^2 \right)$$

VII. For $L^2 \ll L_i^2 \ll L_e^2$, and (1):

$$\delta N_m \approx L/L_i$$

VIII. For $L^2 \ll L_i^2 \ll L_e^2$, and (2):

$$\delta N_m \approx \left(L/L_e + Ld^2/L_i d_i^2 \right)$$

IX. For $L^2 \ll L_i^2 \ll L_e^2$, and (3):

$$\delta N_m \approx \left(Ld^2/L_e d_e^2 + Ld^2/L_i d_i^2 \right)$$

It is evident from the relations in Equation 5.68 that, depending on the values of the coefficients of diffusion in the ionospheric plasma (e.g., on the height $h \equiv z_0$ of the initial irregularity), the dimensions of the irregularity, and the time of the spreading process, its character changes from $t^{-1/2}$ to $t^{-3/2}$. As a result, the nature of the diffusion of the plasma disturbance changes appreciably from *electron-controlled* to *ion-controlled*, and the dependence of the relaxation time t on scales L and d also changes. The latter parameter will be discussed when evolution of the heating-induced inhomogeneities are analyzed in Chapter 6.

Effects of the Inhomogeneous Ionosphere on Diffusion Spreading. Let us compare results of the numerical computation of the system of Equation 5.64, which describes diffusion of weak irregularity in the inhomogeneous ionosphere with the analogous analytical solution (Equation 5.67) obtained from a system (Equation 5.65) for the same irregularity but spreading in the quasi-homogeneous ionosphere without taking into account effects of recombination, which are not so important for the ionospheric heights $h > 160$ km [18,19]. This is carried out to obtain the effects of plasma density gradients, and transport coefficient altitudinal nonmonotonic dependence on the spreading of a weak anisotropic 3-D plasma irregularity in the middle-latitude ionosphere. In Figure 5.19 the results of the numerical calculations of the system (Equation 5.54) are presented in a form of curves of equal values of the normalized density $\delta N_m = \delta N(z,\rho,t)/\delta N(z_0,0,0) = (\delta N_e + \delta N_i)/\delta N(z_0,0,0)$ in an irregularity at $t = 0$ s (a), 5 s (b), 40 s (c), and 100 s (d) for the initial irregularity (Equation 5.46) with characteristic scales $L = 10$ km and $d = 10$ m, and a maximum $\delta N(z_0,0,0)$ situated at $t = 0$ at the altitude of $z_0 = 200$ km. The curves at $t = 5$, 40, and 100 s characterize the change in the form of the initial irregularity ($t = 0$) as it spreads. The dark points show the location of maximum density at each time of spreading. Numbers 1 and 2 at $t = 0$ correspond to densities of $\delta N_m = 0.132$ and 0.0013, with the maximum value $\delta N_{max} = 2$. At $t = 5$ s, $\delta N_{max} = 0.21$, and numbers 1, 2, and 3 near the curves correspond to $\delta N_m = 0.066$, 0.007, and 0.005. At $t = 40$ s, $\delta N_{max1} = 0.025$ (bottom point), $\delta N_{max2} = 0.022$ (top point), and numbers 1, 2, 3, 4, and 5 correspond to $\delta N_m = 0.02$, 0.0175, 0.01, 0.005, and 0.002. At $t = 100$ s, $\delta N_{max1} = 0.0081$ (bottom point), $\delta N_{max2} = 0.0082$ (top point), and numbers 1, 2, 3, and 4 correspond to $\delta N_m = 0.0072$, 0.002, 0.001, and 0.0005. From Figure 5.19 the following features in the spreading of the irregularity are evident. On the whole, the form of the irregularity remains spindle-shaped, as in homogeneous weakly ionized plasma (see previous subsection), but already ($t < 20$ s) a strong asymmetry initially appears in the form of the irregularity along the z-axis (along $\mathbf{B}_0$). The initial irregularity during its spreading is compressed from below and is elongated upward in the upper ionosphere. Thus, at $t = 5$ s its size from the maximum in the direction of increasing z is approximately twice greater than in the opposite direction. This change in the form of the irregularity is related to the intensification of diffusion along $\mathbf{B}_0$ with increasing z but first to the increase of $D_{e\parallel}$ with altitude (see Figure 5.18). In addition, there is a small (by several km) displacement of the maximum of the irregularity upward. Later, at $t = 20$–40 s, a characteristic change is observed in the form of the irregularity. Thus, at an altitude $z = z_{max2}$, whose value is 20–30 km greater than the altitude of the first maximum $z = z_{max1}$, a second maximum in the density appears. Finally, in the course of spreading, the irregularity is split into two plasmoids. At the same time the second maximum moves upward with a velocity increasing with time, as does the irregularity as a whole plasma structure. Thus, at $t = 40$ s, the velocity of the second maximum upwards is about 1 km/s, whereas at $t = 100$ s the velocity equals 2.2 km/s. The velocity of the first maximum upward is lower, on the average being about 80 m/s. The second maximum decays with time more slowly than the first. This follows from Figure 5.20, which shows the change in density at the first maximum (denoted by 1) and at the second maximum after its generation at $t = 20$ s (denoted by 2) as a function of time on the logarithmic scale. It is evident that the rate of spreading of the first maximum is proportional to $t^{-3/2}$, which corresponds to a velocity of spreading at the maximum of the plasma disturbance in a homogeneous weakly ionized plasma in a magnetic field described in the previous subsection. The rate of spreading of the second maximum is proportional to t^{-1}, i.e., much slower. As a result, with time the density in the second maximum becomes equal to or even greater than the density in the first maximum (see, for example, the case $t = 100$ s). Additional analysis of the solutions of the system (Equation 5.64) with a different dependence of the diffusion coefficients on the height z showed that the appearance of the second maximum is related to the nonmonotonic dependence of altitude of the coefficients of transverse diffusion $D_{e\perp}$, $D_{i\perp}$, and primarily, of the coefficient $D_{e\perp}$, i.e., to the nonmonotonic variation of the

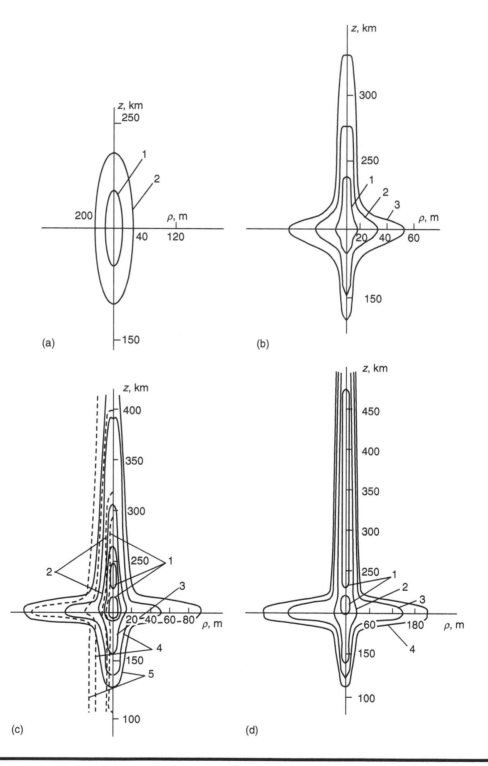

Figure 5.19 Temporal presentation of a shape of small plasma disturbance according to the model of middle-latitude ionosphere.

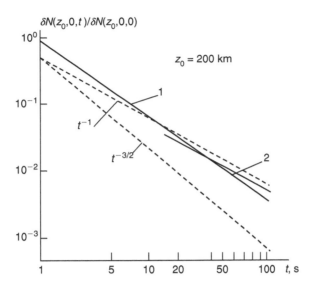

Figure 5.20 Temporal decay of the first (1) and the second (2) maximum of plasma density in the model of weakly ionized plasma (dashed lines) and according to the model of middle-latitude ionosphere (solid curves).

rate of transverse diffusion of electrons with increasing z in the ionosphere (see Figure 5.18). When the monotonic dependences $D_{e\perp}(z)$, $D_{i\perp}(z)$ are given, the second maximum is not formed in the same way as in a homogeneous plasma. However, in this case, the first maximum moves upward with a velocity exceeding several times the velocity under conditions examined here, and its rate of spreading is proportional t^{-1}. What is important to note here is that by introducing additional artificial extrema in the altitudinal dependence of unipolar coefficients, made in Ref. [19], compared to that shown in Figure 5.18, primarily for $D_{e\perp}(z)$ and $D_{i\perp}(z)$, leads to the creation in time ($t > 30$–40 s) of the third and more maxima δN_{max}, as with the second maximum, floating up along $\mathbf{B}_0$, i.e., upward. In this sense we should emphasize that in the case of the polar or equatorial ionosphere, with more complicated nonmonotonic altitudinal dependence of the coefficients of diffusion than that presented in Figure 5.18, we may wait splitting of the initial irregularity in a set of plasmoids elongated along $\mathbf{B}_0$ and penetrating into the upper ionosphere. This will be shown, for example, in Chapter 6 in the context of evolution of the equatorial natural plasma structures called bubbles, and their floating-up penetration from lower to upper ionosphere was observed experimentally.

To ensure that the obtained effects are the pure effects of plasma inhomogeneity along altitude z, in Figure 5.19 at $t = 40$ s, we represent with dashed lines the form of the spreading of the irregularity in the unbounded homogeneous plasma, i.e., for $D(200) = const$ computed using Equation 5.67, which correspond to solid curves computed by using the system of Equation 5.64. Here, $D(200)$ is the value of the transverse and longitudinal components of the diffusion tensor of electrons and ions at the altitude $z = 200$ km in the ionosphere. Comparing the dashed and continuous curves in Figure 5.19 at $t = 40$ s (Figure 5.19c), we see that the spreading of the weak irregularity with the degree of anisotropy $\chi = 10^4$ m/10 m $= 10^3$ in an inhomogeneous ionosphere differs considerably from the spreading of this irregularity in the model of a homogeneous ionospheric plasma. In homogeneous ionospheric plasma the spreading of the irregularity both along and across $\mathbf{B}_0$ is symmetrical relative to the center of the irregularity and has a spindle-shaped form, as in the case of the weakly ionized plasma in the magnetic field described in the previous subsection. The additional features, such as the displacement and splitting of the maximum of the irregularity, do not occur. At the same time, in an inhomogeneous plasma, because of the dependence of the rate of diffusion on altitude z, the speed of spreading along and across $\mathbf{B}_0$ at different altitudes changes considerably, which affects the change in its form (compare the same curves in Figure 5.19c, $t = 40$ s). Thus, for $z < 200$ km, the size of the irregularity along $\mathbf{B}_0$ in an inhomogeneous plasma ($L_{\parallel}^{in}$) is less than that in a homogeneous plasma ($L_{\parallel}^{ho}$) (compare curves 1, 2, 4, and 5). At $z > 200$ km, $L_{\parallel}^{in}$ is likewise less, but at higher altitudes the difference in the dimensions vanishes, and for $t = 40$ s with $z \geq 400$ km $L_{\parallel}^{in}$ exceeds $L_{\parallel}^{ho}$. As the time for the spreading process increases, this effect occurs at lower altitudes. Thus, at $t = 60$ s $L_{\parallel}^{in} > L_{\parallel}^{ho}$ for $z \geq 350$ km; at $t = 100$ s $L_{\parallel}^{in} > L_{\parallel}^{ho}$ for $z \geq 280$ km. The change in an inhomogeneous plasma of the quantity $L_{\parallel}^{in}$ can be explained by the nonmonotonic nature of the increase in the magnitude of the unipolar coefficient $D_{e\parallel}$ with altitude in the ionosphere (see Figure 5.18), which primarily determines the size of the irregularity along $\mathbf{B}_0$. Thus, at $z < 200$ km $D_{e\parallel}(z) < D_{e\parallel}(z_0)$, $z_0 = 200$ km, and at $z > 200$ km $D_{e\parallel}(z)$ exceeds $D_{e\parallel}(z_0)$ insignificantly. Across $\mathbf{B}_0$ with $z = z_0$ in the close zone ($\rho \leq d$) in an inhomogeneous plasma the transverse scale $L_{\perp}^{in}$ of the irregularity is likewise less than the corresponding size in the homogeneous plasma $L_{\perp}^{ho}$ (curves 1, 2), whereas in the far zone ($\rho > d$) it is greater (curves 4 and 5). This fact is also explained well by the nonmonotonic nature of the unipolar coefficient $D_{i\perp}$, as a function of z (see Figure 5.18), which primarily determines the rate of spreading of strongly elongated irregularity across $\mathbf{B}_0$ [16–20]. Thus, for $z > 200$ km $D_{i\perp}(z) < D_{i\perp}(z_0)$, for 130 km $< z < 200$ km $D_{i\perp}(z) > D_{i\perp}(z_0)$, and for $z < 130$ km once again $D_{i\perp}(z) < D_{i\perp}(z_0)$.

An additional analysis of the character of spreading of the first maximum δN_{max1} ($z = z_0$, $\rho = 0$), depending on time t of diffusion spreading, made in Ref. [19], has shown that it depends on the degree of anisotropy of the initial irregularity, or for the constant longitudinal scale $L = const$ on the transverse scale d of the irregularity, as well as on the time of the process of spreading by itself. It is evident from Figure 5.20, where in the logarithmic scale the dependence of $\delta N_{max1}(z = z_0, 0, t)$ on t for $L = 10$ km and $z_0 = 200$ km is shown for different $d = 1$, 10, and 100 m. Here, solid lines correspond to the spreading of the irregularity in the inhomogeneous background plasma with $D = D(z)$ according to Figure 5.18, and the dashed lines correspond to the case of the homogeneous background plasma with $D = D(z_0 = 200$ km$) = const$. The dotted-dashed lines correspond to changes of the value $\delta N_{max1}(z = z_0, 0, t) \propto t^{-1/2} - t^{-3/2}$. The case $d = \infty$ corresponds to a one-dimensional diffusion of the infinitely stretched across $\mathbf{B}_0$ irregularity analyzed in Ref. [19]. It is evident that with the decrease of the scale d or with increase of the time of spreading t, the law of decay of the initial irregularity changes from $t^{-1/2}$ to $t^{-3/2}$. Moreover, the altitudinal inhomogeneity of the background ionospheric plasma does not play a sufficient role in the rate of spreading of the first maximum, nor of the whole irregularity for $d \geq 100$ m and $\chi = L/d \leq 10^2$. Its influence becomes vital for $d \leq 10$ m, and for small d the decay of δN_{max} in the irregular inhomogeneous plasma where $D = D(z)$ accelerates faster compared to the case of homogeneous plasma when $D = D(z_0 = 200$ km$)$ and at large t we get $\delta N_{max} \sim t^{-2}$ (compare solid and dashed lines for $d = 1$ m and $t \geq 40$ s). Again, it is evident that the altitudinal inhomogeneity of the ionospheric plasma significantly influences the character of diffusion spreading of the irregularities strongly stretched along $\mathbf{B}_0$ with a degree of anisotropy greater than $\chi = L/d = 10^2$.

The features of diffusion spreading of the irregularity with a sufficient degree of anisotropy in the inhomogeneous ionospheric plasma obtained during numerical computations can also prove the existence of plasma instabilities associated with gradients of plasma density and with non-monotonic altitudinal dependence of coefficients of unipolar diffusion of electrons and ions of plasma, which generate plasma disturbances and split them into two plasmoids. This is a clear-cut example of striation of plasma structures during a pure diffusion process occurring in the inhomogeneous background plasma with nonmonotonic altitudinal dependence of transport coefficients of electron and ion components of plasma. A numerical analysis carried out in Refs. [17–19] has shown that the effects of recombination are insignificant at altitudes greater than 160 km, and they should only be taken into account at altitudes of the lower ionosphere less than 160 km, where the rate of decay of the irregularity increases when compared to the higher ionosphere ($z > 160$–200 km) and only affects the electronic component of plasma irregularity δN_e. This finally leads to the displacement upward along $\mathbf{B}_0$ of the maximum of the electron component of the irregularity. The influence of the recombination on the ion component of the plasma irregularity δN_i is weaker, and the maximum of the ion component of the irregularity does not change its position in space [19]. The total profile of irregularity concentration, $\delta N = \delta N_e + \delta N_i$, shows the same behavior. This allows the conclusion that the main contribution of recombination is observed below 160–200 km, where the fast decay of δN is usually observed, whereas in the upper ionosphere ($z > 200$ km) the influence of the recombination on the character of the spreading of the irregularity is insignificant.

Following the results obtained in Refs. [7–20], we will now consider the dependence of the character of irregularity spreading from its initial dimensions L (along $\mathbf{B}_0$) and d (across $\mathbf{B}_0$). The analysis of Equation 5.64, without taking into account the effects of recombination for the initial irregularity placed at the height of $z_0 = 200$ km for different parameters L and d, is shown in Figure 5.21 in the form of the dependence of $\delta N(z,0,t_0 - 20\text{ s})$ from the coordinate z (i.e., along $\mathbf{B}_0$). The solid curves correspond to $L = 10$ km and $d = 20$, 40, and 100 m and the dashed ones

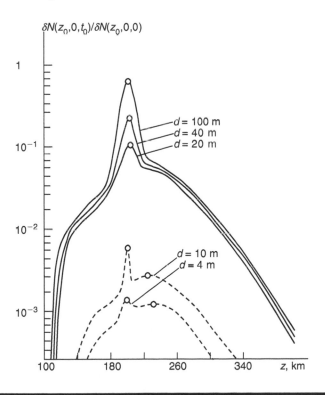

Figure 5.21 Spatial distribution of the initial small plasma disturbance for $L = 1$ km (dashed lines) and for $L = 10$ km (solid curves).

correspond to $L = 1$ km and $d = 4$ and 10 m. Such parameters of the irregularity were selected to obtain the same degree of anisotropy. Thus, $L = 10$ km and $d = 40$ m give the same degree of anisotropy $\chi = 2.5 \cdot 10^2$, as $L = 1$ km and $d = 4$ m, and so on. We denote by circles the maxima at the normalized curves $\delta N(z,0,t_0 = 20 \text{ s})/\delta N(z_0,0,0)$. It can be seen that at time $t_0 = 20$ s after the process of diffusion spreading commenced, the initial plasma irregularity splits onto two plasmoids for the case of $L = 1$ km and $d = 4$ and 10 m, whereas for the same degree of anisotropy but for the larger (by several times and more) transverse dimensions $d = 20$, 40, and 100 m ($L = 10$ km) the effect of splitting has not been observed. The dependence of the rate of spreading of the irregularities elongated along $\mathbf{B}_0$ on the initial dimensions will be investigated further by using the approximate analytical expressions of Equation 5.67 following the corresponding model described by Equation 5.64 or 5.65. In this sense we need to understand how strictly this approximate model describes the process of diffusion spreading of irregularities with finite dimensions both along and across the magnetic field. Therefore, we will now compare both the strict model described by nonlinear inhomogeneous equations (Equation 5.55) and the approximate model described by Equation 5.64 without taking into account the effects of recombination (to have the comparable solutions during numerical computations, as was done in Refs. [18–20]).

The main results of the numerical simulations of the nonlinear inhomogeneous system (Equation 5.55) consisting both of the plasma disturbance density and the normalized potential of the self-consistent intrinsic electric field taking into account initial and boundary conditions are presented in Figures 5.22a through c and Figure 5.23a and b, according to Ref. [20]. In the case of weak plasma disturbance, $\delta N \ll N_0$, for the time of the spreading process T, the values of the total concentration N were close to N_0. This case was investigated for the purpose of comparing the results of the strict model (Equation 5.55) with those of the approximate model (Equation 5.64), which is valid only for the weak plasma perturbations (i.e., for the linear case). The solutions are presented in the form of lines of normalized equal density $\delta N_m = \delta N(z,\rho,t)/\delta N(0,0,0)$ and the additional normalized equal potential $\delta \Phi_m = (\Phi(z,\rho,t) - \Phi_0)/\Phi_0$ (compared with the case of the numerical solution of the approximate model [Equation 5.54] at the moments (a) $t = 5$ s, (b) 20 s, and (c) 40 s, results of which are shown in Figure 5.19). The initial irregularity according to Equation 5.46 has the characteristic scales $L = 10$ km and $d = 5$ m and a maximum situated for $t = 0$ at the altitude $z = 200$ km. As in Figure 5.19, the curves at 5, 20, and 40 s in Figure 5.22 characterize the changes in the form of the initial irregularity as it spreads (the left-hand figures), and also the corresponding changes of equipotential lines in each case (the right-hand figures). The dark points were used to mark the maximum values of density and potential at each moment of time t. When $t = 5$ s, numbers 1–9 correspond to $\delta N_m = 1.3 \cdot 10^{-1}$, $1.2 \cdot 10^{-1}$, 10^{-1}, $8.3 \cdot 10^{-2}$, $2.4 \cdot 10^{-2}$, 10^{-2}, 10^{-3}, 10^{-4}, $7.7 \cdot 10^{-5}$ with the maximum value $\delta N_{max} = 1.37 \cdot 10^{-1}$ at altitude $z_1 = 211$ km. For the potential $\delta \Phi_m$ numbers 1–8 correspond to $\delta \Phi_m = -10^{-1}$, $9.5 \cdot 10^{-2}$, $-8 \cdot 10^{-2}$, $2.5 \cdot 10^{-2}$, $1.5 \cdot 10^{-2}$, 10^{-3}, $8.5 \cdot 10^{-4}$, $4 \cdot 10^{-4}$ with maximum $\delta \Phi_{max} = -1.14 \cdot 10^{-1}$. At the moment of $t = 20$ s there are three maximum values of density perturbations, $\delta N_{max1} = 2.3 \cdot 10^{-2}$ (bottom point; $z_1 = 203$ km), $\delta N_{max2} = 2.85 \cdot 10^{-2}$ (top point; $z_2 = 219$ km), and $\delta N_{max3} = 2.17 \cdot 10^{-2}$ (middle point; $z_3 = 212$ km). The numbers 1–6 correspond to $\delta N_m = 2 \cdot 10^{-2}$, $1.8 \cdot 10^{-2}$, $1.2 \cdot 10^{-2}$, $9 \cdot 10^{-3}$, $3.5 \cdot 10^{-3}$, 10^{-3}. For the potential we have, respectively, $\delta \Phi_{max1} = -2.06 \cdot 10^{-3}$ and $\delta \Phi_{max2} = -2.1 \cdot 10^{-3}$; numbers 1–9 correspond to $\delta \Phi_m = -2 \cdot 10^{-3}$, $-7 \cdot 10^{-4}$, $-2.1 \cdot 10^{-5}$, $-5 \cdot 10^{-7}$, $5 \cdot 10^{-7}$, 10^{-6}, $5 \cdot 10^{-6}$, 10^{-5}, $5 \cdot 10^{-5}$. When $t = 40$ s, there are also three maximum values of density perturbations, $\delta N_{max1} = 1.06 \cdot 10^{-2}$ (bottom point; $z_1 = 205$ km), $\delta N_{max2} = 1.1 \cdot 10^{-2}$ (middle point; $z_2 = 231$ km), and $\delta N_{max3} = 5.4 \times 10^{-3}$ (top point; $z_3 = 242$ km). The numbers 1–6 correspond to $\delta N_m = 10^{-2}$, $7 \cdot 10^{-3}$, $5 \cdot 10^{-3}$, $4 \cdot 10^{-3}$, $3 \cdot 10^{-3}$, 10^{-3}. For the potential we have,

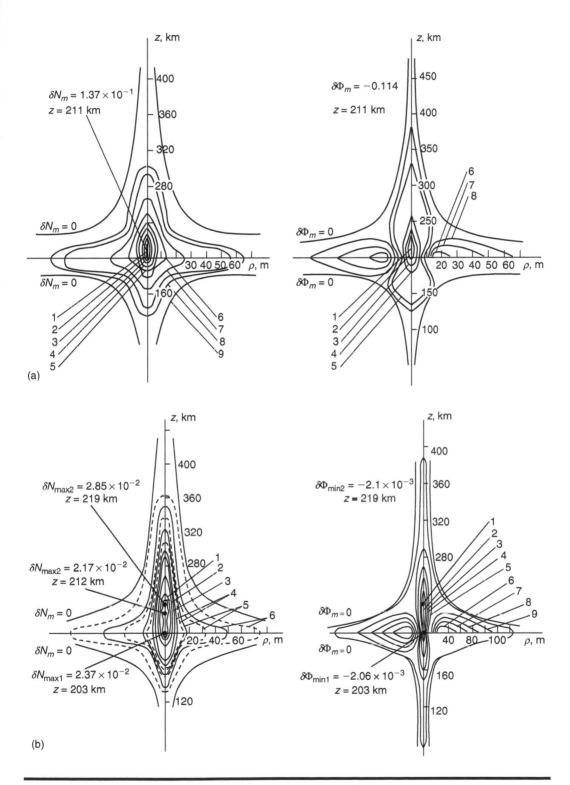

Figure 5.22 Equidense (left panels) and equipotential (right panels) contours during diffusion spreading of strongly anisotropic plasma disturbance according to the model of middle-latitude ionosphere (solid curves), and one of example is according to the model of weakly ionized plasma.

(*continued*)

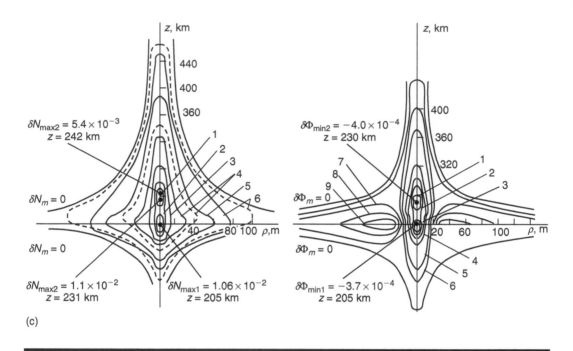

(c)

Figure 5.22 (continued)

respectively, $\delta\Phi_{max1} = -3.7 \cdot 10^{-4}$ and $\delta\Phi_{max2} = -4 \cdot 10^{-4}$; numbers 1–9 correspond to $\delta\Phi_m = -3.3 \cdot 10^{-4}$, $-1.8 \cdot 10^{-4}$, $-7 \cdot 10^{-5}$, $-4.9 \cdot 10^{-6}$, $-2.5 \cdot 10^{-7}$, -10^{-8}, 10^{-10}, 10^{-7}, 10^{-6}. As seen from numerical results illustrated in Figure 5.22a through c, the rigorous approach based on general nonlinear system of inhomogeneous equations of plasma diffusion (Equation 5.55) gives the same new effects (in comparison with the case of a homogeneous plasma) that were observed earlier in Figure 5.19 by using the approximate system of inhomogeneous equations (Equation 5.64). These are the changes in the spindle-shape form of the irregularity (Figure 5.22a), a splitting of the irregularity into two plasmoids (corresponding to the model of the middle-latitude ionosphere presented in Figure 5.18) in the course of spreading at time $t \geq 20$ s, and a floating-up penetration into the upper ionosphere, where the irregularity is compressed from the bottom side and elongated upward (Figure 5.22b and c). However, using the strict model of Equation 5.55, we can investigate the characteristic peculiarities of the spreading of an irregularity in more detail by examining the equipotential lines of the ambipolar electric field inside the disturbed areas. In fact, the curves of surfaces of normalized equal potential depicted at the right side of Figure 5.22 illustrate vividly the physical picture of a spreading process. Thus, at the initial moments (when $t < 20$ s) for a very stretched irregularity, the ions diffuse across $\mathbf{B}_0$ (in the direction of the ρ-axis) and form areas of positive potential. At the same time, the character of irregularity spreading along $\mathbf{B}_0$ (along the z-axis) is determined by the velocity of electron diffusion, which forms areas of negative potential near the maximum of the irregularity (see Figure 5.22). Later, when $t > 20$ s, the areas of positive potential expand both along and across $\mathbf{B}_0$, and larger areas of disturbed plasma are controlled by ions, determining the character and velocity of the irregularity decay in these areas. Along the z-axis and for $\rho < 20$ m, the character of irregularity spreading is controlled by electrons. In Refs. [17,18] it was shown that if the transverse scale of the irregularity exceeds the Larmor ion radius ρ_{Hi} at the considered ionospheric altitudes, then the process of diffusion fundamentally changes its nature. The character of diffusion spreading in the

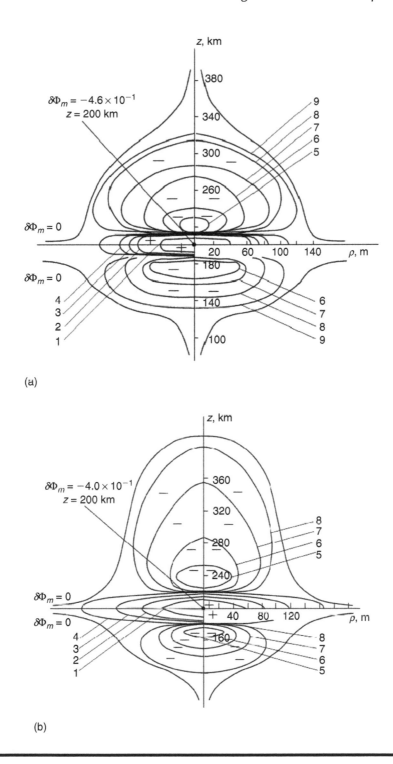

(a)

(b)

Figure 5.23 Same as Figure 5.22, but for isotropic initial plasma disturbance.

maximum of the plasma disturbance is termined by ions. To show the change of character of the spreading, from electron-controlled to ion-controlled, numerical simulation of the strict model (Equation 5.55) for $L = 10$ km and $d = 100$ m ($\chi = L/d = 10^2$) was made. The results of computations in the form of equipotential lines are presented in Figure 5.23 at (a) $t = 5$ s and (b) $t = 20$ s.

Where $t = 5$ s, numbers 1–9 correspond to $\delta\Phi_m = 1.2 \cdot 10^{-1}$, 10^{-1}, $7.5 \cdot 10^{-2}$, $1.8 \cdot 10^{-2}$, $-4.5 \cdot 10^{-3}$, $-2.2 \cdot 10^{-3}$, $-6.5 \cdot 10^{-4}$, $-2 \cdot 10^{-4}$, $-7.8 \cdot 10^{-5}$, with maximum $\delta\Phi_{max} = 4.6 \cdot 10^{-1}$. At the moment of $t = 20$ s, numbers 1–8 correspond to $\delta\Phi_m = 9.5 \cdot 10^{-2}$, $7.3 \cdot 10^{-3}$, $4.2 \cdot 10^{-4}$, $9.6 \cdot 10^{-5}$, $-8.4 \cdot 10^{-3}$, $-2.5 \cdot 10^{-3}$, -10^{-4}, $-6.1 \cdot 10^{-5}$. It can be seen that the character of irregularity decay in the maximum and across $\mathbf{B}_0$ (along the ρ-axis) in the areas $z = 180$–220 km is determined by ions, whereas in the areas $z > 220$ km and $z < 180$ km and near this direction, the diffusion spreading is determined by electrons. Moreover, for the weakly anisotropic irregularities (with $\chi \leq 10^2$), a number of new effects, namely, a change in the shape of the irregularity, a splitting of the irregularity into plasmoids, and a floating-up penetration into the upper ionosphere were not obtained.

This allows us to point out that using the rigorous self-consistent model, we finally unify the problem of diffusion spreading of different kinds of plasma irregularities in the real ionosphere, from the polar cap to the equator, by introducing the real nonmonotonic altitudinal dependence of the transport coefficients of the electron and ion components of plasma obtained for each region. At the same time, the approximate model presented by Equations 5.65 through 5.69, which is valid up to several tens seconds during the process of generation of plasma disturbance (dependent on its degree of anisotropy [17--20]), will allow to obtain simple analytical solutions that have a vivid physical meaning and can directly explain some features observed in systematic ionospheric experiments described in Chapters 6 and 7.

Relaxation of Field Aligned Irregularities. We now return to the solution of the linear problem (Equation 5.65) of relaxation of the strongly stretched along the z-axis (along $\mathbf{B}_0$) weak ($\delta N \ll N_0$) irregularities in the unbounded quasi-homogeneous ionospheric plasma. The process of spreading of such irregularities according to the analysis carried out in Refs. [17–20] is described by Equation 5.67 as a solution to the linear problem of diffusion (Equation 5.65). In deriving Equation 5.67, it was assumed in Refs. [18 and 20] that at initial time $t = t_0 = d^2/8D_{i\perp}$ (for isothermic plasma, $T_e \approx T_i = T$), the density of quasi-neutral particles in the irregularity $\delta N(z,\rho,t_0)$ consists of two equal parts, i.e.,

$$\delta N(z,\rho,t_0) = \delta N_e(t_0) + \delta N_i(t_0) = 2\delta N_0 e^{-\frac{z^2}{L^2}-\frac{\rho^2}{d^2}} \qquad (5.70)$$

One part of δN_e is electronic, which spreads out with the diffusion velocity of electrons (the ions in the irregularity adjust themselves to it according to the Boltzmann law). The second part of δN_i is the ionic component, which spreads out with the velocity of diffusion of ions (the electrons adjust themselves to the irregularity according to the Boltzmann law). The spreading of particles in the two simple one-dimensional (1-D) cases made in Refs. [16,18] using approximate linear model are the following:

(a1) Diffusion along $\mathbf{B}_0$; the irregularity is of infinite size across $\mathbf{B}_0$, $L \ll d \to \infty$.
(a2) Diffusion across $\mathbf{B}_0$; the irregularity is of infinite size along $\mathbf{B}_0$, $d \ll L \to \infty$.

According to the constraint Equation 5.68, the electronic component spreads out more rapidly along $\mathbf{B}_0$ and the ionic component more rapidly across $\mathbf{B}_0$. As with the result in the 1-D case (a1), the irregularity spreads out along $\mathbf{B}_0$ (or along the z-axis) with the velocity of the ions, as is depicted in the upper part of Figure 5.24a. Such cases are characterized for diffusion of the F spread in the ionosphere for which $L \ll d$ (see Chapter 6). In the 1-D case (a2), sketched in the bottom part of Figure 5.24a, the irregularity spreads out across $\mathbf{B}_0$ (along ρ-axis) with the velocity of the electrons.

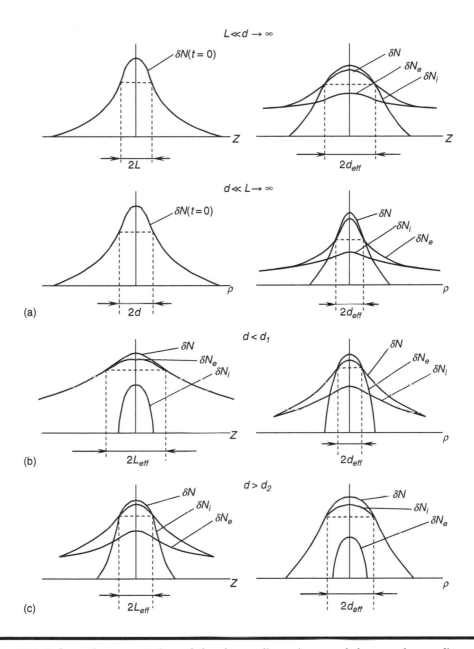

Figure 5.24 Schematic presentation of the shape, dimensions, and degree of spreading of the plasma disturbance of finite dimensions.

In Figure 5.24a the total density δN, the electronic δN_e, and the ionic δN_i parts of the irregularity with the Gaussian profile (Equation 5.56) are shown schematically. It is evident that the characteristic size and the velocity of diffusion of the irregularity along $\mathbf{B}_0$ are determined by the ionic component and by the electronic component across $\mathbf{B}_0$ (as in the case of the absence of magnetic field according to Schottky [1–3]).

However, the pattern of the spreading of the irregularity with finite dimensions along and across $\mathbf{B}_0$ is substantially different in the real three-dimensional (3-D) case. The presence and magnitude of the diffusion flows both along and across the irregularity in each specific case is now important. It was shown in Ref. [18] that depending on the magnitude of the transverse size d of the irregularity, in the 3-D case the character of diffusion spreading changes not only across $\mathbf{B}_0$ but also along $\mathbf{B}_0$. The change in the nature of the diffusion accompanying an increase in the transverse size of the irregularity d is shown schematically in Figure 5.24b for small d, and in Figure 5.24c for large d based on an analysis of the relations (Equation 5.66) and Equation 5.68. It is evident that because of the spreading of the irregularity across $\mathbf{B}_0$ (along ρ-axis), the velocity of diffusion changes at the center of the irregularity and at altitudes close to the maximum. For example, at the level $\delta N_{max} \cdot e^{-1}$ determining the effective dimensions of the irregularity along $\mathbf{B}_0$, $L_{eff} = \sqrt{L^2 + L_e^2}$, and across $\mathbf{B}_0$, $d_{eff} = \sqrt{d^2 + d_i^2}$. For small d ($d < d_1$, the value d_1 will be defined later) the ionic component δN_i of the irregularity has time to diffuse a long distance across $\mathbf{B}_0$ within a short time interval, since $D_{i\perp} \gg D_{e\perp}$. As a result of this, the ion density in the center of the irregularity essentially drops. The characteristic size of the irregularity along and across $\mathbf{B}_0$ is in this case determined by the electronic component δN_e, as shown schematically in Figure 5.24b. For large d ($d > d_2$, the value d_2 will be defined later) there is not enough time for the ionic component in the irregularity to spread out across $\mathbf{B}_0$ and therefore the effective size of the irregularity and the velocity at which it spreads are determined by the ionic component δN_i, as shown schematically in Figure 5.24c. Hence, as the transverse size of the irregularity d increases, the electronic character of the diffusion along and across $\mathbf{B}_0$ (Figure 5.24b) changes into the ionic character (Figure 5.24c). We note that the schemes in Figure 5.24 correspond to estimations made for specific parameters of the upper middle-latitude ionosphere at $z_0 = 200$ km.

We will now analyze the relaxation time (or the lifetime) τ of the weak irregularity defined by the constraint Equation 5.53 in the quasi-homogeneous ionospheric plasma based on the main equations (Equation 5.69) depending on initial dimensions of the irregularity both along L and across d the ambient magnetic field $\mathbf{B}_0$. Thus, the analytical expressions describing the dependence of $\tau = \tau(d)$ following Refs. [18,20] can be expressed as

(a) for a small transverse size $d < d_1$,

$$\tau \approx \frac{L^{2/3} n^{2/3}}{8\sqrt[3]{4} D_{e\|}^{1/3} D_{e\perp}^{2/3}} d^{4/3} \tag{5.71a}$$

(b) for the moderate transverse size $d_1 < d < d_2$,

$$\tau \approx \frac{L^2 n^2}{32 D_{e\|}} \left\{ 1 + \frac{2d}{L} \sqrt{\frac{2D_{e\|}}{n D_{i\perp}}} \right\} \tag{5.71b}$$

(c) for the large transverse size $d > d_2$

$$\tau = \frac{n^2 - 4}{32 D_{i\|}} L^2 \approx \frac{n^2 L^2}{32 D_{i\|}} \tag{5.71c}$$

The dependences $\tau = \tau(d)$ pictured in Figure 5.25 correspond to $n = e$, 5, and 10 for the daytime (D) and the nocturnal (N) ionosphere computed for the parameters of the ionosphere at altitude

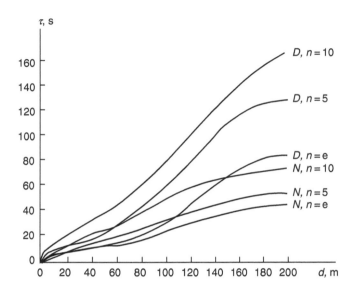

Figure 5.25 **Lifetime dependence versus initial longitudinal dimension of plasma disturbance for diurnal and nocturnal middle-latitude ionosphere.**

$z_0 = 200$ km. It is evident that the relaxation time increases substantially as the transverse size d increases.

Analysis of Equation 5.71a through c gives the following. For small $d (d < d_1)$ $\tau(d) \sim d^{4/3} - d^2$, i.e., relaxation time increases rapidly with d. For moderate $d (d_1 < d < d_2)$ $\tau(d) \sim d^{1/2} - d$, and for large $d (d > d_2)$ τ is actually independent, i.e., the irregularity spreads out as if it is one-dimensional with the velocity of diffusion of ions along $\mathbf{B}_0$ (see Equations 5.71c and 5.69, VII). Thus, the change in the nature of the dependence of $\tau = \tau(d)$ for small and large d is linked to the change in the decay law of the density at the center of the irregularity, which is evident from Equation 5.69: from the *electron character* along and across $\mathbf{B}_0$ when $d < d_1$ to the *ion character* when $d > d_1$. On the transverse scale $d = d_2$, saturation of the spreading process occurs and the relaxation time becomes independent of the transverse size. The character of the decay of the irregularity with $d \geq d_2$ is analogous to the case of stretched-out geomagnetic field irregularities, which are actually observed in the middle-latitude F layer (see Chapter 6).

Following Refs. [18,20], we can now estimate the critical values of transversal scales d_1 and d_2 at which the character of relaxation of the irregularity changes. Using Equation 5.71a through c and real parameters of the middle-latitude ionosphere at altitude $z = 200$ km, we get $d_1 \approx 10-20$ m and $d_2 \approx 100-150$ m for the daytime ionosphere; and $d_1 \approx 3-5$ m and $d_2 \approx 100-120$ m for the nocturnal ionosphere, results that fully coincide with the dependences $\tau = \tau(d)$ shown in Figure 5.25. These estimated values, of course, depend on the altitude of initial irregularity. Figure 5.26 shows the relaxation time τ of irregularities with the longitudinal size $L = 10$ km, depending on their transverse dimension d, calculated for the conditions of the daytime middle-latitude ionosphere at altitudes $z_0 = 200, 300, 400$ km. As can be seen, with increasing altitudes of initial irregularity the *two-step* dependence $\tau = \tau(d)$ also shown in Figure 5.25 becomes a *one-step* dependence $\tau = \tau(d)$ with only one critical scale $d_1 \approx d_2$, at which the

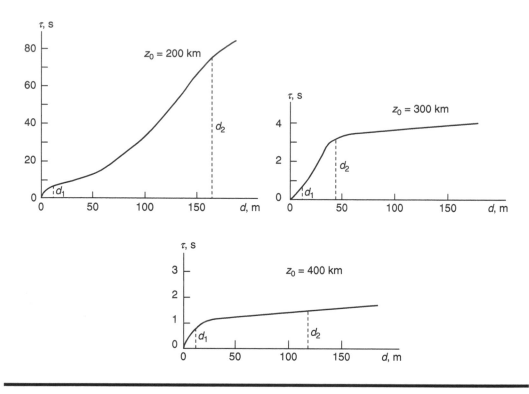

Figure 5.26 Lifetime dependence versus transverse dimension of the initial plasma disturbance for ionospheric altitudes of 200, 300, and 400 km computed according to the model of middle-latitude ionosphere.

relaxation process changes its nature, from electron-controlled to ion-controlled, according to the criteria mentioned earlier.

Hence, the effects discovered in numerical computations carried out according to [17–20] and described previously in the comparative analysis of separate cases of inhomogeneous and quasi-homogeneous ionosphere, such as the change in the form and shape of strongly elongated irregularities in the real ionosphere, splitting of irregularities in two (or more, depending on the complexity of altitudinal dependence of the transport coefficients in the real ionosphere), rise of irregularities, their penetration into the upper ionosphere. Finally, the changes in the character of spreading, from electron-controlled to ion-controlled, with nonmonotonic two-step and one-step dependence of the relaxation time on transverse size of the initial irregularity d. All these features of spreading are caused by the altitudinal inhomogeneity of the ionosphere and by the altitudinal changes of the transport coefficients of the electronic and ionic components of background ionospheric plasma.

5.2 Drift of Plasma Irregularities in the Ionosphere

We assumed in the foregoing that the gas of electrons and ions, i.e., plasma, is at rest relative to the ambient magnetic field B_0 and only diffusion spreading of arbitrary plasma disturbance is observed. In this case, the three-dimensional models of diffusion spreading of any plasma structure were analyzed without considering the origin of its creation. Under real ionospheric conditions, the drift of charged particles relative to $\mathbf{B}_0$ is caused by the wind of the neutral component of plasma with velocity $\mathbf{U}_m$, and by the ambient electric field $\mathbf{E}_0$ and gravitation field with acceleration $\mathbf{g}$, as well as by movements in the crossing electric and magnetic fields, $\mathbf{E}_0 \times \mathbf{B}_0$. Therefore, the results obtained in the previous section are correct for a description of the process of evolution of plasma irregularities in the ionosphere for which the lifetime is shorter than the characteristic time of switching of the drift mechanism through the background ionospheric plasma. Following the numerous theoretical and experimental investigations [6,7,21–37,87–94] carried out for weakly anisotropic large-scale plasma irregularities, such as barium clouds, bubbles, or meteor trails, we have to include both gravity and drift effects and use rigorous theories created both for small- and large-scale plasma disturbances [6,7,89–94], which were then modified in Refs. [49,72,74,81,87,88] for the case of real plasma anisotropic irregularities with various degrees of ionization (compared to background plasma) generated in the ionosphere by different natural and artificial sources. It was shown [6,89–94] that in the presence of ambient electric $\mathbf{E}_0$ and magnetic $\mathbf{B}_0$ fields in the unbounded weakly ionized plasma, the character of plasma disturbance spreading changes greatly compared to the case of diffusion spreading without taking ambient fields into account. Thus, the initial plasma inhomogeneity splits during its movements and spreads into separate plasma structures, caused by the wind of neutral molecules (atoms) and drift of plasma in ambient electrical and magnetic fields and called the *plasmoids* or *strata* in the literature [48–51]. These are stretched along magnetic field lines and attenuate more quickly than when only the diffusion process occurs [41,48,49,70,71]. This new mechanism of spreading and attenuation of plasma disturbances in real time was called a *drift mechanism* [6,87–94]. In Ref. [6] it was consistently analyzed, and then modified in Refs. [87,88,92,94] for real ionospheric conditions, the character of drift spreading of plasma inhomogeneity, its velocity, and the time of relaxation (e.g., the lifetime). It was shown that the phenomenon depends significantly not only on the value and character of the drift of the electron and ion components of plasma (as rigorously shown in Refs. [6,89]), but also on the initial size of the inhomogeneity and its degree of anisotropy $\chi = L/d$ in the ambient magnetic field $\mathbf{B}_0$, as well as on the value of the initial density of plasma disturbance and on the altitude of the inhomogeneity δN ($h \equiv z_0$) with respect to the background ionospheric plasma $N_0(h)$. Here, as before, L and d are the longitudinal and transverse to $\mathbf{B}_0$ scales of inhomogeneity.

Moreover, theoretical and experimental investigations carried out by separate groups of researchers [58–69], describing evolution of artificially injected ion clouds in the ionosphere, have shown that such splitting of initial plasma disturbance on several plasma plasmoids occurred in the first stage of drift spreading. In the next stage, each plasmoid splits into elementary strata and, finally, a cluster of these plasma strata is created in the front edge of the initial plasma irregularity (relative to the direction of the $\mathbf{E}_0 \times \mathbf{B}_0$-drift), each of which is stretched out along the geomagnetic field lines. Unfortunately, during recent decades, only a few realistic numerical models were performed describing the dynamics of strong large-scale plasma disturbances (in comparison with the Debye radius) in the real ionosphere [49,73,74,88], and despite their accuracy, these models described only the initial stage of the evolution of isotropic and weakly anisotropic inhomogeneities embedded into the infinite homogeneous background plasma, but did not explain the more interesting later stages of their evolution, where complicated pictures of plasma striation and cluster strata structures creation were observed.

On the other hand, some numerical models were performed to study the process of drift spreading of plasma inhomogeneities with an arbitrary initial form in the upper ionosphere, taking into account both the influence of recombination observed at the lower E layer, and mutual coupling and interaction of layers E and F [58–63]. In such numerical models of plasma drift, the initial conditions were simplified: the smooth initial inhomogeneity described by the Gaussian distribution was perturbed by a few sinusoidal harmonics. As a result, the striation process was observed at the back edge of the initial inhomogeneity (also relative to the direction of the $\mathbf{E}_0 \times \mathbf{B}_0$-drift). These models were too idealized to be compared to the real active space experiments with rocket injection of ion clouds and beams. At the same time, further performed numerical models [60–65,94–96] gained criteria of drift instability, which is generated by the gradient-drift instability (GDI) causing plasma splitting at the several plasmoids. In Chapter 2 we discussed the main aspects of GDI and showed to what degree of anisotropy of the initial plasma structure (called stretched or field-aligned irregularities (FAI)), the criteria of *drift instability* (and, conversely, *drift stabilization*), theoretically obtained in Refs. [49,64–66,73,74], are valid.

Our aim is to present the reader with the main formulations, functions, and parameters of the problem concerning the drift process of a wide spectrum of ionospheric irregularities independent of the origin of their creation, naturally or artificially, which will then be analysed in more detail in Chapter 6, which describes modern radio and optical methods of study of their spatial and temporal evolution.

5.2.1 A Classical Description of Drift in Unbounded Homogeneous Plasma

In the unbounded homogeneous plasma each disturbance of plasma density occurring in real time moves with the velocity of drift of electrons $\mathbf{V}_{e0}$ and ions $\mathbf{V}_{i0}$, if they are equal. However, in plasma in the presence of the homogeneous magnetic ($\mathbf{B}_0$), electric ($\mathbf{E}_0$), and gravitational ($\mathbf{g}$) fields, electrons and ions of plasma drift along and across $\mathbf{B}_0$ with different velocities:

$$
\begin{aligned}
\mathbf{V}_{e0} &= \mathbf{U}_m - \frac{\hat{\sigma}_e}{eN_e}\left\{\mathbf{E}_0 + \frac{1}{c}\mathbf{U}_m \times \mathbf{B}_0 - \frac{m_e\mathbf{g}}{e}\right\} \\
\mathbf{V}_{i0} &= \mathbf{U}_m + \frac{\hat{\sigma}_i}{eN_i}\left\{\mathbf{E}_0 + \frac{1}{c}\mathbf{U}_m \times \mathbf{B}_0 + \frac{M_i\mathbf{g}}{e}\right\}
\end{aligned}
\tag{5.72}
$$

Here, as in previous chapters, we consider ions of unit charge, where a charge equals the electron charge, i.e., $q = Ze \equiv e$, but with different masses, $M_i/m_e \propto 10^3$; $\hat{\sigma}_e$ and $\hat{\sigma}_i$ are the tensors of conductivity of electrons and ions defined in Chapter 1. As a result of this difference in velocities, the charged particles try to separate and the inner (ambipolar) electric field occurs (as in the case of pure diffusion process), which slows the fast particles and accelerates the slow ones in the corresponding directions to the magnetic field. Finally, because of the influence of the ambipolar field, the initial plasma disturbance moves with the intermediate velocity $\mathbf{V}_a$ called the *ambipolar drift velocity*, and the concentration of the electrons and ions inside this disturbance is always equal; i.e., the quasi-neutrality of plasma ($N_e \approx N_i = N$) occurs. The ambipolar field leads to the spreading of the initial plasma disturbance. The difference between the case of the absence of ambient-sources-caused drift and the case of drift occurrence in plasma is that the plasma perturbation with the wave vector $\mathbf{k}$ not only attenuates because of diffusion, but also propagates in space with the frequency ω according to the law of dispersion

$$
\omega = \mathbf{k} \cdot \mathbf{V}_a(\mu^2) - iD_a(\mu^2)k^2
\tag{5.73}
$$

The phase velocity $\mathbf{V}_a$ of the plasma perturbation (i.e., the wave of plasma density) depends on the angle β between the vectors $\mathbf{k}$ and $\mathbf{B}_0$, introduced in the previous section ($\mu = \cos \beta = \mathbf{k} \cdot \mathbf{B}_0 / kB_0$):

$$\mathbf{V}_a(\mu^2) = \frac{[D_{e\|}\mu^2 + D_{e\perp}(1 - \mu^2)]\hat{D}_i\mathbf{E} - [D_{i\|}\mu^2 + D_{i\perp}(1 - \mu^2)]\hat{D}_e\mathbf{E}}{(D_{e\|} + D_{i\|})\mu^2 + (D_{e\perp} + D_{i\perp})(1 - \mu^2)} \quad (5.74)$$

where $\mathbf{E}$ is the superposition of the ambient, $\mathbf{E}_0$, and the inner self-consistent (ambipolar), $\mathbf{E}_a = -\nabla\varphi$, fields.

The coefficient of ambipolar diffusion $D_a(\mu^2)$ was presented in the previous section both for weakly (Equation 5.11) and strongly (Equation 5.31) ionized plasma. From Equation 5.74 it follows that the different modes (e.g., Fourier harmonics) of plasma disturbance spread with different velocities. This leads to a general spreading of plasma disturbance, analogous to wave packet propagation in the dispersive medium [6,11,89] with the group velocity

$$\mathbf{V}_{gr} = \frac{\partial}{\partial\mathbf{k}}(\mathbf{k} \cdot \mathbf{V}_a(\mathbf{k})) \quad (5.75)$$

and its decay due to diffusion with the coefficient $D_a(\mu^2)$. Both velocities, $\mathbf{V}_a$ and $\mathbf{V}_{gr}$, are functions of $\mu = \cos \beta$; therefore, they are not defined for the case $\mathbf{k} = 0$. As a result, the disturbance of the total number of plasma particles, corresponding to the harmonics $\delta N_\mathbf{k}$ with $\mathbf{k} = 0$, has a different behavior compared with that described in the previous section for a pure diffusion, i.e., in the case of the absence of plasma drift.

Drift in Weakly Ionized Plasma. It was found in the previous section that the existence of the weak plasma disturbance, $\delta N < N_0$, in the unbounded weakly ionized plasma leads to the flow of short-circuit currents through the background plasma with the creation of depletion zones in it. With the increase of plasma disturbance δN (compared to the background plasma N_0), more background particles participate in the forming of such currents—the depletion zones become broad and deep (up to zero). The existence of the drift process leads to a more complicated picture of spreading [6,11,89–91]. At the beginning stage of its evolution the plasma disturbance spreads as in the case of the absence of the drift process, and the picture of spreading is the same as was discussed in the previous section. However, because of the dependence of $\mathbf{V}_a$ and $\mathbf{V}_{gr}$ on the direction of spreading (i.e., on $\mu = \cos \beta$), according to Equations 5.73 and 5.75, the dispersion mechanism of plasma density spreading becomes predominant. This mechanisms lead to new qualitative effects of splitting of the initial quasi-neutral plasma disturbance into three moving plasmoids, each of which has its own orientation to the magnetic field, and higher speed of decay compared to the case of $\mathbf{V}_a = 0$ (absence of drift process). The plasma disturbance δN is maximal along a curve defined by the equation

$$\mathbf{r}_m = \mathbf{V}_a(\mu_m^2)t \quad (5.76)$$

and therefore is called the *curve of extrema*. Here, the parameter μ_m^2 changes from zero to unit. In the absence of collisions between charged particles ($\nu_{ei} \ll \nu_{im} \ll \nu_{em}$), this curve is transformed into a straight line (with the accuracy of the terms of the order $D_{e\perp}/D_{i\perp} \ll 1$ and $D_{i\|}/D_{e\|} \ll 1$). During the process of spreading the curve of extrema (Equation 5.76) moves in the space, conserving its shape, as it is shown from Figure 5.27. In this figure, the curve (Equation 5.76) is presented for three moments of time $t_1 < t_2 < t_3$ in the case where $\mathbf{E} \perp \mathbf{B}_0$, i.e., the curves of extrema are performed in the plane orthogonal to $\mathbf{B}_0$ (see Figure 5.27a) and when $\mathbf{E} \| \mathbf{B}_0$, i.e., the

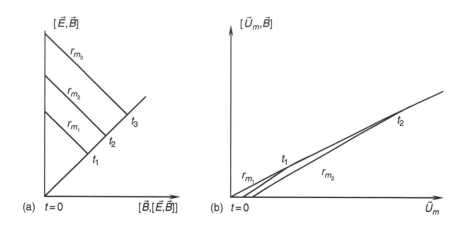

Figure 5.27 Spread of curve of (a) maxima. (Data extracted from Ref. [6].) (b) Its main extrema during drift of weakly ionized point plasma disturbance. (Data extracted from Ref. [89].)

curves of extrema are performed in the plane parallel to $\mathbf{B}_0$ (see Figure 5.27b). In both cases it is evident that the length of the curve $\mathbf{r}_m = V_a\left(\mu_m^2\right)t$ grows linearly with time t. Correspondingly, the maximal value of the plasma density disturbance $\delta N_m(\mathbf{r},t)$ has to be decreased. Hence, the dispersion of drift velocity $\mathbf{V}_a$ (i.e., its dependence on $\mathbf{k}$) leads to the spreading of the plasma disturbance and to the generation of three plasmoids around the points (see Figure 5.28)

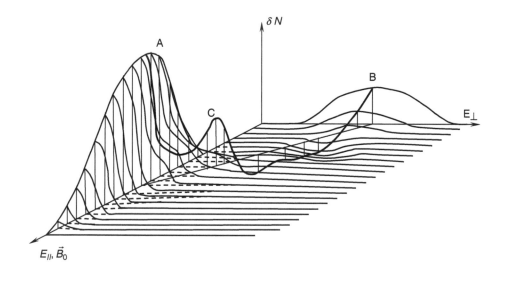

Figure 5.28 Spatial shape of curve of maxima and position of its three main maxima, A, B, and C.

$$\mu_m^2 = 0 \quad (\text{plasmoid } A)$$

$$\mu_m^2 = 1 \quad (\text{plasmoid } B)$$

$$\mu_m^2 = \frac{\mu_0 E_\perp^2}{E_\parallel^2 + \mu_0^4 E_\perp^2} \quad (\text{plasmoid } C)$$

where $\mu_0 = \cos\beta_0 = \sqrt{D_{i\perp}/D_{e\parallel}}$. In Figure 5.28 the schematic picture of the spreading of a weak point disturbance $\delta N(\mathbf{r},0) = n_0\delta(\mathbf{r}) \ll N_0$ is pictured according to Refs. [6,89], where the z-axis is directed along $\mathbf{B}_0$, and $E_\parallel$ and $E_\perp$ are the components of the longitudinal and transverse components (to $\mathbf{B}_0$) of the total electric field (as a superposition of the ambient and intrinsic ambipolar fields), respectively. The number of particles in each of the plasmoids is of the order of the total number of particles n_0 in the initial plasma disturbance. The projection of the curve of extrema (Equation 5.76) at the plane $(\mathbf{E},\mathbf{B}_0)$ is the segment [AB] shown schematically in Figure 5.29. Here, the angle δ is related to the electric field by the expression $\tan\delta = \mu_m^2(E_\perp/E_\parallel)$. When $E_\perp = 0$, $E_\parallel \neq 0$, the mean plasmoid C coincides with point A, and when $E_\parallel = 0$, $E_\perp \neq 0$, it coincides with the point B.

According to Equation 5.72, for the plasmoid A, the components of drift velocity along $\mathbf{B}_0$ and in the direction $\mathbf{E}_\perp \times \mathbf{B}_0$ coincide with the velocity of electron drift $\left(\mathbf{V}_a(\mu_m^2 = 0) \approx \mathbf{V}_{e0}\right)$, and for the plasmoid B, the components of drift velocity transverse to $\mathbf{B}_0$ coincide with the velocity of ion drift $\left(\mathbf{V}_a(\mu_m^2 = 1) \approx \mathbf{V}_{i0}\right)$. This means that the condition of quasi-neutrality does not essentially limit the movements of the plasmoids, which is only possible when accounting for the existence of the depletion zones and for the occurrence of the short-circuit currents through background plasma

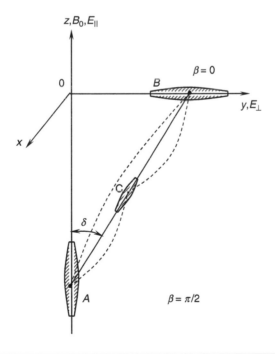

Figure 5.29 Schematic representation and shapes of the three main maxima with respect to ambient electrical and magnetic field directions.

[89,90]. The position of plasmoid C corresponds to reaching of the surface of group velocity $\mathbf{V}_{gr}(\mu^2)$ according to Equation 5.75 to the straight line of extrema (Equation 5.76). Between plasmoids A and C, the plasma disturbance $\delta N_m(\mathbf{r},t)$ is positive at the line $\mathbf{r}_m(t)$ (Equation 5.76), but between plasmoids B and C it is negative (see Figure 5.28). In the particular case of $\mathbf{E} \perp \mathbf{B}_0$ plasmoid C coincides with the *ionic* plasmoid B, and for the case $\mathbf{E} \parallel \mathbf{B}_0$, it coincides with the *electronic* plasmoid A. The concentration in the electronic plasmoid is [6,89,90]

$$\delta N_m\left(V_a\left(\mu_m^2 = 0\right) \cdot t,\, t\right) = \frac{n_0 \Gamma^3(5/4)}{D_{e\perp}^{5/4}(D_{e\parallel}E_\perp)^{1/2} t^{7/4}} \tag{5.77}$$

where $\Gamma(5/4)$ is the Gamma function. This plasmoid is elongated along the magnetic field $\mathbf{B}_0$ with the corresponding scale $\ell_\parallel = (D_{e\parallel}E_\perp t)^{1/2}(D_{e\perp}t)^{1/4}$. Its transverse scale is determined by electron diffusion, i.e., $\ell_\perp \sim (D_{e\perp}t)^{1/2}$. The plasma density inside this plasmoid attenuates as $t^{-7/4}$ and the corresponding disturbance of the potential of ambipolar field is negative and decays in time as $\sim t^{-5/4}$. Because the concentration at the curve of extrema falls as $\sim t^{-2}$ [6,89,90], we can see that the first plasmoid spreads more slowly than the plasma disturbance located at the curve of extrema.

The same tendency was obtained for the ionic plasmoid B [6,89,90]. Here, the concentration inside it equals

$$\delta N_m\left(V_a\left(\mu_m^2 = 1\right) \cdot t,\, t\right) = \frac{n_0}{16\sqrt{2}\pi^{3/2}D_{i\perp}D_{i\parallel}^{3/2}E_\parallel^2 t^{5/2}} \tag{5.78}$$

This plasmoid is stretched across $\mathbf{B}_0$ with the corresponding scale $\ell_\parallel = (D_{i\parallel}t)^{1/2}$ along $\mathbf{B}_0$ and $\ell_\perp = (D_{i\parallel}D_{i\perp}t)^{1/2}E_\parallel t$ across $\mathbf{B}_0$. The concentration in the plasmoid B decreases as $\sim t^{-5/2}$, which is slower than at the curve of extrema ($\sim t^{-2}$). The disturbance of the potential in this point is positive and decays as $\sim t^{-5/2}$. From comparison of the expressions for the plasma concentration in the maximum and accounting for the characteristic scales, it can be found that the number of particles in each plasmoid for these two limited cases corresponds to the total number of particles in the initial disturbance of plasma density n_0.

Both these limiting cases are presented in Figures 5.30 and 5.31, respectively, based on the numerical analysis carried out in Ref. [6] for the case $\mathbf{E} \perp \mathbf{B}_0$ and in Ref. [89] for the case $\mathbf{E} \parallel \mathbf{B}_0$. Following Ref. [6], we point out that the main features occur in the case $\mathbf{E} \perp \mathbf{B}_0$ during drift spreading in the plane XOY (for $\mathbf{z} \parallel \mathbf{B}_0$). In Figure 5.30, the lines of equal concentration are shown for a weak ($\delta N < N_0$) plasma disturbance

$$\delta N(\mathbf{r},t) = n_0\, G(\mathbf{r},t)$$

$$= \frac{n_0}{32\pi^{5/2}t^{3/2}} \int\limits_0^1 \frac{d\mu}{D_a^{3/2}(\mu^2)} \int\limits_0^{2\pi} \left[1 - \frac{\rho^2 \varsigma^2}{2D_a(\mu^2)t}\right] \exp\left\{-\frac{\rho^2 \varsigma^2}{4D_a(\mu^2)t}\right\} d\varphi \tag{5.79}$$

normalized on the concentration in the center of the initial disturbance (Equation 5.14) when only pure diffusion occurs. In Equation 5.79, $\varsigma = \cos(\mathbf{k}, \boldsymbol{\rho})$, $\boldsymbol{\rho} = \mathbf{r} - \mathbf{V}_a(\mu^2)t$, and φ is the azimuthal angle in the same cylindrical coordinate system that was introduced in the previous section for pure

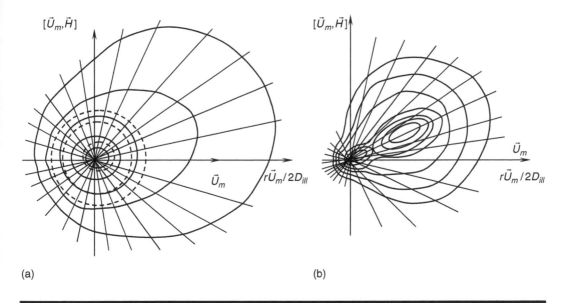

(a) (b)

Figure 5.30 Spatial contours of initial small plasma disturbance during drift spreading and splitting into two plasmoids in the ambient electric and magnetic fields.

diffusion process. The curves in Figure 5.30a are performed for the time $t_1 = 2D_{i\parallel}/U_m^2$, and in Figure 5.30b for $t_2 = 20D_{i\parallel}/U_m^2$, where again U_m is the velocity of the neutral wind in plasma. In the computations, the magnetized plasma was considered, taking $q_H = 50$ and $Q_H = 0.5$ (i.e., $q_H Q_H = 25 \gg 1$). It is evident that for $t < t_1$ the shape of the plasma disturbance is symmetric and determined by a pure diffusion process. At the moment $t = t_1$ the drift begins to influence the spreading process and the shape of the disturbance differs from an ellipsoidal form (here the curves describing a pure diffusion are presented by dashed lines in Figure 5.30a). For $t > t_1$ the form of the disturbance is deformed owing to drift process and is now defined by the dispersion mechanisms of spreading. The initial disturbance splits into two plasmoids, the first moving in the direction of the vector $\mathbf{U}_m$ with comparably small velocity $V_a(0) \ll U_m$. The second plasmoid moves with the velocity $V_a(1)$, which is more of the order higher than the velocity of the first plasmoid in the directions between $\mathbf{U}_m$ and $\mathbf{U}_m \times \mathbf{B}_0$. However, the concentration in the second plasmoid decays with time slower than in the first one, and the transfer of particles from the first plasmoid to the second is observed. During the definite time the concentration in both plasmoids becomes equal. The first plasmoid is strongly stretched along $\mathbf{B}_0$, and the second is in the plane orthogonal to $\mathbf{B}_0$. The electric field created by these plasmoids is close to the dipole with potential $\Phi \sim r^{-2}$. Therefore, the asymptote of the plasma disturbance in the far zone ($r \gg \sqrt{4D_a t}$) is proportional to t/r^4 and caused by the redistribution of the background particles under the control of the disturbed electric field (i.e., by short-circuit currents through the background plasma). The first plasmoids were denoted as electronic, and the second as ionic. However, as we will see in Chapter 6, the speeds and dimensions of these plasmoids do not coincide with real sizes and speeds of the injected electrons and ions in real ionospheric experiments. Introduction of definitions of electron and ion plasmoids in Refs. [6,89,90] was made to simplify a physical picture of the process of spreading, without distortion of the condition of plasma quasi-neutrality. The above approach was proposed in Refs. [6,89,90] to show another similar examples of evolution of the plasma disturbance, particularly, in the limiting

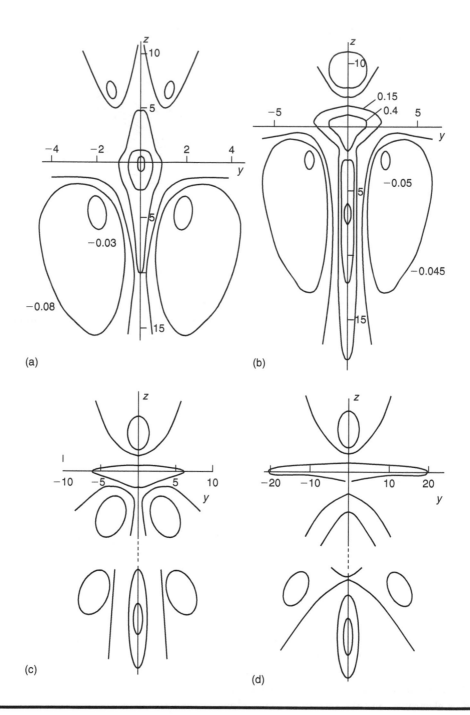

Figure 5.31 Spatial and temporal distribution of initial point plasma disturbance during drift spreading in the ambient electric and magnetic fields for $\tilde{t} = 9$ (a), $\tilde{t} = 10^2$ (b), $\tilde{t} = 9 \times 10^2$ (c), and $\tilde{t} = 10^4$ (d).

case when $\mathbf{E} \parallel \mathbf{B}_0$ presented in Figure 5.31 according to numerical computations made in Refs. [89,90]. In this case, computations were now made also for the case of magnetized plasma, where $q_H = 30$ and $Q_H = 0.3$ (i.e., $q_H Q_H = 9 \gg 1$). The time $\tilde{t}$ in figures was chosen in units $8D_{i\parallel}(D_{e\parallel}|E_{\parallel}|)^2$ and the concentration normalized to the initial concentration $\delta N_d(0,t)$ defined by Equation 5.14 for a pure diffusion process; the distance is presented in units $24D_{i\parallel}D_{e\parallel}E_{\parallel}$. Lines of equal concentration correspond to the relative time $\tilde{t} = 9$ (Figure 5.31a), $\tilde{t} = 100$ (Figure 5.31b), $\tilde{t} = 900$ (Figure 5.31c), and $\tilde{t} = 10^4$ (Figure 5.31d). It is evident that at the initial stage of plasma spreading, for $t \le D_a/V_a^2$ (see Figure 5.31a), the distribution of the concentration δN is determined by pure diffusion. For $t > D_a/V_a^2$ (Figure 5.31b through d) the initial disturbance splits into two plasmoids and the redistribution of the depletion zones of the background plasma defined by the negative numbers near the curves of equal concentration. The same picture as observed for the case $\mathbf{E} \perp \mathbf{B}_0$ is observed in the case $\mathbf{E} \parallel \mathbf{B}_0$. Thus, electrons of the initial point disturbance $\delta N(\mathbf{r},0) = n_0 \delta(\mathbf{r})$ move with the velocity $\mathbf{V}_a(0) = D_{e\parallel}|E_{\parallel}|$. The condition of quasi-neutrality for the corresponding plasmoid strongly stretched along z-axis (or $\mathbf{B}_0$; see Figure 5.31), is provided with the flows of background ions across $\mathbf{B}_0$. Because the velocity of movement and the characteristic sizes of this plasmoid are determined by electrons, and the potential is negative, this plasmoid was defined in Refs. [89,90] as *electronic*. The character of spreading and the characteristic sizes of the second plasmoid, stretched along the y-axis (across $\mathbf{B}_0$; see Figure 5.31), are defined by background ions coming from depletion zones to secure quasi-neutrality in this plasmoid (therefore, the potential in it is positive). In Refs. [89,90], this plasmoid was defined as *ionic*. However, as we will show in Chapter 6, the real injected ions move along the electric field with velocity $\sim D_{i\parallel}|E_{\parallel}|$, whereas the virtual ionic plasmoid does not move along $\mathbf{B}_0$. Hence, movement of plasma disturbance along $\mathbf{B}_0$ (along the z-axis) is the same as for the case of $\mathbf{B}_0 = 0$, where the maximum δN_m is at rest and the scale of plasma disturbance is defined by the ambipolar diffusion. In the transverse directions to $\mathbf{B}_0$ (along the y-axis), the ion plasmoid spreads linearly in time owing to background electrons coming along $\mathbf{B}_0$ from depletion zones, which also move together with the electronic plasmoid.

Drift in Unbounded Plasma with an Arbitrary Degree of Ionization. Here, we consider briefly peculiarities occurring when a small (point) weak plasma disturbance spreads and moves in plasma with arbitrary degree of ionization. As in the previous section, we consider a weak ($\delta N < N_0$) point $\delta N(\mathbf{r},0) = n_0 \delta(\mathbf{r})$ plasma disturbance concentrated in the initial moment in the small volume of the characteristic scale L_0.

In this linear case, drift and diffusion of weak small plasma disturbances can be described by the linear equation obtained from general nonlinear equations presented in Chapter 1, by excluding from them the potential of the ambipolar field (see Refs. [6,89–91]), as done for a purely diffusion process (see Equation 5.48). Finally, the equation of joint drift and diffusion process is as follows:

$$[\nabla\hat{\sigma}_e\nabla + \nabla\hat{\sigma}_i\nabla]\frac{\partial\delta N}{\partial t} + \left[(\nabla\hat{\sigma}_e\nabla)\frac{\partial}{\partial N_0}(N_0\mathbf{V}_{i0}\nabla) + (\nabla\hat{\sigma}_i\nabla)\frac{\partial}{\partial N_0}(N_0\mathbf{V}_{e0}\nabla)\right]\delta N$$
$$- \left[(\nabla\hat{\sigma}_e\nabla)(\nabla\hat{D}_i\nabla) + (\nabla\hat{\sigma}_i\nabla)(\nabla\hat{D}_e\nabla)\right]\delta N = 0 \tag{5.80}$$

Here, all components of tensors of diffusion and conductivity for electrons and ions are as presented in Chapter 1 for the model of the ionospheric plasma as plasma with arbitrary degree of ionization. The solution of such a linearized equation can be obtained by the expansion of the linear solution into the Fourier integral over coordinates, as was done previously for pure diffusion process:

$$\delta N(r,t) = n_0 G(r,t) = n_0 \int \exp\left\{i\mathbf{k}[\mathbf{r} - V_a(\mu^2)t] - D_a(\mu^2)k^2t\right\} d^3k \tag{5.81}$$

Here, $D_a(\mu^2)$ is the coefficient of the ambipolar diffusion defined for the plasma with an arbitrary degree of ionization by expression (Equation 5.41), and the velocity for the ambipolar drift in plasma with arbitrary degree of ionization was obtained in Ref. [88] in the following form:

$$
|\mathbf{V}_a(\mu^2)| = U_m + \eta U_m \left\{ \left[C_3 \mu^2 + \frac{C_4(1-\mu^2)}{A} \right] C_6 - \left[C_3 \mu^2 + \frac{C_5(1-\mu^2)}{A} \right] C_7 \right\}
$$
$$
\times \left\{ C_1 \mu^2 + \frac{C_2(1-\mu^2)}{A} \right\}^{-1}
\tag{5.82}
$$

where $A = (1+p)^2 + q_H^2(1 + 2\eta p - Q_H^2)$, $C_1 = \frac{(1+\eta)}{(1+p)}$, $C_2 = \frac{(1+p)(1+\eta) + Q_H^2 + \eta q_H^2}{A}$

$$
C_3 = \frac{1}{(1+p)}, \quad C_4 = \frac{(1+p+Q_H^2)}{A}, \quad C_5 = \frac{(1+p+q_H^2)}{A}
$$

$$
C_6 = \frac{1}{(1+p)^2} + \frac{(1+p)^2 + q_H^2(2+2p-p^2) + q_H^4(1+Q_H^2+2\eta^2 p)}{A^2} +
$$
$$
+ q_H \frac{2p(1+p) + q_H Q_H(2+p^2+2p\eta^{-1}) + q_H^3 Q_H((1+2\eta p + Q_H^2)}{A^2}
$$

$$
C_7 = \frac{1}{(1+p)^2} + \frac{(1+p)^2 + q_H^2(1+2p+2\eta p^2) + q_H^2 Q_H^2(2+2p+Q_H^2)}{A^2}
$$
$$
+ q_H \frac{(1-p^2) + q_H^2(1+2\eta p + \eta^2 p^2) + q_H^2 Q_H^2((2+2\eta p + Q_H^2)}{A^2}
$$

The dependence V_a/U_m versus the angle θ ($\theta = 90° - \beta$) and for various parameters of ionization $p = \nu_{ei}/\nu_{em} = 0, 1, 5, 20$ is shown in Figure 5.32 for the model of homogeneous middle-latitude ionosphere at the altitude $h = 200$ km, all parameters of which are presented by Figures 1.1 through 1.3. It is evident that the velocity of the ambipolar drift can be less than the velocity of wind of neutrals ($\sim 10^{-2} - 10^{-3}$) for $p \ll 1$, whereas for $p \geq 1$ the effects related to the influence of the magnetic field become weaker, and for $p \gg 1$, the drift spreading goes with velocity close to the wind velocity. This effect depends on the direction of spreading of the plasma disturbance in the ambient magnetic field (on the angle β). Thus, for $\beta \to 90°$, $\theta \to 0$, and for $0 \leq p \leq 1$, the drift velocity is minimal in directions across the magnetic field, whereas for $p \gg 1$, even in this direction, the effects of magnetic field become negligible. For $\beta \to 0°$, $\theta \to 90°$ for arbitrary plasma ionization, the magnetic field does not affect the process of the ambipolar drift.

We now return to the integral (Equation 5.81) and present it in the spherical system of coordinates $\{k,\mu,\psi\}$. Then, following the derivations made in Refs. [6,89–91], we get the result that after a time $t \geq L_0/D_{i0}$, where D_{i0} is the coefficient of ion diffusion in the isotropic plasma for $\mathbf{B}_0 = 0$, the disturbance of plasma concentration δN is described by the following expression:

$$
\delta N(r,\kappa,t) = n_0 G(r,\kappa,t) = \frac{n_0}{8\pi^{5/2} t^{3/2}} \int_0^1 \frac{d\mu}{D_a^{3/2}(\mu^2)} \int_0^\pi [1 - 2B^2] \exp\{-B^2\} \, d\psi
\tag{5.83}
$$

where as before $\boldsymbol{\rho} = \mathbf{r} - \mathbf{V}_a(\mu^2)t$, $\kappa = \boldsymbol{\rho} \cdot \mathbf{B}_0/\rho \cdot B_0$, $\mu = \mathbf{k} \cdot \mathbf{B}_0/k \cdot B_0 = \cos\beta$, and

$$
B^2 = \frac{\rho^2 \left[\mu\kappa + \sqrt{(1-\mu^2)(1-\kappa^2)} \cos\psi \right]}{4D_a(\mu^2)t}
$$

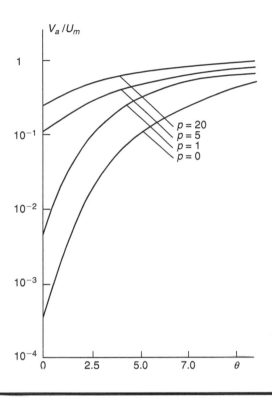

Figure 5.32 Angle distribution of the normalized ambipolar velocity for various degrees of ionization of the background plasma.

Equation 5.83 characterizes the movements and decay of the small weak plasma disturbance in the time domain. In Figure 5.33 the results of numerical analysis of Equation 5.83 are in the form $G(y,\kappa) = G(r,\kappa,t)/G_d(0,t)$, where $G_d(0,t)$ is the profile of plasma concentration in the origin of the coordinate system in the case of the pure diffusion problem with absence of drift movements described by Equation 5.14, depending on the dimensionless parameter $y = U_m^2/D_{i\parallel}$. This parameter was estimated using well-known values U_m and $D_{i\parallel}$ at the corresponding ionospheric altitudes. All parameters were taken according to the model of the middle-latitude ionosphere for altitude $z = 200$ km (see Tables 1.1 through 1.3). From the presented illustrations it is seen that for $p \ll 1$ the velocity of spreading along $\mathbf{B}_0$ (Figure 5.33a; $\beta = 0^\circ$, $\mu = 1$) significantly exceeds the velocity of spreading across $\mathbf{B}_0$ (Figure 5.33b; $\beta = 90^\circ$, $\mu = 0$). This is evident because in the magnetized plasma the character of dispersion spreading of plasma disturbance along $\mathbf{B}_0$ is determined by the electronic component of plasma (i.e., by $D_{e\parallel}$ and $V_{e\parallel}$) and across $\mathbf{B}_0$ by the ionic component (i.e., by $D_{i\perp}$ and $V_{i\perp}$), and $V_{e\parallel} \gg V_{i\perp}$. Even for $p \geq 1$ the effects of the magnetic field are significant, mostly in the direction along $\mathbf{B}_0$, and strong anisotropy is observed if we compare these two limited cases presented, respectively, in Figure 5.33a and b. However, for $p \gg 1$ (more strictly for $\eta p \gg 1$, $\eta = m_e \nu_{em}/M_i \nu_{im}$), all effects related to the influence of the magnetic field discussed earlier become negligible, and the velocity of drift process both along and across $\mathbf{B}_0$ is the same; the anisotropy in the shape of the initial plasma density becomes sufficiently weak. All these features are observed for $y \leq 1$ and $y \geq 1$, when the velocity of the neutral wind U_m is small enough). For $y \gg 1$ (U_m is large) in both cases

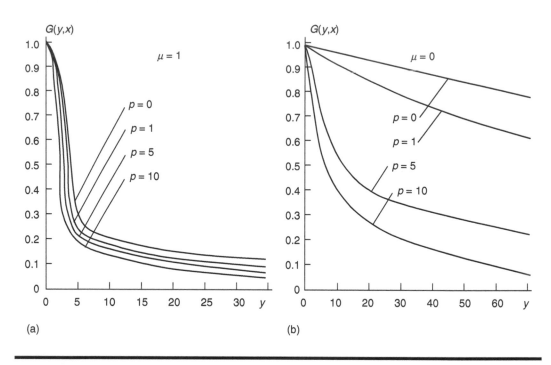

Figure 5.33 Spatial distribution of plasma density profile of small plasma disturbance for various degrees of ionization of the background plasma.

of weakly ($p \leq 1$) and strongly ($p \gg 1$) ionized plasma the influence of the magnetic field becomes insufficient. The weak small disturbance of plasma density spreads isotropy as in the case of the absence of the magnetic field (according to Schottky [1–3]). This effect can be understood if we take into account the fact that for the large wind velocities U_m the neutral molecules (atoms) in their movement interact with the charged particles and force them to move with the same velocity in the direction $\mathbf{U}_m$ [28]. It is also evident that in the presence of drift the influence of electron–ion collisions (i.e., the degree of ionization of plasma) is less compared to their influence in the pure diffusion process of spreading (see previous section). The results, depicted in Figure 5.33, correspond to the case where the vector of wind $\mathbf{U}_m$ is orthogonal to $\mathbf{B}_0$, i.e., $\mathbf{U}_m \perp \mathbf{B}_0$, and the curve of extrema $\rho = \left| \mathbf{V}_a(\mu_m^2) - \mathbf{V}_a(\mu^2) \right| t$ lies in the plane also orthogonal to $\mathbf{B}_0$, i.e., $\kappa(\mu_m^2) \equiv 0$. In this limiting case, as follows from previous discussions, the mean plasmoid C coincides with the second one B (see Figure 5.29). Without loss of generality, we have to point out that in the other limiting case, when $\mathbf{U}_m \| \mathbf{B}_0$ and $\kappa(\mu_m^2) \equiv 1$, the mean plasmoid C coincides with the first plasmoid, A (see Figure 5.29). Here, the same physical picture of the drift spreading of weak small plasma disturbances in the unbounded background plasma with an arbitrary degree of ionization is observed, and the same effects of electron–ion collisions on drift spreading in the strong magnetic field should be taken into account.

The analogous effects are observed in the analysis of time of drift spreading (lifetime) of the plasma disturbance defined earlier by the constraint Equation 5.53. The corresponding expressions, analogous to Equation 5.54b but for the plasma with an arbitrary degree of ionization, were obtained and analysed in Ref. [88]. If according to Ref. [88], we denote t as the time of spreading of initial plasma disturbance caused by drift and diffusion, and τ as the parameter of the pure diffusion, which instead of Equation 5.44a gives the following expression:

$$\frac{\tau}{t_0} = \left\{ \int_0^1 \left[\frac{(1 + 2p)F_2(\mu)}{(1 + p)F_1(\mu)} \right]^{3/2} \right\}^{2/3} \tag{5.84}$$

We can then obtain, following Ref. [88], the lifetime of the first (A) and the second (B) plasmoids at the curve of extrema compared with the lifetime of plasma disturbance in the absence of drift and the magnetic field, $t_0 = L_0^2 n^{2/3}$. Here, as previously in Equation 5.54, $L_0 = [n_0 n / \delta N_m(\mathbf{r}_m, 0)]^{1/3}$ is the characteristic initial size of the plasma disturbance. Thus, for the first plasmoid we get

$$\left(\frac{t}{t_0} \right)_I = \left\{ \frac{\Gamma^3(5/4)[(1 + p)^2 + q_H Q_H(1 + p)]^{5/4}(1 + 2p)^{5/4} C_2}{\pi^{5/4} \eta^{1/2}(1 + p)^{11/4} C_8^{1/2} F_1^{5/4}(\mu_m) x^{1/2}} \right\}^{4/7} \tag{5.85}$$

For the second plasmoid we get

$$\left(\frac{t}{t_0} \right)_{II} = \left\{ \frac{2^{5/4} \Gamma^2(7/6)(1 + 2p)^{7/6} C_1^{2/3} F_2^{7/6}(\mu_m)}{\pi^{5/4} \eta^{2/3}(1 + p)^{7/6} C_8^{2/3} F_1^{7/6}(\mu_m) x^{2/3}} \right\}^{6/11} \tag{5.86}$$

The common expression of the relaxation time of the maximum of plasma disturbance during its spreading caused by drift and diffusion at the curve of extrema, defined by the vector $\mathbf{r}_m = \mathbf{V}_a(\mu_m^2)t$, can be obtained, following Ref. [88], as

$$\left(\frac{t}{t_0} \right)_{\mathbf{r}_m} = 2 \left\{ \frac{(1 + 2p)[(C_1 - C_2)\mu_m^2 + C_2]^2 F_2(\mu_m)}{\mu_m(1 - \mu_m^2)^{1/2} \eta(1 + p) C_8 F_1(\mu_m) x} \right\}^{1/2} \tag{5.87}$$

Here, $x = L_0 U_m / D_{i\parallel}$ is the dimensionless parameter;

$$F_1(\mu_m) = (1 + p)^2 + q_H^2(1 + 2\eta p)\mu_m^2 + q_H^2 Q_H^2 \mu_m^4$$
$$F_2(\mu_m) = (1 + p)^2 + q_H Q_H(1 + p) + q_H^2(1 + \eta p + Q_H^2)\mu_m^2$$
$$C_8 = C_2 C_3(C_6 - C_7) - C_1(C_6 C_4 - C_7 C_5)$$

all other parameters and functions are presented after Equation 5.82. In Figure 5.34 the dependence of $(t/t_0)_{\mathbf{r}_m}$ against μ_m (i.e., the spreading occurring in the direction of the curve of extrema) is presented according to Equation 5.87 computed for $x = 1$ and 10, and for the conditions of plasma with an arbitrary degree of ionization associated with the model of the middle-latitude ionosphere at $z = 200$ km, all parameters of which are described in Tables 1.1 through 1.3. As was expected, for the same value of x (i.e., U_m) the largest lifetime has those plasmoids from a whole spectrum located at the curve of extrema, the direction of which is close to orthogonal direction to $\mathbf{B}_0$. With growth of x (from 1 to 10) or the corresponding wind velocity U_m, for the same parameter of ionization p, the duration of drift spreading decreases by 3–5 times. As parameter p increases, this dependence changes weakly. We explain this phenomenon, following Equation 5.28, by the fact

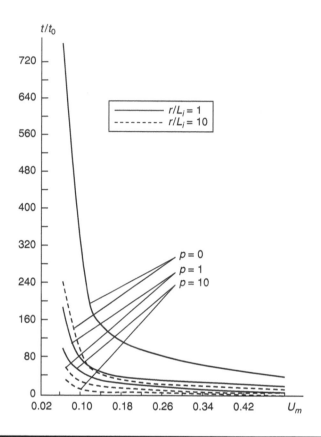

Figure 5.34 **Lifetime of initial small plasma disturbance for various degrees of ionization of the background plasma and at different distances from the maximum of initial disturbance.**

that in their movements, molecules (atoms), because of interactions with plasma electrons and ions, force them to move with the same velocity in the direction $\mathbf{U}_m$, and the influence of the ambient magnetic field on charged particles becomes weaker. In Figure 5.35 we now present the dependences $(t/t_0)_I$ and $(t/t_0)_{II}$ for the first and the second plasmoids, respectively, for the same conditions of ionospheric plasma as for the previous figure. It is evident that the time of drift spreading of the first plasmoid (across $\mathbf{B}_0$ and $\mathbf{U}_m$) exceeds the time of spreading of the second plasmoids (across $\mathbf{B}_0$ and along $\mathbf{U}_m$) by one or two orders of magnitude. This effect depends on the degree of ionization and magnetization of plasma. Thus, with increase of the parameter of ionization p, the value of $(t/t_0)_I$ decreases faster than $(t/t_0)_{II}$ for the same degree of plasma magnetization (see Figure 5.35). However, the additional analysis has shown that for strongly ionized plasma ($p \gg 1$), the lifetime of the first and the second plasmoids becomes equal, and the dispersion mechanism is ineffective; i.e., the influence of the magnetic field on plasmoid spreading becomes negligible. The same tendency (see also Figure 5.34) is evident: with the increase of parameter x (or U_m), the effects of the magnetic field on drift spreading of both plasmoids become weaker.

The critical parameter of ionization p_{cr}, for which this effect is observed, is defined by the equation $(t/t_0)_I = (t/t_0)_{II}$. Estimations made in Ref. [88] for the model of middle-latitude ionosphere give the following: for daytime conditions, $p_{cr} = 110$–120 ($z = 200$ km), $p_{cr} = 340$–350 ($z = 300$ km); for nocturnal conditions $p_{cr} = 180$–190 ($z = 200$ km), $p_{cr} = 425$–440 ($z = 300$ km).

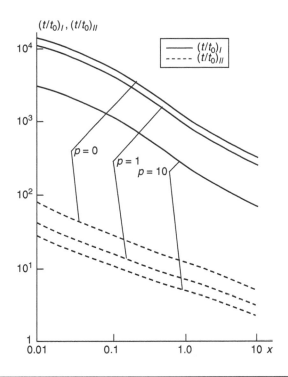

Figure 5.35 **Lifetime of the first and the second maximum of the initial plasma disturbance for various degrees of ionization of the background plasma.**

According to these estimations, we should stress that all previous effects obtained in the real ionosphere cannot be simply interpreted because with increase of the height, simultaneous non-linear growth of degree of magnetization and ionization are observed, and the latter grows sharper than the first one up to altitudes of 500 km (see Figures 1.1 through 1.3 and Table 1.3).

We now consider the question of how the drift generally influences the lifetime of plasma disturbance compared with the case of a pure diffusion process. In Figure 5.36 we presented the dependence t/τ (t corresponds to drift with diffusion, and τ is the pure diffusion) as a function of x (or U_m). Curve 1 corresponds to computations using parameters of the ionosphere at altitude 200 km and curve 2 is for an altitude of 300 km according to Chapter 1. It is seen that the existence of drift leads to a decrease by a factor of 3–5 in the lifetime of the plasma disturbance compared with a situation where the disturbance is in the rest. This effect decreases with altitude i.e., with decrease of altitude from 300 km to 200 km, the ratio t/τ becomes smaller.

However, as follows from Figure 5.35 for $x > 1$, $(t/t_0)_I \gg 1$ and $(t/t_0)_{II} \gg 1$. This means that even taking into account the drift process, the time of relaxation of a weak small plasma disturbance in the ionosphere exceeds the lifetime of that disturbance in the case of absence of ambient magnetic field ($\mathbf{B}_0 = 0$) and drift process ($\mathbf{V}_a = 0$).

5.2.2   Dynamics and Spreading of Arbitrary Irregularity in the Ionosphere

Further modeling of the process of the drift and diffusion spreading of plasma irregularities of comparatively large sizes in the ionospheric plasma has been carried out by several separate groups

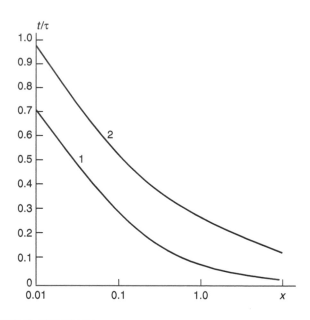

Figure 5.36 **Spatial distribution of the lifetime of the initial plasma disturbance during its drift and diffusion spreading normalized on the lifetime of this plasma disturbance during its pure diffusion spreading, calculated for ionospheric altitudes $h = 200$ km (curve 1) and $h = 300$ km (curve 2).**

of researchers [40,49,58–66,87–88,95–99] using numerical simulation approaches because of the many natural initial and boundary conditions, which limit the use analytical methods. Therefore, only some simplified analytical models concerning inertial drift process and evolution and striation of strong plasma structures in the ionosphere have been developed in recent decades [73,74,87–94]. Thus, in Refs. [40,58–63], the evolution of ion clouds in the ionospheric plasma at altitudes between the E and F layers was investigated numerically, taking into account the natural layers' structure, wind movements, and the process of recombination, mostly occurring in the E region of the ionosphere.

We present the results of their investigations in Chapter 6, where we consider the evolution of artificial ion clouds in the ionosphere and compare results of computations with the corresponding data obtained from numerous rocket experiments and with the corresponding real-time optical observations of dynamics of such plasma structures. We only point out that some of the models under consideration are sufficiently realistic [58–60], because they account for multilayered ionosphere and interactions of the injected plasma structure with bottom and upper layers, as shown in Figure 5.37. As can be seen, the initial plasma cloud lies between two layers; the upper (1) corresponding to the F layer and the bottom one (2) to the E layer. Owing to the interaction between the plasma cloud and the layers, imaginary clouds are generated at these layers with lack of or excess of plasma concentration related to the effect of short-circuit (vortex) current through the background plasma. This vortex current is due to the flow of the electrons between the layers (along the magnetic field $\mathbf{B}_0$ which is usually directed vertically) and the flow of ions between imaginary clouds within each layer, with depletions of density (denoted by "−") and with enhancements of density (denoted by "+"), as shown schematically in Figure 5.37. The direction of the vortex currents is such that the ions and electrons leave the depletion zones and enter zones with

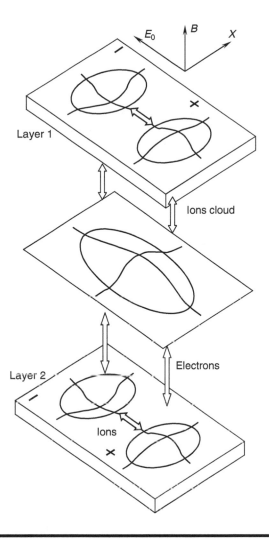

Figure 5.37 Three-layer model of evolution of ion clouds in the ionosphere between the upper F layer and lower E layer.

enhancements of plasma density. It was shown in Refs. [40,58–60] that even in the absence of the wind of neutrals and recombination, but with only the influence of the ambient electric field, the penetration of plasma irregularity in the background plasma occurs, and a significant increase of velocity of spreading of an injection charge (compared to pure diffusion) is observed with an essential deformation of the initial shape and form of the irregularity. The authors of these works related this deformation to the influence of the ambient electric field E_0 and to the multilayered structure of the ionospheric plasma.

In some works [61–63], the consistent numerical analysis of the drift of plasma structures of various dimensions and forms (plate, spherical, and elliptical) was carried out. A new phenomenon was observed experimentally, related to the splitting of the initial plasma perturbation on plasma clumps strongly stretched along B_0, called *strata*, which moved more slowly than the initial mother irregularity. The numerical analyses carried out in Refs. [61–63], and then in Refs. [64–69], show that the splitting in the cloud begins at the front of the cloud when a bifurcation of plasma profile is observed at its back, with further splitting of the initial plasmoid at the set of quasi-stable strata. The authors of these studies related this phenomenon to the $\mathbf{E}_0 \times \mathbf{B}_0$-drift instability described in detail in Chapter 2, and caused, according to Refs. [61–69], by the differences in velocities of ions and electrons in the crossing electric and magnetic fields. The localization of these instabilities (e.g., strata) depends on the gradient of the plasma density inside the large mother cloud, on the vector $\mathbf{E}_0 \times \mathbf{B}_0$, and on the initial velocities of drift of electrons and ions defined by Equation 5.72, and also on the degree of disturbance of the ambient electric field $\delta \mathbf{E}_0$ caused by the injected plasma clouds. In Refs. [61–63] the following criterion of weakness perturbation of the ambient electric field and its influence on movements of initial plasma irregularity in the nondisturbed electric $\mathbf{E}_0$ and magnetic $\mathbf{B}_0$ fields, $\lambda^* = \Sigma_P^{E,F}/\Sigma_P^c \ll 1$, was introduced. Here, Σ_P^c is the Pedersen conductivity along $\mathbf{E}_0$ and across $\mathbf{B}_0$ of the plasma cloud integrated along the height, and Σ_P^E, Σ_P^F are the same conductivities integrated along the z-axis (i.e., along the height), for the E and F layers, respectively. Conversely, the criterion of strong electric field perturbation caused by the injected plasma cloud, $\lambda^* \gg 1$, was introduced in Refs. [61–63]. This criterion is more soft compared with the criterion of ambient electric field perturbation introduced in Refs. [89–91], $\lambda^{**} = \delta N/N_0 \gg 1$, where δN, as before, is the initial plasma concentration inside the irregularity and N_0 the concentration of the background ionospheric plasma. What criterion of plasma perturbation should be taken into account, is the main aspect of further explanation of the observed experiments, because it defines the degree of nonlinearity of the problem of drift evolution of arbitrary plasma disturbance in the ionosphere. Therefore, we will analyze this aspect in detail in Chapter 6, which deals with the evolution of barium clouds in the ionosphere.

Here, we will only note that if the weak initial plasma disturbance ($\delta N/N_0 \ll 1$) occurs in the ionosphere, the vortex currents through background plasma stabilize the $\mathbf{E}_0 \times \mathbf{B}_0$-drift instabilities. However, in the inverse case, $\delta N/N_0 > 1$, when the ambipolar field ($\mathbf{E}_a = -\nabla \varphi$) cannot be fully secured by the vortex currents in the regions above and below the initial plasma disturbance, the movements of plasma disturbance in the total electric field as a sum of the ambient and inner fields ($\mathbf{E} = \mathbf{E}_0 + \mathbf{E}_a$) cause fundamental changes in the form of the irregularity. The inner field in the ionosphere is irregular in space and directed against the gradient of plasma density. Therefore, in the regions with enhancement of plasma concentration, the total field is smaller and these regions move more slowly compared with the depletion ones. This effect leads to bifurcation of the gradients of concentration during irregularity movements. This general result obtained in Refs. [61–63] is clearly seen from illustrations presented in Figure 5.38 of numerical simulations of the nonlinear dynamic of large-scale irregularity in the middle-latitude ionosphere. Here, the moments of time are presented in dimensionless units, t/t_0, where $t_0 = L_0/V_d$, L_0 is the initial size of irregularity, and V_d the magnitude of drift velocity. Along axes, vertical and horizontal, the coordinates are also presented in relation to L_0 units. It is evident that stratification begins at the front of the movements in the direction of $\mathbf{E}_0 \times \mathbf{B}_0$-drift (along the vertical axis), and because of bifurcation of the front profile, several strata are generated and move with the initial irregularity along the vertical axis. Conversely, in some works [35,40,58,59], the generation of plasma waves (oscillations) was observed at the back of the initial irregularity, i.e., opposite the direction of movements, which was explained by the authors taking into account the difference between mobilities of electrons and ions in the ambient electric field. However, in the previously mentioned works, the effects of stratification were generated artificially by the introduction into the numerical model of some artificial quasi-periodical (according to sine and cosine laws) perturbations in the initial Gaussian profile of irregularity.

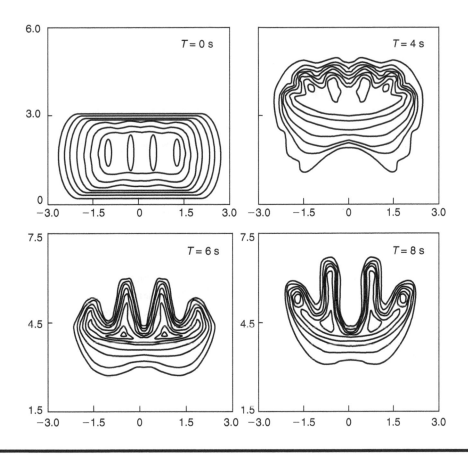

Figure 5.38 Spatial-temporal evolution of the ion cloud according to the three-layer model presented in Figure 5.38.

Some researchers have tried to analyze and resolve the observed contradictions between results of numerical simulations both analytically [89–94] and numerically [70,71]. Thus were resolved some problems of nonlinear drift evolution of strong ($\delta N/N_0 \gg 1$) one- and two-dimensional irregularity in the ionospheric plasma. It was shown that in more realistic cases, owing to spatial inhomogeneity of the total electric field $\mathbf{E}$, the gradient of plasma density increases at the *front* of the irregularity (relative to $\mathbf{E}_0$) with the normal vector along $\mathbf{E}_0$, as well as at the *back* of the irregularity (relative to $\mathbf{V}_d$) with the normal vector orthogonal to $\mathbf{E}_0$ and $\mathbf{B}_0$. During bifurcation of fronts of the irregularity, a sharp increase of plasma density occurs. At the same time, in Refs. [89–91], it was found that in the real ionosphere in the case of drift of irregularities with limited dimensions along $\mathbf{B}_0$, the criterion of $\mathbf{E}_0 \times \mathbf{B}_0$-drift instability development and generation of the corresponding stratified plasma structures in the direction of the $\mathbf{E}_0 \times \mathbf{B}_0$-drift is different from the criterion obtained in Refs. [64–69]. Thus, according to Refs. [89–91] this criterion is defined not only by drift velocity, $\mathbf{V}_d = c(\mathbf{E}_0 \times \mathbf{B}_0)/|\mathbf{B}_0|^2$, and the transverse dimension of the irregularity d, but also by the integration along the height of the Pedersen conductivities for the plasma cloud, Σ_P^c, and the background plasma, Σ_P^0, i.e., $\gamma \geq (V_d/d) \cdot (\Sigma_P^0/\Sigma_P^c)$, whereas in Refs. [61–69] the authors obtained the following criterion $\gamma \geq (V_d/d)$ for the increment of gradient-drift instability (GDI).

In the latter case, the authors analysed the irregularity elongated infinitely along $\mathbf{B}_0$ with the plasma density wave vector $\mathbf{k}$ orthogonal to $\mathbf{V}_d$. As we will show in the Chapter 6, experiments for the injection of ion clouds in the ionosphere have indicated that the stratification (i.e., the growth of instabilities) occurs in the direction of $\mathbf{E}_0 \times \mathbf{B}_0$-drift and that the strata leave the initial irregularity in the direction of $\mathbf{V}_d \sim \mathbf{E}_0 \times \mathbf{B}_0$, as is also seen from Figure 5.38, instead of occurring at the back of the irregularity profile. These experimental observations coincide with the theoretical results obtained numerically in Refs. [64–71] and analytically in Refs. [89–94]. All the contradictions between different researches and their analysis, the theories and experimental data, reveal the involvement of the processes accompanying the evolution of arbitrary plasma disturbance in the real nonregular inhomogeneous ionospheric plasma. Therefore, in Refs. [49,73,74], taking into account effects of real ionosphere on the transport processes occurring at the different altitudes [87,88], a self-consistent model of the nonlinear evolution (drift and diffusion) of the arbitrary plasma irregularity (weak and strong) with arbitrary degree of anisotropy with respect to ambient magnetic field has been developed. Moreover, in Ref. [74] the effects of the degree of ionization, inhomogeneity of background plasma, initial dimensions of the irregularity, and initial perturbation of the ambient electric field on plasma GDI generation were analyzed separately on the basis of results of the classical theory performed in Refs. [6,89–91] and briefly presented in this section. We now introduce the reader to this self-consistent numerical model and compare it with those realistic models used in Refs. [70,71].

In Refs. [49,73,74], the drift of strong plasma disturbances ($\delta N > N_0$) with various degrees of anisotropy in the inhomogeneous ionosphere was investigated both theoretically and numerically on the basis of an ionospheric model that is to a great extent close to conditions in the middle-latitude ionosphere at heights from 100 to 400 km. In these works, the altitude dependence of transport coefficients of charged particles, shown in Figure 5.39, for middle-latitude ionosphere,

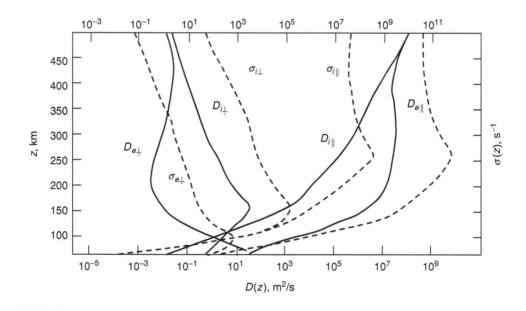

Figure 5.39 Height profiles of the coefficients of diffusion and conductivity obtained in Refs. [16–20] according to the model of middle-latitude ionosphere.

as well as effects of inhomogeneity of background ionospheric plasma, of ambient electric field, and of neutral wind were taken into account. The actual boundary conditions usually used in the middle-latitude ionospheric study [83–86] were also taken into account. It was shown that the smooth diffusion process, which is more important in the initial stage of the spreading, transforms at the later stages into an unstable phase in which the front of the initial Gaussian profile of the irregularity splits into a number of strata (i.e., clusters) stretched along the geomagnetic field $\mathbf{B}_0$. The time of the GDI creation in crossing $\mathbf{E}_0$ and $\mathbf{B}_0$ fields depends on the initial parameters of the irregularity and initial ambient conditions in the ionosphere, such as the ratio of the plasma disturbance δN to the background density N_0, the degree of anisotropy of initial inhomogeneity $\chi = L/d$, the altitude of creation of such an irregularity, and the value of the ambient electric field $\mathbf{E}_0$ in the ionosphere.

As in the foregoing analysis of the smooth diffusion process, we consider next the movement (i.e., simultaneous diffusion and drift spreading) of plasma inhomogeneities with initial dimensions larger than both characteristic electron or ion mean free paths, λ_α, along $\mathbf{B}_0$ and the Larmor's radii, $\rho_{H\alpha}$, across $\mathbf{B}_0$ ($\alpha = e, i$). Here, as previously, we take into account the conditions of potentiality of the internal self-consistent ambipolar field, $\mathbf{E}_c$ ($\mathbf{E}_c = -\nabla\phi$, where ϕ is the potential of the ambipolar field), of plasma quasi-neutrality, $N_e \approx N_i = N$, isothermality, $T_e = T_i = T$, and ambipolarity in the presence of the magnetic field, $\nabla\mathbf{j}_e = \nabla\mathbf{j}_i = \nabla\mathbf{j}$. Under such assumptions, the system of inhomogeneous hydrodynamic equations can be presented in the following manner [49,73,74]:

$$
\begin{aligned}
\frac{\partial N}{\partial t} &= \frac{1}{2}\left(\nabla\hat{\mathbf{D}}_e(z)\nabla + \nabla\hat{\mathbf{D}}_i(z)\nabla\right)N - \frac{1}{2e}\left(\nabla\hat{\boldsymbol{\sigma}}_i(z)\nabla - \nabla\hat{\boldsymbol{\sigma}}_e(z)\nabla\right)\phi \\
&\quad + \frac{1}{2}\left(\nabla\mathbf{V}_{e0}(z) + \nabla\mathbf{V}_{i0}(z)\right)N
\end{aligned}
$$

$$
\begin{aligned}
\frac{1}{e}&\left(\nabla\hat{\boldsymbol{\sigma}}_e(z)\nabla + \nabla\hat{\boldsymbol{\sigma}}_i(z)\nabla\right)\phi \\
&= \left(\nabla\hat{\mathbf{D}}_e(z)\nabla - \nabla\hat{\mathbf{D}}_i(z)\nabla\right)N + \frac{1}{2}\left(\nabla\mathbf{V}_{e0}(z) - \nabla\mathbf{V}_{i0}(z)\right)N
\end{aligned}
\tag{5.88}
$$

Here, as before, $\mathbf{V}_{e0}(z)$ and $\mathbf{V}_{i0}(z)$ are the electron and ion drift velocities; and $\hat{\mathbf{D}}_e(z)$, $\hat{\mathbf{D}}_i(z)$, $\hat{\boldsymbol{\sigma}}_e(z)$, and $\hat{\boldsymbol{\sigma}}_i(z)$, the tensors of diffusion and electric conductivity of electrons and ions, respectively, described in Equation 1.12a and b for plasma with arbitrary degree of ionization and magnetization. Unlike the nonlinear equations used in Refs. [6,89–91], in a system (Equation 5.88) the components of the tensors are functions of altitude z in the ionosphere. Their dependence on ionospheric parameters for the real middle-latitude ionosphere was investigated in Refs. [87,88], and it was found that in the real conditions of the ionosphere the transport coefficients are complicated functions of altitude z (see Figure 5.39) with respect to those in a weakly ionized unbounded homogeneous plasma. Now, if the ambient electric field, $\mathbf{E}_0$, and the neutral wind, $\mathbf{U}_m$, occur in the ionosphere, the drift velocities, $\mathbf{V}_{\alpha 0}(z)$, can be represented as follows ($\alpha = e, i$):

$$
\mathbf{V}_{\alpha 0}(z) = \mathbf{U}_m(z) + \frac{1}{eN}\text{sgn}(e_\alpha)\hat{\sigma}_\alpha(z)\left\{\left[\mathbf{E}_0 + \frac{1}{c}(\mathbf{U}_m \times \mathbf{B}_0)\right] + \text{sgn}(e_\alpha^{-1})m_\alpha\mathbf{g}\right\}
\tag{5.89}
$$

where c is the velocity of light, and e_α and m_α are the charge and mass of the electron and ion components of the plasma inhomogeneity. The last term in Equation 5.89 describes the effect of gravity, an important issue in the process of evolution of large-scale one- and two-dimensional irregularities [87,88], where $|\mathbf{g}| \approx 9.81 \text{ m/s}^2$ is the value of acceleration due to gravity in free space.

The initial density profile of the inhomogeneity, δN, as before, was set to the Gaussian form in the coordinate system z-axis directed along $\mathbf{B}_0$, and the potential $\Phi = (e/T)\phi$ to be constant, i.e.,

$$\delta N(x,y,z,t = 0) = AN_0 \exp\left\{ -\frac{(x^2 + y^2)}{d^2} - \frac{(z - z_0)^2}{L^2} \right\}$$

$$\Phi(\rho,z,t = 0) = \Phi_0$$

(5.90)

The plasma density at the initial moment is equal to

$$N(x,y,z,t = 0) = \delta N(x,y,z,0) + N_0(z)$$

(5.91)

The boundary conditions for the system (Equation 5.88) are as follows [73,74] across and transverse to $\mathbf{B}_0$

$$\delta N(x = \mp x_1,y,z,t) = 0, \quad \delta N(x,y = \mp y_1,z,t) = 0$$

$$\varphi(x = \mp x_1,y,z,t) = 0, \quad \varphi(x,y = \mp y_1,z,t) = 0$$

(5.92a)

along $\mathbf{B}_0$

$$\delta N(x,y,z = z_{1,2},t) = 0, \quad \varphi(x,y,z = z_{1,2},t) = 0$$

$$D_{\alpha\parallel} \cdot \frac{\partial \delta N}{\partial z}\bigg|_{z=z_2} = const$$

(5.92b)

The boundary conditions are chosen according to commonly accepted physical suggestions [83–86] as follows. At the lower ionospheric boundary along the vertical axis, $z = z_1$, as well as at each side in the horizontal plane, $x = \mp x_1$, and $y = \mp y_1$, the density perturbation δN vanishes and the perturbations of electric field potential, created by δN, equal zero at both the upper and lower boundaries and at each side. At the upper boundary, $z = z_2$ and, as follows from Equation 5.92b, either δN vanishes or the charged particle flux is taken to be constant, because the mean free path increases with altitude. The system of equations (Equation 5.88) is nonregular, nonlinear, and inhomogeneous, therefore, as shown in Refs. [49,73,74], it can only be solved numerically. In Ref. [74], the solution was sought in the ionospheric region of $z_1 = 100 \leq z \leq z_2 = 500$ km, $|x| \leq x_1$, $|y| \leq y_1$, $x_1 = y_1 \leq 10$ km. The integration time interval was $t_0 \leq t \leq 100$ s, where, as before, $t_0 = d^2/8D_{i\parallel}$. The steps of numerical computations in time and space were chosen to be variable to justify the stability conditions of the trial-run method and the variable direction method [73,74]. Thus, the sharper the initial shape of the inhomogeneity (i.e., its spatial gradient, $\partial \delta N(x,y,z,0)/dz$), the smaller the step in time, Δt, which was used in the numerical computations according to Ref. [74]. For a step in space, Δz, a new spatial variable $\xi(z)$ was introduced that enabled the investigation in more detail of the spatial zones nearest the initial inhomogeneity, and in less detail (with larger spatial steps) the zones farthest from the inhomogeneity (see details in Ref. [74]).

To obtain the effects of background plasma inhomogeneity, as done in the previous section for pure diffusion spreading, we also consider the unbounded quasi-homogeneous plasma, where the concentration of background plasma is constant, $N_0(z) = N_0(z_0) = const$, and all transport plasma coefficients are constant and correspond to the altitude of the initial maximum $h = z_0$ of plasma disturbance. Using the same approach as for unipolar diffusion of the irregularity with finite dimensions across and along $\mathbf{B}_0$, evaluated in Refs. [16–20], we now obtain the additional effects caused by the drift process, the velocities of which we suggested, for simplicity's sake, to be constant. Using the same equations of diffusion and drift for electron and ion components $(\alpha = e, i)$ of initial plasma perturbation, as Equation 5.65, we can present their solutions for diffusion and drift process as two-dimensional integrals:

$$I_\alpha = \int\int \exp\left\{ -\frac{\beta'^2}{a^2} + ik_\beta(\beta - \beta') - D_{\alpha\beta}k_\beta^2 t \right\} dk_\beta \, d\beta' \tag{5.93}$$

where $\beta = x, y, z$; $\beta' = x', y', z'$; $a = L, d$; $D_{\alpha\beta} = const$ are the coefficients of unipolar diffusion for electrons $(\alpha = e)$ and ions $(\alpha = i)$, which correspond to those depicted in Figure 5.39 for the altitude $h = z_0$ of initial plasma irregularity. Using the Poisson representation for each integral (Equation 5.93)

$$\int_{-\infty}^{\infty} \exp\left[\frac{x^2}{a^2} + i\varsigma x \right] dx = \sqrt{\pi}a \exp\left[-\frac{\varsigma^2 x^2}{4} \right] \tag{5.94}$$

we finally obtain the expression describing the spreading and movement of disturbed plasma inhomogeneity in the uniform, homogeneous background ionospheric plasma:

$$\delta N(\rho, z, t) = \delta N_0 L d^2 \left\{ \delta N_e'(\rho, z, t) + \delta N_i'(\rho, z, t) \right\}$$

$$\delta N_e'(\rho, z, t) = \frac{\exp\left[-\frac{z^2}{(L^2 + L_e^2)} - \frac{x'^2 + y'^2}{(d^2 + d_e^2)} \right]}{(L^2 + L_e^2)^{1/2}(d^2 + d_e^2)} \tag{5.95}$$

$$\delta N_i'(\rho, z, t) = \frac{\exp\left[-\frac{z^2}{(L^2 + L_i^2)} - \frac{x'^2 + y'^2}{(d^2 + d_i^2)} \right]}{(L^2 + L_i^2)^{1/2}(d^2 + d_i^2)}$$

The difference between Equations 5.67 and 5.95 is that in the latter equation new coordinates x' and y' are introduced to take into account effects of drift movements, i.e., $x' = x - \sigma_{e\wedge}E_{0x}^* t / eN_0$ and $y' = y - \sigma_{e\perp}E_{0y}^* t / eN_0$, where $\mathbf{E}_0^* = \mathbf{E}_0 + c^{-1}(\mathbf{U}_m \times \mathbf{b})$ is the ambient electric field in the original frame moving with velocity of the neutral wind $\mathbf{U}_m$, $\mathbf{b} = \mathbf{B}_0 / |\mathbf{B}_0|$; all other parameters were introduced earlier. It is evident that for $|\mathbf{E}_0| = 0$ and $|\mathbf{U}_m| = 0$, Equation 5.95 transforms to Equation 5.67, which describes the pure diffusion spreading of weakly ionized $(\delta N < N_0)$ plasma disturbances in the infinite homogeneous background plasma [18–20].

Before moving to a detailed analysis of the self-consistent 3-D numerical model discussed in Refs. [73,74], we will compare its results with those obtained in the 3-D models discussed in Refs. [70,71], using strong $(A = \delta N / N_0 \gg 1)$, large-scale Gaussian (Equation 5.90) initial perturbations with dimensions: (1) $L = 5$ km and $d = 0.3-5$ km (the corresponding degree of anisotropy $\chi = L/d \approx 1-17$) and $A = 4-100$ [70] injected at the altitudes $h = z_0 = 125, 185, 305$ km, and (2) with $L = 5$ km and $d = 1-5$ km ($\chi = L/d \approx 1-5$) and $A = 25-100$ [71], injected at the altitude $h = z_0 = 160$ km. All computations were made on the assumption of absence of a vertical component (along the z-axis) of the electric field, i.e., for $E_{0\perp} \approx 10$ mV/m and $E_{0\|} \equiv E_{0z} = 0$.

All results of computations in Ref. [70] were presented in the coordinate frame moving with the velocity of drift of initial nonlinear large-scale irregularity in the ambient electric field. Moreover, in Ref. [70], the authors compared their model with the approximate two-layered (2-D) numerical model proposed in Refs. [59,60]. Thus, in Figure 5.40, the isolines of equal concentration for

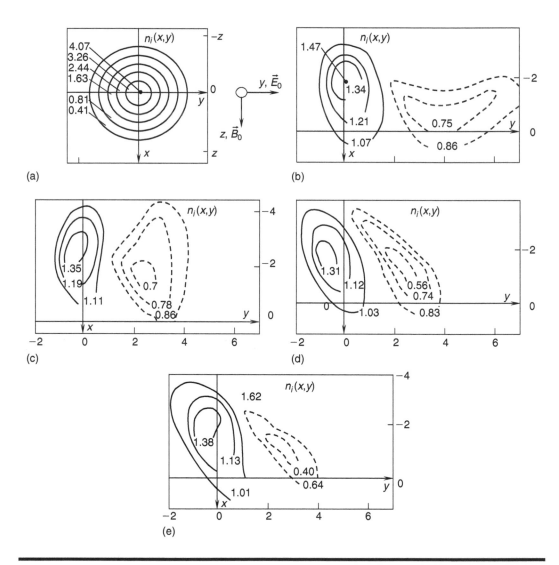

Figure 5.40 2-D profiles of large-scale plasma disturbance (solid contours) and the associated depletion areas of the background plasma (dashed contours) in the space and time domains for $t = 0$ (a); $t = 5$ s; $z_0 = 125$ km (b); $z_0 = 185$ km (c); $z_0 = 305$ km (d); for two-layered model according to Refs. [59,60] (e).

the strong disturbance of the background plasma in the initial moment $t=0$ for $L=d=5$ km (Figure 5.40a) and then for the moment $t=5$ s are presented at altitudes $z_0=125$, 185, 305 km (correspondingly, Figure 5.40b through d), and computed in conditions of a two-layered model [59,60] (Figure 5.40e). The positive profiles of plasma disturbance are presented by solid curves and the depletion regions by dashed curves. It is evident that the shape and density of the disturbance, as well as those for the depletion zones, are not the same at different altitudes. Thus, if at the altitudes of the E layer ($z_0=125$ km; see Figure 5.40b) the irregularity is stretched along the $\mathbf{V}_d \propto \mathbf{E}_0 \times \mathbf{B}_0$ direction in the F_1 layer ($z_0=185$ km; see Figure 5.40c) and in the F_2 layer ($z_0=305$ km; see Figure 5.40d), the rotation of the initial irregularity at the angle $\alpha \approx \tan^{-1}[(b_{e\Lambda} - b_{i\Lambda})/b_{i\perp}]$, estimated in Ref. [70], is observed. In all figures, the vertical x-axis is directed along the $\mathbf{E}_0 \times \mathbf{B}_0$-drift and the y-axis is along the $\mathbf{E}_0$-direction. The z-axis is directed along $\mathbf{B}_0$ orthogonal to the plane of figures. Here, as was introduced at the beginning of this chapter, $b_{e\Lambda}$ and $b_{i\Lambda}$ are the mobilities of electrons and ions in the Hall direction (in the crossing $\mathbf{E}_0 \times \mathbf{B}_0$ direction), and $b_{i\perp}$ is the mobility of ions in the Pedersen direction (in the direction along $\mathbf{E}_0$ and across $\mathbf{B}_0$). Moreover, the perturbed regions at the different altitudes have different shapes and forms and even the different displacements along the x-axis are not so noticeable. It can also be seen that with an increase of the height of injection of the initial irregularity, the decay of plasma concentration inside the regularity is also increased. The same disturbance computed according to the two-layered model (Figure 5.39e) is characterized by a strong elongation along the $\mathbf{E}_0 \times \mathbf{B}_0$ direction and rotation relative to the Hall direction. The comparison between 3-D and 2-D models shows that the two-layer model is a good approximation for the description of 3-D evolution of large-scale strong ionospheric irregularity.

In Ref. [71], the same geometry and coordinate frame was initially taken into account for $b_{i\perp} = 0.5 b_{i\Lambda}$, $b_{e\perp} = 1.25 b_{i\Lambda}$, and $b_{e\perp} = b_{i\parallel} = 0$. In Figure 5.41, the isolines of equal concentration are presented in dimensionless coordinates $\zeta = x/L$, $\eta = y/L$, and $\xi = z/L$ and for dimensionless time $\tau = b_{i\Lambda}E_0 t/L = 1.6$ in the plane $z=0$. It is seen that along $\mathbf{B}_0 \| \mathbf{z}$ the scale of the

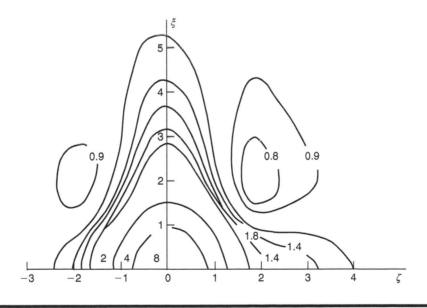

Figure 5.41 Equidensity contours of the large-scale plasma disturbance and the corresponding depletion areas during its drift in ambient electric and magnetic fields.

irregularity in the process of spreading is of the order of the irregularity of the electron, $L_{eff} \sim b_{e\|}^{1/2} (d_{e\perp} E_0 t)^{1/2}$, and across $\mathbf{B}_0$ (in Hall direction) the floating of plasma with the speed of transverse drift of ions is observed. The region where these effects were found is defined by the authors as the *ionic ellipsoid*. Thus, it is again evident that the deformation of the initial irregularity and the floating of the charged particles of plasma occur in the $\mathbf{E}_0 \times \mathbf{B}_0$ direction. The flow of particles in the ionic area is of the order of $N_0 V_0 L^2 (b_{i\perp} b_{iA})^{1/2}$, and the maximum of density in this region is of the order of $N_0 L_{eff}/L = N_0 (b_{e\|}/b_{i\perp})^{1/2}$. Additional analysis of the 3-D model carried out in Ref. [71] gives the floating of plasma particles, electrons and ions, not only in the Hall ($\mathbf{E}_0 \times \mathbf{B}_0$) direction but also in the Pedersen (along $\mathbf{E}_0$ and across $\mathbf{B}_0$) direction. The shapes of the electronic ellipsoid and the ionic ellipsoid and the corresponding depletion regions are quite adequate for the nonlinear spreading of strong large-scale plasma disturbance according to the scheme developed in Refs. [89–91] and presented in Figure 5.42. As is presented schematically in Figure 5.42, the electronic and ionic ellipsoids move in the directions 00_1 and 00_2, respectively, creating regions with dimensions $L_e \sim b_{e\|}^{1/2}(d_{e\perp} E_0 t)^{1/2}$ and $L_i \sim b_{i\|}^{1/3}(b_{i\perp}/b_{e\|})^{1/3}(E_0 t)^{1/3}$. The depletion zones are depicted by an ellipsoid with crossed lines. The authors of Refs. [89–91] showed that the lifetime of the strong large-scale irregularity depends on two physical mechanisms. The longitudinal scale of the injected ions increases according to law $\sim (8D_{i\|} t)^{1/2}$ because of ambipolar diffusion; then in the absence of background plasma, the density of injected ions decays according to the $\sim t^{-1/2}$ law. The existence of background plasma leads to an increase in the decay of the density of injected ions inside the irregularity and their floating in the Hall and Pedersen

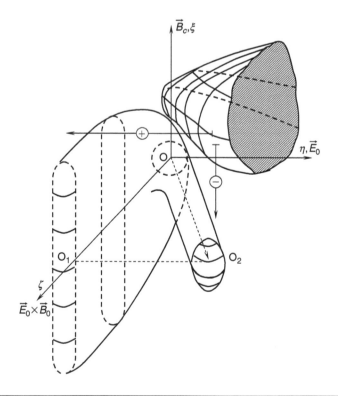

Figure 5.42 Three-dimensional schematic representation of the main extrema and the depletion zones during drift of initial plasma disturbance in ambient electric and magnetic fields.

directions with the creation of ion areas such as 00_1 in Figure 5.42. In the same manner, the electronic areas 00_2 are formed by the injection electrons inside the plasma irregularity, which was initially located in the point O, as follows from Figure 5.42. From this simple scheme the physical meaning of formation of ionic and electronic areas floating from the initial irregularity and of formation of the corresponding depletion zones is evident.

In Ref. [71], the conditions of the initial dimensions of the irregularity were obtained, and using them we can ignore the altitudinal inhomogeneity of the background ionospheric plasma and consider the constant averaged along the height parameters of plasma. Thus, the irregularity with initial transverse d and longitudinal L scales, according to the effect of short-circuit currents, has the real longitudinal scale of $L_{\parallel} \approx d(D_{e\parallel}/D_{i\perp})^{1/2}$. The influence of the longitudinal inhomogeneity of the background plasma depends on relations between AL and $L_{\parallel}$, where as before, $A = \delta N/N_0$. For $AL \ll L_{\parallel}$, we can ignore the inhomogeneity of background plasma as the approximate formulas (Equation 5.95) obtained according to Refs [73,74] are working well. This was also shown in Refs. [70,71], for 300 $m \leq d \leq 1$ km and $AL \leq L_{\parallel}$, where $L_{\parallel} \approx d(D_{e\parallel}/D_{i\perp})^{1/2}$ for the actual altitudes, the 3-D computations are close to those leading from the two-layer (2-D) model [59–61]. If now $AL > L_{\parallel}$, on the spreading of the strongly elongated irregularity ($\chi = L/d \geq 10^2$) the significant influence comes from the gradients of the parameters of the background plasma and the corresponding transport coefficients. This case corresponds to the model proposed in Refs. [73,74].

The main difference between the model proposed in the aforementioned references and the model developed in Refs. [34,73] is that in the former not only was the $\mathbf{E}_0 \times \mathbf{B}_0$-drift taken into account but also were the effects of the longitudinal electric field (even it is smaller than the transverse one, $E_{0\parallel} \ll E_{0\perp}$), of the neutral wind ($\mathbf{U}_m$), and the gravity force ($\sim \mathbf{g}$). The corresponding nonlinear inhomogeneous system for this model with the initial and boundary conditions is described in the foregoing by Equations 5.88 through 5.92. This model was simulated numerically accounting for the transport coefficients variable along height according to Figure 5.39, and the effects of ionization and recombination in the natural boundary conditions (Equation 5.92).

We now return to the comparison of results of the simulation of system (Equation 5.88) with the results of the 3-D models proposed in Refs. [34,73] for the same initial conditions as were taken in their simulations and presented earlier. The results of the numerical analysis of Equation 5.88 are presented in Figure 5.43a for a spherical perturbation in the same conditions as in computations in Ref. [71] and for the following parameters: $A = 10^2$, $h = z_0 = 160$ km, $E_x = E_y = 10$ V/m, $E_z \equiv E_{\parallel} = 0$, $L = d = 5$ km and $t = 5$ s. It is evident the results of computations according to Equation 5.88 for $t \leq 5$ s are in good agreement with the calculations carried out in Ref. [71], where the ionosphere was assumed to be quasi-homogeneous and unbounded; i.e., the parameters were assumed to be constant with altitude. At the same time, the results obtained in Refs. [73,74] differ somewhat from those obtained in Ref. [70] in the case when $L = 5$ km and $d = 0.3$ km ($\chi = L/d \approx 17$) and for $h = z_0 = 300$ km. As indicated in the contours of equal density plotted in Figure 5.43b during the initial phase ($t \leq 5$ s), the spreading is similar to that discussed in Refs. [70,71]. Later, however, the original irregularity splits into two elongated plasmoids for $t \geq 20$ s; they rotate around an axis parallel to $\mathbf{E}_0 \times \mathbf{B}_0$. The second plasmoid moves in the direction $\mathbf{V}_a = (\mathbf{V}_e + \mathbf{V}_i)/2$, which is between $\mathbf{B}_0$ and $\mathbf{E}_0 \times \mathbf{B}_0$. This result is similar to that discussed in Ref. [6] for a small irregularity in which $\mathbf{U}_m \perp \mathbf{B}_0$, and indicates that the initial irregularity splitting is purely dispersive in nature.

Comparison between the proposed self-consistent model in Refs. [73,74] and the tested models created in Refs. [59,60,70,71] and the good agreement between them enabled the authors of Refs. [73,74] to use the developed model based on the algorithm described by Equations 5.88 through 5.92 for investigating the drift-spreading process of the irregularities with various degrees of anisotropy and arbitrary perturbation of plasma density in the irregularities. The examples of simulation of the 3-D drift model are presented for the different cases in Figures 5.44 and 5.45. In Figure 5.44a and b the spatial redistribution of plasma disturbance normalized density, $\delta N(\mathbf{r},t)/N_0$, with the degree of anisotropy of $\chi = L/d = 10$ km/100 m $= 10^2$ for the case of $A = \delta N(0,0)/N_0 = 10^2$ is presented for two moments of time $t_1 = 10$ s (Figure 5.44a) and $t_1 = 30$ s (Figure 5.44b), taking into account the following values of the components of the ambient electric field, $E_{0\parallel} = 10^{-4}$ mV/m and $E_{0\perp} = 10$ mV/m (which is close to real ambient conditions of the middle-latitude ionosphere [83]). It can be seen that for sufficiently large values of the ambient electric field after

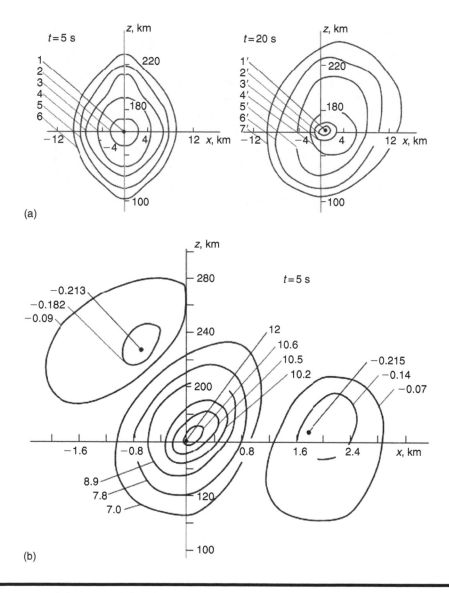

(a)

(b)

Figure 5.43 Spatial-temporal distribution of a large-scale plasma disturbance in ambient electric and magnetic fields.

(continued)

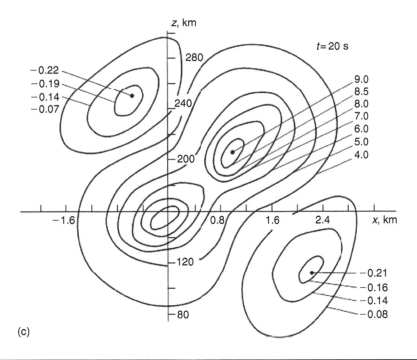

(c)

Figure 5.43 (continued)

the initial stage ($t \geq 10$ s, see Figure 5.44a), a strongly stratified structure is formed both along and across $\mathbf{B}_0$, which grows catastrophically with time and forms a significantly wide negative area of the disturbed background plasma, i.e., depletion region (see Figure 5.44b). The geometry of this depletion region changes during the process of formation of a quasi-periodical structure inside it, the *strata*, and then its dimension also changes in the process of evolution followed by the reconstruction of the irregularity itself. As can be seen from examples presented in Figure 5.44a and b, the process of plasma structure evolution is complicated under real ionospheric conditions and depends on many actual ambient plasma parameters, each of which specifically affects the drift spreading. That is why we will try below to examine each effect independently.

Effect of Longitudinal Component of the Ambient Electric Field. Many investigators exclude this component of the ambient electric field, and this is correspond to the experimental data from direct monitoring of the ionosphere presented by [100,101], according to which $E_{0\parallel} = (10^{-3} - 10^{-5})E_{0\perp}$. We can evaluate $E_{0\parallel} \equiv E_{0z}$ and develop the criterion for its importance in the drift process occurring in the real ionospheric plasma. Taking into account that the field is potential, $\mathbf{E} = -\nabla\Phi$, and that Boltzmann's law is valid, we obtain

$$|\mathbf{E}_{0z}| = \nabla_z \Phi = \nabla_z[(T_e/e)\ln(N/N_0)] \tag{5.96}$$

Its influence on drift spreading is essential if the scale of plasma disturbances generated by E_{0z} is of the same order as that of the diffusion scale of electrons in plasma, i.e., $2D_{e\parallel}E_{0z}t = (4D_{e\parallel}\,t)^{1/2}$. Thus, the critical time of this effect, t_{cr}, can be evaluated as

$$t_{cr} = \left\{ D_{e\parallel}\nabla_z^2[(T_e/e)\ln(N/N_0)]^2 \right\}^{-1} \tag{5.97}$$

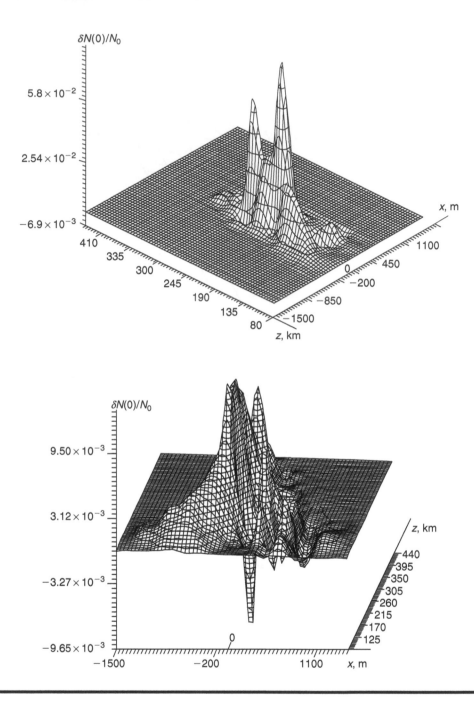

Figure 5.44 Three-dimensional profile of strongly anisotropic moderate plasma irregularity during its drift spreading in ambient electric and magnetic fields in presence of the wind of neutrals in the middle-latitude ionosphere for $t = 10$ s (top panel) and $t = 30$ s (bottom panel).

Now, because $\nabla_z^2[(T_e/e)\ln(N/N_0)]^2 \propto \Delta\Phi/\Delta z \propto \Delta[(T_e/e)\ln(N/N_0)]/\Delta z \propto L_E^{-1}$, we finally obtain

$$t_{cr} \approx L_E^2/D_{e\parallel} \tag{5.98}$$

where L_E is the scale of the vertical component of the ambient field perturbation. For the ionosphere at altitudes $h = 150$–300 km and for $L_E = 10^5$ m, we get $t_{cr} \approx 100$ s.

Thus, in the process of drift spreading of arbitrary irregularities of the ionospheric plasma for which the lifetime exceeds $t_{cr} \approx 100$ s, it is necessary to include the vertical component E_{0z} of the ambient electric field, despite the fact that it is smaller than the horizontal component (parallel to Earth's surface), $E_{0\perp}$, which is what we did in our simulations presented in Figures 5.44 and 5.45,

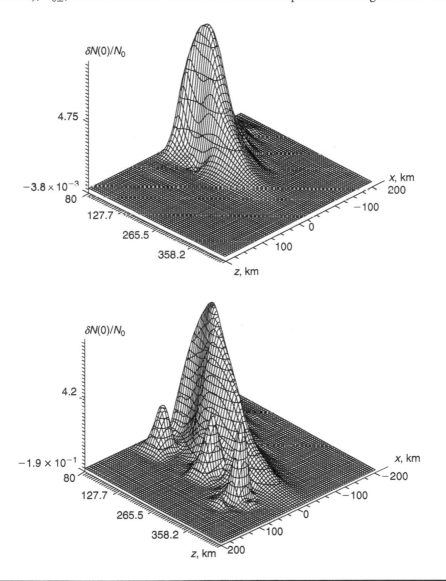

Figure 5.45 Same as Figure 5.44 but for initially isotropic strong plasma irregularity for $t = 30$ s: $E_{0\perp} = 1$ mV/m (top panel) and $E_{0\perp} = 10$ mV/m (bottom panel).

Table 5.1 Decoupling Timescale (t_1) and Layer-Formation Timescale (t_2) as a Function of Transverse Size ($L = 10$ km, $U_m = 0$, $E_0 = 1$ mV/m, and $A = 100$)

d, m	1	5	10	20	50	100	300	10^3	$5 \cdot 10^3$	10^4
t_1, s	0.25	0.57	0.79	1.63	2.25	5.3	8.6	27.5	—	—
t_2, s	0.43	1.09	1.58	2.66	3.88	9.54	25.24	—	—	—

taking in computations the mean value from the data obtained experimentally in [100,101] and mentioned previously.

Effects of Degree of Anisotropy. To understand the role of the degree of anisotropy of an arbitrary irregularity, we now examine a strong ($A = 10^2$) isotropic ($L = d = 5$ km) irregularity (see Figure 5.45a), which is created in the background of inhomogeneous ionospheric plasma in the presence of insufficient ambient electric field ($E_{0\perp} = 1$ mV/m, $E_{0z} = 10^{-4}$ mV/m). It is evident from Figure 5.45a that the velocity of drift spreading is slower than in the previous case of anisotropic initial irregularity (Figure 5.44) and the striation of the initial irregularity into a set of plasmoids (strata) is not observed. The character of drift spreading is isotropic and corresponds to a diffusive smooth expansion. Only at the edges of the Gaussian profile (Equation 5.90) of plasma density inside the irregularity do the local perturbations take place during the process of plasma drift in a weak electric field. This result is in good agreement with results obtained in Ref. [72] for weakly anisotropic irregularities. The same picture was also obtained from the numerical model created in Ref. [71], where the initial stage of evolution of an isotropic irregularity was investigated. The difference between the two results is that the irregularity stretches upward along magnetic field lines (along the z-axis; see Figure 5.45a) owing to a small longitudinal component of ambient electric field. In Figure 5.45a we show only an example of simulation. More studies of different kinds of initial irregularity are presented in Table 5.1, where the dependence of time $t_d \equiv t_1$ and $t_s \equiv t_2$ on the transversal dimension, d, of the initial irregularity is introduced (see Table 5.1). The other parameters of the problem are kept constant for numerical simulations: $L = 10$ km, $E_{0\perp} = 1$ mV/m, $E_{0z} = 10^{-4}$ mV/m, $A = 10^2$. Here, t_d is the time of plasma splitting on several plasmoids owing to dispersive drift spreading (according to Refs. [6,89,90]), and t_s is the time that characterizes the plasma structure creation (i.e., splitting on the grid of strata) due to GDI [74], described in Chapter 2. As can be seen from Table 5.1, with increasing degree of anisotropy of the initial plasma irregularity (with decreasing parameter d), the time of dispersive drift striation t_d and the time of strata structure formation, t_s ($t_s > t_d$ always) decrease. For isotropy irregularities ($L = d$) both effects in the process of plasma irregularity spreading have not been observed.

Effects of the Ambient Electric Field. In Figure 5.45b, the character of spreading is the same as in the case of the strong isotropic plasma irregularity. It is presented for a moment $t = 30$ s after creation of the irregularity for the same conditions shown in Figure 5.54a, but for $E_{0\perp} = 10$ mV/m. The difference between results of simulations shown in Figure 5.54a computed for $t = 30$ s for a weaker electric field ($E_{0\perp} = 1$ mV/m) and those obtained for $E_{0\perp} = 10$ mV/m is evident when comparing the illustrations presented in Figure 5.54a and b. In the latter figure, after $t = 30$ s striation of initial irregularity on the several plasmoids is observed. The comparison between these two figures allows us to conclude that with increasing electric field amplitude (from 1 to 10 mV/m), the striation process grows both along and across $\mathbf{B}_0$. More detailed analysis of the influence of the ambient electric field on the character of plasma irregularity

Table 5.2 Decoupling Timescale (t_1) and Layer-Formation Timescale (t_2) as a Function of Transverse Electric Field ($L = 10$ km, $U_m = 0$, $d = 100$ m, and $A = 100$)

$E_{0\perp}$, mV/m	0.1	1	10	100
t_1, s	—	5.3	3.62	1.24
t_2, s	—	9.54	4.16	1.81

spreading is presented in Table 5.2 as a dependence of the characteristic time t_d and t_s ($t_s > t_d$), introduced earlier, versus the transverse component of ambient electric field, $E_{0\perp}$, for $L = 100$ km, $d = 100$ m, $E_{0z} = 10^{-4}$ mV/m, $A = 10^2$. As can be seen, with increasing amplitude of the ambient electric field, the characteristic times of irregularity splitting due to a dispersive mechanism and further striation due to development of gradient-drift instability (GDI) decrease. On the other hand, for a significantly small ambient electric field these effects have not yet been observed.

Effects of the Inhomogeneity of the Background Plasma. To understand the influence of background ionospheric inhomogeneity on drift spreading, we generate the irregularity in the infinite quasi-homogeneous plasma where the concentration of plasma particles $N_0 \equiv N_0$ ($h - z_0$) is constant and where all transport plasma coefficients presented in Figure 5.39 correspond to the altitude of the initial maximum $h = z_0$ of plasma disturbance. The solution of such a problem is described by Equation 5.85 by taking into account the same initial and boundary conditions (Equations 5.81 and 5.82) that were introduced for the inhomogeneous background ionospheric plasma. It is easy to see that for $E_0 = 0$ and $U_m = 0$ the formulas (Equation 5.95) transform to those describing pure diffusional spreading of weak ($\delta N < N_0$) plasma disturbances in the infinite background plasma that were presented in the previous section in Equations 5.19 and 5.20. Results of numerical calculations of Equation 5.95 are shown in Figure 5.46 as lines of equal concentration, $A = \delta N/N_0 = 10^2$, for the following parameters of numerical simulation, $h = 200$ km, $L = 10$ km, $d = 10$ m ($\chi = L/d = 10^3$), $E_{0\perp}^* = 2$ mV/m (or $E_{0\perp} = 1$ mV/m and $U_m = 30$ m/s according to Refs. [83,100,101]) for two moments of current time, $t_1 = 1$ s (Figure 5.46a) and $t_1 = 5$ s (Figure 5.46b). Here, for $t_1 = 1$ s digits 1–8 correspond to the lines of equal plasma density depicted in Figure 5.46a at the plane (x,y) perpendicular to $\mathbf{B}_0$, $\delta N(x,y,t)/N_0 = 48.3, 17.9, 6.8, 26.7, 20.4, 15, 6, 6.98$, respectively, and for $t_1 = 5$ s digits 1–9 corresponding to the same lines, but for $\delta N(x,y,t)/N_0 = 23.2, 8.91, 3.43 \ 1.95, 23.4, 8.98, 2.51, 1.24, 0.92$, respectively. It can be seen that during drift spreading the initial plasma irregularity moves in the direction between $\mathbf{E}_0$ and $\mathbf{E}_0 \times \mathbf{B}_0$, and splits into two plasmoids owing to pure dispersive drift mechanism only [6,89–91]. Overall, in the regular and homogeneous background plasma, the picture of spreading is symmetrical relative to both Pedersen (along $\mathbf{E}_0$) and Hall (along $\mathbf{E}_0 \times \mathbf{B}_0$) directions and the additional striation on a cluster of strata, observed in Figure 5.44 for the case of the inhomogeneous plasma, is not found in this case. Moreover, we generated the same numerical simulations that are presented in Figure 5.46 for inhomogeneous background plasma with the same inner and outer parameters as their actual complicated altitudinal dependence, also taking into account effects of recombination and its altitudinal dependence (at the lower ionosphere) according to Ref. [73]. Figure 5.47, where the lines of equal density is shown in the (z,x) plane for $t = 5, 10$, and 20 s, clearly indicates that the initial plasma disturbance breaks up into a cluster of stratified plasmoids stretched along the magnetic field $\mathbf{B}_0$, even as little as a few seconds away from the initial time ($t \leq 5$ s). The same

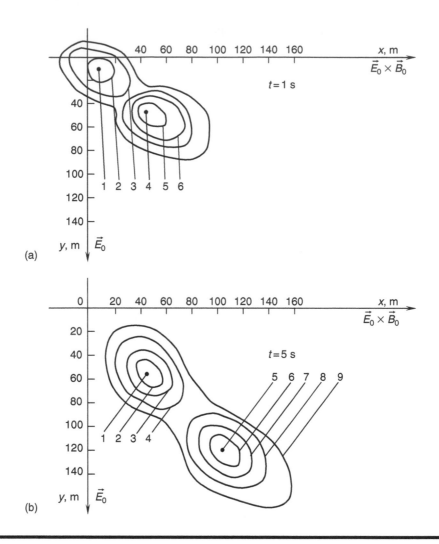

Figure 5.46 Two-dimensional drift model of spreading of weakly anisotropic strong plasma inhomogeneity in the crossing ambient electric and magnetic fields for $t_1 = 1$ s (a) and $t_1 = 5$ s (b).

results were obtained in Refs. [58–63], where the sinusoidal profiles were introduced artificially onto the density profile $\delta N(0,0)$ of initial irregularity in the drift equations. This initial density profile eventually led to further stratification of plasma inside the irregularity. As may also be seen from Figure 5.47, the spreading pattern for perturbation extended highly in the direction along $\mathbf{B}_0$ is quite complex. As mentioned earlier, this pattern depends on a wide variety of parameters describing both the initial irregularity in the background plasma and the background plasma itself. Therefore, we will continue to analyze these effects separately.

Effects of Plasma Disturbance. We now examine effects of plasma disturbance in the initial irregularity that is characterized by the parameter $A = \delta N/N_0$. In Table 5.3 the dependence of the characteristic timescales t_d and t_s $(t_s > t_d)$, introduced earlier with respect to the parameter $A = 10^{-1}–10^3$, is presented for other constant parameters: $L = 10$ km, $d = 100$ m $E_{0\perp} = 1$ mV/m, $E_{0\parallel} = 10^{-4}$ mV/m. It is evident that the larger the parameter A of plasma disturbances, the greater the times of drift spreading characterized by the dispersive mechanism of splitting into two or three plasmoids, t_d, and of plasma striation into cluster of strata, t_s (i.e., the time of GDI creation, as the most probable striation mechanism [64,69]). For $A \geq 10^2$ the processes of plasma splitting and striation are not observed. This case corresponds to that depicted in Figure 5.45a. The self-consistent theoretical analysis carried out in Refs. [73,74] allowed for obtaining the approximate dependence of time of plasma GDI generation and plasma striation onto strat, t_s, as a nonlinear function of the parameter A, the transverse scale d of initial irregularity, and the transverse component of the ambient electric field, $E_{0\perp}$, i.e.,

$$t_s \approx (Ad/E_{0\perp})^k \qquad (5.99)$$

where the exponent k, obtained by fitting the numerical simulation results, ranges over $0.3 < k < 0.6$ [73]. The threshold value of $E_{0\perp}$, needed for striation, can be obtained from the conditions of the $\mathbf{E}_0 \times \mathbf{B}_0$-drift (excitation of GDI) predominance over the diffusive expansion (damping of GDI). This means that the maximum GDI rate, $\gamma_{max} \cong cE_{0\perp}/Bd$ (see Chapter 2 where $\ell_\perp$ was now changed to d) has to exceed the product of the lateral diffusion coefficient $D_\perp$ and the square of the transverse (to $\mathbf{B}_0$) wave number, i.e., $\gamma_{max} > k_\perp^2 D_\perp$. Substituting $k_\perp = 2\pi/d$ and obtaining the Bohm diffusion coefficient, $D_\perp \cong D_{Bohm} \approx cT_e/16\pi eB_0$, following Ref. [91], we get

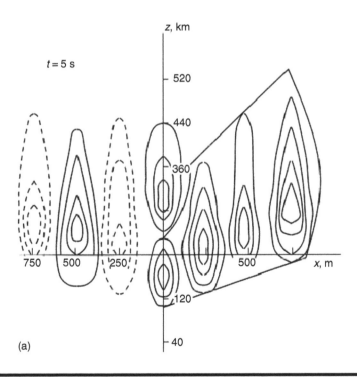

(a)

Figure 5.47 **Equidensity contours spatial and temporal evolution for the large-anisotropic weak plasma disturbance (solid contours) and the corresponding depletion areas (dashed contours) during its drift in ambient electric and magnetic fields in the presence of the wind of neutrals and gravity forces for $t_1 = 5$ s (a)**

(*continued*)

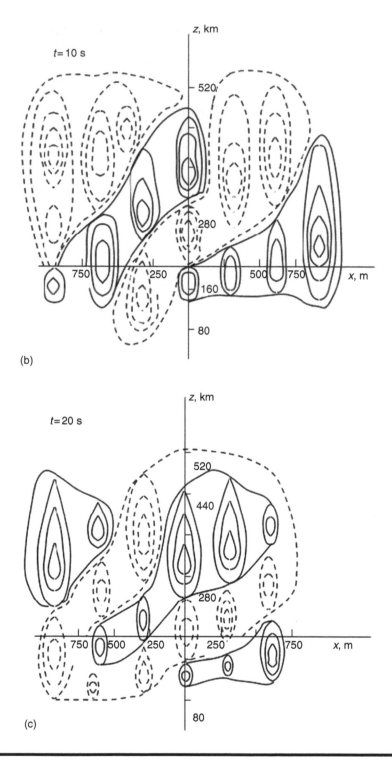

Figure 5.47 (continued) $t_1 = 10$ s (b), and $t_1 = 30$ s (c).

Table 5.3 Decoupling Timescale (t_1) and Layer-Formation Timescale (t_2) as a Function of A ($L = 10$ km, $U_m = 0$, $E_{0\perp} = 1$ mV/m)

$A = \delta N/N_0$	0.1	1	10	100	1000
t_1, s	—	24.9	17.5	5.3	2.85
t_2, s	—	31.5	24.8	9.54	3.37

$$E_{0\perp} \geq E_{0\perp}^{th} \approx T_e/ed \tag{5.100}$$

So, at $d = 100$–300 m and $T_e \approx (0.1$–0.3) eV, for the threshold of $E_{0\perp}$ we get $E_{0\perp}^{th} \approx 1$ mV/m, which can cause the exciting of GDI and the striation of plasmoids into clusters of strata (i.e. the striation mechanisms). We use these estimations in our simulations shown in Figure 5.45.

Effect of Stretching of the Irregularity. In Chapter 2 we briefly presented analysis of the *stretching criterion* for arbitrary plasma irregularity striation due to drift instability based on results obtained in Refs. [64–69] and then modified for the plasma with an arbitrary degree of ionization in Refs. [49,74]. Using estimations made there, we obtain the result that for altitudes of $h = 150$–250 km the striation mechanism becomes effective for striation mechanism for the irregularities with a degree of anisotropy of

$$\chi = L/d \geq \frac{q_H}{(1+p)} \tag{5.101}$$

Taking into account deviations of the parameters of the ionosphere at these altitudes according to Ref. [83] and results presented by Tables 1.1 through 1.3, we get $L/d \geq 10^2$. So, as pointed out earlier, parameters taken for the simulations presented earlier according to Refs. [73,74] are in good agreement with results obtained from constraints (Equations 5.100 and 5.101) according to the phenomenological approach proposed in Refs. [49,74] and described in Chapter 2.

Effect of Ambient Electric Field Perturbations. We will now estimate the perturbation, δE, of the ambient electric field, E_0, generated inside the disturbed plasma region owing to the creation (by injection or by heating) of the plasma irregularity inside it. Using the relations between electric field perturbations and the electric potential, Φ, inside the disturbed region, $\delta E \sim \nabla_\perp \Phi$, and the relationship between the potential of disturbed electric field and plasma density perturbations [91],

$$\begin{aligned}
\Phi^- &\equiv e\varphi^-/T = -\ln(\delta N_e/N_0 + 1) \\
\Phi^+ &\equiv e\varphi^+/T = -\ln(\delta N_i/N_0 + 1)
\end{aligned} \tag{5.102}$$

where $\delta N_e/N_0$ and $\delta N_i/N_0$ are determined by Equation 5.95, we can finally obtain the expressions for the components of the perturbed electric field in the 3-D case, taking into account the finite initial sizes of the plasma disturbances and the real ionospheric parameters at the ionospheric altitude $h = z_0$:

$$\delta E = \left(\delta E_x^2 + \delta E_y^2\right)^{1/2} \tag{5.103}$$

$$\delta E_x = 4E_{0x}\frac{(D_{e\Lambda} - D_{i\Lambda})}{D_{i\Lambda}}\frac{\sqrt{D_{e\parallel}/D_{i\perp}}}{A\chi}\left\{\left[1 + \frac{4\sqrt{D_{e\parallel}/D_{i\perp}}}{A\chi}\right]^2 + \frac{(D_{e\Lambda} - D_{i\Lambda})^2}{D_{i\Lambda}^2}\right\}^{-1} \quad (5.104a)$$

$$\delta E_y = E_{0y}\left\{\frac{(D_{e\Lambda} - D_{i\Lambda})}{D_{i\Lambda}}\left[1 + \frac{4\sqrt{D_{e\parallel}/D_{i\perp}}}{A\chi}\right]^{-2} + 1\right\} \times$$

$$\times \left\{\frac{(D_{e\Lambda} - D_{i\Lambda})^2}{D_{i\Lambda}^2}\left[1 + \frac{4\sqrt{D_{e\parallel}/D_{i\perp}}}{A\chi}\right]^{-2} + 1 + \frac{4\sqrt{D_{e\parallel}/D_{i\perp}}}{A\chi}\right\}^{-1} \quad (5.104b)$$

Here, as before, the signs of $\parallel$, $\perp$, and Λ determine the diffusion coefficients in the $\mathbf{B}_0$ direction, the $\mathbf{E}_0$ (Pedersen) direction, and the $\mathbf{E}_0 \times \mathbf{B}_0$ (Hall) direction, respectively; other parameters were introduced earlier. From Equations 5.103 and 5.104 it follows that for a plasma irregularity weakly stretched along $\mathbf{B}_0$ ($L_{\parallel} \gg A L$), the characteristic scale of the ionospheric plasma altitudinal variations $L_{\parallel}$ is more than the longitudinal scale of the irregularity, the components δE_x and δE_y of field perturbation δE being sufficiently small. In this case of a quasi-homogeneous ionosphere, the magnetic field lines are equipotential. Therefore, in this case, we can assume that the ionosphere is unbounded and homogeneous and can use the models constructed in Refs. [70,71]. In the inverse case, where $L_{\parallel} \ll A L$, i.e., the irregularities are strongly stretched along $\mathbf{B}_0$, the perturbation of the ambient electric field is essential and the aforementioned models do not work. From estimations made in Ref. [74] for the 3-D case described by Equations 5.103 and 5.104 for the case of real middle-latitude ionosphere at altitudes $h = 150$–300 km described in Chapter 1 when $D_{e\parallel}/D_{i\perp} \sim 10^4 - 10^6$, it is evident that the development of gradient-drift instability (GDI) and the essential striation of the initial plasma irregularity occur during the process of spreading of very elongated ($\chi = L/d \geq 10^2$) and strongly perturbed ($A = \delta N(0,0)/N_0 \geq 10^2$) or slightly elongated ($\chi < 10^2$) and weakly perturbed ($A \leq 10$) irregularities.

In the inverse cases we obtain the effects of drift stabilization at the earlier stage of plasma irregularity evolution obtained theoretically for ion clouds in Refs. [67–69] and briefly described in Chapter 2. We return again to a description of the effect of plasma drift stabilization following Ref. [74]. As stated in Refs. [67–69], the criterion of plasma drift stabilization can be presented for isothermic, $T_e \approx T_i = T$, plasma as

$$\frac{2T}{eE_{0\perp}} > \frac{L\chi}{q_H Q_H}\left[1 - V_A\frac{A}{A+2}\right] \quad (5.105)$$

where $V_A \propto c^{-1}[\mathbf{E}_{0\perp} \times \mathbf{B}_0]$ is the drift velocity in the azimuthal direction. We can compare the results of our numerical computations presented in Tables 5.1 through 5.3 with the criterion of drift stabilization (Equation 5.105). Thus, with increasing $\mathbf{E}_{0\perp}$, i.e., the velocity V_A, the threshold of drift stabilization is also increased. Again, we obtain the result that in a strong ambient electric field only development of drift instability can be found, the stabilization effect being absent. This result is clearly seen from results presented in Figures 5.44 and 5.47 and Table 5.2. At the same time, with the increasing parameter of perturbation, A ($A > 10$), the factor $A/(A+2) \to 1$ and as seen from Equation 5.102 for $E_{0\perp} = const$, the threshold of drift stabilization becomes smaller. Therefore, because the stabilizing factor is more strongly manifested for the cases presented in Figure 5.45, we obtained smooth spreading there. We observed the same decrease in the threshold of drift stabilization as that of the degree of anisotropy $\chi = L/d$ (see Equation 5.105). Thus, for weakly anisotropic irregularities, the effect of drift stabilization is strongly manifested, and the foregoing estimations confirm the results of the numerical analysis presented earlier in Figures 5.44 through 5.47 and Tables 5.1 through 5.3. Now, taking into consideration the parameters of the background plasma and the ion cloud itself following Refs. [64–69], we obtain the same fast (of a few seconds) striation and splitting of the initial ion cloud in the strata structure (from Table 5.1 it follows that $t_s = 5$–8 s).

Hence, as follows from results presented in Tables 5.1 through 5.3 and in Figures 5.44 through 5.47, we can, using the self-consistent model described earlier according to Refs. [49,73,74], predict both the process of plasma drift stabilization in the cases of strong ($A > 10$), weakly anisotropic ($\chi \leq 10$) plasma irregularity evolution in a weak ambient electric field ($E_{0\perp} \leq (3$–$4)$ mV/m), and the process of plasma drift instability during the spreading of weak ($A \leq 10$), strongly stretched ($\chi > 10$), plasma irregularity in a strong ambient electric field ($E_{0\perp} > 5$ mV/m). We will use this approach in Chapter 6 for the analysis of optical observations of evolution of ion clouds in Earth's ionosphere.

5.3 Thermodiffusion of Plasma Irregularities in the Ionosphere

The process of diffusion and drift relaxations of plasma structures in the background ionosphere, discussed earlier, mostly characterizes the situation of evolution of isothermal plasma where the temperatures of the charged particles, electrons and ions, are the same. However, most physical processes accompanying the formation and further evolution of plasma structures in the ionosphere are not isothermal, and, first of all, the processes of artificial heating of plasma by powerful short-wave radiation from ground-based facilities or heating of background ionospheric plasma during a magnetic storm and other natural phenomena occur due to solar activity. Investigation of such heating processes is important for the study of both geophysical and radiophysical phenomena, notably, the interaction of powerful radio wave radiation with a cold, dense, partially or fully ionized and magnetized ionospheric plasma (depending on the height of heating and power of the natural or artificial source). Here, the important questions that need to be answered are how electromagnetic waves transfer their energy into the plasma wave oscillation energy, how they create plasma temperature and concentration fluctuations, and finally, how the quasi-periodical grating plasma structures (also defined in the literature as plasma cavities, turbulences, irregularities or inhomogeneities [50–57,75–78]) are generated.

Knowledge of all kinds of plasma structures artificially induced by a heating process or rocket injection at the corresponding altitudes of the ionosphere and their spatial and temporal evolution is necessary to adequately model their effects on propagation of probing radio waves through disturbed regions of the ionosphere and for predicting of propagation characteristics of ionospheric communication links [21,22].

In all these cases, thermodiffusion, a faster process compared to pure isothermal plasma diffusion and drift [75–78,102,103], plays an important role in the formation and further spatial and temporal evolution of striated plasma density structures. Theoretical and experimental studies on the influence of powerful radio waves on ionospheric plasma heating [50–57] indicated the possible development of cavities (in the 1-D case) and quasi-periodical heating-induced ionospheric plasma structures (in the 2-D case) in the regions of heating. However, the possible application of the proposed theories was limited because all these 1-D and 2-D analytical and numerical models of ionospheric plasma heating cannot explain the real spatial redistribution of plasma density during the heating process observed experimentally, namely, plasma striation during its evolution into the grating quasi-periodical structure stretched along the geomagnetic field lines and further nonlinear relaxation of such cluster of strata. In Refs. [75–78], the 1-D and 3-D numerical and analytical models were developed for descriptions of the processes of formation and further evolution of the heating induced plasma irregularities stretched along the geomagnetic field lines as well as the quasi-periodical plasma density grating structure in the inhomogeneous ionospheric background, taking into account the real ionospheric ambient conditions close to those occurring in the middle-ionospheric heating experiment. We describe the results obtained in detail in Chapter 6, where we will discuss such kinds of plasma turbulences. In the following text, we present the formulation of the problem and give the corresponding equations and parameters describing the evolution of such plasma irregularities in the ionosphere. We also present results of modeling of the specific cases of plasma heating (local or resonance) with altitudinal separation of heating sources, following results obtained in Refs. [50–57]. Then, we present the reader the physical picture of formation of the plasma temperature and plasma density profile in the real model of middle-latitude ionosphere at the altitudes of higher–E and lower–F regions of the ionosphere.

5.3.1 General Problem Statement: Functions and Parameters

We now suggest that in a given region of the ionosphere, and at a fixed height $h = h_0$, the ionosphere was heated by a source of heat $Q(z)$. We do not go into detail about the nature of the heater (local, resonance, adaptive, and so on) and will present for future discussions the relations between the coordinate z, which we usually direct along the geomagnetic field $\mathbf{B}_0$, and the height of the initial heating-induced plasma structure, $h = h_0$, $h = z \cos \theta$, $h_0 = z_0 \cos \theta$, where θ is the angle between the vertical axis $\mathbf{h}$ and the geomagnetic field $\mathbf{B}_0$ lined along the z-axis. The point with coordinates ($x = x_0$, $y = y_0$, $z = z_0$) corresponds to the point of initial heating. Here, we consider a more general case occurring in the real ionosphere compared with that described earlier for pure diffusion and drift spreading, directing now the z-axis not along the field $\mathbf{B}_0$, i.e., vertically, but at an angle θ to the real height of the ionosphere h.

According to the classical theory developed in Ref. [102], because of the heating process, the temperature distribution of charged particles becomes nonuniform and inhomogeneous. Because of this, a pressure gradient arises, leading to an inhomogeneous distribution of electron and ion components of plasma density, i.e., disturbances of plasma temperature $T_e = T_{e0} + \delta T_e$ and concentration $N_e = N_{e0} + \delta N_e$ [102]. To describe the process of evolution of perturbed plasma temperature and the corresponding plasma density disturbances for the case of quasi-neutral plasma, $N_e \approx N_i = N$, according to Equation 1.9a, and of ambipolar thermodiffusion, diffusion and drift, $\nabla \mathbf{j}_e = \nabla \mathbf{j}_e = \nabla \mathbf{j}$, according to Equation 1.9b, we based ourselves on the following system of three-dimensional (3-D) equations of plasma transport, introduced earlier in Chapter 1:

a. Equations 1.5 and 1.6, which, together with conditions in Equation 1.9, describe the quasi-neutral and ambipolar diffusion and drift of plasma inhomogeneities in the ambient electric and magnetic fields.
b. Equation 1.10a and b, which describes the energy balance in the ionospheric region of heating. As in Ref. [102], we do not take into account disturbance of the ion temperature T_i; i.e., we neglect Equation 1.10b describing its relaxation. This is possible if the inequality $\delta_{ei} \nu_{ei} \ll \nu_{em}$, which is correct for ionospheric altitudes $h \leq 500$ km, is satisfied (see also Chapter 1 where all parameters of ionospheric plasma components were defined).
c. In Equation 1.10a we introduce two kinds of external sources of heating: the source Q_1 connected with the heating of thermal electrons during their interaction with highly energetic

photoelectrons formed during the process of solar ultraviolet irradiation, and the additional source, Q, connected with powerful radio wave heating. Finally, the total power of the combined natural and artificial sources is $Q_{total} = Q_1 + Q$.

d. In the analysis of Equations 1.5 through 1.10, we took into account the fact that all transport coefficients, described by Equation 1.12a through d, are functions of ionospheric altitude h; i.e., the problem is nonuniform, nonregular, and inhomogeneous.

e. We also consider in our calculations that the frequency of electron–ion collisions, according to Equation 1.3, depends on $T_e(\sim T_e^{-3/2})$, and consequently, the longitudinal coefficient (relative to the geomagnetic field $\mathbf{B}_0$) of electron thermal conductivity $\kappa_{e\parallel} \sim T_e^{5/2}$ (see Equation 1.12d) depends significantly on the perturbations of temperature of electrons. According to expressions of the transverse ($\kappa_{e\wedge}$) and perpendicular ($\kappa_{e\perp}$) coefficients of thermal conductivity given earlier, their temperature dependences are weaker, i.e., $\kappa_{e\wedge,\perp} \sim T_e^{3/2}$ (see Equation 1.12d). We present the corresponding coefficients of thermo-diffusion and thermoconductivity along $\mathbf{B}_0$, denoted by $\parallel$, in Figure 5.48. It is evident that, as with other transport coefficients presented in Figure 5.39, the changes of these transport coefficients in the middle-altitude ionosphere with altitude have a nonmonotonic character.

As shown in Refs. [75,76], a full nonlinear inhomogeneous system of equations (Equations 1.5 through 1.10) is impossible to solve analytically, and therefore, in Refs. [77,78], 1-D and 3-D numerical codes based on the self-consistent nonlinear inhomogeneous system of plasma density and temperature transport with given different heat sources (corresponding to each concrete heating experiment) were developed. To numerically solve such a complicated system for 1-D to 3-D models of thermodiffusion, diffusion and drift, in Refs. [77,78]—as was done in the cases of pure diffusion and drift—the method of trial-run and variable directions was introduced, following Ref. [104]. This was done to obtain stable solutions for an inhomogeneous ionospheric background with complicated height dependence of coefficients of thermodiffusion, conductivity, diffusion, and velocities of drift.

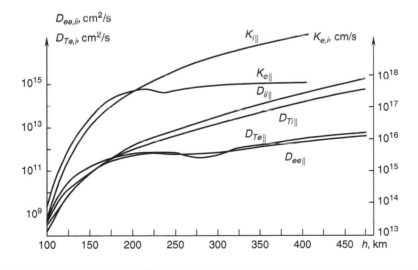

Figure 5.48 Height profiles of the coefficients of thermodiffusion and thermoconductivity according to the model of middle-latitude ionosphere.

We present in the following text results of numerical computations made during a detailed computational analysis of spatial and temporal evolution of various plasma structures independent of the origin of their creation in the inhomogeneous background ionospheric plasma for various methods of heating.

Naturally, to fully describe the total balance of the amount of plasma particles, electrons and different kinds of ions, in the ionospheric background, we have to introduce into Equations 1.5 through 1.10 the ionization and recombination processes, which form the ionization–recombination balance in the ionospheric background. We describe in the following text the main aspects of these phenomena.

5.3.2 One-Dimensional Heating of the E and F Layers of the Ionosphere

First, to understand the main physical aspects of local heating of the ionospheric plasma, we have to consider a nonstationary 1-D problem of the evolution of plasma irregularities highly stretched along $\mathbf{B}_0$, i.e., in a situation close to the real heating experiments carried out in the upper-E and lower-F regions of the ionosphere [50–57]. We consider different kinds of such heating processes, located at fixed altitudes, such as two-source heatings separated along the altitude, resonance and adaptive heating, and their influence on the formation of negative and positive disturbances of plasma density in the field of a powerful radio wave in the upper and lower ionosphere.

Changes of Plasma Temperature and Density in Local Fixed Heating. In the 1-D case a self-matching system (Equations 1.5 through 1.10) of diffusion and thermodiffusion can be presented for the local heating of ionospheric plasma as follows [77]:

$$\frac{\partial N_I}{\partial t} = \frac{\partial}{\partial z}\left[D_{aN||}(z)\frac{\partial N_I}{\partial z} + D_{eT||}(z)\frac{\partial T_e}{\partial z} + D_{iT||}(z)\frac{\partial T_i}{\partial z}\right] + I - R \tag{5.106a}$$

$$\frac{\partial T_e}{\partial t} = \frac{\partial}{\partial z}\left[\frac{\kappa_{e||}(z)}{N_e}\frac{\partial T_e}{\partial z}\right] - \delta_{ei}\nu_{ei}(T_e - T_i) - \delta_{em}\nu_{em}(T_e - T_m) + \frac{2(Q + Q_1)}{3N_e} \tag{5.106b}$$

Here, the transport coefficients of ambipolar diffusion, $D_{aN||}$, thermal diffusion, $D_{\alpha T||}$, $\alpha = e, i$, and thermoconductivity, $\kappa_{e||}$, along $\mathbf{B}_0$ can be simplified from Equation 1.12 as [77]:

$$
\begin{aligned}
D_{aN||} &\approx \frac{D_{ii||} + \eta D_{ee||}}{1+\eta}, \quad D_{eT||} \approx \frac{2 + p(1+\eta)}{(1+p)(1+\eta)}\frac{N_e}{m_e\nu_{em}} \\
D_{iT||} &\approx \frac{\eta + p(1+\eta)}{(1+p)(1+\eta)}\frac{N_e}{M_i\nu_{im}}, \quad D_{ee||} \approx \frac{T_e[1 + \eta p(1 + T_i/T_e)]}{m_e\nu_{em}(1+p)} \\
D_{ii||} &\approx \frac{T_e[p + (1+p)(T_i/T_e)]}{M_i\nu_{im}(1+p)}, \quad \kappa_{e||} \approx \frac{T_e N_e}{m_e\nu_{em}(1+p)}
\end{aligned} \tag{5.107}
$$

All parameters here were introduced in Chapter 1; here we only note that the energy shares transmitted by electrons during their interactions with ions and neutral molecules were estimated in Ref. [102] as $\delta_{ei} \approx 10^{-5}$ and $\delta_{em} \approx 2 \cdot 10^{-3}$, $N_I = [O^+]$ being the concentration of the main ions of monoatomic oxygen that prevails at the altitudes of the F region.

In the right-hand side of the first equation in the coupled system (Equation 5.106) there are terms that take into account ionization (I) and recombination (R) processes in the ionosphere. The coefficient of recombination R is described in Chapter 1 according to analysis of ionization–recombination processes in the ionosphere carried out in Refs. [77,78]. It should also be noted that, in Equation 5.106a, the disturbances of the ion temperature are not taken into account, and this is justified under the condition $\delta_{ei}\nu_{ei} < \nu_{im}$. Using the aforementioned estimations of the parameters made according to Ref. [102], we obtain that this constraint is fulfilled in the height region from E to F layers.

The second equation (Equation 5.106b) describes the energy balance in the ionosphere during the heating process (here, T_e and T_i are expressed, as in Chapter 1 in energetic units). The external (Q_1) and the high-power (Q) source from ground-based facility were described previously.

Furthermore, the plasma irregularities in the ionosphere created by an external heating source Q_1 were calculated in Refs. [77,78] on the background of a natural regular distribution of electron temperature $T_{e0}(z)$ and concentration $N_{e0}(z)$ given according to the model of the middle-latitude ionosphere [105] (see also Chapter 1). Values $T_{e0}(z)$ and $N_{e0}(z)$ were substituted into Equation 5.106, which allowed us to determine the natural source of ionization $J(z)$ and thermal energy $Q_1(z)$, as well as the corresponding values of flows of particles and energy

$$j_0(z)\big|_{z=z_2} = -\frac{1}{M_i \nu_{im}}\left\{\frac{\partial}{\partial z}[P + M_i N_I g \cos\theta]\right\}\bigg|_{z=z_2}$$

$$S_0(z)\big|_{z=z_2} = -\kappa_{e\parallel}\left\{\frac{\partial T_e}{\partial z}\right\}\bigg|_{z=z_2}$$

(5.108)

at the upper boundary, $z_2 \approx 500$ km, of the region of the ionosphere under consideration. Here, $P = (T_e + T_i)N_I$ is the full pressure of the ionospheric plasma; the angle $\cos\theta$ was introduced at the beginning of this section; and $g = 9.81$ m/s^2 is the acceleration due to gravity. We should note that without loss of generality we can take $\theta = 0$ and $h = z$, which is almost correct for auroral conditions and approximately correct for the middle-latitude ionosphere. The values $J(z)$ and $Q_1(z)$, as well as flows $j_0(z)$ and $S_0(z)$, were considered unchanged after switching on the additional source of electron heating Q from a ground-based facility.

In numerical calculations of the system (Equation 5.106) the same boundary conditions as were taken for diffusion and drift and usually used in ionospheric problems [84–86] were introduced: the disturbance of electron concentration δN_e and temperature δT_e is zero at the lower $z_1 \approx 90$ km and upper $z_2 \leq 500$ km boundaries of the considered ionospheric region. The initial conditions correspond to the absence of density and temperature disturbances at the initial time, i.e.,

$$\delta N_e(z, t = 0) = 0, \quad \delta T_e(z, t = 0) = 0$$

(5.109)

It should be noted that in the numerical experiments carried out in Refs. [75–78], the positions of the upper and lower boundaries were chosen in such a way that their changes would not influence the altitudinal variations of $\delta N_e(z)$ and $\delta T_e(z)$.

The external heating source Q connected with a local dissipation of powerful radio wave energy was given in Gaussian form:

$$Q(z) = Q_0 \exp\left\{-B\frac{(z - z_0)^2}{L_T^2}\right\}$$

(5.110)

where Q_0 is the heating amplitude, i.e., the maximum heat intensity at the point $z = z_0$ of the source; parameter B determines the width of the heat profile; and $L_T = \sqrt{\kappa_{e\parallel}/[N_{e0}(\delta_{ei}\nu_{ei} + \delta_{em}\nu_{em})]}$ is the longitudinal (relative to $\mathbf{B}_0$) scale of the penetration of the heat wave front into ionospheric plasma [102]. In calculations in [77], the heating source was taken as being rather narrow at the altitude with $L_T/\sqrt{B} \approx 1$ km. The heat amplitude Q_0 is proportional to the absorbed power of the radio wave incident on the ionospheric layer [76,105]

$$\int_{z_1}^{z_2} Q(z) \, dz = \sqrt{\pi} L Q_0 \approx \frac{\eta PG}{2\pi^2 b_0^2} \tag{5.111}$$

where in the general case, as was determined earlier, $h_0 = z_0 \cos \theta$ is the height of the local maximum; z_1 and z_2 are the lower and upper boundaries of the ionosphere; PG is the equivalent power of the transmitter radiation; the coefficient $0 < \eta < 1$ takes into account both radio wave attenuation in the path of propagation and the portion of the arriving energy, which is later anomalously absorbed and dissipated as heat; and L is the half-width of the region heated by the source (Equation 5.110). For the calculations it is convenient to measure the relative energy $q = Q/N_e$ in kelvins per second (K/s), called the *source capacity*, and from Equation 5.110 we obtain the correlation between parameters of the numerical experiment [77]:

$$Q/N_e \text{ (K/s)} \approx \frac{6.5 \cdot 10^4 \eta PG \text{ (100 MW)}}{L \text{ (km)} h_0^2 \text{ (100 km)} N_e \text{ (10}^6 \text{cm}^{-3})} \tag{5.112}$$

It should be noted that in the given case the source $Q(z)$ is located at the fixed altitude $h_0 = z_0$ during the whole heating process. We compare this case with those where the heating source changes its position and moving into the upper ionosphere simultaneously with the changes of resonance plasma concentration during plasma heating, which is called resonance and adaptive heating [75,77,105].

Figure 5.49 shows the results of the numerical simulations of Equation 5.106 with initial and boundary conditions (Equations 5.108 through 5.110) for $z_0 = 240$ km and $q = 2000$ K/s (Figure 5.49a) and $q = 5000$ K/s (Figure 5.49b) when the time of heating was $\tau_T = 5$ s. In both figures the observed time of ionospheric plasma evolution after switching off the source was taken as $t = 3$ s, i.e., less than the heating time. In all the figures on the left side, the solid lines show the altitudinal (in kilometers) dependence of plasma density $N(z) = N_0(z) + \delta N(z)$ (cm^{-3}). Every heating altitude z_0 corresponds to the plasma density $N(z_0)$, which is indicated on the abscissa axis. The dotted lines indicate the altitudinal dependence of the background ionospheric plasma density $N_0(z)$. In the right-hand panels in Figure 5.49a and b, the altitudinal dependence of the total temperature of electrons $T_e(z) = T_{e0}(z) + \delta T_e(z)$ (solid curves) as well as background plasma temperature $T_{e0}(z)$ (dotted curves) and the altitudinal run of heating source capacity $q(z)$, are presented. From the illustrations it can be seen that, with the growth of heat capacity $q(z)$, the process of plasma ejection in the heat area increases and the negative concentration area, called *cavity*, widens and deepens. Thus, for $q = 2000$ K/s during the heating time $\tau_T = 5$ s, the depletion of plasma in the heating area is about $\delta N(z)/N_0(z) \approx 0.01$; for the heating capacity $q = 5000$ K/s, it is about $\delta N(z)/N_0(z) \approx 0.03$. In Ref. [77] some hypothetical cases of high-energy heating with the heating capacity up to 20,000 K/s were examined (the effective anomalous heating power is up to $\eta PG \approx 250$ MW). It was shown that a considerable heating up of electrons at

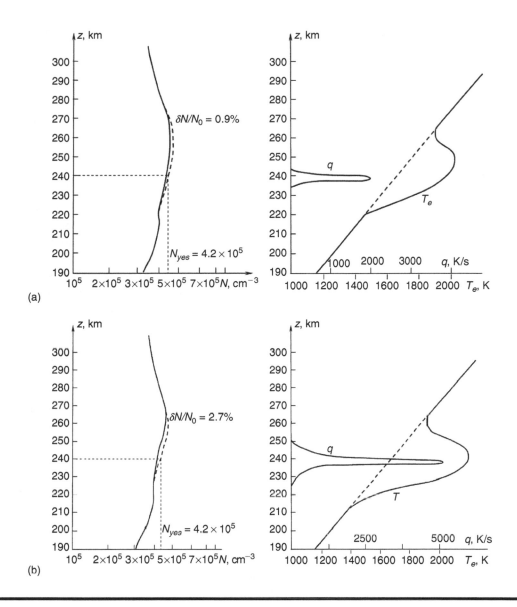

Figure 5.49 Scenario of local heating of the ionospheric plasma at altitude 240 km: the height plasma density profile and the plasma temperature (solid curves), compared with those of the background plasma (dashed curves) for the normalized heating power $q = 2000$ K/s (a) and $q = 5000$ K/s (b).

altitudes $h > 200$ km produces an excess of hydrodynamic pressure, and as a result leads to plasma ejection from areas of the greatest heating. The maximum of the background plasma density decrease can achieve about $\delta N(z)/N_0(z) \approx 0.14–0.15$. It should also be noted that, in the numerical experiment, even when a lower heating power was used, noticeable disturbances of plasma density and temperature are also observed. In particular, as follows from Figure 5.49a and b for anomalous heating power $\eta PG \approx 6$ MW and $\eta PG \approx 25$ MW, respectively, the relative temperature disturbances, some after the maximum 3 s, are about $\delta T_e(z)/T_{e0}(z) \approx 0.12-0.15$. Using Ref. [102], we can evaluate the relationship between plasma density and temperature disturbances

$$\frac{\delta T_e(z)/T_{e0}(z)}{\delta N(z)/N_0(z)} \approx \frac{(1/D_a t + 1/L_N^2)}{(N_e/\kappa_{e\parallel} t + 1/L_T^2)} \tag{5.113}$$

where $L_N^2 = D_a \tau_N$ and $L_T^2 = \kappa_{e\parallel}/N_{e0}(\delta_{ei}\nu_{ei} + \delta_{em}\nu_{em})$ are the characteristic spatial scales of plasma density and temperature disturbances; τ_N is the time of the recombination process; and D_a the coefficient of ambipolar diffusion presented in Section 5.1. The altitude $z_0 = 240$ km is characterized by the fact that the recombination time here is somewhat great, $\tau_N \approx 3-6$ min (according to Ref. [102]), so that, at the altitudes under consideration, of $h > 200$ km, the recombination processes are not manifested. Nevertheless, for altitude $z_0 = 200$ km $\tau_N \approx 30$ s the influence of the recombination process at altitudes less than 200 km is considerable because τ_N is of the same order as the time of heating $\tau_T \approx 5-20$ s. Moreover, the characteristic time, $\tau = (\delta_{ei}\nu_{ei} + \delta_{em}\nu_{em})^{-1} \approx 20-30$ s, during which the stationary distribution of electron temperature can reach the steady state, is also comparable at these altitudes with the full time of the heating process. It can also be seen from the foregoing illustrations that considerable density disturbances reach upward, mainly to the maximum of the nondisturbed F layer of the ionosphere ($h = 270-300$ km). Downward plasma disturbances spread no further than $h = 160$ km. This can be understood because at approximately these altitudes, the character of the recombination process changes radically. In fact, at altitudes greater than 160 km the term describing the recombination losses in Equation 5.106b is proportional to the plasma density. Here, the temperature dependence of the recombination velocity is absent. In contrast, at altitudes less than 160 km, recombination losses become proportional to the square of the plasma density, and the coefficients of dissociative recombination decrease with an increase of temperature (see Chapter 1, where the recombination process is fully described). Therefore, the additional heating at the altitudes less than 160 km produces an increase of plasma density and the creation of plasma enhancements (here according to Ref. [102] the transport processes are insignificant). This effect is seen from Figure 5.50, where the numerical simulations were carried out at an altitude $z_0 = 160$ km for $\eta PG \approx 25$ MW and $\tau_T = 20$ s. It is seen that when the temperature of the ionospheric plasma increases in a somewhat small altitudinal range of 150–180 km, a plasma density increase occurs compared with the background ionospheric plasma.

We will consider now the heating of the ionosphere taking place with the separated sources $Q_1(z)$ and $Q_2(z)$, whose power profiles are exponential [105]:

$$Q(z) = Q_1(z) + Q_1(z) = Q_0 \exp\left\{ -B\frac{(z - z_{01})}{L_T^2} - B\frac{(z - z_{02})}{L_T^2} \right\} \tag{5.114}$$

All parameters, functions, and boundary conditions are the same as described earlier in the case of local heating. Here, z_{01} and z_{02} are the altitudes of the greatest heat intensity for the first (at altitude $z_{01} = 200$ km) and second (at altitude $z_{02} = 240$ km) source, respectively. The plasma density redistribution during such a separated heating along altitude is presented in Figure 5.51 for the normalized heating power of heaters $q_1 = 2000$ K/s and $q_2 = 20,000$ K/s, respectively. The heating time is $\tau_T = 20$ s and the time of observation after switching off the heating sources is $t = 20$ s. It is seen that in the case of separate altitudinal heating, there appear additional plasma pits between source points z_{01} and z_{02}, i.e., a striated plasma structure at the ionospheric altitudes. This fact was

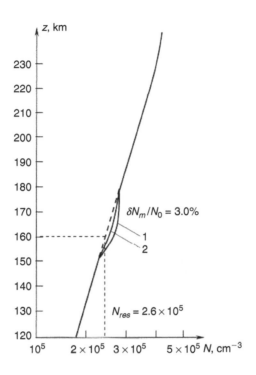

Figure 5.50 The height plasma density profile (solid curves) evolution during local heating of the ionospheric plasma at altitude $h = 160$ km compared to the background plasma (dashed curves).

explained theoretically in Refs. [106,107], taking into account the processes of diffusion and thermodiffusion along the geomagnetic field lines (along altitude h).

We now compare the above results of numerical simulations with those obtained for the regime of *resonance* heating, for which, according to Ref. [75], we consider that the heating source power $Q(z)$ is a function of resonance plasma concentration N_R. It was shown in Refs. [106,107] that the bulk of the energy in the pump radio wave is absorbed near the upper-hybrid resonance (UHR) corresponding to this resonant density, which can be defined as

$$N_R = \frac{\pi m_e}{e^2}\left(f_0^2 - f_{ge}^2\right) \tag{5.115}$$

where f_0 is the frequency of the incident radio wave and f_{ge} is the electron gyrofrequency. The resonant heating source was modeled in the form

$$Q(z) = Q_{max}\exp\left\{-B\frac{N - N_R}{N_R}\right\} \tag{5.116}$$

where the coefficient $B \gg 1$ ensures that the resonance is narrow with respect to the density perturbation. Then the undisturbed ionospheric plasma density profile can be approximated linearly by the function

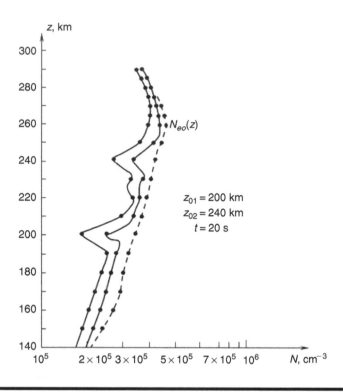

Figure 5.51 Same as Figure 5.50, but for scenario of heating at different altitudes $h = 200$ km and $h = 240$ km.

$$N_{e0}(h) \approx N_R \left(1 + \frac{h - h_R}{L}\right) \qquad (5.117)$$

where the resonance altitude h_R and the range of disturbed ionospheric area L are of some hundreds of kilometers. Strictly speaking and according to Equations 5.116 and 5.117, the heat source follows some specific plasma density, called *resonance density* N_R.

In Figure 5.52, the perturbation of plasma disturbance created at the initial height of the heating source $h_R = 240$ km for effective heating power $\eta PG = 120$ MW is presented. Numerical computations were carried out in Ref. [75] for the frequency of the heating source $f_0 = 8.6$ MHz or, according to the relationship between the frequency of powerful radio wave and resonance plasma density (Equation 5.115), for the value $N_R \approx 9 \cdot 10^5$ cm^{-3}. The dotted curve represents the background undisturbed plasma profile, and the solid curve represents the disturbed plasma profile for the moment when the observed time is equal to the heating time, i.e., $t = \tau_T = 20$ s. As can be seen from Figure 5.52, the location of the heating source is changed according to the change of the resonant plasma density during the heating process and for $t = 20$ s, the difference between the initial and later location becomes about 40–50 km. In the process of floating-up penetration in the upper ionosphere, the reflection point of the wave also moves upward as if penetrating into the depths of the ionosphere [108]. It was shown theoretically in Refs. [106–109] that sufficiently strong resonant heating will lead to the formation of a flat region (a plateau) in the density profile. This feature was also observed from numerical simulations of the problem and shown in

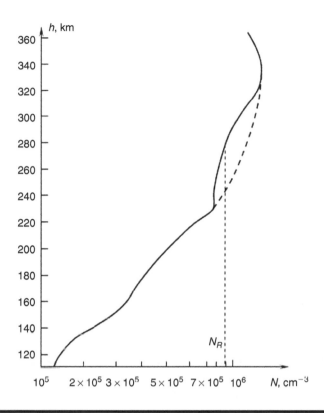

Figure 5.52 Same as Figure 5.50, but for scenario of resonance heating at altitude $h = 240$ km.

Figure 5.52, where such a plateau ranges in altitudes $220 < h < 280$ km. This is the principal difference between the local and the resonance heating (compare Figures 5.49 and 5.50 with Figure 5.52 for plasma density profile altitudinal density).

Finally, we present *adaptive* heating when the heating source is always placed at the minimum of the profile $N(z)$ (the realization of such an experiment will be described in Chapter 6). In Figure 5.53, the disturbed plasma density altitudinal redistribution is presented for the equal heating and observation times, i.e., $\tau_T = t = 30$ s, with effective heating power of $Q_0 = \eta PG \approx 20$ MW. In the computations, both thermodiffusion and changes of recombination were taken into account during the heating process [105]. Concentration $N_1, \ldots, N_5$ in Figure 5.53 is achieved in the minimum of $N(z)$ during the time $t_1 = 1$ s for $N_1 = 4.5 \cdot 10^5$ cm^{-3}; during $t_2 = 5$ s for $N_2 = 4.03 \cdot 10^5$ cm^{-3}; during $t_3 = 10$ s for $N_3 = 2.95 \cdot 10^5$ cm^{-3}; during $t_4 = 16$ s for $N_4 = 2.19 \cdot 10^5$ cm^{-3}; and during $t_5 = 22$ s for $N_5 = 1.89 \cdot 10^5$ cm^{-3}.

Thus, using different kinds of plasma heating, we can control the spatial position of the negative plasma cavity and its depth and shape in the upper ionosphere.

5.3.3   Evolution of 3-D Heating-Induced Irregularity in the Ionosphere

Quasi-periodic grating structures of very elongated plasma irregularities in real conditions of ionospheric heating experiment in the F layer of the ionosphere were observed during and after a

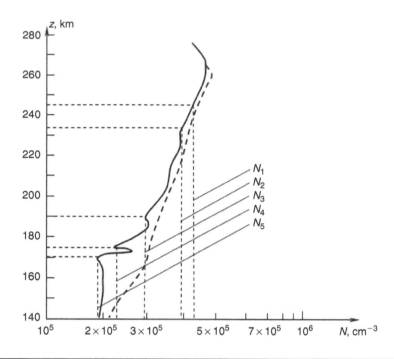

Figure 5.53 Spatial and temporal evolution of ionospheric plasma profile during scenario of the adaptive heating compared to the background plasma (dashed curves).

powerful radio wave had influenced it [50,51,55–57]. Theoretical investigations of the influence of a powerful radio wave on ionospheric plasma heating [50,51,102,110,111] indicate the possible development of quasi-periodical artificial ionospheric turbulence in the region of heating. In Refs. [76,78], the formation and further spatial and temporal evolution of striated plasma density 3-D structures obtained from numerical and analytical computations were examined on the basis of experimentally determined distance–frequency characteristics of back-scattered signals from the ionospheric F region disturbed by heating effect of a powerful radio beam [55] (described also in Chapters 6 and 7 of this book). Following Refs. [76,78], we will analyze in the following text the evolution of the plasma density in the ionospheric F region created in the field of a powerful HF-wave during local heating using 3-D nonstationary hydrodynamics based on a full self-consistent system of transport equations (Equations 1.5 through 1.12) described in Chapter 1, as well as under conditions of the middle-latitude ionosphere described there and by items a–e described earlier. The difference from the problems investigated in Refs. [50,51,102,110,111] is that in Refs. [76,78] the full set of inhomogeneous equations with parameters dependent on the altitude was taken into account. Moreover, in the proposed rigorous self-consistent model, all characteristic frequencies of the plasma, as well as the recombination coefficient and coefficients of conductivity, are functions of electron temperature, i.e., of the heating process. All these aspects are fully described in Chapter 1 and in Section 5.3.1. As before with the local 1-D heating numerical experiment, we can present the amplitude of the plasma heater (Equation 5.110) through the absorbed radio wave power (Equation 5.111 or 5.112) incident on the ionospheric layer.

Inhomogeneous Ionosphere. The nonlinear inhomogeneous system of plasma density and temperature transport with all formulas of transport coefficients (Equations 1.5 through 1.12) with a given heat source (Equations 5.110 through 5.112) and with given boundary and initial conditions (Equations 5.108 through 5.110) was solved numerically using the methods of trial-run and variable directions [19,20,74]. The solution is sought in the region of $z_1 = 100$ km $\leq z \leq z_2 = 400$ km (along $\mathbf{B}_0$) and $|\rho| = \sqrt{x^2 + y^2} \leq 10$ km (across $\mathbf{B}_0$) for the cylindrical symmetry along z-axis (along $\mathbf{B}_0$). The source described by Equation 5.110 with anomalous power $\eta P W \approx 50$–60 MW (which was then chosen according to Equation 5.112 in K/s) was located at the altitude $h = z_0 = 240$ km corresponding to the electron concentration $N_{e0} \approx 4.5 \cdot 10^5$ cm^{-3} (corresponding to the heating experiment described in Ref. [55]; see also Chapter 6) and the time interval of observation was $t_0 \leq t \leq 100$ s, $t_0 = d/D_{i\perp}$, where d and $D_{i\perp}$, as before, are the initial scales of the irregularity and the coefficient of ion diffusion across $\mathbf{B}_0$, respectively. Results of numerical simulations are presented in Figure 5.54a and b as lines of equal concentration in the plane $(x0z)$, i.e., along $\mathbf{B}_0$ (i.e., along the z-axis) and transverse $\mathbf{B}_0$ (i.e., in the $\mathbf{E}_0 \times \mathbf{B}_0$-drift direction) at the moment of switching off the heating source $t = 10$ s (Figure 5.54a) and $t = 20$ s (Figure 5.54b). So, the total time of heating ($\tau_T = 10$ s) and of observation was $t_{total} = 30$ s (in Figure 5.54b). The ambient geophysical parameters were chosen to satisfy the real conditions in the middle-latitude ionosphere at the observed altitudes [16,19,74]: the ambient electric field $E_0 \approx 5$–10 mV/m and the neutral wind velocity $U_m \approx 50$–70 m/s. It can be seen that during the heating process the quasi-periodic plasma structure of induced heating irregularities of plasma density strongly elongated along $\mathbf{B}_0$ with characteristic scales $\ell_\perp \approx 10$–500 m is formed. Because in general equations the processes of drift and diffusion were accounted in the process of thermo-diffusion of plasma structure, we may expect generation of the gradient-drift instability (GDI) described in detail in Chapter 2, which, together with the altitudinal dependence of transport coefficients and external geophysical factors, according to Ref. [74], can be responsible for such a grating structure of irregularities with the degree of anisotropy of $\chi = \ell_\parallel / \ell_\perp \geq 20$–$30$. Together, all factors lead to additional deformation and further striation of irregularities, which grow into separate plasmoids stretched in parallel and transversely to the geomagnetic field lines. It is necessary to note that electron density disturbances with different signs may occur at close height levels. The existence of regions with positive plasma disturbances along the x-axis (along $\mathbf{E}_0 \times \mathbf{B}_0$ direction) is characterized by the influence of boundaries in the process of plasma evolution caused by drift in crossing $\mathbf{E}_0 \times \mathbf{B}_0$ fields [74]. Again, we obtain here a physically evident effect of short-circular currents, which are created by the flow of charge particles through the undisturbed background plasma and which finally create the depletion region surrounding each plasma irregularity, as can be clearly seen from illustrations presented in Figure 5.54.

Quasi-Homogeneous Ionosphere. We now examine the stage of relaxation of a striated plasma structure (SPS) of irregularities strongly elongated along $\mathbf{B}_0$ in the case of homogeneous unbounded ionospheric plasma, the initial Gaussian profile of which in the moment $t = t_0$ can be presented in the same symmetrical form relative to the z-axis (along $\mathbf{B}_0$), as was done in Section 5.1 for pure diffusion and in Section 5.2 for pure drift, but now taking into account a grid consisting of N inhomogeneities in it:

$$\delta N(\rho, z, 0) = \delta N_0 \sum_{j=1}^{N} \exp\left\{ -z^2/\ell_\parallel^2 - \left(\rho \pm \rho_{0j}\right)^2 \Big/ \ell_\perp^2 \right\} \tag{5.118}$$

Here, ρ_{0j} is the transverse (to $\mathbf{B}_0$) coordinate of the center of each irregularity of number j in the SPS; the symbols "+" and "−" correspond to the arrangement of irregularities to the positive and negative directions at the ρ-axis. Now, using the approach evaluated in Ref. [20] for the single

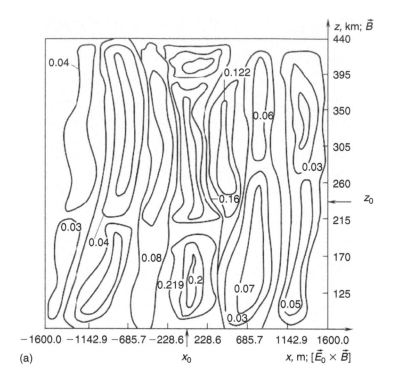

(a)

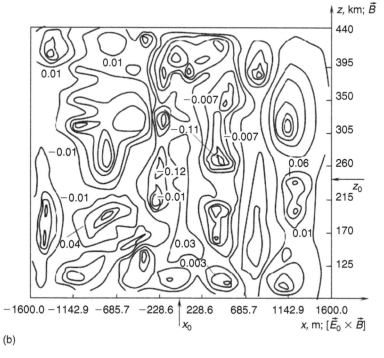

(b)

Figure 5.54 Two-dimensional plot of plasma contours of equal density during (a) 3-D heating of plasma, accounting for all ambient geophysical effects of electric and magnetic field, neutral wind, gravitation, and real transport parameters of the middle-latitude ionospheric plasma, and the same plot during (b) time evolution of heating-induced plasma disturbances.

irregularity and presented in Section 5.1, we can, following Ref. [78], construct the expression describing the stage of SPS spatial and temporal relaxation (using the unified notations of this chapter):

$$\delta N(\rho,z,t) = \delta N_0 \ell_{\|} \ell_{\perp}^2 \sum_{j=1}^{N} \exp\left\{ -\frac{z^2/(\ell_{\|}^2 + L_e^2) + (\rho \pm \rho_{0j})^2/(\ell_{\perp}^2 + d_e^2)}{\sqrt{\ell_{\|}^2 + L_e^2}(\ell_{\perp}^2 + d_e^2)} \right.$$

$$\left. -\frac{z^2/(\ell_{\|}^2 + L_i^2) + (\rho \pm \rho_{0j})^2/(\ell_{\perp}^2 + d_i^2)}{\sqrt{\ell_{\|}^2 + L_i^2}(\ell_{\perp}^2 + d_i^2)} \right\} \tag{5.119}$$

Here, as in previous sections, δN_0 is the concentration of plasma in the maximum of the initial plasma irregularity, $\delta N_0 = AN_0$; N_0 is the concentration of the background ionospheric unbounded plasma; $L_e^2 = 4(1 + T_e/T_i)D_{e\|}t$, $L_i^2 = 4(1 + T_i/T_e)D_{i\|}t$, $d_e^2 = 4(1 + T_e/T_i)D_{e\perp}t$, $d_i^2 = 4(1 + T_i/T_e)D_{i\perp}t$, i.e., are now written for the case of nonisothermal plasma. The numerical analysis of Equation 5.119 in the relative form $\delta\widetilde{N} = \delta N/\delta N_0$ is presented as isodensity lines in Figure 5.55a through c for the case $\ell_{\perp} = 10$ m, $\ell_{\|} = 10$ km, $r_0 = 2\ell_{\perp}$ for single irregularity ($N = 1$), two irregularities in SPS ($N = 2$), and five irregularities in SPS ($N = 5$), respectively. Here,

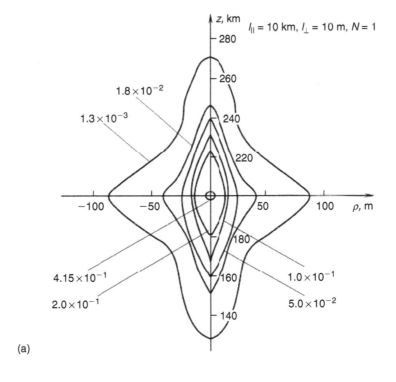

(a)

Figure 5.55 Spatial and temporal evolution of a plasma disturbances: (a) single inhomogeneity

(continued)

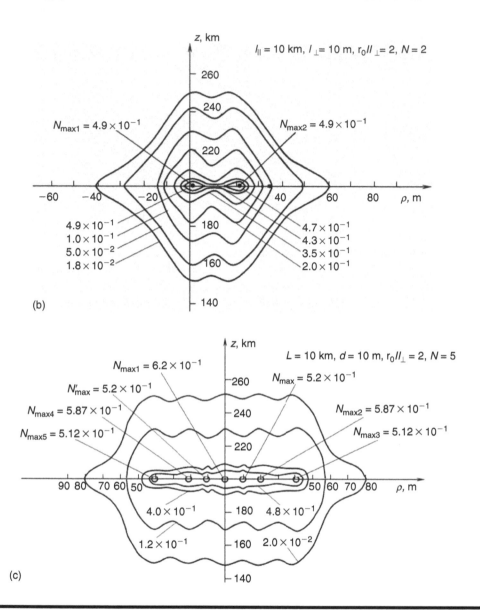

Figure 5.55 (continued) **(b) two inhomogeneities, (c) a grid of five inhomogeneities when the distance between them is less than the gyroradius of plasma ions.**

r_0 is the distance between irregularities in SPS. All transport coefficients were calculated for the fixed altitude $h = 200$ km according to the model of the middle-latitude ionosphere described in Chapter 1. As can be seen from Figure 5.55a, the relaxation of a single irregularity ($N = 1$) at time $t = 10$ s has a symmetrical character relative to the z- and ρ-axes. The concentration at the maximum of the irregularity, $\delta\widetilde{N}(0,0,0)$, attenuates during 10 s from $\delta\widetilde{N} = 2$ to $\delta\widetilde{N} = 0.415$; i.e., plasma concentration decreases by a factor of five. The presence of two irregularities in SPS ($N = 2$; Figure 5.55b) at the same time $t = 10$ s shows a decrease of the attenuation process of each irregularity; the concentration at the maximum falls by a factor ≈ 1.2 only, compared with the larger factor 5 of the $\delta\widetilde{N}$ decay obtained for a single irregularity. The increase of the number of irregularities in the SPS ($N=5$; Figure 5.55c) at the same time $t=10$ s leads to an essential deceleration of the decay of plasma concentration in the center of the SPS ($\rho=0$). We note that relaxation time of the SPS increases by a factor of 1.5 with $N=5$ in comparison with that for a single irregularity. Furthermore, simultaneously with the fundamental deceleration of plasma density attenuation inside the SPS, the appearance of new extrema between the initial irregularities is observed (see Figure 5.55c). Here, at coordinates $\rho_1=-10$ m and $\rho_2=10$ m there appear two local maxima designed by "crests." The concentration at these two maxima exceeds the concentration at the edge irregularities. Thus, $\delta\widetilde{N}$ (10 m) = 0.52, but $\delta\widetilde{N}$ (40 m) = 0.512.

The observed numerical effects depend on the initial distance between irregularities inside the SPS (on its scale r_0). Thus, when $r_0=4\ell_{\perp}$ (see Figure 5.56) the velocity of SPS decay for $N=5$ (five irregularities inside the SPS) is of the same order as that for a single irregularity, and for time $t=10$ s, the relative concentration at its maximum is $\delta\widetilde{N}(0)=0.456$. For the case $r_0=4\ell_{\perp}$ there are no additional extrema between the initial irregularities inside the SPS; the attenuation and spreading are also symmetrical. From comparison between the calculations presented in Figures 5.55 and 5.56, it is seen that the boundary effects do not exist when the plasma density inside the SPS exceeds the plasma concentration outside it. Additional calculations made in Ref. [78] showed that for $r_0 \geq 4\ell_{\perp}$ the SPS spreads analogously to a single irregularity. It becomes clear if, according to Refs. [17–20], we return to Section 5.1 and consider the spreading of a single irregularity strongly stretched along $\mathbf{B}_0$ (with the degree of anisotropy $\chi \geq 10^2$). We will find for this irregularity that the spreading process transverse to $\mathbf{B}_0$ is determined by the transverse ion diffusion (ion-controlled), but the character of attenuation at the maximum (in the center) of this irregularity is electron-controlled. When at the distance (or less) of the Larmor radius of the ions, ρ_{Hi}, additional irregularities are placed near the existing irregularities (as occurs in the SPS with $r_0=2\ell_{\perp}$), the transverse ion diffusion becomes difficult because the ions of each irregularity "screen" the transverse diffusion ion flows from the irregularities adjacent to it inside SPS. In this case, the character of diffusion spreading at the SPS scale $r_0 \leq \rho_{Hi}$ is determined by the polarization

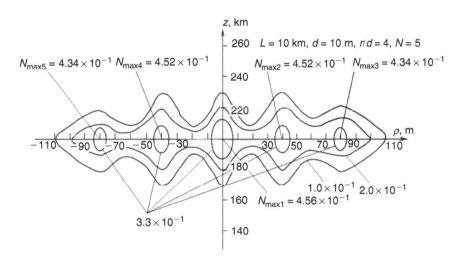

Figure 5.56 Evolution of a grid of plasma disturbances with distance between them exceeding the gyroradius of plasma ions.

field, which decays across the magnetic field lines not with the ion scale $d_i \sim \sqrt{D_{i\perp}}$, as in the case for a single irregularity ($N = 1$) [18–20], but with the electron scale $d_e \sim \sqrt{D_{e\perp}}$. We also stress that when the SPS scale $r_0 \approx \ell_\perp$, the character of plasma density attenuation for $N > 5$ becomes slower, namely, for $t = 10$ s. Moreover, the increase of the number of irregularities in the SPS for $N > 5$ does not lead to an essential change of the character of plasma density attenuation at the SPS center (this fact will also be shown in Chapter 6 during the analysis of the experimental data).

In Chapter 6, we will present the modern methods, radiophysical and optical, based on ground-based, rocket, and satellite facilities, employed in studying the main characteristics of ionospheric irregularities of various origins, artificial and natural, as well as the methods of investigation of geophysical characteristics of the ionosphere based on observations of evolution of such irregularities in the space and time domains.

References

1. Schottky, W., Wadstorm und teorie der positiven saule, *Phys. Zeitchrift.*, Bd. 25, No. 14, 1924, pp. 329–341.
2. Simon, A., Ambipolar diffusion in a magnetic field, *Phys. Rev.*, Vol. 98, 1955, pp. 317–326.
3. Holway, L.H., Ambipolar diffusion in the geomagnetic field, *J. Geophys. Res.*, Vol. 70, 1967, pp. 3635–3645.
4. Gurevich, A.V., Spreading of inhomogeneities in the weakly ionized plasma in the magnetic field (ambipolar diffusion), *J. Experim. Techn. Phys.*, Vol. 44, 1963, pp. 1302–1306.
5. Gurevich, A.V. and Tsedilina, E.E., About character of spreading and form of inhomogeneity in plasma, *Geomagn. Aeronom.*, Vol. 6, 1966, pp. 255–265.
6. Gurevich, A.V. and Tsedilina, E.E., Motion and diffusion of plasma inhomogeneities, *Sov. Phys. Usp.*, Vol. 91, 1967, pp. 609–642.
7. Kaiser, T.R., Pickering, W.H., and Watkins, C.D., Ambipolar diffusion and motion of ion clouds in the Earth's magnetic field, *Planet Space Sci.*, Vol. 17, 1969, pp. 519–551.
8. Polyakov, V.M., About diffusion of charged particles in the *F*-region of the middle-latitude ionosphere, *Geomagn. Aeronom.*, Vol. 6, 1966, pp. 341–351.
9. Rozhansky, V.A. and Tsendin, L.D., Spreading of a small inhomogeneity in the unbounded weakly-ionized plasma in magnetic field, *Physics of Plasma*, Vol. 1, 1975, pp. 944–953.
10. Rozhansky, V.A. and Tsendin, L.D., Spreading of a strong inhomogeneity at the background of weakly ionized plasma in the magnetic field, *Plasma Physics*, Vol. 3, 1977, pp. 382–387.
11. Zhilinsky, A.P. and Tsendin, L.D., Interactive diffusion of partly ionized plasma in the magnetic field, *Sov. Phys. Usp.*, Vol. 131, 1980, pp. 343–385.
12. Rozhansky, V.A. and Tsendin, L.D., Evolution of strong ionospheric inhomogeneities of ionospheric plasma. I., *Geomagn. Aeronom.*, Vol. 24, 1984, pp. 414–419.
13. Rozhansky, V.A. and Tsendin, L.D., Evolution of strong ionospheric inhomogeneities of ionospheric plasma. II., *Geomagn. Aeronom.*, Vol. 24, 1984, pp. 598–602.
14. Rozhansky, V.A. and Tsendin, L.D., Spreading of the shnur of weakly ionized plasma placed under angle to the magnetic field, *Geomagn. Aeronom.*, Vol. 17, 1977, pp. 1002–1007.
15. Rozhansky, V.A. and Tsendin, L.D., *Collisional Transport in a Weakly-Ionized Plasma*, Electroatomizdat, Moscow, 1988.
16. Blaunstein, N.Sh., Diffusion of plasma with arbitrary degree of ionization in the magnetic field (ionospheric case), in book: *Problems of the Cosmic Electrodynamics*, Moscow: Nauka, 1981, pp. 96–119.
17. Blaunstein, N.Sh. and Tsedilina, E.E., Spreading of strongly elongated inhomogeneities in the upper ionosphere, *Geomagn. Aeronom.*, Vol. 24, 1984, pp. 340–344.
18. Blaunstein, N.Sh. and Tsedilina, E.E., Effect of the initial dimensions on the nature of the diffusion spreading of inhomogeneities in a quasiuniform ionosphere, *Geomagn. Aeronom.*, Vol. 25, 1985, pp. 39–44.

19. Blaunstein, N.Sh., Diffusion spreading of inhomogeneities of the middle-latitude ionospheric plasma, in *Investigations of Geomagnetism, Aeronomy and Physics of Sun*, Moscow: Nauka, 1988, Vol. 82, pp. 90–103.

20. Blaunstein, N., Diffusion spreading of middle-latitude ionospheric plasma irregularities, *J. Ann. Geophys.*, Vol. 13, 1995, pp. 617–626.

21. Filipp, N.D., Blaunshtein, N.Sh., Erukhimov, L.M., Ivanov, V.A., and Uryadov, V.P. *Modern Methods of Investigation of Dynamic Processes in the Ionosphere*, Kishinev (Moldova): Shtiintza, 1991.

22. Uryadov, V., Ivanov, V., Plohotnuk, E., Eruhimov, L., Blaunstein, N., and Filipp, N., *Dynamic Processes in the Ionosphere–Methods of Investigations*, TechnoPress, Iasi (Romania), 2006.

23. Frihagen, J. and Jacobson, T., Insitu observations of high-latitude F-region irregularities, *J. Atmos. Terrest Phys.*, Vol. 33, 1971, pp. 519–522.

24. Muldrew, D.B. and Vickrey, J.F., High-latitude F-region irregularities observed simultaneously with ISIS-1 and Chatanika radar, *J. Geophys. Res.*, Vol. 87, 1982, pp. 8263–8267.

25. Hansucher, R.D., Chatanica radar investigation of high-latitude *E*-region ionization, *Radio Sci.*, Vol. 10, 1975, pp. 277–288.

26. Gelberg, M.G., *Irregularities of the High-Latitude Ionosphere*, Novosibirsk: Science, 1986.

27. Fatkullin, M.N., Zelenova, G.L., Kozlov, V.K., Legen'ka, A.D., and Soboleva, T.N., *Empirical Models of the Middle-Latitude Ionosphere*, Moscow: Nauka, 1981.

28. Gershman, B.N., *Dynamics of the Ionospheric Plasma*, Moscow: Nauka, 1974.

29. Filipp, N.D. and Blaunstein, N., Effect of the geomagnetic field on the diffusion of ionospheric inhomogeneities, *Geomagn. Aeronom.*, Vol. 18, 1978, pp. 423–427.

30. Ossakow, S.L. and Chaturvedi, P.K., Morphological studies of rising equatorial spread-F bubbles, *J. Geophys. Res.*, Vol. 83, 1978, pp. 2085–2091.

31. Haerendel, G., Coloured bubbles: an experiment for triggering equatorial spread-F, *ESA-SP* 195, July 1983, pp. 295–298.

32. Balsley, B.B. and Farley, D.T., Radar observation of two-dimensional turbulence in the equatorial electrojet, *J. Geophys. Res.*, Vol. 78, 1973, pp. 7471–7476.

33. Fopple, H. et al., Preliminary experiments for sudy of interplanetary medium, *Planet Space Sci.*, Vol. 13, 1965, pp. 95–114.

34. Fopple, H. et al., Artificial strontium and barium clouds in the upper atmosphere, *Planet Space Sci.*, Vol. 15, 1967, pp. 357–372.

35. Haerendel, G., Lust, R., and Rieger, E., Motion of artificial ion clouds in the upper atmosphere, *Planet Space Sci.*, Vol. 15, 1967, pp. 1–18.

36. Rieger, E., Neuss, H., and Lust, R. et al., High altitude release of barium vapors using Rubis rocket, *Ann. Geophys.*, Vol. 26, 1970, pp. 845–852.

37. Wescott, E.M., Stolarik, J.D., and Heppner, J.P., Electric fields in the vicinity of auroral forms from motion of barium vapor release, *J. Geophys. Res.*, Vol. 74, 1969, pp. 3469–3487.

38. Rosenberg, N.W., Observations of striation formation in barium ion clouds, *J. Geophys. Res.*, Vol. 76, 1971, pp. 6856–6864.

39. Lust, R., Space experiments with barium clouds, *New Scientist.*, Vol. 20, 1072, pp. 154–169.

40. Perkins, F.W., Zabusky, N.J., and Doles, J.H. Deformation and striation of plasma clouds in the ionosphere. 1., *J. Geophys Res.*, Vol. 78, 1973, pp. 697–709.

41. Baxter, A.J., Lower F-region barium release experiments at sub-auroral location, *Planet Space Sci.*, Vol. 23, 1975, pp. 973–983.

42. Glova, I.S. and Ruzhin, Yu.Ya., Cumulative injection of barium at altitude of 170 km. Experiment Spolokh-1, in book: *Dynamics of Cosmic Plasma*, Moscow: Nauka, 1976, pp. 154–174.

43. Antoshkin, V.S. et al., About evolution of barium clouds of high density, *Geomagn. Aeronom.*, Vol. 19, 1979, pp. 1049–1056.

44. Ruzhin, Yu.Ya. and Skomarovsky, V.S., About results of of complex rocket experiment Spolokh-2, in book: *Physical Processes in the Ionosphere and Magnetosphere*, Moscow: Nauka, 1979, pp. 35–45.

45. Dzubenko, N.L., Zhilinsky, A.P., Zhulin, A., and Ivchenko, I.S. et al., Dynamics of artificial plasma clouds in Spolokh experiments: Movement pattern, *Planet Space Sci.*, Vol. 31, 1983, pp. 849–858.

46. Andreeva, L.A., et al., Dynamics of artificial plasma cloud in Spolokh experiments: Clouds deformation, *Planet Space Sci.*, Vol. 32, 1984, pp. 1045–1052.

47. Filipp, N.D., Oraevskii, V.N., Blaunshtein, N.Sh., and Ruzhin, Yu.Ya. *Evolution of Artificial Plasma Irregularities in the Earth's Ionosphere*, Kishinev (Moldova): Shtiintza, 1986.

48. Milinevsky, G.P., Romanovsky, Yu.A., Evtushevsky, A.M., and Savchenko, V.A. et al., Optical observations in active space experiments in investigation of the Earth's atmosphere and ionosphere, *Cosmic. Res.*, Vol. 28, 1990, pp. 418–429.

49. Blaunshtein, N.Sh., Milinevsky, G.P., Savchenko, V.A., and Mishin, E.V., Formation and development of striated structure during plasma cloud evolution in the Earth's ionosphere, *Planet Space Sci.*, Vol. 41, 1993, pp. 453–460.

50. Djuth, F.T., Thide, B., Ierkis, H.M., and Sulzev, M.P., Large-F-region electron-temperature enhancements generated by high-power HF radio waves, *Geophys. Rev. Lett.*, Vol. 14, 1987, pp. 953–956.

51. Duncan, L.M., Sheerin, J.P., and Behnke, R.A., Observation of ionospheric cavities generated by high-power radio waves, *Phys. Rev. Lett.*, Vol. 61, 1988, pp. 239–242.

52. Blaunstein, N.Sh., Boguta, N.I. and Erukhimov, L.M. et al., About the dependence of the development and relaxation time of artificial small-scale irregularities on the diurnal time, *Izv. Vuzov, Radiophyzika*, Vol. 33, 1990, pp. 548–351.

53. Blaunshtein, N.Sh., Erukhimov, L.M., and Uryadov, V.P. et al., Vertical dependence of relazation time of artificial small-scale disturbances in the middle-latitude ionosphere, *Geomagn. Aeronom.*, Vol. 28. No. 4, 1988, pp. 595–597.

54. Bernhardt, P.A., Swartz, W.E., and Kelley, M.C. Spacelab 2 upper atmospheric modification experiment over Arecibo, 2, Plasma dynamics, *Astro. Lett. Communic.*, Vol. 27, 1988, pp. 183–198.

55. Bochkarev, G.S., Eremenko, V.A., and Cherkashin, Ya.N., Radio wave reflection from quasi-periodical disturbances of the ionospheric plasma, *Adv. Space Res.*, Vol. 8, No. 1, 1988, pp. 255–260.

56. Hansen, J.D., Morales, G.J., and Duncan, L.M., Large-scale ionospheric modifications produced by nonlinear refraction of an HF wave, *Phys. Rev. Lett.*, Vol. 65, No. 26, 1990, pp. 3285–3288.

57. Keskinen, M.J., Chaturvedi, P.K., and Ossakow, S.L., Long time scale evolution of high-power radio wave ionospheric heating, 1, Beam propagation, *Radio Sci.*, Vol. 28, No. 5, 1993, pp. 775–784.

58. Volk, H.J. and Haerendel, G., Striation in ionosppheric plasma clouds, *J. Geophys. Res.*, Vol. 76, 1971, pp. 4541–4559.

59. Lloyd, K.H. and Haerendel, G., Numerical modeling of drift and deformation of ionospheric plasma clouds and their interaction with other layers of the ionosphere, *J. Geophys. Res.*, Vol. 78, 1973, pp. 7389–7415.

60. Goldman, S.R., Ossakow, S.L., and Book, D.L., On the nonlinear motion of a small barium cloud in the ionosphere, *J. Geophys. Res.*, Vol. 79, 1974, pp. 1471–1477.

61. Perkins, F.W., Zabusky, N.J., and Doles, J.H., Deformation and striation of plasma clouds in the ionosphere, 2, Numerical simulation of a non-linear two-dimensional model, *J. Geophys. Res.*, Vol. 78, 1973, pp. 711–724.

62. Doles, J.H., Zabusky, N.J., and Perkins, F.W., Deformation and striation of plasma clouds in the ionosphere, 3, Numerical simulation of multilevel model recombination chemistry, *J. Geophys. Res.*, Vol. 81, 1976, pp. 5987–6004.

63. Perkins, F.W. and Doles, J.H. Velocity shear and the $\mathbf{E} \times \mathbf{B}$ instability, *J. Geophys. Res.*, Vol. 80, 1975, pp. 211–214.

64. Drake, J.F. and Huba, J.D., Dynamics of three dimensional ionospheric plasma clouds, *Phys. Rev. Lett.*, Vol. 58, No. 30, 1987, pp. 278–281.

65. Mitchell, H.G., Fedder, J.A., Huba, J.D., and Zalesak, S.T., Transverse motion of high-speed barium clouds in the ionosphere, *J. Geophys. Res.*, Vol. 90, 1985, pp. 11,091–11,103.

66. Huba, J.D. et al., Simulation of plasma structure evolution in the high-latitude ionosphere, *Radio Sci.*, Vol. 23, No. 4, 1988, pp. 503–510.

67. Drake, J.F., Mubbrandon, M., and Huba, J.D., Three-dimensional equilibrium and stability of ionospheric plasma clouds, *Phys. Fluids*, Vol. 31, No. 11, 1988, pp. 3412–3424.

68. Zalesak, S.T., Drake, J.F., and Huba, J.D. Dynamics of three-dimensional ionospheric plasma clouds, *Radio Sci.*, Vol. 23, No. 4, 1988, pp. 591–598.

69. Zalesak, S.T., Drake, J.F., and Huba, J.D., Three-dimensional simulation study of ionospheric plasma clouds, *Geophys. Res. Lett.*, Vol. 17, No. 10, 1990, pp. 1597–1600.

70. Pudovkin, M.I., Golovchanskaya, I., and VKlimov, M.S. et al., About employment of two-layer model for the description of three-dimensional evolution of the strong ionospheric inhomogeneity, *Geomagn. Aeronom.*, Vol. 37, No. 4, 1987, pp. 560–567.

71. Voskoboynikov, S.P., Rozhansky, V.A., and Tsendin, L.D., Numerical modelling of three-dimensional plasma cloud evolution in crossed $\mathbf{E} \times \mathbf{B}$ fields, *Planet Space Sci.*, Vol. 35, 1987, pp. 835–844.

72. Rozhansky, V.A., Veselova, I.Y., and Voskoboynikov, S.P., Three-dimensional computer simulation of plasma cloud evolution in the ionosphere, *Planet Space Sci.*, Vol. 38, No. 11, 1990, pp. 1375–1386.

73. Blaunstein, N.Sh., Tsedilina, E.E., Mirzaeva, L.I., and Mishin, E.V., Drift spreading and stratification of inhomogeneities in the ionosphere in the presence of an electric field, *Geomagn. Aeronom.*, Vol. 30, 1990, pp. 799–805.

74. Blaunstein, N., The character of drift spreading of artificial plasma clouds in the middle-latitude ionosphere, *J. Geophys. Res.*, Vol. 101, No. A2, 1996, pp. 2321–2331.

75. Blaunshtein, N.Sh, Vas'kov, V.V., and Dimant, Ya.S., Resonant heating of the F layer of the ionosphere by a high-power radio wave, *Geomagn. Aeronom.*, Vol. 32, No. 2, 1992, pp. 235–238.

76. Blaunshtein, N.Sh. and Bochkarev, G.S., Modelling of the dynamics of periodical artificial disturbances in the upper ionosphere during its thermal heating, *Geomagn. Aeronom.*, Vol. 33, No. 2, 1993, pp. 84–91.

77. Blaunstein, N., Changes of the electron concentration profile during local heating of the ionospheric plasma, *J. Atmosph. Terrest. Phys.*, Vol. 58, No. 12, 1996, pp. 1345–1354.

78. Blaunstein, N., Evolution of a stratified plasma structure induced by local heating of the ionosphere, *J. Atmosph. Solar-Terrest. Phys.*, Vol. 59, No. 3, 1997, pp. 351–361.

79. Filipp, N.D., Power of H_E-scatter signals, *Izv. Vuzov, Radiophyzika*, Vol. 22, No. 4, 1979, pp. 407–411.

80. Filipp, N.D., *Specular Scattering of Ultra-Short Waves by Middle-Latitude Ionosphere*, Kishinev (Moldova): Shtiintza, 1980.

81. Filipp, N.D. and Blaunstein, N.Sh., Drift of ionospheric inhomogeneities in the presence of the geomagnetic field, *Izv. Vuzov, Radiophyzika*, Vol. 21, 1978, pp. 1409–1417.

82. Yamamoto, M.S., Fukao, S., Woodman, R.F., Tsuda, T., and Kato, S., Mid latitude E-region field-aligned irregularities observed with the MU radar, *J. Geophys. Res.*, Vol. 96, 1991, pp. 15943–15949.

83. Rawer, K., International Reference Ionosphere-IRI 79, *Report VAG-82*, World Data Center A, Boulder, Colo., 1981.

84. Namgaladze, A.A. and Latishev, N.S., Influence of upper boundary conditions on modeling ionospheric parameters, *Geomagn. Aeronom.*, Vol. 19, 1976, pp. 43–49.

85. Deminov, M.G., Kim, V.Yu., and Hegai, V.V., Influence of longitudinal electric fields on the structure of the ionosphere, *Geomagn. Aeronom.*, Vol. 20, 1977, pp. 837–840.

86. Kolesnik, A.G. and Golikov, I.A., Two-dimensional non-stationary model of the ionospheric *F* region, *Geomagn. Aeronom.*, Vol. 24, 1981, pp. 64–70.

87. Blaunstein, N. and Filipp, N.D., Diffusion of strong irregularity of arbitrary configuration in the unbounded ionospheric plasma, in book: *Investigations of the Problems of the Cosmic Geophysics*, Moscow: Nauka, 1982, pp. 25–44.

88. Blaunstein, N. and Filipp, N.D., Drift of plasma with arbitrary degree of ionization in the magnetic field (ionospheric case, in book: *Practical Aspects in Study of the Ionosphere and Ionospheric Radio Propagation*, Moscow: Nauka, 1981, pp. 163–169.

89. Rozhansky, V.A. and Tsendin, L.D., Evolution of strong inhomogeneities of the ionospheric plasma, *Geomagn. Aeronom.*, Vol. 27, 1984, pp. 598–602.

90. Rozhansky, V.A. and Tsendin, L.D., Influence of electric field and wind on spreading of the disturbance of weakly-ionized plasma in the magnetic field, *Geomagn. Research*, Vol. 27, 1980, pp. 93–104.

91. Rozhansky, V.A., Stratification of ionospheric and magnetospheric plasma, *Plasma Physics*, Vol. 7, 1981, pp. 745–751.

92. Rozhansky, V.A. and Tsendin, L.D., Spreding of inhomogeneity of weakly ionized plasma with the current in the magnetic field, *J. Technical Phys.*, Vol. 47, 1977, pp. 2017–2026.

93. Rozhansky, V.A., Perturbed electric fields and evolution of ionospheric inhomogeneities, *Geomagn. Aeronom.*, Vol. 25, No. 6, 1985, pp. 906–911.

94. Voskoboinikov, S.P. et al., Numerical modelling of diffusion of strong disturbance of weakly ionized plasma in the magnetic field, *Plasma Physics*, Vol. 6, 1980, pp. 1270–1276.

95. Vickrey, J.F. and Kelley, M.C., The effects of a conducting *F*-layer on classical F-region cross-field plasma diffusion, *J. Geophys. Res.*, Vol. 87, 1982, pp. 4276–4284.

96. Linson, L.M. and Workman, R. Formation of striations in the ionospheric plasma clouds, *J. Geophys. Res.*, Vol. 75, 1970, pp. 3211–3218.

97. Simon, A., Growth and stability of artificial ion clouds in the ionosphere, *J. Geophys. Res.*, Vol. 75, 1970, pp. 6287–6294.

98. Scannapieco, J. et al., Conductivity ratio effect on the drift and deformation of ionospheric plasma clouds and their interaction with other layers of the ionosphere, *J. Geophys. Res.*, Vol. 79, 1974, pp. 2913–2916.

99. Gatsonis, N.A. and Hastings, D.E., A three-dimensional model and initial time numerical simulation for an artificial plasma cloud in the ionosphere, *J. Geophys. Res.*, Vol. 96, 1991, pp. 7623–7639.

100. Rieger, K., Measurements of electric fields in equatorial and medium altitudes during twilight using barium-ion clouds, *Gerlands Bietr. Geophys.*, Vol. 213, 1980, pp. 243–253.

101. Schultz, S.G., Adams, J., and Mozer, F.S., Probe electric field measurements near a mid-latitude ionospheric barium releases, *J. Geophys. Res.*, Vol. 78, 1973, pp. 6634–6642.

102. Gurevich, A.V., *Nonlinear Phenomena in the Ionosphere*, Springer-Verlag, New York, 1978.

103. Gurevich, A.V. and Milikh, G.M., Artificial airglow due to modifications of the ionosphere by powerful radio waves, *J. Geophys. Res.*, Vol. 102, 1997, pp. 389–394.

104. Roberts, K.V. and Potter, D.E., *Methods of Computational Physics*, Vol 9: *Plasma Physics*, Academic Press, New York, 1970, pp. 335–416.

105. Blaunstein, N.Sh., Local heating of *F*-region of the ionosphere, in book: *Interaction of Radio Waves with the Ionosphere*, Moscow: Nauka, 1990, pp. 51–61.

106. Vas'kov, V.V. and Dimant, Ya.S., Influence of deformation of the regular profile of the ionospheric plasma at anomalous absorption of the powerful radiowave in the resonance region, *Geomagn. Aeronom.*, Vol. 29, 1989, pp. 417–422.

107. Vas'kov, V.V. and Dimant, Ya.S., Excitation of ion-sound density disturbances in the region of resonance of the powerful radiowave, *Geomagn. Aeronom.*, Vol. 30, 1990, pp. 268–274.

108. Vas'kov, V.V. and Gurevich, A.V., Nonstationary phenomena in reflection of the strong wave from the ionosphere, *Geomagn. Aeronom.*, Vol. 25, 1975, pp. 67–73.

109. Vas'kov, V.V. and Gurevich, A.V., Nonliner saturation of disturbances of electron concentration in the F-layer of the ionosphere during affecting by the powerful radiowaves, *Geomagn. Aeronom.*, Vol. 25, 1976, pp. 1112–1115.

110. Vas'kov, V.V. and Gurevich, A.V., Self-focusing and resonance instabilities in the *F*-region of the ionosphere, in book: *Thermal Nonlinear Phenomena in Plasma*, Nauka, Gorky (USSR), 1979, pp. 81–138.

111. Bernhardt P.A., Swartz, W.E., and Kelley, M.C., Spacelab 2 upper atmospheric modification experiment over Arecibo, 2, Plasma Dynamics, *Astro. Lett. Communic.*, Vol. 27, 1988, pp. 183–198.

112. Belikovich, V.V., Benediktov, E.A. Tolmacheva, A.V., and Bakhmet'eva, N.V., *Ionospheric Reseach by Means of Artificial Periodic Irregularities*, Katlenburg-Lindau, Germany, Copernicus GmbH, 2002.

Chapter 6

Modern Radiophysical Methods of Investigation of Ionospheric Irregularities

In recent decades there has been an increased interest in understanding the effects of ionospheric irregularities on radio propagation through the ionosphere in land–satellite communication links performance, as well as in land–ionosphere–land long-distance channel design. For the prediction of the key parameters of such channels, it is important to understand the processes accompanying the generation and further evolution of ionospheric irregularities of various origins, natural or artificial, and to predict how they affect radio signals transmitted through the ionosphere at different operational frequencies. The processes associated with generation of plasma irregularities, such as diffusion, drift, and thermodiffusion, were analysed in detail in Chapter 5.

In the present chapter, modern methods will be described, using ground-based radio and optical facilities as well as satellite facilities developed in recent decades to predict the main parameters of plasma irregularities. These included locations of irregularities in the ionosphere, the degree of plasma perturbations induced by irregularities with respect to the background plasma, the shape and time of relaxation (or lifetime) for future prediction of the key parameters of probing waves sent through the disturbed ionospheric region, such as reflection and scattering, absorption, fading, and so on. To predict these parameters more precisely, different radio and optical methods and the corresponding facilities were used during experimental investigations.

Many modern methods have been devised in recent decades for studying the irregular structure of the ionosphere, equatorial, middle-latitude, and high-latitude, based on conventional ionospheric stations and ionosonde sounding, satellite interferometers and ground-based radar systems, coherent and incoherent, and in situ rocket and satellite probes. To name one, ionosondes with linear frequency modulation of the transmitted signals with the help of which the authors of this book for many years investigated plasma irregularities, naturally and artificially induced in the ionosphere (see Chapters 7 and 8). Using various modern radio methods, signals amplitude and phase oscillations arriving from satellite can be precisely studied, and also the radio wave oscillations coming from stars after passing the disturbed ionosphere.

It is evident, taking into account limiting length of our book and its specific content related to radio propagation through ionospheric radio channels with fading, that we cannot cover all experiments and numerous theoretical explanations of the phenomena under discussion. We also understand that the broad spectrum of geophysical and radiophysical phenomena cannot be covered by this book. We concentrate our effort on only those experimental results that (1) give a systematic description of the corresponding precise experimental observations of natural and artificially induced plasma irregularities; (2) present rigorous of the phenomenon under observation; and (3) present adequate theoretical explanations of the observed phenomena. In our opinion such an approach would attract the attention of the reader to existing problems, how they have been resolved up to the present, and what remains to be done in the future. In other words, we will concentrate our efforts on the analysis of those parameters and characteristics of ionospheric irregularities that finally allow for establishing numerous applications in the process of the performance of the long-distance ionospheric channels, as well as of the land–satellite communication links in different ionospheric latitudes.

We begin with investigations of the artificial irregularities generated during the heating of the E and F layers of the ionosphere by powerful radiation from ground-based HF facilities [1–18]. We will then describe experiments on rocket injection of ion clouds in Earth's ionosphere during active space experiments [19–37]. The main goal of such experiments was mostly to understand and then to estimate the main geophysical factors in the ionosphere, such as the ambient electric and magnetic fields, as well as the neutral winds, vertical and horizontal, in the ionosphere by observing different stages of evolution and dynamics of artificially induced plasma clouds.

In this chapter we will also introduce the experimental study of parameters of natural ionospheric inhomogeneities based on how they are affected by radio propagation through the disturbed ionospheric regions. It should be noted that a wide spectrum of sporadic plasma irregularities has been found experimentally in the high-latitude ionosphere [38–47], the quasi-stationary natural H_E inhomogeneities and meteor trails in the middle-latitude ionosphere [48–51], and the bubbles and sporadic E and F layers in the equatorial electrojet regions [52–61]. All these affect in different ways radio waves passing through disturbed ionospheric regions, causing the radiophysical phenomena such as signal attenuation, absorption, scattering and diffraction, aspects that will be considered in further chapters on ionospheric propagation channels performance (Chapter 7), the corresponding optical and radio devices for investigation of the natural and artificial phenomena occurring in the ionosphere (Chapter 8), and the effects of different types of ionospheric irregularities on these radiophysical phenomena occurring in land–satellite communication (Chapter 9).

6.1 Radio and Optical Methods of Studying Artificially Induced Irregularities in the Ionosphere

6.1.1 Radio Methods for Diagnostics of Heating-Induced Irregularities

We begin with methods of radio diagnostics of formation and relaxation stages of plasma irregularities induced in the ionosphere by powerful short-wave radiation (called *pump wave*) from ground-based facilities that have been developed by numerous investigators [62–83] for study of both geophysical and radiophysical phenomena in the irregular inhomogeneous ionosphere.

Radiophysical Studies of the Disturbed Lower Ionosphere ($z < 80$–100 km). As is well known from the literature [84–89], the effect of cross-modulation occurs in the ionosphere, where the powerful radio wave radiation due to heating ionospheric plasma can change the collisional absorption of the probing waves sent through the disturbed heated region (DHR), because the coefficient of absorption of the probing wave, Γ, in the general formula of wave intensity

$$I = I_0 e^{-\Gamma} \tag{6.1}$$

is related to the collisional frequency of plasma particles, electrons, ions, and neutrals, $\nu_e = \nu_{ei} + \nu_{em}$, as

$$\Gamma = \frac{2\pi e^2}{m_e} \int \frac{N \nu_e(T)}{\omega^2 + \nu_e^2(T)} \frac{dz}{\cos \theta(z)} \tag{6.2}$$

where θ is the angle of incident of the probing wave at the ionospheric layer; the z-axis is directed vertically; all other parameters have been defined in Chapter 1. Hence, owing to the location of the DHR by probing waves, it is possible to obtain information on the parameters of the disturbed plasma region (all details can be found in Refs. [84–87]). Only a short time is needed to establish the quasi-stationary temperature and its relaxation, $(\delta \cdot \nu_e)^{-1} \leq 10^{-3}$ s, thus allowing the use of the new approach of study of altitudinal profiles of $\nu_e(z)$ based on the time tomography. This approach is based on lighting by probing pulse signals of the ionospheric regions that were earlier disturbed by the short pulses of 100–500 μs of powerful radio waves [90]. The probing pulses radiate with a time delay relative to the powerful pulses perturbing the ionosphere at altitude z, while passing through it (in the moment $t = z/c$) crossing DHR during direct propagation and then after reflection from the upper layer of the ionosphere at the different moments of time relative to the moment of heating by the pump wave. Based on the limiting properties of the wave speed, $c = 3 \cdot 10^8$ m/s, this effect allows for obtaining the altitudinal dependence of wave parameters [90] for small periods of changing of T_e and ν_e in the D layer of the ionosphere.

Another method of investigating disturbed ionospheric regions is by using heating-induced irregularities. From Equation 2.1, Chapter 2, the oscillatory speed $\mathbf{v}_\Omega$ of plasma electrons (ions, because plasma is quasi-neutral) in the field of the powerful radio wave, E_Ω, relates linearly to this field, and for the periodically changing electron collision frequency ν_e associated with these oscillatory movements, the alternate electric field can occur at the modulation frequency Ω:

$$\mathbf{j}_\Omega \approx \frac{\nu_{e\Omega}}{\nu_e} eN(\mathbf{V}_{e0} - \mathbf{V}_{i0}) \tag{6.3}$$

where in $\nu_{e\Omega}/\nu_e \sim 5T_e/6T_{e0}$, T_e and T_{e0} are, respectively, the disturbed (after heating) and nondisturbed (before heating) temperature of electrons; all other parameters having been defined in Chapters 1 and 2. The current described by Equation 6.3 leads to the variation of electromagnetic waves with frequency Ω. The conditions of observations of this radiation at frequency Ω depend on conditions of capturing of this radiation by the waveguide Earth-ionosphere and propagation inside it. We will defer a treatment of capturing and further propagation of radio waves within such a waveguide to Chapter 7. Here, we will only mention that conditions for obtaining such a radiation at the ground surface, i.e., of removing it from the waveguide, are defined by this frequency Ω, depending on how the width of the skin layer inside the waveguide has changed. This width controls the level of lower-frequency signal at the upper (ionospheric) bound of the waveguide. For sufficiently low frequencies Ω, the main contribution in the field at the ground surface gives the current momentum of the source, $\int j_\Omega dz$, which is defined by the horizontal dimensions of the disturbed heated region (DHR) and the characteristic height interval, $\Delta z = z_{max} - z_{min}$, of the ionospheric DHR (along the z-axis) [91]. The structure of j_Ω (z) also depends on the relationship between the Hall current ($j_H \sim U_m \times B_0$ or $j_H \sim E_0 \times B_0$) and the Pedersen current ($j_P \sim E_0$). This relation determines changes in the polarization of the receiving radiation [91]. Multiple reflections from the ionosphere (as an upper part of the waveguide) and Earth's surface (as a bottom part of waveguide) of powerful pulses, changing conductivity in the ionospheric current system, cause the series of low-frequency responses at the receiver.

Creation of powerful ground-based facilities operating at low frequencies (LH) for modification of the ionosphere under heating at frequencies close to the gyrofrequency of electrons makes possible continuous diagnostics of middle-latitude current systems in the ionosphere. The artificial radio radiation of low frequencies from the modified (after heating) ionospheric regions can be received successfully, as shown in Figure 6.1, according to experiments carried out in Ref. [66]. Here, the heating of the ionosphere was performed by short pulses ($\tau_1 = 2.5$ μs, Figure 6.1a; $\tau_2 = 0.3$ μs, Figure 6.1b) from the transmitter with effective power of $P = P_0 G = 240$ MW (G is the gain of antenna). Such experiments allow information to be obtained about the DHR using artificial electromagnetic radiation of the ionosphere (ARI) and also the structure of the ambient electric field at altitudes $z < 100$ km.

Another method used in the diagnostics of the lower ionosphere disturbed by a pump-wave heater from ground-based HF radar is the use of artificial periodical irregularities (APIs) created in the field of powerful radio waves. The first to predict the possibility of generation of such heating instabilities in the lower ionosphere was Vilensky, who described and summarized this in a series of joint investigations by him and his colleagues from NIRFI (Gorky, USSR) [62–71]. This methodology, based on numerous systematically prepared experiments, allows for studying a set of important parameters of the lower ionosphere using effects of scattering of probing waves by APIs. For example, in the D region ($z < 70$–80 km), observations of the relaxation time of API allow for finding the coefficient of attractions of electrons to molecules of neutrals after switching off the transmitter. The Doppler measurements allow for searching the vertical motions and displacements of API during their spreading and relaxation (see Section 5.3). Moreover, using different regimes of switching the electromagnetic heater on and off allows for measuring the vertical

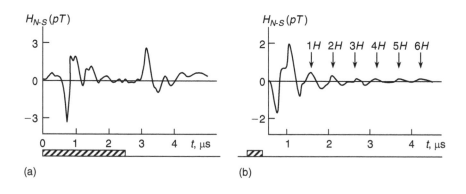

(a) (b)

Figure 6.1 Artificial radio radiation received at low frequencies from the modified ionosphere after its heating by short pulses with durations: (a) $\tau_1 = 2.5$ μs, and (b) $\tau_2 = 0.3$ μs.

component of API movements and also separating wave movements and turbulent mixtures, which are very important when studying recombination processes in the lower ionosphere. At the E layer (at altitudes 90–150 km) the velocity of ambipolar diffusion and drift along $\mathbf{B}_0$ can be estimated on the basis of the observations of the velocity of relaxation of API.

In the F layer, where a collision regime prevails, the nonstationary API excites ion acoustic waves. Their decay is determined by the collisional frequency of ions, ν_{im}, and by the collisionless attenuation at the heating ions. The latter related to the fact that the velocity of ion sound waves, $C_s = \sqrt{(T_e + T_i)/M_i}$, for $T_e \approx T_i$, is close to the thermal velocity of ions, $\nu_{Ti} \approx \sqrt{T_i/M_i}$, where, as mentioned in Chapter 2, for the Maxwellian plasma (i.e., in a steady-state regime), the derivative of the distribution function $\partial f_i(\nu)/\partial \nu$, defining Landau decay, is maximal [84–87]. Therefore, the value of this decay characterizes the ratio T_e/T_i in the ionosphere [84]. Using additional heating of the regular ionosphere by the pump electromagnetic wave, an increase by tens percent of T_e with respect to T_i (due to higher thermomobility; see Chapter 2) allows for separating the regimes of collisionless and collisional absorption because T_i gives information on the collisionless and T_e on collision absorption. Finally, this allows for estimating values ν_{im}, T_e and T_i, because $\nu_{im} \sim \sqrt{T_i}$ [92].

In Figure 6.2, the results of the measurements of electron concentration using the method of resonance backscattering at API, where the angle χ is the angle of inclination with respect to solar radiation, are presented. This method is based on the experimentally obtained fact that the artificial

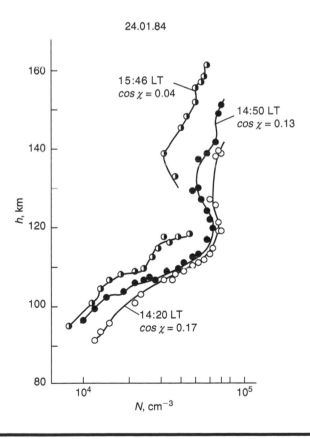

Figure 6.2 Height profile of the ionospheric plasma at lower altitudes measured using methods of backscattering at artificial periodical irregularities (APIs).

grid of irregularities (the characteristic example is modeled in Section 5.3) created by the wave with wavelength $\lambda_0/n_{o,x}$ scatters back a probing wave with the wavelength $\lambda' = \lambda_0/n_{x,0}$. Here, λ_0 is the wavelength in vacuum, and n_o and n_x are the refractive indexes of the ordinary and extraordinary waves, respectively. Therefore, between frequencies of excited wave, f_{ex}, and probing wave, f_{pr}, the following constraint could be satisfied [67]: $f_{ex}n_{o,x}(f_{ex}) = f_{pr}n_{x,o}(f_{pr})$. Taking into account that in the quasi-longitudinal approximation $n_{o,x}$ is proportional to the square of the electron plasma frequency ω_{e0}, and the latter is proportional to concentration of electrons N_e and, finally, using this method of backscattering from API, the electron content in the ionospheric disturbed region can be estimated. So, using the diagnostic of API by probing waves above the sites of the location of heating facilities, the main parameters of the ionosphere in the DHR can be estimated. In Ref. [92] the idea was proposed of using perspective excitation of API elongated along geomagnetic field lines (namely, during vertical excitation in the region of the magnetic equator), which could be more intensive, as well as moving API created by radiation of two waves with close frequencies.

Diagnostics of Large-Scale Heating-Induced Plasma Structures. The possibility of generating the regions with enhancement of electron concentration in the E and lower F layers during the heating of the ionosphere by pump electromagnetic wave was pointed out in Refs. [86,87]. Some studies [64,93] also indicated the possibility of creating a region with depletion of plasma concentration because of thermodiffusion occurring during heating by pump radio wave, and on using this region as a radio lens. Experimental investigations of regions with enhancement and depletion of electron concentration using methods of vertical and oblique radio sounding of the ionosphere were carried out from the beginning of the 1970s to the end of the 1990s [1–11,93–97]. In some of these experiments facilities for incoherent scattering of radio waves were used for studying large-scale plasma disturbances with the horizontal dimensions $L_h \sim \theta_A z$, where z is the height of heating and θ_A the width of the diagram of antenna of power transmitter.

Continuous experiments were carried out in Arecibo, where the powerful beam at $P_0 = 300$ kW with the gain $G = 23$ dB was radiated at the frequency $f_{ex} \approx 3.175$ MHz with the value of power inside the beam of 50 μW/m^2 at the height of the DHR [2–5]. Figure 6.3 shows a strong disturbance of the profile of electron concentration in the F layer (at altitudes 270–290 km) during

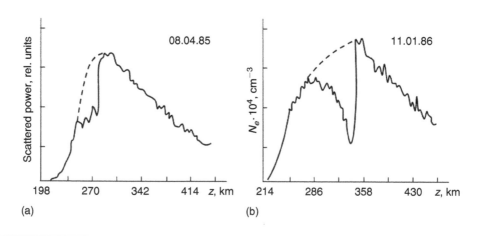

Figure 6.3 Examples of strong perturbations (e.g., depletions) of the background profile of the ionospheric plasma density (dashed curves) occurring in the F layer of the ionosphere obtained during nocturnal periods at the Arecibo observatory for (a) relative power of scattered waves and (b) electron concentration distribution.

heating of the nocturnal ionosphere (with $f_{e0} \approx (3.5 \div 4)$ MHz), the relative power (Figure 6.3a) and the corresponding electron concentration (Figure 6.3b), versus height h. The corresponding undisturbed profiles of power and plasma concentration at altitudes under consideration are shown by dotted curves. It is evident that during local heating a depletion region of plasma concentration (i.e., the negative disturbance) occurs, the effect of which was fully predicted by the self-consistent model of local heating described in Section 5.3. Here, following this model and taking into account the parameters of the ionosphere above Arecibo, as well as the parameters of the experiment presented in Figure 6.3, we present results of numerical simulation of the real heating experiment in Figure 6.4a for the concentration and in Figure 6.4b for the temperature of plasma versus the height of the local heating. It is evident that the computer experiment based on the proposed model is a good predictor of the existence of deep cavity in the nondisturbed profile of plasma density and increase of the temperature of plasma from 750 to 2500 K (in the real experiment the electron temperature was increased from $T_{e0} \approx T_{i0} = 800$ K to $T_e \approx 2400$ K), and decreasing to the nondisturbed value again after switching off the powerful transmitter. The ion temperature changed its value insignificantly during heating. The horizontal dimension of the cavity was compared with the dimension $\sim\theta_A z$, and the cavity was elongated along $\mathbf{B}_0$. Some discrete structures were found inside the cavity moving insignificantly along the gradient of plasma concentration with the effective drift velocity $0 < V_d < 100$ m/s. Diagnostics were carried out by the locator working in pulse regime (with $\tau_{pulse} \approx 52$ µs and with the repetition time of 10 µs), which allows for obtaining the height profiles of plasma density $N(z)$ with resolution of 600 m.

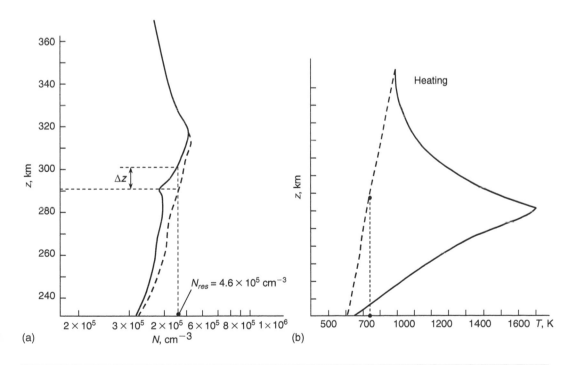

Figure 6.4 Numerical experiment of the local heating of the ionosphere: (a) perturbations of plasma density, and (b) perturbations of the plasma temperature with respect to those of the background ionosphere (dashed curves).

Simultaneously, with the help of another locator, the temperatures T_e and T_i were studied with resolution along an altitude about 45 km and duration time several minutes. When separations with the resonance frequency $f_{ex} \approx 3.175$ MHz were performed, the specific dynamics of the cavity region was observed. Thus, in Figure 6.5 the dynamics of the cavity region and its structure after switching off of the excited wave are presented with separations from $f_{ex} \approx 3.175$ MHz: curve 1 corresponds to separation at 1.2 kHz; curve 2 to 1.7 kHz; curve 3 to 2.2 kHz; curve 4 to 2.7 kHz; curve 5 to 3.2 kHz; and curve 6 to 4.2 kHz. This heating regime can be predicted by the model of the so-called adaptive heating of plasma described in Section 5.3. Here, the extrema of the plasma profile within the cavity following the density corresponds to the current resonance frequency. With increase of the frequency separation (from 1.2 to 4.2 kHz), the corresponding extremum moves downward along the height of the ionosphere.

We should point out that the main heating of electrons can be done either by the acceleration of plasma particles by plasma waves (thermal instabilities; see Chapter 2) or by the creation at the relaxation stage of the polarized (inner) field $E_{p\parallel}$ due to the existence of longitudinal currents or the ambient electrical field E_0. Because amplification of the electric field within the cavity is $E_{p\parallel} \sim E_0(N/N_0)$, and the stationary temperature $T_e \sim E_{p\parallel}^2/\nu_e^2$, and accounting for changes of $\nu_e(T_e)$, we finally arrive at the fact that for $N_0/N \leq 0.3$ the ratio $T_e/T_{e0} \leq 10\nu_{e0}^2/\nu_e^2(T_e)$.

The cavities presented in Figures 6.3 and 6.5 were only observed during nocturnal time, which, according to a theory presented in Section 5.3, can be related to the absence of competition from the ionization–recombination processes against transport processes occurring in the upper ionosphere. At the same time, the altitudinal stratification of the disturbed region (the cavity), as shown in Figure 6.5, was observed and discussed seperately in Ref. [98]. Distortion of the ionograms of vertical sounding during heating with power $P = 100 - 120$ kW, and with gain $G \approx 80$, was found in the earlier morning and later evening hours. Using the special regime of excitation (such as the adaptive regime described in Section 5.3), the possibility of destroying the small-scale low-frequency turbulence and exciting and amplifying the large-scale plasma structures was found during these cycles of experiments [98]. The results obtained showed the possibility of controlling

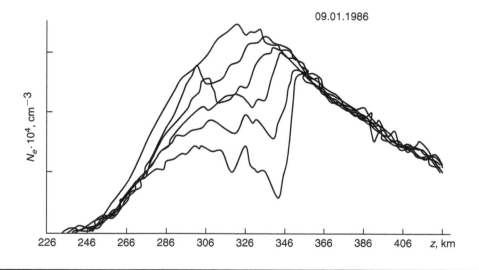

Figure 6.5 Evolution of the depletion area and its structure during the experiment with adaptive changes of the resonance frequency of the powerful transmitter.

the large-scale structure inside the disturbed heated region (DHR) due to change of the resonance frequency of the heater and the time regime of heating process. Performance of such a heating experiment could further the study of heating-induced large-scale ionospheric plasma disturbances. The corresponding theoretical framework of such an experiment, which was described in Section 5.3, allows for realizing all ideas of heating methods during a virtual computer experiment based on the computational self-consistent models presented there.

Measurements of Parameters of Moderate-Scale Artificial Irregularities. The artificial irregularities of the ionosphere of moderate scale 0.1 km $< \ell_\perp < 10$ km can change the structure of the signal reflected from the ionosphere (in the same manner as the large-scale irregularities) during vertical and oblique soundings of the disturbed ionospheric regions, as well as cause changes in phase and amplitude of signals of orbital (LEO and MEO) or geostationary satellites (see details in Chapter 9). Simultaneous observations of amplitude and phase allow for gathering information about the power of moderate-heating-induced irregularities up to the scales of $\ell_\perp = (3 \div 10)$ km, because disturbances of signal phase are not affected by the influence of Fresnel's filtering, which limits occurrence of irregularities with scale $\ell_\perp > \sqrt{\lambda \bar{R}}$ at the variations of signal amplitude. Here, $\bar{R} = R_1 R_2 / (R_1 + R_2)$, where R_1 and R_2 are distances from the transmitter and the receiver to the layer with irregularities [88,99].

Information on spectral characteristics of such irregularities using illuminating of the ionosphere by in situ satellite probes will be discussed in the following text. We now only point out that such observations allow for studying dynamics of artificial irregularities at different altitudes from the source of their excitation (from the levels of reflection of the powerful radio wave from the ionosphere) [79]. Using such a methodology, high velocities ($\nu > (0.5-1)$ km/s) of the vertical transport of the agent responsible for generation of artificial irregularities were observed in Ref. [79].

Similar results can be obtained by using oblique or vertical sounding of the DHRs by signals of different frequencies [70]. The latter relates to the fact that the main contribution to the signal amplitude and phase fluctuations during reflections from the ionosphere introduce irregularities located near the level of wave reflection, first of all, irregularities with $\ell_\perp \geq (1 \div 5)$ km. Therefore, the observations carried out at different frequencies offer the possibility of obtaining altitudinal resolution of the DHR. From the corresponding ionograms of vertical or oblique sounding, the characteristic features of the moderate-scale irregularities can easily be found. Thus, the main characteristic of such turbulences is the phenomenon of diffusivity of the frequency ionograms, defined in the literature as E_{spread} or F_{spread}, depending of the height of heating and observation (see Chapter 7). This phenomenon is related to spread of the pulse signal because of reflection and scattering from irregularities located mainly near the level of reflection of sounding pulses from the ionosphere (see details in Chapter 9). The time of excitation of moderate-scale irregularities increases with an increase in their scale $\ell_\perp$ [78]. Therefore, after studying ionograms at different time intervals after switching the powerful radio transmitter on and off, selective influence of the irregularities of various scales on ionograms of vertical and oblique sounding of the ionosphere at the DHR can be found. As shown during experiments (see detailed analysis in Chapter 7), diffusivity at the ionograms occurs at moments less than 1 min after the commencement of the heating process. However, the occurrence of strong F_{spread} related to the time of excitation exceeding 3–7 min is also observed (see Chapter 7). The observed effect of easy excitation of F_{spread} and increase of its lifetime in the nocturnal ionosphere compared with the daytime ionosphere again demonstrate a correction of the main theory presented in Section 5.3, according to which in the nocturnal ionosphere effects of ionization and recombination are insufficient compared with transport processes. Therefore, in the nocturnal ionosphere all transport processes (thermodiffusion, diffusion, and drift) are easily developed with respect to the daytime ionosphere, where the recombination process depends strongly on the temperature of plasma and decreases with an increase of heating power.

Radio Methods of Studying Small-Scale Artificial Ionospheric Turbulence. We will define small-scale artificial ionospheric turbulence (AIT), which usually occurs near the level of reflection of the excited wave, as the irregularity, the transverse scale of which is described by $\ell_\perp < \lambda_{ex.w}$, where $\lambda_{ex.w}$ is the wavelength of the excited wave in vacuum. As was mentioned earlier and in Chapter 2, the AIT is generated during the transformation of energy of the powerful radio wave into energy of the plasma wave. Finally, this process defines the main method of AIT diagnostics, called the *method of anomalous weakness* (due to anomalous absorption) of the exciting and probing waves. The effect of anomalous absorption is stronger for waves with ordinary polarization according to its behavior in the region of transformation of energy (see Figure 2.2, Chapter 2). The anomalous weakness of the exciting waves, after exceeding the power limit of parametric striction instability (see definition in Chapter 2) of about $P_{eff} \equiv \eta PG \approx 80$ MW, can achieve 70%–80%. At the same time, the anomalous weakness of the probing waves occurs only if their frequency differs from the frequency of exciting waves by not more than a few kilohertz. The fact that the probing waves with frequency $f_{p.w}$, which differs from that of excited waves $f_{ex.w}$ only at several kilohertz, are not displaying any anomalous weakness shows that there is a narrow-height region in the ionosphere where the striction parametric instability can be developed [78]. This experimentally obtained fact was proved by direct measurements of plasma waves using the method of incoherent scattering and an experimentally obtained narrow spectrum of scattering waves. During their further evolution, and after scattering at the heating-induced irregularities and leading to the large-scale striation of the disturbed plasma regions, plasma waves fill large ranges of observation angles and height intervals in the ionosphere (as seen in Figure 6.5). These phenomena cause instability in the region of high-hybrid resonance (HHR) and generation of field-aligned (along $\mathbf{B}_0$) irregularities, which then cause transformation of both exciting waves and probing waves into plasma waves (in the sufficiently wide spectral range – up to hundreds of kilohertz). Times of development of anomalous absorption of the probing waves of various frequencies are determined by the growth of intensity of irregularities of various scales generated at different altitudes. Therefore, the dynamics of the anomalous weakness (absorption) of the probing waves illuminating the disturbed heated region (DHR) provides much information about the spectrum of irregularities formed at the altitudinal range occupied by DHR.

When the intensity of heating-induced irregularities is strong enough, the anomalous absorption can also be observed for the extraordinary probing waves (i.e., waves with extraordinary polarization). In this case the weakness of the wave occurs due to the scattering process, and not by transformation into plasma waves. Diagnostics of AIT using the method of anomalous absorption of the exciting and probing waves provides information on the small horizontal part of DHR (with an area similar to the first Fresnel zone). We mentioned earlier another method using artificial low-frequency radiation of the ionosphere (see Figure 6.1). Observations in Refs. [78,100] of the dynamics of ARI (see also Figure 6.1) in the region of negative and positive ARI frequency deviations from the frequency of the exciting wave, consisting of information on the processes of splitting (and grouping) of plasma waves, and their penetration in the upper ionosphere along altitude caused by photoelectrons, are the problems which in present time do not find their full understanding and physical interpretation.

At the same time, at the Arecibo [1–6] and Tromse [101–103] experimental sites, where powerful radiolocators for diagnostics of artificial plasma disturbances in the ionosphere were used, more complete information about distribution of $N(z)$ and $T_{e,i}(z)$, and about ion acoustic and plasma waves, was obtained having at the same time evident description of the physical phenomena. We did not go into all details presented in Refs. [1–6,101–103].

Another effective method of investigations of AIT, mostly stretched along $\mathbf{B}_0$, is the method of specular scattering of HF and VHF waves, which will be described in more detail in Chapter 7. Now, following Refs. [7,8,80–82], we will direct the reader's attention to radiophysical methods of estimation of parameters of the process of relaxation of AIT, and their lifetime dependence on the initial dimensions of created heating-induced irregularities filling the heating region. Thus, in Figure 6.6a the dependence of the amplitude of scattered probing wave A_s during the short heating of the ionosphere by pulse with period $\tau_{heat} = 0.1$ s, normalized on the value A_{s0} (corresponding to the stationary process) and presented in dB, versus the time of observation after switching off the heater is shown for the different intervals between the further heatings, T_p: curve 1—$T_p = 4.9$ s; curve 2—$T_p = 9.9$ s; and curve 3—$T_p = 19.9$ s. In Figure 6.6b the dependence of the time of achieving maximum value of A_s, denoted by t_m, is shown versus the moments of heating T_p according to Ref. [81]. From Figure 6.6a it is seen that after switching off the heater, the amplitude A_s has a tendency to increase. At the same time, t_m depends on the time of pause between heatings. The velocity of relaxation of A_s also depends on T_p. Thus, from the moments of 1 to 2 s, the fast relaxation of A_s is observed with the coefficient $D \approx 3 \cdot 10^3$ cm²/s, whereas in its further relaxation the velocity of the decay becomes faster and is determined by the coefficient of diffusion $D \sim \ell_\perp^2/(2\pi)^2\tau \approx 1.4 \cdot 10^3$ cm²/s (where τ is the time of diffusion relaxation of A_s defined in Chapter 5, and $\ell_\perp \approx 3.3$ m [81]).

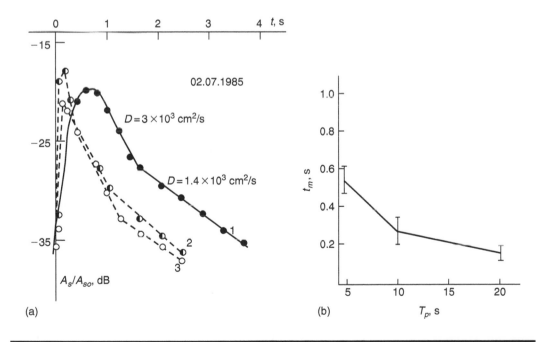

Figure 6.6 (a) Time dependence of the normalized amplitude of the scattered signal for different time intervals between heating pulses, T_p: 1 corresponds to 4.9 s, 2 corresponds to 9.9 s, and 3 corresponds to 19.9 s, and (b) time of the amplitude of scattered signal to achieve its maximum value, t_m, versus T_p.

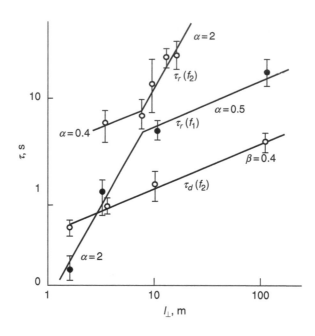

Figure 6.7 Polynomial two-step dependence of the relaxation and excitation time of artificially induced irregularities versus their experimentally observed transversal scales.

The dependence of times of development τ_d and relaxation τ_r of artificial heating-induced irregularities of various scales during stationary heating of the ionosphere using frequency $f_1 = 4.8$ MHz (diurnal time) and $f_2 = 2.95$ MHz (nocturnal time) is presented in Figure 6.7 versus the transverse (to $\mathbf{B}_0$) scale $\ell_\perp$ of irregularities extracted from Ref. [82]. It is seen that the time of development, τ_d, increases continuously with $\ell_\perp$ as $\tau_d \sim \ell_\perp^\beta$, $\beta = 0.4$, whereas the relaxation time has a two-mode character of increasing with $\ell_\perp$ [82]:

$$\tau_d \sim \ell_\perp^\alpha, \quad \alpha = \begin{cases} 2, & \ell_\perp = 1.6 \div 7 \text{ m} \\ 0.5, & \ell_\perp = 8 \div 100 \text{ m} \end{cases} \tag{6.4}$$

Thus, for small $\ell_\perp$ the relaxation time is increased proportionally to $\ell_\perp^2$, whereas for larger irregularities with $\ell_\perp \geq 10$ m, the velocity of relaxation becomes slower and the relaxation time of such irregularities increases sufficiently slowly with $\ell_\perp$, proportional to $\ell_\perp^{1/2}$. Hence, with an increase of pause between heatings (with increase of T_p) during the achievement of the next short heating process, larger irregularities have faded away. Therefore, their influence on excitation and relaxation of irregularities with smaller scales is the evidence of the transformation of energy along the spectrum of turbulence. As shown in Ref. [89], the direction and velocity of this transformation is defined by the initial form of turbulence spectra. Results of experiments [80–82,89] mentioned earlier showed that observations of AIT excited by various regimes of heating, simulated numerically in Section 5.3, can help in finding mechanisms of the formation of turbulence spectra and ability to control such spectra.

We will now return to more precise experiments carried out to estimate relaxation time of AIT and its dependence on initial scales of irregularities, because it in a way contradicts the three-mode dependence of relaxation time versus initial transverse scale $\ell_\perp$ obtained theoretically in Section 5.1, and the two-mode dependence (Equation 6.4) obtained experimentally. Furthermore, differences in dependence of relaxation time for diurnal and nocturnal ionosphere, as well as the height dependence of relaxation time of AIT obtained experimentally, are the subject of the unified theoretical explanations.

The series of experiments was carried out on the radio path Kiev–Gorky–Beltsi (former USSR) in 1985 using the method of back specular scattering of radio waves [7,8] because the transverse scales $\ell_\perp$ responsible for the specular scattering corresponded approximately to the half-wavelength of the probing waves (according to Bragg's law). The heating region was generated using a Sura heating facility located near Gorky and operating at frequencies 4.8–5.8 MHz. The geometry of the path corresponded to conditions close to the back scattering of radio waves. Because the operational frequencies of the probing waves used for diagnostics of artificial irregularities were in the range $f = 11$–23 MHz; the corresponding transverse scales $\ell_\perp$ were equal, $\ell_\perp = \lambda/2 = 6.5$–14 m. The observations were made in diurnal and nocturnal time in different seasons. Thus, the experiments presented in Figure 6.8 were carried out in (1) May 1985 at 19:00–21:00 and (2) November 1985 at (a) 10:00–11:00, (b) 13:00–16:00, and (c) December 1985 at 10:00–11:00 (local time). All figures show the experimental relations of the average relaxation time τ_r versus the transverse scale $\ell_\perp$ (denoted in Chapter 5 by d) of artificial irregularities. It can be seen that τ_r with increase in $\ell_\perp$ grows sharply (as was obtained in Refs. [81,82] and shown in Figure 6.7) according to the mode $\tau_r \sim \ell_\perp^\alpha$, $\alpha = 1.5$–2, for $7 \leq \ell_\perp \leq 9$ m for all seasons and different times, morning, noon, and evening. At the same time the effect of saturation with the mode $\tau_r \sim \ell_\perp^\alpha$, $\alpha = 0.2$–0.5 is observed for $10 \leq \ell_\perp \leq 14$ m. Results of these experimental investigations can easily be understood using theoretical results obtained in Section 5.1 according to Refs. [104,105].

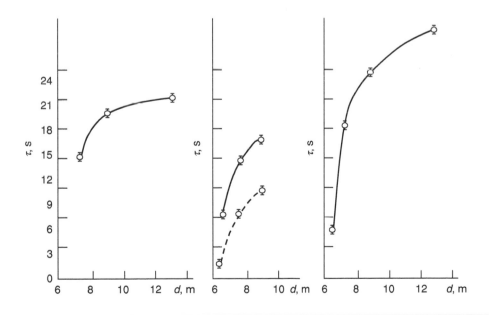

Figure 6.8 Experimental dependence of the relaxation time of heating-induced irregularities (HII) on the experimentally obtained transversal scale, by the receiver located in Beltsi at the middle-latitude radio trace Kiev-Vasil'sursk-Beltsi taking place at (a) 10:30–11:00 (November 1985), (b) 13:00–16:00 (November 1985), and (c) 10:00–11:00 (December 1985).

Thus, according to theoretical predictions for the transverse scale $d < d_1$ the relaxation time is $\tau_r = d^2/(2\pi)^2 D_{a\perp}$, where $D_{a\perp} \approx 2D_{e\perp}$ is the doubled transverse electron diffusion. However, in the case $d > d_1$ the dependence is close to $\tau_r \sim d^{1/2}$, which corresponds roughly with the value $L^2/8D_{i\parallel}$, where $D_{i\parallel}$ is the coefficient of longitudinal ion diffusion. Thus, for large scales d, where the value τ_r depends weakly on d, the main role in relaxation of artificial irregularity is longitudinal ion diffusion. The weak dependence of relaxation time τ_r on d can also be explained by weak dependence of characteristic scale L on d ($L \propto d^{1/4}$), following from the theoretical analysis [104,105].

If we now denote the critical scale d by $\ell_\perp^*$, as done in Refs. [7,8], we can present this critical scale according to experiments carried out in the aforementioned references. Thus, the critical transverse scale $\ell_\perp^*$ and the relaxation time τ_r of artificial irregularities, on which the polynomial dependence of the relaxation time changes from $\tau_r \sim \ell_\perp^{1.5-2}$ for $\ell_\perp < \ell_\perp^*$ to $\tau_r \sim \ell_\perp^{0.5}$ for $\ell_\perp > \ell_\perp^*$, were estimated as follows: for the daytime ionosphere $\ell_\perp^* = 10-25$ m and $\tau_r = 3-10$ s; in the evening time $\ell_\perp^*$ decreases to 6–10 m and $\tau_r = 7-15$ s; and for the nocturnal ionosphere $\ell_\perp^* < 3$ m and $\tau_r = 2-3$ s. In further experiments described in Refs. [7,8], a more detailed study of relaxation time as a function of $\ell_\perp$ in the region $\ell_\perp \approx \ell_\perp^*$ was carried out and a conclusion about the altitudinal character of relaxation time of heating-induced irregularities was drawn. Thus, taking into account estimations presented earlier, we can point out that for $\ell_\perp > \ell_\perp^*$, $\ell_\perp^* = 8-9$ m, and accounting for an expression to the longitudinal of diffusion of ions, $D_{i\parallel} = T_i/M_i \nu_{im}$, it can be found that relaxation time is proportional to ν_{im}, and its altitudinal dependence on z can be determined by the analogous dependence $\nu_{im}(z)$. Following these results, in Refs. [7,8], the experimental altitudinal dependence of the relaxation time (curve 1 in Figure 6.9) was obtained and compared with that dependence obtained using theoretical analysis presented in Section 5.1 based on the aforementioned references (dotted curve 2 in Figure 6.9). The satisfactory agreement between experimental data and theoretical prediction is evident.

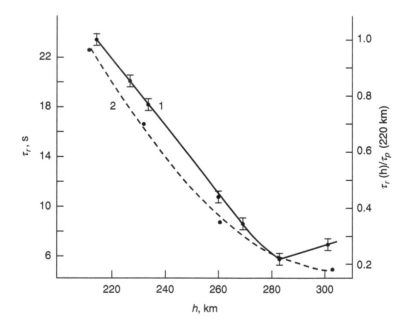

Figure 6.9 Height dependence of the relaxation time of HII obtained experimentally (continuous curve 1) and theoretically (dashed curve 2).

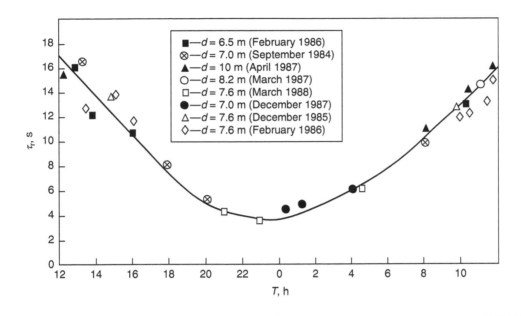

Figure 6.10 Diurnal and nocturnal dependence of the relaxation time obtained experimentally at the middle-latitude radio traces Kiev–Gorky–Beltsi and Moscow–Gorky–Beltsi.

At the next stage, the diurnal dependence of τ_r was obtained experimentally. This dependence is illustrated in Figure 6.10, which shows that τ_r in the altitude region 200–250 km decreases sharply from day to night. As can also be noted, the experimentally obtained diurnal dependence of the relaxation time is the same for any season of the year (see signs indicating various months). All experiments, carried out in the period 1984–1987 [7,8,78–83], showed that, as estimated from experiments, the critical scale $\ell_\perp^*$ is close to the limiting scale $d_{kr} \sim 3(D_{e\perp}/D_{i\parallel})^{1/2}L$ obtained from Equations 5.104 and 5.105, which determines the transition from a longitudinal electron mode to a transverse ion mode of diffusion of weak plasma diffusion [105].

At the same time the theory predicts a three-mode stage of heating-induced irregularities relaxation with two critical transverse scales, $d_1 \equiv \ell_\perp^*$ and $d_2 \equiv \ell_\perp^{**}$, whereas the experiment predicts only two-mode dependence with only one critical scale $d_1 \equiv \ell_\perp^*$. To resolve this contradiction between theory and experiment, in Ref. [106], a grid of irregularities, described in detail in Section 5.3, was proposed. According to this model, the grid or striated plasma structure (SPS) consisting of N separate irregularities spreads in such a manner that only two-mode dependence on $\ell_\perp$ is obtained from a numerical analysis of the problem. Here, as in Section 5.1, for a single irregularity, the lifetime of SPS, τ, was estimated from the level of plasma density attenuation at its maximum until it fell e^{-1} times from its initial value (i.e., when $\delta N(0,0,0) = nN(0,0,\tau)$, $n = e$, 5, etc). Thus, as presented in Figure 6.11, for N changing from 1 to 5, the relaxation time of the SPS at its maximum ($z = 0$, $\rho = 0$) has three-mode dependence with $\ell_\perp^*$ and $\ell_\perp^{**}$ only for $N \leq 2$, whereas for $N \geq 5$ it is analogous to the case of $N = 3$ and $N = 4$, having only a two-mode increase with $\ell_\perp^* \approx \ell_\perp^{**}$. In fact, as follows from illustrations presented in Figure 6.11, the character of attenuation of the SPS with the numbers of irregularities inside $N \geq 5$ and with a transverse scale of each irregularity, $\ell_\perp$, less than the SPS scale, r_0 and less than Larmor ion radius, ρ_{Hi}, emanates from two-stage relaxation process with critical scales $\ell_\perp^* \approx 5-10$ m and $\ell_\perp^{**} \approx 100$ m to one-stage with break point $\ell_\perp^* \approx \ell_\perp^{**} \geq 10$ m (as was found in Ref. [105] for a single irregularity). Now, using

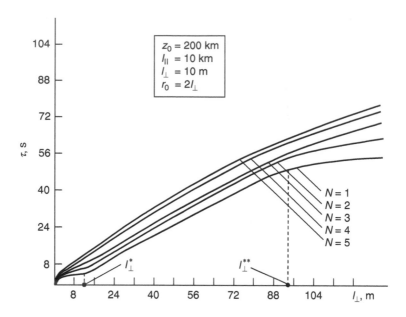

Figure 6.11 Results of numerical simulation of the relaxation time of HII versus their transversal scales for various numbers N of irregularities inside a grid of irregularities.

the results of numerical calculations obtained in Ref. [106] and the peculiarities observed in the process of SPS relaxation (see Figure 6.11), we can successfully explain the two-stage relaxation obtained experimentally (see Figure 6.8) with one break point $\ell_\perp^* \geq 10$ m and with two-mode polynomial dependence of the relaxation time: sharp increase as $\tau_r \sim \ell_\perp^{1.5-2}$ for $\ell_\perp \leq \ell_\perp^*$ and weak increase as $\tau_r \sim \ell_\perp^{0.5}$ with further saturation for $\ell_\perp > \ell_\perp^*$, where $\ell_\perp^* \geq 10$ m. In the regions of small scales before $\ell_\perp^*$, the spreading of the field-aligned heating-induced irregularities across $\mathbf{B}_0$ is "electron-controlled," and with larger scales beyond $\ell_\perp^*$ is "ion-controlled" along $\mathbf{B}_0$ (see Chapter 5). Thus, we have satisfactory agreement between theoretical prediction and experimental data.

The foregoing experimental and theoretical data illustrates possibilities of the method of specular scattering for resolving many problems of radio communication, which will be discussed in detail in Chapter 7. These results need additional experimental investigations for their further confirmation. It is particularly interesting to understand the role of the nonlinear processes of transformation of energy of initial turbulence along the whole spectrum of scales. This aspect can explain contribution of transformation of energy along spectrum of scales to the dependence $\tau_r(\ell_\perp)$ near the "break point" $\ell_\perp^*$. This aspect is also important for concretization of the mechanisms of generation of small-scale artificial plasma turbulence in the ionosphere during its heating by powerful radio waves. It will be interesting in future studies to compare dynamics of artificial turbulences of small scales $(\ell_\perp < \ell_\perp^*)$ and those that are comparable with or larger than $\ell_\perp^*$ in different geophysical situations, and not only for the middle-latitude ionosphere.

6.1.2 Methods of Investigations of Dynamics of Plasma Clouds in the Ionosphere by Rocket Injection

In the first cosmic experiments, with help of rockets and satellites and direct measurements obtained of geophysical parameters and characteristics in the natural conditions, the study of the main properties of ionospheric and magnetospheric plasma was carried out. However, not all natural phenomena and conditions occur often enough to allow for carrying out continuous and regular observations. This led to the idea of re-creating the same situations found in natural conditions in the nearby Earth space which can be modeled (as was done by heating of the ionosphere by the injection of pump electromagnetic waves discussed earlier) to the injection of the artificial ion clouds and beams from rockets into the ionospheric and magnetospheric plasma. Such experiments open the possibility of imitating natural phenomena and the diagnostics of the near-the-Earth plasma. Thus, the artificial cloud can be regarded as a probe to investigate the main characteristics of the space near Earth, such as ambient electrical and magnetic fields, their disturbances, the density and temperature of plasma, its wave properties as well as nonlinear phenomena occurring in plasma under its perturbation.

Experiments for creating artificial plasma structures in the Earth ionosphere began at the end of the 1950s and continued until the middle of the 1990s [19–36,111–117]. Today, extensive experimental data has accumulated and adequate physical models developed, thus allowing the comparison of experimental data with theoretical predictions. The first experiments carried out from the end of the 1960s to the beginning of the 1970s at altitudes 120–170 km [19–21] showed that the more suitable material for the creation of artificial ionized clouds is ions of barium (Ba^+), because their characteristic properties are better for optical and spectral observations compared to other ions (strontium, cadmium, etc.). Then, a cycle of further experiments described in Refs. [22–31] was carried out, which estimated the influence of ambient electric and magnetic fields on the ionized component of the injected cloud and determined the wind of the neutrals in the ionosphere at heights of 120–200 km following observations of the neutral component of the injected cloud.

We will now introduce the reader to the problems that are directly related to investigations of the process of evolution of strong plasma perturbations of the ionosphere created by injections of ion clouds from a special geophysical rocket. It is not so important for the process of creation of such ion clouds, but the methods of investigation of their parameters, as well as characteristics of the background ionospheric plasma during the process of spatial and temporal evolution of clouds.

Nonlinear Evolution of Injected Plasma Clouds in the Ionosphere. Detailed experiments of rocket injections based on visual observations of the evolution of the injected plasma clouds allowing defining a full picture of relative movements of the neutral and ionized components of clouds, and their spreading in the space and time domains, were studied systematically in the United States [19–26] and the former USSR [30–37]. In Ref. [24] the rocket experiment carried out at the height $z = 194$ km at the Florida experimental site in 1967 is described. Detailed observations were carried out at three experimental sites of observation, S_N, S_W, and S_E, during 930 s from the moment of cloud creation. Each site of measurement (see Figure 6.12) is shown in the coordinate system with direction zA to the North, zB to the East, zC vertically upward, zD along $\mathbf{B}_0$, and zE in the direction of the sun. The positions of these observation sites were as follows: S_N is about 100 km in a northerly direction, S_W about 150 km in a westerly direction, and S_E about 150 km in an easterly direction from the region of injection. Site S_E was oriented to the vector $\mathbf{B}_0$, which passes through the point of injection and has a magnetic declination of $\approx 3°$ and a magnetic inclination of $\approx 61.5°$. As can seen from the three groups of isolines on the right of equal optical lightness (or isolines of equal density) observed separately from three experimental sites and the corresponding analysis of experimental data during 930 s after creation, the center of the cloud descends with a velocity of 27 m/s at the height of 168 km (from 194 km) along the lines of geomagnetic field $\mathbf{B}_0$ and forms a columnlike trail with the plasma concentration of $\approx 10^{10}$ cm^{-3}, which was observed at the height of 160–190 km along $\mathbf{B}_0$. During the whole time of the observations, the cloud itself moved across $\mathbf{B}_0$

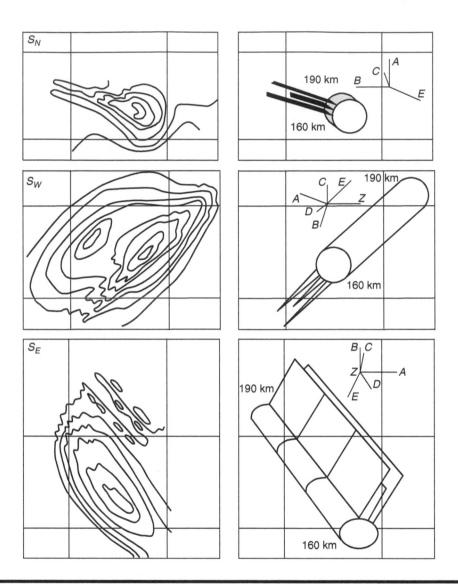

Figure 6.12 Lines of equal lightness inside barium cloud (left panels) and its schematic representation during evolution (right panels) obtained from photos made from point S_N (top panels), S_W (middle panels), and S_E (bottom panel).

with a velocity 35 m/s relative to the neutral wind. Three pictures on the left present isolines of equal density for barium cloud obtained during the numerical modeling of the problem. The main layered structure of the injected cloud formed during the observation time of 930 s is shown separately in Figure 6.13. It can be seen that there are three vertical strata oriented along $\mathbf{B_0}$ floating from the bottom of the ion cloud in the direction of the neutral wind. Estimations showed that the length of strata along $\mathbf{B_0}$ was close to 30 km, their width was about 0.5–0.8 km, and they are placed at the distance of 10 km from the center of the ion cloud. From the sites S_N and S_W the developed striations of cloud into strata, oriented along $\mathbf{B_0}$, were observed precisely. From the site S_E the cloud was observed after 930 s as a bright light sphere with fingerlike layer structures—strata strongly elongated from the center of the cloud. From the site S_N, using azimuth and inclination, the

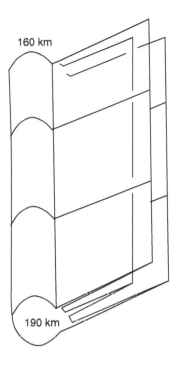

160 km

190 km

Figure 6.13 Modeling of the ion cloud by the cylinder with side strata.

nucleus and three strata were obtained. Crossings of all these coordinates define four lines of the cylinder and three layers parallel to Earth's surface at the point of injection at different altitudes, 160, 170, and 190 km, as shown in Figure 6.13. This simple model of the cloud with strata was chosen to observe the cloud simultaneously from all experimental sites S_N, S_W, and S_E (Figure 6.12). Why is it important? Because, for example, from S_W and S_E through 930 s, the peaks of brightness at heights 165–168 to 185 km were observed; from S_W the bottom ends of strata were better observed, whereas from S_E the upper ends of strata were better observed at altitudes up to 185 km. Despite the absence of a good height resolution at the site S_E, the angle dimensions and thickness of each strata were estimated more accurately.

Additional observations of evolution of the ion cloud (see Figure 6.14) have shown that during $t = 500$ s the ambipolar diffusion across $\mathbf{B}_0$ occurs with the velocity of diffusion, which is only about 10% of the velocity of diffusion of the neutral barium cloud. After 500 s the cloud becomes deformed, as is clearly seen in Figure 6.14. The upper part of the cloud had a Gaussian profile with a radius of about 0.8 km, whereas the bottom part of the cloud, also having a Gaussian profile, had a radius of about 3.7 km. During the time of drift spreading of 500 s, as seen from Figure 6.14, at the fear of the cloud, a striated structure as a cluster of a set of plasma sleeves is generated, each sleeve moving with a velocity of about 10 m/s, i.e., much slower than the barium cloud itself ($\sim$35 m/s) from which they were formed. The dimensions of the created strata are as follows: $L = 30$ km along $\mathbf{B}_0$, the height from the center of the cloud $h = 10$ km, and the width across $\mathbf{B}_0$ $d = 0.6$ km. Accounting for these dimensions, the concentration of barium ions inside the cloud was estimated as $\approx 6 \cdot 10^6$ cm^{-3}, while at the same range of altitudes the concentration of background ionospheric plasma does not exceed $\approx 3 \cdot 10^6$ cm^{-3}; i.e., it is less. Thus, we talk about a strong large plasma structure created in the ionosphere by injection of ion cloud from a rocket.

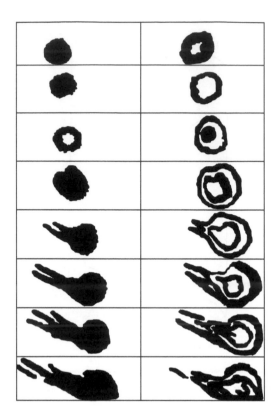

Figure 6.14 Temporal and spatial evolution of plasma density inside the injected cloud (right panel) and its brightness (left panel), obtained experimentally.

The same result was obtained in other experiments [26] whose results are depicted in Figure 6.15, where the dotted line shows the initial position of the injected neutral barium cloud. It is evident that during a definite time the cloud is deformed by the floating of its plasma along $\mathbf{B}_0$, and then at the bottom of the cloud the striation at the plasmoids stretching along $\mathbf{B}_0$ occurs.

In the aforementioned two experiments, the clouds of high density and large dimensions were observed during the limiting time of observation and, unfortunately, the striation and further evolution of strata were not analyzed, because in the problems under consideration the researchers employed a coordinate system moving with the wind of neutrals, and the mechanisms of drift instability generation and the splitting of cloud on several smaller sleeves were not considered.

A more detailed experiment was described in Refs. [29,31]. In Ref. [31] the development of the barium cloud injected at a height of 160 km was registered by an optical device with interference filters allowing observation of the first 100 s of evolution of the neutral central component of the cloud and then of the ion cloud with intermittent returns to observations of the neutral cloud. The complete time of observation was about 12 min, and it was made up to the moment the radiation from ions Ba^+ became lower than the radiation level of the background ionospheric plasma. For an observation time of about 400–500 s, the shape of the neutral cloud deviates from its spherical-symmetrical form (see Figure 6.16a). During observation the fixation of the neutral and ion barium clouds was carried out periodically by fixing their positions in space. After 200 s of barium cloud creation, the displacement of the cloud of ions Ba^+ from the neutral one was about

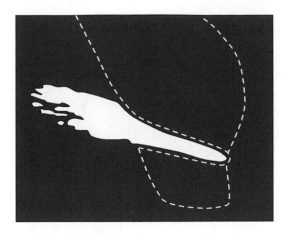

Figure 6.15 Photograph of last stage of evolution of plasma cloud with respect to its initial position (dashed contour).

8 km (Figure 6.16b and c). The distance between the central parts of clouds, neutral and ion, during observations at periods from 190 to 420 s was about 3–6 km (Figure 6.16d). The same picture shown in Figure 6.16 was obtained in Ref. [29]. In both experiments, the profiles of plasma density obtained by using the optical method, both for the neutral and the ion (along $\mathbf{B}_0$) clouds, were close to Gaussian. Thus, the detailed information about evolution and the corresponding profiles of neutral and ion components were obtained [31]. The phenomena observed indicate the following. Initially, the neutral cloud spreads isotropically in the frame moving together with the neutral wind, and the profile of concentration of neutral atoms of Ba observed

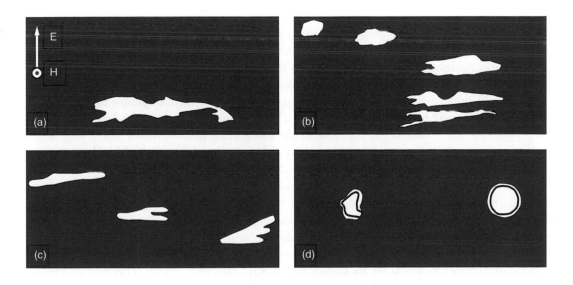

Figure 6.16 Time evolution of the ion cloud and the associated neutral cloud accompanied by splitting and stratification of the initial cloud on several plasmoids arbitrarily oriented to the ambient electric and magnetic fields.

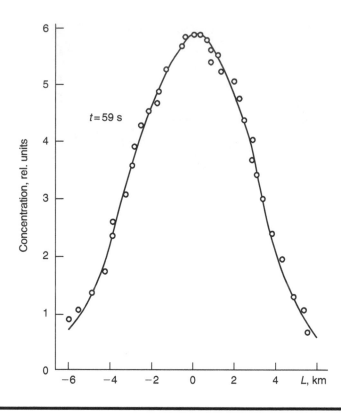

Figure 6.17 Gaussian distribution of plasma density within the ion cloud (curve) as a best fit of experimentally observed data (circles).

experimentally (circles) and predicted theoretically (curve) was close to Gaussian, $e^{-\alpha\tau_i}$, as shown in Figure 6.17. The number of atoms of neutral Ba decreases owing to ionization with the characteristic time $\tau_i \approx 20$ s and with velocity of ionization α found experimentally. Therefore, the dependence of the square of the half-width, d_n^2, of the Gaussian curve is linear (see circles in Figure 6.18), according to which the coefficient of diffusion of neutral atoms of $D_0 \approx 1.3 \cdot 10^8$ cm^2/s was estimated. The initial distribution of ions Ba$^+$ in the ionized component of plasma cloud was also close to Gaussian, with half-width $d_n(\tau_i) \approx 1$ km. The concentration of ions in the center of cloud was $N_c \approx 6 \cdot 10^6$ cm^{-3}, i.e., larger than that of the background plasma. At the initial stage after creation, the spreading of ion clouds with velocity of longitudinal (along $\mathbf{B}_0$) ambipolar diffusion was observed. The longitudinal profile of concentration of ions Ba$^+$ was still Gaussian (see curve with [Ba$^+$]$_{||}$ in Figure 6.19). The linear dependence of d_n^2 for [Ba$^+$]$_{||}$ represented by dots in Figure 6.18 allows for estimating the coefficient of ambipolar diffusion along $\mathbf{B}_0$ which equals $D_{a||}^{(Ba^+)} \approx 4.2 \cdot 10^8$ cm^2/s, and exceeds the previous value D_0 more than three times. If we now assume that $D_{a||}^{(Ba^+)} \approx 2D_0$ [29], which corresponds to conditions of isothermal plasma ($T_e \approx T_i$), then the excess of T_e over T_i in experiments described in Ref. [31] can be more than doubled. The movements of ions Ba$^+$ across $\mathbf{B}_0$ are slower. In Ref. [31] the coefficient of transverse ambipolar diffusion (across $\mathbf{B}_0$) $D_{a\perp}^{(Ba^+)} \approx 1.7 \cdot 10^6$ cm^2/s, was estimated as being about $4.2 \cdot 10^8/1.7 \cdot 10^6 \approx 250$ times slower than in the direction along $\mathbf{B}_0$. In the transverse profile of the ion cloud (Ba$^+$), a fundamental asymmetry, increasing with time and close to the edges of Gaussian profiles of [Ba$^+$]$_\perp$, was observed (compare two bottom curves with points and circles presented in

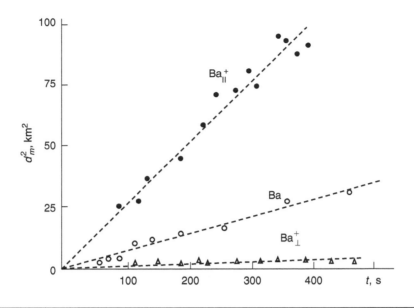

Figure 6.18 Time evolution of square of the diameter of barium clouds along (bold dots) and across (triangles) the ambient magnetic field and of the neutral barium cloud (circles).

Figure 6.19). To understand the effect of this asymmetry of the profile of ions $[Ba^+]_\perp$ in dynamics, we present these two curves separately in Figure 6.20; the first curve for $t = 534$ s denoted by (1) and the second curve for $t = 608$ s denoted by (2). In Ref. [31] the concentration of ions in the ion cloud for $t < 400$ s was estimated to be of the order of $N_c \approx 4 \cdot 10^6$ cm^{-3}, i.e., greater than the concentration of the background ionospheric plasma ($\sim 10^6$ cm^{-3}). For $t > 400$ s a sharp decrease

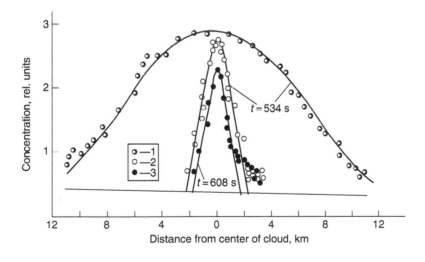

Figure 6.19 Dependence of plasma concentration inside ion clouds on the distance from the center of barium cloud: 1 is for longitudinal distribution of barium ions, 2 and 3 are for transversal distribution of barium cloud observed at 534 and 608 s, respectively.

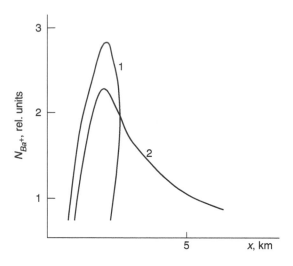

Figure 6.20 Spatial distribution of the neutral barium (curve 1) and ions inside the barium cloud (curve 2).

of the amount of ions in the ion cloud was observed experimentally in Ref. [31]. Because the influence of transverse diffusion on ion cloud spreading begins from the moment of time exceeding $d_n(\tau_i)/D_{a\perp}^{(Ba^+)} \approx 2 \cdot 10^3$ s, i.e., the total time of observation, in Ref. [31] it was assumed that the asymmetry of profile $[Ba^+]_\perp$ shown in Figure 6.20 by curve (2) for $t = 608$ s can only be explained by the effects of the ambient electric field and the wind of neutrals. In the following text, based on theoretical analysis discussed in Section 5.2, we will try to prove this assumption with rigorous theoretical approaches. We will now return to some interesting measurements carried out after the pioneering experiments in barium release and briefly present their results in the observation of spatial and temporal evolution of ion clouds in the ionosphere.

The same phenomena in movements of neutral and ion clouds were observed in Ref. [29], where injection of four clouds with differing number of ions Ba^+ was performed. In three clouds the concentration of ions was large enough, $N_c \approx 7 \cdot 10^6 \div 2 \cdot 10^7$ cm^{-3}, whereas the fourth had a concentration of about $N_c \approx 0.6 \cdot 10^6 \div 2 \cdot 10^6$ cm^{-3}, i.e., of the same order as the concentration of the background ionospheric plasma. The first three ion clouds moved close to the neutral barium (compare circles denoted by (a) for atoms Ba and curves 1–3 together with points denoted by (b) for ions Ba^+ in Figure 6.21), and their velocities of spreading were also close to those for the neutral cloud. This tendency was observed during the entire experiment ($T = 10$ min). As for the fourth ion cloud, from $t = 3$–4 min it sharply changed its direction compared with the neutral cloud (curve 4 in Figure 6.21), and from the beginning its velocity and direction of spreading were different when compared to those for the cloud of neutral atoms Ba.

In parallel with these experiments, in the former USSR, the same barium release experiments were carried out from 1975 onward [30–37]. In light of these investigations, preliminary analysis of observations of evolution of both neutral and ion clouds of barium will be given, and then a physical analysis will be given following results obtained in Section 5.2 and analysis of theoretical works presented in Refs. [122–129]. First, cumulative injections Spolokh 1 and Spolokh 2 were performed at the altitude $z = 170$ km close to the vertical direction ($\propto 2^0 - 5^0$ to $\mathbf{B}_0$).

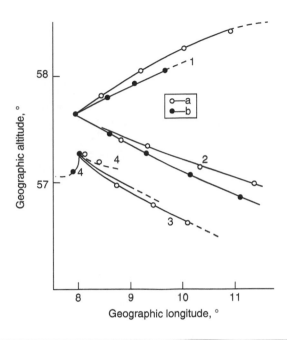

Figure 6.21 **Last stage of evolution of secondary plasmoids generated by the initial barium ion cloud (denoted by circles) and the associated clouds of neutral barium (denoted by points) observed in experiment "Spolokh-1."**

Observations were made during $T = 10$–15 min from three sites (in Spolokh 1) and from two sites (in Spolokh 2) separated 50–300 km from each other. In Figure 6.22 the photos of neutral barium and ion clouds obtained from three sites are denoted by 1–3. Observations were absent in some fragments for all three sites. All photographs correspond to the following moments of time (from top to bottom, respectively): 20, 50, 150, and 400 s. At all optical devices a fast-spreading spherical cloud of Ba was registered, which after 20 s had a radius of about 2 km. After this time, the plasmoid of ions Ba^+ elongated along $\mathbf{B}_0$ was created owing to ionization of the neutral component of barium cloud, which was observed by using special photo filters. At about 130 s both clouds, the neutral and the ion, moved in parallel, and after that the spherical neutral cloud was hardly observed because of its low brightness. After 45 s from this moment, the floating of ions Ba^+ from the main plasmoid in forms of fingerlike structures, strata, elongated along $\mathbf{B}_0$ were clearly observed from experimental site 1 and less so from site 2. Later, the brightness of the main ion plasmoid decreased and strata became more diffusible. At a later stage of ion cloud evolution the brightness of the tail with stratum exceeded that for the main ion plasmoid.

In experiment Spolokh 2, parallel movement of the spherical neutral cloud and the main ion plasmoid 32 s after injection (Figure 6.23a) and at time $t = 100$ s floating of the main plasmoid from the neutral cloud (Figure 6.23b) was also observed. At the later stage of ion cloud evolution (at $t = 15$ min, Figure 6.23b), a tail with strata structure was also observed. Careful optical and spectrometric analysis allowed the distribution of atoms Ba inside the neutral cloud and concentration of ions Ba^+ in the ionized cloud to be obtained. Thus, it was found that concentration in the main ion plasmoid exceeded the concentration of background plasma by more than one order of magnitude. The diffusion coefficient of the neutral component was estimated as $D_0 \approx 8 \cdot 10^8$ cm^2/s (in Spolokh 1) and $D_0 \approx 6 \cdot 10^8$ cm^2/s (in Spolokh 2). During 300–400 s all

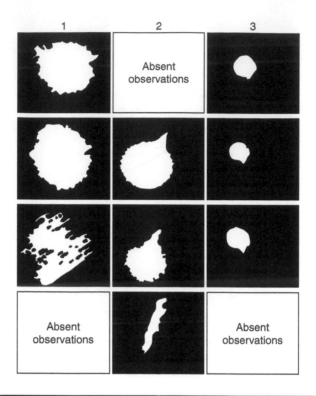

Figure 6.22 Photographs of barium clouds obtained from different observation stations presented at different panels: left, middle, and right. Sometimes, data on brightness of barium clouds was absent.

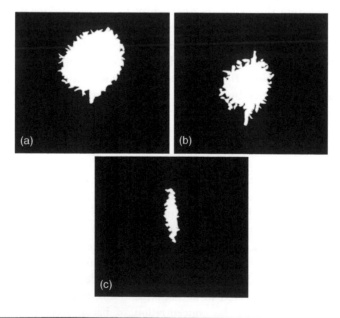

Figure 6.23 Spatial and temporal evolution of ion and neutral barium clouds obtained in experiment "Spolokh-2."

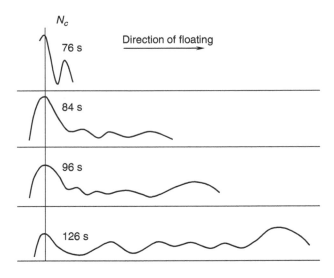

Figure 6.24 Time evolution of the brightness profile of ion clouds in the direction of ion drift.

ions were floated from the main plasmoid to the tail structure. Accounting for the homogeneous distribution of ions inside the ion plasmoid with dimensions $L \approx 20$ km (along $\mathbf{B}_0$), $d \approx 1$ km (across $\mathbf{B}_0$), the average concentration of plasma within the ion cloud was estimated at about $N_c \approx 10^7$ cm^{-3}, which matches with results obtained using radiolocation measurements during these experiments, which yielded $N_c \approx (6.3 \cdot 10^6 \div 1.4 \cdot 10^7)$ cm^{-3}. This value exceeds the concentration of the background ionospheric plasma at altitudes of 160–170 km by about two orders of magnitude. At the same time, estimations made for the tail structure yielded $N_c \approx 2 \cdot 10^5$ cm^{-3}, which is of the same order as the concentration of the background plasma. Stratification of the main ion plasmoid 45 s from the moment of injection obtained experimentally, and floating in the form of strata in the tail can be seen from Figure 6.24, where the time dynamic of the profile of brightness in the direction of floating of ions is shown. As can be seen, the sharp rear of the main ion plasmoid remains stable during the time of observation. From the beginning of the process of stratification, two to three strata leave the main plasmoid. Each stratum has an asymmetrical profile of brightness, smoother than the sharp profile of the main ion plasmoid. From the commencement of this process, the concentration in each leaving strata exceeds that for background plasma. During the process of diffusion, the concentration in both sleeves equalizes. Changes in distance between continuously leaving strata denoted by digits 1–3 are presented in Figure 6.25. It can be seen that the velocity of movement of these three strata is approximately constant, practically equally for these strata structures, and coincides with drift velocity of the floating tail structure in the crossing $\mathbf{E}_0 \times \mathbf{B}_0$ fields. A system of thin, field-aligned plasma filaments, which the authors called *secondary strata*, appeared. These small secondary strata structures with weaker brightness and smaller dimensions are shown by arrows in Figure 6.26. Each filament structure then groups around the primary strata structure. However, their weak brightness observed in this experiment did not allow for obtaining detailed information on filament-structure dynamics.

More detailed observation and investigation of the dynamics of this fine filament structure was carried out in Refs. [33,34,36], and the corresponding interpretation of nonlinear processes accompanying formation and evolution of artificial ion clouds in the inhomogeneous ionosphere was presented in Ref. [37] on the basis of theoretical studies carried out in Refs. [125–129].

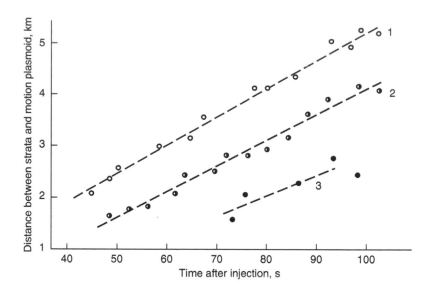

Figure 6.25 Change in time of distances between moving strata (denoted by 1–3) and the main ion plasmoid.

First, we present results of optical observations obtained in Refs. [33,34,36] before entering into a detailed analysis of the problem. In these experiments, ground optical observations of the barium plasma cloud striation process in the middle-latitude ionosphere at altitudes of 155–190 km were made.

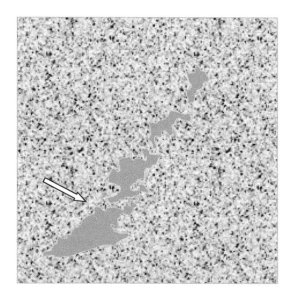

Figure 6.26 Process of stratification of the initial plasma cloud and the periodical tail creation observed experimentally. These periodical strata structures are denoted by an arrow.

Finally, four complex experiments from 1988 to 1990 were carried out using two TV cameras and photo-cameras with a high spatial (~50 m) and temporal (~0.2 s) resolution. In three of the four experiments the striation mechanism was clearly observed, and the fine filament structure of strata elongated close to the magnetic field lines (with deviations of no more than 1°) was investigated in detail. In each release, a similar picture of inhomogeneity evolution was observed. Thus, in the first stage of the cloud's evolution (within about 30 s after each release) when the density of plasma was about two orders of magnitude higher than the background density, the dispersive diffusion stage corresponded to the process described in Section 5.1 according to Refs. [104,105,123,133]. After this time, the cloud began to split into a neutral quasi-spherical and ionic cloud, the latter of which after approximately 50–60 s split into an ionic cluster stretched along $\mathbf{B}_0$. In a further evolution (up to 160 s) in this primary strata, separation of each stratum from the ionic cluster was observed in the $\mathbf{E}_0 \times \mathbf{B}_0$ direction (see Figure 6.27a). After 30 s, i.e., up to 190 s, the striated structure (the grating of primary strata) was replaced with a cluster of $\mathbf{B}_0$-field-aligned plasma filaments (called *secondary strata*) concentrated within the disturbed region with transverse (across to $\mathbf{B}_0$) and longitudinal (along $\mathbf{B}_0$) sizes 5–6 km and 22–25 km, respectively (see Figure 6.27b). The transverse scale, $\ell_\perp$, and longitudinal scale, $\ell_\parallel$, of the separate stratum were

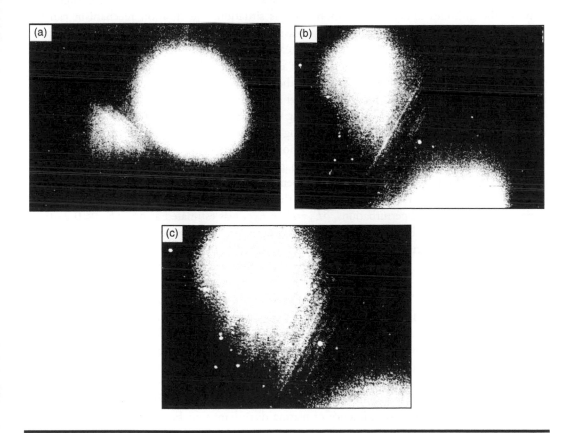

Figure 6.27 Mutual evolution of stratified ion plasma structures and the neutral barium cloud obtained in "Spolokh-2" experiment.

estimated using precise optical methods (the reader can find more in detail about these measurements in Refs. [36,37]) to be within $\ell_\perp \approx 0.1-0.3$ km, and $\ell_\parallel \approx 20-25$ km; i.e., the stretching parameter, $\chi = \ell_\parallel/\ell_\perp$, was about 100. This result is in good agreement with estimations obtained in Section 5.2 based on calculations carried out in Refs. [37,129], where, for the commencement of striation mechanisms, the degree of anisotropy of the initial irregularity has be not less than 100. All strata were oriented along $\mathbf{B}_0$ with an accuracy of $\pm 1^0$ (see Figure 6.27c); i.e., they were field-aligned. The striations took place at the *front*, but not the *back*, edge of the ionic cluster, i.e., in the $\mathbf{E}_0 \times \mathbf{B}_0$ direction. The same results were observed earlier in Refs. [29,31], which contradicted the initial pioneering observations [24–27], where the striation mechanism was observed from the rear of the initial ion cloud.

At the same time the dark areas between bright plasmoids in directions parallel to the geomagnetic field $\mathbf{B}_0$ lines were observed during a detailed optical analysis of experimental data [36]. These peculiarities can be explained by the theoretically predicted and numerically obtained results in Refs. [128,129], taking into account altitudinal inhomogeneity of ionospheric plasma and nonuniform altitudinal dependence of plasma transport coefficients, which can finally accelerate generation of gradient-drift instabilities in plasma during filament-structure creation. Thus, additional estimations of the ion density in each separate strata based on the photometric method [36] yield the value $\delta N_i \approx 5 \cdot 10^5 \div 10^6$ cm^{-3}, i.e., of the same order as concentration of background plasma at the altitudes under consideration. In this situation, the striation mechanism can be developed faster than for the case of strong plasma disturbances' drift spreading. In fact, the characteristic time of new (secondary) stratum creation in the grating structure was estimated as 3–4 s, which was of the same order of magnitude as the time of the splitting of existing (primary) strata. In other words, newly created strata enter the gap between the diffusive ion cluster and the grating structure with the characteristic time of 3–4 s. At the same time, splitting of already-existing (primary) strata takes place inside the strata grating with the same characteristic time of 3–4 s. The lifetime of each separate stratum within the grating was about 30 s. The characteristic diameter of the secondary stratum, moving in the same direction as the whole grating structure, was estimated to be of the order of 100–150 m. Using the estimated drift velocity of separate stratum, $V_d \propto 100$ m/s, and its deviation $\Delta V_d \sim 20$ m/s, which depends on strata luminosity and location in grating (because the boundary stratum spreads faster than inner one), the associated perpendicular component of electric field perturbations, $\delta E_\perp = E_\perp - E_{0\perp}$, was estimated as $\delta E_\perp/E_{0\perp} \approx 0.2-0.3$. All these parameters will be used later to simulate numerically the experimental data described earlier based on the theoretical model developed in Refs. [128,129]. We now explain the evolution of ion clouds through the prism of more realistic theoretical and numerical models referred to in Section 5.2.

Experimental Data via Theoretical Prediction. We now continue the foregoing discussions of nonlinear evolution of artificial ionospheric irregularities created by injection of plasma clouds into the ionosphere, now based on the knowledge obtained in Sections 5.1 and 5.2 regarding processes of diffusion and drift of arbitrary plasma disturbances in the real ionosphere. The developed models of drift referred to there point to the qualitative agreement between observed phenomena and theoretical predictions based on analytical and numerical analysis [112–129]. Thus, for example, it was found that the injected cloud influenced the nondisturbed ambient electric field $\mathbf{E}_0$. Defined in Refs. [123,124], the criterion of influence of plasma disturbance on $\mathbf{E}_0$ supposes a large concentration of ions of barium N_{Ba}^+ in the cloud compared with the concentration of background ionospheric plasma N_0, i.e., $N_{Ba}^+/N_0 \gg 1$. In Refs. [21,27,107], another criterion, less rigorous, of perturbation of the ambient electric field was proposed in the following form: $\lambda^* = \Sigma_p^c/\Sigma_p^0 \gg 1$, where Σ_p^c and Σ_p^0 are the integral Pedersen conductivities of ion cloud and background plasma, respectively, introduced in Section 5.2. We mentioned in this section, following Refs. [123,124], that a fundamental change in the drift process of the strong plasma disturbance occurs only when the depletion region surrounding it becomes deep (up to zero). Therefore, the criterion of disturbance of the ambient electric field proposed in Refs. [21,27,107] was modified for the real ionosphere in Refs. [125–127]. According to the study, the perturbation of $\mathbf{E}_0$ is possible if the full amount of ions in the cloud with scale L, $\int_0^L N_{Ba}^+(z)\,dz$, exceeds the amount of electrons (ions, plasma is quasi-neutral), $\int_0^{z_0} N_0(z)\,dz$, in the plasma column, oriented along $\mathbf{B}_0$ below the cloud, where z_0 is the height of cloud injection. This condition is stronger than $\lambda^* = \Sigma_P^c / \Sigma_P^0 \gg 1$, proposed in Refs. [21,27,107], but weaker than $N_{Ba}^+ / N_0 \gg 1$ (according to Refs. [123,124]). Moreover, the cloud's dimensions are smaller than the characteristic scale of the ionosphere under the cloud H_δ; i.e., $d < L < H_\delta$. Therefore, the parameter of nonlinearity $h = N_{Ba}^+ / N_0$ describing a degree of depletion of regions of the background plasma surrounding the injection cloud (see definitions in Sections 5.1 and 5.2) is much higher that the parameter λ^* introduced in Refs. [21,27,107], taken as a degree of perturbation of the background ionosphere by the injected ion cloud. For the high-density clouds, the ambient electric field $\mathbf{E}_0$ is fully compensated by the inner disturbed electric field $\delta\mathbf{E}$, and the effect of the latter becomes predominant in the movements of injected ions. Therefore, as shown in Refs. [123–125], in the limited case of fully ionized plasma ($N_{Ba}^+ \gg N_0$) when the depletion zones become deep (zero-order), the drift of cloud ions in the crossing $\mathbf{E}_0 \times \mathbf{B}_0$ fields is absent. In the frame related to the neutral wind, the ordinary ambipolar diffusion in the absent of the magnetic field takes place (according to Schottky; see Section 5.1). In this case diffusion of strong ion cloud is isotropic, and the center of the ion cloud coincides with that for the neutral cloud, as was evidently illustrated in Figures 6.22 and 6.23 for the time of observation $t \le 200$ s; in this case $N_{Ba}^+ / N_0 \ge 10^2$, as was estimated from experiments. However, at the edges of the cloud, the electric field $\mathbf{E}_0$ is not fully compensated, and a part of the ions and electrons of plasma cloud are liable to drift in crossing $\mathbf{E}_0 \times \mathbf{B}_0$ fields, i.e., the electrons in the Pedersen (across $\mathbf{B}_0$ and along $\mathbf{E}_0$) direction and the ions in the Hall (across $\mathbf{B}_0$ and across $\mathbf{E}_0$) direction. Thus, they create regions of floating electrons O_1 and floating ions O_2 as illustrated schematically in Figure 5.42 (see Chapter 5). According to this scheme, the injected ions floating in region O_1 create a tail moving in the nondisturbed ambient electric field $\mathbf{E}_0$.

At the *initial stage* of ion cloud evolution ($t < 2$–3 min) the dense plasma cloud moves with the velocity of neutral wind $\mathbf{U}_m$, damping ambipolarly as in the case of $\mathbf{B}_0 = 0$, and only some of the ions float in the crossing $\mathbf{E}_0 \times \mathbf{B}_0$ fields, fully coinciding with the vision of drift spreading following from classical drift theory described in Section 5.2.

At a *later stage* of its evolution ($t > 3$–4 min), the density of plasma within the ion cloud becomes smaller owing to ambipolar diffusion, and we consider the case of moderate nonlinearity, when $N_{Ba}^+ / N_0 \ge 1$ or $N_{Ba}^+ / N_0 \le 1$. Here, the effects of short-circuits currents through the depletion regions and the effects of the background plasma become predominant. Drift spreading of ions occurs faster than in the case of ambipolar diffusion, and most float in an $\mathbf{E}_0 \times \mathbf{B}_0$ direction. The picture of spreading becomes more complicated because in this case, and as has been shown in Refs. [126–129], the shape, form, and orientation of plasma structure in space are changed nonuniformly. Different parts of the ion cloud begin to move with different velocities determined by different physical processes—from ordinary diffusion to the effect of short-circuit currents (see Sections 5.1 and 5.2). However, in this case the problem becomes quasi-linear, and we can again use the results obtained in Section 5.2 following Refs. [122–129]. In this case, a disturbance $\delta\mathbf{E}$ of the ambient electric field $\mathbf{E}_0$ is sufficiently small, and currents of particles caused by this disturbance mostly flow through the background plasma. Here, the movements of the injected ions are determined by faster unipolar diffusion coefficients ($D_{e\parallel}$ along $\mathbf{B}_0$ and $D_{i\perp}$ across $\mathbf{B}_0$), by drift in

$\mathbf{E}_0 \times \mathbf{B}_0$ direction, and by components of velocity of neutral wind, $\mathbf{U}_{m\|}$ along $\mathbf{B}_0$ and $\mathbf{U}_{m\perp}$ across $\mathbf{B}_0$ [128,129]; i.e.,

$$\mathbf{V}_d^{Ba^+} = \mathbf{U}_{m\|} + b_{\|}^{Ba^+} \mathbf{E} + b_{\perp}^{Ba^+} \left(\mathbf{E} + \frac{\mathbf{U}_m \times \mathbf{B}_0}{cB_0} \right) + b_{\Lambda}^{Ba^+} \frac{\mathbf{E}_0 \times \mathbf{B}_0}{B_0} + \frac{\mathbf{U}_{m\perp}}{1 + Q_H^2} \quad (6.5)$$

Here, $b_{\|}^{Ba^+}$, $b_{\perp}^{Ba^+}$, and $b_{\Lambda}^{Ba^+}$ are mobilities of the injected ions in $\mathbf{B}_0$, Pedersen, and Hall directions, introduced in Chapter 5 for regular ions of the ionosphere, all other parameters being the same as in Chapter 1. Because at altitudes $h > 150$ km, ions are magnetized and $(1 + Q_H^2) \gg 1$, they do not move in the direction of $\mathbf{U}_{m\perp}$ but only in the $\mathbf{E}_0 \times \mathbf{B}_0$ direction. Here, in the linear case, described in Refs. [123–127], owing to the dispersion mechanism of spreading of plasma cloud, three plasmoids are created, the first of which associate with the *electronic* (1) and spread with slower velocity across $\mathbf{B}_0$; the second (2) is *ionic*, spreading faster along $\mathbf{B}_0$; and the third (3) spreading in the $\mathbf{E}_0 \times \mathbf{B}_0$ direction (see Figure 6.28, which we present here in a simpler manner than that shown in Figure 5.42). In this figure, the initial position of the cloud after injection is denoted by (0). Now, using results of theoretical prediction, one can easily explain the asymmetry of the density profile of ions Ba^+ across $\mathbf{B}_0$ obtained experimentally in Refs. [29,31] and shown in Figures 6.19 and 6.20. According to the classical presentation of the dispersive mechanism of drift, this phenomenon has an evident physical explanation: the asymmetry of the transverse Gaussian profile of ion concentration in the cloud is caused by transverse movements of injected ions in the $\mathbf{E}_0 \times \mathbf{B}_0$ direction relative to the wind of natural molecules (atoms) occurring in ionospheric plasma. Moreover, because Hall mobilities of electrons and ions are different, the injected cloud in itself cannot move in the $\mathbf{E}_0 \times \mathbf{B}_0$ direction, as this contradicts the condition of quasi-neutrality. This scenario can be secure only by short-circuit currents through the background plasma and the depletion regions occurrence. These circuit currents, denoted by "−" and "+" in Figure 6.28, lead

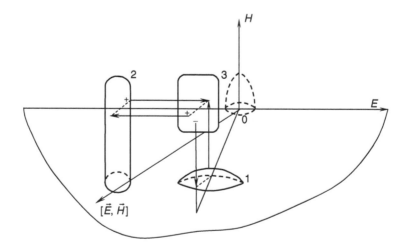

Figure 6.28 Schematic positions and shapes of ion cloud plasmoids with respect to initial neutral cloud position and shape (denoted by dashed contour) in the crossing ambient electric and magnetic fields.

to the creation of a depletion region moving together with plasmoids 1 and 2, corresponding, respectively, to the injected electrons and ions. Such dispersive spreading leads to a fundamental enlargement of the barium cloud, which was observed experimentally [30–36] Moreover, a sharp decrease in the number of ions in the barium cloud after $t \geq 400$ s can obviously be described by the influence of dispersive drift mechanism in the process of plasma ordinary diffusion at the quasi-linear and linear stages of evolution of ion cloud. Thus, in all experiments [29–36], at the stage after $t \geq 400$ s, the observed illumination of the ion cloud was falling fast to the illumination of the background plasma. Now, taking from a theoretical framework the critical time $t_{kr} = d_n(\tau_0)/b_{i\perp}E_\perp^*$ of commencement of the drift process and information about the initial plasma cloud transverse dimension $d_n(\tau_0)$, found from Figure 6.18, we can obtain an estimation of the transverse electric field $E_\perp^*$, defined in Section 5.2 as $E_\perp^* \approx 2-6$ mV/m, which is in good agreement with estimations made during experimental optical observations [31–36].

At the *final stage* of its evolution, the ion cloud transforms into a small-scale plasma structure having a lower density than that of the background ionospheric plasma. During this stage the striation of plasma clouds on the grating structure consisting of clusters of strata stretched along geomagnetic field lines (see Figures 6.12 through 6.14 and Figures 6.26 and 6.27 obtained by different experiments). To explain this phenomenon, various physical mechanisms were analyzed and discussed, and in Ref. [112] three possible types of instabilities, acoustic, drift-dissipative, and gradient-drift, were analyzed analytically; it was found that the more probabilistic source of striation of plasma cloud on strata structure is gradient-drift instability (GDI), discussed in detail in Chapter 2. A first one-dimensional (1-D) theoretical model of ion cloud in which GDI was stressed as a main mechanism of striation into strata was developed in Ref. [108], and then has been generalized for 2-D and 3-D models of plasma cloud in Refs. [114–121]. All authors have pointed out and proved theoretically that GDI in crossing $\mathbf{E}_0 \times \mathbf{B}_0$ fields is the most probabilistic mechanism of plasma structures' striation without entering into the origin of their creation. In Refs. [37,127,128], the phenomenon of fine grating structure creation was analyzed theoretically and experimentally on the basis of precise optical and spectral observation of the process [34–36]. We have described this effect in the foregoing text and in Section 5.2. We now will only reiterate some important aspects of this theory and again present results of the numerical experiment carried out under conditions of one of the experiment described in Refs. [36,37].

Firstly, it was found that the increment of GDI growth (as a limit of striation of plasma plasmoids into grating structure) can be estimated for stretched along $\mathbf{B}_0$ irregularities as $\gamma \approx V_d/\ell_\perp$, where V_d can be estimated from Equation 6.5 and $\ell_\perp$ is the transverse (to $\mathbf{B}_0$) scale of primary strata (plasmoid). To obtain a clearer comparison with experiments carried out in Refs. [35–37] and briefly described previously, in Refs. [37,129] were presented numerical experiments with various kinds of plasma structures related to injected ion clouds. The input parameters of plasma clusters in numerical experiment we want to present now were chosen to be close to those of Experiment 3 described in Refs. [36,37]. Initially, the parameters of the plasma cloud were taken from the barium release phenomena described in Refs. [36,37]: $N_{Ba}^+/N_0 = 100$, the shape of initial plasma cloud was isotropic, $L = 2d = 10$ km. The background electric field (including neutral wind; see Section 5.2) was estimated as $E^* \approx 10$ mV/m. All parameters of the background ionospheric plasma were chosen from the model of middle-latitude ionosphere (real experiments were carried out in such an ionosphere) described in Chapter 1. Presented in Figure 6.29a through c are the isolines of equal density at 10, 30, and 60 s, respectively. Simulations again show the physical picture of the spreading process, described earlier and obtained experimentally. Thus, in the initial stage less than 20 s, the smooth diffusive expansion of the plasma profile is observed. Then, after 20 s, when the degree of anisotropy of the plasma cloud significantly exceeds

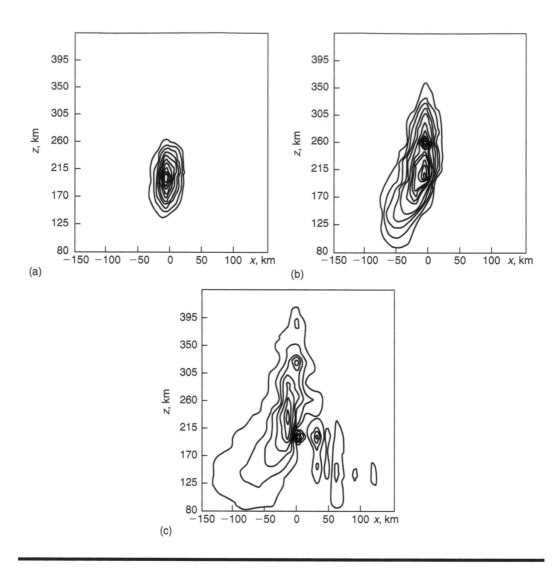

Figure 6.29 Numerical simulation of spatial and temporal evolution of the ion cloud in the crossing ambient electric and magnetic fields in the presence of the wind of neutrals for (a) 10 s, (b) 30 s, and (c) 60 s.

ten (i.e., $L/d > 10$), the front steeping strengthens with subsequent stratification on plasmoids owing to a drift dispersion mechanism in less than 10 s being observed. After that, a striation of plasmoids at strata is clearly seen with a wide spectrum of the strata transverse sizes, $\ell_\perp$, from 100 to 500 m. The drift velocity of each separate stratum varies from 50 to 100 m/s. It is significant that the drift velocities of the extreme strata, as well as the secondary ones in the grating structure, are noticeably higher (by 10–30 m/s) than those of more dense primary strata (1–5 m/s) obtained both numerically and experimentally. It should be noted that the main cloud drifts significantly slower that the strata grating. It is easy to see that the characteristic times of the plasma cloud splitting on plasmoids due to drift mechanism are about $t_d = 20{-}30$ s, and the striated structure formation commences 10–20 s after the splitting process, which is also in good agreement with all the experiments carried out in Refs. [29–36]. These references refer to the characteristic strata sizes, i.e., $\ell_{||} \approx 10-20$ km and $\ell_{\perp} \approx 100-300$ m, as well as to drift velocities of both the strata (50–100 m/s) and the plasma cluster (40–50 m/s).

The interaction between the primary and secondary strata and transformation of energy between these strata are still not clear. To describe this process in terms of nonlinear plasma mode, stable and unstable, coupling and redistribution of energy between them (as carried out in Chapter 2, Section 2.3), to obtain the dependence for the time of striation t_s from the parameters of plasma structures and ambient ionospheric conditions (see Tables 5.1–5.3, Chapter 5), many more numerical simulations and additional experiments for ion cloud releases are needed.

Nevertheless, results presented in this section show that many parameters of artificially induced plasma irregularities can by predicted accurately, so also ambient geophysical characteristics of the ionosphere, by studying spatial and temporal evolution of artificial plasma structures of various origins and at different stages of their evolution, using modern radiophysical and optical methods and the corresponding theoretical and numerical background described in Chapter 5.

6.2   Natural Ionospheric Irregularities and Methods of Their Parameters' Diagnostics

Many aspects of the investigation of artificial plasma irregularities can be directly transformed by studying natural ionospheric irregularities generated by different geophysical and morphological processes occurring in the ionosphere at different altitudes, from the polar cap to the equatorial zone. We will direct the reader's attention to those from a whole spectrum of natural irregularities that are interesting for the purpose of the performance of ionospheric radio channels (Chapters 7 and 8) and for the investigation of their effects on land–satellite radio communication to predict an efficient design of such channels and transfer of information data through them (Chapter 9).

6.2.1   Radio Methods of Diagnostics of Meteor-Induced and H_E Irregularities in the Middle-Latitude Ionosphere

The meteor-induced and the natural field aligned along $\mathbf{B}_0$ irregularities (called H_E irregularities in the literature; see Chapter 7) have been studied at the middle-latitude traces in several regions over the world by separate groups of researchers from the beginning of the 1960s to the beginning of the 1990s [48–51,134–164]. The purpose was to explore the possibility of achieving stable radio channels due to scattering of radio waves by these kinds of plasma disturbances occurring in the ionosphere by meteor trails and by Farley–Buneman (modified two-stream) instability (FBI), respectively, which occur at the lower ionosphere at altitudes 80–150 km. As for sporadic layered structures (denoted by E_s) occurring in the middle-latitude ionosphere, the main mechanism of their generation is associated with gradient-drift instability (GDI) [155–160] and with a combination of GDI and FBI for large- and small-scale irregularities, respectively [160–164].

We will discuss the practical aspects of the creation of such radio links in Chapters 7 and 8. For now, we will consider some actual radio methods for investigating the initial parameters of meteor-induced and H_E irregularities and, first of all, the process of their spreading and the corresponding lifetime that characterizes the efficiency of radio propagation within the ionospheric channel consisting such types of irregularities.

We will begin with meteor trials that participate in long-distance radio wave propagation at heights 80–110 km. In Ref. [144], it was found that the lifetime of meteor-induced irregularities actually depends on orientation of plasma irregularity relative to geomagnetic field lines. Theoretically, in Ref. [130] a strong dependence of diffusion spreading of such irregularities on the degree of orientation with respect to $\mathbf{B}_0$ was obtained. It was also found that only at a deviation of the irregularity at the angle of $\theta = 1°–2°$ from the direction of the magnetic field $\mathbf{B}_0$ does the velocity of spreading increases by two to three orders of magnitude and the corresponding lifetime of such an irregularity decreases by approximately two orders of magnitude. However, in experimental investigations of meteor trails and meteor-induced irregularities carried out in Ref. [144], doubt is cast on such a critical dependence of diffusion spreading (the corresponding ambipolar coefficient of diffusion) on the angle θ between the axis of the irregularity and the vector $\mathbf{B}_0$.

The discrepancy between the experimental [144] and theoretical [130] results may be due to the fact that in Ref. [130] only slightly ionized irregularities (i.e., quasi-linear disturbances with $\delta N \geq N_0$ and $\delta N \leq N_0$) were analyzed without taking into account collisions between charged particles of the weakly ionized background ionospheric plasma. Therefore, in Refs. [50,51] the effects of arbitrary degree of ionization and disturbance of the background ionospheric plasma corresponding to real perturbation and ionization of meteor-induced irregularities were investigated theoretically and experimentally and described in detail in Refs. [134–136]. Analyzing the formula (Equation 5.41) for plasma with arbitrary degree of ionization $p = \nu_{ei}/\nu_{em}$ defined in Chapter 1, and changing this parameter for under-dense to high-dense meteor trails, $p = 1, 10, 100$ according to Refs. [134–136], it was found that the dependence of the character of diffusion spreading depends essentially on the orientation of meteor-induced irregularities only after $\theta \geq 10°$. This result clearly seen from Figure 6.30, where the relative coefficient of ambipolar diffusion (Equation 5.41)

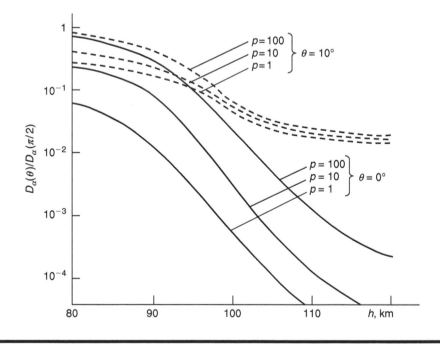

Figure 6.30 Height dependence of the normalized coefficient of diffusion for various degrees of ionization of meteor plasma ($p = 1, 10, 100$) and for different angles of orientation of the meteor trail to the ambient magnetic field, $\theta = 10°$ (dashed curves) and $\theta = 0°$ (solid curves).

normalized at that coefficient calculated for $\theta = \pi/2$, i.e., when the irregularity is oriented perpendicularly to $\mathbf{B}_0$ and diffusion is along $\mathbf{B}_0$, is shown versus the height of the ionosphere. The results indicate that the magnetic field has a significant effect on diffusion only at $h > 90$ km, and mostly on field-aligned irregularities (for $\theta = 0$) with a weak degree of ionization ($p < 10$). With the increase of the angle θ beyond $10°$, the degree of ionization ceases to play an important role in the character of diffusion spreading. The same picture is observed during analysis of lifetime calculated according to formula (Equation 5.54b) of an arbitrary meteor-induced irregularity with parameters of ionization changing from $p = 1$ (weak-dense irregularity) to $p = 100$ (high-dense irregularity). Thus, the relative lifetime according to Equation 5.54, where τ_0 is the lifetime of plasma irregularity in the absence of a magnetic field, is presented (see Figure 6.31) versus the angle θ for height $h = 80$ km (solid curves) and $h = 90$ km (dashed curves). The dot-dash line represents this relation for weakly ionized plasma (when $p = 0$) obtained in Ref. [130] (i.e., for H_E inhomogeneities).

As expected, the magnetic field affects a significant increase in the lifetime of meteor-induced irregularities in the region of angles $0° \leq \theta < 20°$ at altitudes $h \geq 90$ km. In this case, there is a strong dependence on the degree of ionization in the region of angles $0° \leq \theta \leq 5°$, which is most distinct at heights $h \geq 90$ km. Additional analysis carried out in Ref. [50] has shown that the lifetime of meteor trails increases by more than one order of magnitude with respect to the case of the absence of a magnetic field at heights $h > 100$ km within the limited range of angles

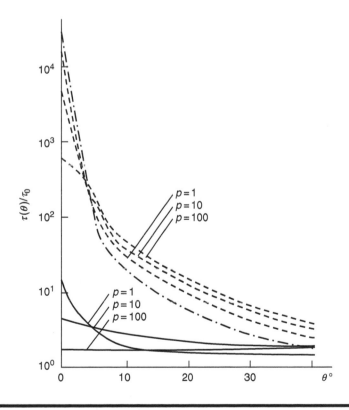

Figure 6.31 Angle dependence of the normalized lifetime of meteor trails in the lower ionosphere, at altitude $h = 80$ km (solid curves) and $h = 90$ km (dashed curves), for various degrees of ionization of meteor plasma: $p = 1$, 10, 100.

$0° \leq \theta \leq 20°$. Considering that on a given radio path described in Chapter 7 according to Ref. [50] and at operating frequency $f = 74$ MHz, $\tau_0 = 0.65$ s, we can for real field-aligned meteor ionizations ($p = 10$–100) at altitudes $h = 100$–110 km obtain the following limits:

$$2.5 \leq \tau(\theta = 0°) \leq 28 \text{ s} \qquad \text{at } h = 100 \text{ km}$$
$$26 \leq \tau(\theta = 0°) \leq 4480 \text{ s} \qquad \text{at } h = 110 \text{ km}$$
$$2.5 \leq \tau(\theta = 5°) \leq 30 \text{ s} \qquad \text{at } h = 100 \text{ km} \tag{6.6}$$
$$17.5 \leq \tau(\theta = 5°) \leq 127 \text{ s} \qquad \text{at } h = 110 \text{ km}$$

whereas for slightly ionized volumetric irregularities (H_E—inhomogeneities or under-dense meteor trails) obtained in Ref. [130] we have strong dependence on the angle θ:

$$\tilde{\tau}(\theta = 0°) \approx 252 \text{ s} \qquad \text{at } h = 100 \text{ km}$$
$$\tilde{\tau}(\theta = 0°) \approx 4000 \text{ s} \qquad \text{at } h = 110 \text{ km}$$
$$\tau(\theta = 5°) \approx 1 \text{ s} \qquad \text{at } h = 100 \text{ km} \tag{6.7}$$
$$\tau(\theta = 5°) \approx 7 \text{ s} \qquad \text{at } h = 110 \text{ km}$$

Hence, we obtained, in particular, that the lifetime of weakly ionized under-dense meteor-induced irregularities or H_E inhomogeneities is much shorter even for small deviations of the angle of their orientation with respect to $\mathbf{B}_0$, at angles $\theta = \pm (1° \div 5°)$, which also follows from the theoretical framework according to Ref. [50].

We now compare theoretical predictions from Ref. [50] with the corresponding experimental data according to measurements carried out at the middle-latitude trace with a radio path of about 1400 km long on frequency $f = 74$ MHz to investigate the scattering effect of meteor-induced irregularities located at the altitudinal range of 80–110 km and of H_E irregularities located at the altitudinal range of 110–150 km [49,50]. As we wrote earlier, full information of how to perform such radio traces will be presented in Chapter 7. Now we will present lifetime of meteor-induced irregularities estimated from measurements by studying the burstlike signals of a duration from several seconds to several minutes, most of which were produced by meteor ionization. In this estimation in Ref. [50] it was taken into account that there is a distinction between the two stages of dissipations of field-aligned meteor trails with two specific mechanisms of diffusion spreading.

In the *first stage*, diffusion takes place in the strongly ionized and disturbed plasma with large velocity of spreading (though relatively smaller than in the absence of the magnetic field). In this stage the scattering burstlike signal from the meteor trail is coherent (i.e., with deterministic oscillations of its strength envelope) with rapid decrease of the amplitude envelope and intense regular fading (with a deep oscillatory character) of the received signal (see Figure 6.32). The corresponding relative time constant of the dissipation of a meteor-induced irregularity at this stage τ_{co}/τ_0 can be estimated by Equation 5.54b for the plasma with an arbitrary degree of ionization and perturbations.

In the *second stage*, the inhomogeneous plasma of the meteor trail becomes weakly ionized and perturbed (as in the case of artificial injected plasma clouds described in the previous section). The velocity of diffusion spreading decreases considerably. The scattering signal is diffuse (i.e., incoherent with random oscillations of its strength envelope), because we now observe scattering from a cluster of small-scale turbulences with a fairly slowly decreasing amplitude envelope and small

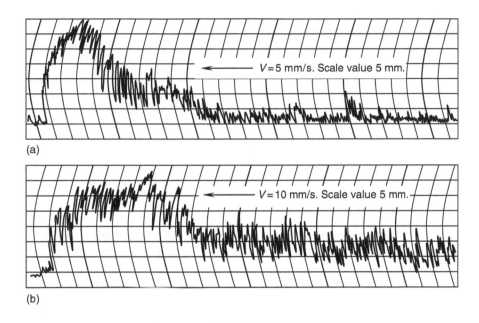

(a)

(b)

Figure 6.32 Example of recorded burstlike CW signal scattered from the meteor-induced irregularity.

irregular fluctuations of the signal (see Figure 6.32). The corresponding relative time constant of the dissipation of a meteor-induced irregularity in this stage, τ_{inc}/τ_0, can be estimated by Equation 5.54c for weakly ionized plasma ($p = 0$). In this phase, regular anisotropic diffusion is determined by the degree of anisotropy of small-scale plasma turbulences (as was observed for heating-induced and injected irregularities mentioned earlier) within the meteor-induced plasma cluster generated during dissipation of the meteor trail of the first phase. In that sense we can stress about the generation by meteor trails a cluster of small-scale irregularities that under the effect of the ambient magnetic field are oriented and extended along $\mathbf{B}_0$ (or close to the lines of force).

According to the fine structure and duration of the scattering signal in the first and second phases, we can now judge the effect of the geomagnetic field on diffusion spreading of meteor-induced irregularities in the ionosphere. We will later discuss another effect of deep fading of the signals caused by drift mechanism and its horizontal gradients along altitudes.

We now estimate the duration of dissipation of meteor-induced irregularities obtained experimentally [50] from all the radio bursts of meteor origin in one of a series of experiments in which only those signals whose duration was more than 5 s were selected, the ratio of the incoherent to coherent part of signal duration was more than unit—i.e., $\tau_{inc}/\tau_{co} > 1$—and the decrease of the signal envelope of which in the first phase was accompanied by a strong regular fading.

The distribution of durations of the burstlike signals is shown in Figure 6.33. The duration of most of these signals is $5 \text{ s} \le \tau \le 160 \text{ s}$, where $\tau = \tau_{co} + \tau_{inc}$. Such a broad spectrum of the time constant of the dissipation process agrees with the results of the theoretical framework [50]. Figure 6.34 shows the dependence of the ratio of the total lifetime of meteor-induced irregularities to the time constant of the dissipation during the first phase of diffusion spreading. It is evident from the presented histogram that the duration of the second phase of dissipation of irregularities is greater (from 3–4 to 10–15 times) than the duration of their first phase.

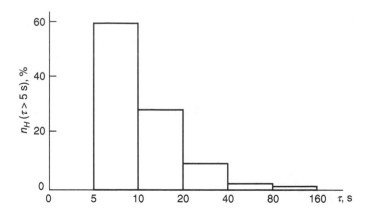

Figure 6.33 **Distribution of durations of the burstlike signals (in percentages) with periods more than 5 s.**

The results of experiments presented in Figures 6.33 and 6.34 are in good agreement with theoretical predictions based on classical theory of diffusion spreading in the real ionosphere with irregularities of arbitrary degree of ionization and perturbation discussed in Section 5.2 on the basis of the corresponding theoretical frameworks [123,133,149–151]. For example, for $p = 10$ and 100 we obtain, respectively: $\tau_{inc}/\tau_{co} \approx 6$ and $\tau_{inc}/\tau_{co} \approx 58$ for $h = 100$–110 km, which coincide quantitatively with results presented in Figure 6.33.

In Refs. [145,146] it was also found that the shape of the meteor trails significantly changes under the effect of drift mechanism of spreading because the horizontal winds predominate at altitudes of 80–110 km, where usually these kinds of ionospheric irregularities are observed using radio methods of location. The speed and direction of the wind in this altitudinal region of the lower ionosphere can vary strongly with insignificant changes in height. One of the reasons for the

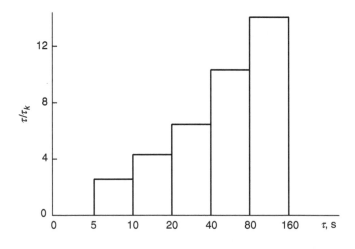

Figure 6.34 **Normalized lifetime of burstlike signals (in percentages) to that of signals, the first phase of decay of which exceeds 5 s.**

change in the shape of meteor trails is the difference in the drift velocities of meteor-induced irregularities at various heights, caused by the geomagnetic field and the differential effect of various height-dependent parameters of the ionosphere. In Refs. [145,146], an approximate formula for the velocity of the ambipolar drift for the particular case at $h \geq 100$ km was found by analyzing the general formula (Equation 5.82). We present it in notations that we used in Chapter 5 as

$$V_a = U_m + \eta U_m \frac{\sin \varphi}{(1+p)^2} \frac{Q_H^2 - q_H^2 \cos^2 \theta}{[1+p)^2 + q_H^2(1 + 2\gamma + Q_H^2)]}$$

$$\times \left[\frac{1+\eta}{1+p} \sin^2 \theta + \frac{(1+p)(1+\eta) + Q_H^2 + \eta q_H^2}{(1+p)^2 + q_H^2(1 + 2\gamma + Q_H^2)} \cos^2 \theta \right]^{-1} \qquad (6.8)$$

Here, a new parameter was introduced with respect to those introduced in Chapter 5, $\gamma = m_e \nu_{ei} / M_i \nu_{em}$; the angle φ is the angle between the direction of neutral wind vector $\mathbf{U}_m$ and the magnetic field vector $\mathbf{B}_0$. Analysis of Equation 6.8 shown in Figure 6.35a and b as the values of relative velocity (to U_m) of the ambipolar drift versus angles θ and φ for the two heights $h = 100$ km (solid lines) and $h = 110$ km (dashed lines) for different parameters of ionization of the meteor trails: $p = 0$ (under-dense), $p = 1$ (moderate-dense), and $p = 10$ (highly-dense). From Figure 6.35a it can be seen that the essential influence of the drift spreading of anisotropic meteor-induced irregularities takes place for angles θ up to 3° at altitudes more than 95 km. The value $0.1 < V_a / U_m < 1$ for meteor ionizations, but for weakly ionized irregularities ($p \ll 1$), such as under-dense meteor trails or H_E inhomogeneities, becomes much smaller. At the same time, as shown in Figure 6.35b, with a decrease of the angle

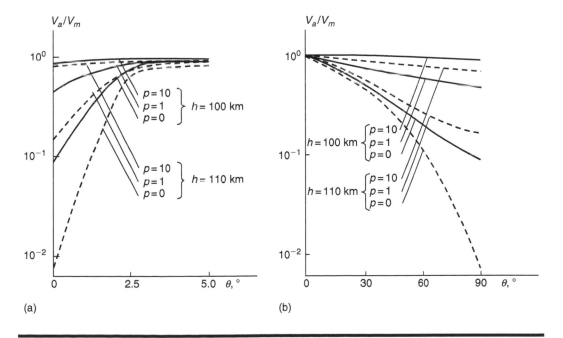

Figure 6.35 Normalized velocity of ambipolar drift of meteor plasma (to the velocity of the neutral wind) versus angles θ (a) and φ (b) for different parameters $p = 0, 1, 10$ and two altitudes of 100 km (solid curves) and 110 km (dashed curves).

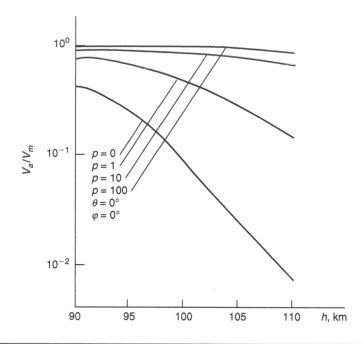

Figure 6.36 The same, as in Figure 6.35, but for $\theta = \varphi = 0°$ and for various altitudes $h = 80–110$ km of the lower ionosphere with degrees of ionization of meteor plasma changed from $p = 0$ to $p = 100$.

φ between $\mathbf{U}_m$ and $\mathbf{B}_0$ the value V_a/U_m increases. In Figure 6.36 the relative dependence V_a/U_m versus the height of meteor trials formation h for $\theta = 0$, $\beta = 90°$, and for different degrees of ionization p of irregularities, is depicted. As can be seen, for the field-aligned meteor-induced irregularities (i.e., for $\theta = \varphi = 0°$) the magnetic field decreases at the order the velocity of their drift spreading elative to wind velocity at altitudes $h \geq 100$ km. This effect increases with increase of the height h and with decrease of the degree of ionization of plasma within the irregularity.

As for H_E irregularities, scattering from which generates quasi-continuous signals of duration from several minutes to several hours with a slow increase and decrease of signal envelope and with irregular fading of signal all the time of observation (see Figure 6.37), most of them are caused by long-lived slightly ionized weakly anisotropic inhomogeneities of the low ionosphere of great extent and has the same sporadic structure as high-latitude and equatorial sporadic structures called *sporadic E layers*. On the basis of experimental studies of a long-distance (about 1400 km) middle-latitude ($\propto 45°$ North Latitude) radio trace in Refs. [49,147], some conclusions about the nature of field-aligned irregularities responsible for foreshortened scattering (at 44 and 74 MHz), called diffusive H_E irregularities, were pointed out. Thus, the daily variation in transmission intensity of quasi-continuous signals of durations from tens of minutes to several hours and their duration and fine structure characterized by fast stochastic fading (a fragment of which is shown in Figure 6.37) indicate that they are produced by long-lived weakly ionized diffusive irregularities of the ionospheric E region of E_{sq} type, consisting of small-scale field-aligned irregularities typical of the equatorial and polar zones [48–51,145–147]. The latter type of sporadic E structures will be discussed later. It should be noted that the diffusion mechanism of these plasma structures is described satisfactorily through the framework of theories concerning diffusion spreading of weakly ionized irregularities in the ionospheric plasma [123,133].

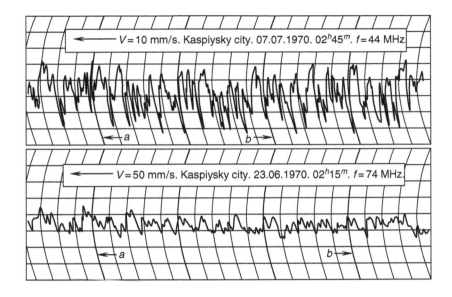

Figure 6.37 **Example of recorded quasi-continuous signal scattered from the H_E irregularity.**

Additionally, in Ref. [147] a correlation was found between the occurrence of sporadic E_s layers with the corresponding critical frequencies (and, consequently, between the blanking frequency $f_b E_s$) and the frequency of occurrence of quasi-continuous signals. The high correlation between them indicates the nature of the quasi-continuous signals, if sporadic irregularities of E_{sq} type, having small blanking frequency, cause great differences between the limiting $f_0 E_s$ frequency and blanking frequency $f_b E_s$, i.e., large $\Delta f E_s = f_0 E_s - f_b E_s$. As found experimentally, quasi-continuous H_E signals are not observed or is a rare phenomenon [49] in which sporadic E_s layers of low values of $\Delta f E_s$ appear. An example of such experimental observations can be seen in Figure 6.38, where (lower part) the limiting frequency $f_0 E_s$ (dashed curves) and blanking frequency $f_b E_s$ (solid curves) as well as the index K of magnetic activity at the area of scattering of radio signals (for detailed description of the radio trace, see Chapter 7) are presented. The circles with a plus sign along the horizontal axis (indicating the local time at middle-latitude radio trace) indicate the time of receiving of quasi-continuous H_E-scatter signals, and the circles with a minus sign indicate the times these signals were absent. The top part of Figure 6.38 gives the magnetograms of the vertical Z and horizontal H components of the geomagnetic field, as well as the corresponding daily pattern of the index of magnetic activity K (arranged inside the lower part of the figure) versus the local time at the radio trace during the period of 18:00 on June 26 to 18:00 on June 27, 1970. A small magnetic storm commenced at 9:00 a.m. and lasted until 21:00 (see decrease of both components of the magnetic field and increase of index K during this period of observation). According to such a combined sketched presentation, the correlation between the occurrence of sporadic irregularities of E_{sq} type and H_E scatter signals can be found, and the role of the magnetic field disturbance (due to large K) and ionospheric disturbances/instabilities (due to $\Delta f E_s$) in the formation of H_E irregularities can also be understood. It can be seen that where $\Delta f E_s = f_0 E_s - f_b E_s \geq 1$ MHz, which mostly occurs in evening time and nocturnal periods, both quasi-continuous signals and sporadic E_s layers are observed successfully, whereas in the morning and daytime periods, when even the blanking frequency exceeds 4–5 MHz but $\Delta f E_s \ll 1$ MHz, the H_E scatter signals were not found.

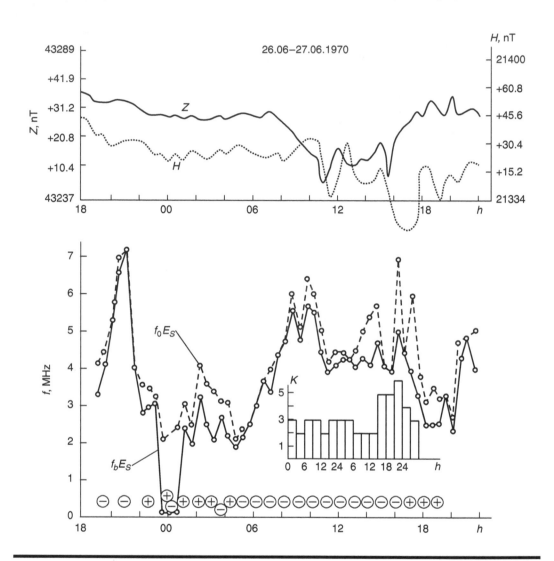

Figure 6.38 Temporal distribution of *Z*- and *H*-components of the ambient magnetic field, in nT (top panel), and temporal distribution of the critical and blanking frequencies in the disturbed ionosphere during magnetic activity determined by the index *K* varied from 2 to 5 during a day. Circles with "+" and "−" indicate correlation or anticorrelation, respectively, between ionospheric perturbations and magnetic field activity, obtained experimentally.

The full statistical analysis, as results of a comparison of 134 simultaneous records of received signals with the corresponding ionograms for various degree of transparency of sporadic layers, is presented in Figure 6.39. Here, the dependence of the coefficient of coincidence of the receiving of the H_E scatter signals, denoted by $\gamma\%$, on the difference $\Delta f E_s$, is shown. This coefficient was computed for each $\Delta f E_s$ as the ratio of the number of H_E scatter signals to the total number of observed sporadic layers (through the corresponding ionograms) with a given difference $\Delta f E_s$. We can see that $\gamma\%$ exceeds 70% for the sporadic layers with $\Delta f E_s > 1$ MHz. So, it follows from the presented statistical analysis that the presence of highly ionized sporadic E_s layers, and consequently, the existence of an even stronger inhomogeneous structure of the ionospheric E region at middle

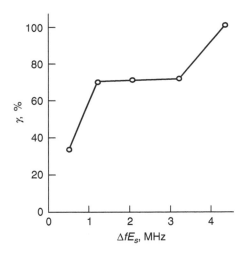

Figure 6.39 Correlation between existence of H_E scatter signals recorded at the receiver (in percentages) and perturbation of the ionospheric plasma (i.e., critical frequency of the sporadic layer F_s).

latitudes, does not produce the scatter of quasi-continuous H_E signals if the difference between the limiting and blanking frequencies is insignificant (i.e., $\Delta f E_s = f_0 E_s - f_b E_s \ll 1$ MHz).

At the same time, statistical analysis carried out in Ref. [147] has found that there is a good correlation (exceeding 70%) between recording of quasi-continuous H_E signals and the presence in the active scattering E region of the ionosphere of sporadic E_s layers with a blanking frequency $f_b E_s < 5$ MHz, but with the difference of the corresponding frequencies $\Delta f E_s > 1$ MHz. Even in the presence of only weakly ionized sporadic E_s layers but with a large $\Delta f E_s$, the presence of such layers and the occurrence of H_E scatter signals coincide. Additionally, it was found that there is no relation between magnetic activity and the occurrence of H_E scatter signals. In other words, no transmission of H_E signals was observed during magnetic storms. At the same time, such signals were recorded frequently (up to 85%–90%) under magnetically quiet conditions several days before and after a storm. Following the results of the analysis of experimental data obtained in Refs. [48–51,145–147] regarding sporadic H_E irregularities causing scattering of quasi-continuous H_E signals (one of the fragments of which is shown in Figure 6.37), we should note that the lifetime of sporadic E_s layers observed at middle latitudes (up to 90–150 min [52–54]) coincides with the duration of quasi-continuous H_E signals (amounting to 120 min at 74 MHz and to 180 min at 44 MHz [145–147]). The fact that such sporadic irregularities are mostly observed in evening and nighttime observation correlates with data obtained by other researchers using other methods of investigation of the E region of the middle-latitude ionosphere who obtained similar features common to middle-latitude sporadic layers and E_{sq} layers of the second type generated by gradient-drift instabilities in the polar and equatorial ionosphere, which we will further discuss in the following text.

Before entering into discussions of results obtained by other researchers, we will show by using methodology proposed in Refs. [145–147] how to estimate drift velocities of such sporadic plasma structures. The comparative analysis carried out in the aforementioned references for highly ionized meteor-induced irregularities and weakly ionized H_E sporadic irregularities showed that for small angles θ, the magnetic field reduces the velocity of drift spreading of H_E irregularities at altitudes $h > 95$ km by a factor 10–100 compared with wind of neutral molecules, whereas at $h < 90$ km the wind completely controls the movement of H_E irregularities. For $\theta > 5°$ the value of V_a/U_m for H_E irregularities is not more than one order of magnitude.

An additional feature, which was mentioned earlier in the context of different drift velocities occurring at different heights of field-elongated meteor-induced irregularities, can be used for the estimation of their real velocity by investigating the Doppler effect [145,146]. In fact, the scattering of radio waves from individual segments of a field-aligned meteor trial at various heights and, therefore, drifting at different velocities are coherent in the first phase of spreading of the meteor trail, as was also mentioned previously. This leads to regular fading of the envelope of the received signal as a result of the Doppler effect. The frequency of this fading, f_D, can be determined from signal records, and this in turn makes it possible to estimate the radial components and the total drift velocity. Following Refs. [145,146], we will determine the effect of the ionospheric wind on the frequency of signal fadings in oblique radio sounding [48,49]. Considering the height dependence of the effect of the magnetic field, we will use various values for the velocity of drift spreading, depending on height. Suppose that the drift velocity at the point B of the active zone of scattering is $\mathbf{V}_{d2}$ (see Figure 6.40). We will now assume that the conditions of specular scattering also occur at the higher point A, where drift velocity is $\mathbf{V}_{d1}$ (Figure 6.40). These points will move by $\mathbf{BB'} = \mathbf{V}_{d2}\,dt$ and $\mathbf{AA'} = \mathbf{V}_{d1}\,dt$, respectively, over time dt. In this case the path difference changes can be found as

$$\Delta s = (KB' - BP) - (NA' - AM) = [(\mathbf{V}_{d2} - \mathbf{V}_{d1})_T + (\mathbf{V}_{d2} - \mathbf{V}_{d1})_R]\,dt \qquad (6.9)$$

Here, we assume that the transmitter, T, and receiver, R, are sufficiently far apart [145]. The frequency of fadings produced by such difference drift can be obtained from the obvious relation $f_D = \lambda^{-1}|\partial\Delta s/\partial t|$, which taking Equation 6.9 into account gives

$$f_D = \lambda^{-1}|(\mathbf{V}_{d2} - \mathbf{V}_{d1})_T + (\mathbf{V}_{d2} - \mathbf{V}_{d1})_R| \qquad (6.10)$$

Without entering into complicated geometrical calculations following from the radio trace described in Refs. [49,145], we present here a histogram of the Doppler frequencies (or of signal envelope

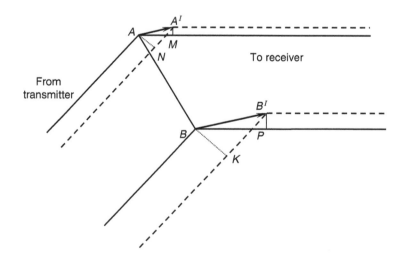

Figure 6.40 The geometry describing the Doppler effect between two points of wave reflection placed at different heights of the strongly elongated meteor trail.

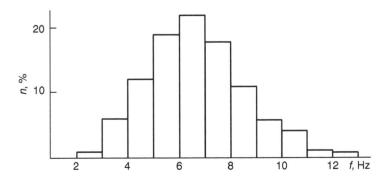

Figure 6.41 Histogram of distribution of echo signals with fading, the Doppler shift of which exceeds 2 Hz.

fadings) of more than 510 meteor radio reflections of a duration within a range of $5 \leq \tau \leq 300$ s and having the ratio of the durations of the incoherent (diffuse) and coherent parts within the range $1 \leq \tau_{inc}/\tau_{co} \leq 15$.

Using experimentally determined limits of the frequency of fadings, 2–12 Hz (see Figure 6.41) and a geometry of the trace described in Ref. [145], we obtain for the difference in the horizontal drift velocities of the scattering centers at various altitudes: (1) 16–106 m/s for north-westerly or south-easterly winds; (2) 9–61 m/s for a wind from the north or south; (3) 12–78 m/s for north-easterly or south-westerly winds; and (4) if we used the minimum fading frequencies, 2–4 Hz, for ionospheric winds from the west to east, we obtain velocity values of 61–122 m/s. These estimates are close to the horizontal velocities of ionospheric winds at heights of $85 < h < 95$ km, because for two-center reflections of field-aligned under-dense meteor trails, the segment of the trail above 100 km is actually not affected by neutral winds. These estimates agree satisfactorily with the data on drift velocity at these heights obtained in other studies [152–154].

The agreement between the estimates of the drift velocity of the meteor-induced irregularities and the H_E sporadic irregularities at the lower ionosphere obtained in Refs. [145,146] and those using other methods [152–154] indicates that the proposed unified explanation of dissipation of these two types of natural ionospheric irregularities at the lower ionosphere due to diffusional and drift spreading under the effect of geomagnetic field and the neutral wind, based on classical theory developed in Refs. [123,133], is vindicated.

Only after one or two decades were more precise methods of investigating drift velocities of meteor trails and sporadic E_s irregularities occurring at the middle latitude ionosphere proposed, based on Doppler shift estimations using the same continuous wave (CW) vertical and oblique sounding of the E region as discussed earlier, and the corresponding radar systems operating in the frequency domain and analyzing the shape and deviations of Doppler spectra of echo-signals recording at the receiver antenna [155–164]. We will discuss in Chapter 7 the experiments and the results of radio propagation performance using scattering from such irregularities. For now we will discuss briefly only two such studies that more accurately and precisely account for the corresponding effects of the Doppler shift of arriving radio waves scattered from the field-aligned irregularities (FAIs) occurring at the E region of the middle-latitude ionosphere, which in our opinion more precisely estimate Doppler shifts and the corresponding drift velocities of FAI movements [162–164]. Thus, in Ref. [162], using the portable radar interferometer arranged as a 50 MHz ($\lambda = 6$ m) Doppler CW backscatter radar located on the island of St. Croix

(17.7°N, 64.8°W), the plasma irregularities with transverse scales of 3 m (corresponding to the half-wavelength for backscattering conditions following from Bragg's law) associated with sporadic E layers were set specifically to find the shape of power spectra and the corresponding Doppler velocities characterizing horizontal drift movements of such irregularities.

These studies were carried out by researchers of Cornell University based on this portable radar interferometer and therefore were called the CUPRI Project. The radar beam pattern maximum was oriented at an azimuth angle of 66.5°W, i.e., over Puerto Rico and slightly to the north of the Arecibo observatory [162]. The corresponding 430 MHz radar assembled at the Arecibo observatory was used to measure the electron density profiles within the ionospheric regions observed by the CUPRI system. Figure 6.42a shows the normalized power spectra of the nighttime echo signals,

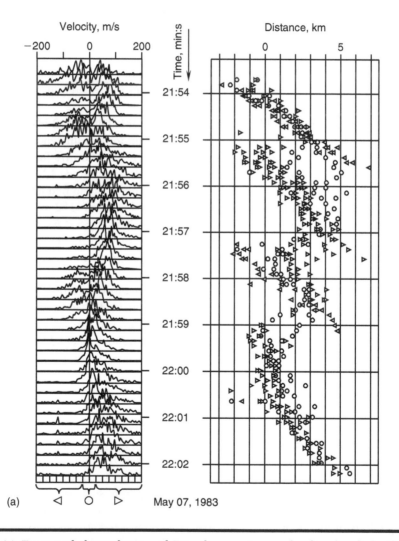

(a) ◁ ○ ▷   May 07, 1983

Figure 6.42 (a) Temporal dependences of Doppler spectrum of echo signals in the Doppler velocity domain (left panel) and temporal displacement (in km) of the corresponding plasma scatterers (right panel) obtained experimentally.

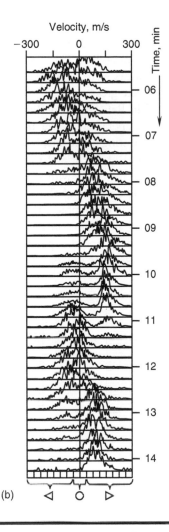

(b)

Figure 6.42 (continued) (b) The same as plotted in the left panel of Figure 6.42a.

where each spectrum is normalized by its own peak value, obtained by radar interferometer from sporadic E_s irregularities located at the range of 251.3 km. The time increases downward. In Figure 6.42a, the periods from 21:55 to 21:57 and from 22:00 to 22:02 (May 7, 1983) are the most interesting because they characterize the broad spread of Doppler frequency (with respect to zero) caused by movements of small-scale irregularities filling the scattered region (see left panel) and separated in space by ranges of 6 km and more (see right panel). According to the figures, the estimated velocities of horizontal drift of plasma irregularities during these two periods can achieve 80–160 m/s. The close results were obtained for the same range of 251.3 km during local nighttime (LT) on August 22, 1983. In Figure 6.42b, where these results are presented, the time is expressed in minutes after 22:00 LT. The authors found that the corresponding FAI occurred at altitudes from only 100 to 130 km when a sporadic E_s layer was present and that this is a pure nighttime phenomenon. The corresponding short-scale irregularities are stretched strongly along geomagnetic field lines with accuracy ±0.5° and less. A theoretical analysis made in Ref. [162] of the corresponding drift instability showed that the short-scale plasma waves (i.e., irregularities) are associated with large-scale plasma waves having wavelengths of 10–12 km and perturbation Doppler velocities around 200 m/s, smaller than those corresponding to the ion acoustic waves that caused modified two-stream (F-B) instability. The measurement results presented in Figure 6.42a and b show that the 3 m irregularities were generated primarily at the peaks and troughs of these waves, where the drift velocities were greatest.

Contradicting the results obtained in Ref. [162], in Refs. [163,164] it was shown that field-aligned irregularities associated with the sporadic E_s layer that ranged from 95 to 120 km along ionospheric altitudes can occur not only during nighttime, but also during daytime. Using a bistatic CW Doppler radar located on the island of Crete and operated at 50.52 MHz in the bistatic regime. The covered area of the E layer (at the center with altitude of 105 km) filled by field-aligned irregularities was about 15×40 km². The details of this circle of experiments are described in Chapter 7. Here, we only point out effects of Doppler shift spectra and the corresponding drift velocities estimations obtained experimentally. Thus, in Figure 6.43a and b the normalized echo signal power spectra, mean Doppler velocities, and the spectrum width are presented for nighttime (22:43 LT, Figure 6.43a) and early-morning time (04:24 LT, Figure 6.43b). Again, we see that the

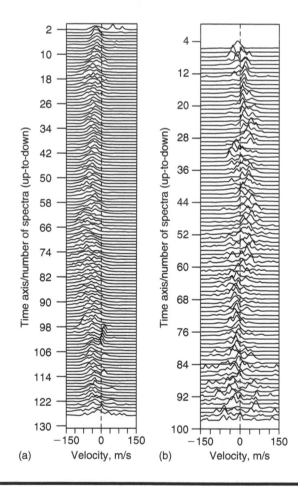

Figure 6.43 The same as plotted in the left panel of Figure 6.42a obtained for (a) nighttime and (b) morning time.

estimated Doppler velocities of 60–120 m/s are less than the ionic acoustic speed range of 300–500 m/s, remaining lower than this threshold of F-B instabilities generation. Nevertheless, the authors of Refs. [163,164] showed that F-B instability can be assumed as the most probable candidate of the source of the associated small-scale field-aligned irregularity generation if instead of simple unit-charged monoatomic ions, the metallic ions are taken into account; this usually occurs in the E layer of the ionosphere. In this case the threshold of ion sound speed becomes significantly smaller because $C_s \sim M_i^{-1/2}$ and with increase of mass of metalling ion (with respect to monoatomic ion) C_s becomes smaller, and the corresponding threshold for F-B instability generation essentially decreases. Thus, the drift velocities obtained experimentally that are shown in Figure 6.43 can be sufficient for exceeding the threshold of FBI and for generation of associated small-scale field-aligned (FAI) irregularities that fully coincide with observed measurements. We will discuss below the type and form of FAI spectra, comparing them with those observed in equatorial and polar regions of the ionosphere.

6.2.2 Radio Sounding of Sporadic Inhomogeneities and Bubbles in the Equatorial Ionosphere

In the equatorial region of the ionosphere there are irregularities of plasma density of various origins, as was expected and found in numerous experiments carried out by engineers and designers of land–ionospheric–land and land–satellite communication links. These irregularities introduce a lot of problems in prediction of key parameters of the corresponding ionospheric channels, such as the path loss, fading, SNR, and so on. In periods of plasma perturbations occurring in the E and F layers of the equatorial ionosphere, signals transmitted and received by satellites and/or ground-based terminals undergo deep and frequent oscillations after passing ionospheric channel, i.e., fading (see definitions in Chapters 3 and 4). Also, strong fading of the transmitted/received information data is observed even at UHF/X-band frequencies, applicable for satellite communications. This causes troublesome problems not only for designers of the ionospheric links, but also for regular users of satellite communication and navigation (called global positioning system, GPS; see Chapter 8), notably, for positioning of any subscriber in areas of service.

Because geophysical processes occurring in regions E and F of the ionosphere differ, and even today a general theoretical approach that could fully predict creation in these regions of different types of plasma instabilities and the associated plasma irregularities, small-, moderate-, and large-scale, is still lacking, we will consider these two regions independently and will present results of experimental investigations of the corresponding irregularities in these two regions separately.

Observations of Irregularities in the Equatorial E layer. We begin by considering the E region of the equatorial ionosphere, which, as was mentioned in Chapter 1, covers a layer of $\pm(15°-20°)$ in latitudes around the geographical dip equator to the north and to the south. In altitudes of E layer of 80–150 km, the electromagnetic processes are related to ionospheric electric fields and currents, even when they can be sufficiently small. Thus, the ambient electric field as was found by in situ spacecraft probe experiments does not exceed a few millivolts per meter (mV/m). Therefore, taking into account measured magnetic field strength of several hundreds and thousands of nanoteslas (nT), we find that the drift velocity of plasma in crossing $\mathbf{E}_0 \times \mathbf{B}_0$ fields is approximately $V_a \approx 80-100$ m/s, i.e., small enough to exceed the ion acoustic or thermal ion speed. We cannot therefore expect generation of drift instabilities in the lower equatorial ionosphere. At the same time, the drift velocity can exceed ion acoustic speed (or even ion thermal velocity), which leads to two-stream (or Farley-Buneman) instability. These kinds of instabilities were analyzed in Chapter 2. Because at the dip equator strength lines of the geomagnetic field are close to the horizontal direction, a large additional vertical electric field (in the Hall direction) occurs with strength of 10–20 mV/m. Moreover, at the center of the equatorial layer, all these phenomena produce the current that in the literature is called the *electrojet*, where the corresponding drift velocities can exceed 500 m/s. Such complicated natural geophysical features and phenomena dictate precise experimental and theoretical investigations of irregularities existing in the equatorial E region of the ionosphere.

Pioneering observations of the E layer of the equatorial ionosphere, and mostly sporadic E irregularities using a vertical sounding of the ionosphere, commenced from all over the world toward the end of the 1940s and beginning of the 1950s. Despite their wide geographical range, these observations were carried out sporadically, with only statements of observed phenomena (see the analysis carried out by Chapman [165]). The study of sporadic irregularities in the equatorial E region was carried out more systematically during the 1958–1959 International Geophysical Year (IGY) at Huancayo, Peru, by researchers from Peru and the United States (Stanford University) using a three-frequency, swept-azimuth backscatter sounder. These studies were then summarized in Refs. [166–169]. The investigations resulted in the discovery of the presence of field-aligned irregularities (FAIs) of a sporadic type in the equatorial ionospheric E layer and the corresponding mechanism for the formation of sporadic E_s irregularities (called E spread or "E_{sq} echoes") [168,169]. Without going into the technical description of the backscatter sounder, we only point out that in these investigations three operational frequencies were used: 12, 18, and 30 MHz. The scattered signal from sporadic equatorial E_s irregularities (E_{sq} echoes) was observed in more than 90% of the daytime observations at 18 and 30 MHz scattered from FAIs placed at altitudes between 100 and 140 km. The typical E_{sq} echo fading rate is illustrated in Figure 6.44, where an A-scope photograph of an E_{sq} echo integrated over approximately 10 pulses is presented. Sporadic occurrence of E_{sq} echoes was observed up to 200 km. The mean-square of deviations of plasma density inside these E_{sq} irregularities, $\langle (\delta N/N)^2 \rangle$, was estimated at between 10^{-3} to 10^{-5}, i.e., weak enough (i.e., of the same order as for middle-latitude ionospheric H_E sporadic irregularities described earlier). These first systematic experiments and theoretical investigations based on backscatter sounding of the ionosphere adequately describe the mechanism of "equatorial-type" sporadic E generation in the same manner as was carried out for the middle-latitude ionosphere, and we will compare them later in this chapter.

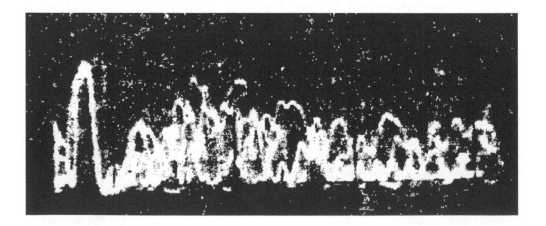

Figure 6.44 An example of recorded typical E_{Sq} echo signal with fading obtained in Ref. [168].

Furthermore, using the Jicamarca radar system located in Peru (2°N, 75°W) and operated at 50 MHz, a detailed analysis of the thin structure of the equatorial region was carried out both in the time and Doppler frequency domains [54,170–174] (see also technical details about the Jicamarca radar in Chapter 8). It was found experimentally that the small-scale irregularities observed by the radar are embedded in the large-scale plasma structures and that the 2-D turbulent plasma eddies create diffused radar echoes at the receiver. On the basis of a theoretical framework of plasma instabilities creation, the researchers assumed that the small-scale irregularities are generated by a 2-D coupling process in which large-scale plasma irregularities, associated with the gradient-drift instability (GDI) due to interactions between unstable and stable modes, generate small-scale irregularities via either the same GDI or the Farley–Buneman (modified two-stream) instability (see Chapter 2 and the referred bibliography there). Analyzing the Doppler spectra of small-scale irregularities, the investigators found an interesting common feature of echo signals received after scattering from such types of plasma irregularities consisting within the equatorial electrojet.

Two characteristic Doppler spectra of echo signals was found for small-scale irregularities located in the lower (E layer) equatorial ionosphere: one is with narrow spectral bandwidths, $\Delta\omega \ll kC_s$, defined as type-1 *irregularity*, and the second is with a wide spectral bandwidth, $\Delta\omega > kC_s$, defined as 2-*type irregularity*. Both are presented schematically in Figure 6.45. Here, as in Chapter 2, C_s is the ion acoustic velocity, and $k = 2\pi/\lambda$ is the wave number (as is well known, according to Bragg's law adapted for backscattering process $\lambda_{irregularity} = \lambda/2$ [54]). To be clear as to what was carried out by researchers, we present here one of the examples of analysis presented in Refs. [54,173]. Thus, in Figure 6.46, an A-scope display (i.e., time-range display of signal strength/intensity) of several thousand superimposed traces of the 50 MHz backscattered signal from electrojet received by the Jicamarca radar. The mean spectrum of the power density of the echo signal, averaged over the entire echoing region and the entire 2 min 40 s period, is shown in Figure 6.47. As can be seen, this spectrum is sufficiently narrow and symmetrical around zero Doppler shift. Therefore, the full coincidence with the spectrum schematically presented in Figure 6.45a is evident.

At the same time, the situation fundamentally changes when altitudinal and time dependence were analyzed in depth. For this purpose, a real signal presented in Figure 6.46, having been sampled and digitized at the five ranges labeled 1 to 5 at the bottom photograph of Figure 6.46, was separated in altitudinal range along the vertical axis (see Figure 6.48) according to the labels, and in

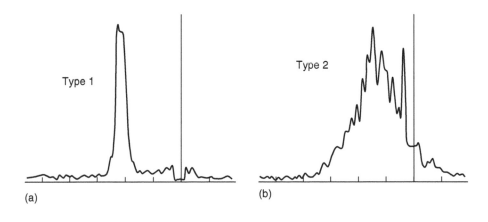

(a) (b)

Figure 6.45 Schematic presentation of spectra of echo signals scattered from irregularities of the first and second types.

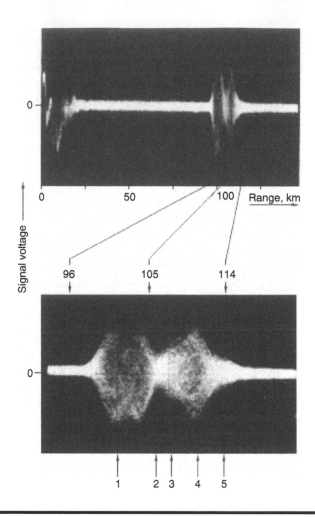

Figure 6.46 The A-scope display of backscattered signals (top panel) and detailed spatial displacement of several of them (bottom panel), positions of which are indicated by solid lines with the numbers. (Extracted from Ref. [54].)

samples of Doppler shifts (-150 to $+150$ Hz) at 10 s intervals along the horizontal axis (Figure 6.48). Moreover, to investigate the dynamics of echo signals in the Doppler shift domain, some of the portions labeled A–D from Figure 6.48 having resolution of $\Delta t = 10$ s were analyzed separately to obtain better time resolution, $\Delta t = 5$ s for a medium row with spectra, and $\Delta t = 2.5$ s for the bottom row with samples of spectra (see Figure 6.49). Detailed analysis of each spectrum demonstrates that the shape can change from the 1-type to the 2-type presented in Figure 6.45a and b, depending on altitude and the time of the process of evolution. Thus, the Doppler shift can vary rapidly with altitude, changing the shape and sign. The bottom row of eight spectra in Figure 6.49 covers a 20 s period with 2.5 s intervals and 5 km total altitudinal range. Type 1 of echo signals presented in Figures 6.45a and 6.47 (and the corresponding irregularities) have been associated with Farley–Buneman instability (FBI) generated in the equatorial plasma (see Chapter 2). According to estimations of Doppler spread, the drift velocities for 2-type echo signals (and corresponding irregularities) presented in Figures 6.45b, 6.48, and 6.49 have been associated

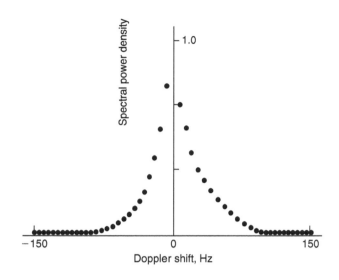

Figure 6.47 An example of spectral power density distribution in the Doppler shift domain, obtained experimentally in Ref. [54].

with drift-gradient waves, i.e., with GDI generated in the equatorial ionospheric plasma. Naturally, the authors of Refs. [54,170–174] worked with the corresponding theoretical frameworks developed before and during the period of their experimental investigations, and the role of large-scale irregularities on small-scale irregularities generation, and conversely, was not fully understood then.

Only after about 10 years, with a new theoretical background that analysed nonlocal effects of wave–wave and wave–particle interactions (see Chapter 2) and with new modern methods of

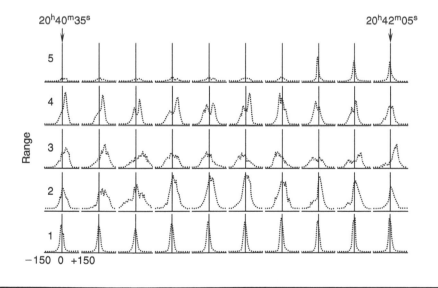

Figure 6.48 Doppler spectra of echo signals distribution separated in the time domain by 10 s intervals.

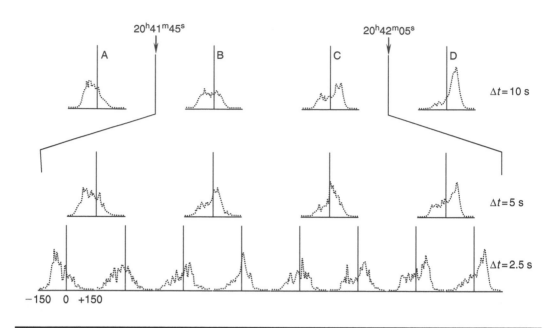

Figure 6.49 Detail presentation of some examples from Figure 6.48, labeled A, B, C, D.

experimental investigations of E layer of equatorial region was a new methodology created to describe features of equatorial electrojet observed experimentally during the latter 20 to 30 years using the Radio Observatory located in Peru. This methodology is based on complex simultaneous observations of spatial, temporal, and Doppler shift evolution of irregular plasma structures using the VHF radar Jicamarca interferometer, the HF backscatter radar operated at 14.2 MHz [55,175,176], and in situ satellite measurements [177–179].

The corresponding experiments and analysis of the theoretical background through the prism of obtained experimental data were carried out in Refs. [55,175–179]. We will briefly follow these excellent works, where full explanations of observed phenomena have been presented based on a self-consistent theoretical framework developed by Sudan and his colleagues and separately by Hamsa and St.-Maurice and their colleagues (see their theoretical background with the corresponding bibliography in Chapter 2).

The complex experiments based on VHF/HF radar [55,175,176] and several rocket in situ measurements [177–179] were unified in a joint Project called "Condor Equatorial Electrojet Campaign" (CEEC) and fully described in the aforementioned references. We will discuss briefly the results obtained there. Major campaign results discussed in these works are the following. First, the simultaneous investigations of kilometer-scale irregularities (large and moderate) were carried out by using radar measurements and rocket detections, and a detailed comparison between data obtained by different methods was made. Moreover, a good coincidence between measured data and theoretical prediction, based on the theoretical background of two-stream instability (or FBI) generation, developed by Sudan and his colleagues, was obtained, taking into account the anomalous diffusion process, nonlocal effects, and coupling between wave modes, large- and small-scale (see Chapter 2). Earlier, in Refs. [175,176], it was found that large-scale irregularities (from few to several kilometers) are more predictable sources of echo signals received both in daytime and nighttime observations of the equatorial E spread layer. Their characteristics can better be explained by using the self-consistent theory of gradient drift instabilities (GDIs) generation described in Chapter 2. Moreover, during their spatial and temporal evolution, the velocity of growth of these plasma perturbations can exceed the ion acoustic velocity; i.e., they are also candidates for FBI generation associated with small-scale irregularities.

Radar observations using a 50 MHz Jicamarca VHF interferometer (see technical details in Chapter 8) of the altitudinal range 90–110 km with a height resolution of 1.75 km have indicated the same results in the Doppler shift domain as presented in Figures 6.48 and 6.49, but performed more precisely with resolution in time (interpulse periods) of 2.25 ms. These results are shown in Figure 6.50a through c for (a) daytime ionosphere at a 96 km altitude and for nighttime ionosphere; (b) at 96 km altitude; and (c) 109.5 km altitude, respectively. As can be seen, for lower ionospheric altitudes (Figure 6.50a and c) the spectra of echo signal distribution in the time domain is the same in both diurnal and nocturnal ionosphere and indicates the existence of irregularities of 1-type, defined in Refs. [170–176] and presented in Figure 6.45a. They also indicate that the electron drift velocity always exceeds the ion acoustic speed, which is enough to create two-stream instability waves (e.g., FBI waves) in the equatorial ionospheric plasma. Thus, the peaks of 1-type of echo signals indicate that the 1-type instable plasma waves (e.g., irregularities) travel radially away from the radar with velocity approximately equal 315 m/s. As for higher altitudes (Figure 6.50c), the spectral characteristics differ completely from those presented for lower altitudes. Here, evidently the 2-type of echo signals occur corresponding to GDI causing large-scale waves (irregularities of type 2) at the top of the equatorial sporadic E layer with oscillatory variations of spectral shapes. The same difference between spectral characteristics along altitudes in the bottom, central, and the top levels of the equatorial E spread layer has been observed from results of radar observations carried out in the space domain, presented in Figure 6.51a (daytime) and Figure 6.51b (nighttime) obtained in Ref. [55]. The intersample gap for the former figure is $\Delta t = 1.44$ s, and that for the latter figure is $\Delta t = 4.32$ s. Again, we see that for both time observations, daytime and nocturnal, the narrow quasi-symmetric-shaped spectra for each of the five consecutive estimates at altitudes less than 100 km (bottom part of layer) and the wide multimode-shaped spectra at the center of the E layer. As for altitudes above 110 km (top part of the E layer), quasi-symmetrical narrow (but not smooth) spectra are observed that at higher altitudes for daytime ionosphere (Figure 6.51a) continuously transform into multimode diffuse spectra, an effect that has not been found for nighttime spectral observations in the Doppler shift domain.

These pictures were also observed by using HF radar, as well as by satellite. Thus, simultaneous observations using the 14.2 MHz HF radar were carried out with a time resolution of 1.44 s at 10 min intervals (during a 2 h period) at an elevation angle of about 20°. The corresponding measured data in the Doppler shift domain (from −50 to +50 Hz) is shown in Figure 6.52. The strong 2-type echo signal spectra are evident, which again allows for pointing out the main role of large-scale plasma irregularities associated with GDI in generation of turbulent small-scale eddies embedded into a large-scale structure. Therefore, some authors, based on results of measurements, have proposed three kinds of the echo-signal power spectra [52]: type 1, type 2, and the mixed type, as shown in Figure 6.53. At the same time, as shown in Ref. [52], the shape of the power spectrum depends on the zenith angle of the radar scattering cross section, which differs for all three types of spectra presented in Figure 6.53. This effect was proved by series of in situ experiments during the corresponding rocket probe measurements of plasma structure and ambient electric field in the equatorial ionosphere [177–179]. Thus, as was shown experimentally by the rocket observations presented in Figure 6.54, the shape of the spectra depends strongly on the elevation angle of observations of the equatorial E spread layer. As seen from illustrations presented in Figure 6.54 for different elevation angles, the 1-type echo-signals (and the corresponding 1-type irregularities)

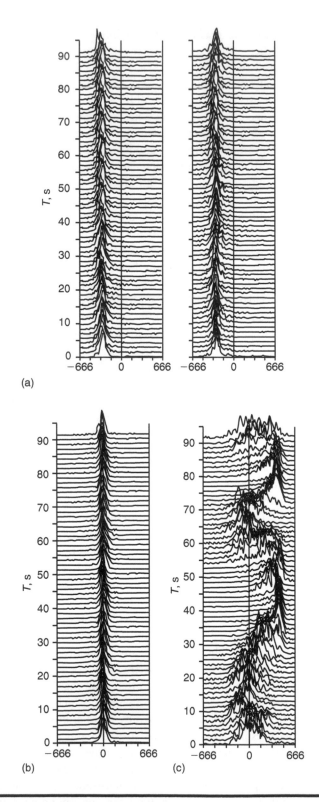

Figure 6.50 Signal power spectra in Doppler shift domain for (a) daytime ionosphere at 96 km, (b) for nighttime ionosphere at 96 km, and (c) for nighttime ionosphere at 109.5 km.

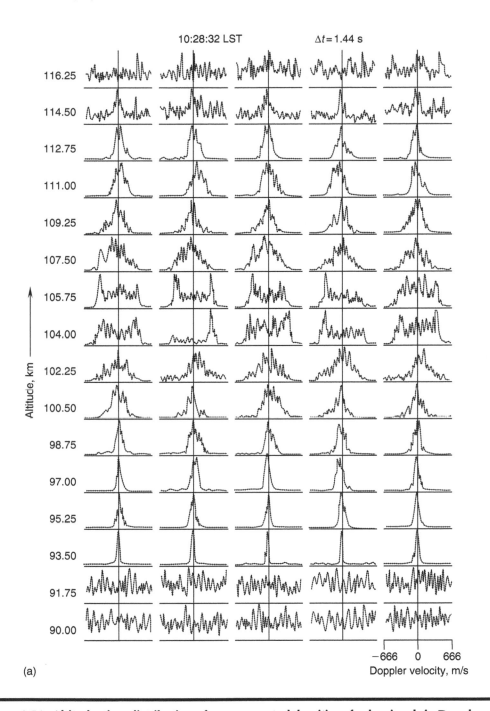

Figure 6.51 Altitude–time distribution of power spectral densities of echo signals in Doppler shift domain (with separation of each spectra in time at 1.44 s along the horizontal axis) for (a) daytime and

(*continued*)

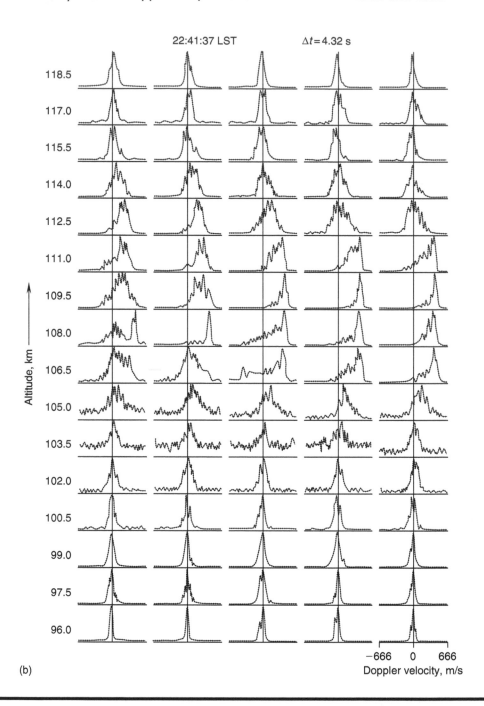

Figure 6.51 (continued) (b) nighttime ionosphere.

dominate for elevation angles smaller than about 35°, whereas high-elevated spectra are usually type 2.

These results indicate that the dynamical spectra of echo signals occurring in the frequency domain can be associated with different localized (along altitude) scattered regions consisting of

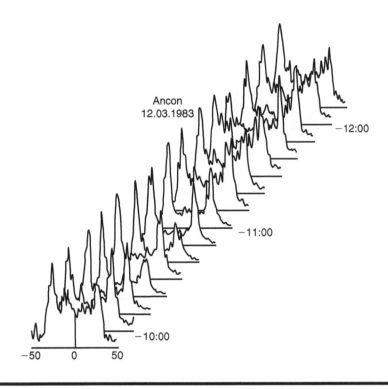

Figure 6.52 Echo signal spectra in joint frequency–time domain obtained in Ref. [52] in Ancon experiment.

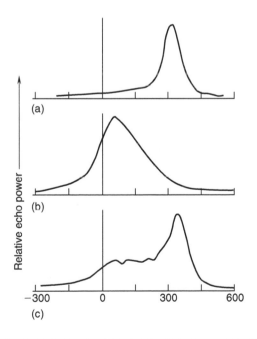

Figure 6.53 Schematic presentation of spectra of echo signals scattered from ionospheric irregularities, divided in Ref. [52] on three types: first (a), second (b), and combined (c).

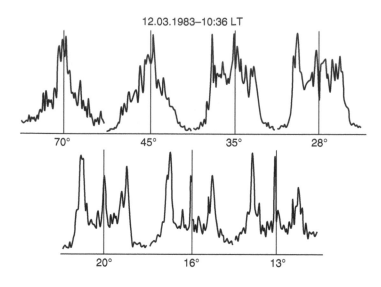

Figure 6.54 Echo signal spectra obtained by rocket experiments in Refs. [177–179] for various elevation angles of the rocket.

turbulent eddies of different scales, from large to small, which finally move across the radar beam with transit time usually greater than the lifetime of this eddies within the turbulent sporadic structure.

Moreover, in Refs. [177–179], the markedly different altitudinal regions of the E layer were found where two kinds of instabilities are generated: from 90 to 105 km, where GDI is the most probabilistic candidate of the field-aligned irregularities (FAIs) creation, and from 103 to 111 km, where FBI is the main source of FAI observed at those altitudes. The corresponding in situ measurements of the altitudinal profile of the plasma concentration inside the inhomogeneous E region (the left panel) and the frequency–height sonogram (the right panel) extracted from Ref. [177] are shown in Figure 6.55. A more detailed examination of the pure two-stream region consisting of F-B instabilities was made in Ref. [178]. In Figure 6.56 frequency–time–altitude dependence is displayed for this region to show the high Doppler shift, which can measure up to several hundreds of hertz and the corresponding drift velocities of several hundreds meters per second which can easily cross the ion acoustic speed threshold for FBI generation in the altitudinal range 106–111 km. The experimentally obtained results were analyzed and summarized in Ref. [179] on the basis of analysis of Figures 6.50 through 6.56, from which it is evident in which altitudinal region the two kinds of instabilities prevail to generate the corresponding plasma irregularities. Additional measurements have shown that for small-scale eddies with scales from 1 m to 15–20 m the plasma density fluctuations (with respect to background ionospheric plasma) do not exceed 1%–2%, whereas for moderate-scale eddies (with scales 30–300 m) these characteristics can achieve 10%–20%. The corresponding fluctuations of ambient electric field $\mathbf{E}_0$ caused by perturbations of plasma density can achieve $\delta E \approx 10-20$ mV/m with corresponding drift velocities $V_d = 100-200$ m/s [179–181], results that were fully predicted by the existing theoretical background concerning generation and evolution of GDI and FBI and the associated plasma waves (e.g., plasma irregularities).

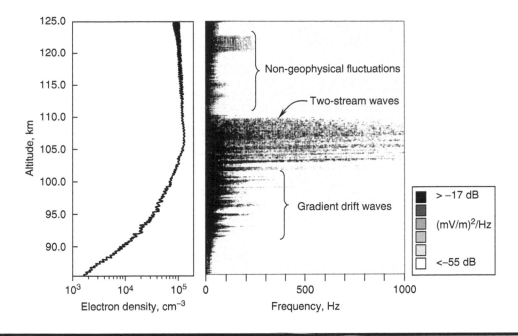

Figure 6.55 Altitudinal profile of electron density obtained experimentally in the lower iono-sphere (left panel), and the corresponding spectral characteristics of plasma waves (instabilities) versus Doppler frequency (horizontal axis) measured in dB (right panel).

Further observations of the equatorial E spread region and the corresponding irregularities filling this region based on the same Jicamarca 50 MHz radar were carried out with unprecedented altitude resolution (of the order of 125–150 m) for altitude–time–power, and Doppler spectra display were reported in Ref. [182]. The authors mostly studied an altitudinal range from 95 to 120 km during the daytime and evening time periods. Because their results are precise, with high resolution of measured data in space, time and frequency domains, we present here further vivid physical results obtained in the aforementioned reference. Thus, we present here only two characteristic plots of altitude–time (Figure 6.57) for echo signal power (or signal-to-noise ratio, S/N, in dB) and for Doppler velocities, V_d, in m/s (Figure 6.58), where the pixel size in Figure 6.57 was about 125 m in the space domain and 1.92 s in the time domain. In Figure 6.58, the black pixels correspond to downward velocities greater than 20 m/s, the light pixels correspond to upward velocities greater than 20 m/s, and velocities of 20 m/s and less are dark. As can be seen in the upper sublayer stretching from 110 to 115 km, the scattered echo signals are strongest. The corresponding irregularities have a tendency to move downward, whereas at the lower sublayer of 95–100 km the tendency to move upward is usually observed and less violent dynamics is exhibited. These two layers are separated by a distinct gap in the echo signals. The tendency of echo signal power and Doppler velocity distribution in a 2-D altitude–time domain observed for midday periods is changed slightly. The corresponding example of these two characteristics is shown in Figure 6.59 for S/N (the left panel) and for V_d (the right panel). It can be seen that in daytime periods more intensive echo signals arrive from the lower sublayer located at altitudes 97–107 km with a tendency of higher downward velocities with respect to the higher layer, which is now sufficiently weak. So, in daytime periods the gradient-drift instability is a more probabilistic

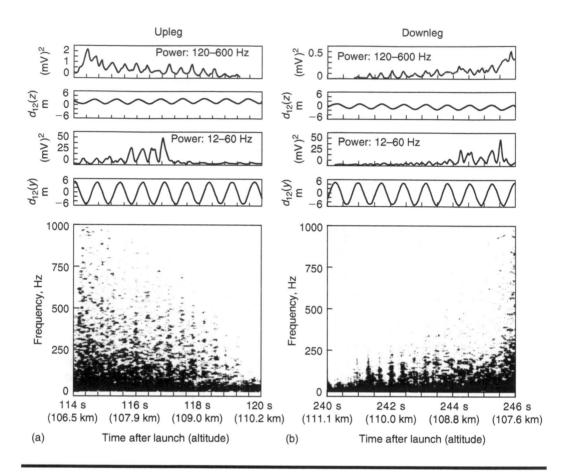

Figure 6.56 Signal power and spatial range distributions along the frequency axis from 120 to 600 Hz (top two panels, respectively) and from 12 to 60 Hz (two further panels, from top to bottom, respectively), and frequency–time–altitude 2-D plot of echo-signals all presented for (a) upward leg and (b) downward leg.

source of irregularities observed in the equatorial E spread layer, whereas in the nighttime periods two types of instabilities are generated, the FBI in the higher sublayer and the GDI in the lower sublayer. This difference is more evident in the echo signal power Doppler spectra altitude–time characteristics, as shown in Figure 6.60. Altitudinal separation of peaks for lower, middle, and upper sublayers with narrow (type 1) symmetrical profiles can clearly be seen at the lower (95–100 km) and upper sublayers (117–120 km) and with a broad spreading shape (corresponding to type 2) at the middle sublayer (107–112 km). The same tendency was observed during entire period of observations in the evening and night periods.

We now return, as was suggested earlier, to the comparison between the sporadic weakly ionized long-term H_E irregularities observed in the middle-latitude E region ionosphere described previously, and the E_{sq} sporadic irregularities observed in the equatorial ionosphere. According to the analysis made in Ref. [147] such middle-latitude irregularities have the same nature as the equatorial sporadic structures associated with the modified two-stream instability. At the same time, comparing Doppler spectra of echo signals from the middle-latitude quasi-regular plasma long-term structures obtained experimentally in Ref. [162] and shown in Figure 6.42 with that for equatorial

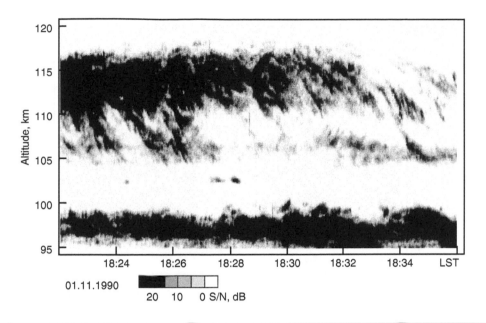

Figure 6.57 **2-D plot of the temporal and altitudinal distribution recorded echo signals, as signal-to-noise (S/N) ratio (in dB).**

sporadic E_{sq} irregularities analyzed experimentally in Refs. [55,181] and shown in Figure 6.50c, we note that the Doppler shift is broad in the equatorial regions and the Doppler velocities for the equatorial plasma irregularities can exceed 350 m/s and more, whereas the same characteristics for

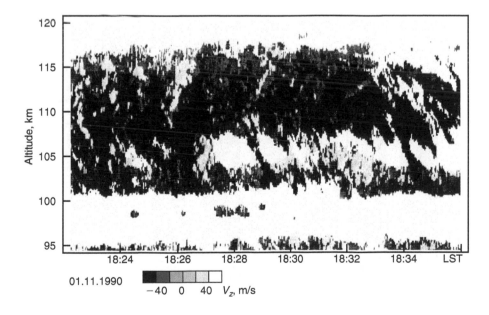

Figure 6.58 **The same as in Figure 6.57, but for Doppler velocity distribution of plasma irregularities in time and altitude domains.**

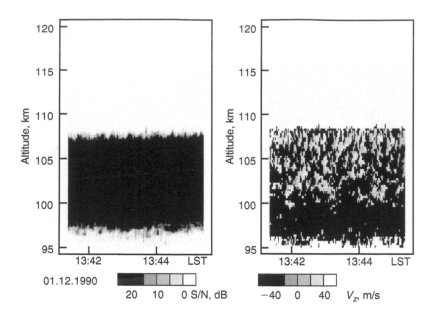

Figure 6.59 Altitude–time 2-D plot for echo signals, as signal-to-noise (S/N) ratio in dB (right panel), and the same plot for Doppler velocities in m/s.

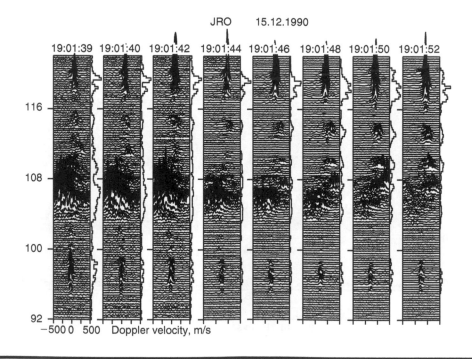

Figure 6.60 Spectral characteristics of the echo signals in height–time 2-D domain with deviation of Doppler velocities from −500 to +500 m/s.

the middle-latitude E layer do not touch even 200–250 m/s. To argue the existence of FBI as a potential source of observed H_E irregularities in the middle latitudes, it was assumed in Refs. [162–164] that if accounting for the metallic ions (namely, Fe^+) instead of monoatomic simple ions, then the corresponding ion acoustic speed threshold becomes several times lower, and the corresponding condition of F-B instability generation can be satisfied with sufficient accuracy also in the middle-latitude E region. Thus, taking $T_e \approx T_i = 400$ K at altitude 120–125 km (middle-latitude conditions; see Chapter 5) and $M_i^{Fe^+} = 56$ amu, we find that the ion acoustic speed is given by $C_s \approx 345$ m/s, which is similar to that observed in the equatorial E region. Hence, it can be suggested that H_E irregularities observed in the middle-latitude ionosphere have the same nature as the equatorial sporadic E_{sq} irregularities and can be associated with the two-stream (or FBI) instability.

We will now describe several modern experimental methods for measuring the electron content and the plasma density profile of the E equatorial region, which recently began to be used in the study of irregular structures of the ionosphere. It is evident that knowledge of plasma density allows for predicting effects of refraction, scattering, and reflection of probing radio waves passing the corresponding irregular layers of the ionosphere (see Chapters 7 to 9). It has been established for many years that the conventional method of measuring electron density profile in the strongly variable E region of the equatorial ionosphere, as a thin rarefied ionospheric layer, by using incoherent scatter radar is generally unacceptable for daytime periods, and rocket probe sounding is the only technique that can provide information about this region of the ionosphere (see Refs. [177–179] and the references contained therein).

In Ref. [183] it was recently reported that even when using the Jicamarca radar system based on an incoherent scatter technique, strong interference of the echo signals arriving at the receiver after scattering from intensive FAIs filling the equatorial E region is observed, which makes it difficult to measure the plasma density profile and the corresponding electron concentration in the ionosphere. Therefore, in Ref. [183], a new remote-sensing technique for measuring plasma (electron) density was proposed and successfully employed, taking into account advantages of the coherent scatter radar based on measurements of the Faraday rotation angle. It is well known that the Faraday rotation of the radio signals passing through ionospheric plasma occurs in proportion to the electron density along the radio path. The differences in the Faraday angle from one radar range gate to the next indicate the plasma density at a given altitude. Using advantages of the Jicamarca bistatic coherent radar operating at 50 MHz, it was shown [183] that this radar can detect signals passing the daytime thin rarefied equatorial E layer with an accuracy of less than 1 rad of the Faraday rotation, i.e., even in the hard conditions when the rotation rate decreases significantly. Figure 6.61, extracted from Ref. [183], represents the geometry of the experiment. A bistatic coherent radar system for measuring the Faraday rotation angle was arranged at Paracas (the transmitter) and Jicamarca (the receiver), Peru. We will not describe this technique in any kind of depth; we will only point out that the scattered ionospheric area was located at altitude of about 100 km. In Figure 6.61 the irregular solid line indicates the Pacific coastline, and the horizontal line bisecting the figure indicates the locus of perpendicularity to $\mathbf{B}_0$ at an altitude of 100 km. At the right-hand side of the figure the radar antenna pattern is shown with spacing of antenna elements at ~1.5 λ for both site terminal antennas. Figure 6.61 shows the effective two-way power pattern in the combined transmitter and receiver arrays. The half-power full beam width of the main antenna lobe was about 6°, and the sidelobes were down by more than 20 dB compared to the main lobe. Using four five-element antennas with ~1.5 λ broadside spacing and sending narrow pulses (3 μs) with high pulse repetition frequency, the authors of Ref. [183] significantly increased the sensitivity of the experiment with minimization of the error in measurements of the Faraday rotation angles. For a whole cycle of experiments both terminal antennas were synchronized using satellite (GPS) receivers.

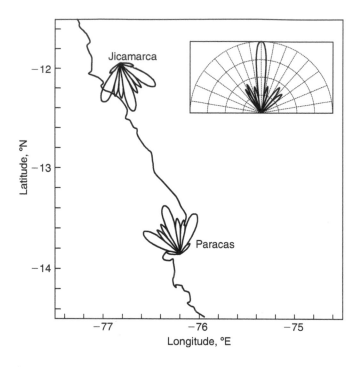

Figure 6.61 Geometry of experiments with the Jicamarca radar. (Extracted from Ref. [183].)

The strongest echo signals received on September 12, 2000, in daytime periods, were plotted in the range–time–intensity (RTI) format in Figure 6.62a. Here, only signals above 5 dB are shown despite the fact that even weak signals below −10 dB were registered accurately. The corresponding electron density profiles computed at about 20 min time intervals are shown in Figure 6.62b based on the Faraday rotation analysis. To compare the new technique proposed in Ref. [183] with that based on a sounding by rocket probe during the Condor campaign (see previous text), the latter measurements of electron concentration in the equatorial E region and the corresponding plasma profile are shown in Figure 6.62c according to Refs. [177–179]. On the basis of the theoretical background that predicted existence of two-stream waves and the cascade wave interaction (from large-scale to small-scale) occurring in the daytime equatorial electrojet, an existence of the echo signals scattered from the topside of the E layer can be observed using bistatic coherent radar technique that record waves with wavelengths at the outer scale of two-stream turbulence [183]. This argument is evident from Figure 6.63, where the altitudinal dependences of (a) S/N, (b) magnitude of correlation (the coherence), (c) the echo signal phase (in radians), and (d) the electron density (in cm^{-3}) are depicted. Figure 6.63d illustrates the density profile, similar to that presented in Figure 6.62c, with a peak density of $\sim 2.0 \cdot 10^5$ cm^{-3} that occurs at an altitude of about 105 km. The observed plateau along altitude in S/N and the coherence (from 95 to 110 km) (Figure 6.63a and b, respectively) fully correspond to the absence of the signal phase deviations (Figure 6.63c) recorded by the radar receiver. Obtained results allow us to point out that this technique, based on measurements of the Faraday rotation angles by backscatter bistatic coherent radar, can be successfully adapted for obtaining F region density profiles when spread F prevents ionogram inversion.

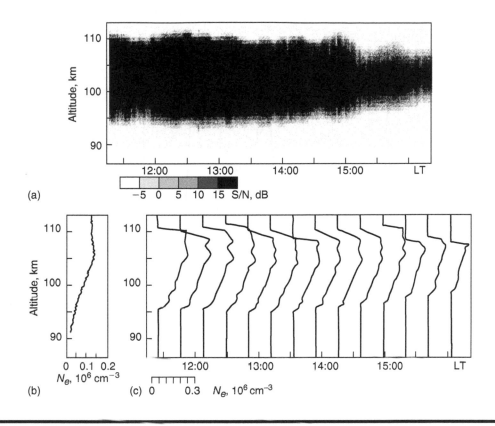

Figure 6.62 Altitude–time plot of scattered signal, as (top panel) (a) S/N ratio (in dB), (b) the plasma density profile along altitude, and (c) time–altitude plot of plasma density observed in *E* region of the ionosphere at altitudes 90–120 km. (Extracted from Ref. [183].)

Several years later, in Ref. [184], the improved technique of measurement of electron density profiles based on incoherent scatter radar (ISR) sounding of the disturbed irregular D to F regions of the equatorial ionosphere was reported and compared with the canonical method based on similar measurements carried out in Ref. [185] based on a Digisonde portable system (DPS). This technique was also validated via the standard "Faraday International Reference Ionosphere" (FIRI) empirical model based on Faraday rotation experiments conducted on flying rockets, which provides electron densities between 60 and 140 km [186], as well as with the analytical Danilov's model [187] as an option of IRI model, but for the D equatorial region (below 90 km). An improved technique is based on the same Jicamarca ISR based on measurements of the Faraday rotation angle of echo signals scattered by the corresponding layers of the irregular equatorial ionosphere. Figure 6.64 shows the electron density profile between 130 and 390 km obtained by ISR (solid line), the same ISR densities obtained using the new Faraday rotation technique (bold point), the DPS ionosonde profile (triangle), and IRI-modeled values based on DSP soundings (square). It can be seen that the agreement between two radar estimate data is excellent and both sets of data correlate well with data obtained by DSP ionosonde, whereas the IRI-modeled profile shows poor agreement with ISR- and DPS-measured values, particularly at the bottom and middle layers of the equatorial F region. At the same time, the modified IRI model (which was called FIRI

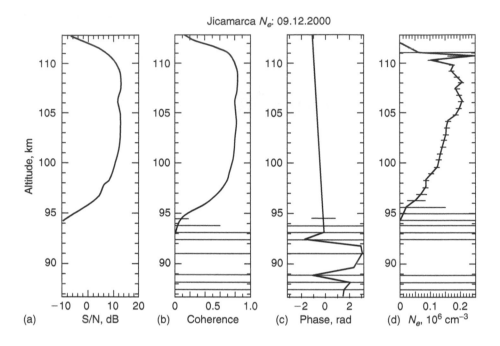

Figure 6.63 Altitudinal distribution of S/N ratio, measured in dB (a), coherence effect (b) measured by use probability of correlation between events and signal phase variations in radians (c), and plasma density (d) in E layer of the ionosphere.

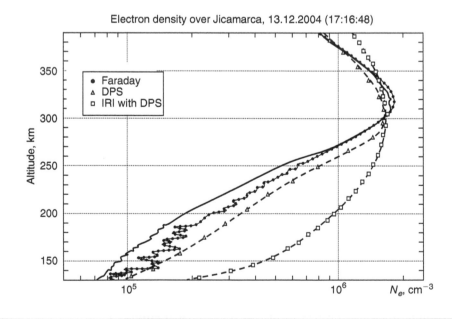

Figure 6.64 Plasma density profile between 130 and 390 km obtained by ISR (solid line), using a new Faraday rotation technique (bold points), investigated DPS ionosonde profile (triangles), and based on IRI model using DSP soundings (squares).

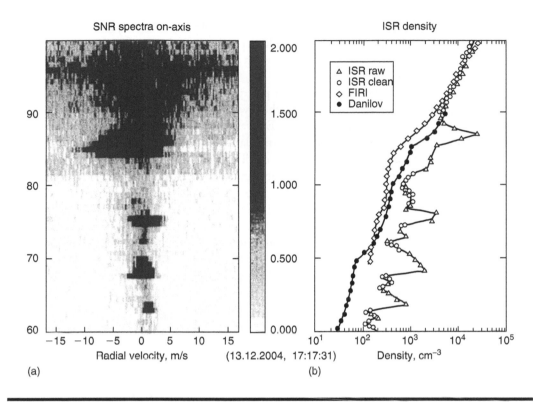

Figure 6.65 **2-D altitude–radial velocity pattern of S/N ratio spectra of echo signals (left panel a) and ISR density altitudinal distribution using different well-known approaches (right panel b).**

earlier) that takes into account the Faraday rotation effects can be a good predictor (together with Danilov's model) of plasma density profile in the equatorial lower E layer and in the D layer, as follows from Figure 6.65. Here, the corresponding FIRI profile is shown by the diamond symbol, whereas profiles obtained by ISR measurements of the Faraday rotation angle in two conditions, raw and clean, are denoted by the triangle symbol and the circle, respectively. Here, the empirical Danilov model is also shown using bold points. Experimental results reported in Refs. [183,184] show increased possibilities in predicting plasma density profiles in all three regions of equatorial ionosphere by using for both coherent and incoherent radar systems a new technique based on Faraday rotation angle estimation, which, as was shown earlier, can be done with satisfactory accuracy.

Observations of Irregularities in the Equatorial F Layer. Before going into the subject of main properties of irregularities of various origins and scales located in the equatorial F region of the ionosphere and of modern methods of studying their main parameters and characteristics important for radio propagation, let us point out some general aspects of the problem. First, according to knowledge obtained in Chapter 1, we should stress that the main difference between the F layer compared with the E layer of the ionosphere is the that region F extends from about 180 to 200 km up to more than 700–800 km, where it continuously transfers to the magnetosphere. Having a thickness more than 500 km, this region of the ionosphere is affected by different physical processes occurring at its bottom part, inside the layer and in the topside of region F. Therefore, it was difficult from the beginning to investigate all processes accompanying formation and further evolution of plasma at different altitudinal ranges of this ionospheric region. As for plasma properties at altitudes of the F layer, we should note that both electron and ion plasma components are magnetized and the background plasma is the plasma with a high degree of ionization, where we should take into account not only interactions between electrons and ions with neutral molecules (atoms), but also collisions between them.

Moreover, at altitudes of the F layer, the transport processes are predominant compared with ionization–recombination processes occurring in the lower E layer of the ionosphere. The ambient conditions are also changed compared with the E layer, because in the equatorial F region the evolution of plasma large- and moderate-scale irregularities associated with various instabilities occurring in the ionosphere is controlled (as in the middle-latitude ionosphere; see previous section) by atmospheric wind, ambient electric, magnetic, and gravitation fields, finally defining the specific dynamics of such irregularities. The small-scale irregularities, as in the E layer, are usually embedded into large-scale plasma structures, and such chaotic (turbulent) structures result in diffuse echo signals (i.e., oscillations of signal amplitude and phase) due to back/forward scattering from turbulent regions consisting of eddies of various origins and with scales varying from dozens of meters to hundreds of kilometers.

Below, following Refs. [188–214], we describe experimental investigations of irregularities occurring in the equatorial F region, covering a layer surrounding the dip equator with a width of ±15° northward and southward, respectively. Then we analyze separately the depletion zones accompanying evolution of large-scale irregularities, called *bubbles* in the literature [56–61,213–215].

Because the geomagnetic field lines lie in the horizontal plane to the ground surface, the magnetic field is perpendicular to the gravity field and to the gradient of plasma density along ionospheric altitudes. Under such conditions, the Pedersen horizontal currents are generated at the bottom side of the F layer, and the effect of short-circuit currents between the highly-conductive E layer and the lower side of the F layer dictates the evolution of irregularities observed experimentally. Large-scale irregularities created at the bottom side of the F layer floating upward penetrate the central and topside of the F region, generating these irregularities of wide spectrum of scales, from moderate (with hundreds of meters to a few kilometers) to small (of dozens of meters to several meters). Here we should note that, despite the fact that some authors have presented another separation on several different kinds of irregularities according to their scales, we divide irregularities into large, moderate, and small scales, because throughout our book, right from the beginning (see Chapter 1), we have compared their scales with characteristic dimensions of plasma components, electrons and ions, such as their free path length, gyroradius in the ambient magnetic field, and so on. Therefore, our differentiation into separate groups of irregularities with their specific spatial and temporal evolution is based on the understanding of physical processes occurring in the ionospheric plasma.

It should also be noted that, as done for irregularities of the equatorial E layer, in this chapter we do not go deeply into the analysis of physical processes accompanying formation of plasma instabilities and associated plasma irregularities. These aspects are fully presented in Chapters 2 and 5. Here, we will draw the attention of the reader to the methodology of modern radio and in situ rocket and satellite methods of observations for predicting the main characteristics of the irregular structure of the equatorial F region that are important for the performance of the corresponding radio propagation links passing through this region of the ionosphere.

The equatorial spread F phenomenon has been studied sporadically for five or six decades. However, the first systematic radar observations of the equatorial F layer were carried out at the end of the 1960s to the end of the 1970s [188–198] and continued through the 1980s [199–205] up to the end of the 1990s [206–214] using HF/VHF radar systems located in different geographical sites of Earth (see details in Chapter 8) combining modern in situ measurements using rocket and satellite physical probes. Such radio sounding of the F region of the equatorial ionosphere together with systematic in situ rocket and satellite probe measurements allowed investigators to create an adequate theoretical background to explain many specific features of plasma irregularities of various origins observed experimentally.

The first systematic experimental studies of large-scale plasma structures in the F spread region of the equatorial ionosphere in the African area were reported in Refs. [188–200] using non-great-circle propagation on the transequatorial path from Lindau (West Germany) to Tsumeb (South West Africa), combined with radar backscatter observations in Nairobi (Kenya). In Lindau ($\approx 51°40'$N, $\approx 10°10'$E), the transmitter sent impulse radio signals at the carrier frequency of 18.21 MHz, which arrived at the receiver located at Tsumeb ($\approx 19°10'$S, $\approx 17°40'$E). This path has a length of 7915 km and crosses the dip equator as shown in Figure 6.66. As is evident, Lindau and

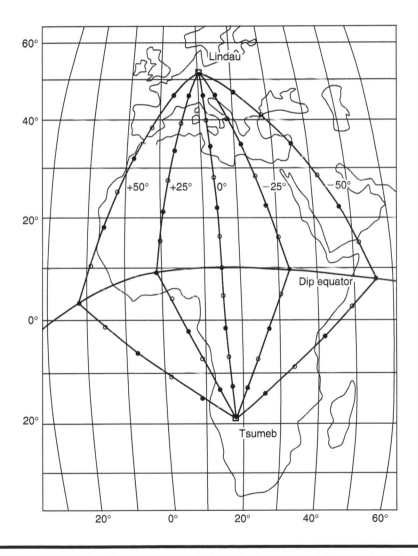

Figure 6.66 Geometry of radio traces passed through the equatorial ionosphere from Tsumeb (Africa) to Lindau (Europe). (Extracted from Ref. [189].)

Tsumeb are located outside the equatorial region. The rotated antenna in the horizontal azimuthal plane covered the region from $\alpha = -50°$ to $\alpha = +50°$ from the line-of-sight direction between ground-based terminal antennas ($\alpha = 0°$). The pulse duration of 200 μs used in this type of ionospheric sounding prevents an exact observation of propagation paths showing azimuthal deviations less than $\Delta\alpha = 10°$ around the great-circle area. Simultaneously, the monostatic back-scatter radar, where the transmitter and receiver were arranged in one location in Nairobi ($\approx 1°30'$S, $\approx 36°50'$E), operating at 27.8 MHz, was used with rotation in azimuth plane antenna to obtained echo signals reflected from the equatorial spread F irregularities.

First, we present results obtained along the radio trace Lindau–Tsumeb. Experiments were carried out according to equatorial spread F conditions when echo signals predominantly occur via F layer scattering. The received propagation modes on non-great-circle paths were characterized by the number of semihops between Earth and the ionosphere, northward and southward from the equator line (Figure 6.66). Using geometry of traces, the mean position of irregularities was found by using information on signal propagation time and azimuth of arriving signals. During experiments, several groups of traces or patches were observed, mostly from 20:00 to 22:00 (UT), showing a wavelike horizontal structure of the area of scattering. The distances between patches were defined by the authors as horizontal scale d_{WE} of wavelike plasma irregularities in the west–east direction. The left panel in Figure 6.67 shows the distribution of d_{WE} obtained from the corresponding sounding of the ionosphere during March 1971. As is evident, the patch size can be a few hundreds of kilometers with a mean value of about 380 km, and the lower and upper boundaries at 210 and 640 km, respectively. At the right panel of Figure 6.67 the distribution of the eastward horizontal velocity (with vertical velocity of -24 m/s, i.e., downward) is presented with the mean value of 110 m/s and the lower and upper boundaries of about 50 and 185 m/s, respectively. The lifetime of irregularity patches, Δt, was determined as the time during which patches were observed at 18.21 MHz. Its distribution is shown in Figure 6.68. It is important to note that the mean lifetime of these irregularities—about 20 min—and the maximum duration exceeding 2 h are in good agreement with the direct backscatter measurements [189,190].

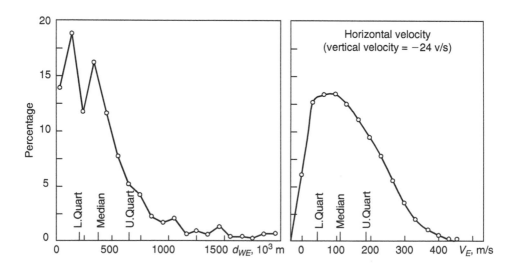

Figure 6.67 Distribution of horizontal scale of wavelike plasma structures, in percentages (left panel), and their horizontal velocity, in m/s (right panel).

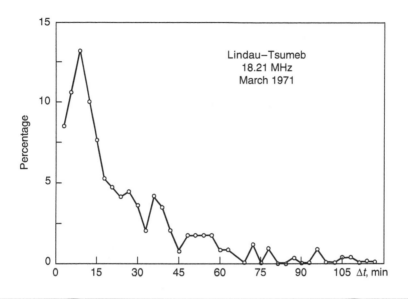

Figure 6.68 Distribution of lifetime of wavelike structures, in percentages, obtained along the trace Lindau–Tsumeb.

To indicate the number of periods of the observed wavelike large-scale plasma structures, in Figure 6.69 the number of simultaneously observed traces (or patches) is shown. It is evident that more patches are observed to the west of the great circle than to the east, giving an indication of sensitivity of irregularity patches and of their tilt as a function of the geographical longitude. Experimentally occurrences of up to 10 irregularly periodical patches were found along a west–east

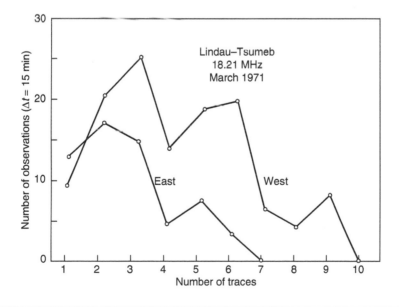

Figure 6.69 Distribution of number of observations of wavelike structures at different radio traces passed through the equator, as shown in Figure 6.66.

horizontal range of about 3000 km with the median length of each period of 350–400 km. Additional analysis of this phenomenon has shown that the observed irregularities are strongly elongated along geomagnetic field lines (the same as was shown for the equatorial E layer; see previous text). So, preliminary observations indicated the existence in the equatorial F region of large-scale quasi-periodical (having wavelike structure) FAIs of plasma density. Therefore, the authors assumed that these irregularities of electron density could be stated as precursors of periodically occurring equatorial spread F irregularities.

Moreover, results obtained by monostatic backscatter radar indicated a high correlation in the occurrence of equatorial large-scale spread F irregularities identified using backscattering echo signals and signals arriving due to bistatic transequatorial propagation through the equatorial region. Further, the in situ satellite probe [191,192] and rocket probe [196–199] direct measurements of ambient plasma parameters in the equatorial zones of the ionosphere proved the existence of the large-scale spread F quasi-periodical (i.e., wavelike) irregularities in the equatorial ionosphere.

However, despite the fact that in Refs. [188–190] an existing close association between the wavelike structures observed experimentally and atmospheric gravity waves occurring in the equatorial zones has been assumed, no theoretical proof was proposed to explain the close correlation between phenomena. This was done by several authors using radar sounding from ground-based facilities placed at different geographical points along the dip equator and by several theoretical works created by analyzing of experimental data [193–195,200–212]. We will describe briefly their results accounting for methods of study of main parameters of these plasma structures.

We start with systematic organized experiments carried out in Peru using the same 50 MHz radar located in Jicamarca [193–195] (see technical details of this radar system in Chapter 8). The same monostatic backscatter radar was used for investigation of spread E irregularities, the technical details of which were given earlier, and, therefore, we will not repeat this description and will instead analyze results of measurements of structure, localization, and dynamics of the spread F irregularities in the equatorial zones. The authors used the same, but modified, more precise technique based on measured data presentation as the range–time–intensity (RTI) radar plots using 5 km pulses. As for a digital power mapping (DPM) technique, 2.5 km pulses were used. Using this modified, very sensitive backscatter radar, the occurrence of spread F echo signals in the space and time domains and their relation to the plasma (electron or ion, plasma is quasi-neutral) density and irregularity drifts was obtained with high accuracy. Thus, Figure 6.70, extracted from Ref. [195], presents spatial and altitudinal distribution of echo signals scattered from spread F irregularities, their relation to electron density ($\log_{10} N_e$) denoted by lines with the corresponding values, electron density gradients ($Sp\ F$) denoted by vertical black lines superimposed on the density contours and indicate areas filling by irregularities, and also present vertical drift velocity plotted around the horizontal axis (around the zero-line). The dashed line defines the maximum altitude of the F region denoted by h_{max}. It is evident that spread F large-scale irregularities occur mostly in the evening time periods, floating both upward (drift velocity is positive) and downward (drift velocity is negative) with vertical drift velocities not exceeding 40 m/s, respectively. The drift velocities have been estimated more precisely from signal power spectra in the Doppler shift domain (using the same methodology that was introduced for estimation of drift velocities of spread E irregularities described earlier).

At the same time, the shape of power spectra (more than 3000 samples of spectral characteristics were obtained) allowed researchers to determine the corresponding types of irregularities occurring in the F region. Thus, in Figure 6.71 a set of selected spectra corresponding to different altitudes and observation times are shown to illustrate the different forms of echo signals. From illustrations presented here it is evident that the shapes of spectra may have a fine form characteristic for the

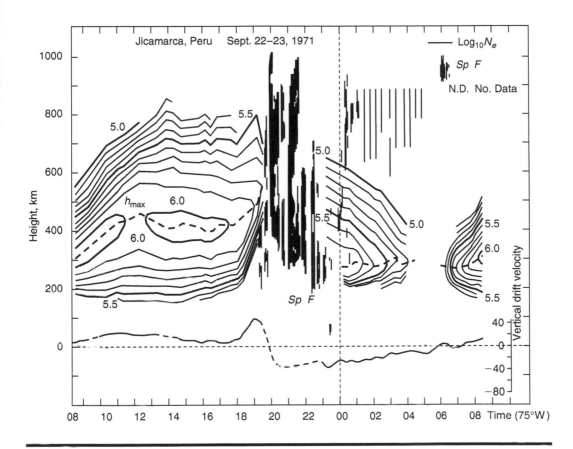

Figure 6.70 Contours of equal electron density, log$_{10}N_e$, and the corresponding scatter pattern in the altitude–time domain. The electron density gradients (*Sp F*) are denoted by vertical bold lines. The time distribution of vertical drift velocity is shown in the bottom part of the panel. The dashed curve defines the maximum altitude h_{max} of the F region. (Extracted from Ref. [195].)

type 1 irregularities with a single narrow peak within the bandwidth or composite structure with several peaks and with broadening within the bandwidth. Single but wide peaks can be associated with type 2 irregularities. Generally speaking, there is no simple separation on the two types of irregularities introduced earlier for the description of equatorial echo signals, and on the corresponding instabilities associated with these types of irregularities. The corresponding fine, narrow, medium and wide spectra are denoted by letters F, N, M, W, as well as by abbreviations WM, MF and so on, as well as their combinations, respectively. Obtained experimental data indicates a more complicated (composite) structure of echo signal power spectra which change with altitude of the F region and evening time period of observation. Thus, the composite spectra with broadening widths are mostly observed for higher altitudes above 400 km, whereas the fine and narrow spectra with narrow widths are recorded at the bottom side of the F region.

Authors do not give a theoretical explanation of these types of spectra; only will outline that they compare the corresponding frequency spectra with power and velocity map pattern. One of the characteristic examples obtained from raw data recording by the Jicamarca backscatter radar on October 28, 1973 is presented in Figure 6.72a (power map pattern) and Figure 6.72b (velocity map pattern). As follows from presented illustrations, in the earlier evening periods the spread F

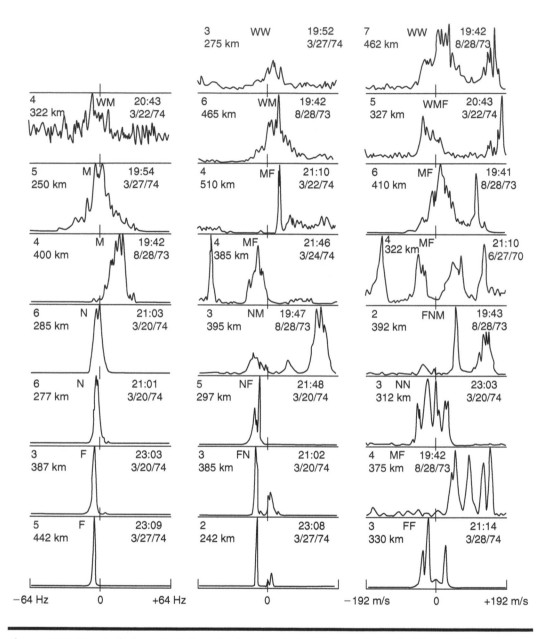

Figure 6.71 Spectral characteristics of echo signals from plasma irregularities of different types scattered from various altitudes versus Doppler shift frequency from −64 to 64 Hz and versus Doppler velocities from −192 to +192 m/s.

irregularities are located at the bottom side of the F region with spectral shape close to type 1 irregularities, and Doppler (phase) velocities do not exceed ±40 m/s. During their further temporal evolution, irregularities float upward to the middle and topside of the F region with velocities exceeding ±(100 ÷ 200) m/s, achieving more complicated composite shapes that can be explained as a mixture of type 1 and type 2 irregularities. Despite the fact that most previous authors stated that the Rayleigh–Taylor instability (see definition in Chapter 2) was the best candidate of

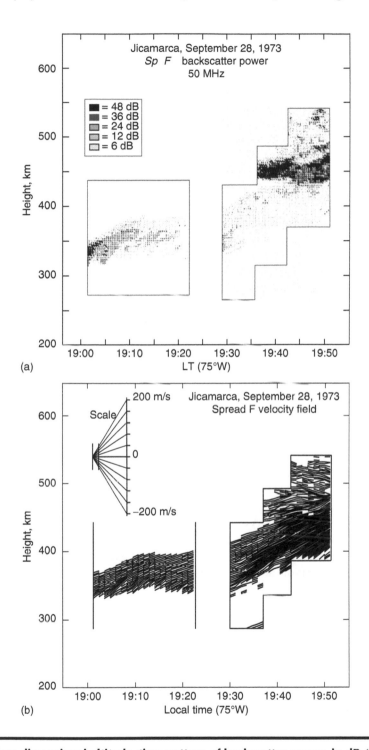

Figure 6.72 Two-dimensional altitude–time pattern of backscatter power, in dB (a), and of drift velocity of the corresponding sporadic irregularities, in m/s. (b) observed experimentally in the ionospheric layer F. (Extracted from Refs. [195,196].)

associated spread F irregularities, the authors of Refs. [193–195] first outlined the two possible candidates of generation of irregularities located at the bottom side of the F region as the gradient-drift instability (GDI) and the drift instability. These irregular plasma structures observed in the spread F region and responsible for strong backscatter from FAIs filling this structure are usually called the *plumes*. As is shown, they float upward from the bottom side to the topside of the F layer. Moreover, the close correlation between the echo signals reflected from the steep bottom side of the F region and existence of steep gradients of plasma density in this altitudinal region was found experimentally. At the same time, in further studies carried out by other authors it was shown that the mechanisms of spread F irregularities generation is more complicated, and these two proposed mechanisms can be taken as an extension of Rayleigh–Taylor instability caused by coupling between the plasma at altitudes of the F region and atmospheric gravity waves.

We will discuss these important experimentally obtained results in the following text, but now we will note again that the authors of Refs. [193–195] more precisely (with respect to preliminary studies) investigated the main features of these plasma structures *plumes*, and, more importantly, related them and their spatial and temporal dynamics to the corresponding depletion plasma areas, called *bubbles*, and their spatial and temporal dynamics. Because this phenomenon is important for the further evolution of the irregular structure of the middle and topside of the F region, we will consider it separately. Finishing the analysis of results obtained in Refs. [193–195] we should point out that the incoherent scatter radar used in the radio observatory of Jicamarca (Peru) allows associating the rapidly growing plumes inside the spread F regions with strong echo signals received after scattering from meter-scale irregularities (according to Bragg's law of backscattering, the scale of the scatterer equals the half-wavelength of radiated radio wave, i.e., $\ell = \lambda/2 \approx 3$ m). At the same time, it has been shown [200–202,204] that this is correct only if the vector of the radiated radio wave is perpendicular to the geomagnetic field lines. When the radar wave vector is only a few degrees away from perpendicularity, wedge-shaped depleted plasma density structures and the collocated plasma bubbles with walls strongly stretched along geomagnetic field lines are usually observed.

The results were obtained using steerable backscatter radar (called ALTAIR radar) located on the island of Kwajalein and operating at 0.96 m wavelength. All technical information on the ALTAIR radar system can be found in Ref. [200]. We will now discuss the main features of the observed irregular plasma structures. A thin structure and dynamics of plumes, and their spatial and temporal evolution, as well as the collocated bubbles, were observed experimentally with the help of the ALTAIR radar system. Thus, in Figure 6.73 sequences of backscatter radio map of echo signals is shown leading the generation and further evolution of secondary plumes A, B, and C. It is evident that during secondary plume generation the continued structuring of the west wall of upwelling is observed. Thus, as shown in Figure 6.73a, the primary plume extends vertically upward, floating up to altitudes of about 800 km. The bubbles collocated with the primary plume also developed upward from the wall. Generation of secondary plumes A (Figure 6.73b), B (Figure 6.73c), and C (Figure 6.73d) is evident during spatial and temporal evolution of the irregular plasma large-scale structure. The dynamics of plumes and the collocated bubbles are controlled by the collisional Rayleigh–Taylor instability, which can generate a gravity-driven equivalent velocity exceeding 180 m/s [202]. Finally, the three secondary plumes shown in Figure 6.73e and f have a mean east–west spacing of 125 km, compared to 400 km for three large upwellings and primary plumes. The west wall continuous structuring was also observed during development and evolution of the three secondary plumes. As was shown experimentally in Ref. [203] by in situ observations of satellite probe measurements, plasma bubbles produce scintillations and Faraday rotation on the polarization plane of radio waves crossing the depletion regions, which are important features for future ionospheric radio link performance.

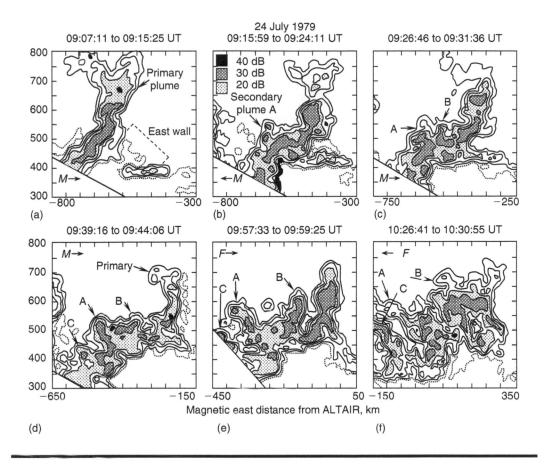

Figure 6.73 Spatial evolution of primary plume (a–d), secondary plumes, A (b–f), B (c–f), and C (d–f), observed by ALTAIR radar system. (Extracted from Refs. [201,202].)

At the same time, in a further study of the phenomena in Ref. [204], it was shown that the collisional Rayleigh–Taylor instability with the growth rate $\gamma_{RT} = g/\nu_e L$ is the primary driving process of plumes and collocated bubbles, where $g = 9.81$ m/s^2 is the gravity field acceleration, $\nu_e = \nu_{em} + \nu_{ei}$ the general collision frequency defined in Chapter 1, and L the plasma density gradient scale length. Existence of the horizontally stratified ionosphere with an increase of the plasma density gradient along altitude, as well as the eastward electric field observed in in situ rocket and satellite experiments [191,192,196–199] after sunset (i.e., exactly in the evening time periods when such echo signals were observed by ALTAIR and Jicamarca radars) are the best candidates for gradient-drift instability generation with the growth rate $\gamma_E = E/BL$ at the further stage of the large-scale irregular plasma structure evolution as an additional source of plasma irregularity generation. Here, E and B are the magnitudes of ambient electric and magnetic fields. In Ref. [204] the effect of a neutral wind was considered as a probable additional candidate for plasma irregularity generation and a destabilizing factor in regions where plasma density contours are tilted at an angle α with the density gradient ∇N partially directed westward. The growth rate of such an additional instability can be estimated via the zonal component of eastward wind U observed experimentally during and after sunset, $\gamma_U = (U/L)\sin\alpha$.

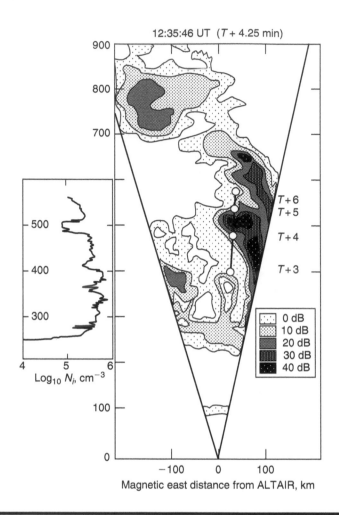

Figure 6.74 Altitude distribution of plasma ion density, $\log_{10} N_i$ (left panel) with corresponding plumes and depletion areas, and the corresponding 2-D pattern of echo signals from these irregular plasma structures (right panel).

Comparison between in situ satellite measurements of equatorial plasma density depletions and simultaneous backscatter power map of meter-scale irregularities ($\ell = 0.96$ m for the ALTAIR radar) are shown in Figure 6.74, where the left panel shows striated enhancement structures of plasma density in the F region filled by irregularities of various dimensions obtained by rocket in situ measurements called Plumex [196–198], whereas the right panel presents the corresponding radio-echo map of areas of observations by ALTAIR radar [200–204], where such irregularities were observed via scattered echo signals. Rocket measurements have found that the upper secondary plumes have to be considered simultaneously with the collocated bubbles that follow the plume dynamics to secure quasi-neutrality of plasma over the large area illuminated by the radar.

Two spectra from electron density measurements of the irregular F region were obtained during the rocket in situ experiments and are shown in Figure 6.75 according to Plumex measurements (see Refs. [198,204]), one near the F layer peak and one above this peak. It is seen that these two

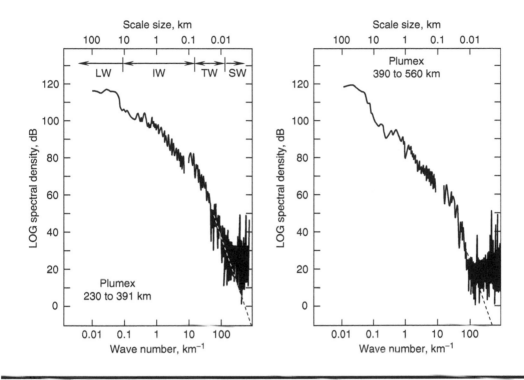

Figure 6.75 Logarithm of the spectral density (in dB) versus the wave number (in km^{-1}) obtained experimentally at altitudes 230–391 km (left panel) and 390–560 km (right panel).

spectra differ both in spectral slope and intensity in the intermediate scale-size range for irregularities with dimensions between 100 m and 20 km (called the intermediate-wavelength, IW). However, both spectra are similar before and beyond this range. The small-scale irregularities (denoted in the left panel as short-wavelength, SW) of 10–100 m have been termed *transitional* whereas those with dimensions above 20 km (called long-wavelength, LW) have been termed *long*. The transition range refers to the small-scale irregularities that are responsible for backscatter at Jicamarca ($\ell = 3$ m) and ALTAIR ($\ell = 0.96$ m) radar observations. Additional studies [200–204] showed that the spectra presented in the left-hand panel of Figure 6.75 varies roughly as k^{-n}, $n = 1.5$–2.0 in the intermediate range, i.e., for the moderate-scale irregularities, whereas for the small-scale irregularities it can vary more sharply, with $n = 2.5$–3 and more (up to $n = 4$–6 [200,201]).

These results also show that introducing effects of turbulent diffusion and wave coupling, i.e., interactions between small-, moderate-, and large-scale irregularities (see Chapter 2), one can explain the experimentally observed complicated spectral distribution plasma density irregularities. Evidence that such processes occur in the equatorial spread F region is shown in Figure 6.76, where spectral distribution of electron density fluctuations for irregularities of various scales filled the bottom side F region. The common form of spectra, k^{-n}, $n = 2 + k$, $k = 1$–2, obtained theoretically for the moderate-scale irregularities, is interrupted by a steep regime. In Ref. [206] three steep gradients and associated gradient-scale lengths were found. The spectral breaks where observed in Figure 6.76 for the inner scales at about 2 km scale and at about 100 m scale, i.e., within the intermediate range. According to Ref. [206] these breaks are directly related to the steep gradients.

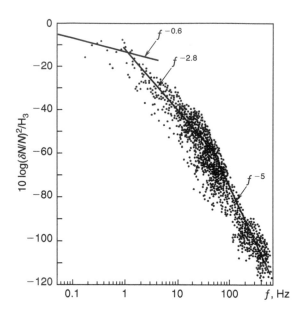

Figure 6.76 Spectral distribution of electron density fluctuations in the frequency domain obtained experimentally during sounding the bottom side F region (bold points) and the corresponding frequency dependence (solid lines) as a best fit of experimental data.

In Refs. [204–206] it was suggested that a vertical structure with the outer scale of hundreds of kilometers can be created by a distortion of the horizontal scale due to vertical uplift, which can be taken as a specific characteristic of equatorial spread F structure. This idea was proved experimentally [206–211] taking into account zonal irregularity drifts and neutral winds (as described in Ref. [204]) based on experiments with Jicamarca and ALTAIR radars combined with in situ satellite probe measurements. Thus, in Figure 6.77 a fine structure of equatorial spread F is shown where the lines correspond to the measurements of echo signal power scattered from different locations of observed areas labeled by A, B, C, and D in Figure 6.78.

The same fine multiple-plume structure was observed during measurements based on the ALTAIR radar observations of the equatorial spread F region (see Figure 6.79). The lines correspond to those shown in Figure 6.80, in which, again, the position of the primary plume, the secondary plumes A, B, and C, as well as the location of east wall are presented. A more complex experiment using the 50 MHz backscatter radar (radar interferometer [207]) at Jicamarca, Peru ($\approx 12°S$, $\approx 76°50'W$), and the observation station of signal oscillations located at Huancayo (HU), Peru, for simultaneous measurements of drifts of 3 m and kilometer-scale irregularities, respectively, was carried out. Parallel, simultaneous measurements of neutral winds by the Fabry–Perot Interferometer (FPI) located at the point A (at Arequipa, $\approx 12°S$, $\approx 75°20'W$) in Peru as shown in Figure 6.81 and measurements in Huancayo during continuous communication with the geostationary satellite FLEETSATCOM operating at 250 MHz were carried out. The corresponding 300 km intersection point of this propagation link is denoted in Figure 6.81 by the point HU/FLT. In this figure, the intersection point denoted by HU/FLT is co-located with the ionospheric region illuminated by Jicamarca radar (denoted by letter *J*); the points A_E and A_W indicate the intersections of the eastern and the western measurements with the 275 km level from the Arequipa station place *A* where 6300 $\mathring{A}$ emission of FPI originated.

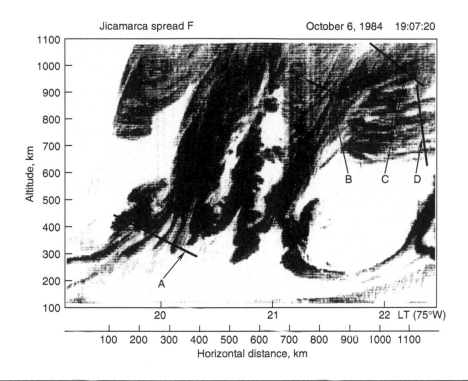

Figure 6.77 **Two-dimensional spatial structure of the sporadic F region is shown, where lines labeled A to D indicate different locations of the observed areas. (Extracted from Ref. [204].)**

In Figure 6.82, the behavior of the 50 MHz backscattered power recorded, as an example, by the Jicamarca radar on October 11, 1988, during the period of 19:50-22:00 LT (Local Time) in the joint altitude–time domain is shown, indicating the temporal variations of altitudinal range and intensity of echo signals scattered from the 3 m plasma irregularities filled equatorial spread F region. From the presented radio pattern of echo signals shown in Figure 6.82, it can be seen that in the initial phase of irregular layer evolution, the irregularities are extended in altitude in the form of three plumes, the central one penetrating at latitudes beyond 800 km. After about 20:20 LT, a steep descent of the spread F region takes place and the irregularities occupy a narrow altitudinal region not more than 100 km in thickness. Thus, during a further period the thickness of the irregular layer with filled irregularities is reduced. Figure 6.83 shows the time variations (scintillations) of signal strength obtained by using the 250 MHz transmission from the FLEETSATCOM satellite which are then received at the Huancayo observatory (see intersection point denoted in Figure 6.82). Thus, the plasma density structure and the corresponding signal scintillations can be obtained simultaneously. It was found in Ref. [207] that such scintillations are caused by drifting irregularities with dimensions ranging from 100 m to 1 km that forward scatter radio signals radiated at 250 MHz.

Comparing now Figures 6.82 and 6.83, one can see that the plume structure ending at 20:20 LT corresponds to the first scintillation signal recorded between 19:40 and 20:20 LT. The second scintillation signal recorded between 20:00 and 21:05 LT correlated with the second backscatter radio map pattern between 20:20 and 20:55 LT. During this period the scintillation index of the signal intensity remains high, having been caused by scattering phenomenon from a relatively thin

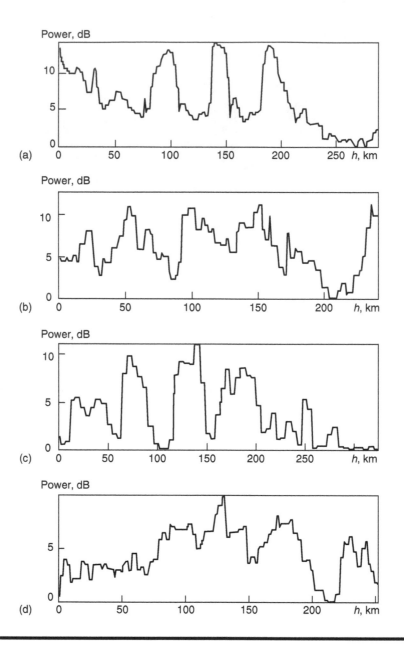

Figure 6.78 The altitudinal profile of plasma density for each location observed experimentally and indicated in Figure 6.77 as A (a), B (b), C (c), and D (d).

layer filled by 3 m FAIs. Decrease of signal intensity oscillations, as well as the amplitude of oscillations, indicates the importance of kilometer-scale irregularities in the further stages of evolution of the irregular F layer because their lifetime is longer than that for the 3-m-scale irregularities [203,207]. Figure 6.84 presents a comparison of results of drift velocity measurements in the east–west direction by scintillation and backscatter Jicamarca radar with the neutral wind measurements in the east–west direction obtained by the FPI at Arequipa. The drift of kilometer-scale irregularities measured by the spaced-receiver scintillation technique [207] are presented by

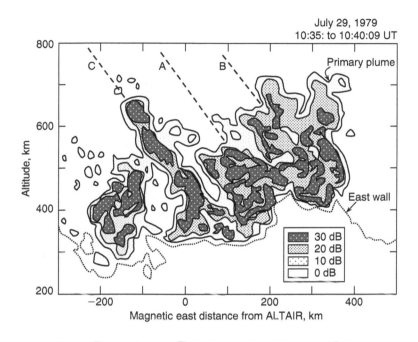

Figure 6.79 Two-dimensional map of scattered signals pattern from irregular structure of F region consisting a primary plume and secondary plumes labeled by A, B, and C, positions of which are indicated by dashed lines.

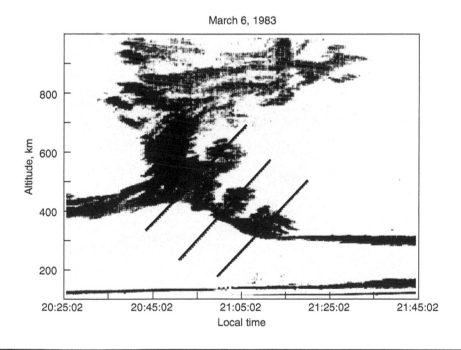

Figure 6.80 The multiple-plume structure corresponded to radio map shown in Figure 6.79, where lines indicate positions of secondary plumes, A, B, and C.

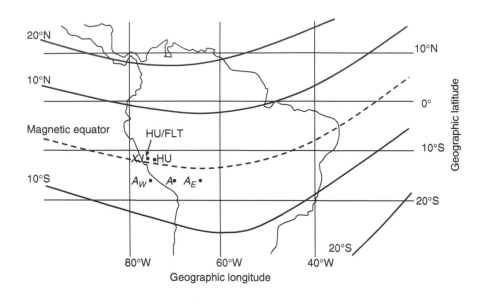

Figure 6.81 Geometry of experiments carried out in Huancayo (Peru) using FPI. (Extracted from Ref. [207].)

the solid curve and show its deviations in time, achieving maximum magnitudes of about 150 m/s between 19:50 and 20:50 LT and decreasing to 50 m/s at midnight. Simultaneous measurements of drift velocities obtained by the Jicamarca radar are indicated by dots in Figure 6.84. Here,

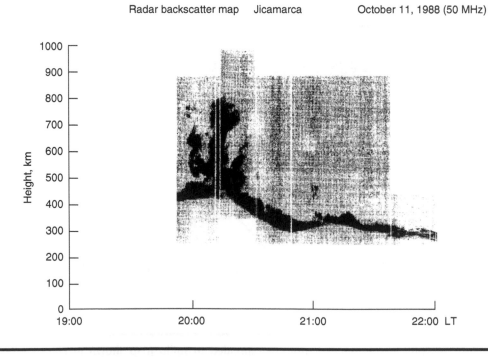

Figure 6.82 Radar backscatter power 2-D height–time map obtained by Jicamarca radar system.

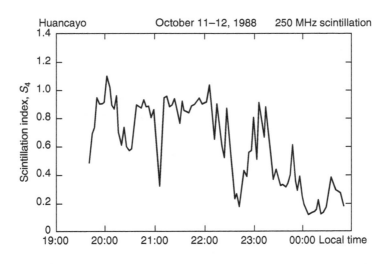

Figure 6.83 Scintillation index deviations in the time domain during experiment carried out by Huancayo FPI system.

the neutral wind flow in the east and the west directions is indicated by vertical segments with letters E and W, respectively. Here also, the best-fit quadratic curve to the neutral wind data during the corresponding night on 11–12 October, 1988, is represented by the dotted curve. It is evident that the maximum of the neutral wind is not correlated with a decrease of zonal drift velocity.

However, it is at least a factor of two higher than the drift velocity obtained from signal scintillations correlation analysis made for kilometer-scale irregularities at about 23:00, and remains

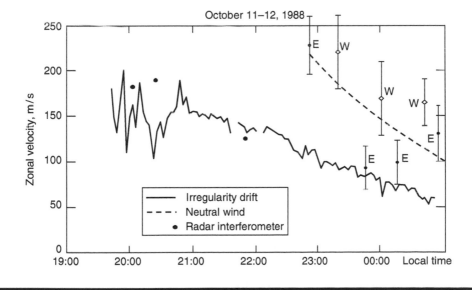

Figure 6.84 Comparison between measurements of drift velocity (solid curve) and neutral wind velocity in the east–west direction (dashed curve). Bold points indicate results of measurements of drift velocity using radar interferometer. Vertical segments indicate standard deviations in measurements of the neutral wind in east (E) and west (W) directions.

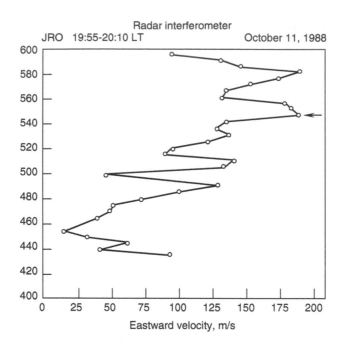

Figure 6.85 Altitudinal profile of the eastward velocity, in m/s, of irregularities observed by radar interferometer. The arrow indicates location of measured point at altitude of 544 km corresponded to that measured at 19:53 LT from Figure 6.84.

50 m/s higher than drift velocity even at 01:00. It is interesting to note that other experiments give the same or an even greater difference between the zonal neutral winds and the drift velocity of observed irregularities filling the equatorial F region, higher by 40% during the equinoxes and by 100% during the winter [203]. Because the obtained data was not sufficient to draw any conclusions, the authors also used in situ obtained satellite data [207,209], results of which are shown in Figure 6.85 in the form of an altitudinal profile of eastward drift velocity (in m/s). The main result shown here is the increase of the zonal drift velocity of irregularities beyond an altitude of about 580 km in the period from 19:55 to 20:10, the onset of the spread F layer. The arrow in Figure 6.85 indicates the selection of an appropriate data point for drift comparison that was replotted in Figure 6.84 at 19:53 LT. A good coincidence between data obtained by different methods is evident.

During recent decades researchers based their effort for investigating the total electron content and plasma density profile during equatorial spread F events on the Jicamarca radar system described earlier. Thus, using the Jicamarca Unattended Long-Term Investigation (JULIA) radar, echo signal scintillations were recorded from two stations, of which the first was located near the magnetic equator (in Ancon) and the second located 12° southward, in Antofagasta, Peru. In investigations of the equatorial F spread, observations of 3 m irregularities forming altitude-elongated plumelike structures were carried out. Thus, Figure 6.86 displays a plot of the magnitude of the coherent echo signals recorded by the JULIA radar during the period 10–11 October, 1998 (top panel) [213]. The bottom panel shows the scintillation index, S4, temporal distribution during the same period measured by the Ancon east and the Antofagasta west receivers. It can be seen that during registration of strong scintillations of signal intensity by a station located in Antofagasta, plasma plumes extending up to altitudes of about 900 km were observed simultaneously by the

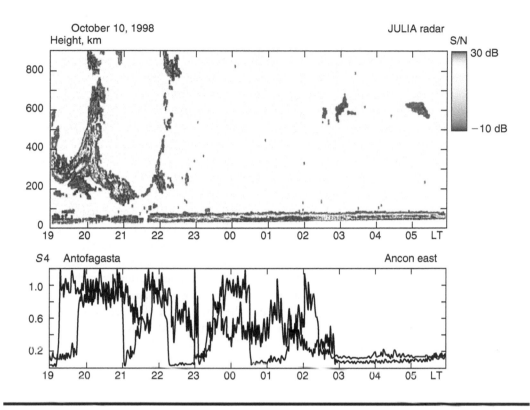

Figure 6.86 **Signal-to-noise height–time distribution of scattered signals, in dB (top panel) and the corresponding time dependence of scintillation index obtained in Ref. [213] by JULIA radar.**

Jicamarca radar. In Ref. [214], results of measurements of the plasma density distribution in the space and time domains by using the same coherent Jicamarca radar are reported. Thus, a mapping of the altitude–time profiles of plasma density is shown in Figure 6.87. It can be seen that above altitudes of 200–250 km, the F region plasma concentration rises during the day just as the F peak also rises in altitude. We should note that below 300 km the corresponding density profile was observed by the same radar, and we described the altitudinal distribution of the plasma density

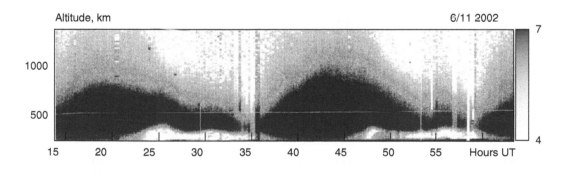

Figure 6.87 **Two-dimensional mapping of plasma density in joint altitude-time domain obtained in Ref. [214].**

profile of the lower F region during analysis of the plasma density profile in the D and E regions (see previous paragraph).

To complete the brief analysis of radio and optical methods of prediction of plasma irregularity parameters filling the spread F layer, we wish to compare irregularities occurring in the thin E layer with those occurring in the thick F layer using high-resolution vertically oriented HF coherent radar system, following the results of precise radio observations described in Ref. [215]. These experiments were carried out at Korhogo in the Ivory Coast ($\approx$9°25′N, $\approx$5°38′W) with continuously varying radiated frequencies in the range 1–30 MHz. Experiments were carried out for several different campaigns (in April, May, July, and October–November) during 1993, the International Equatorial Electrojet Year. The observations were carried out in the daytime, and kilometer-scale irregularities were separated for a sporadic E region with altitude range 100 to 150 km, i.e., greater than the usual width of the equatorial electrojet, as well as from altitudes of the F1 region from 200 to 250 km. We present here only some characteristic results giving information on both groups of irregularities. Figure 6.88 shows the observations carried out on May 27, 1993, for the HF sounding at a frequency of 4.2 MHz from early morning (07:30 LT) to the beginning of the evening (17:30 LT). This sounding frequency usually presents reflections from the F1 region of the ionosphere at altitudes 200–250 km, as well as the large irregular area at 100–160 km, i.e., extending from 100 to 150 km at noon and exceeding the usual width of the equatorial electrojet. From this area individual echoes are detached in the morning and early evening. A detailed altitude–time map is presented for noon (after 14:05 LT) in Figure 6.89 [215]. A weak scattering

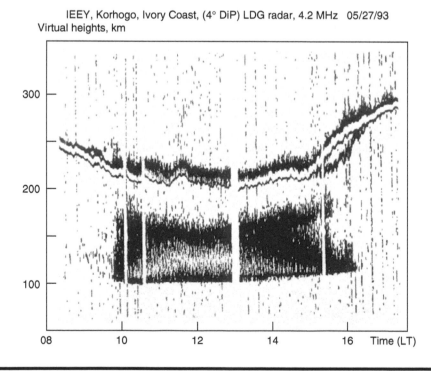

Figure 6.88 Two-dimensional height–time mapping of echo signals distribution scattered from F1 and E layer of the equatorial ionosphere during its HF sounding.

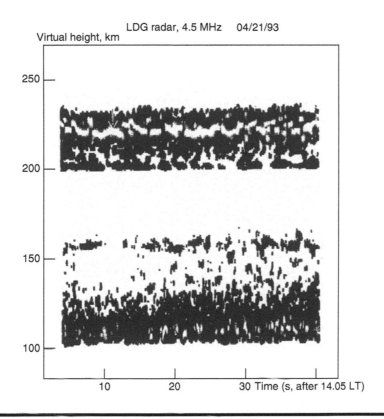

Figure 6.89 The same as in Figure 6.88, but obtained using LDG radar system. (Extracted from Ref. [215].)

altitudinal range around 160 km and below 210 km can be seen, whereas from the range of 100–130 km, typical E_{sq} echo signals, described earlier with respect to the equatorial E region, were recorded as characterized by very fast fading, i.e., fluctuations of signal amplitude and phase. It was also found that reflections arrive at the receiver from easterly and westerly directions in both the E and F regions.

The irregularities responsible for these echo signals and operating frequency 4.5 MHz have dimensions (according to Bragg's rule) of about 33 m, i.e., ten times greater than those of the irregularities observed at 50 MHz backscatter Jicamarca radar. Figure 6.90 shows the Doppler spectra corresponding to the data presented in Figure 6.88, obtained by the sounding of the E region from 99 to 140 km, the intermediate region from 142 to 183 km and the F region from 187 to 240 km. As can be seen, at lower altitudes in the E layer the spectra shape is symmetrical corresponding to the type 1 irregularities, and the symmetry is conserved even at the upper part of the E layer. This symmetry relative to zero is also observed at the intermediate altitudes with wide spreading of spectrum, becoming wider than in the F region where spectra remain broad. What is important is that for the latter two regions, spectra lose the specific shape of the type 2 irregularity. More detailed analysis of experimental data concerning Doppler spread of recorded signal allows for estimating the velocity of the horizontal motion of irregularities filling both regions: for the E region of the order of 100 m/s and for the $F1$ region of the order of 130 m/s.

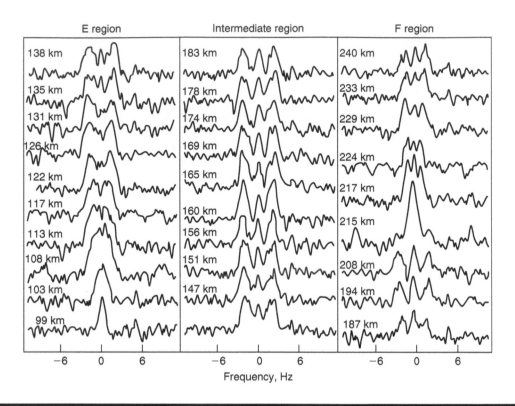

Figure 6.90 Doppler spectra distribution along different altitudes for E region (left panel), intermediate region (middle panel), and F region (right panel).

To explain the results obtained from height–time mapping and Doppler shift spectra shape analysis, the authors proposed several mechanisms by which such irregularities could be generated. Like other authors, they first suggested a gravity wave wind-driven instability as the main candidate of generation of associated FAIs. However, following results presented in Figure 6.90 and the ambient conditions of the experiment, the gravity wave activity was found to be weak. Another mechanism of generation of irregularities proposed during the analysis of the results obtained by using the ALTAIR incoherent radar system (in Kwajalein) is related to the polarization electric field developed in response to the presence of a latitudinal gradient of plasma density [200–202]. In the experiments described in the aforementioned references, the magnetic field lines pass over the magnetic equator at 150 km, and, therefore, the echo signals were observed over Kwajalein at a height of the 150 km. In the current experiment described in Ref. [207], the magnetic field lines cross the magnetic equator at an altitude 120 km. Because the 150 km HF echo signals were observed here simultaneously with a Doppler bilateral symmetry and the extension of the irregularities region upward to the $F1$ region, it is difficult to explain the echo signals arriving from the 150 km region by the same mechanism of latitudinal gradient influence on drift processes described in Refs. [200–202].

This result again allows us to point out (as we did earlier) that the wide spectrum of physical processes occurring in the equatorial E and F regions of the ionosphere and vertical coupling between these two regions finally lead to a very complicated picture of temporal and spatial evolution of irregular plasma structures associated with the spread E and F layers. Therefore, for each feature and each process, the separate combined methods of observation need to be performed, as was done in the study of plasma bubbles associated with plume structures observed in the middle and topside regions of spread F. We will now briefly describe this phenomenon and how it can be studied.

Studies of Equatorial Plasma Bubbles. As noted earlier, the bubbles are the localized plasma depletion regions associated with the plasma enhancements called plumes, which follow the latter in their spatial and temporal evolution to secure conditions of plasma quasi-neutrality at each place of spread F layer filled by disturbed irregularities [56–61,193–195,216–235]. The transverse dimensions (with respect to B_0) of such plasma structures are of the order of tens to hundreds of kilometers. The first systematic radar observation of the equatorial F region [193–195] found that bubbles always collocate the tilted and plumelike plasma structures, and together they filled the whole thickness of the spread F layer, from the bottom side to the topside of its boundaries. From the beginning it was stated [195] and theoretically proved [193,194] that a close relation of bubble structures with Hamilton's mechanism of positive plasma gradients creation due to plasma extension of the gravitational instability (e.g., the Rayleigh–Taylor instability described in Chapter 2) is responsible for the chaotic motion taking place whenever a heavier fluid rests on top of a lighter one.

In these works, to explain the existence of regions with negative gradients of plasma density, i.e., the bubbles observed experimentally at the topside of the equatorial F region, a three-fluid phenomenological model was proposed. According to this model, the lightest fluid is located at the bottom side of the F region, and the second, the heaviest, is at its middle part (denoted by crests), and the fluid with the moderate density is located at its topside (denoted by straight crossing lines), as shown in Figure 6.91, where dynamics of bubblelike plasma structure, having cylindrical shape with the axis directed along B_0, is sketched schematically. As can be seen from this sketch of a

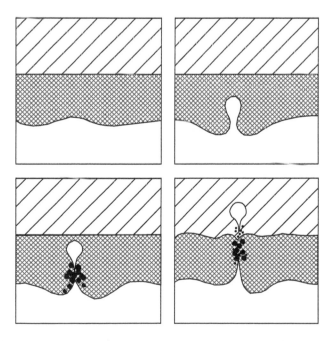

Figure 6.91 Schematic presentation of dynamics of bubblelike structure in the upper ionosphere.

three-fluid model of the region F, the bubble forming at the bottom side of the F spread having lower density floats upward, penetrating into the topside layer of the F region, as most gravitationally stable its part. In the coordinate system moving with the $\mathbf{E} \times \mathbf{B}_0$ drift, a vertically floating upward bubble with its wake having any configuration produces in its evolution a plumelike vertical structure. To explain the phenomenon of tilting of plumelike structure in Refs. [193–195], it was assumed that the bubble could move eastward with a velocity not exceeding the background $\mathbf{E} \times \mathbf{B}_0$ drift velocity. When the neutral atmosphere moves eastward faster than $\mathbf{E} \times \mathbf{B}_0$ drift, the bubble will move more slowly than the background plasma. However, this difference between velocities of the neutral and ionized medium was not observed experimentally. Therefore, another mechanism was assumed to explain the tilt phenomenon of plumes, taking into account a sheared drift in the ionosphere, assuming the east–west drift to be faster at the bottom side of the F layer compared with that at the topside. We note that the same mechanism was proposed during observation of the barium cloud evolution (see previous text). These were not systematic evaluations of the problem.

The more rigorous models based on linear and nonlinear collisional Rayleigh–Taylor instability type generated in plasma were performed to explain processes of creation and further evolution of bubbles in the topside of the spread F layer [56–58,218–235]. Some rigorous numerical models, taking into account many experimentally observed natural features occurring in the equatorial F spread, have finally shown stable evidence of the existence of upward-floating bubbles, as vertically elongated depleted plasma density regions extending upward to the topside ionosphere and accompanying evolution of plumelike plasma enhancement plasma structures. We will not go deeply into the mathematical aspects of the problem and the corresponding physical models. We will present the most interesting experiments using simultaneous ground-based radar and satellite observations carried out for finding parameters of plasma structures that could be important for predicting the propagation characteristics of probing radio waves passing the F region of the equatorial ionosphere, which is important both for land–ionospheric and land–satellite communication.

More systematically, plumelike and bubblelike spatial and temporal evolution were analyzed [218–235] using the ground-based radars located at Jicamarca and Kwajalein described earlier and comparing the measured data with those obtained during satellite observations,. We present only those results that give full presentation of the phenomena under consideration. Thus, we focus on experiments carried out by using the ALTAIR backscatter radar located at Kwajalein operating simultaneously at 155.5 and 415 MHz [217,218] combined with in situ measurements carried out by satellite probes [61,215,219,226,227]. Thus, in Refs. [217,218], comparing the observed data obtained by different methods, the location of bubbles relative to plumelike plasma density enhancement structures, their dynamics, shape, density profile, and velocity of movements were reported. The in situ measurements carried out by satellites having a low inclination orbit over island Kwajalein at altitudes ranging from 372.5 to 374.5 km have shown, during all the sets of experiments, full depletion (of about 99%) of regions collocated the plume structures, which were defined by researchers as typical bubbles [227]. Thus, in Figure 6.92 the ion-density profiles obtained by in situ probe measurements during the satellite trajectory passing the ALTAIR radar during a short-term period from 10:41 to 10:43 UT (which corresponds to 21:51–21:53 LT) on August 1978. The satellite trajectory is presented in Figure 6.92 by solid lines with bold points with 16 s intervals between points directed from west to east near the zero magnetic aspect ($B_\perp$) curve.

The ALTAIR radar scan was close in time to the moments of satellite trajectory to cross the magnetic meridian at about 21:51 to 21:53 LT (10:41-10:43 UT). The authors of Refs. [217,218] extrapolated spatially the measured ion density data along the magnetic meridian into a plane

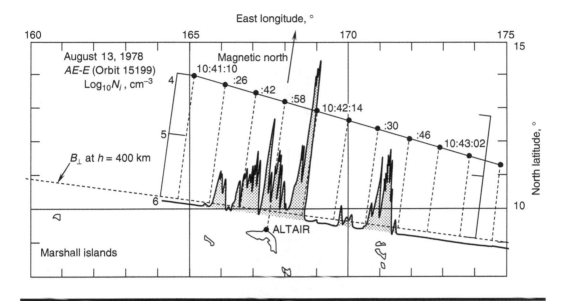

Figure 6.92 Plasma ion density longitudinal–latitudinal distribution measured by satellite, the trajectory of which is shown by straight line. The numbers near each bold point along this trajectory indicate the current local time of satellite at each point. (Data extracted from Ref. [227].)

scanned by the radar. Thus, in Figure 6.92 the ion density variations were plotted along the ALTAIR radar zero magnetic aspect curve, shown by the dashed line placed at altitude of 400 km. The authors have identified ion-density variations of interest in terms of four bubbles denoted by letters A to D (see the topside of Figure 6.93). The monotonic increase of depth of depletion areas from A to D is evident, with bubble D depleted by more than two orders of magnitude from the background plasma, i.e., more than the other (A to C) bubbles.

The relations between plasma bubbles and the collocated plumes obtained by ALTAIR radar during the same period are shown in Figure 6.93. The top panel presents the ion-density variations measured by the satellite during its trajectory (inverted from those presented in Figure 6.92), whereas the bottom panel presents the corresponding radar backscatter radio map. The path of the satellite is superimposed above this radio map corresponding to the height of the satellite above Earth's surface. The contours of the equal strength corresponding to the strongest echo signals (of 40–50 dB) can be associated with plumelike structures observed by the radar. Thus, two plume structures are clearly seen at the time of the satellite pass. The major plume extends upward at an altitude exceeding 700 km and the minor one appears next to the west side of the major plume. This plume pair was associated by the authors with the large-scale upwelling that has an east–west dimension of approximately 400 km (or even more).

The relationship of bubbles with collocated plumes was determined by the vertical lines between the upper and bottom panels, as shown in Figure 6.93. Bubble A corresponds to a weak patchy backscatter region located to the west of the minor plume, which was interpreted in Refs. [215,216] by the absence of plasma structures within bubble A. Bubbles B and C were more structured. Bubble D corresponds to the strongest backscatter region in the neck of the major plume. Based on experimental observations, the two-dimensional shape of the plasma-depleted region was obtained (see Figure 6.94). The different character and structure of each bubble can be seen. Thus, bubble D corresponds to the primary wedge that is usually generated from near the

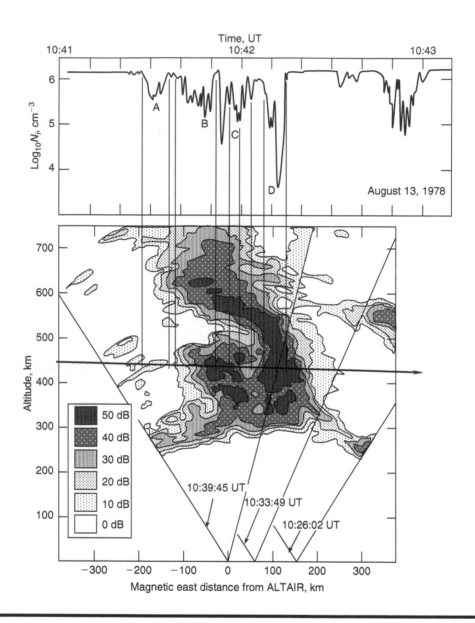

Figure 6.93 Fluctuation of plasma density inside bubbles, indicated by A to D in the time domain (top panel) and the 2-D spatial distribution of echo signals scattered from these corresponding depletion areas (bottom panel). Vertical lines between panels indicate the spatial position of each bubble, from A to D, at the 2-D radio map. (Data extracted from Refs. [215,216].)

crest of the large-scale upwelling denoted by a lowest dot-dashed contour of constant plasma density placed at the bottom side of the F region. Bubbles B and C can be interpreted as bifurcated segments of a single wedge that developed from the west wall of the large-scale altitude modulation (see details in Ref. [61]). Finally, bubble A can be associated with a pinched-off bubble (called a dead bubble [226]), which according to Ref. [226] is separated from a low-density bottom side of the F region and does not move upward any longer (see Figure 6.95, where schematic presentation extracted from Ref. [221] of evolution of depletion regions is shown).

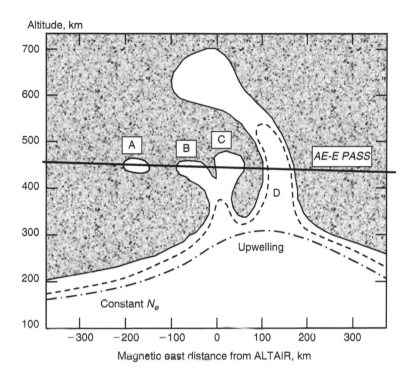

Figure 6.94 Spatial evolution of bubbles, from A to D, presented by contours of equal plasma densities (solid contours). The large-scale upwelling is indicated by the dot-dashed curve, and the initial stage of bubble D is indicated by the dashed curve. (Extracted from Ref. [221].)

In Figure 6.95 bubble (a) observed in the early evening drifts upward during the observation period caused by an eastward background electric field (see the corresponding arrow depicted below), whereas bubble (b) has two directions of floating, both upward and downward, caused by background zonal electric field turning westward (see the corresponding arrow depicted below). The combination of downdrifting and updrifting velocities due to the existence of eastward and westward electric field, as shown in case (c), leads to the pinching off of the upper part of the bubble. As observed experimentally from satellite measurements [59,61], such a plasma structure appears to be collapsing inward with a velocity of about 20–40 m/s. At the same time, the authors of Refs. [217,218] associated bubble A with the fragments of bubbles C and B, which can be separated by accounting the west–east $\mathbf{E} \times \mathbf{B}_0$ drift occurring in the equatorial ionosphere at altitudes of the F region.

Later, in Refs. [219–235], a set of experiments carried out in different sites over the world around the magnetic dip equator using a ground-based radar system both separately from and combined with satellite radio and optical observation of the spread F region, as well as the corresponding theoretical background, was reported regarding observations and a further description of phenomena accompanying generation and evolution of bubblelike structures and the collocated plumelike structures occurring in the equatorial ionosphere. As most of the works alluded to report similar results based on radar and in situ observations of bubble structure dynamics and similar theoretical explanation of the generation, due to Rayleigh–Taylor instability, of these plasma structures, we will not mention any of them in our further discussions of the problem. Instead, we direct the reader's attention to modern methods of studying thin plasma

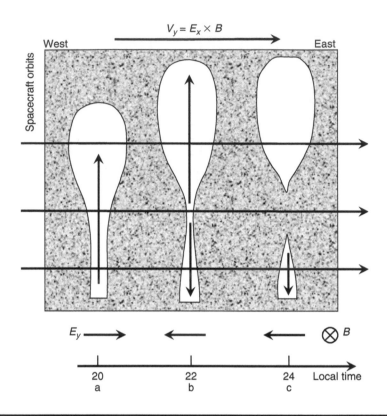

Figure 6.95 **Schematic presentation of different scenarios of evolution of bubble observed from moving spacecraft (its orbits are indicating by bold horizontal arrows). (Extracted from Ref. [221].)**

structure within bubbles using precise satellite probes, electromagnetic and optic observations reported during the last 10 years or so [227–235].

Thus, in Refs. [227–230], a systematic cycle of measurements of plasma density and the structure inside bubbles and the collocated enhancements of plasma density (plumes) as well as their altitudinal location, velocity, and direction of drift in the ambient electric and magnetic fields were reported. The corresponding first satellite for these purposes was operated in the Republic of China and therefore has the abbreviation ROCSAt-1. This satellite was launched into a 600 km circular orbit with low inclination oriented at an angle of 35°. Equatorial plasma structures were measured over several months in 1999, using an electromagnetic probe operating at 1024 Hz. During a cycle of experiments, two adjacent equatorial bubble structures were examined: (1) one characterized by the bulk plasma upward flow during its evolution, as predicted theoretically and shown schematically in Figure 6.95a, and (2) the second, in which the bulk plasma flow is small or characterized by a stagnated structure (see Figure 6.95c). A thin irregular structure and active bubble dynamics were systematically observed in this cycle of measurements. We will present only the more interesting results characterizing the bubblelike structure in the equatorial F region. Thus, during launching on March 28, 1999, a special electromagnetic probe operating at 1024 Hz registered two depletion areas labeled A and B in Figure 6.96a and b. In the first figure, a full observation time period, 02:55 to 03:45 UT, is presented, and its expanded plot for the 2 min time period (03:25 to 03:27 UT) is shown in the second figure. As can be seen at the bottom panel in Figure 6.96a, both bubble structures have depletions in density of about two orders of magnitude with respect to the density of the background ionospheric plasma. The size of depletions A and B can approach about 240 km each in the horizontal direction. The two velocity components V_z (in the vertical plane) and V_y (in the horizontal plane) measured in the satellite ROCSAT-1 coordinate system are plotted in the top and middle panel, respectively. It can be seen that the component $(-V_z)$ is positive upward, and V_y is positive to the right in the flight path of the satellite, which is inclined southwards. A detailed analysis of bubbles A and B carried out for the 2 min time period and presented in Figure 6.96b shows their rough structure caused by turbulent mixing movements of small-scale irregularities (with scales up to 10–15 m) embedded into a large-scale bubblelike structure with dimensions up to several tens of kilometers, which were generated, according to the theoretical framework presented in Refs. [225,226], these bubblelike structures can be finally created by the coupling of the atmospheric gravity waves and Rayleigh–Taylor instability and finally creating bubblelike structures.

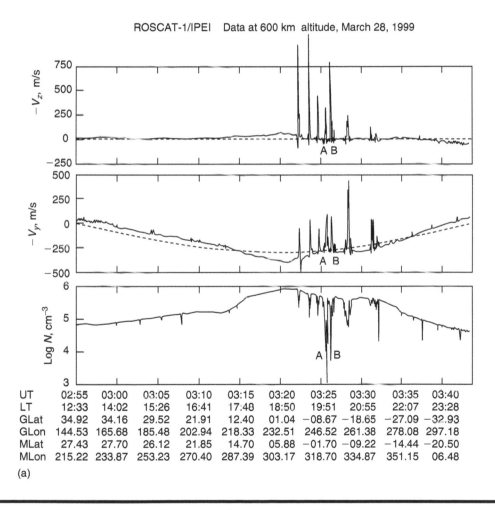

Figure 6.96 Time dependence of the vertical (top panel) and horizontal (middle panel) velocity of depletion zones labeled by A and B, and plasma density variations in these zones (bottom panel) observed by satellite ROSCAT during whole period of satellite observation (a) and during specific time period from 19:51 to 20:16 LT and

(*continued*)

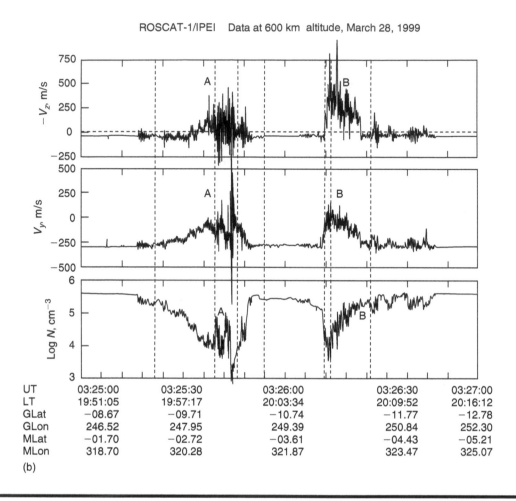

ROSCAT-1/IPEI   Data at 600 km altitude, March 28, 1999

Figure 6.96 (continued) **(b) when these regions were clearly observed.**

What is interesting to point out is that the results obtained during these experiments fully agree with data collected over more than 13 years (from 1989 to 2002) by the polar-orbiting satellites according to a special campaign called "Defense Meteorological Satellite Program" (DMSP), results of which were reported in Refs. [231,232], and comparison with data obtained during ROCSAT-1 observations was reported in Ref. [233]. We present here only a few characteristic results obtained during these two programs comparing results obtained at the same time periods of observations of the spread F region. Thus, in Figure 6.97, following Ref. [233] are presented simultaneously the northeasterly trajectory of the satellite ROCSAT-1 observed during periods from 13:03 to 13:21 UT (corresponding to the evening magnetic local time sector from 18:00 to 24:00 MLT) and comparing measurements of four DMSP satellites labeled F11, F12, F14, and F15 whose orbits crossed the magnetic equator near 19:15, 20:15, 20:30, and 21:06 MLT, respectively.

Figure 6.98a shows plasma density measured by satellite ROCSAT-1 during the time periods when the satellites F11 and F14 (F12 and F15) intersected ROCSAT-1 trajectory close to 100°E at 12:49 and 13:57 UT (12:49 and 13:30 UT), respectively, plotted as a function of UT (universal time), GLon (geographic longitude), MLat (magnetic latitude), and MLT (magnetic local time). It can be seen that all satellites discovered strong depletion regions with plasma density decreasing by

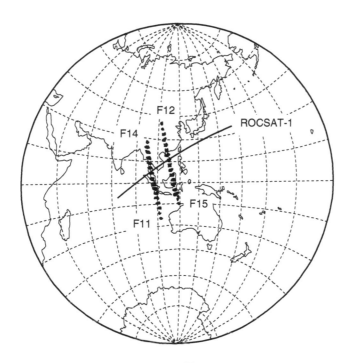

Figure 6.97 Geometry of joint experiments carried out by satellite ROCSAT-1 and DMSP satellites labeled by F12 to F15. Trajectory of ROCSAT is indicated by solid line, whereas trajectory of other satellites is indicated by bold points. (Extracted from Ref. [233].)

more than one order of magnitude compared to the background plasma, i.e., from 10^6 cm^{-3} to $7 \cdot 10^4$ cm^{-3}. At the same time, Figure 6.98b and c shows the plasma depletion regions during this interval obtained by measurements carried out by DMSP satellites F12 and F15. As is evident, both satellites observed strongly irregular bubblelike structures with decrease of plasma density inside them from $(3 \div 5) \cdot 10^5$ cm^{-3} to a minimum $\sim 9 \cdot 10^4$ cm^{-3}. At the same time, two other satellites F11 and F14 did not detect the existence of the bubblelike structures near the magnetic equator.

Statistical analysis carried out in Ref. [233] allows for estimating the occurrence of the equatorial plasma bubble structures systematically studying a whole cycle of satellite observations. Thus, in Figure 6.99, the occurrence rate of bubbles (in %) for March and April 2000 (diamonds) and 2002 (triangles) is plotted against magnetic latitude λ. The Gaussian best fit to experimental data, represented by a dashed curve, is found as an exponential function of λ, $P(\lambda) = 16.5\% \cdot \exp\{-\lambda^2/2\sigma^2\}$, with the standard deviation of magnetic latitude, $\sigma = 8°$. It was found that the satellite ROCSAT-1 passing the magnetic equator detected more than 90% of bubblelike structures occurring in the equatorial region at attitudes of about 600 km within the $\pm 16°$ region around a dip equator, with a maximum of more than 15% of occurrence of bubbles during the whole cycle of measurements.

We will now show an interesting comparison between two satellite campaigns using different methods of bubble-structure observation, optical and electromagnetic. Thus, we compare the surveyed data by satellite ROCSAT-1 and the data obtained in the IMAGE satellite campaign using a far ultraviolet (FUV) probe for nighttime imaging of equatorial bubbles by measuring their brightness signatures [234,235]. The IMAGE satellite was launched in March 2000 into a highly elliptical orbit. Without going deeply into the technical details of the experiment and description of

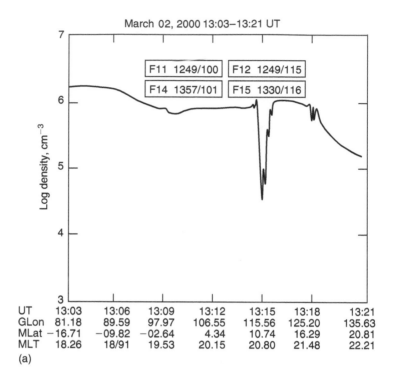

(a)

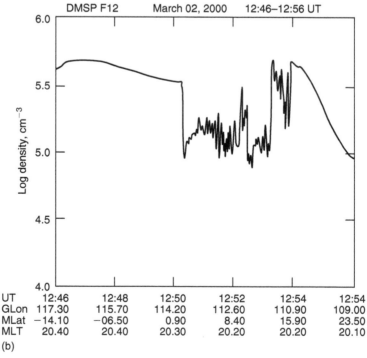

(b)

Figure 6.98 Complicated structures of depletion zones observed in Ref. [233] by satellites during period of 13:03–13:21 UT (a) 12:46–12:56 UT (b).

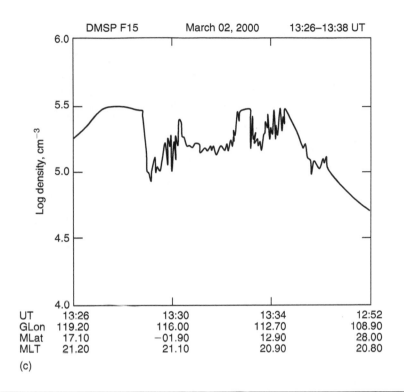

Figure 6.98 (continued) 13:26–13:38 UT (c).

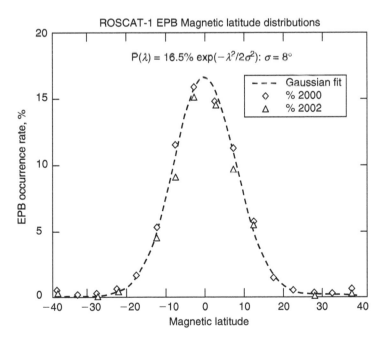

Figure 6.99 Occurrence rate of bubbles (in percentages) obtained experimentally in March and April 2000 (diamonds) and in 2002 (triangles); dashed line is the best Gaussian fit to experimental data.

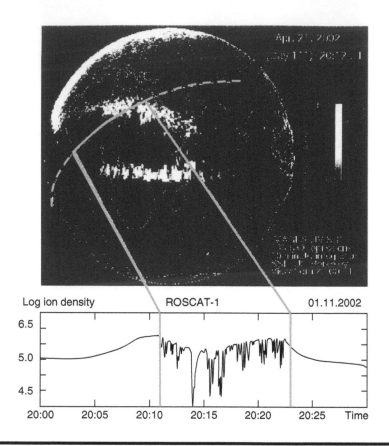

Figure 6.100 Indication of existence of bubble phenomena obtained by using optical method (top panel) and electromagnetic in situ probe (bottom panel) obtained by satellite ROCSAT-1, trajectory of which presented at top panel by solid-dashed curve. (Extracted from Ref. [234].)

the FUV optical probe, we present in Figure 6.100 a comparison of two different methods, the optical (top panel) and the electromagnetic discussed previously (bottom model). The image obtained by the FUV sensor is mapped onto a projection of geographic coordinates (top panel), where the trajectory of ROCSAT-1 is also mapped on the FUV images. The solid line at the top panel covers the time interval when ROCSAT-1 detected equatorial plasma bubbles (EPB), whereas the dashed line represents the satellite trajectory without detecting bubbles. In the bottom panel, plasma density measured by satellite ROCSAT-1 at a height of 600 km is shown, where the vertical lines specify the period during which ROCSAT-1 observed EPB corresponding to images at the top panel. Again, we see a strong (one order of magnitude) drop of plasma density within bubble structures, from background plasma of $\sim 5 \cdot 10^5$ to $\sim 5 \cdot 10^4$ cm^{-3}. The occurrence of at least seven distinct plasma bubbles is evident, which seems to be structured periodically.

Analyzing results of observations presented in this paragraph obtained by different methods, radar, electromagnetic, and optical, we observe in every case a complicated picture of plasma structures evolution in the spread F equatorial region (similar to that observed in the spread E equatorial region). Again, we should point out that to explain the entire mechanism of plasma structures, plumes, and bubbles generated in the equatorial F region, the researchers need to perform more systematic and complex experiments, including the ground-based facilities operating at the wide spectrum of frequencies, from HF to UHF, combined with in situ rocket and satellite measurements, based on electromagnetic and optical imaging of plasma irregularities, which will finally give rise to new theoretical models based on analytical and numerical studies of these complicated phenomena.

6.2.3 Observations of Irregularities in the High-Latitude Ionosphere

We will now analyze the modern radiophysical methods of studying plasma irregularities occurring in the high-latitude ionosphere, which is usually divided into two regions, the auroral zone ionosphere and the polar cap. We will consider investigations of irregular structures of both these regions, concentrating our attention on the processes occurring separately in the E and F regions, as was done in the previous subsection dealing with the equatorial ionosphere, because, as with the equatorial ionosphere, different physical processes accompany the spatial and temporal evolutions of the corresponding irregularities filling these two regions. Furthermore, different methods of observations, as well as the different theoretical models, are used to analyze the dynamics of these two regions of the ionosphere. Before going deeper into the subject, we should note that the geophysical ambient conditions are strongly varied in the high-latitude ionosphere compared with those observed in the middle-latitude and equatorial ionosphere, and that many observations carried out in the last four to five decades contradict one another as well as the corresponding theoretical frameworks. Therefore, we will not compare results of measurements and the corresponding theoretical models to discuss how to resolve all these contradictions.

We will instead show in chronological order the perspective radio and optical methods based on a ground-based facility coupled with rocket and satellite probe observations of perturbed regions of the ionosphere filled by various irregularities, to better understanding the complicated physical processes and phenomena accompanying generation and further evolution of these irregularities in the high-latitude ionosphere for future understanding of their reaction to probing radio wave propagation through the disturbed ionospheric regions. This is why, in our further discussions, we divide the corresponding analysis into three subjects. The first aspect relates to the study of morphology of the high-latitude ionosphere, its plasma content, temperature, ambient electric and magnetic fields, and winds of neutral gas (all existing in both quiet and perturbed conditions). The second aspect deals with generation of plasma irregularities of a wide spectrum of dimensions, methods of observations of their dynamics, spatial location along altitudes, and their lifetime. As for the third subject, we will briefly discuss instabilities associated with the corresponding plasma irregularities, pointing out their specific reaction to radio waves propagating through the perturbed irregular ionospheric regions. The two latter aspects will be discussed simultaneously by referring to the experiments carried out and the corresponding measured data obtained using separate or jointly used modern radiophysical methods of the study of high-latitude ionosphere.

Study of E Region Irregularities in the High-Latitude Ionosphere. In the high-latitude ionosphere E region plasma density irregularities are completely controlled by the ionospheric currents and plasma drifts in crossing $\mathbf{E}_0 \times \mathbf{B}_0$ fields, because in high latitudes the ambient magnetic field is directed close to the vertical axis of Earth's surface, and therefore plasma irregularities are field-aligned (oriented along $\mathbf{B}_0$) and affected by the ambient electric field transverse to $\mathbf{B}_0$ and also by the neutral wind with the horizontal component (across $\mathbf{B}_0$) exceeding its vertical component (along $\mathbf{B}_0$).

Radar investigations of the E region of high-latitude ionosphere began even before the Second World War and are continuing even now. Extensive measurements of the auroral and polar zones of the ionosphere were carried out in the mid-1960s and 1970s. More systematical investigations of the effects of ambient phenomena (solar activity, high-energy particles of the cosmic rays, magnetic activity, and so on) on ionospheric plasma parameters deviations were carried out [39,236,237] using simultaneous Chatanika, Alaska, incoherent scatter radar, auroral optical camera, and photometric observing data (see technical details about these systems in Chapter 8). All ground-based facilities were arranged at the research center located near Chatanika (magnetic latitude 65.7°, geographic latitude 65°N). This radar system based on a unique technique allows for obtaining electron density profiles during quiet to strongly perturbed ambient conditions occurring in the high-latitude ionosphere at altitudes of 90–120 km. We have to note that there is a principal difference in the thickness of the auroral and equatorial E layers; the latter extends along altitudes up to 150 km and even higher, whereas the former extends only up to 120–130 km. In Refs. [121,122], not only was the plasma content measured, but also the variations of plasma disturbances in time, $\Delta N/\Delta t$, whose daytime variation Is given by $\Delta N/\Delta t \approx 3.7$ cm^{-3}sec^{-1}. More detailed analysis using an ionosonde located at Anchorage, Canada, showed the parameters $\Delta N/\Delta t \approx 18.7$ cm^{-3}s^{-1} and $N_{max}/N_{min} \approx 4.0$ of E layer plasma density and its critical frequency deviations as reactions of E region to total solar eclipse. It was also found in Refs. [121,122] that the largest effects observed in the auroral E region occurred during the solar eclipse, when the obstruction of the solar disk was about 100% at the altitude $h = 100$ km at the observation station. Then, in Ref. [39], the effects of auroral particle precipitations due to fast-moving high-energy particles falling at the ionospheric altitudes of the E region were analysed. Such particles could cause visible auroral emission. It was noted during observations made in Ref. [39] that whenever the discrete auroral arc passes directly over a vertical ionosonde position, the sporadic E layer maximal frequency (fE_s) increases by several megahertz [121,122].

During the next two decades of investigations of the high-latitude E spread phenomena and a wide spectrum of irregularities observed experimentally, researchers focused their attention on different kinds of instabilities as being the potential sources of large- and small-scale plasma density irregularities, which were responsible for the auroral echo signals obtained experimentally at different high-latitude experimental sites throughout the world (see Refs. [41–47,238–247] and the bibliographies therein). We note only some that, in our opinion, provide more detailed information about the shape and dimensions of irregularities and their degree of perturbation (compared with background plasma), Doppler spectra, and the corresponding velocities of the observed plasma waves (i.e., irregularities), their spatial localization, dynamics, and lifetime. More interesting in our opinion is a cycle of combined complex experiments carried out in Alaska [238–244], using the Chatanika incoherent scatter radar described earlier briefly (see details in Chapter 8), and the 398 MHz phased array radar and 1210 MHz backscatter radar located at Homer, Alaska (59.72°N, 151.53°W) The geometry of the complex experiment is presented in Figure 6.101, where the magnetic aspect contours for an altitude of 110 km are shown by solid curves, changing from 0° to $+6°$, and the L-shell contours by dashed curves extending from $L = 5$ to $L = 9$. The bold-boundary sector with the crossing fan-shaped lines indicates the working area of the Homer radar. During the operating period of the Chatanika radar, the Homer phased array (with scanning in azimuth domain ±20° by the narrow beams with width of 2.3°) covered this sector with Chatanika near the center of this sector. The first experiments were carried out at the operating frequency of 398 MHz. The creation of the radio coverage map observed by the Homer radar was correlated with optical observations using the combined all-sky camera coverage from Chatanika, the operated area of which is shown in Figure 6.101 by the crosshatched circle. Without going into the technical details of experiments described in detail in Refs. [238–244], we should point the essential effect of the ambient electric field occurring at altitudes of the E region with values of 30 to 40 mV/m when auroral echoes occur. Moreover, a good correlation between occurrence of the strong electric field and echo signals recording was found experimentally. At the same time, the plasma density at observed altitudes

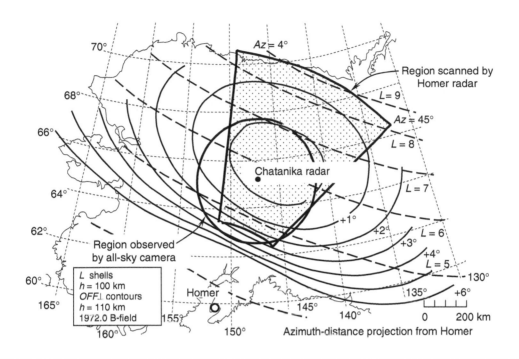

Figure 6.101 Geometry of the complex experiment using radar systems located at the Chatanika and Homer and All-sky optical camera. (Extracted from Ref. [234].)

from 110 to about 120 km varied from $1.5 \cdot 10^6$ to $5 \cdot 10^4$ cm^{-3} in the daytime, with a further increase to $2 \cdot 10^5$ cm^{-3} in the midday periods.

The corresponding plasma irregularities were observed above the Chatanika ionospheric region using both optical and radio observations, mostly in the morning periods (06:00 to 10:00 UT), and no nighttime (between 02:00 to 03:00 UT) not being observations were made. Further analysis of the experimentally obtained data allows for finding a short-term correlation between the auroral echo signal strength, ambient electric field, and plasma density of the observed irregularities. As an example, in Figure 6.102 are shown the results of measurements of these three parameters during a joint campaign carried out in the early morning (from 05:40 to 06:10 UT). In the bottom panel, the time dependence of the northward electric field component, measured at an altitude of 171 km, is depicted together with echo signal amplitude time deviations (solid curve), and the corresponding noise level (dashed curve). For the north–south electric field component, the scale in millivolts per meter is given on the right ordinate. The top panel of Figure 6.102 represents altitudinal distribution of plasma density within the high-latitude E region (from 95 to 170 km), where all numbers near lines of equal electron density are related to $1.0 \cdot 10^5$ e/cm^3 by timing each number to this value.

It can be seen that both curves at the bottom panel show satisfactory agreement (except for 2 min between 06:00 and 06:02 UT) and between these two curves. The development of a strongly irregular plasma structure in the E region, mostly occurs in periods when the echo signals have higher amplitude and the corresponding electric field exceeds 30 mV/m. The existence of two discrete radar peaks at this 2 min period is related by the researchers to the two arcs observed visually by the combined all-sky optical camera which, according to their assumption, reached their position just north of Chatanika passing the radar observation field. To understand more precisely the spatial distribution of the large-scale irregularities, such as arcs and blobs, extending in the high-latitude

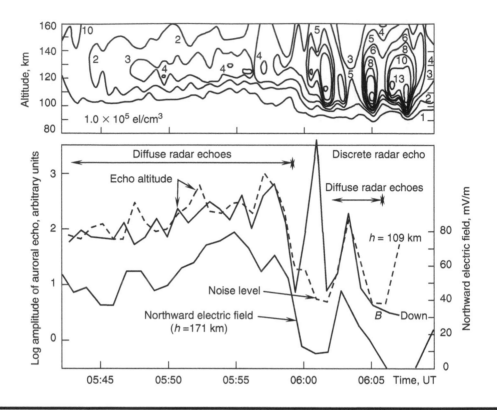

Figure 6.102 Plasma density distribution in joint height–time domain, where numbers indicate amount of electrons per cm³ timing on 10^5 (top panel), and time distribution of diffuse (dashed curves) and discrete (solid curve) radar echo signals (bottom panel). The bottom solid curve in this panel describes time dependence of northward electric field strength in mV/m.

ionosphere up to thousands and hundreds kilometers, respectively, and to differentiate their effect on sounding waves with respect to effects of irregularities with scales less than 10–100 m and even less than 1 m, a 1210 MHz ($\lambda \approx 0.25$ m, $\ell_\perp \approx 0.12$ m, according to Bragg's law) with backscatter radar at Homer, was used [243]. The main goal of this experiment was to find the close relation (according to the suggestion by the authors) between small-scale irregularities and the associated two-stream (or F-B) instability. We will return to this aspect after more precise investigations of spectral characteristics of the observed echo signals and their relations to the types of the corresponding instabilities as potential sources of associated plasma irregularities.

Now, we present the possibilities of Homer radar to obtain radio response of the irregular structures of auroral ionosphere (called *radio aurora*), such as arcs and blobs. The 1210 MHz auroral echo signals recorded during the early morning periods in magnetic activity (Figure 6.103a) and relative magnetic quiescence (Figure 6.103b), taken 30 min later than in the previous figure, are shown on the geographical map of Alaska. The areas consisting of echo signals (i.e., the corresponding irregularities scattering the recorded echo signals) are shown as cross-hatched regions. All parameters shown in Figure 6.103a and b are the same, as in Figure 6.102. It is evident that using multi-beam steering antenna operating at 398 MHz ($\lambda \approx 0.75$ m, $\ell_\perp \approx 0.37$ m) together with 1210 MHz ($\lambda \approx 0.25$ m, $\ell_\perp \approx 0.12$ m), a thin, irregular plasma small-scale structure can be observed, as well as the boundaries of the large-scale structures such as arcs and blobs.

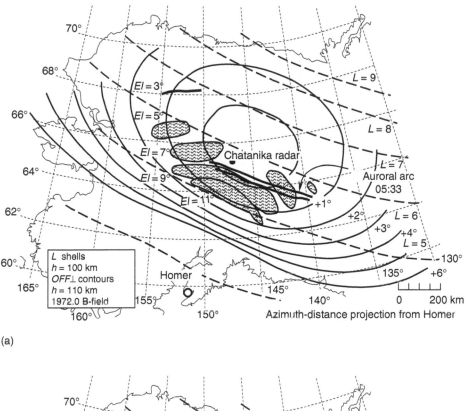

(a)

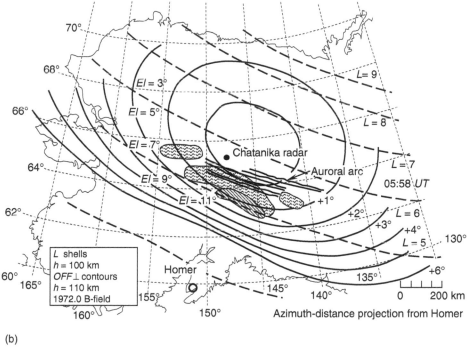

(b)

Figure 6.103 Contours of scattered areas surrounding the Chatanika radar system observed during the magnetic active period (a) and magnetic quite period (b). (Extracted from Ref. [240].)

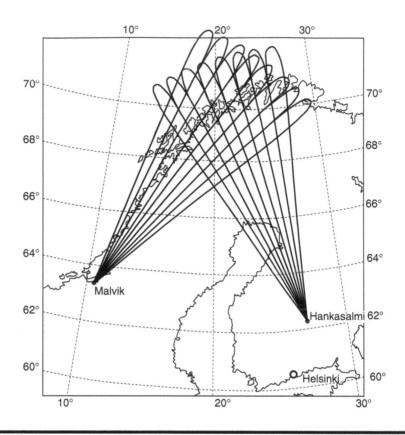

Figure 6.104 Geometry of experiments carried out in northern Scandinavia. (Extracted from Ref. [245].)

Systematic observations of the spread F irregular layer and irregularities that filled this region of the high-latitude ionosphere were performed with the Scandinavian Twin Auroral Radar Experiment (STARE) in Sweden [245–247] consisting of two bistatic pulsed backscatter radars operating at a frequency of 140 MHz (i.e., observing irregularities with scales $\ell_\perp = \lambda/2 \approx 1$ m), which have a common observation region over northern Scandinavia, as shown in Figure 6.104. The multiple beams receiving a pattern produced by the phased array antennas, where the ends of each beam indicate the approximate E region horizon, give a spatial resolution of 20 km × 20 km. The backscatter signal intensity was analysed for various days and for different ranges of the velocities of movements of observed irregularities, from 300 to 350 and from 800 to 850 m/s. Statistical analysis of signal intensity distribution for these two velocity ranges found the same behavior for the 6 days of observations. This allowed the authors to compute the mean backscatter intensities, shown in Figure 6.105 along the vertical axis, for a wide velocity interval from 50 to 1000 m/s with a step of 50 m/s, and a broad scanning angle interval from 10° to 90° with a step of 10° (horizontal axis). It can be seen from the presented illustration that the minimum in the backscattered signals is achieved at 90°, i.e., where a velocity independence is observed. Along a whole velocity range, this parameter has a tendency to increase toward parallel propagation. Moreover, below the ion acoustic velocity ($C_s \approx 300$–350 m/s) this increase does not exceeded 5 dB, whereas above C_s (for drift velocities $V_d \approx 500$–550 m/s) for parallel propagation this difference can reach 20 dB, corresponding

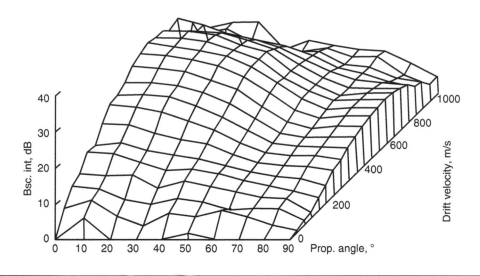

Figure 6.105 Backscattered signal intensity angle-drift velocity 3-D mapping obtained in Ref. [247] by using numerical computations.

to a density fluctuation of 10 times compared with the background plasma density. For higher V_d (more than 650–700 m/s), corresponding to a strong ambient electric field varying between 20 and 30 mV/m, the maximum of the backscatter signal intensity decreases weakly. The authors of Refs. [45–47] explain the observed results by the influence of small-scale irregularities (i.e., short-wavelength plasma waves) moving with drift velocities less than C_s during the process of destabilization of the long-wavelength plasma wave (i.e., the large-scale irregularities) via the wave–wave interactions using the cascade turbulent theory (see Chapter 2). According to Refs. [245–247], the small-scale irregularities causing backscatter signals at the receiver are produced by the two-stream (F-B) instability. At the same time, to explain backscatter signal behavior for drift velocities higher than C_s, the authors artificially introduced an additional damping process. As in the case of experiments described previously, the obtained experimental results again did not offer an adequate theoretical explanation of observed phenomena and did not answer the question of what type of instability is responsible for this behavior of the scattered signals.

To understand the role of different types of instabilities as the potential sources of the resonant aurora radar echo signals (i.e., the corresponding irregularities responsible for these signals), we should analyze the echo signal spectral distribution in the Doppler shift domain that was reported by several separate research groups in the 1980s and 1990s in the United States and Canada [43–47,248–251] based on HF and UHF radars and on the radar interferometer; in Alaska [252] based on the same Chatanika backscatter radar; and in Norway [253]. All the investigators found that, according to spectral characteristics of the echo signals scattered from the E spread, plasma irregularities observed in the high-latitude ionosphere could be classified in the same manner as irregularities were classified as occurring in the middle-latitude and equatorial ionosphere (depicted in Figure 6.45 as types 1 and 2). Finally, according to experimental results, the observed echo signals and corresponding irregularities were divided into two classes, types 3 and 4, the Doppler spectra of which are shown in Figure 6.106. These types of spectra (as will be discussed in the following text) were found only at high latitudes of the ionosphere, where they were usually observed after midnight and when ionospheric plasma was strongly perturbed. Before going into

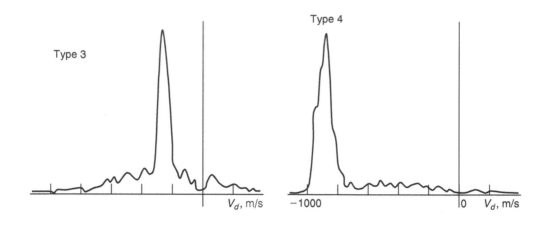

Figure 6.106 Schematic presentation of spectral characteristics of irregularities of 3rd and 4th occurring in the high-latitude ionosphere.

details, we should note that up to now a general physical framework that could interpret these spectra does not exist. Moreover, there are many contradictions between the theoretical background and the observations. Nevertheless, obtained experimental results allow for stating that both spectra have short lifetimes, and the corresponding irregularities have a mixed structure localized in space, where small-scale irregularities are embedded into the moderate- and large-scale irregularities with a high dynamic of coupling processes between irregularities with transformation of energy along the wide spectral waveband (i.e., scales).

We will now analyze possible explanations for the observed phenomena occurring in the high-latitude irregular E region following experiments carried out and described in Refs. [43–47,248–253]. First, it is evident from spectra presented schematically in Figure 6.106 that both of them are narrow with a different Doppler shift. Thus, if the maximum of the type 3 echo signal corresponds to Doppler shift velocity of 200–250 m/s, which is small compared with the speed of ion acoustic waves, it cannot be considered as associated with F-B (two-stream) instability (see Chapter 2). At the same time, the maximum of the type 4 echo signal is usually located at the much higher Doppler shift velocities achieving 800–900 m/s, i.e., exceeding the threshold of the F-B instability, but we can associate the observed type 4 irregularities with two-stream instabilities occurring in the E region high-latitude ionosphere. To prove these statements, we will show some interesting results from references under consideration referring the reader to works [43–47,248–252], where the complex experiments and detailed theoretical interpretation of the observed phenomena were reported.

Systematic observations of resonant auroral echo signal by the backscatter radar operating at 50 MHz and located northeast of Ithaca, New York (76.5°W, 42.5°N), were reported in Ref. [248]. This radar and methods of observation are very similar to those used in the Alaska campaign described earlier, and therefore we do not go further into the technical description of the experiment. We will show the geometry of the experiment and discuss some of the more interesting results of measurements of the Doppler spectra and dynamics of the irregularities responsible for auroral echo signals observed by this radar system. Thus, in Figure 6.107 we present a 2-D latitude–longitude map of the location of the Ithaca radar with respect to *L*-lines, where the radar's main beam direction is shown by a solid line and the range (in kilometers) covered by this radar

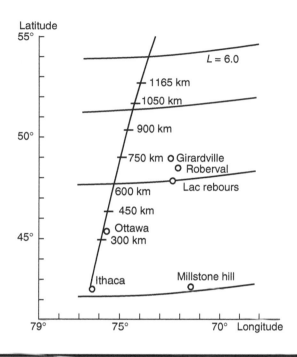

Figure 6.107 A 2-D latitude–longitude map of the location of the Ithaca radar with respect to *L*-lines (e.g., magnetic meridian lines). The radar main beam direction is shown by a solid line and the numbers along this line indicate range (in km) covered by this radar pattern. (Extracted from Ref. [248].)

pattern. The Doppler shift spectra of echo signals, i.e., the corresponding phase (Doppler) velocities of irregularities responsible for the radar echoes, were measured with a frequency resolution of 7.81 Hz, equivalent to a phase velocity resolution (in conditions of direct visibility, LOS) of 23.4 m/s, taking into account that at 50 MHz radiation, a Doppler shift of 1 Hz corresponds to a velocity of 3 m/s. As was expected, two types of echo signal spectra against the Doppler shift were observed. We present only two typical variations of the Doppler spectra (1) in the time domain for a particular range of arriving scattered echoes of 398 km (Figure 6.108a), and (2) in the joint time and range domains (Figure 6.108b). We see, from Figure 6.108a, temporal variations of the shape of Doppler spectra from type 3 (i.e., narrow with a small shift) to type 4 (i.e., broad with a sufficiently high shift), whereas in Figure 6.108b the nearly constant Doppler shift corresponding to a 3-like spectra is always observed with narrow width, which, as shown in Ref. [248], was equal or smaller than the typical equatorial type 1 echo widths observed from the spread E region (see the previous subsection). The authors of Ref. [248] found that the direction of irregularities responsible for these radar echo signals is usually in nearly north–south, with velocities varying in a wide range from 200 to 300 m/s (corresponding to Doppler shifts of 70–90 Hz) to sometimes even exceeding 800 m/s. The authors explained such spectra shifts and the corresponding phase velocities of generated irregularities by the existence of unstable ion cyclotron waves and stated them to be a potential source of radar echo signal observed from the auroral E region of the ionosphere.

At the same time, observations obtained by Johns Hopkins University, Maryland, [249], using HF radar located in Goose Bay, Labrador, operating from 10.5 to 12.5 MHz, and based on studying the E region scatter pattern during nighttime periods, produced estimated Doppler velocities and spectral widths less than those observed in Ref. [248]. In Ref. [249] the Doppler velocity variations were investigated experimentally against the angle α between the radar wave vector **k** and the contour of the magnetic meridian parameter L, introduced earlier, i.e., $V_{Doppler} = V_{||} \cos \alpha + V_{\perp} \sin \alpha$. Here, $V_{||}$ is the plasma irregularities drift velocity parallel to the L contour, and $V_{\perp}$ the drift velocity perpendicular to the L contour. On the basis of the obtained values of $V_{||}$ and $V_{\perp}$, -161 and -24 m/s, respectively, and finally, $V_{Doppler}$ varying from 160 to 220 m/s (see Figure 6.109a,) the authors suggested that the drift velocity of plasma irregularities is almost aligned with the L contour and eastward directed. However, sometimes drift velocities of the E spread irregularities achieved -500 m/s (see Figure 6.109b) and more (even up to 800 m/s; see Figure 6.109c). The sensitivity effects of the aspect angle of the radar orientation with respect to the magnetic meridian contours (from 0° to 10°) and its variation effects on intensity, phase velocity, and altitudinal location of 34 cm E region high-latitude irregularities were studied in Ref. [250]

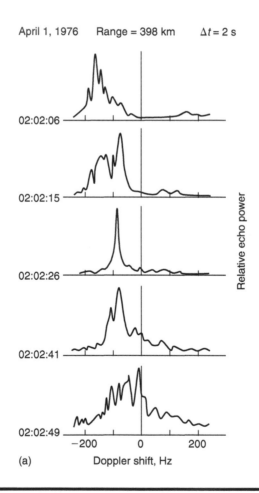

(a)

Figure 6.108 Typical variations of the Doppler spectra in the time domain for a particular range of arriving scattered echoes from ionosphere at 398 km (a).

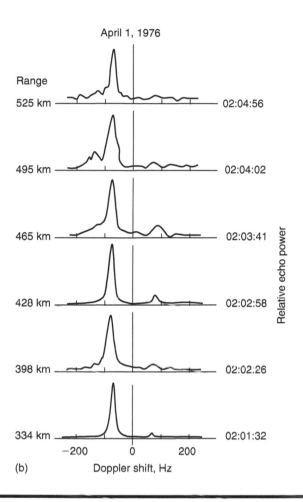

(b)

Figure 6.108 (continued) **The same variations in the joint time and range domains (b).**

based on the systematic observations carried out by Millstone Hill, Massachusetts, 440 MHz backscatter radar (see technical description in Chapter 8). It investigated both small (less than 3°) and sufficiently large (more than 3°) aspect angles. It was found that for the first case the typical echo signals associated with Farley–Buneman waves (i.e., two-stream instability) are usually recorded at the receiver and their phase velocities are consistent with the ion acoustic velocity and located mostly in the altitudinal range 110 to 115 km.

At the same time, for the second case of large aspect angles, the observed phase velocities of 100–200 m/s were less than the ion acoustic velocity and were observed mostly at altitudes around the peak at 107 km. We again see that the processes of plasma irregularity formations cannot be described by only one general unstable mechanism; many factors, including conditions of the radar experiment, should be taken into account. Therefore, more precise observations and the corresponding modern methodology are needed to investigate different types of echo signals and the associated types of plasma irregularities. This was done in further radar and satellite investigations, separate and joint, in the past two decades [44–47,253–258]. Thus, detailed studies of type 4 echo

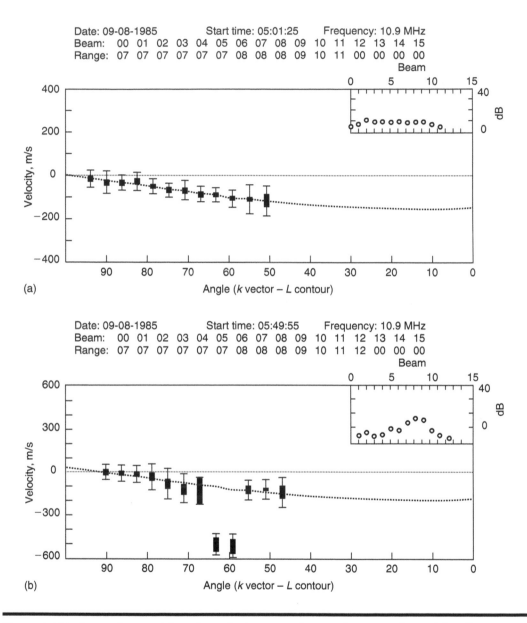

Figure 6.109 Dependence of Doppler velocity on angle α to the magnetic meridian line parameter L obtained at 05:01:25 UT (a), at 05:49:55 UT (b), and 05:51:35 UT (c).

signals scattered from small-scale irregularities localized in strongly unstable plasma regions in the auroral ionospheric E layer were investigated by two groups of researchers from Cornell and Saskatoon (Canada) universities, based on the two 50 MHz backscatter CW radars developed at the University of Saskatchewan (Canada).

The results of joint radar experiments were reported in detail in Refs. [44,47], which name their radar experiment SAPPHIRE, the acronym for Saskatchewan Auroral Polarimetric PHased array Ionospheric Radar Experiment. Again, we do not go deeply into the technical details of these

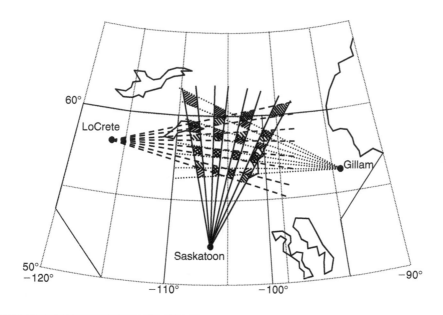

Figure 6.110 Configuration of the experiment SAPPHIRE projected on a geographical map of central Canada. (Extracted from Ref. [46].)

experiments, but only present in Figure 6.110 the SAPPHIRE experimental configuration projected on a geographical map of central Canada. In the previous experiments, the radar echo signal Doppler spectra related to the corresponding type 4 irregularities, being one of the numerous events shown in Figure 6.111, were investigated systematically. According to the typical spectra presented here, we can point out that the type 4 echoes usually have two close peaks, one of which is associated with the primary stable plasma waves with phase velocities constantly observed near −500 m/s and the second with the secondary unstable waves with phase velocities achieving −900 to −1000 m/s (see also Figure 6.112, where several 4-type echo signal spectra are shown).

More detailed experiments for identification of different types of echo signals on types 3 and 4 were reported in Ref. [47]. The spectrograms obtained in further SAPPHIRE experiments and presented in the joint time–frequency domains are plotted in Figure 6.113. According to these spectrograms, type 3 spectral signatures were selected and shown in Figure 6.114 with the existence of localized (with peaks near ±(50–100) Hz), narrow-width (small Doppler shift of −55 Hz; see Figures 6.113 and 6.114), and high-amplitude echo signal spectra, caused by the corresponding 3-*m* (*f* = 50 MHz) irregularities localized in the strongly unstable regions in the E layer consisting of large-amplitude plasma waves (i.e., instabilities). As for the type 4 echo signals, they were observed more rarely than those of type 3, but have their own specific features of the Doppler spectra (see Figure 6.115) characterized by highly shifted narrow peaks, shown in Figure 6.115 by shaded areas and by the existence of additional peaks.

To complete this review of the more interesting international experiments carried out for the investigation of the strongly unstable (and therefore highly irregular) E region of the high-latitude ionosphere, we will show briefly results obtained in northern Europe (Norway and Sweden [253,256], Iceland [257]), and at the Antarctic zone [254].

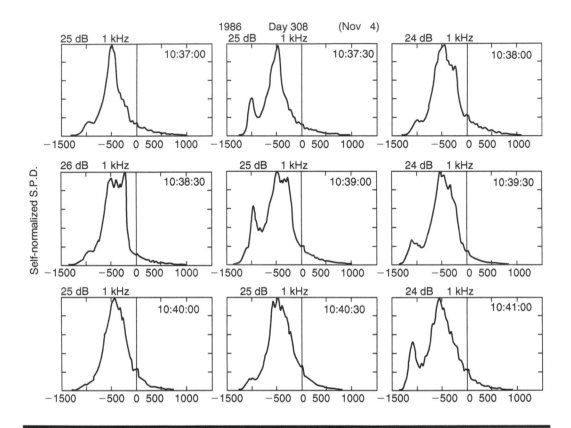

Figure 6.111 Radar echo signal Doppler spectra obtained experimentally in Refs. [44,47] related to the corresponding type 4 irregularities.

Thus, in Refs. [253,256], the same complicated spectral characteristics of echo signals were observed and reported simultaneously using a combined Cornel University Portable Radar Interferometer (CURPI) operated at VHF-band ($f = 46.9$ MHz) and the European Incoherent Scatter (EISCAT) radar, described in the previous section. The dependence of the phase velocity of E region small-scale ($\ell_\perp = \lambda/2 \approx 3.2$ m, according to Bragg's law) irregularities on perturbations of the ambient electric field was investigated systematically and with more care than other investigators. We show in Figure 6.116 one of the characteristic examples of the radio trace range dependence (in kilometers) of the altitudinal localization (left-side panel), Doppler spectra (middle-side panel), values and west–east velocities mapping (right-side panel) of irregularities, as scatterers of the observed echo signals. As can be seen from the presented illustrations, the observed echoes appear mostly at the altitudinal range of from about 115 to 135 km, where the complicated shapes of Doppler spectra are observed covering both 3-type (roughly from about 115 to 120 km and from about 125 to 130 km) and 4-type (roughly from about 120 to 125 km and from 130 to 135 km) echo signals. The arrows represented the apparent drift velocities of the scatterers, indicating their direction as being mostly to the west from the east. As for the lower altitudes from 100 to 115 km, the stable 3-type echo spectra are rarely observed (see the left-side panel). The largest arrows correspond to drift speed of the scatterers exceeding 600 m/s, i.e., the threshold of two-stream instability generation at the observed altitudes.

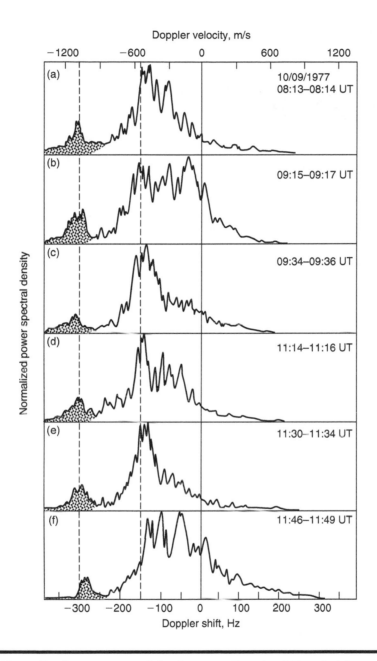

Figure 6.112 Normalized power spectral density versus Doppler shift varied from −300 to +300 Hz and Doppler velocity varied from −1200 to +1200 m/s. Shadow regions indicate complex form of the obtained spectra.

It is also interesting to present here the results of a three-year (1995–1997) systematic observation of auroral coherent echo signals observed at the Antarctic Syowa station at 50 and 12 MHz. The corresponding experiment configuration with the radar beams overlapping pattern (shadow zones) is shown in Figure 6.117. In this experiment, the Super Dual Auroral HF radars operated at 12 MHz (east and south), labeled SD, and the Communication Laboratory 50 MHz

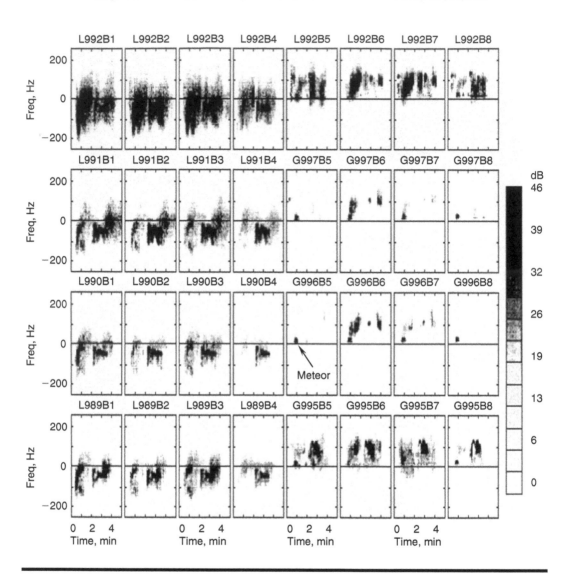

Figure 6.113 Frequency–time 2-D mapping of echo signals obtained during the experiment SAPPHIRE. (Extracted from Ref. [47].)

radar are presented in the Polar Anglo-American Coordinate System (PACE) based on the magnetic parallels, called L shell directions (see above), and shown in Figure 6.117 by solid curves. The bold white curve represents the line of zero off-perpendicular (called *aspect*) angles, Ψ, over the scan at an altitude of 110 km. In the PACE coordinate frame, the lines of equal magnetic latitudes are shown by the bold curves, with numbers denoting magnetic meridians within the angle range $\Lambda = 65°$ to $\Lambda = 75°$. Figure 6.118 shows the azimuthal distribution of echo signal power and Doppler velocity for CRL 50 MHz radar (Figure 6.118a and b) and SD 12 MHz radar (Figure 6.118c and d) for different aspect angles varying from $\Psi = -2°$ to $\Psi = +2°$ (for CRL radar) and at the same range, with an additional curve corresponding to $\Psi = +5°$ (for SD radar), shown by white bold curves. The contour steps for the power were 5 dB and for the velocity 100 m/s. As can be seen from Figure 6.118a and b, the echo signals observed at 50 MHz have power and velocity maxima close to

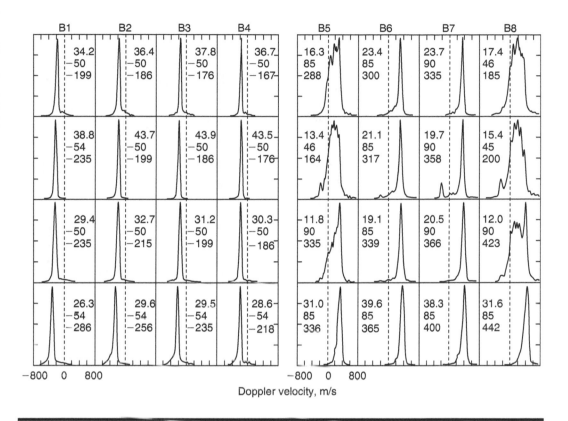

Figure 6.114 Spectral characteristics of echo signals In the joint time–frequency domain obtained during the experiment SAPPHIRE.

the line of the zero aspect angle and extended in the azimuth domain from 85° to 110°, i.e., for the observations approximately along the L shells (i.e., magnetic parallels), and with minimal power at an azimuth of about 130°, i.e., roughly perpendicular to the magnetic L shells. At an azimuth of about 150°, local enhancement of plasma density was observed. Velocities shown in Figure 6.118b have maximal magnitudes exceeding 700 m/s in the range from 85° to 110°, and decreasing at larger azimuthal angles. SD radar data plotted in Figure 6.118c and d shows the same minimum in power across the magnetic L shells, but with more regularly distributed power over the azimuthal range (Figure 6.18c), and with maximal magnitudes of drift velocities of 300–400 m/s localized at the azimuthal range from 80° to 130°. The results were in good agreement with those obtained over the Iceland East using the SD radar system developed in Iceland, operating between 8 and 20 MHz, and reported in Ref. [257].

The obtained statistical data of monthly power (in dB), velocity (in m/s), and spectral width (in m/s) distributions is presented in Figure 6.119 as a range–time plot of each summarized parameter. It can be seen that the strongest echo signals are usually observed in the earlier evening and premidnight periods, where the corresponding velocities of the scatterers correspond to, or even exceed, the ion acoustic speed C_s, with peaks close to ±350 m/s. Nevertheless, small peaks near ±700 m/s were rarely observed, indicating an existence of type 4 irregularities. Sometimes, during these experiments a mixed type of echo signal was observed.

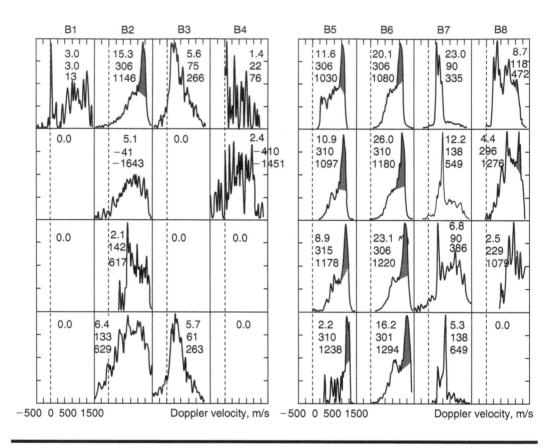

Figure 6.115 Doppler spectra of echo signals in Doppler velocity domain at the ranges from −500 to 1500 m/s.

The complicated structures of the observed Doppler spectra of both types of echo signals and the wide range of velocities, less and more than the ion acoustic speed C_s, were estimated for the corresponding small-scale irregularities occurring in the high-latitude E region, and stimulated for creation of new theoretical frameworks and modifications of existing theoretical background concerning generation of the intercharge instabilities, such as GDI and FBI, and so on, in the strongly unstable ionospheric plasma regions. Thus, in Ref. [251], based on results of measurements made in Refs. [44–47,248–250], the possibility was suggested of combining the nonlocal nonlinear wave–wave and wave–particle interactions (i.e., primary large-scale stable waves and secondary unstable small-scale waves) with elements of the cascade turbulent theory. We discussed in detail the theoretical aspects of these new modern nonlinear theories in the last section of Chapter 2. Such new approaches give a satisfactory explanation for the existence of types 3 and 4 small-scale plasma irregularities embedded into the unstable large-scale plasma regions (structures), which together form the observed two-peak narrow-band and several-peak broad-band spectral characteristics of the echo signals scattered from spread E high-latitude mixed highly unstable plasma region.

Study of F-Region Irregularities in the High-Latitude Ionosphere. As mentioned before, the F layer of the ionosphere, high-, middle- and low-latitude, including the equatorial zones, and occupying a

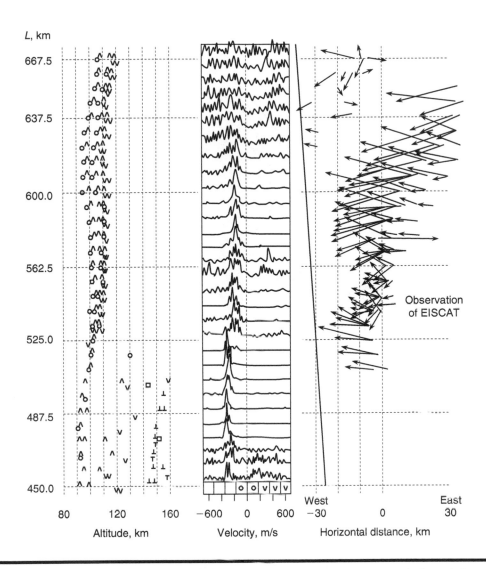

Figure 6.116 Examples of altitudinal localization of the radio trace range *L* (in km) (left panel), Doppler spectra (middle panel), and west–east mapping of directed vectors of Doppler velocities (right-side panel) of irregularities. (Extracted from Refs. [253,256].)

larger region at the altitudinal range, of about 500–600 km, than the E layer, and being closer to the magnetosphere and near Earth's space environment, can be affected by many external factors such as the high-energy particle precipitation, solar wind, magnetosphere–ionosphere coupling, as well as by internal factors, such as convection of currents due to coupling between high conductive E and F regions, and so on. Naturally, it is obvious that a wider spectrum of plasma structures, large, moderate, and small, was observed experimentally by using ground-based, rocket, and satellite radio and optical facilities. Moreover, interactions between different plasma structures (i.e., plasma waves of various wavelengths) lead to a complicated physical picture of their spatial and temporal evolution. Thus, a situation with a full description of physical processes accompanying this evolution is more complicated because some of the observations and the corresponding

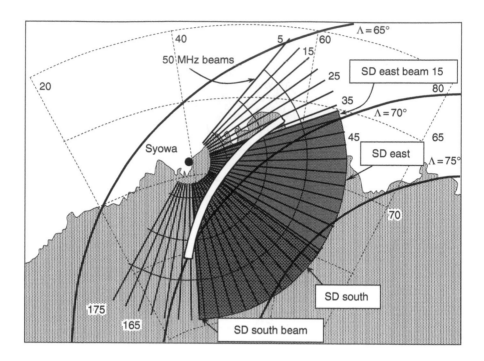

Figure 6.117 The experiments configuration carried out by Antarctic Syowa station with the radar beams overlapping pattern (shadow zones). Here, the Super Dual Auroral HF radars is labeled by SD, and the Communication Laboratory 50 MHz radar is represented by solid curves. (Extracted from Ref. [257].)

theoretical descriptions of the observed processes sometimes contradict one another. Therefore, in our further description of high-latitude F region irregularities, we briefly analyze the main plasma structures observed in the high-latitude F region, their sources of generation, and some of their characteristics, such as density, shape, and dimensions, spatial localization within the F region, spectral properties, velocities, and direction of movements, i.e., those characteristics that are important for radio propagation through the auroral and polar outward ionosphere, on which we will concentrate in Chapter 7.

Entering into a subject, it should be noted that during the last 40–50 years, researchers of the ionosphere selected much experimental data regarding morphology, structure, and dynamics of the high-latitude F region, and a series of systematic works were published mostly during the 1980s and the beginning of the 1990s [38–40,259–271]. Below we follow only some of them [259–262], where the main aspects mentioned at the beginning of this subsection were summarized.

As mentioned in Refs. [259–262], the plasma structures occurring in the high-latitude ionospheric F region can be classified on large-scale (about 10–100 km), such as polar cap patches, auroral and subauroral blobs, and sun-aligned arcs, with dimensions ranging over tens to several hundreds kilometers, and moderate-scale structures, such as plumes (i.e., enhancements) and bubbles (i.e., depleting plasma structures), with dimensions exceeding 1 km, and small-scale structures with dimensions less than 1 km (up to dozens of meters). We will show in Chapter 9 for land–satellite communications that the latter irregularities are more problematic because they determine fast fading inside the channel, i.e., strong random oscillations of the recorded radio

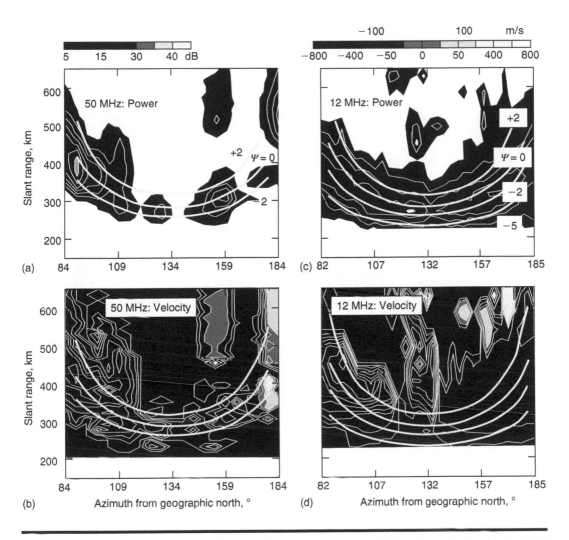

Figure 6.118 The azimuthally distributed echo signal power and Doppler velocity for CRL 50 MHz radar (left panels (a) and (b)) and SD 12 MHz radar (right panels (c) and (d)) for different aspect angles varying from Ψ = −2° to Ψ = +2° (for CRL radar) and at the same range with additional curve corresponding Ψ = −5° (for SD radar), shown by white bold curves. (Extracted from Ref. [257].)

signals and, therefore, loss of information sent through ionospheric radio communication links. This is the reason why we will mostly discuss characteristics of small-scale irregularities via the prism of experiments carried out in the F region of the high-latitude ionosphere.

The patches are localized enhancements of the background plasma density observed at the polar cap with horizontal dimensions ranging from several hundred to thousands kilometers (and therefore called the *largest-scale* [259–262]). They are spatially collocated with enhanced kilometer-scale irregularities [263–269]. An example of patches is shown in Figure 6.120 [267], where measurements of variations in the total electronic content (TEC) using dual-frequency transmissions from the global positioning satellite (GPS) are presented. Multiple patch structure can be seen from presented illustrations during a 5 h period, with a sufficiently complicated structure within each

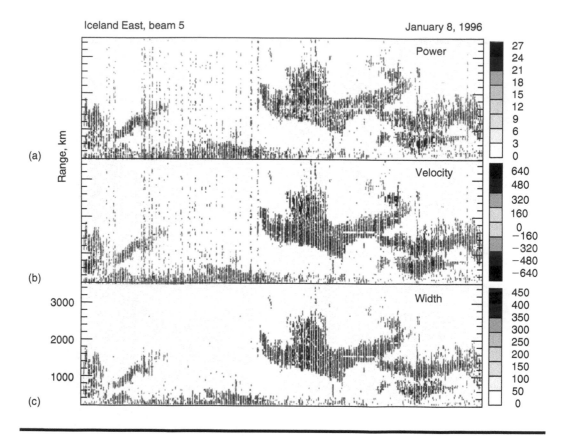

Figure 6.119 The monthly distribution of echo signal power (a) in dB, Doppler velocity (b) in m/s, and spectral width of echo-signals (c) in m/s, obtained in the Iceland East experiments. (Extracted from Ref. [257].)

patch. Taking into account that one TEC unit used as the standard unit equals 10^6 el/cm^2, we can estimate that the largest TEC variation can exceed 20 TEC units. Converting this value to the plasma density, we found that peaks of plasma density in these patches, equaling $8 \cdot 10^5$ el/cm^2, can exceed those for the background ionospheric plasma by factors of eight to ten. It was experimentally obtained [263–267] that the observed patches in the polar cap are the solar-produced plasma structures occupying not only the polar cap but also extending to the auroral and subauroral zones.

Radar observations of the irregular structure of the high-latitude F region have found the existence of irregular regions with widths of less than 100 km that are spatially correlated with 100-km-scale enhancements of plasma density called *blobs* [259–262]. Radar and additional satellite measurements indicated that blobs are characterized by steep gradients of plasma density transverse to the ambient geomagnetic field $\mathbf{B}_0$. Most observations of the variations in TEC can also be explained as blob-produced plasma structures. At the same time, in Ref. [38], commenting on the results of rocket measurements, it was suggested that these variations in TEC can be explained by effects of spatial collocation of intense small-scale irregularities with large-scale gradients of plasma density. Using a theoretical framework concerning convective plasma processes occurring in the polar cap, as well as interchange instability, some authors have attempted to explain the generation of such kilometer-scale plasma structures, which earlier, during the 1960s, many authors connected with particle precipitation (see Refs. [259–262] and the referred literature therein).

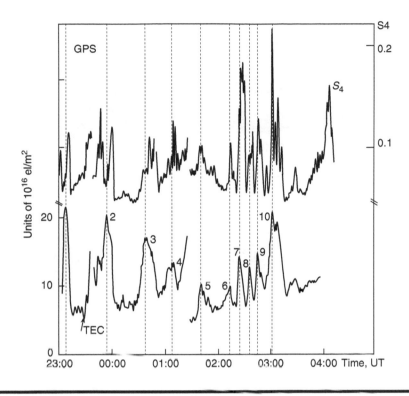

Figure 6.120 Total electron content (TEC) time distribution obtained by GPS satellite experiment. (Extracted from Ref. [267].)

We refer the reader to these systematically collected studies, where full information about these plasma structures will be found.

As an example, the generation of bloblike structures in the high-latitude ionosphere embedded in the ionospheric plasma of the F region produced by solar activity is clearly seen from Figure 6.121, where the altitude-distance 2-D mapping of contours of equal density is presented for different time periods of observation using the Chatanika backscatter radar mentioned earlier in the previous subsection. The numbers near each contour indicate plasma density compared to 10^5 cm^{-3}. Figure 6.121a and b obtained in postmidnight periods shows the dynamics of blobs and their floating and penetration upward along altitudes from the bottom- to the topsides of the F region, ranging over areas of hundreds of kilometers. At the same time, in the earlier-morning periods, blobs either decayed (Figure 6.121c) or were sufficiently weak (Figure 6.121d) [40,269]. In Figure 6.122, three panels of 2-D plasma density distribution at three altitudes, 360 km (bottom panel), 400 km (middle panel), and 440 km (top panel), are presented [266]. The same tendency in the dynamics of plasma bloblike structures with penetration upward and localization at the top level of the F region take place, as was observed in Figure 6.121. According to Ref. [270], a plot of contours of northward plasma irregularities drift velocity is presented in Figure 6.123 in the distance-latitude 2-D format, i.e., in the same format as Figure 6.122. In the area of observation the plasma density inside irregular plasma structures was eight times that for the background plasma.

Now, we present the combined experiment using Chatanika backscatter radar measurements together with satellite observations of the irregular high-latitude F region [40]. In Figure 6.124 we present a geographical map of the position of Chatanika radar in Alaska together with the ISIS-1

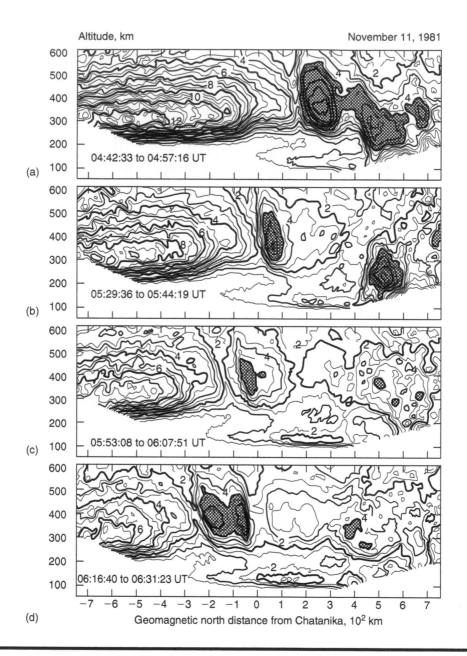

Figure 6.121 Two-dimensional mapping of equidensity contours of ionospheric plasma at altitudes of region F. (Extracted from Ref. [269].)

satellite trajectory during observations carried out at about 11:45 UT on February 20, 1981. The invariant latitudes at the satellite height of about 700 km are given by a bold line. The numbers along the satellite path indicate universal time (UT). The satellite was within the radar pattern sector from 11:43 to 11:46 UT.

Thus, in Figure 6.125a, the electron density 2-D altitude-latitude contour map obtained with Chatanika radar on February 20, 1981, is shown according to Ref. [40], between 11:30

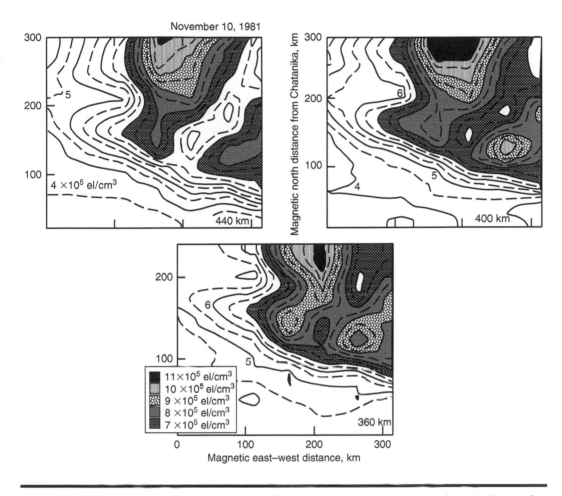

Figure 6.122 **The three panels of 2-D plasma density distribution in the space domain for 440 km (top panel), 400 km (middle panel), and 360 km (bottom panel). (Extracted from Ref. [266].)**

and 11:45 UT. All contours are separated by $0.2 \cdot 10^5$ cm^{-3} and enhanced contours indicating irregular structures of the F and E regions, are given at 0.6, 1.2, and $2.4 \cdot 10^5$ cm^{-3}. A coincident ISIS-1 satellite trajectory is shown by a solid straight line, with numbers indicating the plasma density in units 10^5 cm^{-3} obtained from the ionograms along the satellite path. Figure 6.125b presents the same situation in the ionosphere, but observed simultaneously in the period of 11:40 to 11:57 UT by ground-based and satellite facilities. It is evident that the same 2-D altitudinal distributions of plasma density as depicted by Figures 6.121 and 6.122 are clearly seen from Figure 6.125a and b. Obtained data measured by ISIS-1 satellite allows for estimating electron density profiles for these two events, shown in Figure 6.125a and b, which are presented in Figure 6.126 and denoted by 1 and 2, respectively.

It is interesting to note that the first graph corresponds to the event occurring before an energetic electron precipitation, and the second to the period during and after this event. It can be seen that because of high-energy electron precipitation the height of the F layer peak drops downward from 325 to 300 km, with an increase of its maximum by about 15%, and with the

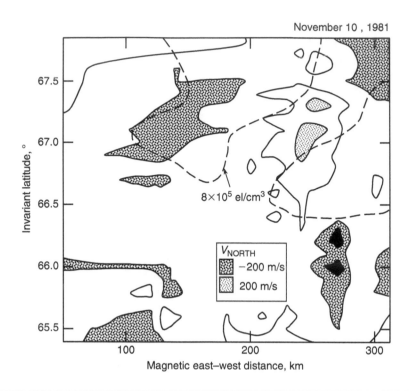

Figure 6.123 Joint 2-D distribution of plasma density and Doppler velocities. (Extracted from Ref. [270].)

tendency of increasing plasma density by about 50% at a lower height (about 250 km). At the same time, a decrease in plasma density (i.e., depletion areas generation) is observed between 350 and 600 km (compare graphs 1 and 2 in Figure 6.126).

During the last 10 years and using the Supper Dual Radar Network (SuperDARN) based on HF radars, large-scale ionospheric (with horizontal dimensions of $L = 300$ km) convection studies were carried out and then summarized in Ref. [271]. In these investigations of solar–terrestrial interactions, for us the most interesting aspects are estimations of the ambient electric fields and the corresponding convective movements of plasma at the topside of the F layer, allowing for the estimation of drift velocities responsible for creation of the unstable high-latitude irregular F spread regions. Thus, in Figure 6.127, we present only part of the convection pattern, obtained by the Goose Bay–Stokkseyri radar pair (described earlier in this section). Here according to Ref. [271], a velocity-divergent structure of the unstable irregular large-scale plasma F region is shown, giving large velocities of amplitude about 100–200 m/s and with a large divergence of the order of about 7 MHz, which is about one order of magnitude greater than the typical values observed by satellite observations (<1 MHz) [258]. The typical scale of this unstable F region is of the order of 200 km.

Completing this subsection of most precise experimental observations of unstable irregular high-latitude F region, we should point out that despite numerous experimental and theoretical investigations, most of which are systematically prepared and argued and can be found by scanning the literature published of the last two to three decades, some aspects of morphology, structure, inhomogeneous plasma content, and interactions between irregularities of various origins and

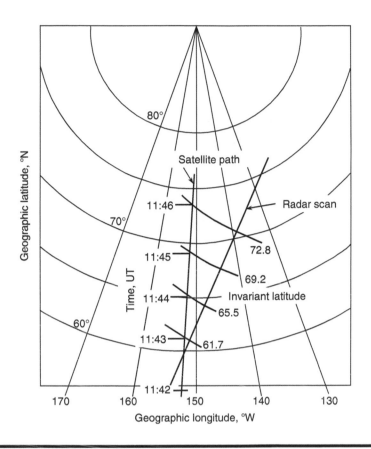

Figure 6.124 Position of Chatanika radar in Alaska and the satellite path (bold straight line) in joint experiment. (Extracted from Ref. [40].)

dimensions of the upper high-latitude ionosphere, and particularly of its F region, require additional experimental and theoretical investigations.

As for new theoretical frameworks, we suggest that some new ideas be combined with the well-known theoretical background described in Chapters 2 and 5. Thus, for example, together with the direct cascade theory of Kolmogorov (see Chapter 2), according to which the stable large-scale organized plasma structures generate unstable small-scale plasma waves, consider also the inverse cascade mechanism, in which large-scale plasma structures can be generated by the inverse energy cascade, i.e., by transferring energy from small-scale to large-scale plasma waves. In such inverse cascade motion, as shown in Ref. [272] (see also bibliography therein), the small-scale plasma structures can be triggered by viscous dissipative processes occurring in the ionospheric plasma or by disturbances of the ambient electric and magnetic fields usually observed in the high-latitude and equatorial ionosphere. These processes lead to the formation of anomalous energy process discussed in Chapter 2 using another mechanism of wave–wave and wave–particle interactions. As mentioned in Ref. [272], a new dissipative process usually occurs in the ionospheric plasma in the presence of the nonvanishing helicity of the velocity field of plasma (as a multi-component fluid), and finally due to the appearance of the helical dissipative mechanism. The formation of the stable large-scale "quasi-regular" plasma structures from the energy of "dissipative-type" unstable small-scale "sporadic" plasma structures, exciting in such nonequilibrium regime,

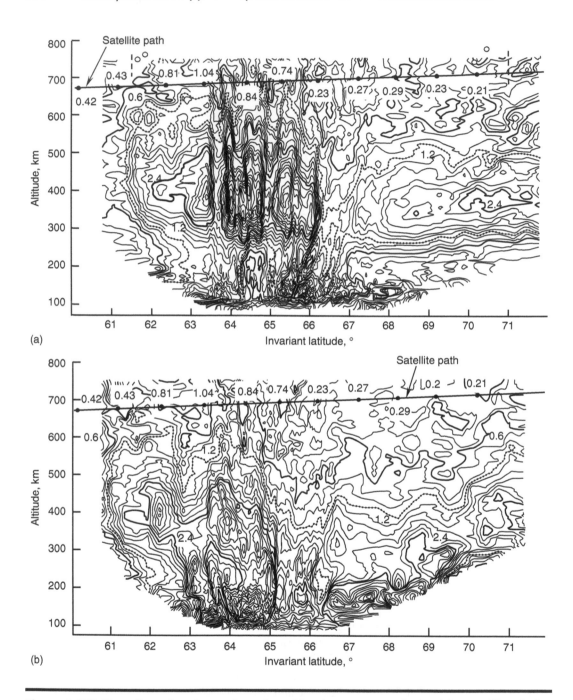

Figure 6.125 Two-dimensional mapping of equidensity contours obtained by satellite during 11:30–11:45 UT (a) and 11:45–11:57 UT (b). The trajectory of the satellite is shown by a straight bold line. (Extracted from Ref. [40].)

may become an important mechanism accompanying generation of the wide spectrum of plasma turbulences, large, moderate and small, observed experimentally in the disturbed nonregular high-latitude or equatorial ionospheric plasma. In other words, the experimental results need to be explained by new theoretical ideas and approaches. We did not introduce the reader to this "exotic"

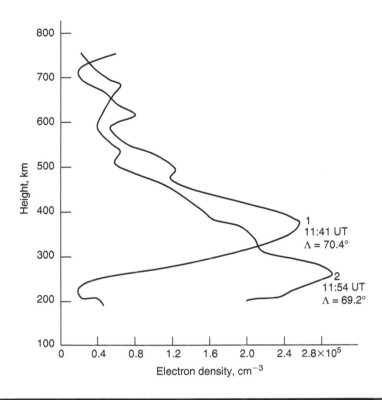

Figure 6.126 Height profile of plasma density corresponding to picture presented by 6.125a (a), and to picture presented by Figure 6.125b (b).

theory, not due to its complexity and limitations of space, but as this excellent theoretical framework nowadays is not supported by experimental evidence.

We present now another example to show how novel theories created during the past decade can be successfully introduced to describe natural phenomena occurring in the ionospheric plasma on the basis of observations of recent experiments employing the novel methodology of measurements in joint space, time, and frequency domains and the corresponding modern ground-based and satellite-assembled radars and radio interferometers. This aspect is closely related to the natural phenomena occurring during interactions of meteoroids with the lower ionosphere at the E and upper D regions described in Section 6.2.1, based on theoretical background developed from the 1970s to 1980s of the last century, as well as on the adequate (existing at that moment) measurements carried out using possibilities of current ground-based facilities described in this section. Despite the objective limitations of these experiments and the corresponding theories, the three main stages of meteor trail evolution in the ionospheric plasma, based on physical principles of meteor plasma dynamics [134–136] and on the basic concepts of plasma diffusion and drift [104–106,122–129], were described in Refs. [49–51]. The first stage is observed when meteor plasma is super-dense (with respect to the ionospheric background plasma) and therefore is not sensitive to the background ionospheric plasma behavior. The second stage occurs in enough dense meteor plasma where strong interaction with background plasma is observed, leading to generation of a strong depletion region surrounding the meteor trail. In the third (latter) stage of meteor trail evolution, the concentration of plasma ions in it is the same as that for the background plasma, and

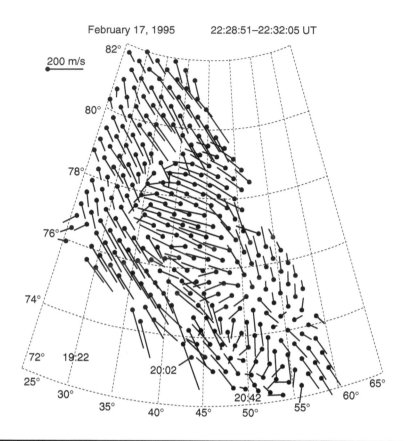

Figure 6.127 Drift velocities vectors convection in the space domain. Scale of 200 m/s allows information to be obtained about deviations of drift velocity of plasma irregularities, their value, and direction during convection. (Extracted from Ref. [271].)

the low-density ionospheric plasma contains a mixture of meteor ions and ions of the background ionospheric plasma. As was pointed out in Refs. [48–51], using a phenomenological approach, in this stage of meteor trail dissipation the Farley–Buneman instability may play a main role in the process of the later stage of meteor trail evolution and in generation of field-aligned H_E small-scale irregularities observed experimentally at the lower ionosphere (see Section 6.2.1).

Only in the recent theoretical studies [273–278] have the roles of Farley–Buneman and other instabilities on meteor plasma later-stage evolution been rigorously examined, both analytically and numerically. The anomalous diffusion and long-term nonlinear evolution of under-dense meteor plasma, described self-consistently in Refs. [273–279], were observed experimentally during the recent decade using powerful and sensitive Arecibo and Jicamarca radars and satellite radio interferometers [279–283], based on a novel methodology of measurements (with respect to that described in Section 6.2.1 according to Refs. [48–51]) for simultaneous observation of specular (coherent) and diffuse (incoherent) radio echoes in the space, time, and frequency domains. However, in the theoretical studies carried out in Refs. [273–278], the authors have ignored the fact that the meteor ions have heavier masses compared to those of background plasma (and therefore, different mobilities). As was shown in earlier studies concerning plasma inhomogeneities evolution [122–129], at the initial stage of penetration of the high-density plasma structure, injected into the background ionospheric plasma at altitudes of $h = 80$–120 km and having ions with mobilities different from those of background ions, two profiles of plasma density disturbances are observed—the stable ambipolar plasmoid (according to Schottky [1–4]) and the plasmoid moving with the velocity of injected ions (see also Chapter 5, where diffusion of plasma in absence and presence of the magnetic field is discussed). The occurrence of the latter plasmoid can be explained by the fact that due to difference between the mobilities, the background plasma cannot be fully compensated (or screened) by the disturbances of the injected plasma. Furthermore, if the injected (meteor) plasma is the complicated multicomponent plasma, as usually predicted by the physics of meteoroids [134–136], consisting of electrons and $(k - 1)$ species of ions, then the amount of plasma disturbances propagating in the plasma (i.e., the profiles) increases and can reach the level of $(k - 2)$ plasmoids, which was observed experimentally [48,49]. Moreover, accounting for the effects of ambient electric field and strong neutral wind, which currently exist in real E layer of the ionosphere (and were taken into account in Refs. [51,145,146]), leads to generation of additional irregularities arbitrary oriented with respect to magnetic field lines. The array of plasmoids, elongated along $\mathbf{B}_0$ and having arbitrary, small to large, angles with geomagnetic field lines were observed experimentally using recent radio observations carried out by Arecibo and Jicamarca ground-based radar systems [279–282]. So, the authors of Refs. [273–278] have simplified their theoretical models taking ions of meteor plasma to be equal those in background plasma. On the other hand, in Refs. [273–278] the processes of recombination, which are sufficiently strong at the altitudes of the lower ionosphere (see Chapter 1), were ignored as the authors found a small rate of recombination for heavy ions with respect to light ions of the ionospheric background plasma. We found here some contradiction between the basic statements of the problem stressed by the authors in Refs. [273–278]. Therefore, we do not present here these novel approaches, hoping that recent and future observations of meteor trails carried out by ground-based radar systems may give additional "push" to the additional improvements of the novel self-consistent models and self-similar numerical codes introduced in the aforementioned references.

We should point out again that the pioneering ideas and concepts generated during the 1970s and 1980s and experimental observations by using radio methods of meteor trails made during these periods (see Section 6.2.1), obviously have spawned novel theoretical and numerical approaches [273–278], as well as new technical possibilities based on modern radiophysical methods and the corresponding radar system (such as, for example, the highly sensitive Arecibo and Jicamarca radars, described in detail in Chapter 8).

At the end of this chapter, we suggest that for future experiments studying natural phenomena occur in the Earth's ionosphere, a more intensive combination of different methods, optical and radio, as was done in investigations of artificial irregularities, such as ion-injected and heating-induced, created mostly in the middle-latitude ionosphere (see Section 6.1). This should be done taking into account that the morphology and structure of the high-latitude ionosphere are more complicated compared with those of the middle-latitude and equatorial ionosphere, which have more common features with respect to more unstable and irregular high-latitude ionosphere affected by external cosmic sources, such as high-energy particle precipitation, solar wind; it is also closer to the magnetosphere and may quickly react on magnetosphere–ionosphere interaction and coupling. Such complex combined experiments can give an impetus to new theoretical studies and new theoretical ideas.

References

1. *Radio Sci.*, Special issue, Vol. 9, No. 11, 1974, pp. 881–1090.
2. Fejer, J.A., Gonzales, C.A., and Ierkic, H.M. et al., Ionospheric modification experiments with the Arecibo heating facility, *J. Atmos. Terr. Phys.*, Vol. 47, 1985, pp. 1165–1179.
3. Djuth, F.T., Thide, B., Ierkis, H.M., and Sulzev, M.P., Large-*F*-region electron-temperature enhancements generated by high-power HF radio waves, *Geophys. Rev. Lett.*, Vol. 14, 1987, pp. 953–956.
4. Rietveld, M.T. and Stubbe, P., Ionospheric demodulation of powerful pulsed radio waves: A potential new diagnostic for radar suggested by Tromso heater result, *Radio Sci.*, Vol. 22, No. 6, 1987, pp. 1084–1090.
5. Duncan, L.M., Sheerin, J.P., and Behnke, R.A., Observation of ionospheric cavities generated by high-power radio waves, *Phys. Rev. Lett.*, Vol. 61, 1988, pp. 239–242.
6. Bernhardt, P.A., Swartz, W.E., and Kelley, M.C., Spacelab 2 upper atmospheric modification experiment over Arecibo, 2, Plasma dynamics, *Astro. Lett. Communic.*, Vol. 27, 1988, pp. 183–198.
7. Blaunstein, N.Sh., Boguta, N.I., and Erykhimov, L.M. et al., About the dependence of the devepment and relaxation time of artificial small-scale irregularities on the diurnal time, *Izv. Vuzov, Radiophyzika*, Vol. 33, 1990, pp. 548–351.
8. Blaunshtein, N.Sh., Erukhimov, L.M. Uryadov, V.P. et al., Vertical dependence of relaxation time of artificial small-scale disturbances in the middle-latitude ionosphere, *Geomagn. Aeronom.*, Vol. 28, No. 4, 1988, pp. 595–597.
9. Bochkarev, G.S., Eremenko, V.A., and Cherkashin, Ya.N., Radio wave reflection from quasi-periodical disturbances of the ionospheric plasma, *Adv. Space Res.*, Vol. 8, No. 1, 1988, pp. 255–260.
10. Wong, A.Y. and Brandt, R.G., Ionospheric modification—An outdoor laboratory for plasma and atmospheric science, *Radio Sci.*, Vol. 25, 1990, pp. 1251–1267.
11. Hansen, J.D., Morales, G.J., and Duncan, L.M., Large-scale ionospheric modifications produced by nonlinear refraction of an HF wave, *Phys. Rev. Lett.*, Vol. 65, No. 26, 1990, pp. 3285–3288.
12. Filipp, N.D., Blaunshtein, N.Sh., Erukhimov, L.M., Ivanov, V.A., and Uryadov, V.P., *Modern Methods of Investigation of Dynamic Processes in the Ionosphere*, Kishinev (Moldova): Shtiintza, 1991.
13. Keskinen, M.J., Chaturvedi, P.K., and Ossakow, S.L., Long time scale evolution of high-power radio wave ionospheric heating, 1, Beam propagation, *Radio Sci.*, Vol. 28, No. 5, 1993, pp. 775–784.
14. Rietveld, M.T., Turunen, E., Matveinen, H., Goncharov, N.P., and Pollari, P., Artificial periodic irregularities in the auroral ionosphere, *Ann. Geophys.*, Vol. 14, 1996, pp. 1437–1453.
15. Djuth, F.T., Groves, K.M., and Elder, J.H. et al., Measurements of artificial periodic inhomogeneities at HIPAS observatory, *J. Geophys. Research*, Vol. 102, 1997, pp. 24.023–24.025.
16. Rietveld, M.T. and Goncharov, N.P., Artificial periodic irregularities from the Tromso heating facility, *Advances in Space Research*, Vol. 21, 1998, pp. 693–696.
17. Ureadov, V., Ivanov, V., Plohotniuc, E., Eruhimov, L., Blaunstein, N., and Filip, N., *Dynamic Processes in the Ionosphere–Methods of Investigations*, Iasi (Romania): TechnoPress, 2006.
18. Belikovich, V.V., Benediktov, E.A., Tolmacheva, A.V., and Bakhmet'eva, N.V., *Ionospheric Research by Means of Artificial Periodic Irregularities*, Berlin: Copernicus GmbH, 2002.
19. Fopple, H. et al., Preliminary experiments for sudy of interplanetary medium, *Planet Space Sci.*, Vol. 13, 1965, pp. 95–114.
20. Fopple, H. et al., Artificial strontium and barium clouds in the upper atmosphere, *Planet Space Sci.*, Vol. 15, 1967, pp. 357–372.
21. Haerendel, G., Lust, R., and Rieger, E., Motion of artificial ion clouds in the upper atmosphere, *Planet Space Sci.*, Vol. 15, 1967, pp. 1–18.
22. Wescott, E.M., Stolarik, J.D., and Heppner, J.P., Electric fields in the vicinity of auroral forms from motion of barium vapor release, *J. Geophys. Res.*, Vol. 74, 1969, pp. 3469–3487.

23. Rieger, E., Neuss, H., and Lust, R. et al., High altitude release of barium vapors using Rubis rocket, *Ann. Geophys.*, Vol. 26, 1970, pp. 845–852.

24. Rosenberg, N.W., Observations of striation formation in barium ion clouds, *J. Geophys. Res.*, Vol. 76, 1971, pp. 6856–6864.

25. Lust, R., Space experiments with barium clouds, *New Scientist.*, Vol. 20, 1972, pp. 154–169.

26. Perkins, F.W., Zabusky, N.J., and Doles, J.H., Deformation and striation of plasma clouds in the ionosphere. 1, *J. Geophys Res.*, Vol. 78, 1973, pp. 697–709.

27. Lloyd, K.H. and Haerendel, G., Numerical modeling of drift and deformation of ionospheric plasma clouds and their interaction with other layers of the ionosphere, *J. Geophys. Res.*, Vol. 78, 1973, pp. 7389–7415.

28. Zabusky, N.J., Doles, J.H., and Perkins, F.W., Deformation and striation of plasma clouds in the ionosphere, 2, Numerical simulation of a non-linear two-dimensional model, *J. Geophys. Res.*, Vol. 78, 1973, pp. 711–724.

29. Baxter, A.J., Lower F-region barium release experiments at sub-auroral location, *Planet Space Sci.*, Vol. 23, 1975, pp. 973–983.

30. Glova, I.S. and Ruzhin, Yu.Ya., Cumulative injection of barium at altitude of 170 km. Experiment Spolokh-1, in book: *Dynamics of Cosmic Plasma*, Moscow: Nauka, 1976, pp. 154–174.

31. Antoshkin, V.S. et al., About evolution of barium clouds of high density, *Geomagn. Aeronom.*, Vol. 19, 1979, pp. 1049–1056.

32. Ruzhin, Yu.Ya. and Skomarovsky, V.S., About results of of complex rocket experiment Spolokh-2, in book: *Physical Processes in the Ionosphere and Magnetosphere*, Moscow: Nauka, 1979, pp. 35–45.

33. Dzubenko, N.L., Zhilinsky, A.P., Zhulin, A., Ivchenko, I.S. et al., Dynamics of artificial plasma clouds in Spolokh experiments: Movement pattern, *Planet Space Sci.*, Vol. 31, 1983, pp. 849–858.

34. Andreeva, L.A. et al., Dynamics of artificial plasma cloud in Spolokh experiments: Clouds deformation, *Planet Space Sci.*, Vol. 32, 1984, pp. 1045–1052.

35. Filipp, N.D., Oraevskii, V.N., Blaunshtein, N.Sh., and Ruzhin, Yu.Ya., *Evolution of Artificial Plasma Irregularities in the Earth's Ionosphere*, Kishinev (Moldova): Shtiintza, 1986.

36. Milinevsky, G.P., Romanovsky, Yu.A., Evtushevsky, A.M., Savchenko, V.A. et al., Optical observations in active space experiments in investigation of the Earth's atmosphere and ionosphere, *Cosmic. Res.*, Vol. 28, 1990, pp. 418–429.

37. Blaunstein, N.Sh., Milinevsky, G.P., Savchenko, V.A., and Mishin, E.V., Formation and development of striated structure during plasma cloud evolution in the Earth's ionosphere, *Planet Space Sci.*, Vol. 41, 1993, pp. 453–460.

38. Frihagen, J. and Jacobson, T., *In-situ* observations of high-latitude F-region irregularities, *J. Atmos. Terr. Phys.*, Vol. 33, 1971, pp. 519–522.

39. Hansucher, R.D., Chatanica radar investigation of high-latitude E-region ionization, *Radio Sci.*, Vol. 10, 1975, pp. 277–288.

40. Muldrew, D.B. and Vickrey, J.F., High-latitude F-region irregularities observed simultaneously with ISIS-1 and Chatanica radar, *J. Geophys. Res.*, Vol. 87, 1982, pp. 8263–8267.

41. Fejer, B.G. and Providakes, J.F., High-Latitude E-region irregularities: New results, *Physica Scripta*, Vol. T18, 1987, pp. 167–178.

42. Gelberg, M.G., *Irregularities of the High-Latitude Ionosphere*, Novosibirsk: Science, 1986.

43. Haldoupis, C.A., A review of radio studies of auroral E region ionospheric irregularities, *Ann. Geophys.*, Vol. 7, 1989, pp. 239–258.

44. Haldoupis, C.A., Sofko, G.J., Koehler, J.A., and Danskin, D.W., A new look at type 4 echoes of radar aurora, *J. Geophys. Res.*, Vol. 96, 1991, pp. 11,353–11,362.

45. Koehler, J.A., Sofko, G.J., and Mu, J., Type-III echoes of radar auroral observational aspects, *Ann. Geophys.*, Vol. 13, 1995, pp. 2–9.

46. Haldoupis, C.A., Sofko, G.J., Hussey, G.C., and Mu, J., An overview of type-3 radar auroral research: Basic observational properties and new interpretation propositions, *Ann. Geophys.*, Vol. 13, 1995, pp. 10–24.

47. Haldoupis, C.A., Kochler, J.A., and Sofko, G.J. et al., Localized and strongly unstable plasma regions in the auroral E region ionosphere and implification for radar experiment, *J. Geophys. Res.*, Vol. 100, 1995, pp. 7771–7782.

48. Filipp, N.D., Power of H_E-scatter signals, *Izv. Vuzov, Radiophyzika*, Vol. 22, No. 4, 1979, pp. 407–411.

49. Filipp, N.D., *Specular Scattering of Ultra-Short Waves by Middle-Latitude Ionosphere*, Kishinev (Moldova): Shtiintza, 1980.

50. Filipp, N.D. and Blaunstein, N., Effect of the geomagnetic field on the diffusion of ionospheric inhomogeneities, *Geomagn. Aeronom.*, Vol. 18, 1978, pp. 423–427.

51. Filipp, N.D. and Blaunstein, N.Sh., Drift of ionospheric inhomogeneities in the presence of the geomagnetic field, *Izv. Vuzov, Radiophyzika*, Vol. 21, 1978, pp. 1409–1417.

52. Ierkic, H.M., Fejer, B.G., and Farley, D.T., The dependence of zenith angle of the strength of 3-meter equatorial electrojet irregularities, *Geophys. Res. Lett.*, Vol. 7, 1980, pp. 497–500.

53. Yamamoto, M.S., Fukao, S., Woodman, R.F., Tsuda, T., and Kato, S., Mid latitude E-region field-aligned irregularities observed with the MU radar, *J. Geophys. Res.*, Vol. 96, 1991, pp. 15943–15949.

54. Balsley, B.B. and Farley, D.T., Radar observations of two-dimensional turbulence in the equatorial electrojet, *J. Geophys. Res.*, Vol. 78, 1973, pp. 7471–7478.

55. Kudeki, E., Fejer, B.G., Farley, D.T., and Hanuise, C., The Condor equatorial electrojet compaign: Radar results, *J. Geophys. Res.*, Vol. 92, 1987, pp. 13,561–13,577.

56. Ossakow, S.L. and Chaturvedi, P.K., Morphological studies of rising equatorial spread-F bubbles, *J. Geophys. Res.*, Vol. 83, 1978, pp. 2085–2091.

57. Haerendel, G., Coloured bubbles: an experiment for triggering equatorial spread-F, *ESA-SP*-195, July 1983, pp. 295–298.

58. Maruyama, T. and Matuura, N., Longitudinal variability of annual changes in activity of equatorial spread F and plasma bubbles, *J. Geophys. Res.*, Vol. 89, 1984, pp. 10,903–10,912.

59. Aggson, T.L., Maynard, N.C., Hanson, W.B., and Saba, J.L., Electric field observations of equatorial bubbles, *J. Geophys. Res.*, Vol. 97, 1992, pp. 2997–3009.

60. Laasko, H., Aggson, T.L., Herrero, F.A., Pfaff, R.F., and Hanson, S.B., Vertical neutral wind in the equatorial F-region deduced from electric field and ion density measurements, *J. Atmos. Terr. Phys.*, Vol. 57, 1995, pp. 645–651.

61. Aggson, T.L., Laasko, H., Maynard, N.C., and Pfaff, R.F., In-situ observations of bifurcation of equatorial ionospheric depletions, *J. Geophys. Res.*, Vol. 101, 1996, pp. 5125–5132.

62. Vilenskii, I.M., Izraileva, N.P., Plotkin, V.V., and Freiman, M.E., *Artificial Quasiperiodic Inhomogeneities in the Lower Ionosphere*, Novosibirsk: Science, 1987.

63. Belikovich, V.V., Benediktov, E.A., and Itkina M.A. et al., Scattering of radio waves by periodic artificial ionospheric irregulatities, *Radiophys. Quantum Electron.*, Vol. 20, 1978, pp. 1250–1253.

64. Belikovich, V.V., Benediktov, E.A., Getmantsev, G.G., Ignat'ev, Y.A., and Komrakov, G.P., Scattering of radio waves from the artificially perturbed F region of the ionosphere, *JETF Lett.*, Vol. 22, 1975, pp. 243–244.

65. Belikovich, V.V., Benediktov, E.A., Goncharov, N.P., and Tolmacheva, A.V., Diagnostics of the ionosphere and neutral atmosphere at E-region heights using artificial periodic inhomogeneities, *J. Atmos. Solar-Terr. Phys.*, Vol. 59, 1997, pp. 2447–2460.

66. Budilin, A.V., Getmantsev, G.G., and Kapustin P.A. et al., Localization of heights of nonlinear currents responsible for the low-frequency radiation in the ionosphere, *Izv. Vuz., Radiophyzika*, Vol. 20, No. 1, 1977, pp. 83–89.

67. Belicovich, V.V., Benedictov, E.A., and Terina, G.I., Diagnostics of the lower ionosphere by method of resonance scattering of radio waves, *Sov. Phys. JETP*, Vol. 48, No. 11, 1986, pp. 1247–1253.

68. Belicovich, V.V., Benedictov, E.A., and Getmantsev, G.G. et al., Ionospheric electron density measurement using radio wave scattering from artificial plasma inhomogeneities, *Radiophys. Quantum Electron.*, Vol. 21, 1979, pp. 853–854.

69. Belicovich, V.V., Benedictov, E.A., Gol'tsova, Y.K., Komrakov, G.P., and Tolmacheva, A.V., Determination of ionospheric parameters in the F-region by means of resonance scattering, *Izv. Vuz., Radiophyzika*, Vol. 29, 1986, pp. 131–138.

70. Belicovich, V.V., Benedictov, E.A., and Gol'cova, Yu.C. et al., Definition of ionospheric parameters in the F-region by method of resonance scattering, *Izv. Vuz., Radiophyzika*, Vol. 29, 1986, pp. 131–138.

71. Mityakov, N.A., Editor, *About Possible Intensification Mechanisms of Periodic Plasma Grid Meshes in the Ionosphere*, Moscow: Science, 1978.

72. Vas'kov, V.V., Gurevich, A.V., and Karashtin, A.N., Self-focusing instability under the oblique incidence of radio waves on the ionosphere, *Geomagn. Aeronom.*, Vol. 16, 1976, pp. 322–325.

73. Vas'kov, V.V. and Gurevich, A.V., Resonance instability of small-scale plasma perturbations, *Sov. Phys. JETF*, Vol. 46, 1977, pp. 903–912.

74. Vas'kov, V.V. and Gurevich, A.V., Nonlinear resonance instability of plasma in the reflection region of ordinary electromagnetic wave, *Sov. Phys. JETF*, Vol. 42, 1975, pp. 91–103.

75. Vas'kov, V.V. and Gurevich, A.V., Saturation of self-focusing instability for radio wave beams in plasma, *Sov. J. Plasma Phys.*, Vol. 3, 1977, pp. 185–194.

76. Bochkarev, G.S., Eremenko, V.A., and Lobachevskii, L.A. et al., Nonlinear interaction of decametre radio waves at close frequencies in oblique propagation, *J. Atmos. Terr. Phys.*, Vol. 44, No. 12, 1982, pp. 1137–1141.

77. Bochkarev, G.S., Eremenko, V.A., and Lobachevskii, L.A. et al., Nonlinear interaction of radiowaves in oblique propagation, in: *Effects of Artificial Modification of the Earth's Ionosphere, Moscow: Science*, 1983, pp. 53–71.

78. Erukhimov, L.M., Metelev, S.A., Mytyakov, N.A., Myasnikov, E.N., and Frolov, V.L., The artificial ionospheric turbulence, *Izv. Vuz., Radiophyzika*, Vol. 31, 1987, pp. 208–225.

79. Erukhimov, L.M., Mityakova, A.E., and Myasnikov, E.N. et al., About spectrum of artificial ionospheric irregularities at the different altitudes, *Izv. Vuz., Radiophyzika*, Vol. 20, 1977, pp. 1814–1818.

80. Erukhimov, L.M., Komrakov, G.I., and Frolov, V.L., About spectrum of smale-scale part of artificial ionospheric turbulence, *Geomagn. Aeronom.*, Vol. 20, 1980, pp. 1112–1114.

81. Erukhimov, L.M., Korovin, A.V., Mityakov, N.A., Mityakova, E.E., and Nasyrov, A.M., About diffusion of small-scale artificial inhomogeneities in the upper ionosphere, *Izv. Vuz., Radiophyzika*, Vol. 25, 1983, pp. 360–362.

82. Belicovich, V.A., Erukhimov, L.M., and Zuzin, V.A. et al., About times of evolution and relaxation of artificial small-scale irregularities, *Izv. Vuz., Radiophyzika*, Vol. 31, 1988, pp. 251–256.

83. Erukhimov, L.M., Metelev, S.A., and Razumov, D.V., Diagnostics of ionospheric irregularities due to artificial radio radiation, *Izv. Vuz., Radiophyzika*, Vol. 31, 1988, pp. 1301–1308.

84. Ginsburg, V.L., *Propafation of Electromagnetic Waves in Plasma*, Moscow: Nauka, 1964.

85. Alpert, Ya.L., *Propagation of Electromagnetic Waves in the Ionosphere*, Moscow: Nauka, 1967.

86. Gurevich, A.V. and Shvartzburg, A.B., *Nonlinear Theory of Radio Wave Propagation in the Ionosphere*, Moscow: Nauka, 1973.

87. Gurevich, A.V., *Nonlinear Phenomena in the Ionosphere*, New York: Springer-Verlag, 1978.

88. Gershman, B.N., Erukhimov, L.M., and Yashin, Yu.Ya., *Wave Phenomena in the Ionosphere and Cosmic Plasma*, Moscow: Nauka, 1984.

89. Likhter, Ya.L., Gul'el'mi, A.V., Erukhimov, L.M., and Mikha'lova, M. *Wave Diagnostics of the Near-the-Earth Plasma*, Moscow: Nauka, 1989.

90. Babichenko, A.M., Mityakov, S.N., and Mironenko, L.F., Cross-modulation studies of dynamic structure of the artificial perturbances in the D-layer, in book: *Modification of the Ionosphere by Powerful Radio Radiation*, Moscow: Nauka, 1986, pp. 124–129.

91. Beliaev, P.P., Kotik, D.S., Mityakov S.N. et al., Generation of electromagnetic signals of combined frequencies, *Izv. Vuz., Radiophyzika*, Vol. 31, 1987, pp. 248–267.

92. Mityakov, N.A., Excited temperature scattering of electromagnetic waves in ionospheric plasma, *Izv. Vuz., Radiophyzika*, Vol. 24, No. 6, 1981, pp. 671–674.

93. Blioh, G.B. and Bruhovetsky, A.S., Focusing of radio waves by artificially created ionospheric lense, *Geomagn. Aeronom.*, Vol. 9, 1968, pp. 545–549.

94. Bochkarev, G.S., Pahotin, V.A., and Rudika, L.V. Propagation of HF-signals through the artificially excited region of the ionosphere, in book: *Diffractional Effects of Short Waves*, Moscow: Nauka, 1981, pp. 38–44.

95. Bahmet'eva, N.V., Benedictov, E.A., and Bochkarev, G.S. et al., Some effects of influence of artificial excitation of the ionosphere on the short waves propagation, in book: *Effects of Artificial Heating by Powerful Radio Radiation of the Earth's Ionosphere*, Moscow: Nauka, 1983, pp. 45–48.

96. Bahmet'eva, N.V., Benedictov, E.A., and Ivanuk, V.A. et al., Influence of the artificial ionospheric turbulence on the statistical characteristics of amplitude of reflected radio waves from the ionosphere, *Effects of Artificial Heating by Powerful Radio Radiation of the Earth's Ionosphere*, Moscow: Nauka, 1983, pp. 71–79.

97. Gurevich, A.V., Milih, G.M., and Shluger, I.S., Rotational amplification of the crossing modulation of radio waves in the ionosphere, *Geomagn. Aeronom.*, Vol. 1, 1977, pp. 79–82.

98. Erukhimov, L.M., Kovalev, V.Ya., and Kurakin, E.P., Results of first experiments for excitation of artificial ionospheric turbulence with help of heating facilities Gissar, *Izv. Vuz., Radiophyzika*, Vol. 28, 1985, pp. 658–662.

99. Ritov, S.M., Kravtsov, Yu.A., and Tatarsky, V.I., *Introduction in the Statistical Radiophysics*, Moscow: Nauka, 1978, Part 2.

100. Grach, S.M., About electromagnetic radiation of artificial plasma turbulence in the ionosphere, *Izv. Vuz., Radiophyzika*, Vol. 28, 1985, Vol. 684–693.

101. Providakes, J.E., Farley, D.T., and Fejer, B.G. et al., Observations of auroral E-region plasma waves and electron heating with EISCAT and VHF radar interferometer, *J. Atmos. Terr. Phys.*, Vol. 50, 1988, pp. 339–356.

102. Rietveld, M.T., Collins, P.N., and St-Maurice, J.-P., Naturally enhanced ion acoustic waves in the auroral ionosphere observed with the EISCAT 933 MHz radar, *J. Geophys. Res.*, Vol. 96, 1991, pp. 19,291–19,305.

103. Rietveld, M., Kohl, H., Kopka, H., and Stubbe, P., Introduction to ionosphere heating at Tromso, 1, Experimental overview, *J. Atmos. Terr. Phys.*, Vol. 55, 1993, pp. 577–599.

104. Blaunstein, N.Sh. and Tsedilina, E.E., Effect of the initial dimensions on the nature of the diffusion spreading of inhomogeneities in a quasiuniform ionosphere, *Geomagn. Aeronom.*, Vol. 25, 1985, pp. 39–44.

105. Blaunstein, N., Diffusion spreading of middle-latitude ionospheric plasma irregularities, *J. Ann. Geophys.*, Vol. 13, 1995, pp. 617–626.

106. Blaunstein, N., Evolution of a stratified plasma structure induced by local heating of the ionosphere, *J. Atmosph. Solar-Terr. Phys.*, Vol. 59, No. 3, 1997, pp. 351–361.

107. Volk, H.J. and Haerendel, G., Striation in ionosppheric plasma clouds, *J. Geophys. Res.*, Vol. 76, 1971, pp. 4541–4559.

108. Linson, L.M. and Workman, R., Formation of striations in the ionospheric plasma clouds, *J. Geophys. Res.*, Vol. 75, 1970, pp. 3211–3218.

109. Scannapieco, J. et al., Conductivity ratio effect on the drift and deformation of ionospheric plasma clouds and their interaction with other layers of the ionosphere, *J. Geophys. Res.*, Vol. 79, 1974, pp. 2913–2916.

110. Rieger, K., Measurements of electric fields in equatorial and medium altitudes during twilight using barium-ion clouds, *Gerlands Bietr. Geophys.*, Vol. 213, 1980, pp. 243–253.

111. Schultz, S.G., Adams, J., and Mozer, F.S., Probe electric field measurements near a mid-latitude ionospheric barium releases, *J. Geophys. Res.*, Vol. 78, 1973, pp. 6634–6642.

112. Simon, A., Growth and stability of artificial ion clouds in the ionosphere, *J. Geophys. Res.*, Vol. 75, 1970, pp. 6287–6294.

113. Gatsonis, N.A. and Hastings, D.E., A three-dimensional model and initial time numerical simulation for an artificial plasma cloud in the ionosphere, *J. Geophys. Res.*, Vol. 96, 1991, pp. 7623–7639.

114. Doles, J.H., Zabusky, N.J., and Perkins, F.W., Deformation and striation of plasma clouds in the ionosphere, 3, Numerical simulation of multilevel model recombination chemistry, *J. Geophys. Res.*, Vol. 81, 1976, pp. 5987–6004.

115. Perkins, F.W. and Doles, J.H., Velocity shear and the $\mathbf{E} \times \mathbf{B}$ instability, *J. Geophys. Res.*, Vol. 80, 1975, pp. 211–214.

116. Drake, J.F. and Huba, J.D., Dynamics of three dimensional ionospheric plasma clouds, *Phys Rev. Lett.*, Vol. 58, No. 30, 1987, pp. 278–281.

117. Mitchell, H.G., Fedder, J.A., Huba, J.D., and Zalesak, S.T., Transverse motion of high-speed barium clouds in the ionosphere, *J. Geophys. Res.*, Vol. 90, 1985, pp. 11,091–11,103.

118. Huba, J.D. et al., Simulation of plasma structure evolution in the high-latitude ionosphere, *Radio Sci.*, Vol. 23, No. 4, 1988, pp. 503–510.

119. Drake, J.F., Mubbrandon, M., and Huba, J.D., Three-dimensional equilibrium and stability of ionospheric plasma clouds, *Phys. Fluids*, Vol. 31, No. 11, 1988, pp. 3412–3424.

120. Zalesak, S.T., Drake, J.F., and Huba, J.D., Dynamics of three-dimensional ionospheric plasma clouds, *Radio Sci.*, Vol. 23, No. 4, 1988, pp. 591–598.

121. Zalesak, S.T., Drake, J.F., and Huba, J.D., Three-dimensional simulation study of ionospheric plasma clouds, *Geophys. Res. Lett.*, Vol. 17, No. 10, 1990, pp. 1597–1600.

122. Pudovkin, M.I., Golovchanskaya, I.V., Klimov, M.S. et al., About employment of two-layer model for the description of three-dimensional evolution of the strong ionospheric inhomogeneity, *Geomagn. Aeronom.*, Vol. 37, No. 4, 1987, pp. 560–567.

123. Rozhansky, V.A. and Tsendin, L.D., Evolution of strong inhomogeneities of the ionospheric plasma, *Geomagn. Aeronom.*, Vol. 27, 1984, pp. 598–602.

124. Rozhansky, V.A. and Tsendin, L.D., Influence of electric field and wind on spreading of the disturbance of weakly-ionized plasma in the magnetic field, *Geomagn. Research*, Vol. 27, 1980, pp. 93–104.

125. Rozhansky, V.A. and Tsendin, L.D., *Transport Phenomena in Partially Ionized Plasma*, Philadelphia: Taylor and Francis, 2001.

126. Voskoboynikov, S.P., Rozhansky, V.A., and Tsendin, L.D., Numerical modelling of three-dimensional plasma cloud evolution in crossed $\mathbf{E} \times \mathbf{B}$ fields, *Planet Space Sci.*, Vol. 35, 1987, pp. 835–844.

127. Rozhansky, V.A., Veselova, I.Y., and Voskoboynikov, S.P., Three-dimensional computer simulation of plasma cloud evolution in the ionosphere, *Planet Space Sci.*, Vol. 38, No. 11, 1990, pp. 1375–1386.

128. Blaunstein, N.Sh., Tsedilina, E.E., Mirzaeva, L.I., and Mishin, E.V., Drift spreading and stratification of inhomogeneities in the ionosphere in the presence of an electric field, *Geomagn. Aeronom.*, Vol. 30, 1990, pp. 799–805.

129. Blaunstein, N., The character of drift spreading of artificial plasma clouds in the middle-latitude ionosphere, *J. Geophys. Res.*, Vol. 101, No. A2, 1996, pp. 2321–2331.

130. Kaiser, T.R., Pickering, W.H., and Watkins, C.D., Ambipolar diffusion and motion of ion clouds in the Earth's magnetic field, *Planet Space Sci.*, Vol. 17, 1969, pp. 519–551.

131. Rozhansky, V.A. and Tsendin, L.D., Spreading of the cord of weakly ionized plasma placed under angle to the magnetic field, *Geomagn. Aeronom.*, Vol. 17, 1977, pp. 1002–1007.

132. Polyakov, V.M., About diffusion of charged particles in the F-region of the middle-latitude ionosphere, *Geomagn. Aeronom.*, Vol. 6, 1966, pp. 341–351.

133. Gurevich, A.V. and Tsedilina, E.E., Motion and diffusion of plasma inhomogeneities, *Sov. Phys. Usp.*, Vol. 91, 1967, pp. 609–642.

134. Kasheev, B.L., Lebedenetz, V.N., and Lagutin, M.F., *Meteor Phenomena in the Earth's Atomosphere*, Moscow: Nauka, 1967.

135. Bondar, B.G. and Kasheev, B.L., *Meteor Communication*, Kiev: Tekhnica, 1968.

136. Fialko, E.I., *Some Problems in Radiolocation of Meteors*, Tomsk, USSR: Tomsk University Press, 1971.

137. Heritage, J.L., Wesbrod, S., and Fay, W.J., Evidence for a 200-Megacycles per second ionospheric forward scatter mode associated with the Earth's magnetic field, *J. Geophys. Res.*, Vol. 64, 1959, pp. 1235–1244.

139. Heritage, J.L., Fay, W.J., and Bowen, E.D., Evidence that meteor trails produce a field-aligned scatter signals at VHF, *J. Geophys. Res.*, Vol. 67, 1962, pp. 953–961.

140. Kuriki, I., Some observations of the VHF pulse wave scattered from the ionosphere, *J. Radio Res. Labs* (Japan), Vol. 10, 1963, pp. 69–81.

141. Kuriki, I., Measurements of direction of scattering points on VHF anomalous scattered propagation, *J. Radio Res. Labs* (Japan), Vol. 12, 1965, pp. 1–14.

142. Kuriki, I., On the scattering propagation caused by the field aligned irregularities, *J. Radio Res. Labs* (Japan), Vol. 13, 1966, pp. 57–74.

143. Kuriki, I., Further experimental study of scattering propagation caused by the field aligned irregularities in VHF band, *J. Radio Res. Labs* (Japan), Vol. 14, 1967, pp. 57–77.

144. Wotkins, C.D., James, R., and Nicholson, T.F., Further studies of the effect of the Earth's magnetic field on meteor trails, *J. Atmos. Terr. Phys.* Vol. 33, 1971, pp. 1907–1916.

145. Filipp, N.D. and Gleybman, Ye.Ya., One of the reasons for radio signal fadings upon scattering by magnetically oriented inhomogeneities of the lower ionosphere, *Geomagn. Aeronom.*, Vol. 19, 1979, pp. 302–306.

146. Blaunshtein, N.Sh. and Filipp, N.D., Drift of plasma with arbitrary degree of ionization in the magnetic field (ionospheric case), in book: *Practical Aspects in Study of the Ionosphere and Ionospheric Radio Propagation*, Moscow: Nauka, 1981, pp. 163–169.

147. Filipp, N.D., Comparison of the characteristics of ionospheric irregularities of type H_E at middle latitudes and of type E_{sq} in the polar and equatorial zones, *Geomagn. Aeronom.*, Vol. 21, 1981, pp. 466–469.

148. Bowman, G.G. and Hajkowicz, L.A., Small-scale ionospheric structures associated with mid-latitude spread-*F*, *J. Atmos. Terr. Phys.*, Vol. 53, 1991, pp. 4447–4457.

149. Blaunstein, N.Sh., Diffusion of plasma with arbitrary degree of ionization in the magnetic field (ionospheric case), in book: *Problems of the Cosmic Electrodynamics*, Moscow: Nauka, 1981, pp. 96–119.

150. Blaunstein, N.Sh., Diffusion spreading of inhomogeneities of the middle-latitude ionospheric plasma, in *Investigations of Geomagnetism, Aeronomy and Physics of Sun*, Moscow: Nauka, Vol. 82, 1988, pp. 90–103.

151. Blaunstein, N., Diffusion spreading of middle-latitude ionospheric plasma irregularities, *J. Ann. Geophys.*, Vol. 13, 1995, pp. 617–626.

152. Rybtsov, L.N. and Solovey, B.G., Vertical wind distribution in the meteor region, *Geomagn. Aeronom.*, Vol. 13, 1973, pp. 804–806.

153. Kashcheyev, B.L., Verba, N.D., Kal'chenko, B.L., and Nechitaylenko, V.A., Characteristics of the method of determining the vector of an atmospheric wind of arbitrary orientation in the meteor zone, *Geomagn. Aeronom.*, Vol. 11, 1971, pp. 146–148.

154. Zhalkovskaya, L.V., Some results of a study of inhomogeneities in the lower ionosphere, *Geomagn. Aeronom.*, Vol. 10, 1970, pp. 276–277.

155. Tanaka, T. and Venkateswaran, S.V., Characteristics of field-aligned *E*-region irregularities over Iioka (36°N), Japan–I, *J. Atmos. Terr. Phys.*, Vol. 44, 1982, pp. 381–393.

156. Tanaka, T. and Venkateswaran, S.V., Characteristics of field-aligned *E*-region irregularities over Iioka (36°N), Japan–II, *J. Atmos. Terr. Phys.*, Vol. 44, 1982, pp. 395–406.

157. Kato, S., Ogawa, T., and Tsuda, T. et al., The middle and upper atmosphere radar: First results using partial system, *Radio Sci.*, Vol. 19, 1984, pp. 1475–1484.

158. Yamamoto, M., Fukao, S., and Woodman, R.F. et al., Mid-latitude *E* region field-aligned irregularities observed with MU radar, *J. Geophys. Res.*, Vol. 96, 1991, 15,943–15,949.

159. Yamamoto, M., Fukao, S., Ogawa, T., Tsuda, T., and Kato, S., A morphological study on mid-latitude *E*-region field-aligned irregularities observed with the MU radar, *J. Atmos. Terr. Phys.*, Vol. 54, 1992, pp. 769–778.

160. Yamamoto, M., Komoda, N., and Fukao, S. et al., Spatial structure of the *E* region field-aligned irregularities revealed by the MU radar, *Radio Sci.*, Vol. 29, 1994, pp. 337–347.

161. Tsunoda, R.T., Fukao, S., and Yamamoto, M., On the origin of the quasi-periodic radar backscatter from midlatitude sporadic *E*, *Radio Sci.*, Vol. 29, 1994, pp. 349–365.

162. Riggin, D., Swartz, W.E., Providakes, J., and Farley, D.T., Radar studies of long-wavelength associated with mid-latitude sporadic *E* layers, *J. Geophys. Res.*, Vol. 91, 1986, pp. 8011–8024.

163. Haldoupis, C. and Schlegel, A 50-MHz radio Doppler experiment for midlatitude E region coherent backscatter studies: System description and first results, *Radio Sci.*, Vol. 28, 1993, pp. 959–978.

164. Haldoupis, C., Schlegel, K., and Farley, D.T., An explanation for type 1 radar echoes from the midlatitude E-region ionosphere, *Geophys. Res. Lett.*, Vol. 23, 1996, pp. 97–100.

165. Chapman, S., The equatorial electrojet as detected from the abnormal electric current distribution above Huancayo, Peru, and elsewere, *Arch. Meteorol., Geophys., und Bioklimatol.*, *A*, Vol. 4, 1951, pp. 368–390.

166. Knecht, R.W., An additional linear influence on equatorial E_s at Huancayo, *J. Atmos. Terr. Phys.*, Vol. 14, 1959, pp. 348–349.

167. Gates, D.M., Preliminary results of the National Bureau of Standards radio and ionospheric observations during IGY, *J. Research NBS*, Vol. 63D, 1959, pp. 1–14.

168. Egan, R.D., Anisotropic field-aligned ionization irregularities in the ionosphere near the magnetic equator, *J. Geophys. Res.*, Vol. 65, 1960, pp. 2343–2352.

169. Peterson, A.M., Egan, R.D., and Pratt, D.S., The IGY three-frequency backscatter sounder, *Proc. IRE*, Vol. 47, 1959, pp. 300–315.

170. Balsley, B.B., Some characteristics of non-two-stream irregularities in the equatorial electrojet, *J. Geophys. Res.*, Vol. 74, 1969, pp. 2333–2347.

171. Balsley, B.B. and Farley, D.T., Radar studies of the equatorial electrojet at three frequencies, *J. Geophys. Res.*, Vol. 76, 1971, pp. 8341–8351.

172. Farley, D.T. and Balsley, B.B., Instabilities in the equatorial electrojet, *J. Geophys. Res.*, Vol. 78, 1973, pp. 227–239.

173. Farley, D.T., Fejer, B.G., and Balsley, B.B., Radar observations of two-dimensional turbulence in the equatorial electrojet, 3, Night-time observations of type 1 waves, *J. Geophys. Res.*, Vol. 83, 1978, pp. 5625–5634.

174. Fejer, B.G., Farley, D.T., Balsley, B.B., and Woodman, R.F., Vertical structure of the VHF backscattering region in the equatorial electrojet and the gradient drift instability, *J. Geophys. Res.*, Vol. 80, 1975, pp. 1313–1321.

175. Kudeki, E., Farley, D.T., and Fejer, B.G., Long wavelength irregularities in the equatorial electrojet, *Geophys. Res. Lett.*, Vol. 9, 1982, pp. 684–687.

176. Kudeki, E.D. and Fawcett, C.D., High resolution observations of 150 km echoes at Jicamarca, *Geophys. Res. Lett.*, Vol. 20, 1993, pp. 1987–1990.

177. Pfaff, R.F., Kelley, M.C., Kudeki, E., Fejer, B.G., and Baker, K.D., Electric field and plasma density measurements in the strongly driven daytime equatorial electrojet, 1, The unstable layer and gradient drift waves, *J. Geophys. Res.*, Vol. 92, 1987, pp. 13,578–13,596.

178. Pfaff, R.F., Kelley, M.C., Kudeki, E., Fejer, B.G., and Baker, K.D., Electric field and plasma density measurements in the strongly driven daytime equatorial electrojet, 2, Two-stream waves, *J. Geophys. Res.*, Vol. 92, 1987, pp. 15,597–15,612.

179. Pfaff, R.F., Rocket observations of the equatorial electrojet: current status and critical problems, *J. Atmos. Terr. Phys.*, Vol. 53, 1991, pp. 709–728.

180. Ecklund, W.L., Carter, D.A., and Balsley, B.B., Gradient drift irregularities in mid-latitude sporadic *E*, *J. Geophys. Res.*, Vol. 86, 1981, pp. 858–862.

181. Tsunoda, R.T., Enhanced velocities and a shear in daytime E_{sq} over Kwajalein and their relationship to 150 km echoes over the dip equator, *Geophys. Res. Lett.*, Vol. 21, 1994, pp. 2741–2744.

182. Swartz, W.W. and Farley, D.T., High-resolution radar measurements of turbulent structure in the equatorial electrojet, *J. Geophys. Res.*, Vol. 99, 1994, pp. 309–317.

183. Hysell, D.L. and Chau, J.L., Inferring E region electron density profiles at Jicamarca from Faraday rotation of coherent scatter, *J. Geophys. Res.*, Vol. 106, 2001, pp. 30,371–30,380.

184. Chau, J.L. and Woodman, R.F., D and E region incoherent scatter radar density measurements over Jicamarca, *J. Geophys. Res.*, Vol. 110 (A12), 2005, A12314 (pp. 1–7).

185. Reinisch, B.W., New techniques in ground-based ionospheric sounding and studies, *Radio Sci.*, Vol. 21, 1986, pp. 331–341.

186. Fredrich, M. and Torkar, K.M., FIRI: A semiempirical model of the lower ionosphere, *J. Geophys. Res.*, Vol. 106, 2001, pp. 21,409–21,418.

187. Danilov, A.D., New ideas on the D-region modeling, *Adv. Space Res.*, Vol. 25, 2000, pp. 5–14.

188. Rottger, J., Wave-like structures of large-scale equatorial spread-F irregularities, *J. Atmos. Terr. Phys.*, Vol. 35, 1973, pp. 1195–1206.

189. Kelleher, R.F. and Rottger, J., Equatorial spread-F irregularities observed at Nairobi and on transequatorial path Lindau–Tsumeb, *J. Atmos. Terr. Phys.*, Vol. 35, 1973, pp. 1207–1211.

190. Kelleher, R.F. and Skinner, N.J., Studies of F region irregularities at Nairobi, 2. By direct backscatter at 27.8 MHz, *Ann. Geophys.*, Vol. 2, 1971, pp. 195–200.

191. McClure, J.P. and Hanson, W.B., A catalog of ionospheric F region irregularity behavior based on Ogo 6 retarding potential analyse data, *J. Geophys. Res.*, Vol. 78. 1973, pp. 7431–7440.

192. Dyson, P.L., McClure, J.P., and Hanson, W.B., In situ measurements of the spectral characteristics of F region ionospheric irregularities, *J. Geophys. Res.*, Vol. 79, 1974, pp. 1497–1502.

193. Farley, D.T., Balsley, B.B., Woodman, R.F., and McClure, J.P., Equatorial spread F: Implications of VHF radar observations, *J. Geophys. Res.*, Vol. 75, 1970, pp. 7199–7216.

194. McClure, J.P. and Woodman, R.F., Radar observations of equatorial spread F in a region of electrostatic turbulence, *J. Geophys. Res.*, Vol. 77, 1972, pp. 5617–5621.

195. Woodman, R.F. and LaHoz, C., Radar observations of F region equatorial irregularities, *J. Geophys. Res.*, Vol. 81, 1976, pp. 5447–5466.

196. Szuszczewicz, E.P., Tsunoda, R.T., Narcisi, R.S., and Holms, J.C., Coincident radar and rocket observations of equatorial spread F, *Geophys. Res. Lett.*, Vol. 7, 1980, pp. 537–541.

197. Szuszczewicz, E.P., Tsunoda, R.T., Narcisi, R.S., and Holms, J.C., PLIMEX II: A second set of coincident radar and rocket observations of equatorial spread F, *Geophys. Res. Lett.*, Vol. 8, 1981, pp. 803–807.

198. Kelley, M.C., Pfaff, R., and Baker, K.D. et al., Simulteneous rocket probe and radar measurements of equatorial spread F: Transitional and short wavelength results, *J. Geophys. Res.*, Vol. 87, 1982, pp. 1575–1588.

199. Valladares, C.E., Hanson, W.B., McClure, J.P., and Cragin, B.L., Bottomside sinusoidal irregularities in the equatorial F region, *J. Geophys. Res.*, Vol. 88, 1983, pp. 8025–8042.

200. Tsunoda, R.T., Baron, M.J., Owen, J., and Towle, D.M., ALTAIR: An incoherent scatter radar for equatorial spread studies, *Radio Sci.*, Vol. 14, 1979, pp. 1111–1119.

201. Tsunoda, R.T. and White, B.R., On generation and growth of equatorial backscatter plums, 1. Wave structure in the bottomside F layer, *J. Geophys. Res.*, Vol. 86, 1981, pp. 3610–3618.

202. Tsunoda, R.T., On the generation and growth of equatorial backscatter plumes, 2. Structuring of the west walls of upwellings, *J. Geophys. Res.*, Vol. 88, 1983, pp. 4869–4874.

203. Basu, S. and Basu, S., Equatorial scintillations: advances since ISEA-6, *J. Atmos. Terr. Phys.*, Vol. 47, 1985, pp. 753–768.

204. Kelley, M.C., Equatorial spread F: recent results and outstading problems, *J. Atmos. Terr. Phys.*, Vol. 47, 1985, pp. 745–752.

205. LaBelle, J. and Kelley, M.C., The generation of kilometer wave irregularities in equatorial spread, *J. Geophys. Res.*, Vol. 91, 1986, pp. 5504–5512.

206. Kelley, M.C. and Hysell, D.L., Equatorial spread-F and neutral atmospheric turbulence: A review and a comparative anatomy`, *J. Atmos. Terr. Phys.*, Vol. 53, 1991, pp. 695–708.

207. Basu, S., Basu, S., and Kudeki, E. et al., Zonal irregularity drifts and neutral winds measured near the magnetic equator in Peru, *J. Atmos. Terr. Phys.*, Vol. 53, 1991, pp. 743–555.

208. LaBelle, J. and Lund, E.J., Bispectral analysis of equatorial spread *F* density irregularities, *J. Geophys. Res.*, Vol. 97, 1992, pp. 8643–8651.

209. Hysell, D.L., Kelley, M.C., Swartz, W.E., and Farley, D.T., VHF radar and rocket observations of equatorial spread *F* on Kwjalein, *J. Geophys. Res.*, Vol. 99, 1994, pp. 15,065–15,085.

210. Hysell, D.L., Kelley, M.C., Swartz, W.E., Pfaff, R.F., and Swenson, C.M., Steepened structures in equatorial spread *F*, 1. New observations, *J. Geophys. Res.*, Vol. 99, 1994, pp. 8827–8840.

211. Farley, D.T. and Hysell, D.L., Radar measurements of very small aspect angles in the equatorial ionosphere, *J. Geophys. Res.*, Vol. 101, 1996, pp. 5177–5184.

212. Blanc, E., Mercandalli, B., and Hongminou, E., Kilometric irregularities in the *E* and *F* regions of the daytime equatorial ionosphere observed by a high resolution HF radar, *Geophys. Res. Lett.*, Vol. 23, 1996, pp. 645–648.

213. Valladares, C.E., Basu, S., and Groves, K. et al., Measurement of the latitudinal distributions of total electron content during equatorial spread *F* events, *J. Geophys. Res.*, Vol. 106, 2001, pp. 29,133–29,152.

214. Venkatraman, S., Heelis, R.A., and Hysell, D.L., Comparison of topside equatorial parameters observed from DMSP, Jicamarca, and another model of the ionosphere (SAMI2), *J. Geophys. Res.*, Vol. 110 (A1), 2001, A01307, pp. 1–9.

215. Kelley, M.C., Haerendel, G., and Kappler, H. et al., Evidence for a Rayleigh Taylor type instability and upwelling of depleted density regions during equatorial spread *F*, *Geophys. Res. Lett.*, Vol. 3, 1976, pp. 448–451.

216. Ott, E., Theory of Rayleigh–Taylor bubbles in the equatorial ionosphere, *J. Geophys. Res.*, Vol. 83, 1978, pp. 2066–2070.

217. Tsunoda, R.T., Livingston, R.C., McClure, J.P., and Hanson, W.B., Equatorial plasma bubbles: vertically elongated wedges from the bottomside *F* layer, *J. Geophys. Res.*, Vol. 87, 1982, pp. 9171–9180.

218. Hudson, M.K., Spread *F* bubbles: Nonlinear Rayleigh–Taylor mode in to dimensions, *J. Geophys. Res.*, Vol. 83, 1978, pp. 3189–3198.

219. Aggson, T.L., Burke, W.J., and Maynard, N.C. et al., Equatorial bubbles updrafting at supersonic speeds, *J. Geophys. Res.*, Vol. 97, 1992, pp. 8581–8590.

220. Huang, C.-S. and Kelley, M.C., Nonlinear evolution of equatorial spread *F*, 3. Plasma bubbles generated by structured electric field, *J. Geophys. Res.*, Vol. 101, 1996, pp. 303–313.

221. Hanson, W.B. and Bamgboye, D.K., The measured motions inside equatorial plasma bubbles, *J. Geophys. Res.*, Vol. 89, 1984, pp. 8997–9008.

222. Laakso, H., Aggson, T.L., and Pfaff, R.F., Downdrafting plasma flow in equatorial bubbles, *J. Geophys. Res.*, Vol. 99, 1994, pp. 11,507–11,515.

223. Sekar, R. and Raghavarao, R., Critical role of the equatorial topsede *F* region on the evolutionary characteristics of the plasma bubbles, *Geophys. Res. Lett.*, Vol. 22, 1995, pp. 3255–3258.

224. Sekar, R., Suhasini, R., and Raghavarao, R., Evolution of plasma bubbles in the equatorial *F* region with different seeding conditions, *Geophys. Res. Lett.*, Vol. 22, 1995, pp. 885–888.

225. Chaturvedy, P.K. and Ossakow, S.L., Nonlinear theory of the collisional Rayleigh–Taylor instability in equatorial spread *F*, *J. Geophys. Res.*, Vol. 83, 1978, pp. 4219–4228.

226. Zalesak, S.T. and Ossakow, S.L., Nonlinear equatiorial spread *F*: Spatially large bubbles resulting from large horizontal scale initial perturbations, *J. Geophys. Res.*, Vol. 85, 1980, pp. 2131–2141.

227. Basu, S., McClure, J.P., Hanson, W.B., and Aarons, J., Coordinate study of equatorial scintillations and in situ and radar observations of nighttime *F* region irregularities, *J. Geophys. Res.*, Vol. 75, 1980, pp. 5119–5128.

228. Kil, H. and Heelis, R.A., Equatorial density irregularity structures at intermediate scales and their temporal evaluations, *J. Geophys. Res.*, Vol. 103, 1998, pp. 3981–3989.

229. Chang, Y.-S., Chiang, W.-L., and Ying, S.-Y. et al., System architecture of the IPEI payload on ROCSAT-1, *Terr. Atmos. Oceanic Sci.*, Vol. 10, 1999, pp. 7–18.

230. Su, S.-Y., Yeh, H.C., and Heelis R.A. et al., The ROCSAT-1 IPEI preliminary results: Low-latitude ionospheric plasma and flow variations, *Terr. Atmos. Oceanic Sci.*, Vol. 10, 1999, pp. 787–804.

231. Yeh, H.C., Su, S.-Y. Heelis, R.A., and Wu, J.M., The ROCSAT-1 IPEI preliminary results: Vertical ion drift statistics, *Terr. Atmos. Oceanic Sci.*, Vol. 10, 1999, pp. 805–820.

232. Su, S.-Y., Yeh, H.C., and Heelis, R.A., ROCSAT 1 ionospheric plasma and electrodynamics instrument observations of equatorial spread F: An early transitional scale result, *J. Geophys. Res.*, Vol. 106, 2001, pp. 29,153–29,159.

233. Burke, W.J, Huang, C.Y., and Valladares, C.E. et al., Multipoint observations of equatorial plasma bubbles, *J. Geophys. Res.*, Vol. 108 (A5), 2003, A51221, pp. 1–9.

234. Burke, W.J., Gentile, L.C., Huang, C.Y., Valladares, C.E., and Su, S.-Y., Longitudinal variability of equatorial plasma bubbles observed by DMSP and ROCSAT-1, *J. Geophys. Res.*, Vol. 109 (A12), 2004, A12301, pp. 1–12.

235. Lin, C.S., Immel, T.J., Yeh, H.-C., Mende, S.B., and Burch, J.L., Simultateous observations of equatorial plasma depletion by IMAGE and ROCSAT-1 satellites, *J. Geophys. Res.*, Vol. 110 (A6), 2004, A06304, pp. 1–11.

236. Hunsucker, R.D. and Bates, H.F., Survey of polar and auroral region effects on HF propagation, *Radio Sci.*, Vol. 4, 1969, pp. 347–365.

237. Baron, M.J. and Hunsucker, R.D., Incoherent scatter radar observations of the auroral zone ionosphere during the total solar eclipse of July 10, 1972, *J. Geophys. Res.*, Vol. 78, 1973, pp. 7451–7460.

238. Greenwald, R.A., Ecklund, W.L., and Balsley, B.B., Auroral currents, irregularities, and luminosity, *J. Geophys. Res.*, Vol. 78, 1973, pp. 8193–8202.

239. Ecklund, W.L., Balsley, B.B., and Greenwald, R.A., Doppler spectra of diffuse radar aurora, *J. Geophys. Res.*, Vol. 78, 1973, pp. 4797–4806.

240. Tsunoda, R.T., Presnell, R.I., and Leadabrand, R.L., Radar auroral echo characteristics as seen by a 398-MHz phased-array radar operated at Homer, Alaska, *J. Geophys. Res.*, Vol. 79, 1974, pp. 4709–4717.

241. Greenwald, R.A., Ecklund, W.L., and Balsley, B.B., Diffuse radar aurora: Spectral observations of non-two-stream irregularities, *J. Geophys. Res.*, Vol. 80, 1975, pp. 131–141.

242. Greenwald, R.A., Diffuse radar aurora and gradient drift instability, *J. Geophys. Res.*, Vol. 79, 1974, pp. 4807–4809.

243. Wang, T.N.C. and Tsunoda, R.T., On crossing field two-stream instability in the auroral plasma, *J. Geophys. Res.*, Vol. 80, 1975, pp. 2172–2182.

244. Tsunoda, R.T. and Presnell, R.I., On a threshold electric field associated with the 398-MHz diffuse radar aurora, *J. Geophys. Res.*, Vol. 81, 1976, pp. 88–96.

245. Greenwald, R.A., Weiss, W., Nielsen, E., and Thomson, N.R., STARE: A new radar auroral backscatter experiment in northen Scandinavia, *Radio Sci.*, Vol. 13, 1978, pp. 497–500.

246. Haldoupis, C. and Sofko, G., Doppler spectrum of 42 MHz CW auroral backscatter, *Can. J. Phys.*, Vol. 54, 1976, pp. 1571–1584.

247. Haldoupis, C., Nielsen, E., and Ierkic, H.M., STARE Doppler spectral studies of westward electrojet radio aurora, *Planet Space Sci.*, Vol. 32, 1984, pp. 1291–1300.

248. Fejer, B.G., Reed, R.W., Farley, D.T., Swartz, W.E., and Kelley, M.C., Ion cyclotron waves as a possible source of resonant auroral radar echoes, *J. Geophys. Res.*, Vol. 89, 1984, pp. 187–194.

249. Villain, J.P., Greenwald, R.A., Baker, K.B., and Ruohoniemi, J.M., HF radar observations of *E* region plasma irregularities produced by oblique electron streaming, *J. Geophys. Res.*, Vol. 92, 1987, pp. 12,327–12,342.

250. Foster, J.C., Tetenbaum, E., DelPozo, C.F., St-Maurice, J.-P., and Moorcroft, D.R., Aspect angle variations in intensity, phase velocity, and altitude for high-latitude 34-cm *E* region irregularities, *J. Geophys. Res.*, Vol. 97, 1992, pp. 8601–8617.

251. St-Maurice, J.-P., Prikryl, P., and Danskin, D.W. et al., On the origin of narrow non-ion-acoustic coherent radar spectra in the high-latitude *E* region, *J. Geophys. Res.*, Vol. 99, 1994, pp. 6447–6474.

252. Hall, G.E. and Moorcroft, D.R., Doppler spectra of the UHF diffuse radio aurora, *J. Geophys. Res.*, Vol. 93, 1988, pp. 7425–7440.

253. Prividakes, J.E., Farley, D.T., and Fejer, B.G. et al., Observations of auroral *E*-region plasma waves and electron heating with EISCAT and VHF radar interferometer, *J. Atmos. Terr. Phys.*, Vol. 50, 1988, pp. 339–356.

254. Kustov, A.V., Igarashi, K., and Andre, D. et al., Observations of 50- and 12-MHz auroral coherent echoes at the Antarctic Syowa station, *J. Geophys. Res.*, Vol. 106, 2001, pp. 12,875–12,887.

255. Prikryl, P., James, H.G., and Knudsen, D.J. et al., OEDIPUS-C topside sounding of a structural auroral *E* region, *J. Geophys. Res.*, Vol. 105, 2000, pp. 193–204.

256. Maeda, S., Nozawa, S., Ogawa, Y., and Fujiwara, H., Comparative study of the high-latitude *E* region ion and neutral temperatures in the polar cap and the auroral region derived from the EISCAT radar observations, *J. Geophys. Res.*, Vol. 108 (A8), 2005, A08301, pp. 1–16.

257. Lacroix, P.J. and Moorcroft, D.R., Ion acoustic HF radar echoes at high latitudes and far ranges, *J. Geophys. Res.*, Vol. 106, 2001, pp. 29,091–29,103.

258. Vickrey, J.F., Rino, C.L., and Potemra, T.A., Chatanika/TRIAD observations of unstable ionization enhancements in the auroral region, *Geophys. Res. Lett.*, Vol. 7, 1980, pp. 789–793.

259. Hanuise, C., High-latitude ionospheric irregularities: A review of recent radar results, *Radio Sci.*, Vol. 18, 1983, pp. 1093–1141.

260. Szuszczewicz, E.P., Theoretical and experimental aspects of ionosphere structure: A global perspective on dynamics and irregularities, *Radio Sci.*, Vol. 21, 1986, pp. 351–387.

261. Tsunoda, R.T., High-latitude *F* region irregularities: A review and synthesis, *Rev. Geophys.*, Vol. 26, 1988, pp. 719–760.

262. Kintner, P.M. and Seyler, C.E., The status of observations and theory of high-latitude ionospheric and magnetospheric plasma turbulence, *Space Sci. Res.*, Vol. 41, 1985, pp. 91–104.

263. Weber, E.J. and Buchau, J., Polar cap *F* layer auroras, *Geophys. Res. Lett.*, Vol. 8, 1981, pp. 125–128.

264. Buchau, J., Reinisch, B.W., Weber, E.J., and Moore, J.G., Structure and dynamics of the winter polar cap *F* region, *Radio Sci.*, Vol. 18, 1983, pp. 995–1004.

265. Buchau, J., Weber, E.J., Anderson, D.N. et al., Ionospheric structures in the polar cap: Their origin and relation to 250-MHz scintillations, *Radio Sci.*, Vol. 20, 1985, pp. 325–334.

266. Rino, C.L., Livingston, R.C., and Tsunoda, R.T. et al., Recent studies of the structure and morphology of auroral zone *F* region irregularities, *Radio Sci.*, Vol. 18, 1983, pp. 1167–1175.

267. Weber, E.J., Klobuchar, J.A., and Buchau, J. et al., Polar cap F layer patches: Structure and dynamics, *J. Geophys. Res.*, Vol. 91, 1986, pp. 12,121–12,132.

268. Kelley, M.C., Vickrey, J.F., Carlson, C.W., and Torbert, R. On the origin and spatial extent of high-latitude *F* region irregularities, *J. Geophys. Res.*, Vol. 87, 1982, pp. 4469–4478.

269. Vickrey, J.F., Rino, C.L., and Potemra, T.A., Chatanika/TRIAD observations of unstable ionization enhancements in the auroral F layer, *Geophys. Res. Lett*, Vol. 7, 1980, pp. 789–792.

270. Tsunoda, R.T., Haggstrom, I., Pellinen-Wannberg, A., Steen, A., and Wannberg, G., Direct evidence of plasma density structuring in the auroral F region.

271. Andre, R. Villain, J.-P. Krassnosel'skikh, V., and Hanuise, C., Super dual aroral radio network observations of velocity-divergent structures in the F region ionosphere, *J. Geophys. Res.*, Vol. 105, 2000, pp. 20,899–20,908.

272. Branover, H., Eidelman, A. Golbraikh, E., and Moiseev, S., *Turbulence and Structures: Chaos, Fluctuations, and Helical Self-Organization in Nature and the Laboratory*, San Diego: Academic Press, 1999.

273. Dyrud, L.P., Oppenheim, M.M., and Vom A.F., Endt, The anomalous diffusion of meteor trails, *Geophys. Res. Lett.*, Vol. 28, 2001, pp. 2775–2778.

274. Oppenheim, M.M., Dyrud, L.P., and Ray, L., Plasma instabilities in meteor trails: Linear theory, *J. Geophys. Res.*, Vol. 108, A2, 2003, 1063, 14 pages.

275. Oppenheim, M.M., Dyrud, L.P., and Vom A.F., Endt, Plasma instabilities in meteor trails: 2-D simulation studies, *J. Geophys. Res.*, Vol. 108, A2, 2003, 1064, 13 pages.

276. Oppenheim, M.M. and Dimant, Ya., Meteor induced ridge and trough formation and the structuring of the nighttime *E*-region ionosphere, *Geophys. Res. Lett.*, Vol. 33, L24105, 5 pages, 2006.

277. Dimant, Y.S. and Oppenheim, M.M., Meteor trail diffusion and fields: 1. Simulations, *J. Geophys. Res.*, Vol. 111, A12312, 14 pages, 2006.

278. Dimant, Y.S. and Oppenheim, M.M., Meteor trail diffusion and fields: 2. Analytical theory, *J. Geophys. Res.*, Vol. 111, A12313, 15 pages, 2006.

279. Chapin, E. and Kudeki, E., Radar interferometric imaging studies of long-duration meteor echo observed at Jicamarca, *J. Geophys. Res.*, Vol. 99, 1994, pp. 8937–8949.

280. Chapin, E. and Kudeki, E., Plasma-wave excitation on meteor trails in the equatorial electrojet, *Geophys. Res. Lett.*, Vol. 21, 1994, pp. 2433–2436.

281. Zhou, Q.H., Mathews, J.D., and Nakamura, T., Implications of meteor observations by the MU radar, *Geophys. Res. Lett.*, Vol. 28, 2001, pp. 1399–1402.

282. Hocking, W.K., Experimental radar studies of anisotropic diffusion of high altitude meteor trails, *Earth Moon Planets*, Vol. 95, 2004, pp. 671–679.

283. Janches, D. and Chau, J.L. Observed diurnal and seasonal behavior of the micrometeor flux using the Arecibo and Jicamarca radars, *J. Atmos. Terr. Phys.*, Vol. 67, 2005, pp. 1196–1210.

Chapter 7

Performance of Radio Communication in Ionospheric Channels

7.1 Absorption of Radio Waves in the Regular Ionosphere

We will first analyze the effects of the irregular ionosphere on radio propagation, considering it, first, as a gaseous quasi-homogeneous continuum consisting of plasma with regular layered structures that are changed only in magnitude of the total plasma content. In the absence of inhomogeneities of plasma density, radio waves passing through the quasi-homogeneous regular ionosphere as layered plasma medium are absorbed because of collisions between electrons and ions and neutral molecules (atoms). Then, the intensity I of radio wave can be defined as [1,2]

$$I(k) = I_0 \exp\left\{-2k \int \kappa \, ds\right\} = I_0 \exp\left\{-2\frac{\omega}{c} \int \kappa \, ds\right\} \tag{7.1}$$

where I_0 is the intensity of incident wave; κ, the coefficient of absorption; and c, k, and ω, the light speed, wave number, and wave angular frequency, respectively, introduced in previous chapters. Integration in Equation 7.1, occurs along the wave radio path (i.e., the wave trajectory). For frequencies of interest, $\omega \gg \omega_H$ and $\omega > \omega_{pe}$, the coefficient of absorption can be presented by the following expression:

$$\kappa = \frac{1}{2} \frac{\omega_{pe}^2 (\nu_{ei} + \nu_{em})}{\omega\left[\omega^2 + (\nu_{ei} + \nu_{em})^2\right]} \tag{7.2}$$

All parameters, such as the plasma angular (Longmire) frequency, ω_{pe}, the electron gyrofrequency, ω_H, and collision frequencies between electrons (e) and plasma ions (i) and neutrals (m),

$\nu_e = \nu_{ei} + \nu_{em}$, were defined in Chapter 1. The magnitude of absorption, denoted by A_κ, can be defined according to Equation 7.1 in decibels (dB) as

$$A_\kappa = 10 \log_{10} \frac{I_0}{I(k)} = \frac{4.3}{c} \int \frac{\omega_{pe}^2 (\nu_{em} + \nu_{ei})}{\omega^2 + (\nu_{em} + \nu_{ei})^2} \, ds \qquad (7.3)$$

This parameter fully characterizes ionospheric radio traces and can usually be estimated by measuring radiometric absorption at the fixed frequencies and by knowledge of the frequency dependence of the coefficient of absorption κ. Thus, taking two different frequencies, ω_1 and ω_2, it can be inferred from Equation 7.3 that the ratio of the absorption coefficients at these two frequencies is a function only of the relation between these frequencies, i.e.,

$$\frac{A_{\kappa 1}}{A_{\kappa 2}} = \frac{\displaystyle \int \frac{\omega_{pe}^2 (\nu_{em} + \nu_{ei})}{\omega_1^2 + (\nu_{em} + \nu_{ei})^2} \, ds}{\displaystyle \int \frac{\omega_{pe}^2 (\nu_{em} + \nu_{ei})}{\omega_2^2 + (\nu_{em} + \nu_{ei})^2} \, ds} \qquad (7.4)$$

In Refs. [1–3], the approximate polynomial function describing the frequency dependence of absorption $A_\kappa(\omega)$ was derived for separate intervals $\Delta\omega$. The power parameter s of this polynomial function was changed at the range $0 < s < 2$ (depending on the ratio $(\nu_{em} + \nu_{ei})/\omega$). This power parameter s can be determined experimentally in measurements of radiometric absorption at two separate frequencies, and using the following expression:

$$s = \frac{\log_{10} \dfrac{A_\kappa(\omega_1)}{A_\kappa(\omega_2)}}{\log_{10} \dfrac{\omega_1}{\omega_2}} \qquad (7.5a)$$

Then, the standard deviation of this parameter equals [1–3]:

$$\delta s = s \left\{ \left[\frac{\delta A_\kappa(\omega_1)}{A_\kappa(\omega_1)} \right]^2 + \left[\frac{\delta A_\kappa(\omega_2)}{A_\kappa(\omega_2)} \right]^2 \right\}^{1/2} \qquad (7.5b)$$

The value was found in Ref. [4] to be $\delta s \leq 0.2$, with errors of measurements of $\delta A_\kappa(\omega) < 0.1$ dB and $\delta A_\kappa(\omega_1) \geq 0.1$ dB, $\delta A_\kappa(\omega_2) \geq 0.1$ dB. Additional spectral analysis of distribution of radiometric absorption has shown that in most events (>75%) the power spectral parameter s is changed inside the expected interval of $0 < s < 2$ [4,5].

The effects of absorption of probing radio waves have been observed during excitation in the ionosphere of small-scale artificial periodic irregularities (APIs) [6,7]. Effects of absorption associated with natural plasma irregularities located in the F region of the ionosphere were observed experimentally using incoherent scatter radar and explained theoretically in Ref. [8]. Absorption of wave energy at the API was explained theoretically in Refs. [9,10] by resonance interaction of reflected probing waves with heating-induced plasma periodical turbulences created by powerful ground-based radio facilities. This phenomenon was explained by conversion of probing radio wave into plasma wave (Z-mode), and finally, by absorption of this Z-mode in the ionosphere. The same effect of probing radio wave absorption occurs in the polar ionosphere when small-scale plasma turbulences are created in natural conditions [3]. Here, radio waves are absorbed during interaction with natural inhomogeneities of the ionospheric plasma having dimensions $\ell < 2\pi v_{Te}/\omega_H$ (see Chapter 2), and v_{Te} and ω_H are the thermal velocity and gyrofrequency of plasma electrons, respectively. As was noted in Ref. [3], the additional absorption of probing radio waves passing through the high-latitude ionosphere is observed at frequencies of 15–30 MHz (i.e., the events with $s > 2$) caused by collective collisions between charged particles, electrons, and ions. However, this effect has tendency to weaken within a short period of several seconds and, finally, the spectral distribution of radiometric absorption becomes sharper [10].

As for the method of estimation of absorption proposed in Ref. [8] that is based on observations of F region natural plasma structures using incoherent scatter radar, to explain this effect the authors used a formula other than Equation 7.3 for estimations of absorption of radio waves in the irregular ionosphere, derived from the Appleton–Hartree magneto-ionic theory:

$$A_K \approx \frac{4.58 \cdot 10^{-5}}{\omega^2} \int N_e (\nu_{em} + \nu_{ei}) \, ds \qquad (7.6)$$

which is valid only if $\omega \gg \omega_H$ and $\omega \gg \nu_{ei} \geq \nu_{em}$, i.e., for conditions valid only for the ionosphere at altitudes greater than the altitudes of the D and E layers. Using incoherent scatter radar (IRIS) operated at 38.2 MHz using 49 beams, and the proposed method of evaluation of plasma density through measurements of limited frequency of the F layer, f_0F_2, the magnitude of probing radio wave absorption was confirmed. It was found that the absorption usually cannot exceed values of about 0.2–0.3 dB on days without magnetic disturbances, sometimes increasing up to 0.5 dB during small magnetic disturbances. As we showed in the previous chapter, strong magnetic storms can increase this value significantly (up to several dB).

7.2 Radio Scattering Caused by Plasma Irregularities of Various Origins in the Ionosphere

In the previous chapter we discussed plasma irregularities, both large and small, with various degrees of perturbations compared with regular background ionospheric plasma, and with a different degree of anisotropy relative to the geomagnetic field $\mathbf{B}_0$. Such kinds of irregularities are observed experimentally in the ionosphere at all latitudes, longitudes, altitudes [11–84]. Plasma irregularities associated with various plasma instabilities caused by natural geophysical phenomena, such as meteor trails, neutral winds, gravity waves due to atmosphere–ionosphere coupling, magnetic storms caused by solar activity, etc., or introduced in the ionosphere artificially by heating of plasma by powerful radio waves or plasma cloud injection, are usually stretched along the geomagnetic field $\mathbf{B}_0$ with a degree of anisotropy $\ell_\parallel/\ell_\perp \geq 10$, where $\ell_\parallel$ and $\ell_\perp$ are the longitudinal and transverse (to $\mathbf{B}_0$) initial dimensions of plasma irregularities, respectively. The character of radio signals scattered from such anisotropic plasma irregularities of the ionosphere, which are oriented along the geomagnetic field $\mathbf{B}_0$, is not the same at the different geographic latitudes [25–42]. This phenomenon can be explained by the strong ionization factors occurring in the high-latitude ionospheric regions and the spatial selectivity of scattered signals from field-aligned plasma irregularities in the middle-latitude ionospheric regions. Why is this question so important? It is important for designers of long-range radio communication channels passing through ionospheric layered structures, D, E, and F, consisting of irregularities with a wide spectrum of scales and orientations with respect to radio waves of various frequencies passing through ionospheric layers. To predict conditions of propagation in such ionospheric channels, it is important to predict location, height, shape, and concentration of such irregularities to understand their influence on scattering of radio signals, back and forward (oblique). Scattering effects allow prediction of final fading of signals, signal-to-noise ratio and, finally, loss of information sent through such ionospheric channels, both for long land–ionosphere–land and satellite communications.

In this section, we consider the effects of scattering of probing radio waves by small- and moderate-scale plasma irregularities, for which dimensions $\ell_\perp$ across $\mathbf{B}_0$ range from several tens of centimeters to several meters, i.e., $\ell_\perp \approx \lambda$ for HF/VHF-wave bandwidth. We consider the cases of backscattering and forward scattering extending up to 180°. First, we begin by discussing some pioneering experiments carried out from the end of the 1950s to the beginning of the 1990s in different geographical places of the world [45–84]. Results of these experiments allowed researchers to finally create satisfactory theoretical models of specular reflection and scattering from plasma irregularities of various origins occurring in the ionospheric plasma at different latitudes [11–16].

7.2.1 Experimental Investigations of Backscattering and Forward Scattering in Region E of the Ionosphere

During the cycle of systematic and well-organized experiments carried out in pulse regimes in the United States at 206 MHz and at 106 MHz, it was found that most of the scattered pulse signals arriving at the receiver from heights of the E layer of the ionosphere ($h = 105$–115 km) were caused by field-aligned plasma irregularities strongly affected by the geomagnetic field, which finally formed an array of plasma columns with radius $r \leq \lambda$ across $\mathbf{B}_0$ and with lengths $L \gg \lambda$ along $\mathbf{B}_0$ [45,46]. Therefore, the authors denoted such types of signals as H_E-type, and defined the corresponding scattering phenomena as H_E-scattering. At the same time [47–52], all researchers analyzing scattering of radio waves from plasma irregularities of the E layer middle-latitude ionosphere have used the same abbreviations to stress the effect of the geomagnetic field on the character of evolution of such anisotropic plasma structures. We will use the same terminology in our further discussions, following the corresponding analysis made in Refs. [15–18], which describe evolution of such kinds of plasma irregularities and their effects on backscattering and forward scattering.

In the first cycle of pioneering experiments on forward scattering from the middle-latitude E layer plasma H_E irregularities [45,46], a power pulse transmitter of high power was used, working at 206 MHz and then at 106 MHz. Experiments were carried out simultaneously in several states of the south-west of the United States (see Figure 7.1). The transmitter was located south of the state of Texas. The signals were registered simultaneously in Arizona and California in such a manner that the receiver antennas were located at different contours of specular reflection (denoted in Figure 7.1 by dashed curves). These contours define the geometrical positions at ground surface where it is possible to observe radio signals reflected specularly from field-aligned E layer plasma H_E irregularities at the two active points of the receiver's location.

During this cycle of experiments, the difference between H_E-scattering and pure meteor scattering was found and analyzed. We present only one record of a typical pure meteor burst-type signal (Figure 7.2a) compared with two kinds of H_E-scattering signals: the burst-type H_E-scattering signal (Figure 7.2b and c) and the quasi-regular continuous H_E-scattering signal (Figure 7.2d). It was concluded that the burst-type H_E-scattering signals come from a region located between altitudes of the 100 and 120 km within the E layer and fully correlated with sporadic meteor trails, both diurnally and seasonally. Thus, it can be seen from Figure 7.2a that typical pure meteor signals are characterized by fast growth and the approximately exponential

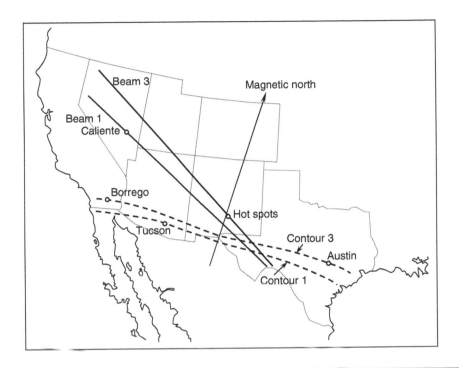

Figure 7.1 Geometry of experiments carried out in the south-west of the United States.

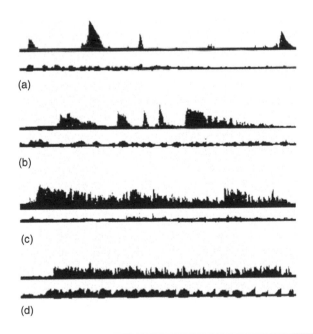

Figure 7.2 Typical record of pure meteor burstlike signal (a), quasi-regular H_E-scattered signals (b and c), and quasi-continuous signal from the sporadic E_s layer (d).

smooth decay of the envelope of scattering signal. The time of attenuation of such burst-type signals is sufficiently small (a few seconds), which is consistent with meteor-head-echo lifetime obtained during observations of pure meteor reflections from the ionosphere and described in Refs. [52–55]. In Figure 7.2b, both meteor-scattering and H_E-scattering signals are shown, from which the fundamental difference between them can be seen. Thus, the H_E-scattering signals have a much longer "lifetime" (from several tens of seconds to several minutes) and a significantly higher degree of "roughness" caused by fast fading (see Figure 7.2c). As for quasi-regular signals, they can also be characterized as "rough," but with a constant envelope of amplitude of fading (an average); they were observed over periods of several tens of minutes. The same results were obtained by the use of the frequency of 106 MHz. The diurnal and seasonal correlation of H_E-scattering signals with those scattered from sporadic meteors allow the researchers to assume that the meteor trails are the main factors causing burst-type H_E-scattering signals [45,46]. The main question remaining after these experiments was: Why is the time of relaxation of the H_E-scattering signals larger than that for pure meteor-scattering signals and why is the fading of H_E-scattering stronger? The authors of Refs. [45,46] have not analyzed the "thin" structure of H_E-scattering signals; they only established their existence and noted differences compared to pure meteor signals. However, large-scale spaced antenna measurements indicated the existence of field-aligned irregularities with lengths of the order of 25 meters embedded within a scattering volume with a horizontal scale of 10 km.

The same statements without any physical explanation of the effects of H_E-scattering observed in the E layer of the middle-latitude ionosphere were stressed in Refs. [47–50]. A cycle of experiments at frequency $f = 56$ MHz were carried out in Japan at two traces of 959 km (Kokubunji–Yamagawa) and 935 km (Kokubunji–Hoshika), as shown in Figure 7.3, to investigate radio scattering from field-aligned plasma irregularities generated at the E layer of the ionosphere. What is interesting here is that most of the signals of a quasi-continuous type (see Figure 7.4, which shows the same range–time records at Yamagawa as shown in Figure 7.2d) were recorded with relaxation time from several minutes to several hours. During these experiments, vertical sounding of the ionosphere was carried out simultaneously in the regions close to zones of specular scattering. What is important to mention here is that researchers did not find any correlation between specular H_E-scattering of pulse radio signals and the existence of sporadic E_s layers in the ionosphere. So, the main difference between the two cycles of experiments is that in Japan more quasi-continuous signals were recorded than those of the burst-type observed in the United States. At the same time, in Refs. [47–50], it was found that the observed field-aligned irregularities contributed much to the scattering of radio waves at altitudes 100–120 km and have at least the scale of $L \approx 30$–50 m in a direction parallel to $\mathbf{B}_0$ and of $T \approx 3$–5 m perpendicular to $\mathbf{B}_0$; i.e., the degree of anisotropy of $L/T \geq 10$. The scales and the corresponding degree of anisotropy of field-aligned irregularities obtained during experiments contribute most of the scattering propagation in the E ionospheric region around Japan.

The positive contribution of these pioneering experiments to our knowledge of plasma inhomogeneities occurring in the E region of the middle-latitude ionosphere that are responsible for the scattered propagation in HF/VHF band is that they fully differentiated the active zone of specular H_E-scattering from zones of meteor activity. This fact allows comparing specular H_E-scattering signals with those obtained from pure meteor reflections. At the same time, many peculiarities of signals scattered by field-aligned plasma irregularities were not analyzed in Refs. [45–50], in particular, the "thin" structure of scattered signals, classification of quasi-continuous and burst-type H_E-scattering signals, their nature. This can be done only following detailed theoretical analysis of plasma dynamics and plasma instabilities generation, based on the models

Figure 7.3 Geometry of experiments carried out in Japan.

of real plasma structures and ambient factors occurring in the inhomogeneous ionosphere. This was done only after one or two decades, both experimentally [15,16,51,52] and theoretically [17,18,56–58]. We will discuss these investigations later.

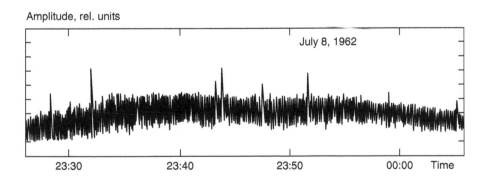

Figure 7.4 An example of joint distance–time record of signal registered during experiments carried out in Japan.

We will describe, first, results of experiments carried out during the early 1970s in the middle-latitude ionosphere of the former USSR at the radio trace with a stationary transmitter and the moving receiver using a continuous regime of radiation of signals operating at 44 and 74 MHz. The main purpose of the experiments was to investigate the "thin" structure of H_E-scatter signals compared with pure meteor-scatter signals [53–55]. We briefly mentioned this cycle of experiments in Chapter 6 in connection with methods of investigation of the main parameters of these types of irregularities. Now we will examine these experiments in greater detail. In the first cycle of experiments, two transmitters were operated at 44 and 74 MHz, each with an output power of 2 kW and with an orientation to the direction of the great circle of 12° and to the position of the receiver R of 24°, respectively, as shown in Figure 7.5. At the receiver, two channels were used: one for registration of H_E-scattering signals and the other for recording of only pure meteor-scattering signals, using the common receiving antenna with width of antenna pattern of 60°. The corresponding geometry of trace is shown in Figure 7.5. The distance between the transmitter and the receiver was about 1500 km. Two different channels allowed for the recording of only H_E-scattering signals in the corresponding channel but which were not observed in the meteor channel. It was found that the main difference between the two types of signals, the H_E-type and the meteor-type, is the form of their envelope, the time of decay, and the character of their fading. As in Refs. [45,46], the same features that differentiate these two types of signal were found, but in Refs. [15,16,51,52] a detailed classification of H_E-scattering signals was carried out. Thus, it was found that most of the several thousands of recording signals can be divided into two basic types, according to the form of their amplitude envelope, their statistical character, and term of fading: the *burst-type* H_E-scattering signals with time of decay of about tens of seconds to several minutes, and *quasi-continuous* H_E-scattering signals with time of decay about several minutes to several hours. In Figures 7.6 through 7.8 are shown typical signals recorded during forward scattering from the E layer of the ionosphere for the radio trace of investigation shown in Figure 7.5.

The burst-type H_E signals are close to those of meteor I–III types [53–55] but have a longer relaxation time and other "thin" statistical features regarding fading phenomena. According to the character of growth and decay of the signal envelope, they can also be divided into three types. Thus, the I-type of burst is characterized by a sharp front and by exponential decay of the amplitude envelope, as shown in Figure 7.6. Usually, the character of fluctuations (i.e., fading)

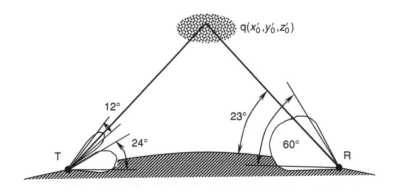

Figure 7.5 Geometry of middle-latitude radio trace. (Extracted from Ref. [16].)

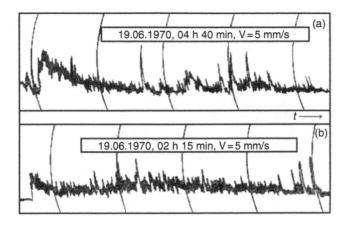

Figure 7.6 **I-type of burstlike signal recorded during experiments carried out at middle-latitude radio trace according to Refs. [16–18].**

of most of these signals was not changed from the beginning to the end of the "lifetime." This effect can be explained by the influence of weakly ionized meteor trails, whose trajectories are close to the directions of geomagnetic field strength lines [16,53–55]. The type-II of burst is characterized by a sharp front but a very slow decay of the amplitude envelope with terms of several to tens of minutes. A typical signal is presented in Figure 7.7a. In Figure 7.7b, the record of the beginning of the type-II of burst signal (of the 13-min lifetime) is shown. As mentioned in Refs. [15–18,51–55], such a signal can be caused by high-ionized meteor trails, the lifetime of which is controlled by the

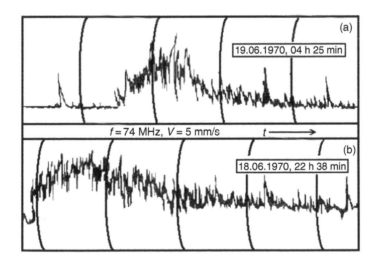

Figure 7.7 **II-type (a) and III-type (b) of burstlike signal recorded during experiments carried out at middle-latitude radio trace according to Refs. [16–18].**

geomagnetic field, which significantly decreases the rate of their diffusion in radial directions to $\mathbf{B}_0$ [56–58] (see also results described in Chapter 5). The signals of this type were rarely recorded owing to low probability of observation of high-density (i.e., highly ionized) meteor trails in the ionosphere of Earth [54,55]. The type-III bursts are those signals whose envelope slowly increases until they achieve a maximum level, and then slowly decreases until they achieve a minimum level (see Figure 7.7b). Such bursts were often observed during the entire cycle of experiments carried out at different frequencies. Most decrease asymmetrically, as shown in Figure 7.7b. This effect was explained [17,18] by the influence of the geomagnetic field on the process of evolution of meteor plasma at the stage when it passes from highly ionized (high-density) to weakly ionized (weak-density), and fully elongated along $\mathbf{B}_0$. As was shown in Ref. [17] (see also discussions in Chapter 5), the lifetime of weakly ionized plasma structures stretched along $\mathbf{B}_0$ is longer than that of highly ionized plasma structures. Therefore, the relaxation time in the stage of signal envelope decay is much larger than the signal envelope initial growth.

As was found during current experimental observations (see also Refs. [47–50]), the existence of continuous signals was observed over a long time exceeding 10 min and continuing up to several hours. Such signals are characterized by a sharp fast fading with deep signal amplitude oscillations, and with a very slow (nonexponential) decay of signal level. In Figure 7.8, the typical recording quasi-continuous signal is shown at the beginning, when the amplitude growth is usually observed and then during the steady-stage term, when amplitude is approximately constant (Figure 7.8a), and at the time of slow decay (Figure 7.8b). The time of growth and decay was selected to be shorter because during experiments, many signals of this type were registered in the first and last stages of more than 5 to 6 min. Sometimes, the time of growth was of the same order as the time of decay [16]. In the steady-state regime the amplitude fading of quasi-continuous signals has a stationary character. The existence of such long-term signals observed at the middle-latitude ionosphere in different geographic regions was explained in the same manner as was for the high-latitude ionosphere, i.e., by existence of turbulent movements owing to gradients of neutral wind and the generation of two-stream instability in the ionospheric plasma (see Chapter 2).

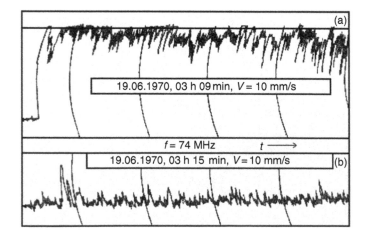

Figure 7.8 An example of quasi-regular H_F-scattered signal recorded during experiments carried out at middle-latitude radio trace according to Refs. [16–18].

Following the theoretical investigations described in Refs. [15,16], geometric computations of the specular contour of forward scattering for the concrete radio trace shown in Figure 7.5 was carried out to find the active zone of H_E-scattering as well as the active heights of localization of H_E plasma irregularities.

The specular-scattered waves from elongated anisotropic ionospheric irregularities create a conic-shaped pattern with the origin at the point of reflection Q as shown in Figure 7.9. This pattern, crossing Earth's surface, creates at the ground surface zones of points where the scattered signals can be received successfully. The axis of the conic-shaped pattern coincides with the longitudinal axis of plasma inhomogeneity; we denote by θ the angle between this axis and the pattern direction. The corresponding zone is defined by this specular contour consisting of specular-scattered waves at the ground surface. The main task of the research is to find the geometry of such scattered zones at the ground surface, i.e., the geometrical distribution of points of the specular scattering for fixed positions of the transmitting and the receiving antennas. These parameters are important for link budget design inside the ionospheric communication links operated mostly at HF/VHF-frequency bands. In Figure 7.5, x'_0, y'_0, and z'_0 denote the coordinates in the system related to positions of the transmitter (T) and the receiver (R). The system (X,Y,Z) is related to the center of Earth, where axis Z passes through the geographic poles, the plane XOZ coincides with the plane of zero-meridian, and the plane XOY is placed at the equatorial plane. Then, the latitude and the longitude of the specular contour will be determined by angles φ and ψ, respectively. So, for our task we need to use the initial parameters such as the coordinates of the transmitter $T(\varphi_1, \psi_1)$ and receiver $R(\varphi_2, \psi_2)$. Then, we calculate points that simultaneously lie at the conic pattern and the sphere of Earth with the radius R_e. We will not enter into these complicated geometrical derivations, which are based on elements of analytical geometry, and will only refer the reader to the original study [15]. One result of the computation [15,16] for the middle-latitude radio trace of 1500 km described earlier with coordinates of the transmitter, $\varphi_1 = 46°$ and $\psi_1 = 30°$, and receiver, $\varphi_2 = 47°$ and $\psi_2 = 48°$, is presented in Figure 7.9. Here, positions of the

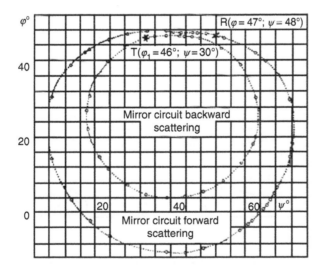

Figure 7.9 Conic-shaped pattern created by mirror-scattered waves from anisotropic ionospheric irregularities.

terminal antennas are denoted by "stars." Using the same geometrical approach described in Refs. [15,16], we can estimate the active zone of H_E-scattering and the heights of localization of H_E plasma irregularities, as this scattering has a strong directive character. Then, using the approach proposed in Refs. [59,60], we present here results of computations of two middle-latitude radio traces. The first is of 1130 km with coordinates of the transmitter, $\varphi_1 = 46°29'$, $\psi_1 = 30°36'$ and the receiver, $\varphi_2 = 45°26'$, $\psi_2 = 42°20'$, respectively. The coordinates of the second trace of 1500 km was described earlier. Figure 7.10a shows the active zone and the altitudinal range for the first radio trace, and Figure 7.10b is for the second radio trace.

In Ref. [61] was proposed the close criterion of estimation of the active zone of scattering from the plasma irregularities observed at the high-latitude ionosphere, mostly appearing in polar caps. Following results presented in Figure 7.10, for some middle-latitude traces, the active zone of H_E-scattering from field-aligned plasma irregularities and that for meteor ionizations can be overlapped. Moreover, it was proved experimentally that the burstlike signals presented in Figures 7.6 and 7.7 have a meteor-induced nature as the corresponding plasma irregularities generated by meteor trails occur in the middle-latitude ionosphere at altitudes from 90 to 130 km.

The foregoing method of investigation of field-aligned irregularities of the ionospheric E region is based on continuous wave (CW) sounding in the ionosphere using back- and forward (oblique) scattering of probing radio waves.

Now, we present a cycle of systematic measurements carried out at the end of the 1970s in Japan, the United States and also in Australia (i.e., at the middle latitudes toward the South Pole) using different kinds of ground-based facilities, such as HF oblique radar, multifrequency iono-sondes and interferometers, and working not only in the mutual range–time domain but also in the frequency domain (strictly speaking, in the Doppler spread domain) [64–84], which allows the researcher to obtain the main characteristics of field-aligned middle-latitude E layer irregularities using modern radio methods. Following these works, we briefly present the main results that allow one to predict experimentally the main characteristics of E region irregularities of the middle-latitude ionosphere, which is important for predicting radio traces that can be performed

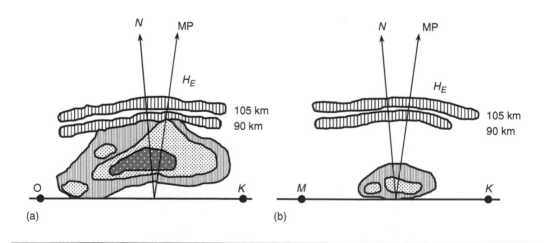

Figure 7.10 Position of active zones of mirror scattering for the first radio trace (a) and the second radio trace (b).

owing to back and oblique scattering from field-aligned sporadic plasma structures of various origins and having a wide spectrum of dimensions.

We start with drift measurements of the lower E layer carried in the United States at middle latitudes (Billerica, Massachusetts) using a pulse radar system operating at 2.35 and 2.89 MHz that produced pulses of 100 μs length with a repetition rate of 50 Hz during transmitting time [65]. The samples of the original data recording at the receiver after scattering from the ionosphere are plotted in Figure 7.11a and b, observed over 150 s. It can be seen from this example that a rapid amplitude fading and large change in phase occurs, with a strong correlation between them. Whenever a deep fading occurs, the phase changes rapidly, owing to the interference of rays' multiple reflections from different layers of the E region. The rigorous correlation analysis made in Ref. [65] showed rapid oscillations in both the autocorrelation and cross-correlation functions, as well as essential broadening of the signal power spectrum. This precise correlation analysis coincided with the experimentally observed existence in the E region of the ionosphere of wavelike plasma structures strongly elongated along magnetic field lines. The same conclusions were made [66] on investigating plasma structures responsible for rapid fading of radio signals reflected from E layer of the middle-latitude ionosphere (Bribie Island, 152°E, 27°S) using radar operating at a frequency of 1.98 MHz. The daytime ionospheric E layer was analyzed under small, moderate, and high disturbed conditions. The same strong correlation between deep rapid amplitude fading and large deviations of phase of the recording signal arriving at the receiver was found. We present in Figure 7.12 one of the example obtained during the moderate disturbance of the ionosphere. The author analyzed the role of gravity waves as well as the possibility of turbulence in the neutral air motions in creation of small short-lived plasma structures. It was outlined that the main role in plasma small isotropic and anisotropic plasma structure creation is played by turbulent motions of neutral gas at altitudes 90–100 km, the result which agrees with results of previous aforementioned works. At the same time, in Ref. [66] were observed other types of discrete scatters, which form presumably sporadic E clouds, owing to their long lifetime. Variations of such long-lived discrete sporadic E clouds is shown in Figure 7.13 [66]. Additional analysis of phase deviation of the signals obtained from short-lived plasma structures allows for finding velocities and direction of movements of such type of irregularities. Thus, in Figure 7.14, the data divided into 15 min intervals (see bottom and top figures) is shown. The maximum negative and maximum positive Doppler shift corresponds to negative (opposite) and positive (toward) maximum velocity of discrete reflectors of about ±150 m/s. Usually, during 7 days of observations these discrete reflectors were observed at altitudes of 95–115 km, mostly near 100 km with velocities varied in the range 30–60 m/s. These results were compared with those obtained in Ref. [64] during investigations of the E layer at Adelaide (Australia, 139°E, 35°S) and good coincidence was found between obtained results concerning deep fading of recorder signals and large phase shift, allowing estimation of the velocity of moving field-aligned plasma structures in the range 50–100 m/s.

More detailed information about large-scale plasma sporadic structures (called sporadic E clouds or blobs), which form sporadic E layer, was presented in detail [67,68] and then summarized in a special issue of Whitehead [69], in which results obtained in Refs. [67,68] were compared with other studies carried out all over the world, including those mentioned earlier. We recommend that the reader study this issue.

As for the experiments [67,68], several types of irregularities, which form sporadic E_s layer, were observed there using the same Bribie Island radar (see earlier text and also Ref. [66]), operating at 3.84 and 5.8 MHz. First, detailed information about the horizontal structure of the E_s layer was obtained and compared with the well-known wind-shear theory (which was successfully used for

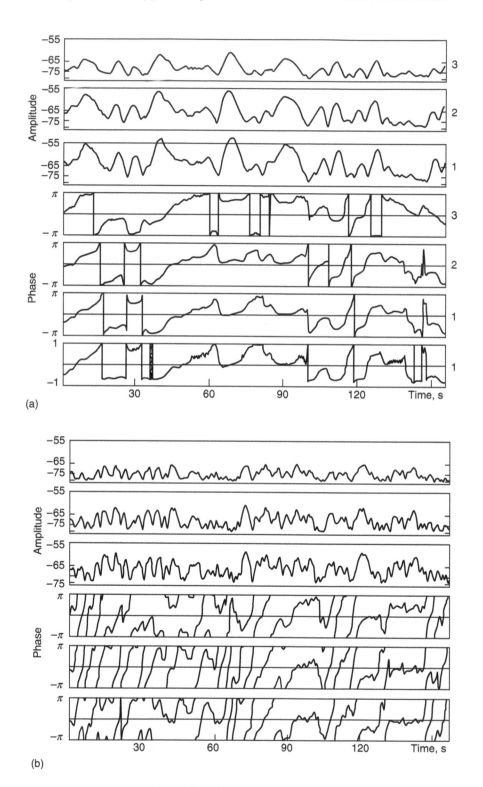

Figure 7.11 Temporal dependence of amplitude and phase of scattered signals recorded at 2.35 MHz (a) and 2.89 MHz (b). (Extracted from Ref. [66].)

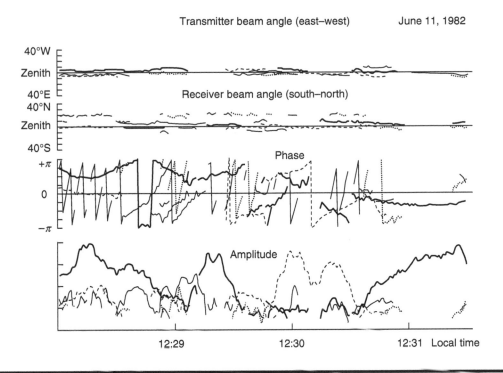

Figure 7.12 Temporal dependence of amplitude (bottom panel) and phase (middle panel) and spatial-temporal distribution of scatterers observed from the receiver and transmitter antennas (top panel) obtained on June 11, 1982. (Extracted from Ref. [66].)

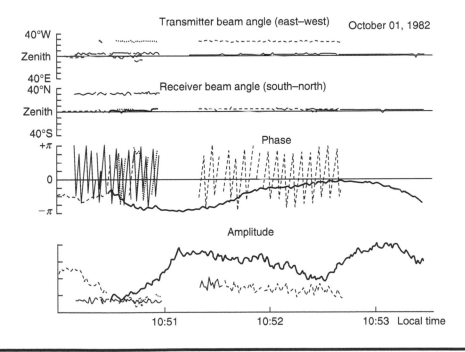

Figure 7.13 The same as Figure 7.12, but obtained on October 1, 1982.

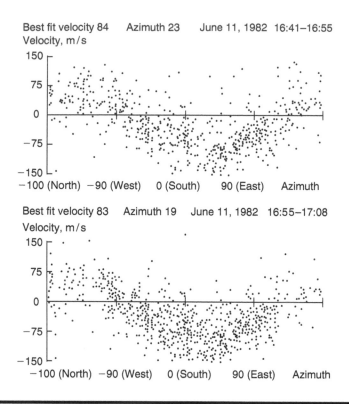

Figure 7.14 Azimuthally distributed Doppler velocities of observed irregularities during experiment carried out on June 11, 1982.

artificial plasma cloud evolution; see Chapters 5 and 6). The authors created a model of sporadic E cloud in the form of anisotropic ellipses with changing dimensions: the length is from 5 to 50 km, and the thickness is from 0.5 to 5 km, elongated along geomagnetic field lines. These plasma structures, according to Doppler measurements, moved in different directions with velocities (we do not present here numerous complicated illustrations in Refs. [67,68]) varying from 20 to 100 m/s. Isolated clouds due to wind-shear effect collected in the clusters moved approximately with the same velocity. The latter finally create large horizontal (several hundreds of kilometers) and thin vertical (several tens of kilometers) sporadic E_s layer in the middle-latitude ionosphere.

In Ref. [70], based on digital ionosonde observation of sporadic E blobs consisting of the E_s layer at Halley ($\approx 27°W$, $\approx 75°S$), rapid fading of radio waves was found accounting for Doppler shift and the corresponding velocities of movements of such plasma structures. The ionosonde was operated at 4.5 MHz, and each ionogram was recorded over every 12 min. The experimentally obtained Doppler shift for different locations of echo signals coming from different blobs located within the sporadic E_s layer (denoted by A to E) is shown in Figure 7.15, where the power spectra derived are shown for each sample for sets of six overlapping time series of 1 min duration separated by 12 min. It can be seen again, as simultaneously was obtained for the south (toward from Equator to South Pole) middle latitudes [66–69], that blobs (clouds) formed in the E_s layer have different velocities and move in various directions, whereas the cluster moves at the same velocity of about 10 m/s.

In Refs. [71,72] were reported backscatter radar observations carried out over Iioka, Japan (35.7°N and geomagnetic latitude 29.6°N, 140.8°E). Observations were made over one year

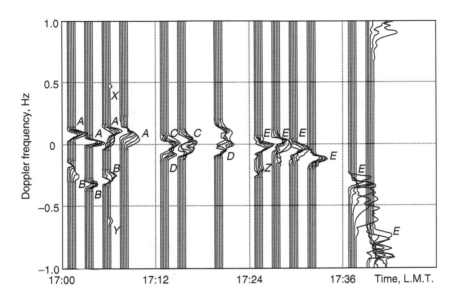

Figure 7.15 Temporal variations of Doppler shift frequencies of echo signals arriving from different blob structures located within E_s sporadic layer, labeled by A to E. (Extracted from Refs. [71,72].)

(December 1973–November 1974) at 25.54 MHz (i.e., the observed irregularities for backscatter regime have scales of about $\ell_{ir} = \lambda/2 \approx 6$ m) using a 100 μs single pulse. The recording signals were obtained in the form of range–azimuth–intensity (B-scan) and as range–time–intensity (RTI) displays. Continuous observations showed that coherent echo signals sent to the ionosphere return from plasma density irregularities aligned along geomagnetic field lines occurring mostly in the nighttime and associating with the sporadic E_s layers located at altitudinal range of 100–110 km. So, they identified the echo signals with the presence of E_s layers. Thus, in Figure 7.16 the central positions of field-aligned echoes are represented by open circles obtained from the B-scan records (i.e., in the azimuth domain). The solid curves denote three different altitudes of origin for the echo signals: 90, 100, and 150 km. For altitude $z = 100$ km the reflection points for slightly off-perpendicular incidence with aspect angles of ±5° are also shown. In addition, the perpendicular points for the altitudes 90 and 150 km are also indicated by dashed curves in Figure 7.16. As can be seen, echo signals are absent when the slant range is less than 170 km and for the slant ranges between 300 and 400 km. Data obtained from RTI-records allows one to estimate the direction and speed of these field-aligned irregularities of the order 50–60 m/s, by using Doppler shift spectra, as shown in Figure 7.17, which illustrates temporal and spatial variations of the motion of irregularities obtained from Doppler shift spectra. The 3 dB spectral widths are denoted in Figure 7.17 by the horizontal bars, and the observed directions are indicated for individual data points. Without entering into details of measurements, we only will outline that according to Doppler shift obtained from RTI-records, the horizontal drift velocities of the irregularities were estimated during two-day continuous measurements. During the entire cycle of measurements, the average velocity of horizontal drift varied from 54 to 65 m/s, the results which are close to those discussed earlier using other methods of Doppler shift estimation. The broadening of Doppler spectra obtained in these experiments was rather large, of about 10–12 Hz. In further cycle of experiments described in Ref. [72], the researchers used a multifrequency regime of sounding of the ionosphere (4–64 MHz). Using statistics of the

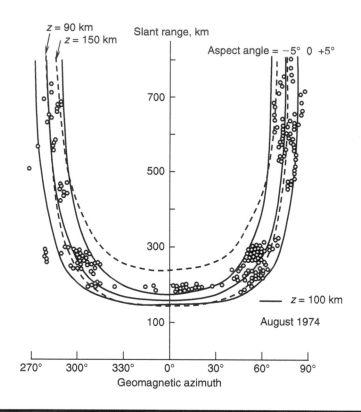

Figure 7.16 Range–azimuth distribution of echo signals scattered from field-aligned irregularities denoted by circles. Solid lines correspond to the best fit obtained for altitude of 100 km, dashed lines correspond to those for altitudes of 90 and 150 km. (Extracted from Refs. [71,72].)

corresponding ionogram recorded by the receiving station, the position of field-aligned backscatter echoes (FAEs) was found as shown in Figure 7.18. It can be seen that FAE appears at the range 380 km in the frequency interval 8–27 MHz. As vertical incident F layer echoes and the obliquely backscattered E_s layer were also recorded, they are also presented in a reproduction and sketch of backscattered ionograms obtained at the Iioka experimental site. According to RTI measurements the Doppler shift spectra and their broadening were found to be changing from +3 to +13 Hz and from 15 to 25 Hz, respectively. The precise measurements of these two parameters allow estimating velocities of horizontal drift of field-aligned irregularities and the direction of their movements. Thus, in Figure 7.19, extracted from Ref. [72], in upper panel (a), the observed westward component of drift velocity obtained from B-scan records is represented by dots versus the local time at the station. The solid curve gives estimated average values. Here, horizontal bars represent the same average velocities but obtained using Doppler shift estimations via RTI-scan records. The vertical bars give standard deviations from these average values. The bottom panel (b) gives the local time variations (in percentages) of the occurrence probability of FAEs. So, again we see the predominant westward motion of field-aligned irregularities associated with the E_s layer with velocities of movement varied from 40 to 100 m/s, the result which coincides previous radio observations of the sporadic middle-latitude E regions. Moreover, systematic statistics obtained in Refs. [71,72] allow one to conclude that the E_s-scattering echo signals are mostly observed in summer nights.

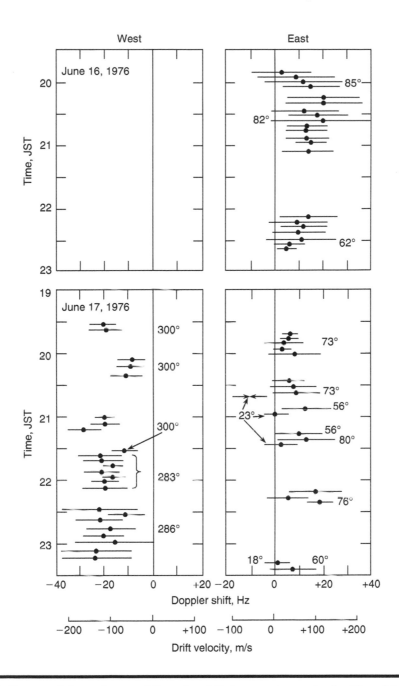

Figure 7.17 Temporal dependence of Doppler shift frequencies and velocities obtained during experiments carried out on June 16, 1976 (top panels) and June 17, 1976 (bottom panels).

After 15 years another group of researchers from Japan using a monostatic and then a multi-beam pulsed (MU) backscatter radar located at Shigaraki (34.9°N, 136.1°E, geomagnetic latitude 25.0°N) and operated at the VHF-band has investigated carefully spatial structure of sporadic E layer field-aligned irregularities using simultaneously a theoretical framework trying to explain

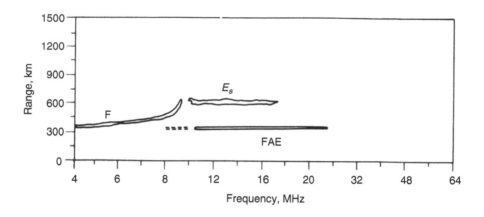

Figure 7.18 Ionogram obtained by sounding of the ionosphere in the frequency interval from 4 to 64 MHz, according to Ref. [72].

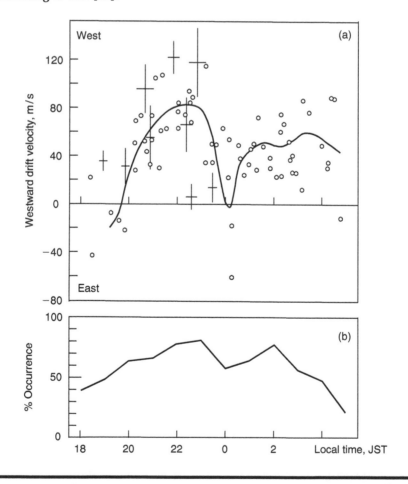

Figure 7.19 Temporal distribution of drift velocity westward component (top panel) and the time occurrence (in percentages) of the corresponding irregularities observed in the ionosphere during experiments described in Ref. [72].

generation of such irregularities by the effect of gravity waves generated in the lower ionosphere (see Refs. [73–78] and the corresponding bibliographies therein). Using the same range–time–intensity (RTI) record, as used by the foregoing researchers, the range–time radio map of echo signals (measured in dB) after backscattering from the corresponding sporadic irregularities was obtained, and one example is presented in Figure 7.20. A very complicated striated structure of the cluster of small-scale irregularities covering altitudinal range from 90 to 120 km can be seen from the shown illustration. The main features of obtained data during previous measurements that can be outlined based on results presented in Refs. [73–75] are the following. The small-scale irregularities with the transverse scale of 3 m (because for a backscattering process $\ell_{ir} = \lambda/2 \approx 3$ m) spread in directions perpendicular both to drift velocity $\mathbf{V}_d$ and to ambient geomagnetic field $\mathbf{B}_0$, mostly in westward direction from the radar. The researchers explain the striation mechanism by generation of gravity instability (see Chapter 2). The observed irregularities were closely correlated with the occurrence of the sporadic E layer mostly in nighttime periods. Even more precise experiments with high temporal, range and altitudinal resolutions (see Refs. [76–78] and the publications mentioned therein) were based on the same MU backscatter radar operated at 46.5 MHz. To obtain echo signals from 3-m-scale field-aligned irregularities in the E region, the multibeam regime was used when an antenna with diameter of 103 m was steered northward from vertical direction to zenith angles of 50°–60° to achieve orthogonal intersection with $\mathbf{B}_0$ at an altitude of 100 km, as is shown from Figure 7.21, where antenna beams for the multibeam observations are shown by thin solid lines (i.e., 12 lines denote 12 beams). Finally, the range of observation was changed from 140 to 293 km. We will not enter details of measurements, but will only show main results obtained in these systematically and precisely performed experiments. A characteristic RTI plot of echo signals backscattered from field-aligned irregularities, shown in Figure 7.22, shows again that the cluster of such irregularities with very complicated striated quasi-periodic structure spreading at the wide

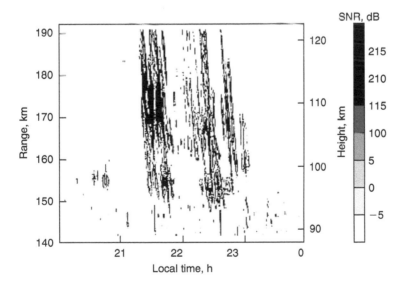

Figure 7.20 Spatial-temporal distribution of echo-signals by measuring of signal-to-noise ratio (SNR) at the receiver. (Extracted from Ref. [75].)

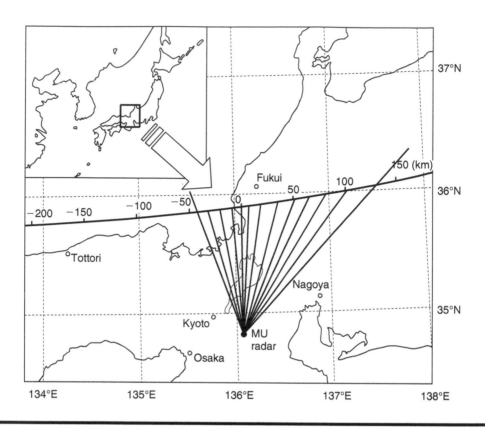

Figure 7.21 Location of the MU multibeam radar used in experiments carried out in Japan and geometry of experiments. (Extracted from Refs. [75,76].)

altitudinal range of 30–40 km (from 100 to 140 km). Here, all echoes appear from a direction perpendicular to the geomagnetic field $\mathbf{B}_0$. The observed picture of striation strongly depends on the direction of the corresponding beams and orientation with respect to the vertical axis. Thus, a different picture (see Figure 7.23) was observed when the antenna beam was directed geographically northward with a zenith angle of 51.0°. As can be seen, very intense echo signals from field-aligned irregularities occur throughout the duration of observations. Echoes are distributed surrounding altitude of about 100–105 km, with narrow spreading along altitudes (only up to 110 km); also, striated structures were not so clear, as follows from Figure 7.22. In both experiments, the movements of sporadic plasma structures were usually in westward directions with velocities 120 to 160 m/s [75,76]. We should outline that the researchers [73–78] have obtained a lot of statistics concerning daily and seasonal variations of scatter phenomena from the E layer that explains the probability of occurrence of echo signals in the time, space (altitude and range), and Doppler shift domains. The main conclusion of the authors of these studies is that gravity waves can decrease the threshold of gradient-drift instabilities (GDIs), modulating and accelerating such kinds of instabilities, and finally generating the associated field-aligned plasma small-scale structures assembled during their motion into striated structures—clusters, which move with lesser velocity

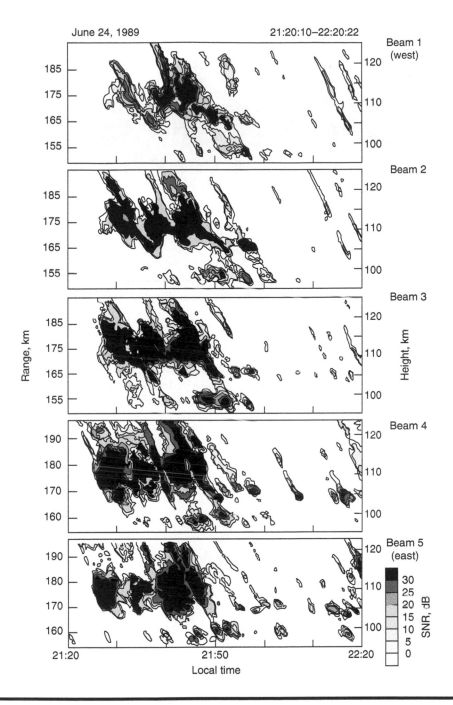

Figure 7.22 Example of 2-D spatial-temporal map of SNR of echo signals (in dB) recorded for different beams from 1 to 5 directed to the west and east. (Extracted from Refs. [76,77].)

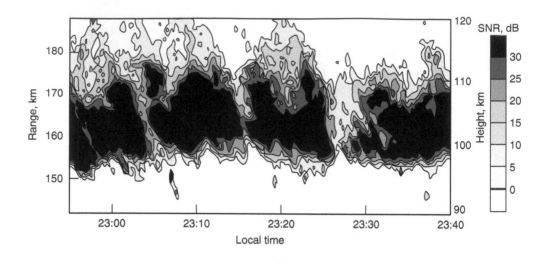

Figure 7.23 The same as in Figure 7.22, but for the beam directed to the north.

(of 50–60 m/s) than each separate stratum (100 m/s and more). The same results were obtained by a previous group of researchers [51,52,66–69] not so systematically investigating large-scale plasma blobs/clouds formed sporadic E layer in the middle-latitude ionosphere by measuring of SNR (in dB) at the receiver (see Figure 7.24). At the same time, most of the aforementioned works could

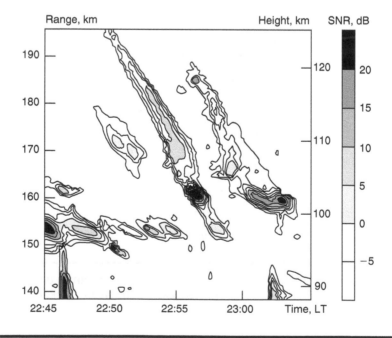

Figure 7.24 The 2-D mapping of echo signals by measuring of SNR (in dB) at the receiver.

not explain the occurrence of backscatter echo signals in the morning and daytime periods, the location of FAIs below 5–10 km than the E_s sporadic layer, the predominant role of gravity waves in formation of quasi-periodic (called *wavelike*) plasma structures, and so on.

Two separate research groups from Europe and the United States [79–81,82–84] tried to resolve problems that could not be satisfactorily explained. We described these studies in Chapter 6 regarding the investigation of types and geophysical characteristics of E layer irregularities occurring in the middle-latitude ionosphere, and concerning the main sources of their generation. Here we will present, as was done with other aforementioned works, results of investigation of location, height, movements, and "power" of such kinds of irregularities using different kinds of backscatter and oblique radars; i.e., we will point out those characteristics of middle-latitude E layer irregularities that will help us to understand "thin" scattering effects caused by such irregularities.

In Refs. [79–81], attention was mainly on obtaining information about the distribution of areas consisting of field-aligned irregularities that form sporadic E layers in the middle-latitude ionosphere, their shape and dimensions, direction and velocity of movements, lifetime, by using CW bistatic radar performed not only for studying high-resolution Doppler spectra (described in Chapter 6), but also for analyzing detailed dynamic changes and short-term characteristics of the scattering signals. The difference between these experiments and earlier ones is that in Refs. [79–81], as was done in experiments described in Refs. [15–18], using a CW regime, the transmitter and receiver were ranged at a distance of about 100 km, whereas for backscatter experiments the authors of Refs. [67–78] used a monostatic system in which the transmitter and receiver were assembled in a joint system. The transmitter described in Refs. [79–81] was operated in the range 50–52 MHz ($\lambda \approx 6$ m), and the receiver is in the range 48–52 MHz. So, observed irregularities had a transverse scale $\ell_{ir} = \lambda/2 \approx 3$ m, in accordance with the rule that in the configuration of oblique scattering, the signal, scattered by those irregularities as plasma waves, propagates with a wave vector of value $k_{ir} = 2\pi/\ell_{ir}$ that relates to the radar wave vector value $k_r = 2\pi/\lambda$ and scattering angle θ_s (see Figure 7.5) as $k_{ir} = 2k_r \sin(\theta_s/2)$. So, for $\theta_s = 180°$, we immediately get $k_{ir} = 2k_r$ or $\ell_{ir} = \lambda/2$. As was mentioned in Chapter 6, the circle of experiments was carried out on the island of Crete, Greece (35°N and 28°N geomagnetic latitude; 24°E), and was called the "Sporadic E Scatter (SESCAT) Experiment." The geographic map presented in Figure 7.25 shows the geometry of these experiments, with the transmitter and the receiver separated by about 100 km. As is seen, the center of the scattering ionospheric region is located at altitude of 105 km, denoted by the solid straight line corresponding zeros magnetic aspect angle. In Figure 7.25, at the left bottom side, some geomagnetic field parameters are summarized, such as invariant magnetic latitude along the line passing the center of scattered area, $\Lambda = 30.8°$; the magnetic field magnitude, $B = 0.43$ Gauss; the magnetic dip angle, $I = 52.5°$; and the magnetic declination, $D = 2.5°$. The latter parameter defines the line-of-sight (LOS) with respect to geomagnetic north at the east direction, which equals 5.5°. So, we can note that experiments were carried out (and, consequently, scatter signals were observed) close to the geomagnetic north–south direction and therefore measured meridian motions of observed irregularities. According to measurements, and the corresponding antenna radiation pattern, in Refs. [79–81] was estimated the scattering area for associated magnetic aspect angles. This azimuthally extent of the scattering area is shown in Figure 7.26 as a family of the contour lines spaced at 3 dB gain intervals. The corresponding shaded area is also shown in Figure 7.25. Computations presented in Ref. [79] give estimations of the scattering area; its extent is about 20 km in longitude and of about 45 km in latitude. Using bistatic radar, the echo signals, "weak" and "strong", reflected from the ionospheric meteor trails were investigated during first circle experiments, both "weak" and "strong" backscatter signals. What is important to note here is that

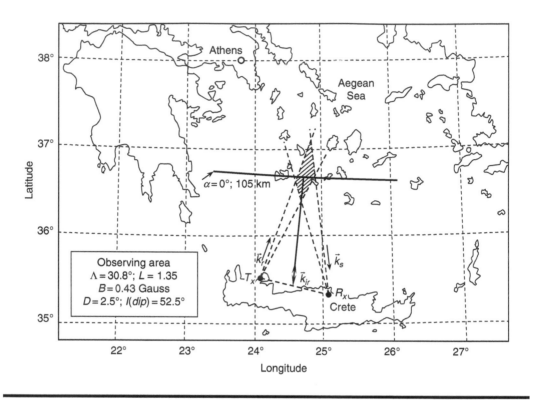

Figure 7.25 Geometry and positions of the transmitter and receiver during SESCAT experiments. (Extracted from Ref. [79].)

the authors of Ref. [79] reported the first clearly observed midday event of daytime echo signals caused by middle-latitude E_s scatter phenomenon. Thus, the upper part of Figure 7.27 evidently illustrates this event, occurring at 14:05 local time. Here, the spectrogram of the E_s event is shown against local time. The corresponding velocities of movements of irregularities responsible for the backscattering were estimated through the measured Doppler shifts in the spectrum, where 1 Hz corresponds to a velocity of 3.11 m/s. So, the movements of irregularities responsible for such a sporadic signal occur at the radial velocity range of ±100 m/s. In the bottom part of Figure 7.27 one of the remaining events occurring only during nighttime is shown. This phenomenon starts with the short-lived burst-type signal having narrow spectrum width and concentrated at the range of ±30 m/s. After several minutes a long-lived continuous signal (of at least 40 min duration) was observed that associated with sporadic E_s scatter phenomenon from large-scale weakly ionized irregularities of the ionosphere. More detailed investigation of burstlike scatter signals also in earlier morning and nighttime, such as shown in the top and bottom parts of Figure 7.28, having the same structure as the burst signal presented in Figure 7.27, as well as another short-lived signal (with duration up to several minutes), having different velocity of movements (in Figure 7.28, the spectrum gives a radial velocity range of 150–200 m/s and achieves even ±250 m/s), allowed the authors to associate these burstlike signals with the meteor scatter signal. So, the obtained results are in good agreement with results obtained 20 years earlier at the middle-latitude trace passing the Black Sea [15–18]. What is important to point out is that the authors of Refs. [15–18] (one of the coauthors of this work was involved in these studies) also observed clearly meteor-scatter (burstlike) signals during midday periods.

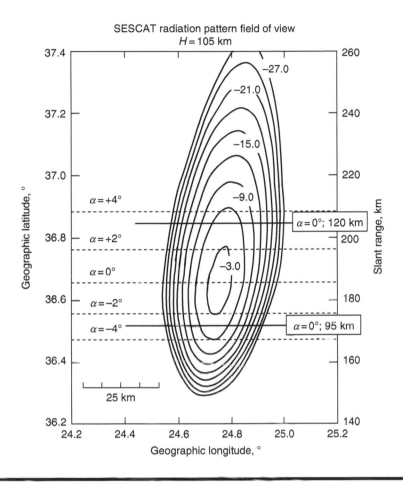

Figure 7.26 Spatial mapping of the active scattered zones spaced by 3-dB-gain intervals for various associated magnetic aspect angles $\alpha = -4°$ to $\alpha = +4°$. (Extracted from Ref. [80].)

In contradiction with these results, the authors of Refs. [82–84] observed E_s scatter phenomenon only at nighttime. The authors from Cornell University, using a portable radar interferometer (CUPRI), i.e., a 50 MHz Doppler radar system operating also under a CW regime over several months (spring–autumn) on the island of St. Croix (17.7°N, 64.8°W) near Puerto Rico to study irregularities of nighttime E_s sporadic layers and the corresponding instabilities causing such irregularities (see Figure 7.29). Association of E_s spread irregularities and the corresponding instabilities were discussed in detail in Chapter 6 using results obtained in Refs. [82–84]. As we mentioned in this chapter, for us it is very important to obtain information about localization and "power" of such irregularities for future estimation of radio-scattering effects. Therefore, in Figure 7.30, we present the estimated electron profiles of irregularities responsible for strong backscattering of echo signals with duration of several hours observed at altitudes 95–116 km on the night of May 1983 as a more physically vivid example of obtained backscatter records. Figure 7.30 indicates the increase of ionospheric plasma density during the evening hours from about 17:00 to about 23:00 (local time). As is seen, from about 17:00 to about 19:00, the irregularities of the sporadic E_s layer responsible for scatter echo signals are located around 115–116 km, with a

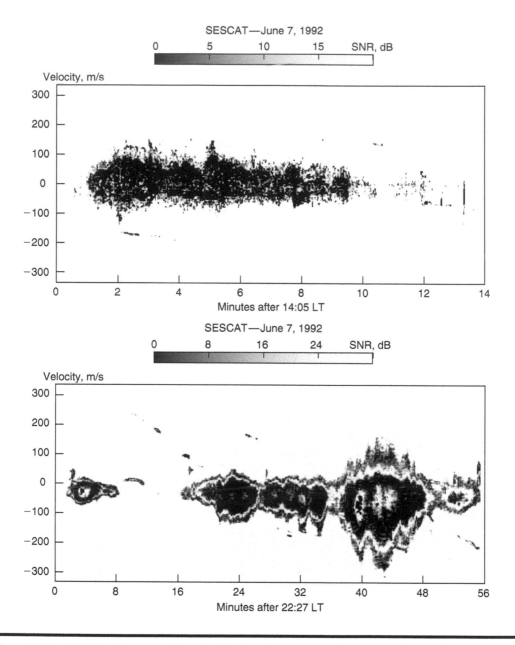

Figure 7.27 Velocity–time mapping of echo signals by measuring SNR (in dB) at the receiver in experiments carried out on June 7, 1992. (Extracted from Ref. [80].)

maximum density of about $5 \cdot 10^5$ cm^{-3}. Then the irregularities moved slightly down, reaching at 21:50 an altitude of about 105 km. After this time and up to the end of observations at 23:00, the maximum of plasma concentration tended to increase up to 10^6 cm^{-3} and more. What is interesting to note is that after about 19:30 the effect of the F layer was evident, but accompanied by the fast decay of "power" of irregularities from $5 \cdot 10^4$ to $5 \cdot 10^3$ cm^{-3}, i.e., to the threshold where scattering signals are recorded at the level of noise.

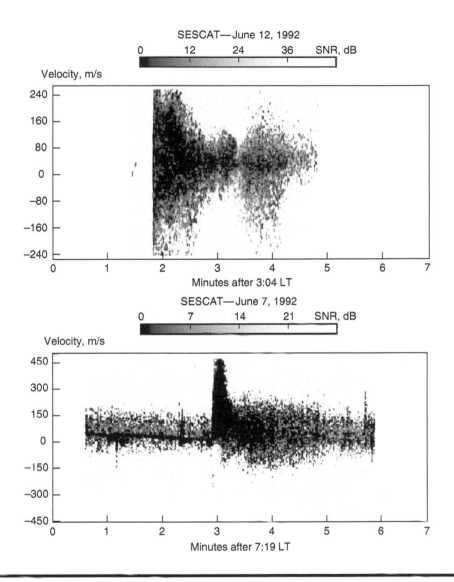

Figure 7.28 The same as Figure 7.27, but measured on June 12, 1992.

Observed experimental data evidently showed that there are two types of irregularities created in the E region of the middle-latitude ionosphere, meteor-induced irregularities and field-aligned irregularities, which many researchers denoted as H_E irregularities, which form sporadic layers in this region of the ionosphere. Now we put a question: how do these kinds of irregularities scatter radio waves passing through the E region of the ionosphere?

Therefore, in the following text we will estimate the power of radio signals scattered in forward directions from anisotropic plasma irregularities in the middle-latitude ionosphere. First, we will present the classical Booker's theory of backscattering from small-scale magnetic-field-oriented plasma irregularities, which was developed to investigate the radio aurora in polar regions of the ionosphere. Then, we will present a modified concept that can extend Booker's results to the case of

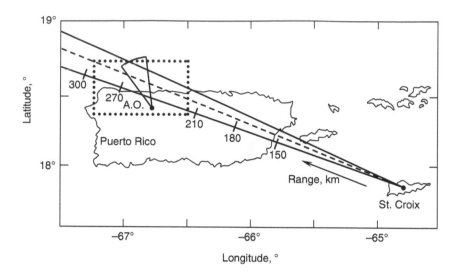

Figure 7.29 Geometry and radar location during CUPRI experiment. (Extracted from Ref. [82].)

forward scattering from anisotropic ionospheric irregularities of various origins and will verify the obtained results via the corresponding experiments carried out at middle-latitude regions of the ionosphere described earlier according to Refs. [15,16].

7.2.2 Theory of Backscattering of Radio Waves from Anisotropic Plasma Irregularities

The theory of backscattering of radio waves from anisotropic moderate-scale and small-scale ionospheric irregularities was created by Booker [11–14] for the case of the auroral ionosphere. We briefly present his theory below and will then expand it to describe forward scattering at wide range of angles without entering into the origin of generation of irregularities, because the obtained results were adapted for different kinds of ionospheric plasma structures. Thus, this theory fully describes scattering from natural ionospheric irregularities, such as H_E type and meteor type [15–18], as well as from quasi-periodical plasma structures artificially induced by heating of the ionosphere by powerful waves [19–22].

In Figure 7.31, the geometry of scattering of radio wave at a wide range of angles is presented. The origin is arranged at point O inside the volume V containing scatterers. The transmitter is located at the point P_1, and the receiver is at point P_2.

The field $E_0(\mathbf{r})$ is defined by vector $\mathbf{r}_T$ from the transmitter of power W, which transmits energy using an omnidirectional antenna and can be presented as [11,12]

$$E_0(\mathbf{r}_T) = \sqrt{\frac{ZW}{2\pi}} \frac{e^{i\mathbf{k}\mathbf{r}_T}}{r_T} \qquad (7.7)$$

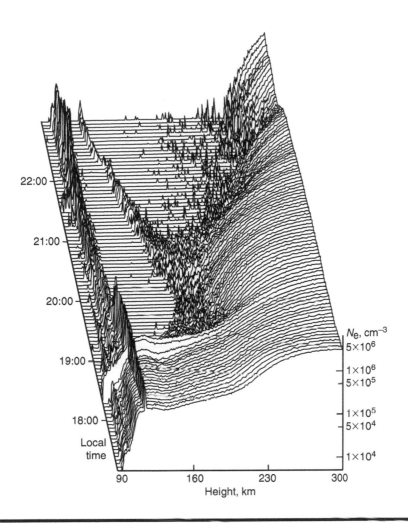

Figure 7.30 Altitude–time distribution of plasma density obtained during sounding of the ionosphere by using backscatter radar system. (Extracted from Ref. [84].)

where Z is the electrical impedance of plasma at the heights under consideration. Owing to fluctuations of plasma permittivity, $\delta\varepsilon = \varepsilon - \varepsilon_0$, the electric moment that occurs within unit volume dv equals

$$\mathbf{M}_E = \delta\varepsilon \cdot \mathbf{E}_0 \tag{7.8}$$

The radio wave, scattered by this elementary volume, arrives at the point P_2 of the receiver. Its strength is [11–14]:

$$|\delta\mathbf{E}| = \frac{k_0^2 \sin\chi}{4\pi} |\mathbf{M}_E| \frac{e^{-i\mathbf{k}_2(\mathbf{R}_2 - \mathbf{r})}}{|\mathbf{R}_2 - \mathbf{r}|} dv \tag{7.9}$$

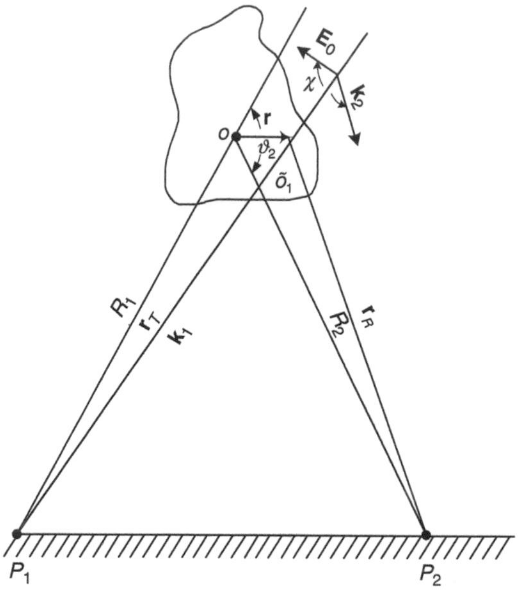

Figure 7.31 The geometry of forward scattering from the irregular ionosphere.

where χ is the angle between the vector of ambient electric field $\mathbf{E}_0$ and the wave vector of the scattered wave $\mathbf{k}_2$ (see Figure 7.31). The total field from the whole volume V of area of scattering, containing plasma irregularities, can be expressed as [11–14]:

$$E \equiv |\mathbf{E}| = \frac{k_0^2 \sin \chi}{4\pi} \sqrt{\frac{ZW}{2\pi}} \int \delta\varepsilon(\mathbf{r}) \frac{e^{i(\mathbf{k}_1 \mathbf{r}_T - \mathbf{k}_2 \mathbf{r}_R)}}{|\mathbf{r}_R \cdot \mathbf{r}_T|} dv \qquad (7.10)$$

In Equation 7.10, the case of isotropic antenna was considered with the spatial angle of 2π, i.e., with the antenna gain equaling 0 dB [11–14]. Then, the average intensity of the scattered signal at the receiving antenna is

$$\langle I \rangle \equiv \langle |\mathbf{E}|^2 \rangle = ZW \frac{k_0^4 \sin^2 \chi}{32\pi^3} \int \langle \delta\varepsilon(\mathbf{r}_R) \cdot \delta\varepsilon(\mathbf{r}_T) \rangle \times \frac{e^{i(\mathbf{k}_1 \mathbf{r}_T - \mathbf{k}_2 \mathbf{r}_R - \mathbf{k}_1 \mathbf{r}'_T + \mathbf{k}_2 \mathbf{r}'_R)}}{\mathbf{r}_R \cdot \mathbf{r}_T \cdot \mathbf{r}'_R \cdot \mathbf{r}'_T} dv\, dv' \qquad (7.11)$$

As shown in Refs. [11–14], if plasma density fluctuations are uniformly distributed within the volume V containing small-scale plasma irregularities, then

$$\langle \delta\varepsilon(\mathbf{r}_R) \cdot \delta\varepsilon(\mathbf{r}_T) \rangle = \int U_\varepsilon(\mathbf{K}) \cdot e^{-i\mathbf{K}\Delta\mathbf{r}} d^3K \qquad (7.12)$$

Here, $U_\varepsilon(\mathbf{K})$ is the power spectrum of perturbations of plasma permittivity, $\mathbf{K}^{-1}$ describes the spatial scale of such perturbations, and $\Delta\mathbf{r} = \mathbf{r} - \mathbf{r}'$. Further theoretical analysis can be simplified by suggestion that the maximal linear scale $L_{\max}$ of the volume V is smaller than distances, R_1 and R_2. At the same time, it is larger than characteristic scales of plasma density (e.g., permittivity [11–14]) perturbations. Under such assumptions, the expression describing the average intensity of scattered waves reduces to

$$\langle I \rangle = I_0 V \frac{k_0^4 \sin^2 \chi}{16\pi^2 R_2^2} \int U_\varepsilon(\mathbf{K}) \, \mathrm{d}^3 K \int_{-\infty}^{\infty} \frac{e^{i\{[\mathbf{K} - (\mathbf{k}_1 - \mathbf{k}_2)]\Delta \mathbf{r}\}}}{|\mathbf{r}_R \cdot \mathbf{r}_R|} \, \mathrm{d}v$$

$$= I_0 V \frac{k_0^4 \sin^2 \chi}{R_2^2} U_\varepsilon(\mathbf{k}_1 - \mathbf{k}_2) \tag{7.13}$$

where I_0 is the wave intensity within the volume V of scattering.

Now, following Refs. [11–14], we introduce the well-known parameter defined in the literature as the *differential* (also called *effective* or *radar*) *cross section*, denoted by $\sigma(\vartheta_2, \chi, k_0)$ [11–17], where the corresponding angles of scattering are shown in Figure 7.31. This parameter determines the power of the wave scattered by the unit volume over the unit spatial angle for the incident wave with the unit field intensity, i.e.,

$$\sigma(\vartheta_2, \chi, k_0) = \frac{k_0^4 \sin^2 \chi}{4\pi^2} U_\varepsilon(\mathbf{k}_1 - \mathbf{k}_2) \tag{7.14}$$

For a given angle of wave scattering, ϑ_2, the absolute value of wave vectors difference, $|\mathbf{k}_1 - \mathbf{k}_2|$, equals

$$|\mathbf{k}_1 - \mathbf{k}_2| = k_0(1 - \cos \vartheta_2) \tag{7.15}$$

From Equation 7.15 it follows that the backscattering occurs from plasma density perturbations (e.g., turbulences) with scales $\ell = \lambda/2$. Because

$$k_0^4 = \frac{(2\pi)^4}{\lambda^4} = \frac{\omega^4}{c^4} \tag{7.16}$$

and

$$\langle (\delta\varepsilon)^2 \rangle = \frac{\omega_{pe}^4}{\omega^4} \left\langle \left| \frac{\delta N}{N_0} \right|^2 \right\rangle \tag{7.17}$$

the coefficient in Equation 7.14 before the spectral function $U_\varepsilon(\mathbf{K})$ does not depend on wave frequency. The frequency dependence of the differential cross section defined by Equation 7.14 leads from the normalized spectrum of plasma permittivity perturbations, $\delta\varepsilon$, i.e., from $U_\varepsilon(\mathbf{K})/\langle(\delta\varepsilon)^2\rangle$.

If plasma permittivity perturbations are distributed according to Gaussian law in the Cartesian coordinate system (x, y, z) with the z-axis oriented along the geomagnetic field lines ($\mathbf{z} \| \mathbf{B}_0$), i.e.,

$$\rho_\varepsilon(x, y, z) = \exp\left\{ -\frac{x^2 + y^2}{l_\perp^2} - \frac{z^2}{l_\|^2} \right\} \tag{7.18}$$

then

$$U_\varepsilon(2\mathbf{k}_0)/\langle(\delta\varepsilon)^2\rangle = (2\pi)^{3/2} l_\perp^2 l_\parallel e^{-k_0^2(l_\perp^2 + l_\parallel^2 \sin \Psi_s)} \tag{7.19}$$

where $\Psi_s = \pi/2 - \beta_s$, β_s is the angle between the ambient magnetic field $\mathbf{B}_0$ and the direction of scattering, called the angle of specular scattering [15–18]. Substituting now the expression Equation 7.14 in Equation 6.14, we can finally obtain the frequency dependence of the effective cross section

$$\sigma(f) = \exp\left(-\frac{f^2}{f_0^2}\right) \tag{7.20}$$

As was shown in Ref. [23], such an exponential frequency dependence, determined by Equation 7.20, was observed during experiments mentioned in Ref. [14], and can be obtained only by using a polynomial correlation function of plasma permittivity fluctuations $\delta\varepsilon$, i.e., for

$$\rho_\varepsilon(x,y,z) \cong \frac{1}{\left((2\pi)^{-1} + \alpha_1^2 x^2 + \alpha_2^2 y^2 + \alpha_3^2 z^2\right)^2} \tag{7.21}$$

where α_1, α_2, α_3 are the coefficients of anisotropy of small-scale plasma irregularities along the corresponding axis. Further analysis of the intensity of scattering signals from small-scale plasma irregularities, obtained without assumption $L_{max} \ll R_1$, R_2, was carried out by Booker [11–14].

It was shown that the obtained results, based on special Legendre's polynomial functions, cannot be used for analysis of scattering processes from small-scale irregularities elongated along the geomagnetic field $\mathbf{B}_0$.

In the following subsection, we present a more general approach to forward scattering from moderate-scale and small-scale anisotropic plasma irregularities, following results described in Refs. [15–18].

7.2.3 Forward Scattering of Radio Waves from Anisotropic Irregularities

Let us consider the geometry of the same problem as presented in Figure 7.31, where again the isotropic transmitter antenna located at the point P_1 radiates a linear-polarized wave that achieves region of volume V in the ionosphere consisting of plasma irregularities of dielectric permittivity ε. Here, we unify all notations made by Booker with those made in Refs. [15–18]. The electric field at the arbitrary point O within the scattered volume can be defined by the same expression (Equation 6.9). Again, owing to fluctuations of the plasma permittivity, $\delta\varepsilon = \varepsilon - \varepsilon_0$, the electric momentum $\mathbf{M}_E = \delta\varepsilon \cdot \mathbf{E}_0$ occurs within the unit volume dv. This electrical momentum creates at the point of the receiver antenna, P_1, the field potential

$$\delta\Pi = \frac{1}{4\pi} |\mathbf{E}_0| \frac{\delta\varepsilon}{\varepsilon} \frac{e^{-ikR_1}}{R_1} dv \tag{7.22}$$

Then the full potential at the receiver created by plasma permittivity irregularities, $\delta\varepsilon/\varepsilon$, within volume V is

$$\Pi = \left(\frac{ZW}{32\pi^3}\right)^{1/2} \int \frac{\delta\varepsilon}{\varepsilon} \frac{e^{-i(\mathbf{k}_1\mathbf{R}_1+\mathbf{k}_2\mathbf{R}_2)}}{|\mathbf{R}_1 \cdot \mathbf{R}_2|} dv \qquad (7.23)$$

Making the same assumption, $V^{1/3} \ll R_1, R_2$, as postulated by Booker, we can replace $|\mathbf{R}_1 \cdot \mathbf{R}_2|$ in Equation 7.23 by $|\mathbf{r}_T \cdot \mathbf{r}_R|$, i.e.,

$$\Pi = \left(\frac{ZW}{32\pi^3}\right)^{1/2} \frac{e^{-i(\mathbf{k}_1\mathbf{r}_T+\mathbf{k}_2\mathbf{r}_R)}}{|\mathbf{r}_T \cdot \mathbf{r}_R|} \times \int \frac{\delta\varepsilon}{\varepsilon} e^{-i(\mathbf{k}_1(\mathbf{R}_1-\mathbf{r}_T)+\mathbf{k}_2(\mathbf{R}_2-\mathbf{r}_R))} dv \qquad (7.24)$$

Now, if for the arbitrary point $O(x,y,z)$ we introduce the unit vectors of the incident wave $\mathbf{e}_1(l_1,m_1,n_1)$ and of the scattered wave $\mathbf{e}_2(l_2,m_2,n_2)$, we obtain for the integral the following formula:

$$I = \int \frac{\delta\varepsilon}{\varepsilon} e^{-i|\mathbf{k}|((l_2-l_1)x+(m_2-m_1)y+(n_2-n_1)z)} dx\,dy\,dz$$
$$= F[|\mathbf{k}|(l_2 - l_1), |\mathbf{k}|(m_2 - m_1), |\mathbf{k}|(n_2 - n_1)] \qquad (7.25)$$

where $F\lfloor|\mathbf{k}|(l_2 - l_1), |\mathbf{k}|(m_2 - m_1), |\mathbf{k}|(n_2 - n_1)\rfloor$ is the Fourier transform of function $\frac{\delta\varepsilon}{\varepsilon}(x,y,z)$. Here, we denoted, as in Refs. [15,16], the integral in Equation 7.24 by I, and assumed $|\mathbf{k}_1| = |\mathbf{k}_2| = |\mathbf{k}|$. Taking into account now the fact that $F[l,m,n] \times F^*[l,m,n]$ is the Fourier transform of function $\frac{\delta\varepsilon}{\varepsilon}(x,y,z) \times \frac{\delta\varepsilon^*}{\varepsilon}(X + x, Y + y, Z + z)$, we finally get:

$$I^2 = V\left\langle\left|\frac{\delta\varepsilon}{\varepsilon}\right|^2\right\rangle K[|\mathbf{k}|(l_2 - l_1), |\mathbf{k}|(m_2 - m_1), |\mathbf{k}|(n_2 - n_1)] \qquad (7.26)$$

where $K\lfloor|\mathbf{k}|(l_2 - l_1), |\mathbf{k}|(m_2 - m_1), |\mathbf{k}|(n_2 - n_1)\rfloor$ is the Fourier transform of the normalized spatial correlation function $\rho(x,y,z)$ of fluctuations of the relative dielectric permittivity of the ionospheric plasma, $\delta\varepsilon/\varepsilon$. Denoted now by χ, the angle between the directions of scattered wave $\mathbf{k}_2$ and the vector of electric field $\mathbf{E}_0$ (see Figure 7.31), we get the electric field at the receiver:

$$E = k^2 \sin\chi \cdot \Pi = \left(\frac{ZW}{32\pi^3}\right)^{1/2} k^2 \sin\chi \frac{e^{-i(\mathbf{k}_1\mathbf{r}_T+\mathbf{k}_2\mathbf{r}_R)}}{|\mathbf{r}_T \cdot \mathbf{r}_R|} \cdot I \qquad (7.27)$$

Then the density of the total power of the scattered wave (called the magnitude of Poynting vector) at the receiver equals [15,16]:

$$S = \frac{1}{2}|\mathbf{E} \times \mathbf{H}^*| = \frac{|\mathbf{E}|^2}{2Z} = \frac{W}{64\pi^3} k^4 \sin^2\chi \frac{1}{|\mathbf{r}_T \cdot \mathbf{r}_R|^2} \cdot |I|^2 \qquad (7.28)$$

From Equation 7.26, one can obtain a more general expression for the effective (or differential) cross section than that described by Equation 7.14, which describes forward scattering of radio waves from ionospheric volume V consisting plasma irregularities as follows:

$$\sigma(\chi,\lambda) = \frac{\pi^2 \sin^2\chi}{\lambda^4}\left\langle\left|\frac{\delta\varepsilon}{\varepsilon}\right|^2\right\rangle K[|\mathbf{k}|(l_2 - l_1), |\mathbf{k}|(m_2 - m_1), |\mathbf{k}|(n_2 - n_1)] \qquad (7.29)$$

Here, the character of plasma irregularities is defined by function K[·]. The dependence of σ from trace geometry and its geographic position is determined by the function K[·] and the angle χ.

Let us consider, as in Refs. [11–18], that volume V consists of the field-aligned plasma irregularities strongly stretched along $\mathbf{B}_0$ with scales $\ell_\parallel$ and $\ell_\perp$ along and across $\mathbf{B}_0$, respectively (e.g., *anisotropic*). Radar measurements carried out at the middle-latitude ionosphere to investigate scattering effects caused by natural inhomogeneous plasma structures have shown that the form of the normalized correlation function of fluctuation of plasma permittivity $\delta\varepsilon$ is closer to the Gaussian distribution (Equation 6.18) than to a polynomial distribution (Equation 7.21) [15–26]. The same tendency was shown by measurements made in the high-latitude inhomogeneous ionosphere [27–31] and in some radar observations in equatorial regions of the ionosphere [32–38], as well as in active experiments of heating of plasma by powerful radio waves [6–8,39–42]. From the problem sketched in Figure 7.32, it is seen that

$$\left[(l_2 - l_1)^2 + (m_2 - m_1)^2 + (n_2 - n_1)^2\right]^{1/2} = 2\sin\frac{\vartheta}{2} \qquad (7.30)$$

where ϑ is the angle of wave scattering.

In Figure 7.32 the following geometrical notations are introduced: T is the point of the transmitter; R the point of the receiver; TQ the incident wave at the ionospheric layer; QR the wave scattered under angle ϑ; $\mathbf{e}_1(l_1,m_1,n_1)$ and $\mathbf{e}_2(l_2,m_2,n_2)$ are the unit vectors of incident and scattered waves, respectively; the point C is the middle of the segment $|\mathbf{AB}|$; QS the direction of plasma irregularity; and the point C lies at the projection of QS at the plane $(\mathbf{e}_1,\mathbf{e}_2)$ containing both incident and scattered waves. For the case of mirror scattering (with angle ψ [15–18]) angles, $\angle SQA = \angle SQB$ and SQ $\perp$ AB. For the case of specular scattering, the following constraint is valid: $n_2 - n_1 = 0$.

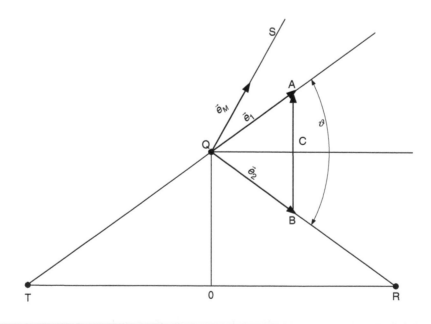

Figure 7.32 Geometry of the problem of forward scattering.

For the case where there is deviation from specular scattering, the angle between lines SQ and AB equals $\pi/2 + \psi$ and

$$|n_2 - n_1| = 2 \sin \frac{\vartheta}{2} |\sin \psi| \tag{7.31}$$

Substituting Equations 7.30 and 7.31 in Equation 7.29, after straightforward derivations, we get:

$$\sigma(\chi,\lambda) = \frac{2^{3/2} \pi^{7/2} \sin^2 \chi}{\lambda^4} \left\langle \left| \frac{\delta \varepsilon}{\varepsilon} \right|^2 \right\rangle \ell_\perp^2 \ell_\parallel$$

$$\times \exp \left\{ -2k^2 \left[\ell_\perp^2 \sin^2 \frac{\vartheta}{2} + \left(\ell_\parallel^2 - \ell_\perp^2 \right) \sin^2 \frac{\vartheta}{2} \sin^2 \psi \right] \right\} \tag{7.32}$$

In plasma, there is a relation between the normalized plasma permittivity $\varepsilon/\varepsilon_0$ and plasma wavelength λ_N [1,2]: $\varepsilon/\varepsilon_0 = 1 - \lambda^4/\lambda_N^4$, where λ is the wavelength of the radio wave. Then,

$$\left\langle \left| \frac{\delta \varepsilon}{\varepsilon} \right|^2 \right\rangle \cong \left(\frac{\lambda}{\lambda_N} \right)^4 \left\langle \left| \frac{\delta N}{N_0} \right|^2 \right\rangle \tag{7.33}$$

and

$$\sigma(\chi,\lambda) = \frac{2^{3/2} \pi^{7/2} \sin^2 \chi}{\lambda_N^4} \left\langle \left| \frac{\delta N}{N_0} \right|^2 \right\rangle \ell_\perp^2 \exp \left\{ -2k^2 \ell_\perp^2 \sin^2 \frac{\vartheta}{2} \right\}$$

$$\times \ell_\parallel \exp \left\{ -2k^2 \left(\ell_\parallel^2 - \ell_\perp^2 \right) \sin^2 \frac{\vartheta}{2} \sin^2 \psi \right\} \tag{7.34}$$

Equation 7.34 is a general equation for an effective cross-sectional description for the case of forward scattering from the ionospheric volume containing anisotropic plasma irregularities. From this general case two special cases can be obtained that have practical significance for radar echo observations. Thus, for plasma irregularities strongly stretched along $\mathbf{B}_0$, when $\ell_\parallel \gg \ell_\perp$, Equation 6.34 reduces to

$$\sigma(\chi,\lambda) = \frac{2^{3/2} \pi^{7/2} \sin^2 \chi}{\lambda_N^4} \left\langle \left| \frac{\delta N}{N_0} \right|^2 \right\rangle \ell_\perp^2 \exp \left\{ -\frac{8\pi^2 \sin^2 \frac{\vartheta}{2}}{\lambda^2} \ell_\perp^2 \right\}$$

$$\times \ell_\parallel \exp \left\{ -\frac{8\pi^2 \sin^2 \frac{\vartheta}{2}}{\lambda^2} \psi^2 \ell_\parallel^2 \right\} \tag{7.35}$$

For the other critical case of backscattering investigated by Booker [11–14], when $\chi = \pi/2$ and $\vartheta = \pi$, we get

$$\sigma_B(\chi,\lambda) = \frac{2^{3/2} \pi^{7/2} \sin^2 \chi}{\lambda_N^4} \left\langle \left| \frac{\delta N}{N_0} \right|^2 \right\rangle \ell_\perp^2 \exp \left\{ -\frac{8\pi^2}{\lambda^2} \ell_\perp^2 \right\} \times \ell_\parallel \exp \left\{ -\frac{8\pi^2}{\lambda^2} \psi^2 \ell_\parallel^2 \right\} \tag{7.36}$$

If we now change the angle ψ to Ψ_s introduced above in Booker's model, we obtain the same results obtained by Booker, which proves the results of derivations obtained in Refs. [15,16] investigating the general case of forward scattering.

We now will estimate the propagation effects of radio waves for forward scattering and backscattering. For this purposes, we will analyze the ratio σ/σ_B defined by Equation 7.35 for forward scattering, and by Equation 7.36 for backscattering, respectively, and the role of geometrical parameters of radio traces (ϑ, χ, ψ) in changes of scattered energy for both phenomena of radio scattering from the stratified plasma structures in the ionosphere. As follows from Equation 7.35, the ratio σ/σ_B is proportional to $\sin^2 \chi$, i.e., it depends on the type of radio wave polarization [16]. Furthermore, in the middle- and high-latitude ionosphere, the active region of ionospheric scattering is localized at the northern directions from a great circle (as well as localized at the southern directions from a great circle for the low-latitude ionosphere). Therefore, the constraint $\chi < \pi/2$ is valid in many cases of ionospheric radio scattering [11–16]. This fact leads to a decrease of σ for forward scattering compared with σ_B for backscattering. Furthermore, the existence in Equation 7.35 of the exponential term

$$A(\ell_\perp) = \ell_\perp^2 \exp\left\{ -\frac{8\pi^2 \sin^2 \frac{\vartheta}{2}}{\lambda^2} \ell_\perp^2 \right\} \tag{7.37a}$$

leads to an increase of differential coefficient of scattering σ compared with σ_B. In Figure 7.33, the ratio $A(\ell_\perp)/A_0(\ell_{0\perp})$ is shown against the radii of transverse correlation $\ell_\perp$ (for forward scattering) and $\ell_{0\perp}$ (for backscattering, when $\sin \frac{\vartheta}{2} = 1$), where

$$A_0(\ell_{0\perp}) = \ell_{0\perp}^2 \exp\left\{ -\frac{8\pi^2}{\lambda^2} \ell_{0\perp}^2 \right\} \tag{7.37b}$$

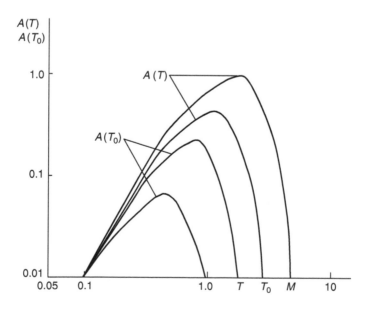

Figure 7.33 Dependence of the exponential terms $A(\ell_\perp)$ from Equation 7.37a and $A_0(\ell_{0\perp})$ from Equation 7.37b on the radii of transverse correlation for forward scattering and backscattering, respectively.

During the computations we used the data of experiments described in Refs. [15,16,45–52], which were carried out in Japan at frequency $f_1 = 56$ MHz ($\lambda_1 = 5.36$ m), in the former USSR at $f_2 = 44$ MHz ($\lambda_2 = 6.8$ m) and $f_3 = 75$ MHz ($\lambda_3 = 4.1$ m), and in the United States at $f_4 = 206$ MHz ($\lambda_4 = 1.45$ m) and $f_5 = 106$ MHz ($\lambda_5 = 2.83$ m). It is seen from presented illustrations that $A(\ell_\perp) > A_0(\ell_{0\perp})$ always at the same wavelength.

From constraints $dA(\ell_\perp)/d\ell_\perp) = 0$ and $dA_0(\ell_{0\perp})/d\ell_{0\perp}) = 0$, we can now find the optimal values of transverse radii of correlation that give maximal contributions in the power of forward and backscattering, respectively:

$$\ell_\perp^{(opt)} = \frac{\lambda}{2^{3/2}\pi \sin\frac{\vartheta}{2}} \quad \text{and} \quad \ell_{0\perp}^{(opt)} = \frac{\lambda}{2^{3/2}\pi} \tag{7.38}$$

Thus, for the frequencies mentioned earlier, we can easily derive that

$$\ell_\perp^{(opt)} \ (44 \text{ MHz}) = 1.9 \text{ m} \quad \text{and} \quad \ell_{0\perp}^{(opt)} \ (44 \text{ MHz}) = 0.77 \text{ m}$$

$$\ell_\perp^{(opt)} \ (56 \text{ MHz}) = 1.5 \text{ m} \quad \text{and} \quad \ell_{0\perp}^{(opt)} \ (56 \text{ MHz}) = 0.61 \text{ m}$$

$$\ell_\perp^{(opt)} \ (74 \text{ MHz}) = 1.1 \text{ m} \quad \text{and} \quad \ell_{0\perp}^{(opt)} \ (74 \text{ MHz}) = 0.45 \text{ m}$$

$$\ell_\perp^{(opt)} \ (106 \text{ MHz}) = 0.77 \text{ m} \quad \text{and} \quad \ell_{0\perp}^{(opt)} \ (106 \text{ MHz}) = 0.31 \text{ m}$$

$$\ell_\perp^{(opt)} \ (206 \text{ MHz}) = 0.4 \text{ m} \quad \text{and} \quad \ell_{0\perp}^{(opt)} \ (206 \text{ MHz}) = 0.165 \text{ m}$$

It is also clearly seen that for $\ell_\perp < \ell_\perp^{(opt)}$ and $\ell_\perp > \ell_\perp^{(opt)}$, functions $A(\ell_\perp)$ and $A_0(\ell_{0\perp})$ are sharply decreased. Taking into account Equation 6.38, one can estimate the increase of differential coefficient of scattering σ compared with σ_B, due to the term $A(\ell_\perp)$

$$\frac{\sigma}{\sigma_B} \sim \left(\frac{\ell_\perp^{(opt)}}{\ell_{0\perp}^{(opt)}}\right)^2 \approx 5\text{--}10 \tag{7.39}$$

for radio traces described in Refs. [15,16].

Now, let us estimate effects of deviations from the specular scattering, which is described by the angle ψ. These deviations from ψ are determined by the term $\sin^2 \frac{\vartheta}{2}$. Existence of this term in both exponents in Equation 7.35 significantly increase the effective area of scattering σ compared with σ_B; i.e.,

$$\frac{\sigma}{\sigma_B} \sim \frac{\exp\left\{-8\pi^2 \frac{\ell_\parallel^2}{\lambda^2} \psi^2 \sin^2 \frac{\vartheta}{2}\right\}}{\exp\left\{-8\pi^2 \frac{\ell_\parallel^2}{\lambda^2} \psi^2\right\}} \approx 10\text{--}15 \tag{7.40}$$

for radio traces described in Refs. [15,16]. Moreover, deviations of scattered radio wave from the angle ψ lead to a decrease of sensitivity of specular scattering both for forward scattering and backscattering. The width of the pattern of scattering (i.e., the maximum deviation $\Delta\psi$ from ψ), which is determined by decrease of scattered energy in e^{-1} times in the dominator and denominator of expression Equation 7.40, can be found from the corresponding expressions of differential cross sections, σ and σ_B, respectively. Finally, we get the width of the pattern of scattering:

For forward scattering

$$\Delta\psi_F = \frac{1}{2^{2/3}\pi \sin\frac{\vartheta}{2}}\frac{\lambda}{\ell_{\parallel}} \tag{7.41}$$

For backscattering

$$\Delta\psi_B = \frac{1}{2^{2/3}\pi}\frac{\lambda}{\ell_{\parallel}} \tag{7.42}$$

It can be seen that the width of the pattern of scattering (also called the effective width of diagram of irradiation from H_E irregularities [15]) for forward scattering differs compared with the same value for backscattering. This difference is determined by the term $\left(\sin\frac{\vartheta}{2}\right)^{-1}$. Thus, for the radio trace shown in Figure 7.3 and for $\alpha = \ell_{\parallel}/\lambda = 10, 3.16, 1$, we have for deviations $\Delta\psi$: $\Delta\psi_F = 3°24'$, $10°24'$, $32°40'$ for forward scattering, and $\Delta\psi_B = 1°18'$, $4°06'$, $12°57'$ for backscattering. For the same conditions in the ionosphere (i.e., for the same scales $\ell_{\parallel}$ and $\ell_{\perp}$), the width $\Delta\psi$, as follows from Equation 7.41, is directly proportional to wavelength λ and decreases with the increase of scattering angle ϑ. When $\ell_{\parallel}$ and $\ell_{\perp}$ are of the same order of magnitude, irregularities become isotropic, and the sensitivity to the specular scattering becomes weaker. Then, the effective cross section σ from Equation 7.35 becomes a function of only the angle ϑ, angle χ, conditions of the ionosphere (the ratio $\delta N/N_0$), and the average scale of isotropic irregularities ($\ell_{\parallel} = \ell_{\perp} = \ell_0$).

7.2.4 Power of H_E-Scatter Radio Signals

Using the knowledge of the effective cross section, one can easily obtain expressions for power of H_E-scattering signals using the following well-known relations between these main parameters of scattered signal [11–16]:

$$P_R = P_T\left(\frac{\lambda}{4\pi}\right)^2 \int\limits_V \frac{G_T G_R \sigma}{r_T^2 r_R^2}\,dv \tag{7.43}$$

The geometry of the radio trace and all notations are presented in Figure 7.31; G_T and G_R are the gain of the transmitter and the receiver antenna, respectively. The effective cross section of scattering, σ, is defined by Equation 7.34 or 7.35 for the case of forward scattering, and by Equation 7.36 for the case of backscattering. In our further description of the problem, we deal with forward scattering from anisotropic plasma irregularities following results obtained in Refs. [15–18].

For sufficiently long radio paths, when the working volume of the active zone of scattering is not too large (i.e., $V^{1/3} \ll r_T, r_R$), the product $r_T^2 r_R^2$ in the denominator of the integral (Equation 7.43) can be assumed to be constant and replaced on r^4, where r is the direct distance between the ground terminal antennas, the transmitter and the receiver, and the center of the projection of the active zone at the ground surface. If so, the integral (Equation 7.43) can be reduced to

$$P_R = P_T \left(\frac{\lambda}{4\pi}\right)^2 \int\limits_V \frac{G_T G_R \sigma}{r^4} \, dv \qquad (7.44)$$

Now we return to Equation 7.35 and take into account the fact that the factor $A(\ell_\perp)$ described by Equation 7.37a has a maximum for $\ell_\perp^{(opt)}$ described by Equation 7.38. As this factor decreases sharply for values of $\ell_\perp$ differing from $\ell_\perp^{(opt)}$, we can assume that irregularities of such transverse scales to $\mathbf{B}_0$ are mostly involved in the scattering process for which $\ell_\perp = \ell_\perp^{(opt)}$. In this case, the factor $A(\ell_\perp) = e^{-1}$.

Taking the foregoing into account and introducing the notation $\alpha = \ell_\parallel / \lambda$ introduced earlier, we finally get

$$P_R = P_T \frac{\lambda^5}{32e\sqrt{2\pi}} \frac{\alpha}{r^4} \left\langle \left(\frac{\delta N}{N_0}\right)^2 \right\rangle \int\limits_V \frac{G_T G_R \sin^2 \chi}{\sin^2 \frac{\vartheta}{2}} \exp\left(-8\pi^2 \alpha^2 \psi^2 \sin^2 \frac{\vartheta}{2}\right) dv \qquad (7.45)$$

For the practical use of this formula, one has to know the angle parameters ψ, ϑ, and χ at each point of the working volume of scattering. The corresponding computations based on the elements of analytical geometry are described in detail in Ref. [15], on the basis of geometrical and geophysical parameters of one of the middle-latitude radio traces. Using these parameters as well as the parameters of terminal antennas, one can estimate the effective volume of the active zone and other parameters having practical meaning for ionospheric radio link design.

Thus, the working volume can be found by numerical integration of Equation 7.45 by plotting the lines of the equal quantities $\gamma = \psi \sin(\vartheta/2) = const$ at the plane of the corresponding experiment, as shown in Figure 7.34. Here, the values of the angle γ are presented at the left of each line. The numbers in dB, which are given at the right side of the map, indicate the decrease of the received signal power due to the exponential factor $\sim \exp(-8\pi^2 \alpha^2 \gamma^2)$. The numbers in the first column correspond to $\alpha^2 = 10$, those in the second column correspond to $\alpha^2 = 100$ (irregularities with a larger degree of anisotropy). It is clearly seen that for $\alpha^2 = 10(\ell_\parallel = 3.16\lambda)$ integration in Equation 7.45 can be limited within the working volume by the isolines $\gamma = \pm 3°$, whereas for irregularities with a larger degree of anisotropy ($\alpha^2 = 100$, $\ell_\parallel = 10\lambda$), this volume lies between isolines $\gamma = \pm 1°$; i.e., it is much smaller. Figure 7.34 also shows the limitations of the working volume of scattering by effects of directivity of both terminal antennas having sufficiently narrow spatial radiation pattern. The ellipses E_T and E_R represent the intersection of the antenna diagrams by the horizontal plane plotted at the center of the active zone at heights of 100–105 km above Earth's surface. Experiments carried out at other geographical radio traces [11–14,45–50], as well as theoretical studies [16–18], have shown that the active zone of H_E-scattering is at altitudes from 100 to 110 km. So, in our future estimations, we can take a value of about 5–7 km for the effective height of the working volume. With this small height, the comparison with the linear dimensions of the active region allows us to assume that the working volume can be presented as a cylindrical body, strongly stretched along geomagnetic field lines.

Let us now estimate the accuracy of the foregoing theoretical analysis for description of scattering by meteor-induced H_E irregularities, the recorded signals from which have the *quasi-continuous* and *burstlike* nature (see Figures 7.6 and 7.8). As shown in Refs. [15,16], Equation 7.45

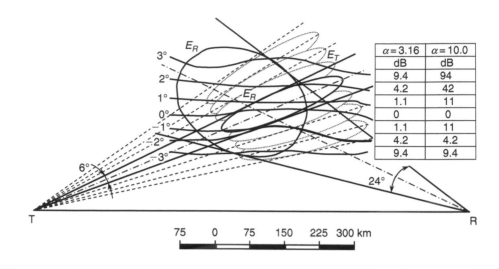

$\alpha = 3.16$ dB	$\alpha = 10.0$ dB
9.4	94
4.2	42
1.1	11
0	0
1.1	11
4.2	4.2
9.4	9.4

Figure 7.34 Schematic presentation of the geometry of localization of the active zone of forward scattering overlapped by ellipses of two antenna pattern. In table the received scattered echo signals (in dB) are presented for different meridians of position of elongated irregularities (indicated by solid curves and associated numbers from $-3°$ to $+3°$) and for two degrees of their anisotropy, $\alpha = 3.16$ and $\alpha = 10.0$.

is common for both types of signals. At the same time, it allows estimating different parameters of such types of irregularities and obtaining important information about the active zone of scattering, the characteristics of which are important for prediction of long-path radio channels on the basis of effects of radio scattering from nonregular inhomogeneous ionospheric plasma.

First, we will analyze Equation 7.45 for *quasi-continuous* signals, taking into account relations between the background plasma density N_0 and plasma frequency f_N (or plasma wavelength λ_N):

$$N_0 = \frac{f_N^2}{80.8} = \frac{9 \cdot 10^{16}}{80.8 \cdot \lambda_N^2} \tag{7.46}$$

Then, we can obtain the part of scattered power from element of volume ΔV, taking into account expression for the effective cross section (Equation 7.35):

$$\Delta P_R = P_T \frac{G_T G_R}{4\sqrt{2\pi}r^4} \frac{\lambda^2 \sin^2 \chi}{\lambda_N^4} \left\langle \left(\frac{\delta N}{N_0}\right)^2 \right\rangle \ell_\perp^2 \ell_\| \exp\left(-\frac{8\pi^2 \alpha^2 \gamma^2 \ell_\|^2}{\lambda^2}\right) \exp\left(-\frac{8\pi^2 \ell_\perp^2 \sin^2 \frac{\vartheta}{2}}{\lambda^2}\right) \tag{7.47}$$

As before, we assume that the main scattering effects can be obtained from irregularities with $\ell_\perp = \ell_{\perp opt}$, for which the last exponent in Equation 7.47 has a decay of e^{-1}. Finally, instead of a strict equation (Equation 7.45) for the total scattered power, we can obtain the approximate formula, which is more convenient for future estimations of parameters of the active zone of scattering. Substituting the parameter $\alpha = \ell_\|/\lambda$ in Equation 7.47 and taking the sum of all elements ΔV of the active zone of scattering, we can finally derive the power of H_E-scattering, replacing Equation 7.45 by the following expression:

$$P_R = P_T \frac{G_T G_R}{32e\sqrt{2\pi}} \frac{\alpha\lambda^5}{r^4\lambda_N^4} \left\langle \left(\frac{\delta N}{N_0}\right)^2 \right\rangle \sum_V \frac{\sin^2\chi}{\sin^2\frac{\vartheta}{2}} \exp\left(-8\pi^2\alpha^2\gamma^2\right)\Delta V \tag{7.48}$$

To estimate the effective zone of H_E-scattering, we use the geometry of the trace described earlier (see Figure 7.5) and the isolines $\gamma = \psi \sin(\vartheta/2) = const$, as shown in Figure 7.34. These lines, presented for two degrees of anisotropy of plasma irregularities, $\alpha = 3.16$ and $\alpha = 10$, allow us to estimate, together with the intersections of ellipses E_R and E_T, the effective "working" zone of H_E-scattering.

Furthermore, if powers P_R and P_T can be experimentally measured and the plasma wavelength obtained according to the corresponding ionograms of the ionospheric E region obtained by the ionospheric stations or ionosondes, the preceding equation (Equation 7.48) allows us to estimate the relative root-mean-square (*rms*) deviations of plasma (e.g., electron) density. This main parameter of ionospheric irregularities is important because it determines the intensity of scattering of radio waves from the inhomogeneous ionospheric plasma. Thus, from Equation 7.48, we get:

$$\left\langle \left(\frac{\delta N}{N_0}\right)^2 \right\rangle \approx \frac{217.6 \cdot r^4}{\lambda^5 P_T} \frac{U_R^2}{R_{in}} \frac{\lambda_N^4}{\alpha} \left[\sum_V \frac{\sin^2\chi}{\sin^2\frac{\vartheta}{2}} \exp\left(-8\pi^2\alpha^2\gamma^2\right)\Delta V \right]^{-1} \tag{7.49}$$

where U_R is the input voltage at the receiver and R_{in} the input resistance at the receiver. For determining elements ΔV of the active zone of scattering, it has to be divided on elements limited by the isolines $\gamma = \psi \sin(\vartheta/2) = const$ and the overlapping area of two antenna diagrams, as shown in Figure 7.34. Then, the product of area of each element on the height $\Delta h = 5$–7 km, mentioned earlier and obtained experimentally in Refs. [11–16], has to be derived and introduced in a total sum.

In experiments described in Ref. [15] the sensitivity toward the position of scattering plasma structures during H_E-scattering was examined. In this case, the receiver antenna has the constant pattern determined by ellipse E_R limited by the solid lines, and the transmitter antenna scanned in the azimuth domain (every 3° in azimuth, as shown in Figure 7.34). During the corresponding step of scanning, the azimuthal dependence of the signal power recorded by the fixed receiver can be measured. The dashed lines in Figure 7.34 give the position of ellipse E_T corresponding to several positions (i.e., steps) of the transmitter antenna during its scanning in the azimuth domain. To estimate the sensitivity of specular scattering to the position of the antenna pattern (e.g., ellipse), we have to determine for each position the following sum:

$$S(\varphi) = \sum_V G_T G_R \frac{\sin^2\chi}{\sin^2\frac{\vartheta}{2}} \exp\left(-8\pi^2\alpha^2\gamma^2\right)\Delta V \tag{7.50}$$

and compare it with the sum $S(0)$. Here, φ is the azimuthal angle of scanning from direction toward the center of the volume ΔV of the active zone of H_E-scattering defined by $\varphi = 0°$ with the corresponding sum $S(0)$.

The ratio of these two sums (which actually describes the ratio of the corresponding powers P_φ and P_0) is shown in Figure 7.35 for a volume limited mainly by the terminal antennas ellipses and by lines $\gamma = const$, as shown in Figure 7.34. As can be seen from Figure 7.35, the computed

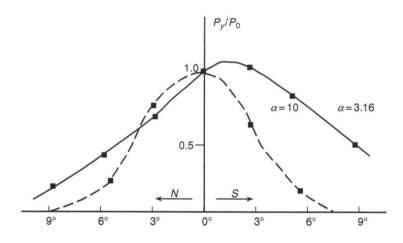

Figure 7.35 Longitudinal distribution of scattered power for two degrees of anisotropy of field-aligned irregularities, $\alpha = 3.16$ and $\alpha = 10.0$.

curves do not give any reason to expect a marked sensitivity toward the position of the scattering region for the radio path shown in Figure 7.5, because the half-power width of the graphs is $\Delta\varphi = 8°$ for $\alpha = 10$ and $\Delta\varphi = 14°$ for $\alpha = 3.16$.

From these experiments, the plasma density perturbation within the active zone of scattering was estimated. Thus, Table 7.1 shows the corresponding computations of the relative *rms* of plasma perturbations within irregularities estimated for experiments carried out at $f = 74$ MHz for $P_T = 4$ kW, $G_T = 23.4$ dB, $G_R = 17.4$ dB, $R_{in} = 75$ Ω (ohm), and $r \approx 700$ km, described earlier and shown in Figure 7.34, for different input voltages at the receiver (in μV), corresponding to the levels of the received quasi-continuous signals varied from 0.1 to 1.2 μV. Because the plasma wavelength at the E region of the middle-latitude ionosphere is usually $\lambda_N = 75$–100 m [1–3,16,52], estimated values of $(\delta N/N_0)^2 \approx 10^{-7}$–$6 \cdot 10^{-5}$ are more realistic. These estimations are close to those of $(\delta N/N_0)^2 \approx 2.1 \cdot 10^{-6}$–$2.9 \cdot 10^{-6}$ obtained during ionospheric experiments carried out at $f = 56$ MHz on the middle-latitude radio traces in Japan [47–50] for the same solar activity conditions under assumptions that $\alpha = \ell_\parallel/\lambda = 5$, 15.8. At the same time, estimations made in the equatorial regions of the ionosphere [37,38] showed that usually $(\delta N/N_0)^2 \approx 10^{-5}$–$6 \cdot 10^{-3}$ and in the polar ionosphere, it is also high enough, i.e., $(\delta N/N_0)^2 \approx 6 \cdot 10^{-4}$–$10^{-3}$ [29,31,40]. So, in equatorial and polar regions of the ionosphere the relative *rms* of plasma density of field-aligned H_E irregularities are approximately higher than that obtained for middle-latitude H_E irregularities by two to three orders of magnitude.

As for *burst-type* signals, which are most likely of meteor origin, presented in Figures 7.6 and 7.7, we should deal with a volume ΔV_F, where $(\Delta V_F)^{1/3}$ is the half width of the Fresnel zone that covers the scattered points of ΔV_F and the receiver. The volume ΔV_F can be estimated from Equation 7.48 as [15]

$$\Delta V_F = \left[\frac{\lambda r_T r_R}{(r_T + r_R)(1 - \cos^2\beta \sin^2\phi)} S \right]^{1/2} \qquad (7.51)$$

Here, S is the total cross section of a plasma inhomogeneity, β is the angle between the direction of the inhomogeneity and the TQR plane, and ϕ is a half-angle TQR (see Figure 7.5). As for cross

Table 7.1   RMS of Plasma Disturbances verses Plasma Wavelength and Voltage at the Input of the Receiver

| $\lambda_N,\ m$ | $\alpha = 3.16$ | | $\lambda_N,\ m$ | $\alpha = 10$ | |
	$U_R,\ \mu V$	$\overline{(\Delta N/N)}^2$		$U_R,\ \mu V$	$\overline{(\Delta N/N)}^2$
100	0.1	$2.3 \cdot 10^{-9}$		0.1	$3.3 \cdot 10^{-8}$
	0.5	$5.8 \cdot 10^{-7}$	100	0.5	$8.2 \cdot 10^{-7}$
	1.2	$3.3 \cdot 10^{-6}$		1.2	$4.8 \cdot 10^{-6}$
75	0.1	$7.2 \cdot 10^{-9}$		0.1	$1.0 \cdot 10^{-9}$
	0.5	$1.8 \cdot 10^{-7}$	75	0.5	$2.6 \cdot 10^{-7}$
	1.2	$1.0 \cdot 10^{-6}$		1.2	$1.5 \cdot 10^{-6}$
60	0.1	$3.0 \cdot 10^{-9}$		0.1	$4.3 \cdot 10^{-9}$
	0.5	$7.4 \cdot 10^{-8}$	60	0.5	$1.2 \cdot 10^{-7}$
	1.2	$4.3 \cdot 10^{-7}$		1.2	$6.4 \cdot 10^{-7}$
50	0.1	$1.4 \cdot 10^{-9}$		0.1	$2.1 \cdot 10^{-9}$
	0.5	$3.7 \cdot 10^{-8}$	50	0.5	$5.3 \cdot 10^{-8}$
	1.2	$2.1 \cdot 10^{-7}$		1.2	$3.1 \cdot 10^{-7}$
30	0.1	$1.8 \cdot 10^{-10}$		0.1	$2.6 \cdot 10^{-10}$
	0.5	$4.9 \cdot 10^{-9}$	30	0.5	$6.5 \cdot 10^{-9}$
	1.2	$2.6 \cdot 10^{-8}$		1.2	$3.9 \cdot 10^{-8}$

section S, it is reasonable to take the area of the circle whose diameter equals the optimal radius of the cross-correlation, i.e., the $\ell_{\perp opt}$. Now taking into account Equation 6.46, we can represent the factor $\langle (\delta N/N_0)^2 \rangle / \lambda_N^4$ in the form $8 \cdot 10^{-31} \cdot \langle (\delta N)^2 \rangle$. Using this relation, we can reduce Equation 7.48 to the more simple formula allowing estimation of the linear plasma (e.g., electron) density in an inhomogeneity in the case of a burst-type signal. With allowance for the foregoing, Equation 6.48 takes the form

$$P_R = P_T \frac{G_T G_R}{2.65 \cdot 10^{32}} \frac{\alpha \lambda^5}{r^4} \langle (\delta N)^2 \rangle \frac{\sin^2 \chi}{\sin^2 \frac{9}{2}} \exp\left(-8\pi^2 \alpha^2 \gamma^2\right) \Delta V_F \qquad (7.52)$$

From Equation 7.52, one can now estimate $\sqrt{\langle (\delta N)^2 \rangle}$, which is the quantity of linear electron density in the inhomogeneity for $\sqrt{\langle (\delta N)^2 \rangle} > N_0$ [15]. At the same time, electron density can be determined from the well-known formula of the signal power reflected from under-dense meteor trails [62,63]

$$P_R = P_T \frac{G_T G_R \lambda^3 q^2 r_e^2 \sin^2 \chi}{16\pi^2 r_T r_R (r_T + r_R)(1 - \cos^2 \beta \sin^2 \phi)} \qquad (7.53)$$

from which it is easy to find the linear density of electrons q and also the volume electron density $q_0 = q/V_0$, where V_0 is the volume of a column of unit meteor trail length and r_e is the radius of electron in meteor plasma. The comparable analysis of computations of linear electron density computed from Equation 7.52 for meteor-induced burst-type signals and that obtained from Equation 7.53 are presented in Table 7.2. It is clear that $\sqrt{\langle (\Delta N)^2 \rangle}$ is very close to q_0, mostly for $\alpha = \ell_{\parallel}/\lambda = 300$, which for $f = 74$ MHz ($\lambda = 4$ m) gives the length of the half-width of the Fresnel zone; i.e., it validates the theory of forward scattering proposed earlier for estimation of meteor-induced H_E irregularities of the lower ionosphere at altitudes of region E.

The method of estimation of degree of anisotropy of such irregularities [15,45,46,64] is based on comparison of the received signal level decay (in dB) for any azimuthal deflection of the transmitting antenna determined by an angle φ from the direction toward the center of the active scattered zone, denoted by $W(\varphi)$, with the signal decay (in dB) directly obtained from the center of the active zone, denoted by $W(0)$. Then, the parameter of anisotropy equals

$$\alpha = \frac{\ell_{\parallel}}{\lambda} = \frac{\sqrt{W(\varphi) - W(0)}}{4\pi \sqrt{5 \cdot \log 2 [\gamma^2(\varphi) - \gamma^2(0)]}} \tag{7.54}$$

For the middle-latitude radio trace described earlier and for $f = 74$ MHz, for changes in recorded level from $W(\varphi) = 3$ dB to $W(0) = 1.6$ dB we get, following Equation 7.54, that $\alpha = 7.4$ for the geometrical parameters of experiment shown in Figure 7.34. Operating at an other frequency $f = 44$ MHz, we get $\alpha = 3.63$. From experiments carried out at 206 MHz and described in Refs. [45,46], changes in signal power from 5 to 40 dB were registered, which finally give us $2 < \alpha < 9$, i.e., close to those obtained in Ref. [15]. Results obtained in radio traces in Japan operating at 56 MHz have shown that $5.9 < \alpha < 9.3$, which were also close to those obtained in the aforementioned reference.

It was discovered in all the cycles of experiments carried out in three different middle-latitude geographical zones of Earth—Japan, the United States, and the former USSR—that the transverse dimensions of field-aligned irregularities $\ell_{\perp}$ was close to $\ell_{\perp opt}$, at which the function $A(\ell_{\perp})$ is maximal. Decrease of this function from its maximum value by a factor of e allows us to estimate limits of $\ell_{\perp}$. Thus, for the experiments described in Refs. [15,16], carried out at $f = 74$ MHz, we get 0.4 m $< \ell_{\perp} < 2$ m. Moreover, experiments showed that irregularities responsible for H_E-scattering have more or less the same degree of anisotropy, approximately

Table 7.2   Average Plasma Density verses Degree of Anisotropy α of Meteors and Comparison with Volume Plasma Density q_0 of Meteors

α $\diagdown$ U_R, μV	3.16	10	20	300	
	$\sqrt{\overline{\Delta N^2}}$, e/m^3				q_0, e/m^3
0.1	$4.8 \cdot 10^{12}$	$2.65 \cdot 10^{12}$	$1.52 \cdot 10^{12}$	$4.8 \cdot 10^{11}$	10^{12}
0.5	$2.63 \cdot 10^{13}$	$1.33 \cdot 10^{13}$	$7.6 \cdot 10^{12}$	$2.43 \cdot 10^{12}$	$5.3 \cdot 10^{12}$
1.2	$5.6 \cdot 10^{12}$	$3.1 \cdot 10^{13}$	$1.84 \cdot 10^{13}$	$1.84 \cdot 10^{12}$	$1.3 \cdot 10^{13}$

within $5 \leq \ell_{||}/\ell_{\perp opt} \leq 18$ for irregularities responsible for quasi-continuous signals, and within $20 \leq \ell_{||}/\ell_{\perp opt} \leq 30$ for those responsible for burst-type signals, i.e., with a higher degree of anisotropy. Hence, the foregoing longitudinal and transverse scales of field-aligned E layer irregularities at middle latitudes obtained in experiments at 26, 44, 74, and 206 MHz [44,47,56] suggest that scattering is frequency selective. The active zone of H_E-scattering apparently contains a whole spectrum of small-scale anisotropic irregularities of different dimensions. Irregularities of given scales, depending on wavelength of radiated signals, are more involved in the irradiation process than those with other scales differ from mentioned earlier.

Hence, we can stress that the statistical character of *quasi-continuous* signals (see Figure 7.8) and *burst-type* signals (see Figures 7.6 and 7.7) is properly described by the theory of forward scattering of radio waves from local anisotropic irregularities of the E region of the ionosphere.

7.2.5 Forward and Backscattering from Irregularities of the F Region of the Ionosphere

The continuous and systematic experiments regarding investigation of scattering effects, back and forward, from small-scale field-aligned plasma irregularities of the F region of the ionosphere caused by natural phenomena [25–41] and artificially induced by heating of ionospheric plasma by powerful radio waves [85–94], were carried from the end of the 1970s to the middle of the 1980s by several groups of researchers at different geographical latitudes and longitudes of Earth. Then, during recent decades, theoretical models that describe the effects of such irregularities on radio scattering and propagation through the ionosphere were developed [19–22,95–102]. Using methods of radiolocation of small-scale anisotropic plasma irregularities by CW (or narrowband) and pulse (or wideband; see definitions in Chapter 3) sounding of the ionosphere in the HF/UHF frequency band, the main properties and parameters of such field-aligned irregularities, the active zones of their localization, as well as other effects of radio scattering, were analyzed theoretically and experimentally. In the following text, we briefly present the main results discussed in Refs. [95–102] regarding scattering of radio waves from anisotropic plasma structures of the region of the ionosphere that is strongly elongated along the geomagnetic field lines.

In Ref. [95], it was shown that forward scattering and backscattering of radio waves by anisotropic plasma irregularities of various origins can be determined by the following relation of their transverse scale $\ell_\perp$ (to $\mathbf{B}_0$), the angle of scattering θ, and the wavelength λ:

$$\ell_\perp = \lambda/2 \sin(\theta/2)$$

It determines the specular sensitivity, i.e., the fact that energy of irradiation is concentrated in the narrow angle interval inside the cone of specular reflection. This definition follows from expressions of effective cross section of radio scattering described by Equations 7.34 through 7.36. In Refs. [95,100], for Gaussian-type field-aligned irregularities, the effective cross section was presented in the same form as in Equations 7.34 through 7.36, i.e., as

$$\sigma(\gamma,\lambda) = \frac{k^4 \ell_\perp^2 \ell_{||} \langle (\delta\varepsilon)^2 \rangle}{16\pi^{1/2}} \exp\left\{ -\left(k^2 \ell_{||}^2 \sin^2 \frac{\theta}{2} \right) \left(\cos^2 \gamma + \frac{\ell_\perp^2}{\ell_{||}^2} \sin^2 \gamma \right) \right\} \tag{7.55}$$

where $k = \omega/c = 2\pi/\lambda$, $\langle (\delta\varepsilon)^2 \rangle = \frac{f_N^4}{f^4} \left\langle \left(\frac{\delta N}{N_0} \right)^2 \right\rangle$, $\cos \gamma = (\cos \psi - \cos \chi)\left(2 \sin \frac{\theta}{2}\right)^{-1}$.

As before, angle θ is the angle of scattering; the angles χ and ψ are the angles between wave vectors of incident and scattered waves and the direction of $\mathbf{B}_0$ in the region of scattering, respectively, as shown in Figure 7.31. From Equation 7.55 it follows that for $\lambda \ll \ell_{\parallel}$ and $\ell_{\perp} \gg \ell_{\parallel}$, the energy of the scattered radiation is concentrated in the area of the cone of specular reflection, where $\psi \approx \chi$, and the angular width of such a cone can easily be estimated by the following expression: $\Delta\psi = \lambda/\pi\ell_{\parallel} \sin\psi$. As is mentioned in Refs. [95–97,100], this peculiarity of radio scattering from plasma irregularities strongly stretched along $\mathbf{B}_0$ can be usefully used for over-horizon HF/UHF radio communication and for the excitation of waveguide modes propagation with a low degree of attenuation.

The same contour of the active zone of intersections of specular cone with the Earth's surface can be obtained using the theoretical approach proposed in Ref. [100], which uses the geometry of scattering presented in Figures 7.31 and 7.32. Thus, in Figures 7.36 through 7.39 are shown the specular contours at the ground surface as generated by specular scattering from the scattering volume ΔV at the altitude z_c of the scattered area above Earth's surface for different values of the scattered angle ψ (denoted by different numbers): 1 corresponds to $\psi = 87°$, 2 to $\psi = 93°$, 3 to $\psi = 92°$, 4 to $\psi = 90°$ (which corresponds to the contour of backscattering), 5 to $\psi = 88°$, 6 to $\psi = 91°$, and 7 to $\psi = 89°$. At the axes, the corresponding geographic coordinates are denoted, the latitude φ (corresponding to middle latitudes in the territory of the former USSR) and the longitude λ (corresponding to eastern longitudes in the territory of the former USSR) of the points of intersection of scattered rays with the Earth's surface. Figure 7.36 was computed for an experiment that corresponds to radiated frequency of the probing waves $f = 10$ MHz and which deals with scattering observed at the height $z_c = 200$ km; Figure 7.37 is for $f = 25$ MHz. The same contours of specular scattering from field-aligned irregularities of the F region occurring in the middle-latitude ionosphere were observed in experiments carried out at $f = 25$ MHz when the main scattering effects were observed at the altitudes surrounding $z_c = 230$ km (Figure 7.38) and $z_c = 260$ km (Figure 7.39). Also taking effect of refractions described in Ref. [100], other shapes of

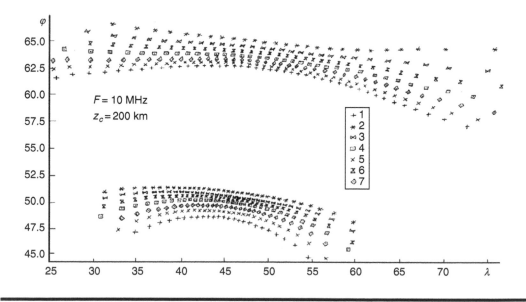

Figure 7.36 Longitudinal–latitudinal distribution of irregularities responsible for mirror scattering from the ionosphere at $z = 200$ km operating at a frequency of 10 MHz.

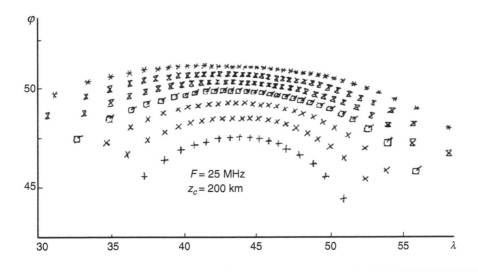

Figure 7.37 The same as Figure 7.36, but for a frequency of 25 MHz at $z = 200$ km.

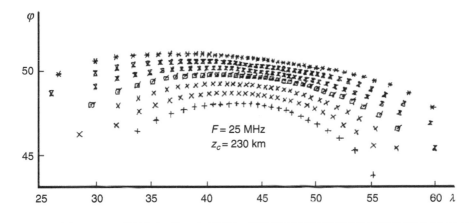

Figure 7.38 The same as Figure 7.37, but for $z = 230$ km.

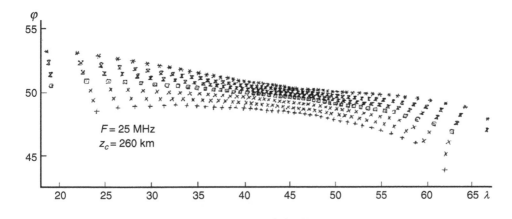

Figure 7.39 The same as Figure 7.37, but for $z = 260$ km.

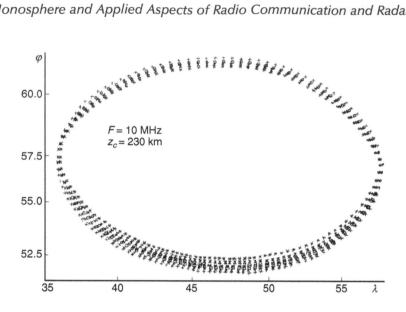

Figure 7.40 The same as Figure 7.36, but for $z = 230$ km.

contours of the active zone can be obtained, as shown in Figures 7.40 and 7.41, computed for $f = 10$ MHz, $z_c = 230$ km and $z_c = 260$ km, respectively.

Special experiments for observation of forward sounding of the F region of the middle-latitude ionosphere after its heating by powerful radio waves at the trace Moscow–Gorky–Kiev were carried out in April 1984; the results were published in Refs. [95–98] and summarized in Ref. [100]. The ionosphere was perturbed by the vertical pump waves using ground-based facilities located near Gorky (now Nizhny Novgorod). During the operation of the transmitter, the probing radio was located near Moscow and the receiver located near Kiev. During the entire working time of the high-power transmitter (i.e., the heater), additional signals were observed at the receiver, which

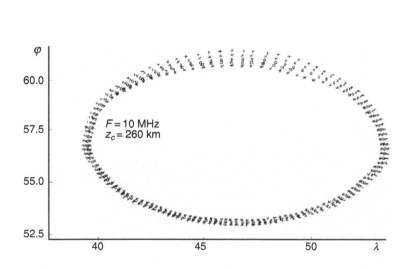

Figure 7.41 The same as Figure 7.36, but for $z = 260$ km.

were caused by specular scattering of probing radio waves at the small-scale anisotropic heating-induced plasma irregularities. Also, taking into account real ionograms of vertical sounding of the disturbed ionosphere after heating above Gorky, as well as ionograms of the nondisturbed ionosphere Moscow and Kiev, the numerical computation of conditions of specular scattering of probing radio waves radiated at $f = 16$–20.5 MHz were carried out. In computations, the scattered area was modeled by a disk with a 50 km radius and a thickness of 20–30 km having a center above the point of the transmitter of the pump waves (the heater) at the ground. The horizontal dimensions of the area of scattering in the ionosphere above the center of heating were determined by the width of the antenna pattern of the heating transmitter at the height of the induced wave; the vertical dimensions were obtained from experimental data according to Ref. [98]. As a result, the geographic coordinates of points satisfying conditions of specular scattering of corresponding radio rays were obtained. On the basis of computations of the time of group propagation of scattering rays, τ_{gr}, from transmitter to the scattered area and then from the scattered area to the receiver, the spread (diffusion) of receiving signals due to scattering in the time domain, $\Delta\tau_{gr}$ (or the frequency shift Δf in the frequency domain), was estimated. In Figure 7.42, the results of computations of the active zone of radio signals recorded at Earth's surface are presented in the same manner as was shown in Figures 7.36 through 7.41. Here, following pictures from the left-side to the right-side both at the top and at the bottom of Figure 7.41, we have respectively: $f = 16$ MHz, $\Delta\tau_{gr} = 0.4$ μs ($\Delta f = 2.5$ kHz), $f = 17$ MHz, $\Delta\tau_{gr} = 0.5$ μs ($\Delta f = 2.0$ kHz), $f = 18$ MHz, $\Delta\tau_{gr} = 0.6$ μs ($\Delta f = 1.66$ kHz), $f = 19$ MHz, $\Delta\tau_{gr} = 0.6$ μs ($\Delta f = 1.66$ kHz), $f = 20$ MHz, $\Delta\tau_{gr} = 0.5$ μs ($\Delta f = 2.0$ kHz), and $f = 20.5$ MHz, $\Delta\tau_{gr} = 0.3$ μs ($\Delta f = 3.33$ kHz). The computed bandwidth of the specular-scattered signals from $f = 16$ MHz to $f = 20.5$ MHz and the large spread (shift) of scattering signals of $\Delta\tau_{gr} = 0.6$ μs at $f = 18$ MHz to $f = 19$ MHz fully correspond to experimental data [95–98].

Another cycle of experiments to find the stratified structure of the disturbed ionosphere after its heating by powerful radio waves was carried out during the middle of the 1980s. The interpretation, analysis and theoretical background of the observed effects were described in Refs. [19,21,99,101]. Let us briefly introduce the reader to the observed results. The one-hop middle-latitude radio path

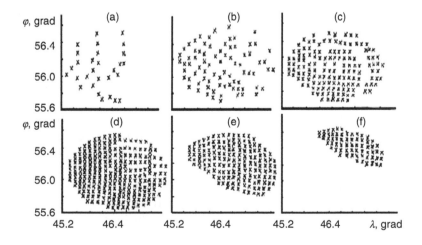

Figure 7.42 Longitudinal–latitudinal mapping of the active zone of backscattering.

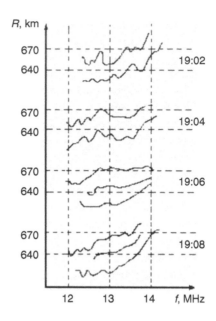

Figure 7.43 Ionograms of backscattering obtained in Ref. [21], from heating-induced irregularities at the middle-latitude ionosphere.

with range 610–680 km with the heating transmitter installation at the midpoint of the trace and using the transmitter of probing radio waves operating at the frequency band of 12–14 MHz (i.e., near the maximum useful frequency of the F layer, $f_{F_2} \approx 13$ MHz) was performed to investigate backscattering of radio waves from heating-induced artificial ionospheric irregularities [101]. The observed structure of the scattered signals was complicated by the increase of the duration of heating, the two to three traces on the range–frequency characteristic diagram, as shown in Figure 7.43. To explain the observed striated structure of the disturbed plasma after the heating process, in Ref. [102] the problem was solved for the diffraction and reflection of waves from an artificially disturbed region of the ionosphere, the plasma density and, thus, the dielectric permittivity $\varepsilon(h)$, the height dependence of which was modeled according to results obtained in Ref. [19,21] as

$$\varepsilon(h) = 1 + \delta \sin Kh \qquad (7.56)$$

where δ is the amplitude of the inhomogeneity fluctuations and K is the inverse scale of the irregularity. The wave-vector of the probing wave was oriented to be normal to the geomagnetic field $\mathbf{B}_0$. The solution yielded a reflection coefficient for a strongly oscillatory monochromatic wave. Generalization of the solution to the short trapezoidal radio pulse used in experiments [101] allows for the calculation of the envelope of the signal diffracted from small-scale anisotropic ionospheric irregularities, taking into account results obtained in Ref. [19,21]. In Figure 7.44, the shape of the signal envelope after diffraction and reflection is shown. As can be seen, the shape of the signal is strongly irregular, and the number of maxima corresponds to the number of traces observed experimentally and shown in Figure 7.43. So, using results obtained in Ref. [19,21] as well as those described in Chapter 6 regarding the problem of reflection and diffraction of radio signals from artificially disturbed regions of the upper ionosphere [102] yields a satisfactory agreement with the results of the experimental studies of backscattered signals described in Ref. [101].

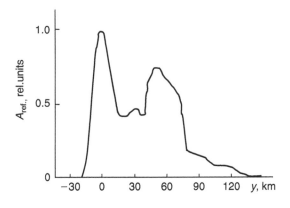

Figure 7.44 The shape of signal envelope obtained in Ref. [21] during experiments.

Now, using the same approach presented in Section 7.2.4, we can, as was done in the previous sections, estimate the parameters of the field-aligned small-scale plasma irregularities observed at the upper ionosphere. Thus, the knowledge of the effective altitudes (about 220–310 km) of the backscattering at small-scale field-aligned natural irregularities obtained in experiments carried out in the F layer of the subpolar ionosphere [98] at a frequency band of 11–20 MHz, as well as geomagnetic latitudes of the region of observation ($\varphi \approx 57°-58.5°$), the angles of incidence of scattered signals ($\approx 11° \pm 3°$), and the delay of these signals ($\tau \approx 7-10$ μs), allows estimation of the horizontal and vertical scales of the volume of scattering. In experiments described in Ref. [98], the vertical scale was estimated in the same manner as in Section 7.2.3 by the decay in e^{-1} time of the exponential (Gaussian) term in the expression of the signal power at the receiver for longitudinal scales $\ell_\parallel$ (to $\mathbf{B}_0$). In this case the range of the corresponding angles is $\Delta\psi \approx \Delta\theta = \lambda/\pi\ell_\parallel$. The horizontal scale of the volume of scattering was determined by the transmitter antenna pattern $\Delta\phi \approx 15°$. Finally, the effective value of backscattering was estimated as $V_{eff} \approx 3 \cdot 10^5$ km^3.

Also, taking into account that $\alpha = \ell_\parallel/\lambda = 5$ and $\ell_\perp = \lambda/2$ (for the case of backscattering), the deviations of plasma density with respect to the background plasma were also finally estimated as $\left[\langle(\delta N/N_0)^2\rangle\right]^{1/2} \approx 1.7 \cdot 10^{-3}$, i.e., twice as high as that obtained for the middle-latitude ionosphere (see Section 7.2.4).

At the same time, experiments carried out for the heating-induced anisotropic irregularities generated at the F region of the middle-latitude ionosphere [94] show the same order perturbations of the background ionospheric plasma inside such irregularities, $\left[\langle(\delta N/N_0)^2\rangle\right]^{1/2} \approx 10^{-2}-10^{-3}$, for irregularities with transverse scales $\ell_\perp \approx 10-25$ m. The range of altitudes occupied by the small-scale artificial irregularities is concentrated near the critical height of reflection of the powerful heating wave and for such irregularities is about $\Delta h \approx 20-30$ km [94].

7.3 Long-Distance Radio Propagation through the Ionospheric Channels

In recent decades modification of the ionosphere by powerful ground-based HF facilities has attracted much attention, not only for purely scientific geophysical purposes but also for the problems of artificially channeling of probing radio waves using artificially or naturally generated plasma irregularities by capturing radio waves into the ionospheric waveguides or implementing the multi-hop ionosphere–land communication channel [100–103]. As shown in Chapter 6, the heating of the real inhomogeneous ionospheric plasma by the powerful HF waves leads, in the E and F regions of the ionosphere, to excitation of heating-induced plasma instabilities (e.g., waves; see Chapter 2) at frequencies in the vicinity of plasma resonances [85–103]. In this case, most of the radio wave energy is absorbed and transformed into energy for the plasma wave oscillations [90–93], and finally leads to intensive local heating of plasma electrons or ions (as plasma is quasi-neutral; see Chapter 1) owing to the collision dissipation of this energy of radio wave. As a result, increase (enhancements) or decrease (depletions) of plasma density in the heating regions is observed experimentally [85–90].

The same effects and similar processes are observed using ionosondes or radars at high latitudes [105–107] and at the magnetic equator [32–38], which are related to the heating of background plasma electrons by highly energetic photoelectrons or cosmic rays [29–31,39–41]. For efficient performance of wireless communication links associated with long-distance radio propagation through ionospheric channels, created both artificially and naturally, it is important to be able to predict the influence of ionospheric heating processes on broadband propagation. Mentioned in Refs. [29–31,39–41] motivated to use the approach based of ray tracing simulation in the presence of heated-induced (i.e., perturbed) plasma media.

In the following text, we will first describe the effects of radio wave propagation through the heating plasma regions containing irregular plasma density structures and profiles generated by parametric plasma instabilities (see Chapter 2) during heating of ionospheric plasma by powerful radio waves or containing moderate- and small-scale plasma structures generated by different kinds of instabilities at the E and F layers of the natural ionosphere. Then, we will discuss briefly the capture of probing wave energy into the ionospheric waveguide caused by the multilayer structure of the regular ionosphere. Finally, we will analyze the so-called channeling of these waves at extremely long distances owing to the multiple-hopping effect.

To more clearly understand the aforementioned processes, we need to asses the effects of losses on the ray paths in the absorbing plasma media completely described in Refs. [108–117] for the naturally stratified layered structure of the ionosphere as an anisotropic media, and then modified for the case of inhomogeneous ionospheric plasma containing heating-induced plasma irregularities [118,119].

7.3.1 Wave Propagation in the Ionosphere Disturbed by Powerful Radio Waves

In this section, our main goal is to assess for the reader the behavior of radio waves resulting in significant changes of the plasma density height profiles during the resonance heating process using ray presentation. Here, following Ref. [119], we will present briefly the extended model of ray tracing (with respect to the familiar Hamilton equations [115–118]) to examine the influence of losses on ray paths for HF/UHF propagation through the E and F layers of the natural ionosphere, taking into account their real plasma density profiles and the corresponding collision frequencies (see Chapter 1). Despite the fact that for low losses and smooth plasma height profiles the problem is understood quite well [110–118], new effects were found [119] for irregular profiles generated in the case of heating, namely, significant increase (or decrease, depending on the height of heating, see Chapter 6) of plasma density and the corresponding irregular perturbations of its height profile, facilitating reflections of probing radio waves passing such disturbed regions at frequencies transcending the HF regime of 3–30 MHz, up into the 1 GHz range, and, finally, changes of the ray trajectories (creating one-hop and more radio paths) and of the field strength of probing radio waves.

As the redistribution of plasma density and the changes in its profile during resonance heating, local and adaptive, described in Chapters 5 and 6, finally lead to a very complicated physical processes, the analytical investigations of the problem are limited by the complexity of the wave propagation through such a stratified ionospheric structure and by complicated physical processes accompanying resonance heating of plasma. All this leads to increasingly complex dispersion relations between plasma characteristics and parameters, and to very complicated wave equations describing the ray-tracing effects. Therefore, for analysis of the problem we use a numerical model and the corresponding ray equation formalism based on the extended Hamilton equations by introducing the dispersion relation $F(\mathbf{k},\omega,\mathbf{r},t) = 0$ in the complex plane, that is, considering whether the propagation vector $\mathbf{k} = (k_x,k_y,k_z)$ and the angular frequency ω may be complex [118,119]:

$$\mathbf{v} = \frac{d\mathbf{r}}{dt} = -\frac{\partial F/\partial \mathbf{k}}{\partial F/\partial \omega}$$

$$\frac{d\mathbf{k}}{dt} = \frac{\partial F/\partial \mathbf{r}}{\partial F/\partial \omega} + i\boldsymbol{\beta} \tag{7.57}$$

$$\frac{d\omega}{dt} = -\frac{\partial F/\partial t}{\partial F/\partial \omega} + i\mathbf{v} \cdot \boldsymbol{\beta}$$

Here, $\mathbf{r} = (r_x,r_y,r_z)$ is the local vector in the space domain, t is the time, which also serves to define a parameter along the ray trajectory, and $\boldsymbol{\beta}$ the vector that guarantees a real group velocity $\mathbf{v}$ along the ray trajectory defined as

$$\boldsymbol{\beta} = -\left[\text{Re}\left(\frac{\partial \mathbf{v}}{\partial \mathbf{k}} + \frac{\partial \mathbf{v}}{\partial \omega} \cdot \mathbf{v}\right)\right]^{-1} \cdot \text{Im}\left[\frac{\partial \mathbf{v}}{\partial \mathbf{k}} \cdot \frac{\partial F/\partial \mathbf{r}}{\partial F/\partial \omega} - \frac{\partial \mathbf{v}}{\partial \omega} \cdot \frac{\partial F/\partial t}{\partial F/\partial \omega} + \frac{\partial \mathbf{v}}{\partial \mathbf{r}} \cdot \mathbf{v} + \frac{\partial \mathbf{v}}{\partial t}\right] \tag{7.58}$$

The expression (Equation 7.58) follows from the constraint

$$d(\text{Im}(\mathbf{v}))/dt = 0 \tag{7.59}$$

which states that along the ray trajectory the incremental imaginary part of the group velocity is equal to zero. Thus, the model presented in Refs. [108,109] and described by Equations 7.57 and 7.58 extends the real Hamilton equations into the multivariate complex space but adds a constraint to force ray trajectories and the group velocity into the real part of the complex space $(\mathbf{r},t)$, and to retain the analcites of the dispersion equation. In other words, the constraint (Equation 7.59) guarantees that the ray trajectories are always present in joint spatial-temporal domain, and consequently also for the associated group velocity because the propagation vector and the frequency are assumed to be complex. Earth's magnetic field in such a model was approximated by a localized dipole field with the dipole located near the center of Earth. Under these conditions, the Appleton–Hartree (sometimes called Appleton–Lassen [116,117]) dispersion relation, obtained for the cold, nonisothermal, collisional, magnetized ionosphere, i.e., for real ionospheric conditions, occurs in the E and F regions of the ionosphere (see Chapters 1 and 6). It can be presented as [118,119]:

$$F = k^2 - \frac{\omega^2}{c^2}\left[1 - \frac{X}{(1 - iZ) - \frac{Y_\parallel^2}{2(1-X-iZ)}} \pm \sqrt{\frac{Y_\parallel^4}{4(1 - X - iZ)^2} + Y_\perp^2}\right] = 0 \tag{7.60}$$

where $X = \omega_N^2/\omega^2$, $Z = \nu/\omega$, $Y_{\parallel} = |\mathbf{Y}| \cos\psi$, and $Y_{\perp} = |\mathbf{Y}| \sin\psi$ are the longitudinal and transverse components of the normalized geomagnetic field vector $\mathbf{Y}$ with respect to the direction of $\mathbf{k}$, and $|\mathbf{Y}|$ is the magnitude of vector $\mathbf{Y}$. The electron plasma frequency denoted by ω_N can be defined as $\omega_N^2 = 4\pi e^2 N_0/(m_e\varepsilon_0)$, the gyrofrequency denoted by ω_{eH} defined in Chapter 1, and m_e and e are the mass and the charge of electron component of plasma, respectively. Here, $\nu = \nu_{em} + \nu_{ei} = \nu_{em}(1+p)$ is the collision frequency related to losses, $p = \nu_{ei}/\nu_{em}$, ε_0 the permittivity of plasma, N_0 the background plasma density, and ψ the angle between vectors $\mathbf{k}$ and $\mathbf{Y}$. The absolute value of the vector $\mathbf{Y}$ can be defined from knowledge of its components in the right-handed Cartesian system with the z-axis in the direction of the North geomagnetic pole, for which angle $\theta = 0$ (see Figure 7.45, where the angle θ is subtended by the vector $\mathbf{v}$ of group velocity and the z-axis) as [118,119].

$$|\mathbf{Y}| = \sqrt{Y_{\parallel}^2 + Y_{\perp}^2} = \frac{|\mathbf{M}|}{\omega}\sqrt{\left(-\frac{3xz}{r^5}\right)^2 + \left(-\frac{3yz}{r^5}\right)^2 + \left(-\frac{r^2 - 3z^2}{r^5}\right)^2} \qquad (7.61)$$

where
$\mathbf{M} = \frac{\mu_0|e|\mathbf{H}}{m_e}$ is Earth's magnetic dipole moment
μ_0 the permeability of the medium
$\mathbf{H}$ the magnetic field strength vector

The derivatives in the ray equations (Equations 7.57 and 7.58) are quite complicated to analytically solve for the real model of ionospheric plasma, parameters of which vary with height and conditions of experiment (see Chapters 1 and 6). The corresponding data for numerical computation of Equation 7.57 was obtained from the corresponding solutions of the heating problem and from the tables presented in Section 5.3. Then the obtained data was converted with a quasi-parabolic approximation to smooth functions describing the various plasma profiles after its heating [99–103]. These quasi-parabolic profiles also serve to integrate the ray equations using suitable mathematical tools and the corresponding numerical code [118,119]. The variations of ray

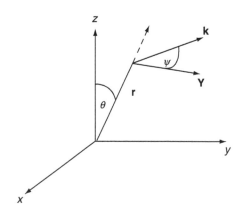

Figure 7.45 Geometry of the problem of radio propagation through the disturbed ionosphere according to Ref. [119].

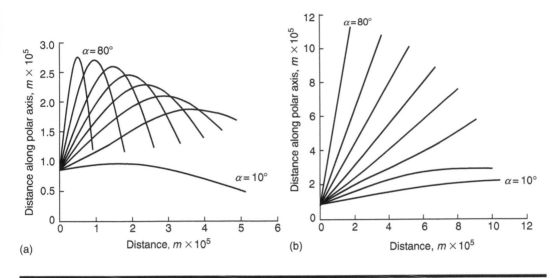

(a)   (b)

Figure 7.46 Trajectories of radio waves for various initial grazing angles α and for $f = 10$ MHz (a) and $f = 30$ MHz (b).

paths according to frequency, grazing angle, collision frequency, and plasma density profile before, during, and after heating process were examined for the E and F layers of the ionosphere.

Now let us present computations of the ray trajectories for the quasi-regular non-disturbed ionosphere with the corresponding smooth altitudinal profile of background plasma density. The results are shown in Figure 7.46a and b for different frequencies for the ordinary mode of probing waves ($f = 10$, 30 MHz), and for various grazing angles α ($\alpha = 90° - \theta$, $\theta = 10°$, $20°, \ldots, 80°$). In the figures, the frequency of probing radio waves passing through the ionosphere is at the same level or higher than the maximum usable frequency (MUF) of the investigated ionospheric layers for the ordinary mode; i.e., for $f < 20$ MHz [119]. As can be seen from illustrations, in the quasi-regular nondisturbed ionosphere for probing waves of 10 MHz, i.e., less than 20 MHz, and for small grazing angles $\alpha = 10°-30°$, the only one-hop propagation is observed because of reflection from the ionospheric quasi-regular plasma layers. With decreasing frequencies of the probing waves (from 10 to 30 MHz), the possibility of obtaining propagation channels with higher grazing angles (up to $\alpha = 50°-60°$) increases (see Figure 7.46a). This effect is recognized in commonly used propagation ionospheric channels for frequencies of probing waves that are close to the MUF for each ionospheric layer. After a virtual experiment with the local resonance heating of plasma described in details in Section 5.3, the same ray trajectories derived from the foregoing equations can be obtained. Results are shown in Figure 7.47a through c for the same variations of grazing angles of probing radio waves with an angular step $\Delta\alpha = 10°$, and for different frequencies $f = 30, \ldots, 400$ MHz, and $f = 1$ GHz, respectively. In computations, the resonance frequency was $f_0 = 8.6$ MHz, which according to results presented in Section 5.3 corresponds to the value of resonance plasma density $N_R = 9 \cdot 10^{11}$ m^{-3}, i.e., to the initial height of the local heating source $h_R = 240$ km. This disturbed plasma density profile due to the heating process was obtained in Section 5.3, and the disturbed collision frequency $\nu_{ei} \approx T_e^{-1/2}(z)N_e(z)$ was also obtained from results presented in previous figures. For the case of local resonance heating, we use for computations the ray-tracing

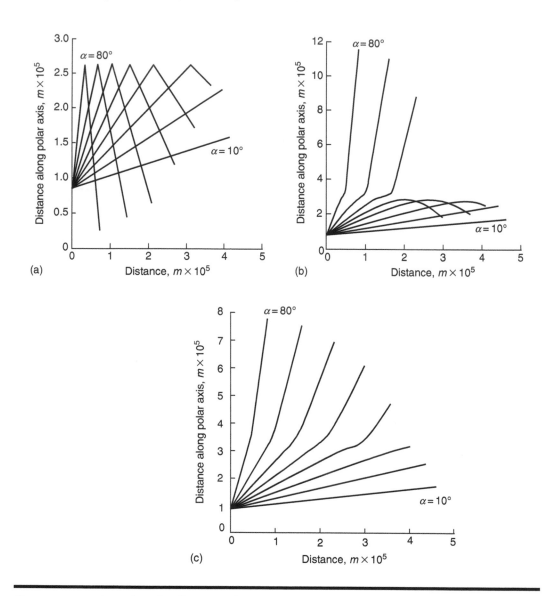

Figure 7.47 Trajectories of radio waves for various initial grazing angles α and for $f = 10$ MHz (a), $f = 20$ MHz (b), and $f = 30$ MHz (c).

model (Equations 7.57 through 7.61) and the following quasi-parabolic approximation for plasma electron density [99,119]

$$N_e = N_{e\max}\left[1 - \left(\frac{r - r_m}{h_m}\right)\right], \quad r_b < r < r_m + h_m \tag{7.62}$$

where N_e and $N_{e\max}$ are the regular and the maximum density of electrons in plasma, respectively; r is the radial distance from Earth's center; r_m is the value of r where $N_e = N_{e\max}$; r_b is the value of r at the layer base, i.e., at $r_m - h_m$; and h_m is the layer semithickness. The corresponding data is shown in Table 7.3. For the case of a plane stratified ionosphere, Equation 7.62 enables a smooth

Table 7.3 Electron Density $N(h)$

Altitude h (km)	$N_0(h) \cdot 10^{11}$ (m^{-3})	$N(h)/N_0(h)$
135	2.05	1
140	2.2	1
145	2.7	1
150	2.95	1
155	3.05	1
160	3.2	1
165	3.55	1
170	3.7	1
175	3.8	1
180	4.0	1
185	4.2	1
190	4.4	1
195	4.7	1
200	5.05	1
205	5.3	1
210	5.8	1
215	6.0	1
220	6.5	1
225	7.0	1
230	7.5	1
235	8.2	0.92
240	9.0	0.85
245	9.5	0.83
250	10.0	0.82
255	11.0	0.75
260	12.0	0.68
265	12.5	0.67
270	13.0	0.67

(continued)

Table 7.3 (continued)   Electron Density N(h)

Altitude h (km)	$N_0(h) \cdot 10^{11}$ (m^{-3})	$N(h)/N_0(h)$
275	13.5	0.67
280	14.0	0.67
285	14.5	0.68
290	15.0	0.69
295	15.5	0.72
300	16.0	0.75
305	16.1	0.81
310	16.2	0.86
315	16.4	0.92
320	16.5	0.97
325	16.2	1
330	16.0	1

transition between the quasi-parabolic layers, providing continuous gradients at the boundaries. The height of each layer was chosen by taking into account the changes appearing in the data. The electron collision frequency (i.e., losses) has been approximated by an exponential function

$$\nu = 2.074 \cdot 10^6 \cdot \exp\left(-4.8 \cdot 10^{-2} \cdot h\right), \text{ Hz} \qquad (7.63)$$

and presented in Table 7.4, where p_0 and p are the regular and perturbed parameters of ionization of the ionospheric plasma.

From Figure 7.47a through c, by using a modified profile of the ionospheric plasma during its heating, larger ranges of one-hop propagation can be obtained for probing waves with frequencies close to MUF ($f < 20$ MHz) of the undisturbed ionosphere for the wide range of grazing angles α. Furthermore, the existence of a heating-induced ionospheric region at the wide range of ionospheric altitudes allows using high-frequency probing waves for one-hop propagation at long distances owing to reflections from the disturbed ionospheric sporadic layers. Thus, for frequencies $20 < f < 400$ MHz of probing waves, efficient one-hop propagation for distances more than 500 km and for wide range of grazing angles $10° < \alpha < 40°-50°$ can be achieved (see Figure 7.47a and b). From Figure 7.47c, by using local resonance heating of the ionospheric F layer, a long-distance propagation channel can be implemented owing to reflections from disturbed ionospheric regions for higher-frequency bands, up to the UHF/L band for a sufficiently wide range of grazing angles $10° < \alpha < 30°$. These results can clearly be seen from the computations presented in Figure 7.48a and b for $\alpha = 10°$ and $\alpha = 45°$, respectively, for frequencies of probing waves greater than MUF, i.e., for $100 < f < 1000$ MHz. Thus, for small grazing angles, $\alpha \leq 10°$, all waves in this frequency band reflect from the disturbed region at altitudes of the ionospheric F layer, and one-hop propagation to large distances occurs (see Figure 7.48a). With an increase of the grazing angles,

Table 7.4   Electron Collision Frequency and Parameters of Ionization

Altitude			
h (km)	ν_{em} (s^{-1})	p_0	p
110	$1.34\ 10^4$	$7.1 \cdot 10^{-2}$	$6.8 \cdot 10^{-2}$
120	$5.29 \cdot 10^{-3}$	$9.35 \cdot 10^{-2}$	$8.5 \cdot 10^{-2}$
130	$2.41 \cdot 10^{-3}$	$1.63 \cdot 10^{-1}$	$1.4 \cdot 10^{-1}$
140	$1.26 \cdot 10^{-3}$	$2.8 \cdot 10^{-1}$	$2.5 \cdot 10^{-1}$
150	$7.32 \cdot 10^2$	$5.3 \cdot 10^{-1}$	$4.8 \cdot 10^{-1}$
160	$4.57 \cdot 10^2$	$6.8 \cdot 10^{-1}$	$6.2 \cdot 10^{-1}$
170	$2.94 \cdot 10^2$	1.2	1
180	$1.97 \cdot 10^2$	1.78	$9 \cdot 10^{-1}$
190	$1.37 \cdot 10^2$	2.19	1.1
200	98.6	2.93	1.47
210	76.9	3.85	1.92
220	53.8	4.73	2.36
230	41.7	6.5	3.25
240	30.5	8.4	2.38
250	24.1	13.3	3.81
260	18.7	18.1	7.1
270	15.9	25.4	5.75
280	12.5	33.4	7.46
290	9.8	38.8	9.2
300	7.81	45.1	12.1
310	6.58	56.4	18.6
320	5.45	65.2	24.3
330	4.19	73.9	33.6
340	3.62	82.5	41.2

for example, $\alpha \geq 45°$ (see Figure 7.48b), only probing waves with frequencies less than 500 MHz create the one-hop propagation channels at distances longer than 500–600 km. Probing waves with $f \geq 500$ MHz penetrate into the ionosphere without returning to the ground surface and slightly change their paths because of refraction in the upper ionosphere.

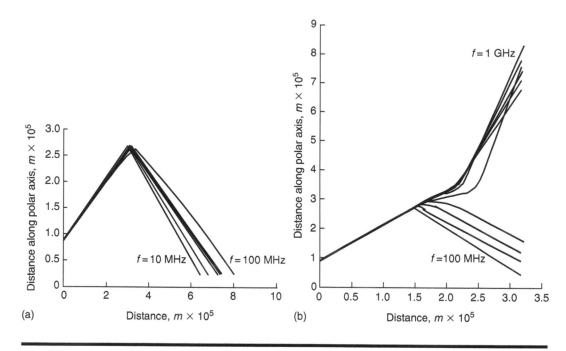

Figure 7.48 Trajectories of radio waves for various frequencies from 10 to 100 MHz and $\alpha = 10°$ (a) and from 100 MHz to 1 GHz, $\alpha = 45°$ (b).

We will also present the effects of adaptive resonance heating of the F layer of the ionosphere, because this heating regime is closer to the heating effects observed in the natural polar ionosphere at high latitudes owing to high-energetic photoelectron and cosmic ray interaction with cold background plasma [104–107]. Such a regime of heating gives similar but weaker effects, as will be shown in the following text. One of our main goals is to examine this regime with a view to predicting probing wave propagation through polar and auroral regions of the ionosphere over a wide frequency band and at various grazing angles. The ray tracing model described by Equations 7.57 through 7.61 is still valid but for the adaptive plasma density and collision frequency it is necessary to use different quasi-parabolic layers, as was shown in Tables 7.5 and 7.6. In this case the

Table 7.5 Electron Density for Adaptive Resonance Heating

Altitude h (km)	$N_b(h) \cdot 10^{11}$ (m^{-3})	$N(h)/N_0(h)$
135	2.05	0.85
140	2.2	0.81
145	2.7	0.68
150	2.95	0.65
155	3.05	0.61
160	3.2	0.56

**Table 7.5 (continued)   Electron Density
for Adaptive Resonance Heating**

Altitude h (km)	$N_b(h) \cdot 10^{11}$ (m^{-3})	$N(h)/N_0(h)$
165	3.55	0.5
170	3.7	0.59
175	3.8	0.57
180	4.0	0.7
185	4.2	0.71
190	4.4	0.5
195	4.7	0.72
200	5.05	0.83
205	5.3	0.88
210	5.8	0.86
215	6.0	0.85
220	6.5	0.83
225	7.0	0.89
230	7.5	0.84
235	8.2	0.83
240	9.0	0.86
245	9.5	0.82
250	10.0	0.88
255	11.0	0.85
260	12.0	0.82
265	12.5	0.88
270	13.0	0.92
275	13.5	0.96
280	14.0	0.99
285	14.5	1
290	15.0	1
295	15.5	1
300	16.0	1
305	16.1	1

**Table 7.6 Electron Collision Frequency
and Parameters of Ionization for Adaptive
Resonance Heating**

Altitude			
h (km)	ν_{em} (s^{-1})	p_0	p
110	$1.34 \cdot 10^4$	$7.1 \cdot 10^{-2}$	$5.1 \cdot 10^{-2}$
120	$5.29 \cdot 10^{-3}$	$9.35 \cdot 10^{-2}$	$5.8 \cdot 10^{-2}$
130	$2.41 \cdot 10^{-3}$	$1.63 \cdot 10^{-1}$	$7.5 \cdot 10^{-2}$
140	$1.26 \cdot 10^{-3}$	$2.8 \cdot 10^{-1}$	$1.1 \cdot 10^{-1}$
150	$7.32 \cdot 10^2$	$5.3 \cdot 10^{-1}$	$1.7 \cdot 10^{-1}$
160	$4.57 \cdot 10^2$	$6.8 \cdot 10^{-1}$	0.21
170	$2.94 \cdot 10^2$	1.2	0.35
180	$1.97 \cdot 10^2$	1.78	0.46
190	$1.37 \cdot 10^2$	2.19	0.54
200	98.6	2.93	0.96
210	76.9	3.85	1.2
220	53.8	4.73	1.28
230	41.7	6.5	1.85
240	30.5	8.4	2.5
250	24.1	13.3	4.18
260	18.7	18.1	5.1
270	15.9	25.4	8.3
280	12.5	33.4	11.8
290	9.8	38.8	15.9
300	7.81	45.1	18.1
310	6.58	56.4	22.6
320	5.45	65.2	28.5
330	4.19	73.9	36.9
340	3.62	82.5	41.8

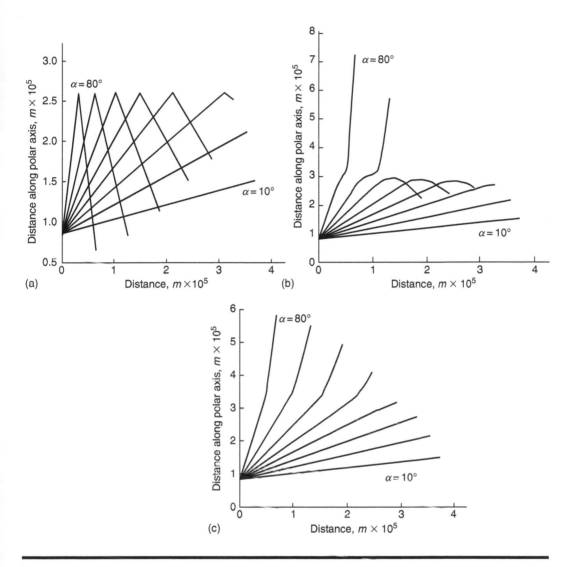

Figure 7.49 The same as Figure 7.47, but for frequencies 100 MHz (a), 300 MHz (b), and 500 MHz (c).

height h was divided into different quasi-parabolic layers according to the data. The electron collision frequency (losses) has been approximated by an exponential function

$$\nu = 2.7 \cdot 10^6 \cdot \exp\left(-4.8 \cdot 10^{-2} \cdot h\right), \text{ Hz} \qquad (7.64)$$

The results of computations are presented in Figure 7.49a through c in the same manner as Figure 7.26a through c for various grazing angles of probing waves and their frequencies. It is clear that with the increase of probing wave frequency for a significantly wide range of grazing angles, $\alpha < 30°$, the ray trajectories of one-hop propagation are larger and can reach 1000 km and more. This effect is clearly seen from the results of ray tracing simulations presented in Figure 7.50a and b for sufficiently small ($\alpha \leq 10°$) and large ($\alpha \geq 45°$) grazing angles, respectively. From the results presented, it follows that for small grazing angles one can predict

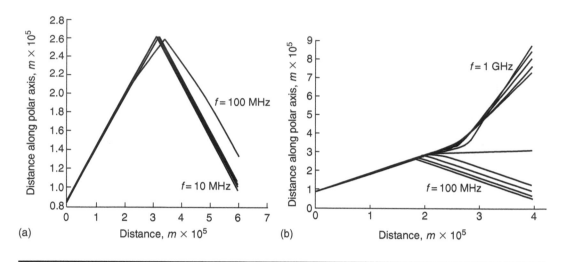

(a)   (b)

Figure 7.50 The same as Figure 7.48, but for $\alpha < 10°$ (a) and $\alpha > 45°$ (b).

a guiding effect for long distances up to 1000 km and more for sufficiently high frequencies, up to 800–1000 MHz. This effect can be observed in the disturbed ionosphere affected by natural or artificial heater, but cannot be completely observed in a nondisturbed quasi-regular ionosphere. With the increase of grazing angles of probing waves passing through the modified ionospheric regions, for example, $\alpha \geq 45°$ (see Figure 7.50b), these effects become weaker, and a guiding effect can be observed for frequencies larger than MUF but smaller than 400–500 MHz. Waves with higher frequencies penetrate to the upper ionosphere with a slight refraction effect.

Various examples shown earlier substantiate the fact that the trajectory of radio waves through the ionosphere itself depends on the absorption and the plasma (i.e., electron) concentration gradients involved. It appears that for the strong plasma depletions (or cavities) generated in the disturbed ionosphere because of the process of forcing out plasma during a relatively short heating time, Earth's magnetic field does not have an appreciable effect on the radio wave trajectories. A more important parameter for radio propagation at extremely long distances that is caused by reflections from disturbed ionospheric regions is the grazing angle and operational frequency of the ground-based facilities. Thus, for different kinds of heating processes, local or adaptive, it has been found that with increase of grazing angles for the same operational frequency of probing waves or with increase of operational frequency for the same grazing angles, the effect of channeling of one-hop modes in the ionosphere–earth waveguide channel becomes more and more unrealistic.

Finally, we can outline that because the heating regimes presented earlier and discussed more in detail in Chapters 5 and 6 are close to real-world natural conditions of the high-latitude ionosphere that correspond to natural nonstationary thermal processes caused by ambient sources, we can predict and explain the changes in probing wave trajectories and guiding effects at long distances in wide frequency bands, up to the L band, which are used in communication with atmospheric missiles, platforms, and space vehicles reentering the higher atmosphere, as well as in satellite communication channels for a wide range of grazing angles of probing waves passing through the ionosphere.

We will now briefly describe and summarize the effects of channeling of waves inside the ionospheric layered plasma waveguides due to capturing of probing waves by naturally or artificially induced plasma irregularities, following results obtained and described in Refs. [120–127].

7.4 Capturing of Radio Waves into the Ionospheric Layered Waveguides

From the earlier stages of study of the ionosphere by radio methods, it was found [120,121] that the regular layered plasma structure of the ionosphere allows, in principle, for guiding low-frequency radio waves at the extremely long distances around Earth. In other words, the layered quasi-regular ionosphere works for such waves as the waveguide where radio waves (called *waveguide modes*) reflect from the upper and lower boundaries and therefore can be channeled owing to multiple reflections at long distances. In the literature this effect was defined long-range ionospheric radio propagation [120–127]. It was shown both theoretically and experimentally that to obtain the guiding effect for probing radio waves propagating inside such plasma layered waveguides, it is necessary for the guiding modes to satisfy the following constraints [120,121]:

$$\omega \cdot \sec \beta_{kr} < \omega_e(z_1)$$
$$\omega \cdot \sec \beta_{kr} < \omega_e(z_2) \tag{7.65}$$

where ω is the angular frequency of the probing wave; ω_e the frequency of background plasma oscillations, all defined in Chapter 1; and z_1 and z_2 are the lower and upper boundaries of the plasma waveguide shown in Figure 7.51. Constraints (Equation 7.65) govern the periodic reflection from the bottom to top walls of waveguide modes during propagation inside the guiding plasma-layered structure. What is important to mention here is that these waveguide modes can propagate at long distances with sufficiently low attenuation of about 5–10 dB for each 1000 km [120,121].

If we assume initially that the regular layered ionosphere is spherically symmetric, containing layers with horizontal dimensions (with respect to $\mathbf{B}_0$) larger than the wavelength, the radio waves transmitted at an angle α_0 from the ground transmitter with frequencies satisfying constraints (Equation 7.65) cannot achieve the critical height z_c located in the middle axis of the waveguide

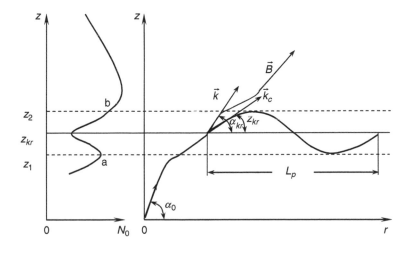

Figure 7.51 Altitudinal profile of plasma density (left panel) and geometry of capturing of probing wave in the layered ionosphere (right panel).

under angles β_{kr}. This also holds if we consider the transmitter assembled at the satellite. Capturing of probing waves energy in the ionospheric waveguide can only be achieved if rays arriving at the waveguide under angles $\alpha_0 > \beta_{kr}$ are oriented in such a manner as to achieve the angle α_{kr} inside the waveguide according to constraints (Equation 7.65), as shown in Figure 7.51. This additional displacement of the angle α from α_0 to α_{kr} can be achieved owing to refraction of radio waves at the horizontal gradients of the ionospheric plasma density or due to moderate- and small-scale irregularities causing scattering effects of the probing radio waves.

As mentioned earlier, such plasma irregularities can be induced either by natural phenomena (meteor trails, magnetic storms, etc.) or by modification of the ionosphere using ground-based facilities for heating the ionosphere or by injection of plasma beams or clouds by rockets [120–127]. Thus, as shown [120–127] both theoretically and experimentally, the heating of the ionosphere by powerful radio waves at the altitudes of resonances of plasma stimulates creation of moderate-scale and small-scale plasma irregularities, which then can work as "lenses" for capture of radio waves into the ionospheric waveguide and channeling their energy at long distances. Our further analysis will base itself on the theory developed in Refs. [120–124].

We must now stress that the corresponding theory is correct only if spatial gradients of parameters of plasma waveguide are smooth enough, i.e., when

$$\frac{d(\ln \xi)}{dx} \cdot \frac{\Lambda_\xi}{2\pi} \ll 1 \tag{7.66}$$

where ξ is the parameter of the plasma waveguide and Λ_ξ is the scale of spatial scintillations of probing waves inside the layered waveguide between Z_1 and Z_2 (see Figure 7.51).

As Λ_ξ generally is larger than 100 km in the channel between the E and F layers, the characteristic scale of the horizontal gradients of the parameters inside the plasma waveguide can exceed 500 km [120,121]. In other words, the horizontal gradients of plasma density are more important than vertical refraction of radio waves in the inhomogeneous layered ionosphere mentioned earlier.

Following Refs. [120,126], we assume that radio waves arriving at and entering the ionosphere at an angle α_0 cross the waveguide along the path denoted ab in Figure 7.51. The virtual midaxis of the waveguide is placed at the height z_{kr}. Then, the angle between the passing ray and the corresponding axis is denoted α_{kr} (see Figure 7.51). The ray trajectory inside the waveguide is characterized by the angle $\beta_{kr}(z_{kr})$ and if so, the boundaries of the capture ray are defined by the following condition [120,121]:

$$\varepsilon(z) = \varepsilon(z_{kr}) \cos^2 \beta_{kr}, \quad z_1 < z_{kr} < z_2 \tag{7.67}$$

Scattering of radio waves at the small-scale plasma irregularities described in previous sections leads to the turning of reflected waves relative the vector k_{kr}. All rays for which the angle of scattering is larger than the difference $\alpha_{kr} - \beta_{kr}$ will be seized inside the ionospheric waveguide. The angle β_{kr} can be determined from the following relationship:

$$\varepsilon(z_1) = \varepsilon(z_{kr}) \cos^2 \beta_{kr} \tag{7.68}$$

For small β_{kr}, from Equation 7.68 it follows that

$$\beta_{kr} = \left[\frac{\omega_e^2(z_1) - \omega_e^2(z_{kr})}{\omega^2} \right] \tag{7.69}$$

According to the Booker theory, described earlier, the angle of scattering θ_s for each ray can be found from the angle of the incident ray with coordinates $(\alpha_0, 0)$ and the angle of scattered ray with coordinates (α, φ), where φ is the angle in the azimuth (e.g., horizontal) plane measured from the axis **x**. The effective cross section of scattering according to the Born approximation can be presented in another form as was done previously following Booker theory [120,121], i.e.,

$$\sigma(\theta_s, \chi, \mathbf{k}_0) = \frac{k_0^4 \sin^2 \chi}{4\pi^2} U_\varepsilon(\mathbf{k}_1 - \mathbf{k}_2) \tag{7.70}$$

Assuming $K \equiv |\mathbf{k}_1 - \mathbf{k}_2| = 2k_0 \sin^2(\theta_s/2)$, we can rewrite Equation 7.70 as

$$\sigma(\theta_s, \chi(z), z) = \frac{k_0^4 \sin^2 \chi(z)}{4\pi^2} U_\varepsilon(K) \tag{7.71}$$

Now, taking into account Equation 7.71, we can by following Equation 120 obtain the coefficient of capture of radio waves inside the waveguide, which now will propagate along the specific trajectory of guiding mode characterized by the angle β_{kr}:

$$\Gamma_0(\alpha_0, \beta_{kr}, \varphi) = \frac{\pi}{2} \int_{z_1(\beta_{kr})}^{z_2(\beta_{kr})} k_0^4 \frac{\sin^2 \chi(z)}{\sin \alpha(z)} U_\varepsilon(K(z)) \, dz \tag{7.72}$$

Then, the total coefficient of capture of radio waves as a sum of waveguide modes (i.e., rays) can be defined by [120,121]:

$$\Gamma_\Sigma(\alpha) = - \int_{-\beta_{kr}}^{\beta_{kr}} \cos \beta'_{kr} \, d\beta'_{kr} \int_0^{2\pi} \Gamma_0(\alpha, \beta'_{kr}, \varphi) D(\varphi, \alpha') \, d\varphi \tag{7.73}$$

where

$$D(\varphi, \alpha') = \begin{cases} 1, & -\beta_{kr} < \alpha' < \beta_{kr} \\ 0, & |\alpha'| > \beta_{kr} \end{cases}$$

Equation 7.73 determines the coefficient of capture of radio waves into a layered ionospheric channel with frequencies less than the maximum useful frequencies (MUF). From the preceding equation it follows that the possibility of channeling radio waves in the ionospheric layered waveguide is determined by the intensity $U_\varepsilon(K)$ of small-scale plasma irregularities of various origins, artificially or naturally generated in the ionosphere, having the outer dimensions $\ell < \frac{\omega}{c}$, where c is the light velocity.

According to investigations presented in Refs. [120–126], the optimal capture of radio waves with frequencies of 10–20 MHz occurs when, at the ionospheric altitudes of the waveguide, plasma irregularities exist with horizontal dimensions (to geomagnetic field $\mathbf{B}_0$) $\ell_{||} = 100 - 500$ m. It follows from the aforementioned examples that such irregularities can be artificially generated during heating of the ionosphere by powerful radio wave [122–127] or naturally created by meteor trails [15–18]. Because the relative plasma disturbances of such plasma irregularities are of the order

$$\sqrt{\left\langle\left(\frac{\delta N}{N_0}\right)^2\right\rangle} \approx 10^{-3} - 10^{-2} \tag{7.74}$$

the coefficient of capturing of probing radio waves into the layered ionospheric waveguide with frequencies ranging from 10 to 20 MHz lies in the limits of $10^{-4} - 10^{-2}$ [120–127].

7.5 Partial Scattering of Radio Waves in the D Region of the Ionosphere

The method of partial diffuse reflection (called *scattering*) is one of the main methods of radio sounding of the D layer of the ionosphere [128–134], which allows for determining the profile of the electron content of ionospheric plasma, the wind velocity and characteristics of plasma irregularities generated at the lower ionosphere at altitudes $h = 60–90$ km by different kinds of plasma instabilities (see Chapter 2). This method is based on vertical sounding of radio signals at frequencies of 2–6 MHz, i.e., more than the plasma frequency of the D region of the ionosphere. The total reflection of these signals occurs at regions E and F of the ionosphere, but weak signals are also simultaneously registered from altitudes of 60–100 km, which result from diffuse reflection of the incident radio wave with the coefficient of reflection of approximately $10^{-3} - 10^{-7}$ at random plasma irregularities of the D layer.

If the sounding signal is a pulse signal, as shown in Figure 7.52, it propagates to the height $h = 0.5c\tau$, where the irregularity is located, and reflects from this irregularity with the coefficient of reflection, $R_{o,x}$, reflecting from Earth's surface. Here, τ is the time of pulse propagation, and c the

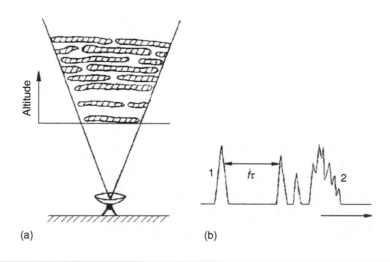

(a)   (b)

Figure 7.52 Schematic presentation of radio sounding of the lower layered ionosphere (a) and presentation of linear modulated signal used for such sounding (b).

light velocity; indexes o and x correspond, respectively, to the regular and nonregular modes of the signal. The wave-field amplitude, $A_{o,x}(h)$, depends mostly on $R_{o,x}$ and on the coefficient of absorption, $\Pi_{o,x}(h) = \exp(-2k_0 \int_0^h \kappa_{o,x} dh')$, of radio wave during its propagation below the level of reflection h [128,129], i.e.,

$$A_{o,x}(h) = R_{o,x}\Pi_{o,x}(h) \tag{7.75}$$

where $\kappa_{o,x}$ is the index of absorption of the regular and nonregular signal modes, $k_0 = \omega/c = 2\pi/\lambda$, $\omega = 2\pi f$, and f is the carrier frequency. The same relation as in Equation 7.75, between the parameters of the signal and those for ionospheric plasma, is accurate also for frequency-modulated continuous signals, called linear-frequency-modulated (LFM) waves in the literature [100,130]. The reflection coefficient in the model of consideration can be found by the Fresnel formula [128,129]:

$$R_{o.x} \cong \frac{\delta n_{o,x}}{2} \approx \frac{1}{2}\frac{\partial n_{o,x}}{\partial N}\delta N \tag{7.76}$$

where $\delta n_{o,x}$ is the fluctuation of the refractive index for two types of signal modes, and δN the fluctuation of the plasma density (plasma is quasi-neutral, and $N_e \approx N_i = N$). If we now assume that the reflected signal is random owing to diffuse reflection (i.e., due to random character of plasma density fluctuations), then the standard deviation of the signal amplitude equals

$$\langle A_{o,x}^2(h)\rangle = |R_{o.x}|^2\Pi_{o,x}^2(h) \tag{7.77}$$

where $|R_{o.x}|^2 = \left|\frac{\partial n_{o,x}}{\partial N}\right|^2 \langle(\delta N)^2\rangle$. In this case, the ratio $A^2 = \langle A_x^2\rangle/\langle A_o^2\rangle$ equals:

$$A^2(h) = R^2(h)\Pi^2(h) \tag{7.78}$$

Here, $R^2(h) = |R_x|^2/|R_o|^2$ and

$$\Pi^2(h) = \exp\left(-4k_0\int_0^h \left(\frac{\kappa_x - \kappa_o}{N}\right) N\, dh'\right) \tag{7.79}$$

The use of the foregoing formulas for descriptions of radio wave propagation in the D layer of the ionosphere contradicts the use of the well-known classical magneto-ionic theory [131–135], according to which the refractive index in the quasi-longitudinal approximation can be presented in the following form:

$$n_{o,x}^2 = 1 - \frac{\omega_e^2}{\omega\nu_{em}}\left[X^{\pm} \cdot C_{3/2}(X^{\pm}) + i\frac{5}{2}C_{5/2}(X^{\pm})\right] \tag{7.80}$$

where $X^{\pm} = (\omega \pm \omega_H \cos\alpha)/\nu_{em}$; ω_e is the Longmire frequency of electrons; ω_H the gyrofrequency of electrons; ν_{em} the frequency of the electron–neutral collisions (see definitions in Chapter 1);

α the angle between the vertical axis to the ground surface and the direction of the Earth's magnetic field $\mathbf{B}_0$; and

$$C_p(X) = \frac{1}{p!} \int_0^\infty \frac{\varepsilon^p e^{-\varepsilon}}{\varepsilon^2 + X^2} \, d\varepsilon \tag{7.81}$$

In Ref. [125], the integral functional $C_p(X)$ of different orders p was approximated by simple sums, which allows one to obtain $R^2(h)$ and $\frac{\kappa_x - \kappa_o}{N}$, respectively, and finally, using Equation 7.78, to define the profile $N(h)$ [100,135]:

$$N(h) = \frac{N}{2\log_{10} ek_0(\kappa_x - \kappa_o)\Delta h} \left\{ \log_{10}\left[\frac{R(h_2)}{R(h_1)}\right] - \log_{10}\left[\frac{A(h_2)}{A(h_1)}\right] \right\} \tag{7.82}$$

where h_1 and h_1 are two close heights at which the corresponding plasma parameters were measured, $\Delta h = h_2 - h_1$, $h = (h_1 + h_2)/2$. Now, if in Equation 7.79 the integral is less than the unit, which has satisfied conditions of the lower D layer of the ionosphere at altitudes $h < 65$ km, then $\Pi^2(h_1) \approx 1$ and $A(h_1) \approx R(h_1)$. This result allows for obtaining the frequency of electron–neutral collisions [as $R(h)$ depends on $\nu_{2m}(h)$] through the measurements of the field amplitude of partially reflected signals at altitudes less than 65 km.

As follows from foregoing, the ratio of amplitudes of the partially reflected signals, ordinary and nonordinary, are functions not only of plasma density N but also of the collision frequency ν_{em}. Therefore, measuring of N has to be done simultaneously with the knowledge of ν_{em}. It is a very complicated problem [128–131], but can be resolved using knowledge of the average phase difference, $\langle\Delta\Phi\rangle$, between two magneto-ion components, i.e.,

$$\langle\Delta\Phi\rangle = 2k_0 \int_0^h \langle\mu_x - \mu_o\rangle \, dh' + \Delta\Psi_n \tag{7.83}$$

where μ_o and μ_x are the real parts of the reflection index of the ordinary and nonordinary waves, respectively, and $\Delta\Psi_n$ is the difference of absolute values of the reflection coefficients R_x and R_o. The corresponding derivations of all these parameters can be achieved using Equations 7.76 and 7.79. The deviations $\Delta\Psi_n$ can be obtained by deriving the difference of phases of two magneto-ion components of the pulse or continuous LFM signal recorded using two receiving antennas separated in space (i.e., using space diversity).

The same method, using two spatially separated receiving antennas, was arranged in Ref. [136] to determine winds and motions in the D layer of the ionosphere. Following Ref. [136], we can assume that the field of the partially reflected radio waves creates a diffraction picture at the ground surface where can be differentiated dark and the light zones separated by lines of equal illumination, as shown in Figure 7.53. Owing to movements of plasma irregularities with velocity $\mathbf{V}$, the lines of equal illumination, i.e., the diffraction picture, will move along Earth's surface with velocity $\mathbf{U}$, passing through the receiving points placed in the tips of the triangular area defined by the coordinates x_0 and y_0, as shown in Figure 7.53, where positions of receiving antennas are denoted by "crests." Then, the coefficient of cross-correlation of the partly reflected signals at the basis x_0 and y_0 can be found as [100]:

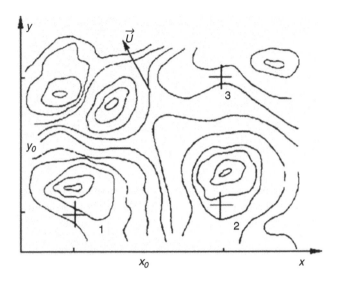

Figure 7.53 Movement of diffraction picture in the space domain relative to the transmitter and receiver indicated by "crests" and labeled by numbers 1, 2, and 3.

$$\rho_A(x_0,\tau) = \frac{\langle A_1(t) \cdot A_2(t)\rangle - \langle A_1(t)\rangle \cdot \langle A_2(t)\rangle}{\sqrt{\langle A_1^2(t)\rangle - \langle A_1(t)\rangle^2} \cdot \sqrt{\langle A_2^2(t)\rangle - \langle A_2(t)\rangle^2}}$$

$$\rho_A(y_0,\tau) = \frac{\langle \tilde{A}_1(t) \cdot \tilde{A}_2(t)\rangle - \langle \tilde{A}_1(t)\rangle \cdot \langle \tilde{A}_2(t)\rangle}{\sqrt{\langle \tilde{A}_1^2(t)\rangle - \langle \tilde{A}_1(t)\rangle^2} \cdot \sqrt{\langle \tilde{A}_2^2(t)\rangle - \langle \tilde{A}_2(t)\rangle^2}} \tag{7.84}$$

Here, $A_{1,2}(t)$ and $\tilde{A}_{1,2}(t)$ are the amplitudes of signals received at the bases x_0 and y_0, respectively, at the first and the second antennas, denoted in Figure 7.53 by crests 1 and 2 for $\rho_A(x_0,\tau)$, and by 2 and 3 for $\rho_A(y_0,\tau)$. The maxima of these coefficients will be displaced in time at τ_x at the base x_0 and at time τ_y at the base y_0 with respect to the point at the origin of the coordinate system. It was found that because of the spherical nature of radio waves, diffraction picture moves along the ground surface with velocity $\mathbf{U}$ exceeding twice the velocity $\mathbf{V}$ of movements of plasma irregularities [100,136]. The movement of irregularities is considered to be horizontal at the heights of reflection of probing partially reflected radio waves. Then, the knowledge of the directional diagram of the antenna, θ_A, and the central carrier frequency, f_c (see Figure 6.34), allow for calculating the Doppler-shift frequencies located at the range of $\Delta f_D = 4f_c|\mathbf{V}|\sin\theta_A/c$. The corresponding velocities $\mathbf{V}$ can be found by using of the geometry of experiment and results obtained from measurements of the velocity $U = |\mathbf{U}|$ of the diffraction picture:

$$V = \frac{U}{2} \frac{x_0 y_0}{\sqrt{x_0^2 \tau_y^2 + y_0^2 \tau_x^2}} \tag{7.85}$$

The knowledge of the cross-correlation between the signal's amplitudes measured at the basis points of the antenna location allows for defining the parameters of the inhomogeneities in the D layer of

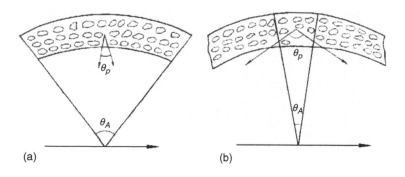

(a)   (b)

Figure 7.54 Geometry of backscattering depending on the angle of antenna diagram θ_A and the angle of the pattern of scattering θ_p.

the ionosphere using the method of partial reflections of radio waves. Thus, the spatial radius of correlation $\rho_A(r,\tau)$, where r is the distance between receiving antennas and τ the time displacement of arriving waves, allows estimation of spatial radius r_k and temporal radius τ_k, which can be defined as the distance and time at which the value of $\rho_A(r,\tau)$ is decreased $e^{-1} \approx 0.38$ times [137,138]. The study of dependences of r_k and τ_k from the angle θ_A of the antenna pattern allows for estimating parameters of inhomogeneities scattering the probing waves in the D region. For these purposes the two main functions have to be defined: the angle spectrum of correlation function of fluctuations of dielectric permittivity $\Delta\varepsilon$, $\Phi_{\Delta\varepsilon} \sim \exp\{-\theta^2/\theta_p^2\}$, and the angle spectrum of antenna diagram, $K^2 \sim \exp\{-\theta^2/\theta_A^2\}$. Here, θ_p is the width of angle spectrum of back-reflected signals, $\theta_p = \lambda/\pi\ell$, measured at the signal strength level of $e^{-1} \approx 0.38$, where ℓ is the horizontal dimension, and θ_A was defined previously (see Figure 7.54). Two cases can occur. The first is shown in Figure 7.54a, where $\theta_A^2 \gg \theta_p^2$. The spatial radius of amplitude correlation simply equals ℓ—i.e., $r_k = \frac{\lambda}{\pi\theta_p} = \ell$—and the temporal scale equals $\tau_k = \ell/2|\mathbf{V}|$. The second case is presented in Figure 7.54b, where $\theta_A^2 \ll \theta_p^2$ (i.e., the diagram of the antenna is narrower than the angle spectrum of backscattering), the spatial radius of amplitude correlation equals $r_k(\varphi) = \lambda/\pi\theta_A$, and the temporal scale is $\tau_k = \lambda/2|\mathbf{V}|\pi\theta_A$. In the last case, for an anisotropic diagram of antenna pattern, i.e., for $K^2 = K^2(\theta,\varphi)$, the diffraction picture will also be anisotropic and $r_k(\varphi) = \lambda/\pi\theta_A(\varphi)$. Here φ is the angle in the azimuth domain,

In the case, where $\theta_A^2 \approx \theta_p^2$, the foregoing formulas can be changed [137,138]:

$$r_k = \frac{\lambda}{\pi\theta_{\textit{eff}}}, \quad \tau_k = \frac{\lambda}{2\pi|\mathbf{V}|\theta_{\textit{eff}}} \qquad (7.86)$$

where $\theta_{\textit{eff}}^2 = \theta_A^2\theta_p^2/(\theta_A^2 + \theta_p^2)$.

Hence, from the corresponding experiments [136–138] in which the antenna diagrams are the same as the angle spectrum of backscattered wave, the horizontal dimensions of the inhomogeneities can be defined for the case when their horizontal scale exceeds the vertical scale, i.e., it has a form of a disk. Strictly speaking, for study of different situations occurring in the experiments on backscattering of radio waves from the D region of the ionosphere, it is useful to use several receiving antenna systems with different angles of antenna pattern θ_A.

7.6 Methods of Design of Ionospheric Radio Propagation Channels

The increasing demand for ionospheric short-wave radio channel design requires the strict prediction of propagation characteristics of such channels. Solutions to this complicated problem depend on various physical processes occurring in the ionospheric plasma. According to discussions in previous chapters, there are many factors and features, artificial and natural, that influence the ionospheric propagation of HF/VHF waves. Thus, sporadic E_s and F_s layers play an important role; their existence leads to the screening of the regular layers, E and F, of the ionosphere. At the same time, these stratified layers cause above-horizon waveguide propagation at the extremely long distances of short and sometimes ultrashort waves at frequencies significantly exceeding the maximum useful frequencies (MUFs) of the F_1 and F_2 layers of the ionosphere. Significant influence of short-wave propagation in the ionosphere is observed because of the existence of ionospheric irregularities of various natures and different scales. The large-scale irregularities (with scales of $\ell \gg \lambda = 10-100$ km) mostly respond to refraction of radio waves passing thorough the ionospheric channel, whereas the moderate-scale (with scales of $\ell > \lambda = 0.1-1$ km) and small-scale irregularities (with scales of $\ell \le \lambda = 1-100$ m) cause multipath phenomena owing to multiple scattering and diffraction of radio waves. These effects cause creation of a diffuse structure of ionograms during oblique and vertical soundings of the ionosphere by ionosonde; as they also lead to strong fading of radio signals. At the same time, as was shown earlier, they "work" as lenses that, due to capture of wave energy, can take the waveguide ionospheric modes into the ionospheric waveguide and then, after traversing a long distance within it, can direct the guiding waves (i.e., modes) back to the ground-based facilities. As the nature of the generation of these plasma irregularities is very complicated and depends on many natural factors—such as solar activity; penetration of high-energy particles into the polar ionosphere; and creation of magnetic storms, mostly in the high-latitude ionosphere, which then penetrate into the lower latitudes; electrojet into the equatorial ionosphere; and so on—it is very complicated to take these physically different factors into account and create the adequate theoretical models of ionospheric radio channels. As has been mentioned by many investigators [139–145], significant progress in the investigation of ionospheric HF/UHF (up to several GHz) radio propagation is due to the development of modern experimental methods of diagnostics of ionospheric channels. One of the most effective of these is based on modern methods of wideband sounding of the ionosphere using signals of complicated structures and having low energy, such as the signals with linear frequency modulation (LFM) [146–150]. The use of LFM ionosondes in regimes of oblique sounding of the ionosphere allows researchers to obtain information regarding the main characteristics of short-wave ionospheric channels defined by the distance–frequency (DF) characteristic and the amplitude–frequency (AF) characteristic, obtained in the form of corresponding ionograms (see the examples given later).

Below, we briefly present results of experimental investigations of LFM sounding of the middle-latitude ionosphere at the different traces of the former USSR, in which the authors of this book have been successfully involved.

The Radio Trace of Alma-Ata–Gorky. Experiments on LFM sounding of the ionosphere at the radio trace of about 2700 km between two cities, Alma-Ata (Kazakhstan) and Gorky (now is Nizhny Novgorod, Russia), were carried out during June and July of 1987 using the frequency band 5–22 MHz with the speed of frequency deviations of 300 kHz/s. For the control of the ionospheric environment during each experiment, ionograms of vertical sounding were registered in the ionosphere above the places of the localization of the transmitter and the receiver. Thus, in Figure 7.55a through h, the measured average (averaging over all probing modes) diurnal

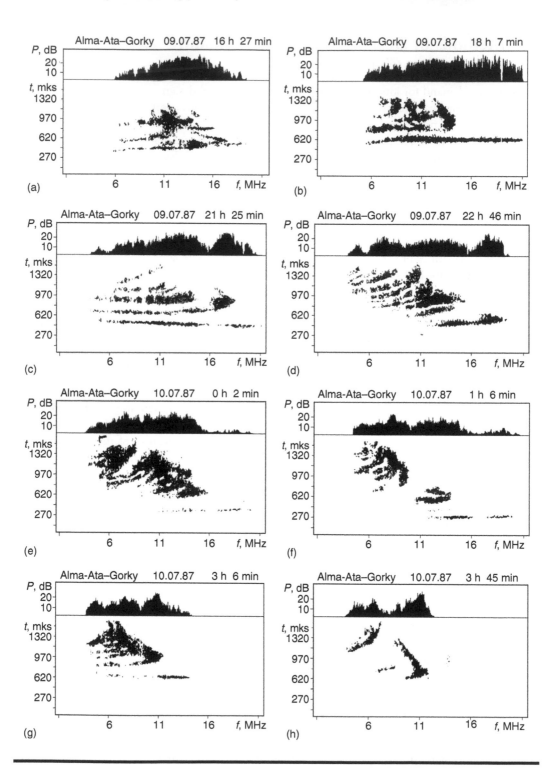

Figure 7.55 Amplitude–frequency (upper panels) and distance–frequency characteristics of signals scattered by the irregular ionosphere at radio trace Alma-Ata–Gorky obtained during several days of observation for (a) $t = 16{:}27$, (b) $t = 18{:}07$, (c) $t = 21{:}25$, (d) $t = 22{:}46$, (e) $t = 00{:}02$, (f) $t = 01{:}06$, (g) $t = 03{:}06$, **and** (h) $t = 03{:}45$.

distance–frequency (DF) and amplitude–frequency (AF) characteristics are presented to illustrate the time dependence of parameters of the short-wave ionospheric communication channel. From these illustrations a great difference can be seen between diagrams obtained at different times, i.e., the change of DF and AF characteristics of the channel with time, and a strong multiray structure of the receiving signals reflected from the real ionosphere. Here, for July 9, 1987: Figure 7.55a corresponds to 16 h 27 min of Moscow time, Figure 7.55b to 18 h 07 min, Figure 7.55c to 21 h 25 min, Figure 7.55d to 22 h 46 min; for July 10, 1987: Figure 7.55e to 00 h 02 min, Figure 7.55f to 01 h 06 min, Figure 7.55g to 03 h 06 min, Figure 7.55h to 03 h 45 min. Using vertical and oblique soundings, it was found that the main influence on mode structure of the reflected signals is observed from the sporadic E_s layer, which was registered in more than 80% of ionograms. In oblique ionograms the mode $2E_s$ was registered mostly at frequencies exceeding the MUF of F_2 at 2–7 MHz. In the evening and the night, the focusing effect of radio signals was observed experimentally, which closely correlates with the existence of moving ionospheric perturbations (MIP) at the F_2 layer of the ionosphere, described in the previous chapters. Taking into account the speed of such plasma perturbations of $V \approx 100$ m/s, we finally find that the characteristic dimensions of MIP generated in the ionospheric F layer by internal gravitation waves [100] are $L \approx V \cdot T = 0.1$ (km/s) $\cdot 1800$ (s) $= 180$ km.

The existence of such plasma structures explains the strong diffusion of reflected signals caused by F scattering of radio waves from natural middle-latitude irregularities and the multiray phenomena. The maximum average number of rays registered at frequencies $f \approx (0.7-0.8)$ *MUF* F_2, as follows from Figure 7.55a through h, is $n \approx 4$, with time delays of $\Delta \tau \approx 500$ μs between distant rays. Single-ray propagation was observed only at frequencies $f > MUF$ F_2 and related to the mode $2E_s$.

The Radio Trace of Khabarovsk–Gorky. Experimental investigations of propagation of LFM signals at the trace Khabarovsk–Gorky of 5830 km was carried out during March–April 1988 at the frequency band of 5–26 MHz with speeds of frequency deviations of 300 kHz/s. In Figure 7.56a through h, the ionograms of DF and AF characteristics are shown for different diurnal times: for 21 March, 1988: Figure 7.56a is for 07 h 11 min, Figure 7.56b for 9 h 55 min, Figure 7.56c for 10 h 50 min, Figure 7.56d for 11 h 07 min, Figure 7.56e for 13 h 45 min, Figure 7.56f for 14 h 17 min, Figure 7.56g for 14 h 41 min, and Figure 7.56h for 15 h 14 min. For the control of the ionospheric conditions during oblique sounding of the ionosphere in real time, vertical sounding of the ionosphere was also used at the points of location of ground-based terminals. The figures illustrate different DF and AF characteristics as main parameters of the short-wave ionospheric channel. Thus, the difference between ionograms, fast fading of most signals, and a strong multiray phenomenon can be seen from the ionograms. Deep and strong deviations of the received signals are related to the screening effects of E_s sporadic layer for the higher F_1 and F_2 layers of the ionosphere. The same effects were observed during the current experiment. Thus, the same multiple splitting of the ray of Pedersen (up to six rays) was observed in addition to magneto-ion splitting, which is determined in Refs. [151–153]. This effect can be explained by stratification of the ionosphere at the proximity of maximum of F_2 layer owing to modulation of background plasma electrons by sinusoidal MIP. The interpretation of such periodic effects, which was found to explain the observed focusing effect of radio signals, was the same as before, that it is caused by moving ionospheric perturbations (MIPs). At the ionograms of oblique sounding, we can observe that radio tracks occur parallel to the mode $3F_2$ with the delay time of 50–100 μs with respect to the main trail (see Figure 7.56b). Sometimes, the additional track overlaps the main mode. This leads to diffusion of the trail of mode $3F_2$. At other times in the nocturnal ionosphere (22–04 h), the probing signals were not observed. We can suppose that the chief influence on short radio wave propagation at long distances is due to variations of plasma density distribution along the radio path. If so, in nocturnal time, a strong negative gradient of the plasma concentration can occur.

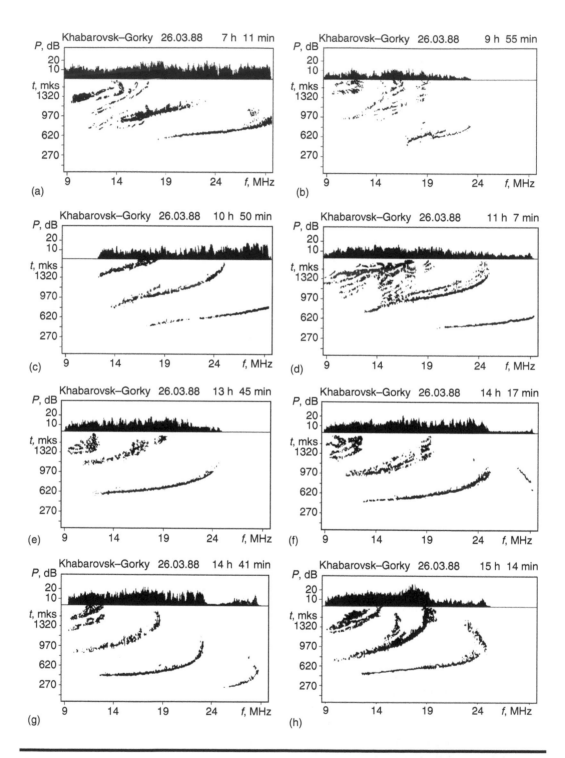

Figure 7.56 The same as Figure 7.55, but obtained at radio trace Khabarovsk–Gorky (a) $t = 07{:}11$, **(b)** $t = 09{:}55$, **(c)** $t = 10{:}50$, **(d)** $t = 11{:}07$, **(e)** $t = 13{:}45$, **(f)** $t = 14{:}17$, **(g)** $t = 14{:}41$, **and (h)** $t = 15{:}14$.

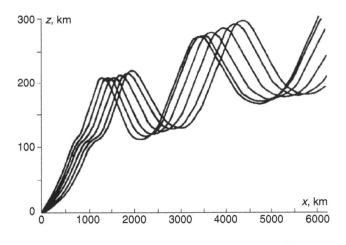

Figure 7.57 Trajectory of radio waves propagation in the ionospheric multilayered waveguide.

Owing to refraction of probing waves caused by these gradients (see Chapter 6), the trajectory of rays appears from Earth's surface, and capture of the radio wave energy into the ionospheric waveguide is also observed (see Figure 7.57). This assumption was proved by computations of the ray trajectories based on the real ionospheric parameters. As an example, in Figure 7.58 we present results of computations for carrier frequency of 12 MHz. Transfer of signals' energy back to Earth's surface can be made by using either refraction of radio waves at the large-scale gradients of plasma density or scattering of radio waves from local ionospheric irregularities, natural and artificially induced [151]. Results of investigations have shown that parameters of the probing signals and the character of the ray of Pedersen depend on the level of magnetic activity. It was found that during magnetic storms with growth of perturbations of the geomagnetic field, the role of ionospheric irregularities on ionospheric radio propagation becomes predominant (see results in Chapters 6 and 8).

Radio Trace Beltsi (Moldova)–Dushanbe (Tajikistan). The oblique sounding of the ionosphere at two traces of 3380 and 7120 km, respectively, was carried out during October and November of 1991 (see the corresponding traces presented in Figure 7.59) [154]. The main goal of these experiments was to find diurnal and nocturnal ranges of working frequencies in the one-mode regime of radio wave propagation, $\Delta f = f_{max,nF_2} - f_{max,(n+1)F_2}$. Radiation and recording of LFM signals were carried out at the frequency range of 1–30 MHz with the recording rate and frequency scanning of $df/dt = 483.3$ kHz/s. The power of the transmitting signals was 200 W. Registration of

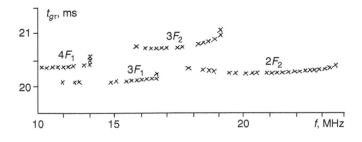

Figure 7.58 Time–frequency ionogram computed according to the model of multilayered iono-sphere.

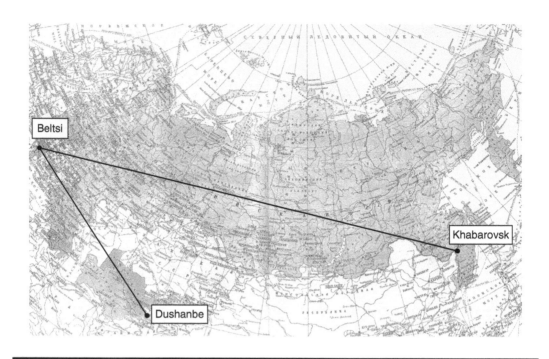

Figure 7.59 Geometry of radio traces Beltsi–Khabarovsk and Beltsi–Dushanbe.

signals was made in the nonstop regime (i.e., 24 h). Registration of one ionogram was done during the period of $t = 1$ min. For each hour, between 5 and 15 ionograms were registered. So, during two months (November–December) more than 2500 ionograms were registered. In all ionograms, the multimode character of radio propagation was observed. Frequency ranges of the one-mode regime of radio wave propagation for the traces Beltsi–Dushanbe and Beltsi–Khabarovsk (Figure 7.59) were defined and the following dependences were analyzed: (a) the time dependence of maximum working frequency of radio signal for different propagation modes; (b) the time dependence of minimum working frequency of radio signal for different propagation modes; (c) the time dependence of frequency range for the one-mode propagation of radio waves; (d) the diurnal and nocturnal amplitude–time dependence for different frequencies, and so on.

The corresponding ionograms of DF–AF characteristics of probing signals are shown in Figure 7.60 for the trace Beltsi–Dushanbe, where P is the power of the receiving signal, t is the time of arrival, and f is the signal operating frequency. The three modes of LFM signal propagation are clearly seen, which correspond to different arrival times at the receiver. For example, for the frequency $f = 18$ MHz the signal of mode $3F_2$ was registered at the decoder at 400 μs later than the mode $1F_2$, and 250 μs later than the mode $2F_2$. Furthermore, at the ionograms can be distinguished the region of the multimode propagation of radio signal denoted as region A ranging from 14.5 to 22 MHz and the range of one-mode propagation denoted as region B ranging from 22 to 25.5 MHz. The median value of the amplitude of receiving signal was about 34 μV at the frequency band from 15 to 27 MHz. In Figure 7.61a through c, the time dependences of modes $1F_2$, $2F_2$, and $3F_2$ are presented, respectively, for maximum f_{max,nF_2} (circles) and minimum f_{min,nF_2} (squares) working frequencies.

Thus, from Figure 7.61a it follows that for mode $1F_2$ the corresponding frequencies achieve values of $f_{max,1F_2} = 27$ MHz and $f_{min,1F_2} = 16$ MHz during the time interval from 8:30 to 14:30.

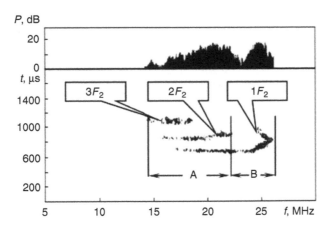

Figure 7.60 Amplitude–frequency characteristic (top panel) and time–frequency ionogram obtained at the radio trace Beltsi–Dushanbe.

In the local time interval of 6:00–9:00 the rate of growth of $f_{max,1F_2}$ from 16 to 27 MHz is about 3.7 MHz/h, and from 14:00 to 17:00, the decrease of $f_{max,1F_2}$ from 27 to 13 MHz takes place at a rate of about 4.7 MHz/h. At the same time, the rate of growth (in the time interval 6:00–9:00) and decrease of $f_{min,1F_2}$ (in the time interval 14:00–17:00) is about 1.3 MHz/h. The same tendency can be found for the mode $2F_2$ (see Figure 7.61b). Thus, the corresponding frequencies achieve values of $f_{max,2F_2} = 21$ MHz and $f_{min,2F_2} = 14$ MHz at the time interval from 9:30 to 14:30. During morning time, between 6:00 and 9:00, the rate of growth of $f_{max,2F_2}$ is about 3.0 MHz/h, and from 14:00 to 17:00 $f_{max,2F_2}$ decreased at a rate of about 3.6 MHz/h. The frequency $f_{min,2F_2}$ grows at a rate of 3.0 MHz/h (in the time interval 6:00–9:00) and is damped in the time interval 14:00–17:00 at a rate of about 3.6 MHz/h. As for the mode $3F_2$, during periods of 8:30–13:30 its maximum and minimum working frequencies, $f_{max,3F_2} = 17$ MHz and $f_{min,3F_2} = 10$ MHz, are constant. The frequency $f_{max,3F_2}$ has the tendency to increase in the morning period of 7:00–9:30 at a rate of 2.0 MHz/h and to damp during time interval 14:00–16:00 at a rate of 2.4 MHz/h (Figure 7.61c). The frequency $f_{min,3F_2}$ has a tendency to increase at a rate of 0.9 MHz/h and to damp at a rate of 0.8 MHz/h during the same time intervals as $f_{max,3F_2}$.

Now we will analyze the character of one-mode radio propagation. Thus, in Figure 7.62, the time dependence of the frequency range, $\Delta f = f_{max,1F_2} - f_{max,2F_2}$, of the one-mode radio propagation along the radio trace, the geometry of which differs significantly from the usually observed latitudinal "big-circle" radio propagation. At the trace under consideration, the one-mode character of radio signal propagation occurs in the period 7:00–15:00, where the range of working frequencies achieves the maximum value of $\Delta f = 5$ MHz, and therefore can be utilized for the implementation of ionospheric radio channel for radio communication for such traces, which have a complicated geometry of radio propagation.

Radio Trace Beltsi (Moldova)–Khabarovsk (Russia). This trace is longer than others because its length is about 7120 km, and such traces were not earlier used for radio communication through the ionosphere. In Figure 7.63 for the radio trace Beltsi–Khabarovsk the corresponding ionograms of DF–AF characteristics of probing signals are shown. The same results as were obtained for the shorter Beltsi–Dushanbe trace can be clearly seen from Figure 7.64a through c for the longer trace Beltsi–Khabarovsk. Only one peculiarity can be found: there are no realized modes $1F_2$ and $2F_2$ at

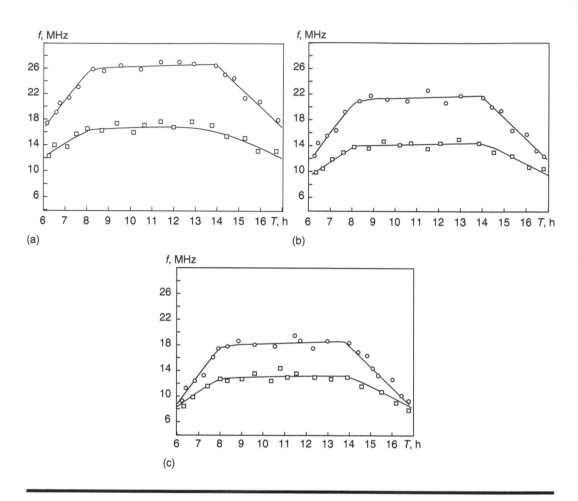

Figure 7.61 The time dependence of $1F_2$ mode (a), $2F_2$ mode (b), and $3F_2$ mode (c) for maximum (circles) and minimum (squares) working frequencies of the nF_2 layer, respectively, at radio trace Beltsi–Dushanbe.

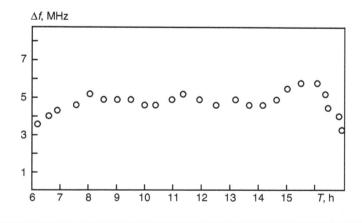

Figure 7.62 Diurnal variations of the one-mode radio propagation characteristic.

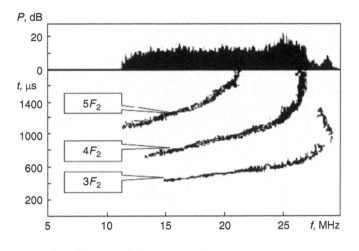

Figure 7.63 Amplitude–frequency characteristic (top panel) and time–frequency ionogram obtained at the radio trace Beltsi–Khabarovsk for modes $3F_2$, $4F_2$, and $5F_2$.

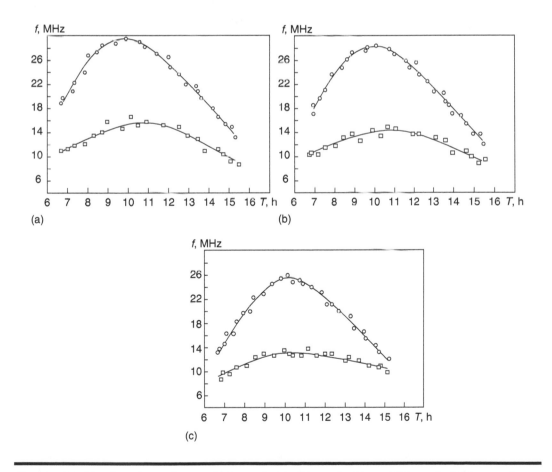

Figure 7.64 The time dependence of $3F_2$ mode (a), $4F_2$ mode (b), and $5F_2$ mode (c) for maximum (circles) and minimum (squares) working frequencies of the nF_2 layer, respectively, at radio trace Beltsi–Khabarovsk.

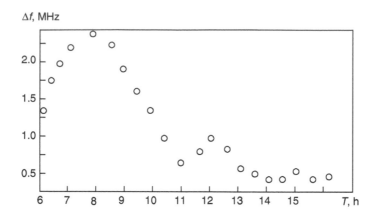

Figure 7.65 The same as in Figure 7.62, but for radio trace Beltsi–Khabarovsk.

the longer traces; instead, there are new modes, $4F_2$ and $5F_2$, which correspond to four- and five-mode radio propagation. We leave it to the reader to analyze simultaneously all three modes, their growth, and damping for different time intervals with different rates. We will only mention that the time intervals, where the corresponding working frequencies for multimode propagation are constant, are lesser than those observed for the shorter radio trace. Furthermore, for the longer trace the time dependence of the frequency interval, $\Delta f = f_{max,3F_2} - f_{max,4F_2}$, of the one-mode radio propagation along the radio trace, shown in Figure 7.65, has a tendency to achieve its maximum value of $\Delta f = 2.0$–2.5 MHz in a very narrow time interval (from 7:00 to 9:00). In other daytime intervals, the one-mode character of radio propagation is very difficult to implement. So, the radio communication channel Beltsi–Khabarovsk is not as stable as that for trace Beltsi–Dushanbe.

The obtained results allow for outlining the essential role that plasma irregularities of various origins play in forming radio propagation ionospheric channels and in prediction of the key parameters of radio signals passing such ionospheric channels.

References

1. Al'pert, Ya.L., Ginzburg, V.L., and Feinberg, E.L., *Propagation of Radio Waves*, Moscow: Gostechizdat, 1953.
2. Al'pert, Ya.L., *Propagation of Electromagnetic Waves in the Ionosphere*, Moscow: Nauka, 1973.
3. Gel'berg, M.G., *Irregularities of the High-Latitude Ionosphere*, Novosibirsk: Nauka, 1986.
4. Belikovich, V.V., Benediktov, E.A., and Itkina, M.A. et al., Frequency dependence of anomalous absorption of cosmic radio radiation in the ionosphere in the region of polar cups, *Geomagn. Aeronom.*, Vol. 9, 1969, pp. 485–490.
5. Hodga-Akhmetov, I.L., About frequency dependence of the cosmic noise level in the periods of anomalous absorption of II type, *Geomagn. Aeronom.*, Vol. 5, 1965, pp. 348–350.
6. Belikovich, V.V., Benediktov, E.A., Tolmacheva, A.V., and Bakhmet'eva, N.V., *Ionospheric Research by Means of Artificial Periodic Irregularities*, Berlin: Copernicus GmbH, 2002.
7. Mityakov, N.A., Rappaport, V.O., and Trakhtengertz, V.Yu., About scattering of regular wave near the point of reflection at the small-scale irregularities, *Izv. Vuzov, Radiophyzika*, Vol. 18, 1975, pp. 1273–1278.

8. Wang, Z., Resenberg, T.J., Stauning, P., Basu, S., and Crowley, G., Calculations of riometer absorption associated with *F* region plasma structures based on Sondre Stomfjord incoherent scatter radar observations, *Radio Sci.*, Vol. 29, 1994, pp. 209–215.

9. Chen, C.H. and Fejer, J.A., Effect of small ionospheric irregularities on radio wave absorption, *Radio Sci.*, Vol. 10, 1975, pp. 167–181.

10. Jones, T.B., Robinson, T., Stubbe, P., and Kopka, H., Frequency dependence of anomalous absorption caused by high power radio waves, *J. Atmos. Terr. Phys.*, Vol. 46, 1984, pp. 147–153.

11. Booker, H.G. and Gordon, W.E., A theory of radio scattering in the ionosphere, *Proc. IRE*, Vol. 38, 1950, pp. 400–412.

12. Booker, H.G., A theory of scattering by non-isotropic irregularities with applications to radar reflection from the Aurora, *J. Atmos. Terr. Phys.*, Vol. 8, 1956, pp. 204–221.

13. Booker, H.G. and Majidi, A.G., Theory of refractive scattering in scintillation phenomena, *J. Atmos. Terrestr. Phys.*, Vol. 43, 1981, pp. 1199–1214.

14. Booker, H.G., Application of refractive scintillation theory to radio transmission through the ionosphere and the solar wind and to reflection from a rough ocean, *J. Atmos. Terrestr. Phys.*, Vol. 43, 1981, pp. 1215–1233.

15. Filipp, N.D., Power of H_E-scatter signals, *Izv. Vuzov, Radiophyzika*, Vol. 22, No. 4, 1979, pp. 407–411.

16. Filipp, N.D., *Specular Scattering of Ultra-Short Waves by Middle-Latitude Ionosphere*, Kishinev (Moldova): Shtiintza, 1980.

17. Filipp, N.D. and Blaunstein, N.Sh., Effect of the geomagnetic field on the diffusion of ionospheric inhomogeneities, *Geomagn. Aeronom.*, Vol. 18, 1978, pp. 423–427.

18. Filipp, N.D. and Blaunstein, N.Sh., Drift of ionospheric inhomogeneities in the presence of the geomagnetic field, *Izv. Vuzov, Radiophyzika*, Vol. 21, 1978, pp. 1409–1417.

19. Bochkarev, G.S., Eremenko, V.A., and Cherkashin, Ya.N., Radio wave reflection from quasi-periodical disturbances of the ionospheric plasma, *Adv. Space Res.*, Vol. 8, No. 1, 1988, pp. 255–260.

20. Filipp, N.D., Oraevskii, V.N., Blaunshtein, N.Sh., and Ruzhin, Yu.Ya., *Evolution of Artificial Plasma Irregularities in the Earth's Ionosphere*, Kishinev (Moldova): Shtiintza, 1986.

21. Blaunshtein, N.Sh. and Bochkarev, G.S., Modeling of the dynamics of periodical artificial disturbances in the upper ionosphere during its thermal heating, *Geomagn. Aeronom.*, Vol. 33, No. 2, 1993, pp. 84–91.

22. Belikovich, V.V., Benediktov, E.A., Tolmacheva, A.V., and Bakhmet'eva, N.V., Ionospheric *Research by Means of Artificial Periodic Irregularities*, Berlin: Copernicus GmbH, 2002.

23. Pogorelov, V.I. and Tomashuk, L.Ya., About correlation function of fluctuations of aurora's ionization, *Geomagn. Aeronom.*, Vol. 13, 1973, pp. 1122–1123.

24. Leadabrand, R.L., Larson, A.G., and Hodges, J.C., Preliminary results on the wavelength dependence and aspect sensitivity of radar echoes between 50 and 3000 MHz, *J. Geophys. Res.*, Vol. 72, 1967, pp. 3877–3887.

25. Bowman, G.G. and Hajkowicz, L.A., Small-scale ionospheric structures associated with mid-latitude spread-F, *J. Atmos. Terr. Phys.*, Vol. 53, 1991, pp. 4447–4457.

26. Yamamoto, M.S., Fukao, S., Woodman, R.F., Tsuda, T., and Kato, S., Mid latitude *E*-region field-aligned irregularities observed with the MU radar, *J. Geophys. Res.*, Vol. 96, 1991, pp. 15943–15949.

27. Fejer, B.G. and Providakes, J.F., High latitude *E*-region irregularities: New Results, *Physica Scripta*, Vol. T18, 1987, pp. 167–178.

28. Robinson, T.R., Towards a self-consistent non-linear theory of auroral backscatter, *J. Atmos. Terr. Phys.*, Vol. 48, 1986, pp. 417–422.

29. Tsunoda, R.T., High-latitude *F* region irregularities: A review and synthesis, *Rev. Geophys.*, Vol. 26, 1988, pp. 719–760.

30. Prividakes, J.E., Farley, D.T., and Fejer, B.G. et al., Observations of auroral *E*-region plasma waves and electron heating with EISCAT and VHF radar interferometer, *J. Atmos. Terr. Phys.*, Vol. 50, 1988, pp. 339–356.

31. Haldopis, C., A review of radio studies of auroral *E* region ionospheric irregularities, *Ann. Geophys.*, Vol. 7, 1989, pp. 239–258.

32. Farley, D.T., Balsley, B.B., Woodman, R.F., and McClure, J.P., Equatorial spread *F*: Implications of VHF radar observations, *J. Geophys. Res.*, Vol. 75, 1970, pp. 7199–7216.

33. Balsley, B.B. and Farley, D.T., Radar observation of two-dimensional turbulence in the equatorial electrojet, *J. Geophys. Res.*, Vol. 78, 1973, pp. 7471–7476.

34. Woodman, R.F. and LaHoz, C., Radar observations of *F* region equatorial irregularities, *J. Geophys. Res.*, Vol. 81, 1976, pp. 5447–5466.

35. Tsunoda, R.T., Backscatter measurements of 11-cc equatorial spread irregularities, *Geophys. Res. Lett.*, Vol. 7, 1980, pp. 848–853.

36. Hysell, D.L., Kelley, M.C., Swartz, W.E., and Farley, D.T., VHF radar and rocket observations of equatorial spread F on Kwjalein, *J. Geophys. Res.*, Vol. 99, 1994, pp. 15065–15085.

37. Swartz, W.E. and Farley, D.T., High resolution radar measurements of turbulent structure in the equatorial electrojet, *J. Geophys. Res.*, Vol. 99, 1994, pp. 309–317.

38. Blanc, E., Mercandalli, B., and Hongminou, E., Kilometric irregularities in the E and F regions of the daytime equatorial ionosphere observed by a high resolution HF radar, *Geophys. Res. Lett.*, Vol. 23, 1996, pp. 645–648.

39. Frihagen, J. and Jacobson, T., In-situ observations of high-latitude *F*-region irregularities, *J. Atmos. Terr. Phys.*, Vol. 33, 1971, pp. 519–522.

40. Hansucher, R.D., Chatanica radar investigation of high-latitude *E*-region ionization, *Radio Sci.*, Vol. 10, 1975, pp. 277–288.

41. Muldrew, D.B. and Vickrey, J.F., High-latitude *F*-region irregularities observed simultaneously with ISIS-1 and Chatanica radar, *J. Geophys. Res.*, Vol. 87, 1982, pp. 8263–8267.

42. Ossakow, S.L. and Chaturvedi, P.K., Morphological studies of rising equatorial spread-*F* bubbles, *J. Geophys. Res.*, Vol. 83, 1978, pp. 2085–2091.

43. Haerendel, G., Coloured bubbles: An experiment for triggering equatorial spread-*F*, *ESA-SP*-195, July 1983, pp. 295–298.

44. Blaunstein, N.Sh. and Tsedilina, E.E., Effect of the initial dimensions on the nature of the diffusion spreading of inhomogeneities in a quasi-uniform ionosphere, *Geomagn. Aeronom.*, Vol. 25, 1985, pp. 39–44.

45. Heritage, J.L., Wesbrod, S., and Fay, W.J., Evidence for a 200-Megacycles per second ionospheric forward scatter mode associated with the Earth's magnetic field, *J. Geophys. Res.*, Vol. 64, 1959, pp. 1235–1244.

46. Heritage, J.L., Fay, W.J., and Bowen, E.D., Evidence that meteor trails produce a field-aligned scatter signals at VHF, *J. Geophys. Res.*, Vol. 67, 1962, pp. 953–961.

47. Kuriki, I., Some observations of the VHF pulse wave scattered from the ionosphere, *J. Radio Res. Labs* (Japan), Vol. 10, 1963, pp. 69–81.

48. Kuriki, I., Measurements of direction of scattering points on VHF anomalous scattered propagation, *J. Radio Res. Labs* (Japan), Vol. 12, 1965, pp. 1–14.

49. Kuriki, I., On the scattering propagation caused by the field aligned irregularities, *J. Radio Res. Labs* (Japan), Vol. 13, 1966, pp. 57–74.

50. Kuriki, I., Further experimental study of scattering propagation caused by the field aligned irregularities in VHF band, *J. Radio Res. Labs* (Japan), Vol. 14, 1967, pp. 57–77.

51. Filipp, N.D., Gleibman E.Ya., and Shliakhovo', A.M., Specular scattering of short radio waves by HE irregularities of middle-latitude ionosphere, in *Scientific Issue of Teaching University in Math. and Phys.*, Beltsi (Moldova), Vol. 12, 1969, pp. 3–17.

52. Filipp, N.D., *Scattering of Radio Waves by Anisotropic Ionosphere*, Kishinev (Moldova): Shtiintza, 1974.

53. Fialko, E.I., *Some Problems in Radiolocation of Meteors*, Tomsk, USSR, Tomsk University Press, 1971.

54. Bondar, B.G. and Kasheev, B.L., *Meteor Communication*, Kiev: Technika, 1968.

55. Kasheev, B.L., Lebedinetz, V.N., and Lagutin, M.F., *Meteor Phenomena in the Atmosphere of the Earth*, Moscow: Nauka, 1967.

56. Filipp, N.D., Blaunshtein, N.Sh., Erukhimov, L.M., Ivanov, V.A., and Uryadov, V.P., *Modern Methods of Investigation of Dynamic Processes in the Ionosphere*, Kishinev (Moldova): Shtiintza, 1991.

57. Blaunstein, N., Diffusion spreading of middle-latitude ionospheric plasma irregularities, *J. Ann. Geophys.*, Vol. 13, 1995, pp. 617–626.

58. Blaunstein, N., Changes of the electron concentration profile during local heating of the ionospheric plasma, *J. Atmos. Terr. Phys.*, Vol. 58, No. 12, 1996, pp. 1345–1354.

59. Millman, G.H., The geometry of the Earth's magnetic field at ionospheric heights, *J. Geophys. Res.*, Vol. 64, 1959, pp. 717–726.

60. Millman, G.H., Field-aligned ionization scatter geometry, *J. Geophys. Res.*, Vol. 74, 1969, pp. 900–911.

61. Finney, J.W., Smith, E.K., Tveten, L.H., and Watts, J.M., IGY observations of *F*-layer scatter in the far East, *J. Geophys. Res.*, Vol. 64, 1959, pp. 403–405.

62. Kasheev, B.L., Lebedenetz, V.N., and Lagutin, M.F., *Meteor Phenomena in the Earth's Atmosphere*, Moscow: Nauka, 1967.

63. Bondar, B.G. and Kasheev, B.L., *Meteor Communication*, Kiev: Tekhnica, 1968.

64. Goodwin, G.L. and Thomas, J.A., Field aligned irregularities in the E_s region, *J. Atmos. Terr. Phys.*, Vol. 25, 1963, pp. 707–719.

65. Pfister, W., The wave-like nature of inhomogeneities in the E-region, *J. Atmos. Terr. Phys.*, Vol. 33, 1971, pp. 999–1025.

66. Jones, K.L., Structures responsible for rapid fading of medium frequency radio reflections from the daytime E-layer, *J. Atmos. Terr. Phys.*, Vol. 46, 1984, pp. 1179–1191.

67. From, W.R., Sporadic E movement followed with a pencil beam high frequency radar, *Planet Space Sci.*, Vol. 31, 1983, pp. 1397–1407.

68. From, W.R. and Whitehead, J.D., E_s structure using HF radar, *Radio Sci.*, 1986, pp. 309–312.

69. Whitehead, J.D., Recent work on mid-latitude and equatorial sporadic-E, *J. Atmos. Terr. Phys.*, Vol. 51, 1989, pp. 401–424.

70. Astin, I. and Thomas, L., Rapid fading of radio waves reflected from sporadic-E ionization, *J. Atmos. Terr. Phys.*, Vol. 53, 1991, pp. 99–104.

71. Tanaka, T. and Venkateswaran, S.V., Characteristics of field-aligned E-region irregularities over Iioka (36°N), Japan. I, *J. Atmos. Terr. Phys.*, Vol. 44, 1982, pp. 381–393.

72. Tanaka, T. and Venkateswaran, S.V., Characteristics of field-aligned E-region irregularities over Iioka (36°N), Japan. II, *J. Atmos. Terr. Phys.*, Vol. 44, 1982, pp. 395–406.

73. Kato, S., Ogawa, T., and Tsuda, T. et al., The middle and upper atmosphere radar: First results using partial system, *Radio Sci.*, Vol. 19, 1984, pp. 1475–1484.

74. Woodman, R.F., Yamamoto, M., and Fukao, S., Gravity wave modulation of gradient drift instabilities in mid-latitude sporadic E irregularities, *Geophys. Res. Lett.*, Vol. 18, 1991, pp. 1197–1200.

75. Yamamoto, M., Fukao, S., and Woodman, R.F. et al., Mid-latitude E region field-aligned irregularities observed with MU radar, *J. Geophys. Res.*, Vol. 96, 1991, 15,943–15,949.

76. Yamamoto, M., Fukao, S., Ogawa, T., Tsuda, T., and Kato, S., A morphological study on mid-latitude E-region field-aligned irregularities observed with the MU radar, *J. Atmos. Terr. Phys.*, Vol. 54, 1992, pp. 769–778.

77. Yamamoto, M., Komoda, N., and Fukao, S. et al., Spatial structure of the E region field-aligned irregularities revealed by the MU radar, *Radio Sci.*, Vol. 29, 1994, pp. 337–347.

78. Tsunoda, R.T., Fukao, S., and Yamamoto, M., On the origin of the quasi-periodic radar backscatter from midlatitude sporadic E, *Radio Sci.*, Vol. 29, 1994, pp. 349–365.

79. Haldopis, C. and Schlegel, K., A 50-MHz radio Doppler experiment for midlatitude E region coherent backscatter studies: System Description and first results, *Radio Sci.*, Vol. 28, 1993, pp. 959–978.

80. Schlegel, K. and Haldopis, C., Observation of the modified two-stream plasma instability in the midlatitude E region ionosphere, *J. Geophys. Res.*, Vol. 99, 1994, pp. 6219–6226.

81. Haldopis, C., Schlegel, K., and Farley, D.T., An explanation for type 1 radar echoes from the midlatitude E-region ionosphere, *Geophys. Res. Lett.*, Vol. 23, 1996, pp. 97–100.

82. Riggin, D., Swartz, W.E., Providakes, J., and Farley, D.T., Radar stadies of long-wavelength associated with mid-latitude sporadic E layers, *J. Geophys. Res.*, Vol. 91, 1986, pp. 8011–8024.

83. Kelley, M.C., Riggin, D., and Pfaff, R.F. et al., Large amplitude quasi-periodic fluctuations associated with midlatitude sporadic *E* layer, *J. Atmos. Terr. Phys.*, Vol. 57, 1995, pp. 1165–1178.

84. Huang, C.-M., Kudeki, E., Franke, S.J., Liu, C.H., and Rottger, J., Brightness distribution of midlatitude *E* region echoes detected at Chung-Li VHF radar, *J. Geophys. Res.*, Vol. 100, 1995, pp. 14,703–14,705.

85. Djuth, F.T., Thide, B., Ierkic, H.M., and Sulszer, M.F., Large-F-region electron-temperature enhancements generated by high-power HF radio waves, *Geophys. Rev. Lett.*, Vol. 14, No. 9, 1987, pp. 953–956.

86. Basu, S., Basu, S., Stubbe, P., Kopka, H., and Waaramaa, J., Daytime scintillations induced by high-power HF waves at Tromso, Norway, *J. Geophys. Res.*, Vol. 92, 1987, pp. 11149–11157.

87. Duncan, L.M., Sheerin, J.P., and Behnke, R.A., Observation of ionospheric cavities generated by high-power radio waves, *Phys. Rev. Lett.*, Vol. 61, No. 2, 1988, pp. 239–242.

88. Hansen, J.D., Morales, G.J., and Duncan, L.M., Large-scale ionospheric modifications produced by nonlinear refraction of an HF wave, *Phys. Rev. Lett.*, Vol. 65, No. 26, 1990, pp. 3285–3288.

89. Keskinen, M.J., Chaturvedi, P.K. and Ossakow, S.L., Long time scale evolution of high-power radio wave ionospheric heating, 1, Beam propagation, *Radio Sci.*, Vol. 28, No. 5, 1993, pp. 775–784.

90. Gurevich, A.V. and Migulin, V.V., Investigations in the USSR of non-linear phenomena in the ionosphere, *J. Atmos. Terr. Phys.*, Vol. 44, No. 12, 1982, pp. 1019–1024.

91. Vas'kov, V.V. and Dimant, Y.S., Excitation of ion-sound density disturbances in the region of powerful radio wave resonance, *Geomagn. Aeronom.*, Vol. 30, No. 2, 1990, pp. 268–274.

92. Blaunshtein, N.Sh, Vas'kov, V.V., and Dimant, Ya.S., Resonant heating of the F layer of the ionosphere by a high-power radio wave, *Geomagn. Aeronom.*, Vol. 32, No. 2, 1992, pp. 235–238.

93. Blaunshtein, N.Sh., Erukhimov, L.M., Uryadov, V.P., Filipp, N.D., and Tsyganash, I.P., Vertical dependence of relazation time of artificial small-scale disturbances in the middle-latitude ionosphere, *Geomagn. Aeronom.*, Vol. 28. No. 4, 1988, pp. 595–597.

94. Erukhimov, L.M., Metelev, S.A., Myasnikov, E.N., Mit'akov, N.A., and Frolov, V.L., The artificial ionospheric turbulence, *Izv. Vuzov, Radiofizika*, Vol. 30, 1987, pp. 208–225.

95. Erukhimov, L.M., Mitugin, S.N., and Uryadov, V.P., To the question of radio wave propagation in the ionospheric waveguide channel, *Izv. Vuzov, Radiophysika*, Vol. 18, No. 9, 1975, pp. 1297–1304.

96. Belenov, A.F., Bubnov, V.A., and Erukhimov, L.M. et al., About parameters of artificial small-scale ionospheric irregularities, *Izv. Vuzov, Radiophysika*, Vol. 20, No. 12, 1977, pp. 1805–1809.

97. Boguta, N.M., Vovk, V.Ya., and Maximenko, O.I. et al., Influence of refraction at the frequency dependence of occurrence of short-wave signals during specular scattering of radiowaves at the artificial ionospheric irregularities, *Geomagn. Aeronom.*, Vol. 28, No. 1, 1988, pp. 152–154.

98. Boguta, N.M., Maximenko, O.I., Uryadov, V.P., and Tsiganash, I.P., Back scattering of waves of decameter bandwidth by irregularities of *F*-layer of the sub-polar ionosphere, *Izv. Vuzov, Radiophysika*, Vol. 30, No. 11, 1987, pp. 1399–1401.

99. Blaunstein, N., Evolution of a stratified plasma structure induced by local heating of the ionosphere, *J. Atmos. Solar-Terr. Phys.*, Vol. 59, No. 3, 1997, pp. 351–361.

100. Uryadov, V.P., Ivanov, V., Plohotnuk, E., Blaunstein, N., and Filipp, N., *Dynamic Processes in the Ionosphere—Methods of Investigation*, Iasi, Romania: Technopress, 2006 (in Romanian).

101. Bahkmet'eva, N.V., Benediktov, E.A., and Bochkarev, G.S. et al., Change of range-frequency characteristics of oblique sounding in condition of artificial disturbance of the upper ionosphere, *Geomagn. Aeronomy*, Vol. 25, No. 2, 1985, pp. 233–238.

102. Bochkarev, G.S., Eremenko, V.A., and Ignat'ev, Yu.A. et al., The modeling of back-scattering signals from an artificially disturbed regions of the upper ionosphere, *Izv. Vuzov, Radiofizika*, Vol. 30, No. 4, 1987, pp. 482–489.

103. *Radio Science* (special issue), 1974, Vol. 9, 1974, pp. 881–1090.

104. Marubashi, K., Structure of topside ionosphere in high latitudes, *J. Radio Res. Lab.*, Vol. 81, 1970, pp. 7175–7181.

105. Kelley, J.D. and Wickwar, V.B., Radar measurements of high-latitude ion composition between 140–300 km altitude, *J. Geophys. Res.*, Vol. 86, 1981, pp. 7617–7626.

106. Mehta, N.E. and D'Angelo, N.D., Cosmic noise absorption by *E*-region plasma waves, *J. Geophys. Res.*, Vol. 85, 1980, pp. 1779–1782.
107. Grebowsky, S.M., Taylor, H.A., and Linsay, J.M., Location and source of ionospheric high latitude trough, *Planet Space Sci.*, Vol. 31, No. 1, 1983, pp. 99–105.
108. Croft, A.T. and Hoogasian, H., Exact ray calculations in a quasi-periodic ionosphere with no magnetic field, *Radio Sci.*, Vol. 3, No. 1, 1968, pp. 69–74.
109. Jones, R.M., Ray theory for lossy media, *Radio Sci.*, Vol. 5, 1970, pp. 793–801.
110. Suchy, K., The velocity of wave packet in an anisotropic absorbing media, *J. Plasma Phys.*, Vol. 8, No. 1, pp. 33–51.
111. Suchy, K., Ray tracing in an anisotropic absorbing medium, *J. Plasma Phys.*, Vol. 8, No. 1, pp. 53–56.
112. Suchy, K., The propagation of wave packet in inhomogeneous anisotropic media with moderate absorption, *Proc. IEEE*, Vol. 62, 1974, pp. 1571–1577.
113. Bennett, J.A., Complex rays for radio waves in the absorbing ionosphere, *Proc. IEEE*, Vol. 62, 1974, pp. 1577–1585.
114. Bennett, J.A., Non-uniqueness of 'real ray' equation when the ray direction is complex, *Proc. IEEE*, Vol. 65, 1977, pp. 1599–1601.
115. Censor, D., Ray propagation and self-focusing in non-linear absorbing media, *Phys. Rev.*, Vol. A18, 1978, pp. 2614–2617.
116. Censor, D. and Gavan, J., Wave packets, group velocity and rays in lossy media revised, *IEEE Trans. Elect. Compat.*, Vol. 31, No. 3, 1980, pp. 262–272.
117. Dyson, P.L. and Bennett, J.A., Exact ray path calculations using realistic ionosphere, *IEEE Proc. H*, Vol. 139, No. 5, 1992, pp. 407–413.
118. Sonnenschein, E., Censor, D., Rutkevich, I., and Bennett, J.A., Ray tracing in an absorptive collisional and anisotropic ionosphere, *J. Atmos. Solar-Terr. Phys.*, Vol. 59, No. 16, 1997, pp. 2101–2110.
119. Sonnenschein, E., Blaunstein, N., and Censor, D., HF ray propagation in the presence of resonance heated ionospheric plasmas, *J. Atmos. Solar-Terr. Phys.*, Vol. 60, 1998, pp. 1605–1623.
120. Gurevich, A.V. and Tsedilina, E.E., *Extremely-Long-Range Propagation of Short Waves*, Moscow: Science, 1979.
121. Gel'berg, M.G., *Inhomogeneities of High-Latitude Ionosphere*, Novosibirsk: Science, USSR, 1986.
122. Krasnushkin, P.E., *Method of Normal Waves to the Problem of Long Radio Communication*, Moscow: Moscow University Publisher, 1947.
123. Borisov, N.D. and Gurevich, A.V., To the theory of short radio waves in the horizontal-inhomogeneous ionosphere, *Izv. Vuzov, Radiophyzika*, Vol. 19, 1976, pp. 1275–1284.
124. Erukhimov, L.M., Matugin, S.N., and Uryadov, V.P., To question of radiowave propagation in the ionospheric waveguide channel, *Izv. Vuzov, Radiophyzika*, Vol. 18, 1975, pp. 1297–1304.
125. Gurevich, A.V., Gurevich, L.M., and Kim, V.Yu. et al., Influence of scattering on seizure of radio waves in the ionospheric channel, *Izv. Vuzov, Radiophyzika*, Vol. 18, 1975, pp. 1305–1316.
126. Erukhimov, L.M. and Trahtengertz, V.Yu., About some effects of scattering of radio waves in the ionosphere, *Geomagn. Aeronom.*, Vol. 9, 1969, pp. 834–841.
127. Mit'akov, N.A., Rapoport, V.O., and Trahtengertz, V.Yu., About scattering of an ordinary wave near the point of reflection at the small-scale irregularities, *Izv. Vuzov, Radiophyzika*, Vol. 18, 1975, pp. 1273–1278.
128. Benediktov, E.A., Grishkevich, L.V., and Ivanov, V.A., Simultaneous measurements of electron concentration and collision frequency of electrons in D-layer of the ionosphere using method of partial reflections, *Izv. Vuzov, Radiophyzika*, Vol. 15, No. 5, 1972, pp. 695–701.
129. Benediktov, E.A., Grishkevich, L.V., and Ivanov, V.A. et al., Usefulness of the method of back scattering of radio waves for investigation of inhomogeneities of ionization and their motions in D-layer of the ionosphere, *Geomagn. Aeronom.*, Vol. 14, No. 4, 1974, pp. 645–650.
130. Rinnert, K., Schleger, K., and Kramm, R., A partial reflection experiment using the FM-CW technique, *Radio Sci.*, Vol. 11, No. 12, 1876, pp. 1009–1018.
131. Sen, H.K. and Wyller, A.A., On the generation of the Appelton-Hartree magnetoionic formulas, *J. Geophys. Res.*, Vol. 65, No. 18, 1960, pp. 3931–3950.

132. Belrose, J.S. and Burke, M.J., Study of the lower ionosphere using partial reflection, *J. Geophys. Res.*, Vol. 69, No. 13, 1964, pp. 2799–2818.

133. Benson, R.F., The quasi-longitudinal approximation in the generalized theory of wave radio absorption, *J. Res. Nat. Bur. Standards*, Vol. 689, No. 2, 1964, pp. 219–223.

134. Geishkevich, L.A., Ivanov, V.A., and Fedoseeva, T.N., About usefulness of 'quasi-longitudinal approximation in the method of partial redlections, *Geomagn. Aeronom.*, Vol. 11, No. 2, 1971, pp. 348–349.

135. Hara, E.H., Approximations to the semiconductor integral Cp(X) and Dp(X) for use of the generalized Appelton–Hartree magnetoionic formulas, *J. Geophys. Res.*, Vol. 68, No. 14, 1963, pp. 4388–4389.

136. Fraser, G.J., The measurement of atmospheric winds and altitudes of 64–120 km using ground-based radio equipment, *J. Atmos. Sci.*, Vol. 22, No. 2, 1965, pp. 217–218.

137. Benediktov, E.A., Grishkevich, L.V. and Ivanov, V.A. et al., Investigation of diffraction picture occurs at the Earth's surface during back scattering of radio waves by inhomogeneities of the lower ionosphere, *Izv. Vuzov, Radiophyzika*, Vol. 17, No. 6, 1974, pp. 798–807.

138. Vinsent, R.A., The interpretation of some observations of radio wave scattered from the lower ionosphere, *Austral. J. Phys.*, Vol. 26, 1973, pp. 815–827.

139. Fenwick, R.B. and Barry, G.H., Sweep frequency oblique ionospheric sounding at medium frequencies, *IEEE Trans. Broadcast.*, Vol. BC-12, No. 1, 1966, pp. 25–27.

140. Barry, G.H. and Fenwick, R.B., Oblique chirp sounding, *Agard Conf. Proc.*, Vol. 13, 1969, pp. 487–490.

141. Fenwick, R.B. and Lomasney, J.M., Test of monostatic FM-CW vertical-incidence sounder, Radioscience Lab. Stanford Univ., Stanford, California, *Report SU-SEL*-68-077, No. 144, 1968.

142. Barry, G.H. and Fenwick, R.B., HF measurements using extended chirp-radar techniques, Radioscience Lab. Stanford Univ., Stanford, California, *Report SU-SEL*-65-056, No. 103, 1965.

143. Barry, G.H., A low-power vertical-incidence ionosonde, *IEEE Trans. Geoscience Electronics*, Vol. GE-9, No. 2, 1971, pp. 86–89.

144. Whitehred, J.D. and Kantarizie, E., Errors in the measurement of vertical height using a phase ionosonde, *J. Atmos. Terr. Phys.*, Vol. 7, 1967, pp. 1483–1488.

145. Fenwick, R.B., Oblique chirpsounders: The HF communications test set, *Communic. News*, Feb. 1974, pp. 32–33.

146. Lynch, J.T., Fenwick, R.B., and Villard, O.G., Measurement of best time-delay resolution obtainable along east–west and north–south ionospheric paths, *Radio Sci.*, Vol. 7, No. 10, 1972, pp. 925–929.

147. Ivanov, V.A., Frolov, V.A., and Shumaev, V.V., Sounding of the ionosphere by continuous LFH-radiosignals, *Izv. Vuzov, Radiophyzika*, Vol. 29, No. 2, 1986, pp. 235–237.

148. Erukhimov, L.M., Ivanov, V.A., and Mitiakov, N.A. et al., Sounding of the artificially disturbed ionosphere by the LFM-ionosound, in book: *Propagation of Radio Waves in the Ionosphere*, Moscow: Science, 1986, pp. 80–86.

149. Zinichev, V.A., Ivanov, V.A., Frolov, V.A., and Shumaev, V.V., The use of LFM-method of vertical sounding for investigations of modification of upper layers of the ionosphere, *Izv. Vuzov, Radiophyzika*, Vol. 29, No. 5, 1986, pp. 629–631.

150. Sharadze, Z.C., Kvavadze, N.D., Liadze, Z.L., and Masashvili, N.V., Moving ionospheric disturbances (MID) and the phenomenon of F-scattering at the middle-latitude ionosphere, *Geomagn. Aeronom.*, Vol. 26, No. 1, 1986, pp. 144–147.

151. Kravtsov, Yu.A., Tinin, M.V., and Cherkashin, Yu.N., About possible mechanisms of excitation of ionospheric wave channels, *Geomagn. Aeronom.*, Vol. 19, No. 5, 1979, pp. 769–787.

152. Sazhin, V.I. and Tinin, M.V., About long propagation by the ray of Pedersen, *Geomagn. Aeronom.*, Vol. 15, No. 4, 1975, pp. 748–749.

153. Tinin, M.V., Role of the ray of Pedersen and the directed waves related with this ray in propagation along the parabolic layer, in the book: *Investigations in Geomagnetism, Aeronomy and Physics of the Sun*, Moscow: Science, 1973, pp. 157–166.

154. Plohotniuc, E.F. and Pascaru, M.D., Sounding of the ionosphere by LFM signals, Scientific Anales of University A. Russo, Vol: Mathematics, Physics, Techniques, Beltsi, State University of A. Russo, Vol. 20. 2004, pp. 52–61.

Chapter 8

Optical and Radio Systems for Investigation of the Ionosphere and Ionospheric Communication Channels

As was shown in Chapters 2 and 5, the propagation of radio waves through the quickly changing plasma density in the nonregular ionosphere during its perturbations is considerably affected, and ultimately the quality and efficiency is decreased of wireless communications, land–satellite, and satellite–satellite, including the positioning of any subscriber, stationary or moving, located in areas of service. An increase in the efficiency of short-wave wireless communication, as was shown in Chapters 6 and 7, is impossible without continuous observation of nonlinear processes occurring in the perturbed ionospheric plasma and without taking into consideration radio traces parameters, i.e., the key parameters of ionospheric communication link.

The parameters and dynamical processes occurring in the nonregular ionosphere are possible to investigate using modern radiophysical methods based on different devices and radio systems, such as optical devices, incoherent scatter radars, Global Navigation Satellite Systems (GNSS), super-DARN, and ionosondes/digisondes, briefly described in Chapters 6 and 7.

This chapter deals mainly with the operation characteristics of linear-frequency-modulated (LFM) radio facilities, called LFM ionosondes, which have been successfully used by the authors of this book for detailed study of the peculiarities of ionospheric propagation of LFM radio signals through the middle-latitude ionosphere under natural and artificial (e.g., man-made) conditions. The main goal of such signals processing was to study the problems concerning radio wave dispersion and short-wave wireless communications through the ionosphere.

Before describing the main aspects of LFM-signals propagation and the operational characteristics of LFM ionosondes, we shall briefly summarize here the existing modern optical and radio systems that were successfully used for investigation of the main nonlinear characteristics of the ionospheric plasma, its various instabilities, and the corresponding irregularities of plasma density at various latitudes of the ionosphere, which were described in detail in Chapters 6 and 7, without entering into a technical description of devices and systems.

Here, we draw the reader's attention to the possibilities of each corresponding device or system, again without going into technical details and schematic descriptions (which can be found in the referred literature) and describing only their operation characteristics and possibilities to investigate natural and artificial ionospheric phenomena. This is done to show the advantages and disadvantages of LFM ionosondes in regard to existing modern devices and systems.

8.1 Devices and Systems for Diagnostics of Ionospheric Phenomena

8.1.1 Optical Devices

Optical devices, such as photometric and spectrometric devices, digital all-sky cameras, and TV complexes, were used to investigate the atmospheric and ionospheric parameters and processes, which, as was described in Chapter 6, were successfully used separately, or with radar systems and ionosondes, when operated regularly in many places over the world, such as Alaska, Arecibo, Arequipa, Fairbanks, New Zealand, Peru, Puerto Rico, Russia, and the North and South Poles [1–16].

For example, at Arecibo Incoherent Scatter Radar System (AISRS), an optical set of equipment includes two tilting-filter photometers, an Ebert–Fastie spectrometer with 1 m focal length, two pressure-scanned Fabry–Perot interferometers, each having 6 in. diameter etalon plates, and lidars. The set of optical devices used in investigations of the atmosphere and the ionosphere are presented in Table 8.1.

Table 8.1 Optical Instruments Used to Investigate Atmospheric Parameters and Processes

Instrument	Observed Item	Altitude	Mode
Photometer (tiny ionospheric photometer)	The airglow lines of various gases in the atmosphere, such as OH, O_2, O, O^+, N, N_2^+, H, He, and others	Different altitudes of ionosphere	Horizontal distribution
Fabry–Perot interferometer	Horizontal and vertical wind velocity, and temperature at airglow layers	Upper mesosphere, lower thermosphere (when quiet: 85, 95, 250 km; when active: 85, 120, 250 km)	Horizontal distribution, night (new moon phase)
Fourier-transform infrared spectrometer	Trace constituents	Troposphere, lower stratosphere (10–30 km)	Vertical distribution, day
Rayleigh lidar	Winds, temperature	Stratosphere, lower mesosphere (30–80 km)	Vertical distribution, night

Table 8.1 (continued)   Optical Instruments Used to Investigate Atmospheric Parameters and Processes

Instrument	Observed Item	Altitude	Mode
Multiwavelength lidar	Aerosol, cloud	Upper troposphere, lower stratosphere (5–40 km)	Vertical distribution, night
All-sky camera	Luminosity of airglow layer, atmospheric waves	Upper mesosphere, lower thermosphere (when quiet: 85, 95, 250 km; when active: 85,120, 250 km)	Horizontal distribution, night (new moon phase)
Aurora Web camera	Aurora live image	Operate at different altitudes of ionosphere	Day and night

Photometers, combined with a set of optical filters, are used to measure the intensities of airglow emissions of the visible and near-infrared part of the optical spectrum [1,2]. Specific filters are available at Arecibo to measure the airglow lines of various gases in the atmosphere, such as OH, O_2, O, O^+, N, N_2^+, H, He, and Na.

Tilting-filter photometers assembled at AISRS have the following parameters: a single channel with bandwidth (at the level of 3 dB) of 0.3 to 1.0 nm (depending on choice of interference filter), and a programmable filter tilt controlled by stepping motors with a 10° tilt range or approximately 2.5 nm scan range with field-of-view varied between 0.25° and 5.0°, using selective field stops.

Spectrometers are used to measure spectral blends of airglow emission bands at medium to high spectral resolution [3,4]. The Ebert–Fastie spectrophotometer arranged at AISRS has the following parameters: 1 m focal length with a bandwidth varied between 0.02 to 1.0 nm. A programmable wavelength scanning takes place via stepping motors with maximum scan range limited to 100 nm (anywhere between roughly 300 and 900 nm) with variable field-of-view, which varies between 0.1° and 9.0°.

Interferometers are generally used to measure Doppler temperature and winds that originate in the E and F regions of the ionosphere, or to measure the spectral distribution and temporal variation of the hydrogen geocorona [5–7]. Fabry–Perot interferometers (minimum two are available) at AISRS have the following parameters: each interferometer is with 1.2 m focal length and 0.15 m clear apertures with typical bandwidth of 0.001 nm and free spectral range of 0.01 nm. The wavelength change takes place via pressure scanning using pistons and choice of scanning gas of Ar, CO_2, or SF_6. The field-of-view depends on the choice of aperture size, but it is typically 0.25° for 630 nm observations. It also has the He–Ne frequency stabilized laser for linewidth calibration and thermal control of etalon and prefilter.

The optical laboratories at AISRS have one Doppler–Rayleigh Lidar and two resonance fluorescence lidars. The Doppler–Rayleigh lidar is used to measure the Doppler shifts and widths of the spectrum of the laser light that is broadened and backscattered from the atmosphere and lower ionosphere from about 15 to 70–80 km of altitude and have the following parameters: Nd/YAG-based laser transmitter with the 24 W average power (with 100 MW in its peak) operating at 532 nm and at frequency of 40 Hz with the pulse width of 6 ns. It has the 80 cm diameter Cassegrain telescope receiver. All the other details can be found in Refs. [8–16].

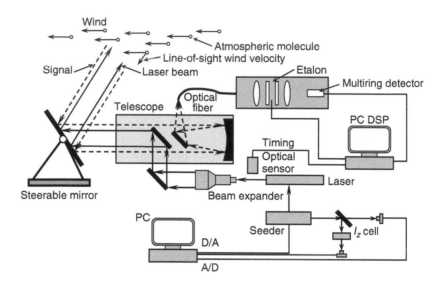

Figure 8.1 Scheme of the Rayleigh–Doppler lidar.

The corresponding scheme of this lidar is presented in Figure 8.1. Temperature profiles observed by Rayleigh lidar (shadow areas) and theoretical results (dotted curves) obtained with CIRA-86 model are presented in Figure 8.2.

Resonance fluorescence lidars (alexandrite-laser based and dye-laser based) can measure various metallic species of the upper atmosphere and lower ionosphere between about 70 and 115 km altitude. An alexandrite laser transmitter can be set between 720 and 800 nm with doubled frequency output between 360 and 400 nm. It has a 3 W average power (with 0.5 MW at the power peak) operating at 770 nm with operational frequency of 28 Hz and the pulse width of 200 ns. A dye laser transmitter is tunable between about 300 and 800 nm depending on the dye/solvent used, as well as on the sets of doubling and mixing crystals. It has 4 W average power

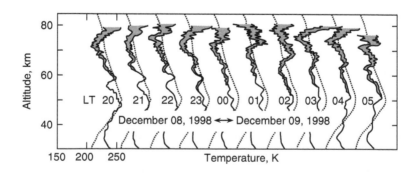

Figure 8.2 Temperature profiles at Poker Flat (average scale: 4 km).

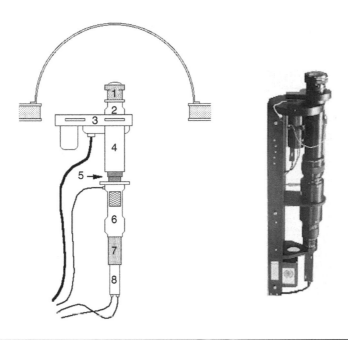

Figure 8.3 The schematic diagram of all-sky camera: 1. Mamiya all-sky lens. 2. Telecentric collimating lens. 3. Filter wheel assembly from Keo Consultants. 4. Positive diopter lens and tube. 5. Canon 50 mm f/1.2 lens. 6. Varo image intensifier tube assembled with refocusing optics. 7. Zoom lens. 8. Pulnix TM-745 CCD camera with NTSC video output.

(with 20 MW peak power) operating at 589 nm with a working frequency of 40 Hz and a pulse width of 5 ns.

Both lidar systems employ 80 cm diameter, pointable Cassagrain telescopes and gated photo-multiplier tube detectors with low-altitude choppers. The results obtained by Doppler–Rayleigh lidar and resonance fluorescence Lidar are presented in Refs. [8–16].

Simultaneous imaging observations of several airglow layers are very important to study vertical propagation of atmospheric gravity waves, which is one of the main aspects in bubbles-structure generation in the equatorial layer F (see Chapter 6).

Moreover, simultaneous observations of wind, temperature, and airglow images are needed to study the dynamics of short-period waves [17]. Figure 8.3 presents a schematic diagram and original view of one of the all-sky camera system. The all-sky imagers that are used in the new Finnish all-sky cameras are manufactured by KeoConsultants. Each imager has telecentric and nonvignetting optics, and the field of view of the fish-eye lens is 180°. The technical specifications of all-sky camera systems are presented in Table 8.2.

The filter wheel can accommodate seven narrow bandwidth interference filters. In normal operation one filter holder is left free to be able to acquire nonfiltered images. Every station has three filters: green, blue, and red.

The faint images are intensified before the final image is acquired by the CCD camera (b&w) and digitized by the frame grabber card of the station computer. This intensification allows shorter exposure times with less expensive CCD cameras, and typically an exposure takes 500 ms. Examples

Table 8.2   Technical Specifications of All-Sky Camera Systems

Item	Description
Fish-eye lens	Canon 15 mm/F2.8
Additional optics	Telecentric lens elements
Filter wheel	7-position filter wheel for 2″ filters
Filters	Interference filters, wavelengths 557.7, 427.8, and 630.0 nm (BW 2.0 nm)
Intensifier lens	Canon 85 mm/F1.2
Image intensifier	Varo 25 mm MCP Gen II Image Intensifier Model 3603
Reimaging optics	Canon 100 mm/F2
CCD camera lens	Fujinon 25 mm/F0.85
CCD camera	Pulnix 765E, 756(H)x581(V)

of the auroral images, representing observations along a north–south meridian versus latitude and time periods, are presented in Figure 8.4a. This particular data example is taken from a test operation of the new Finnish digital all-sky camera operated at Kiruna Observatory, Sweden. Figure 8.4b is an optical intensity map of the red auroral light at wavelength 630 nm emitted by oxygen atoms.

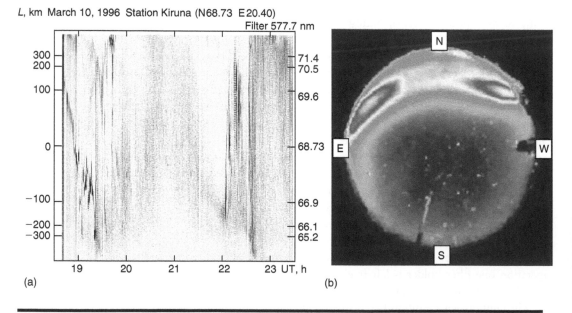

(a)   (b)

Figure 8.4   (a) Example of an auroral Keogram, representing auroral observations along a north–south meridian versus latitude and time (the scale is inverted, i.e., dark zones represent auroral emissions). (b) The false-color intensity map of the red auroral light at wavelength 630 nm emitted by oxygen atoms.

8.1.2 Incoherent Scatter Radars

For ionospheric remote sensing from the ground, the most refined and most modern equipment at present time is probably the *incoherent scatter radar*. Incoherent scatter radar (ISR) is a technique for detecting and studying remote targets by transmitting a radio wave in the direction of the target, observing the reflection of the wave and providing the direct information about electron densities, line-of-sight (LOS) drift velocities, the height of ionospheric layers and ratio electron and ion temperatures, T_e/T_i [18–35]. They also provide indirect measurements of a number of additional parameters, the most reliable of which is the neutral wind below ionospheric altitudes of 130 km. The placements of incoherent scatter radars over the world are presented in Figure 8.5.

Initially, ISR systems were located at Arecibo and at Jicamarca. The Arecibo [18–20] and Jicamarca radars are both monostatic in nature (the transmitter and receiver are co-located, see Chapters 6 and 7).

The Jicamarca Radio Observatory was built in 1960–1961 by the Central Radio Propagation Laboratory of the National Bureau of Standards (see Figure 8.6). The first incoherent scatter measurements at Jicamarca were made in 1961. The Jicamarca Radio Observatory is the premier scientific facility in the world for studying the equatorial ionosphere. As was mentioned in Chapters 6 and 7, the 49.92 MHz incoherent scatter radar is the principal facility of the observatory. The radar antenna consists of a large square array of 18,432 half-wave dipoles arranged into 64 separate modules of 12 × 12 crossed half-wave dipoles. Each linear polarization of each module can be separately phased, and the modules can be fed separately or connected in almost any desired fashion. The isolation between the linear polarizations is very good, at least 50 dB, which is important for certain measurements.

An additional antenna module with 12 × 12 crossed dipoles was built in 1996. It is located 204 m to the west of the west corner of the main antenna and increases the lengths of the available interferometer base line to 564 m.

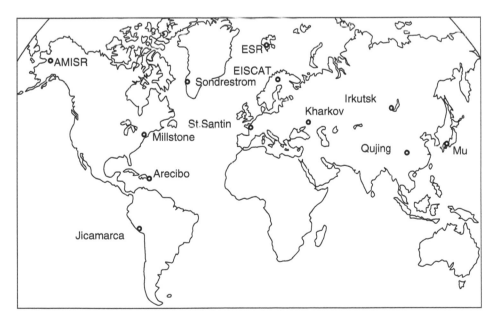

Figure 8.5 World incoherent scatter radars placements.

Figure 8.6 Jicamarca Radio Observatory.

There are three additional 50 MHz array antennas with steering up to $+/-70°$ of zenith angles in the east–west direction only. Each consists of 4×2 half-wave dipoles mounted a quarter wavelength above a ground screen.

The JULIA radar shares the main antenna of the Jicamarca Radio Observatory. As was mentioned in Chapter 6, JULIA is the abbreviation which stands for Jicamarca Unattended Long-term investigations of the Ionosphere and Atmosphere. It has an independent PC-based data acquisition system and makes use of some of the exciter stages of the Jicamarca radar along with the main antenna array. Because this system does not use the main high-power transmitters, it can operate for a long period of time. With a pair of 30 kW peak power pulsed transmitters driving a 290×290 m modular antenna array, JULIA is a formidable MST/coherent scatter radar. It is uniquely suited for studying day-to-day and long-term variability of equatorial plasma irregularities and neutral atmospheric waves, which until now have only been investigated episodically or in campaign mode (description of the corresponding experiments see in Chapter 6).

The Jicamarca Radio Observatory has the following measurement capabilities [21–23]:

1. From all existing ISR-type radars, the Jicamarca radar gives the most accurate measurements of drift velocity and electric field in the equatorial ionosphere (see Chapters 6 and 7). This is because of its unique equatorial geometry. Pointing perpendicular to the magnetic field makes it possible to measure line-of-sight (LOS) drift velocities with accuracy of the order of 0.5 m/s. Vertical F-region plasma drifts measured with such an accuracy allows obtaining information about zonal (e.g., equatorial) electric field with accuracy of about 12 μV/m. By studying the variation of drift velocity with altitude, up to 800–1000 km (or perhaps even higher), it is possible to study the electrodynamics of the entire low-latitude ionosphere, up to the anomaly latitudes, because of the electric field maps along the geomagnetic field lines.

2. The Jicamarca radar also has the unique capability to probe the ionosphere up to very high altitudes. Because of the long radar wavelength, the incoherent scatter is not affected by

problems of the Debye length at low electron densities, and usable signals can be obtained from altitudes of 5000 km and higher, giving densities and perhaps temperatures (but not drifts, as the beam cannot be simultaneously pointed perpendicular to **B**).

3. Absolute F-region measurements of electron density are performed by the Jicamarca Radar using Faraday rotation. Electron and ion temperatures and ion compositions are obtained with a double pulse technique that generates the signal auto-correlation function (ACF). Pulses are transmitted using orthogonal polarization to reduce clutter.

4. The Jicamarca radar is the most sensitive radar in the world. Thus, it is capable of probing even the "gap" region near 45–50 km, partly because of its long wavelength and partly because it has the largest power-aperture product compared with any other VHF radars.

The construction of the Arecibo Radio Observatory began in the summer of 1960. At present, the Arecibo Observatory has selected a data on ionospheric parameters, processes and phenomena over the 1966–2007 periods. We do not enter into a detailed description of Arecibo Radio Observatory because of the wide spectrum of devices and systems arranged there. The reader can find all technical details in the corresponding literature [18–20].

The next generation of incoherent scatter radars (ISR) comprise some of the most advanced radar systems in the world. The best among these are the Millstone Hill and Sondre Strømfjord [27] radars in the American sector and the EISCAT radars (the European Incoherent SCATter Association) in Northern Scandinavia [24–26,32,33]. Their potential regarding observations of thin structure of disturbed irregular ionosphere was shown in Chapters 6 and 7.

The Millstone Hill Radar System consists of two antennas of 25 and 68 m radius, the operating frequencies for which are 1295 and 440 MHz, respectively. The 68 m dish is fixed in the vertical direction, whereas the 25 m dish has full steerability in the azimuth and elevation domains. The peak powers are 3 MW for the 68 m dish and 4 MW for the 25 m dish.

Further, incoherent scatter radar, having a 27 m receiver dish and operating at a frequency of 1300 MHz, was developed at Stanford Research Institute, California, during the 1960s. This radar, also a steerable monostatic-pulsed facility like the smaller of the Millstone Hill system, was subsequently moved to Chatanika, Alaska, where it carried out some groundbreaking observations of the high-latitude ionosphere during the 1970s [27–30], before being moved again to Sondre Strømfjord, Greenland, where it is presently located. The detailed observations of the auroral ionosphere obtained by using this radar system were described in detail in Chapter 6.

The EISCAT Radars, located in Northern Scandinavia and on the Svalbard Archipelago, are currently at the leading edge of incoherent scatter system development. EISCAT is an international collaboration involving the research groups from France, Germany, Finland, Norway, Sweden, the United Kingdom, and Japan (which joined this association in 1996). EISCAT operates using three independent incoherent scatter radar systems: a tristatic UHF "mainland" system, with a transmit/ receiver system located at Tromsø, Norway, and receiver-only facilities located at Kiruna, Sweden, and Sodankylä, Finland.

A new monostatic UHF radar (the EISCAT Svalbard Radar or ESR [26]) is operated close to Longyearbyen, Svalbard. In addition, a monostatic VHF radar is located at Tromsø. All of these radars are pulsed systems, capable of a very wide range of different transmitter modulations and with the most advanced signal processing capabilities for any system of this kind. The mainland UHF system commenced operations in 1981, and the first results from the VHF were obtained in 1985, which is described in detail in Chapters 6 and 7. The EISCAT Svalbard Radar began operations on March 15, 1996. The ESR is UHF radar operating at frequencies around 500 MHz with a 10 MHz bandwidth. The antenna is a Cassegrain-fed dish, similar to the dishes used by the mainland UHF radar system. All signal processing operations are performed digitally, so that much of the analogue hardware found for other types of ISR is not required for the ESR.

A mainland UHF radar system operates at frequencies close to 933 MHz, with 16 different frequencies being available at 0.5 MHz intervals. The system comprises three fully steerable dishes of 32 m diameter. Transmission, reception, and signal sampling are controlled with microsecond accuracy and a dedicated digital correlator exists to form ACFs.

A VHF-radar antenna is a cylindrical in the cross-section plane with the dimensions 120 × 40 m. This antenna can only be moved in the elevation domain. Transmission and reception can also be carried out independently on the two halves of the antenna, allowing the VHF to provide a dual beam capability. The operating frequency of the VHF is 224 MHz, and the bandwidth availability is the same as in the UHF-radar (16 frequencies separated by 0.5 MHz).

The new system, known as EISCAT–3D, will retain the unique and powerful multistatic configuration of the mainland EISCAT UHF-system. Phased-array technology will be employed throughout. The design goals include more precise temporal and spatial resolution of the observed data, an extension of the instantaneous measurements of full-vector ionospheric drift velocities from a single point to the entire altitude range of the radar, and built-in interferometric capabilities. For optimal performance in conditions of low-density plasma, and for the middle-latitude ionosphere, a frequency in the high VHF band (~240 MHz) will be used. The facility will provide high-quality ionospheric parameters measured in real time, as well as near-instantaneous response capabilities for researchers who need data to study unusual and unpredictable disturbances and phenomena occurring in high- and middle-latitude ionosphere. The geographic coordinates of incoherent scatter radars are given in Table 8.3.

Table 8.3 Geographic Coordinates of Incoherent Scatter Radars

Station	Latitude	Longitude	Altitude	Invariant Latitude*	Country
Arecibo	18.34500°	293.24700°	0.00000 km	32.17857°	Puerto Rico
Eiscat tromsøradar	69.58300°	19.21000°	0.03000 km	66.40458°	Troms in Norway, Kiruna in Sweden, and Sodankyl in Finland (Scandinavia)
Svalbard radar (esr)	78.09000°	16.02000°	0.43400 km	74.87426°	Scandinavia
Irkutsk	52.17000°	104.45000°	0.45500 km	45.89960°	Russia
Jicamarca	−11.95000°	283.13000°	1.50000 km	13.90181°	Peru
Kharkov	50.00000°	36.20000°	0.00000 km	45.75379°	Ukraine
Millstone	42.61950°	288.50827°	0.14600 km	53.40967°	United States
Mu	34.80000°	136.10000°	0.00000 km	24.51836°	Japan
Sondre stromfjord	67.00000°	309.00000°	0.00000 km	73.17249°	Greenland

* Calculations of the invariant latitude are unstable near the equator.

To complete the survey of incoherent scatter radars currently operating around the world, we should mention the two systems in the former Soviet Union at Kharkov and Irkutsk, which are capable only of simple long-pulse operation.

The long-term ISR observations provide an extremely valuable data about the ionosphere. The EISCAT Scientific Association, as an international research organization operating with three geophysical research incoherent scatter radar systems, with an ionospheric heater located in Northern Scandinavia, has a data basis for the 1997–2007 period (these heating experiments were briefly discussed in Chapters 6 and 7). The EISCAT-Tromsø Radar system collected data during the period of 1984–2007; the St. Santin radar system (France, 44.6 N, 2.2 E), during 1973–1986; the Millstone Hill radar, from 1970 to 2007; the Arecibo radar, from 1966 to 2007, the Shigaraki middle and upper atmosphere radar (MU radar, Japan, 34.8 N, 136.1 E) from 1986 to 2003 [27], and the Sondrestrøm radar during 1990–2007 [31–35].

The scientific purpose of the measurements is to determine the radar operating modes. Different transmitted pulse schemes are used, depending on the need for range resolution, signal strength, time resolution, and frequency resolution. The rapid steering capability of the radar antenna allows to measure not just range, but also latitude and local time. Because of the many data-taking options, there are many ways to display data.

Finishing this subsection, we present below only a few examples of the results obtained with Sondrestrøm radar (see also details on the Web site: http://isr.sri.com/iono/issdata.html).

Thus, Figure 8.7 is a typical display of data from a 120° elevation scan in the plane of the magnetic meridian. The vertical axis shows altitude from ground level to 600 km and the horizontal axis shows 1000 km of ground range from the south (at the left of the image) to the north (at the right of the image). From left to right and from top to bottom, this figure shows color-shaped electron density, electron temperature, and ion temperature and ion velocity in LOS conditions.

Figure 8.8 shows the radar data from a single 3 min integration with the antenna pointing parallel to the local magnetic field. Some of the basic parameters derived from radar records are electron density (N_e), ion drift velocity (V_i), and electron and ion temperatures (T_e and T_i). All of these quantities are obtained as a function of distance along the radar beam.

In Figure 8.9, the variation of electron density is shown as a function of altitude and time for a 2 h period in March of 1992. Because of the large scale heights in the F region, a 48 km pulse was used. This provided greater signal strength and higher time resolution.

The variations in electron density (Figure 8.10a) and ion drift velocities (Figure 8.10b) as a function of latitude and time for a fixed altitude (300 km) over a 24 h period of observations during May, 1996 demonstrate the steering capability of the antenna and show measurements as a function of latitude. The radar sits under a doughnut-shaped band of data at 74° invariant latitude. The data is displayed north and south of the radar as the site rotates with Earth.

8.1.3 SuperDARN

The SuperDARN network (super dual auroral radar network [36]) is an international system for the studying of the Earth's upper troposphere and ionosphere, and their connections with the magnetosphere and outer space surrounding the Earth. Currently, it consists of a set of radars in the northern and in the southern hemispheres (see Table 8.4). The geometry of the corresponding traces of the SuperDARN international system in the northern and southern hemispheres is shown in Figure 8.11.

The construction of all the radars is roughly identical, with some minor differences in antenna design to accommodate the physical conditions at the site (see Figure 8.12). Each of the radars has

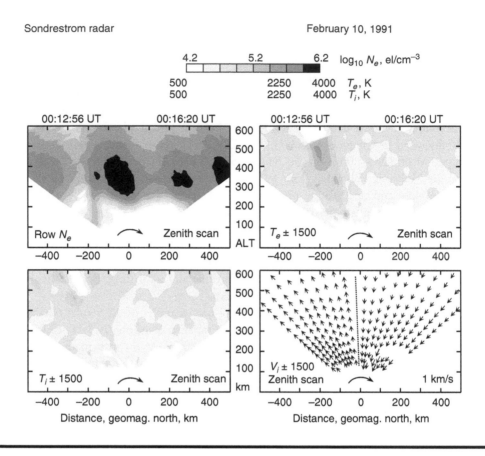

Figure 8.7 Data from a 120 ft elevation scan in the plane of the magnetic meridian.

two arrays of antenna towers, the primary array consists of sixteen towers, and the secondary, interferometer array, consists of four towers. A phasing matrix attached to the antenna array is used to form the beam and to electronically steer the radar into one of 16 different beam directions. The radar transmits a short sequence of pulses in the HF-band and samples the echo signals arriving from the ionosphere (see details of the corresponding experiments in Chapter 6).

A sequence of pulses, referred to as a multipulse sequence, is carefully designed to allow the Doppler characteristics of different targets to be determined at multiple ranges by using the auto-correlation function (ACF) of the received samples. Many sequences are transmitted and the calculated ACFs integrated over a period of several seconds to minimize the effect of noise. The final average ACF is then used to calculate back-scattered power, spectral width, and Doppler velocity of the plasma density irregularities in the nonregular ionosphere. In a standard operating mode, a multipulse sequence (of seven pulses) is transmitted and sampled to resolve 75 ranges with a 45 km separation [37–42].

The operation of a radar is controlled by the radar operating system (ROS), which is responsible for controlling the radar hardware, data processing, and data analysis and storage. A radar control program defines the overall mode of the radar, including the operating frequency, integration period, range separation, and the beam pattern used.

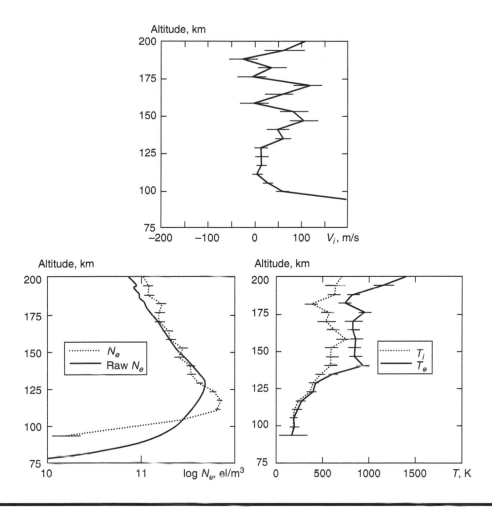

Figure 8.8 The electron density (N_e), ion velocity (V_i), and electron and ion temperatures (T_e and T_i) as a function of distance along the radar beam.

By combining the LOS measurements from a number of radars, the SuperDARN system can produce a two dimensional (2-D) pattern of the ion drifts. The advantages of the system are: an increasingly good coverage in both hemispheres, although the southern hemisphere coverage is less developed; direct, accurate measurement of an important coupling parameter; and the rapid temporal coverage. The greatest weakness of the system is that many of its problems are exacerbated during times of high geomagnetic activity (e.g., magnetic storms). Therefore, SuperDARN radars should be an important component of validation but should be used in conjunction with other data sources.

8.1.4 The Global Positioning System in Investigations of the Ionosphere

The global positioning system (GPS) is the autonomy functional global navigation satellite system (GNSS). It consists of up to 24 medium Earth-orbiting satellites in six different orbital planes, with

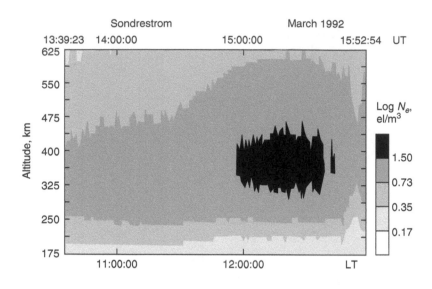

Figure 8.9 The variation of electron density as a function of altitude and time for a 2 h period in March of 1992.

the exact number of satellites varying as older satellites are retired and replaced. These satellites are traveling at speeds of roughly 7,000 mi/h. Transmitter power is only 50 W or less. Operational since 1978 and globally available since 1994, GPS is currently a most utilized satellite navigation system in the world.

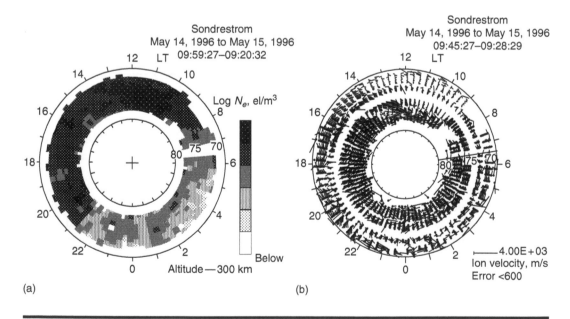

Figure 8.10 (a) This clock dial plot displays the variations in electron density as a function of latitude and time for a fixed altitude (300 km) over a 24 h period in May of 1996. (b) The ion drifts velocities 2-D mapping for the same time period, as in Figure 8.10a.

Table 8.4 SuperDARN Network

Station	Geographic Coordinates	AAGM Magnetic	Status of Work	Location	Principal Investigator
Northern Hemisphere					
Rankin Inlet	62.82° N, 93.11° W	72.96° N, 28.17° W	24 h/d	Nanavut, Canada	Institute of Space and Atmospheric Studies University of Saskatchewan, Saskatoon, Canada
King Salmon	58.68° N, 156.65° W	57.43° N, 100.51° E	24 h/d	Alaska	Communications Research Laboratory, Tokyo, Japan
Kodiak	57.60° N, 152.2° W	57.17° N, 96.28° W	24 h/d	Kodiak Island, Alaska	Geophysical Institute University of Alaska Fairbanks
Prince George	53.98° N, 122.59° W	59.88° N, 65.67° W	24 h/d	British Columbia, Canada	Institute of Space and Atmospheric Studies University of Saskatchewan, Saskatoon, Canada
Saskatoon	52.16° N, 106.53° W	61.34° N, 45.26° W	24 h/d	Saskatoon, Canada	Institute of Space and Atmospheric Studies University of Saskatchewan, Saskatoon, Canada
Kapuskasin	49.39° N, 82.32° W	60.06° N, 9.22° W	24 h/d	Ontario, Canada	Johns Hopkins Applied Physics Laboratory, Laurel, Massachusetts
Goose Bay	53.32° N, 60.46° W	61.94° N, 23.02° E	24 h/d	Goose Bay, Canada	Johns Hopkins Applied Physics Laboratory, Laurel, Massachusetts
Stokkseyri	63.86° N, 22.02° W	65.04° N, 67.33° E	24 h/d	Stokkseyri, Iceland	LPCE/CNRS Orleans, France
Þykkvybaer	63.86° N, 19.20° W	64.59° N, 69.65° E	24 h/d	Þykkvybaer, Iceland	Department of Physics, University of Leicester, England

(*continued*)

Table 8.4 (continued)   SuperDARN Network

Station	Geographic Coordinates	AAGM Magnetic	Status of Work	Location	Principal Investigator
Hankasalmi	62.32° N, 26.61° E	59.78° N, 105.53° E	24 h/d	Hankasalmi, Finland	Department of Physics, University of Leicester, England
Wallops Island	37.93° N, 75.47° W	30.63° N, 75.52° E	24 h/d	Wallops Island, Virginia	Johns Hopkins Applied Physics Laboratory, Laurel, Massachusetts
Hokkaido	43.53° N, 143.61° E	38.14° N, 145.67° W	24 h/d	Hokkaido, Japan	Solar-Terrestrial Environment Laboratory, Nagoya University, Japan
Southern Hemisphere					
Halley	75.52° S, 26.63° W	61.68° S, 28.92° E	24 h/d	Halley Station, Antarctica	British Antarctic Survey High Cross, Cambridge, England
Sanae	71.68° S, 2.85° W	61.52° S, 43.18° E	Began operation in February, 1997	Sanae, Antarctica	School of Physics, University of KwaZulu-Natal, Durban, South Africa
Syowa South	69.00° S, 39.58° E	55.25° S, 23.00° E	24 h/d	Syowa, Antarctica	National Institute of Polar Research, Tokyo, Japan
Syowa East	69.01° S, 39.61° E	55.25° S, 22.98° E	24 h/d	Syowa, Antarctica	National Institute of Polar Research, Tokyo, Japan
Kerguelen	49.35° S, 70.26° E	58.73° S, 122.14° E	24 h/d	Kerguelen Island	LPCE/CNRS, Orleans, France
Tiger	43.38° S, 147.23° E	55.31° S, 133.36° W	24 h/d	Tasmania	Department of Physics, La Trobe University Bundoora, Australia
Tiger Unwin	46.51° S, 168.38° E	55.15° S, 106.54° W	24 h/d	Unwin, New Zealand	Department of Physics, La Trobe University Bundoora, Australia

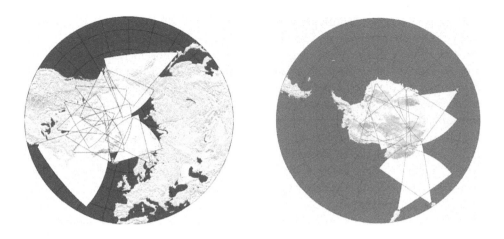

Figure 8.11 Geometry of the radio paths for northern and southern hemispheres of the Super-DARN international system.

Developed by the United States Department of Defense, it is officially named NAVSTAR GPS. GPS was originally intended for military applications, but then, from 1983, the system became available for civilian use.

GPS has become a widely used aid to navigation worldwide, and a useful tool for map-making, land surveying, commerce, and scientific uses. GPS also provides a precise time reference used in

Figure 8.12 The Halley Bay station.

many applications including scientific study of earthquakes and synchronization of telecommunications networks. A GPS receiver must be locked on to the signal of at least three satellites to calculate a 2-D position (latitude and longitude) and track any movements of subscribers. With four or more satellites in view, the receiver can determine the 3-D position of any subscriber (e.g., latitude, longitude, and altitude). Once the user's position has been determined, the GPS unit can calculate other information, such as speed, bearing, tracking, trip distance, distance to destination, sunrise and sunset time, and more.

GPS satellites transmit the following radio signals: military—1227.6 MHz, civilian L1—1575.42 MHz, civilian L2—1227.60 MHz, nuclear burst detection L3—1381.05 MHz, and telemetry on 2227.5 MHz (see details at http://www.tbs-satellite.com/tse/online/prog_gps_freq.html). Beginning from around 2008, civilians will have access to three GPS signals: L1—1575.42 MHz, L2—1227.60 MHz, and L5—1176.45 MHz [43].

A GPS radio signal contains three different bits of information: a *pseudorandom code, ephemeris data,* and *almanac data.* The *pseudorandom code* is simply a code that identifies which satellite transmits information.

Ephemeris data tells the GPS receiver where each GPS satellite should be at any time throughout the day. Each satellite transmits ephemeris data showing the orbital information for that satellite and for every other satellite in the system.

Almanac data, which is constantly transmitted by each satellite, contains important information about the status of the satellite ("healthy" or "unhealthy"), current date, and time. This part of the signal is essential for determining a position. The main factors that can degrade the GPS signal and thus affect the accuracy of subscriber positioning include the following:

- *Ionosphere* and *troposphere group delays* due to effects of multipath occurring in the ionosphere (see Chapters 7 and 9). The GPS uses a special algorithm that calculates an average amount of delay to partially correct for this type of error.
- *Signal multipath* fading [44], which increases the travel time of the signal, thereby causing errors (see definitions in Chapter 4).
- *Receiver clock errors* occur because a ground-based receiver's built-in clock is not as accurate as the atomic clocks onboard the GPS satellites. Therefore, it may have very slight timing errors.
- *Orbital errors* are also known as *ephemeris errors*; these are inaccuracies of the satellite's reported location.
- *Number of satellites visible* means that the more satellites a GPS receiver can "watch," the better the accuracy. Some strong plasma irregularities may partly block signal reception, causing position errors or possibly no position reading at all. Moreover, ground-based obstructions (buildings, hills, sea, soil, etc.) can fully block records from satellites. Therefore, GPS units typically do not work in indoor, underwater, or underground environments. As will be shown in Chapter 9, the same difficulties with recording of signals occur during such natural disasters as magnetic storms.
- *Satellite geometry* is referred to the relative position of the satellites at any given time. Ideal satellite geometry exists when the satellites are located at wide angles relative to each other. Poor geometry results when the satellites are located in a line or in a tight grouping.

Typical GPS positioning accuracy was estimated as 15 m [43,44]. Typical GPS accuracy of user position is 3–5 m and the typical wide area augmentation system (WAAS) position accuracy is less than 3 m. Of course, these estimations are correct only in outdoor environments and for regular and nondisturbed ionospheric and atmospheric conditions. Existence of natural or man-made disturbances of the ionosphere decreases essentially the accuracy of subscriber positioning. Thus, the accuracy of the original GPS, which was subject to accuracy degradation under the Selective Availability Program (in the United States), is 100 m.

Similar satellite-based differential systems were developed in other countries. In Asia, there is the Japanese Multi-Functional Satellite Augmentation System (MSAS), whereas Europe has the Euro Geostationary Navigation Overlay Service (EGNOS). Eventually, GPS users around the world will have access to precise position data using these and other compatible systems. The Russian GLONASS is a global navigation satellite system in the process of being restored to full operation. The European Union's Galileo positioning system is a next-generation of GNSS in its initial deployment phase, scheduled to be operational in 2010. China has indicated it may expand its regional Beidou navigation system into a global system at the same time. India's IRNSS, a next-generation GNSS is in the developmental phase and is scheduled to be operational only around 2012.

GPS ionospheric sounding is a powerful tool for remote sensing of the ionosphere. GPS radio signals L1 and L2 have provided a unique opportunity to study short-scale-length variations in total electron content (TEC) along the signal path in the presence of ionospheric irregularities and scintillations caused by these irregularities (GPS-based ionospheric measurement can measure TEC variations smaller than 10^{-2} TEC units) [45–47]: isolated ionospheric disturbances [48]; simulation of the ionospheric disturbances caused by earthquakes, explosions, cyclones, and tsunamis [49,50]; variations of atmospheric water vapor [51]; ground complex permittivity [52]; processors of horizontal and vertical crustal deformation [53]; influences of the ionosphere on satellite communications and satellite measurements, and so on. Effects of magnetic storms will be discussed separately in Chapter 9 as the most evident example of influence of natural disasters on land–satellite communication.

Hajj and Romans (1998) demonstrated the use of GPS in obtaining profiles of electron density and other geophysical variables such as temperature, pressure, and water vapor in lower ionosphere. This work presents a set of ionospheric profiles obtained from GPS/MET with the Abel inversion technique. The effects of the ionosphere on the GPS signal during occultation, such as bending and scintillation, was also examined [45]. Electron density profiles obtained from GPS/MET are compared with the ones obtained from the Parameterized Ionospheric Model (PIM) and with ionosonde and incoherent scatter radar measurements. Statistical comparisons of $NmF2$ values obtained from GPS/MET and ionosondes indicate that these two types of measurements agree with accuracy of about 20%.

A latitudinal-distributed network of GPS receivers has been operating within Colombia, Peru, and Chile; sufficient latitudinal separation for measuring of the absolute TEC at both crests of the equatorial anomaly is presented in Ref. [47]. The network also indicated the latitudinal extention of GPS scintillations and TEC depletions. We present here only a characteristic example of measurements carried out in [47]. Thus, Figure 8.13 shows TEC depletions measured at several sites, using signals from three different GPS satellites. On September 7, 2001, TEC depletion was observed at all stations except Antuco. In addition, the receiver at Santiago suffered a loss of signals caused by the strong fading which was likely associated with TEC depletions. Figure 8.13 (top graph) shows the passage of a TEC depletion, 35 TEC units (10^{16} el/m2) in depth, detected by the receiver at Bogota between 20:00 and 20:20 LT. Below the TEC curve is displayed with the scintillation index (called an S4-index; see definitions in Chapter 9), which was calculated on-line using the signal received from each of GPS satellites. The gray shadowing indicates the times when the S4-index is above the noise level, and the satellite elevation is above 20°. The TEC depletion illustrated by the top graph is accompanied by high levels of GPS scintillations (S4 = 0.5) on both the east and west walls of the depletion.

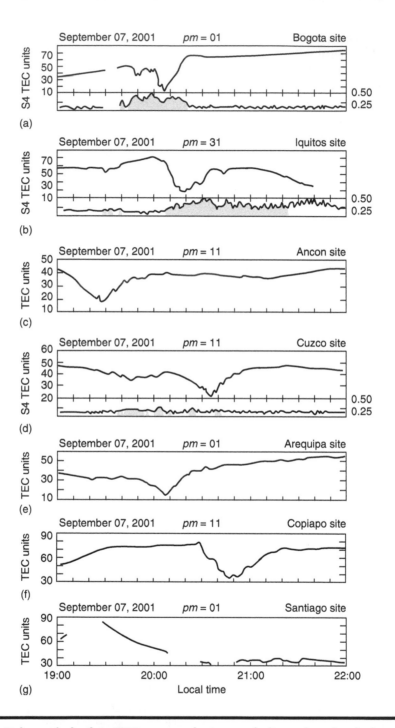

Figure 8.13 Values of absolute TEC measured at several sites using signals from three GPS satellites. The additional subpanel plotted below the Bogota, Iquitos, and Cuzco TEC data corresponds to the GPS scintillation index S4.

The next graph (from the top) illustrates the same TEC depletion that was detected at the Bogota receiver 11 min earlier. At the Iquitos receiver, strong values of the scintillation index (S4 = 0.5) were observed only on the west wall and weaker scintillations on the east side (S4 = 0.25) of the depletion. Due to smaller angle of observation, the receiver at Iquitos observed the same TEC depletion that was obtained at Bogota receiver with an apparent wider width. The authors of Ref. [47] used the transit time of TEC depletions between the Ancon receiver and Cuzco receiver of about 65 min, separated by a 550 km range in the magnetic east–west direction. They also estimated a 140 m/s zonal drift.

Scintillations at Cuzco receiver (the third graph from the top) were of less intensity, due to a smaller density that commonly prevails near the magnetic equator when the anomaly is fully developed. At Arequipa receiver, the lowest value within the TEC depletion was observed at 20:08 UT. This is the same time that the minimum TEC was detected at Bogota receiver (see the top graph), which is located 225 km westward and several hundred kilometers northward from Arequipa receiver. This apparent discrepancy can be explained if we allow the bubbles to tilt westward at altitudes above the F-region peak (effects of bubbles are described in detail in Chapter 6). The westward tilt will make the part of the TEC depletions that extend to higher latitudes to appear at slightly later times in a way similar to the plasma plumes seen with coherent radars. The Copiapo receiver (the second graph from the bottom) detected a 30-TEC-unit depletion spanning between 20:30 and 21:10 LT.

8.2 Diagnostics of the Nonregular Ionosphere by LFM Ionosondes

8.2.1 Developments of LFM Ionosondes in the Historical Perspective

In the recent decades in different regions of the world, active scientific studies of the ionospheric physics and ionospheric radio propagation were carried out with local ground-based networks, called *LFM Ionosondes*, which also have been started intensively for the selection of the optimal ionospheric radio channels (see Chapter 7). This relates to the fact that LFM ionosondes have advantages compared to pulse ionosondes, such as high defense against noise and high electromagnetic compatibility (the radiated power is only about 2 to 100 W), as well as smaller dimensions of the facility.

Special signals with LFM were first used for the purpose of radiolocation [54] and then, in 1954, for the sounding of the ionosphere in regions with high absorption, about 80–90 dB. Later, in 1962, such signals were used for investigation of the *D*-layer of the ionosphere [55–57].

Theoretical aspects of radiation and recording of pulse LFM signals have found their application after creation of methods and approaches dealing with generation of LFM signals. The first tests showed immediately high efficiency of LFM ionosondes for the vertical and oblique sounding of the ionosphere [61]. In Ref. [62], the experimental ionospheric sounding, vertical and/or oblique, is presented using the frequency range from 600 kHz to 2 MHz. The equipment employs a linear frequency sweep transmission generated by direct synthesis from a frequency standard. An identical sweep is used for demodulation at the receiver. Sounding records were obtained at night-time periods along 2000 km radio paths. Radiated power was ranged from 25 to 250 W. A good time delay resolution, together with suppression of interfering signals, makes it possible ionospheric sounding with a greatly improved quality of transmission of any information at large distances using low energy of the transmitting signals.

For applied studies during 1960s, special systems were performed, operating continuous LFM signals [63]. The study of the ionosphere with these systems allowed to obtain a set of original results described in Refs. [64,65]. Further, great success in creation of LFM ionosondes was achieved by Barry Research Co. in the United States, the typical models of ionosondes of which for vertical sounding (VS) and oblique sounding (OS) of the ionosphere, VOS−1 [66], RCS−2 [67], and RCS−4 [68], allowed to obtain ionograms of high quality using very low radiation power, of about 5–10 W only. At the present time, this company produces LFM transivers XCS-6 and TST 4280.

During the period from 1980 to 2000, in different regions of the world, high-effective systems were realized for investigation of the ionosphere, as well as developments of new methods of diagnostics of short-wave ionospheric radio channels, including prediction of the working frequencies (called *maximum useful frequencies*, MUF, see Chapter 7) of radio communication links and also service of short-wave over-horizon radars [57]. Modification of the standard LFM ionosonde of Barry Research Co. (see Refs. [71,72]) by introducing multichannel recording with digital registration and signal processing allowed creation of a monostatic ionosonde of vertical sounding with registration of arriving angles, with estimation of polarization and Doppler speed of displacement points of wave reflection.

A bistatic LFM ionosonde was constructed in England to study dispersion distortions of wideband signals on short-length radio traces [73,74]. A set of experiments at the one-hop traces in the polar and equatorial ionosphere was carried out to investigate the effects of ionospheric irregularities on radio scattering phenomena [75].

TCI/BR Communications has carried out a global experiment using 16 LFM ionosondes to provide and secure radio communication at the short-frequency band. They used data obtained from 30 radio traces of oblique sounding [76]. The same European project was made at two radio traces using BR Communications ionosondes [77,78].

For the study of "thin" ionospheric effects in the propagation of short-wave radio signals, the research groups Defense Evaluation and Research Agency (DERA, United Kingdom) created a high-quality ionosonde for oblique sounding, called Improved Radio Ionospheric Sounder (IRIS) [79,80]. In 1992, at South-West Research Institute (San Antonio, Texas), a small multichannel interferometer was developed on the basis of an LFM ionosonde for measurements of group delay and angles-of-arrival in the azimuthal and elevation local planes [81]. Over-horizon radars of short-wave frequency band also used LFM signals for ionospheric diagnostics and to test the accuracy of ionospheric models, as well as conditions of the ocean surface and ocean flows [82–85].

The Australian Defence Science and Technology Organisation created in Australia a wide net of ionosondes of oblique sounding to investigate the low-latitude ionosphere and to ensure a stable operation of an over-horizon short-wave radio locator [86–88]. A net of LFM ionosondes of the IPS-71 type, produced by the Australian company KEL Aerospace Pty. Ltd. and situated on the territory of Australia, includes ionosondes for vertical and oblique sounding of the ionosphere. Such ionosondes have the velocity of frequency changes of 500 kHz/s (see definitions below). Studies of the peculiarities of the low-latitude ionosphere above the Asian-Pacific-Ocean region [88] are carried out using LFM ionosondes developed by Australian Defense Science and Technology Organization (DSTO).

In Ref. [87], a technique is presented for using the measurements of the time-varying narrow-band (10 kHz) component-transfer function of a high-frequency (HF) frequency-modulated (FM) continuous wave, transmitted at 15.085 MHz, and its decomposition into propagation modes. One event occurring at a 5244 km transmission path, which exhibits flat fading (see definitions in

Chapter 4) within the 10 kHz bandwidth, is analyzed and found to exhibit severe phase distortion due to multipath. A component-transfer function for an individual propagation mode was also obtained in Ref. [87] using 2-D filtering of the signal in the joint Doppler frequency (DF) and time delay (TD) domain, resulting in a significant reduction in phase distortion.

In the former Soviet Union, an ionosonde was first constructed for vertical sounding (VS) of the ionosphere that used quasi-continuous FM signals [57,89]. Then, ionosondes for VS of the ionosphere were created on the basis of synthesizers of continuous LFM signals and their different modifications for VS and OS of the ionosphere [90–96]. With their help, the studies of frequency effects of modification of the ionosphere with powerful short-wave signals were made, as well as long-term investigations of the conditions of propagation of continuous LFM signals at radio traces of various orientations and lengths [97–100]. Now, in Russia, an experimental net of LFM ionosondes is operated for oblique sounding (OS) of the ionosphere, the software–hardware methodology of which was developed by several groups [57,98–100], including a group working at Moldavian Radio Observatory [97,100].

To secure a stable communication at the short-wave band between countries of NATO, a global net of LFM ionosondes over the world was arranged based on the system AN/TRQ-35(V) (Tactical Frequency Sounding System), which includes 77 transmitters [57]. The AN/TRQ-35(V) is an ionospheric sounding system that easily operates with ionospheric propagation statistics on the real-time basis. The system is used to minimize outages related to unpredictable changes of ionospheric characteristics and conditions. It is intended to improve frequency management and assignments of frequencies for HF communications systems and, finally, results in more effective and efficient utilization of the HF spectrum, producing more reliable HF communication with improved grade of service (GoS).

Depending on the method of utilization of LFM ionosondes, it is possible to divide them into LFM ionosondes of vertical sounding (VS), oblique sounding (OS), and back-oblique sounding (BOS) [57]. Ionosondes of VS are used mostly to monitor the ionosphere above the local place of the ground-based diagnostic system, as well as to study physical processes occurring during natural and artificial perturbations of the ionosphere [101]. The ionosondes of OS are used mostly to study the ionosphere along the trace of short-wave radio propagation under various geographic conditions in adaptive communication systems and to secure frequency during selection of the optimal radio channel [82,101,104]. Ionosondes of BOS are used to study propagation conditions in the ionosphere [83–85] and to investigate conditions of sea surface in large spatial regions, as well as in the systems of frequencies secured in short-wave radio communication channels and over-horizon short-wave radars [86].

8.2.2 Operating Principles of LFM Ionosondes

At present, there is no strict self-consistent theory of propagation of extremely wideband signals of LFM ionosondes in the ionospheric plasma as a dispersive medium. Only a few approximate approaches of mathematical description of operational functions of LFM ionosonde were developed, accounting for specific features of processing of registered signals by the receiver, as well as based on well-known behavior of narrowband signals in the ionosphere [57]. The influence of dispersion distortions in the ionosphere on the structure of the complicated signals (including LFM-signals) during VS and OS of the ionosphere was analyzed using theoretical frameworks described in Refs. [100,107–110].

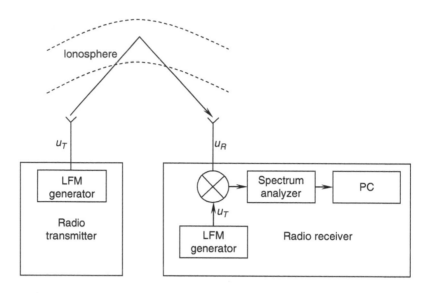

Figure 8.14 Functional scheme of the LFM ionosonde.

The basic principle of LFM ionosonde operation can be explained by using a simple functional scheme shown in Figure 8.14. Let us suppose that at the transmitting channel of an ionosonde the transmitter radiates a continuous signal with linear modulation of its frequency:

$$u_T(t) = u_0 \cos(\omega_S t + \beta t^2/2), \qquad (8.1)$$

where u_0 is the signal amplitude, $\omega_S = 2\pi f_S$ is the initial circular frequency, $\beta = 2\pi \dot{f}$ is the velocity of changes of the circular frequency. Then, the current radiated frequency, $f = d\,(\omega_S t + \beta \cdot t^2/2)/2\pi\,dt$, changes according to linear low at the range from f_S to f_E with the velocity which has fixed values in the limits from 25 kHz/s to 1 MHz/s depending of the regime of sound and the problem that is being solved. Usually, $f_S \sim$ 2–3 MHz and $f_E \sim$ 15–30 MHz, and therefore the radiated signal has period of several minutes and occupies a frequency bandwidth of several tens megahertz, i.e., it is extremely wideband.

We also suppose that the LFM signal formed by the transmitter arrives at the input of the radio channel, containing the receiving–transmitting antenna-fider devices; the input of the receiver and the land ionosphere waveguide have all elements of the channel coinciding with each other in the whole frequency band $[f_S, f_E]$. Let us assume for simplicity that in Equation 8.1 $u_0 = 1$. In the assumption of linearity of the characteristics of the radio channel, a signal at its input can be presented as a convolution of the radiated signal $u_T(t)$ and the pulse characteristic of the channel $h(\tau)$:

$$u_R(t) = \int_0^\infty h(\tau)u_T(t - \tau)\,d\tau. \qquad (8.2)$$

The signal $u_R(t)$ passes the input circuits of the receiver and is then affected by first processing procedure using the method of its compression [1]. Compression of the spectrum is realized by the consistent filter at the receiver, in which the signal $u_R(t)$ is produced with the signal of the basic generator shifted with respect to radiated generator at the time t_0 and having the time dependence presented by Equation 8.1. The result of this product contains low-frequency (difference between frequencies) and high-frequency (sum of frequencies) components. The latter component can be then filtered with low-frequency filter.

From the continuous signal of low frequency, using the corresponding analysis, the sets with period T_D are selected by the "window" $w(t)$ with time duration around their centers of $t_k = t_0 + (k - 1/2)T_D$, $k = \overline{1, N}$. We should note that the window in the time domain with time period T_D corresponds to the segment of the input LFM signal with the frequency band of $\Delta f_D = \dot{f} \cdot T_D$. Therefore, the frequency band of this segment has to be not larger than the frequency band of the receiver's selector.

Then the selected sets of the initial LFM signal enter at the input of the spectrum analyzer. The result of its work is described by the Fourier transform, where the spectrum of the k-selected set has the form:

$$S_k(\Omega) = S_k(F) = \frac{1}{4\pi} \int\limits_{-\infty}^{\infty} \int\limits_{0}^{\infty} w(t - t_k)h(\tau) \cos\left[\frac{\beta}{2}(\tau - t_o)^2 - \beta t(\tau - t_o)\right] e^{i\Omega t}\, d\tau\, dt, \qquad (8.3)$$

where $\Omega = 2\pi F$.

The form of the window $w(t)$ in the device is realized in such a manner that its Fourier transform $W(F)$ has a narrow band, about of 1–10 Hz (in most cases $\sim$1 Hz) with total elimination of the side loops. The selected sets with time period, T_D, correspond to segments of the LFM signal with basis $B_D = \Delta f_D \cdot T_D$ in the frequency domain. For the case of $\dot{f} = 10^5$ Hz/s and $T_D = 1$ s, we get $\Delta f_D \cong 100$ kHz and $B_D \cong 10^5$.

Expressing the pulse characteristic of the radio channel $h(\tau)$ via its Fourier transform, called transferred function $H(\omega)$, the following approximate expression of the registered spectrum at the input of the spectrum analyzer can be obtained [111]:

$$S_k(\Omega) = \frac{\pi}{2} e^{i\psi} \int\limits_{-\infty}^{\infty} H(\omega)B(\beta(t_k - y_0) + \omega + \omega_H)e^{-i\omega y_0}\, d\omega, \qquad (8.4)$$

where $\psi = \beta(y_0^2 - t_0^2)/2 - \omega_S(y_0 - t_0)$, $y = \Omega/\beta - (\tau - t_0)$, $y_0 = \Omega/\beta + t_0$, $B(\beta(t_k - y_0) + \omega + \omega_S)$, and the Fourier transform of the function equals

$$b(y) = e^{i\beta y^2/2} W(\beta y). \qquad (8.5)$$

The term $b(y)$ is defined as a range of the argument values where the function differs essentially from zero. According to Equation 8.5, it is close to the term describing spectrum of the window of selections, $W(\beta y)$. In this case, it is also supposed that the lower limit of the frequency band Ω_0, measured by the analyzer exceeds the frequency band of the window spectrum. In these definitions, the expression for spectrum of the k-selected set is the same as that for the response of the radio channel against any equivalent signal $b(y)$ with the spectrum $B(\omega)$.

According to Equation 8.5 the time duration of the pulse signal $b(y)$ is defined by the frequency resolution of the spectrum analyzer $\delta\Omega$ and by the velocity of frequency changes according to the formula $\delta\tau \cong \delta\Omega/\beta$.

Usually, the time duration of the equivalent pulse is of the order of 10^{-5}–10^{-4} s, and the band of its spectrum is of the order of 10–100 kHz. In the short-wave range in most cases such a spectrum can be considered narrowband. In the conditions mentioned here the scheme of modeling the characteristics of continuous LFM signals when sounding the ionosphere is reduced to a simple problem of propagation with narrowband signals. It is well known [112,113] that when sounding the ionosphere with narrowband pulse signals, several responses of the radio channel are registered by the receiver with the frequency band equal to the frequency band of the signal (in conditions of full coincidence).

As a rule, $H(\omega)$ can be presented as a sum of transferred functions $H_l(\omega)$, corresponding to different modes of propagation:

$$H(-\omega) = H^*(\omega) = \sum_l H_l^*(\omega) = \sum_l |H_l(\omega)|e^{-i\omega P_l(\omega)/c}, \tag{8.6}$$

where $P_l(\omega)$ is a phase trajectory of the l-th mode at frequency ω.

Let us expand $H_l^*(\omega)$ at the proximity of the frequency

$$\omega_k = \omega_S + \beta(t_k - y_0), \tag{8.7}$$

which is a central frequency in the spectrum $B(\omega)$. Within the limits of the examined band, the value $|H_l(\omega)|$ can be taken as constant, and the function $P_l(\omega)$ as a smooth function. Therefore, we can limit the Taylor expansion of the signal phase by using only its second-order term:

$$\frac{\omega P_l(\omega)}{c} = \frac{\omega_k P_l(\omega_k)}{c} + (\omega - \omega_k)\tau_l + (\omega - \omega_k)^2\frac{\tau_l'}{2}, \tag{8.8}$$

where

$$\tau_l = \left.\frac{d[\omega P_l(\omega_k)]}{cd\omega}\right|_{\omega_k} \tag{8.9}$$

is the group delay of the l-th mode at the frequency ω_k, and

$$\tau_l' = s_l/2\pi = \left.\frac{d\tau_l}{d\omega}\right|_{\omega_k} \tag{8.10}$$

After submission of Equation 8.8 into the right part of Equation 8.6, the expression for the output spectrum $S_k(\Omega)$ can be obtained in the following form [114]:

$$S_k(\Omega) = S_k(F) = \frac{\pi}{2}e^{i(\psi+\omega_k y_0)}\sum_l H_l^*(\omega_k)q(y_0 - \tau_l), \tag{8.11}$$

where $q(y_0 - \tau_l)$ is the inverse Fourier transform of $Q(\omega_k - \omega) = B(\omega_k - \omega)e^{-i\frac{(\omega-\omega_k)^2\tau_l'}{2}}$.

From these current derivations it follows that without accounting for frequency dispersion (i.e., for the second-order term in the expansion (Equation 8.8)), the nondistorted signals are registered at the output of the spectrum analyzer. This case corresponds to propagation of various modes during the sounding of the ionosphere with narrowband pulse with envelope $b(y_0)$.

Actually, the pulse envelope is the faction $W(\beta y_0)$, since the power in the exponent of expression 8.11 for the case of consideration is very small. Therefore, the form of pulses is defined by the form of the window $w(t)$ of the spectrum analyzer. The position of the center of l-th signal at the axis Ω is determined from the constraint $y_0 = \tau_l$ and corresponds to the frequency

$$\Omega_l = \beta(\tau_l - t_0) \text{ or } F_l = \dot{f}(\tau_l - t_0) \tag{8.12}$$

Expression 8.12 gives the relation between time delay of the registered signal of mode l and the variable of the analyzer Ω (or the difference F), and permits the interval of group delays of interest to coincide with the working band of the analyzer $[F_0, F_{max}]$ by using the corresponding selection of the parameter t_0.

To obtain record of all modes of the signal with delays $\tau_1, \ldots, \tau_n$, it is necessary that the frequency band of the receiver exceeds the value $\Delta F = \dot{f}(\tau_n - \tau_1)$. For example, if $(\tau_n - \tau_1) = 5$ ms and $\dot{f} = 100$ kHz/s, then $\Delta F = 500$ Hz. This means that for the given velocity of rearrangements of the frequency at the receiver, its working frequency band of 500 Hz corresponds to the time window, which equals 5 ms.

The spectrum analyzer divides the band of the receiver into m equal subbands of value δF. The latter characterizes errors in estimating the time of group delay, which is defined by the time of the equivalent pulse, $\delta\tau$, and has to be smaller than the expected minimum difference of signal delays for different modes of propagation.

By using fast Fourier transform for derivation of the signal spectrum at the difference frequency, the corresponding quadrature components should be outlined. Using these components, the amplitude and phase spectra can be calculated. The amplitude spectrum is used for performance of the corresponding ionograms. Index k in formula 8.11 determines the current frequency ω_k, at which the characteristics of LFM signal of mode l are calculated. Changes of the index k from 1 to N, correspond to the frequency of the LFM signal running along the whole frequency band used for sounding.

As an illustration, in Figure 8.15 the experimental spectrum amplitude $|S_k(\tau)|$ is presented during sounding on the trace Ioshkar–Ola–Balti (described in Chapter 7) on December 10, 1991. The group path $P = c\tau$ is arranged along the axis of delays. The registered signals correspond to bottom and top rays of one-, two- and three-hope modes of propagation. The graphical imaging of the cross-section of $|S_k(\Omega)| = |S_k(\tau)|$ at the given level presents an actual ionogram of oblique sounding. The dependence $|S_k(\tau)|$ at the time t_k presents the amplitude relief of the signal at the current frequency f_k for all the registered modes of propagation.

In a number of studies [93,107,115], the radiated continuous LFM signal was performed in the frequency domain (unlike the above described approach operating with signal in the time domain). For this case, the method of stationary phase was used, giving a good approximation for the real spectrum.

In our further discussions we suppose that the obtained solution of the problem of propagation of the harmonic signal in the ionosphere is found and known. Then, it defines the transfer function of the ionosphere containing the amplitude-frequency characteristic (AFC) and phase-frequency characteristic (PFC) for the wave frequency under consideration. If so, the product of radiated signal and the transfer function of the ionosphere finally gives the spectrum of the signal at the input of the receiver, and the Fourier transform of this product gives us the desired signal.

Further mathematical derivations of the signal can be done according to the scheme described above. As a result of such an approach, we can obtain analytical dependences for the amplitude spectrum on difference frequency accounting for dispersion in propagation of radio signal in the ionosphere.

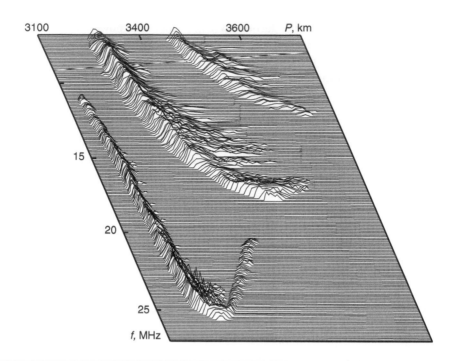

Figure 8.15 Amplitude of the spectra of the output signal at the LFM spectral analyzer at the oblique trace Ioshkar–Ola–Baltsi.

Concluding this paragraph, we will consider a question of stability of LFM ionosonde against noises with respect to the pulse ionosonde described in [91,100]. The main sources of noises in the short-wave frequency band are the operating radio broadcasting stations. In Figure 8.16a according to Ref. [116] the frequency band of the pulse receiver is shown (curve 1).

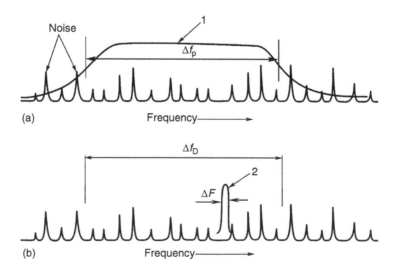

Figure 8.16 Influence of the stationary noises inside the frequency band of: (a) pulse ionosonde, (b) LFM ionosonde.

We should stress that the band of the pulse ionosonde has to be not less than $\Delta f_p \sim 1/T_p$, where Δf_p and T_p are the band of radio frequencies and the time duration of the sounding pulse, respectively. For typical values of $T_p = 50$ μs we easily estimate $\Delta f_p \sim 20$ kHz, i.e., the band of a pulse ionosonde is much bigger than the band ΔF of an LFM ionosonde. All noises from radio stations can be hit simultaneously inside the bandwidth of the impulse ionosonde, and practically cannot be rejected from this wide bandwidth.

The receiver of an LFM ionosonde has much less frequency band ΔF (curve 2, Figure 8.16b) and selects signals from radio stations continuously, step by step, transforming them into impulse noise, the spectral density of which is sufficiently small and which can be rejected by several technical methods, namely, either by frequency rejecting at the input filter of the receiver, or by time rejecting of the signal at difference frequency.

If, for example, at the input of the receiver the noise signal arrives in a form of:

$$u_n(t) = u_0 \cos \omega_n t, \tag{8.13}$$

then, after product operation with the signal (Equation 8.1) in the receiver, this signal has the form:

$$u_n(t) = u_0 \cos\left((\omega_n - \omega_S)t - \frac{\beta}{2} \cdot t^2\right). \tag{8.14}$$

At the same time, the difference signal during the time T can be assumed to be harmonic, that is, its power is concentrated on the element δF. During the time T_D the noise power P_n is distributed along the band Δf_D in such a manner that the spectral density of its power equals:

$$N_n = \frac{P_n}{\Delta f_D}. \tag{8.15}$$

Hence, at the element δF will be concentrated the noise density recorded at the output of the receiver, which equals:

$$(P_n)_{exit} = \frac{P_n}{\Delta f_D} \cdot \delta F = \frac{P_n}{B_D}, \tag{8.16}$$

where $B_D = \Delta f_D/\delta F = \Delta f_D T_D$ is the basic parameter of the signal with noise. In this case, the ratio of signal power to noise power at the input and output of the receiver is described by the following formula [117]:

$$\left(\frac{P_{sign}}{P_n}\right)_{exit} = B_D \left(\frac{P_{sign}}{P_n}\right)_{ent}. \tag{8.17}$$

Thus, during broadcasting activity and the influence of noise arriving at the input of LFM receiver, we see, according to (8.17), that the signal-to-noise ratio (SNR) at the output of the receiver exceeds the SNR at the input of the receiver in B_D times (for $B_D \gg 1$), which allows to decrease significantly the transmitted power of the LFM ionosonde. It should be pointed out that using the narrowbeam antenna grid with the dimensions of 2.5 km, the ionograms of oblique sounding can be easily obtained at power of radiation of the order of units of milliwatt [118,119] (see the results of experiments discussed at the end of Chapter 7).

Further, we shall formulate briefly the main requirements to the parameters of an LFM ionosonde used to investigate the ionosphere and to diagnose short-wave communication ionospheric channels. The transmitting part of an LFM ionosonde can be built on the basis of industry-produced broad-band power amplifiers and elaborated program-run synthesizers of LFM signals [66–72]. The main requirement for the transmitting part is the spectral purity of the formed LFM signal and the frequency stability of the LFM signal synthesizer reference generator, which must be not worse than 10^{-8} Hz. (Brief instability leads to the appearance of mistakes in LFM signal synthesis and to deviation from linear law which drives the transmitter; long instability leads to deviation of frequency absolute values of the forming LFM signal.) Frequency range of an LFM ionosonde is determined by the length of usable radio-routes. In years with high solar activity, maximum-suitable frequencies can reach 40 MHz; therefore, an LFM ionosonde should have a frequency range not less than $2 \div 40$ MHz. Radiation power of a power amplifier should not be too big so as not to interfere with other short-wave communication systems.

Usually, a transmitter's power should not exceed 100 W. The requirements for the receiver bandwidth of an LFM ionosonde are contradictory. On one hand, it should be big enough to transmit without distortion all the signal propagation modes; on the other hand, an increase of the bandwidth leads to an increase of noise that gets into the receiver tract and to its masking of the useful signal. The width of the receiving tract band pass is closely connected with the speed of LFM signal frequency change and with the range of the signal mode group delay. The necessary range of group delay is calculated taking into account radio-route length with some reserve, which compensates the influence of the instabilities of the receiving and transmitting tracts, temporary lack of synchronization of the timescales of the receiving and transmitting parts, and daily and annual variations of group delays under the influence of the Sun radiation. When sounding on short and middle radio-routes, the calculated range of group delays does not exceed 3–4 ms. Thus, when using an LFM signal with the frequency change speed $df/dt = 100$ kHz/s, the receiving tract transition band should be not less than 500 Hz.

When working in an automatic regime on different radio routes with a large difference in the levels of received signals, insufficient dynamic range of the system can influence the quality of received ionograms. A signal that exceeds the boundary of a dynamic range may cause distortions in nonlinear circuits of the receiving tract (frequency changers, amplifiers). The problem can be solved by using variable step-switching attenuators connected at the input of the radio-receiving installation.

An LFM ionosonde should work on both long (more than 3000 km) and short (up to units in kilometers) radio traces. This requires the use of at least two antenna systems for long and short radio traces and of antenna commutators for their quick switching. To ensure the possibility of receiving signals of the world net of LFM ionosondes, the receiving antennas should have a circle radiation pattern in a horizontal plain. The required exactness of synchronization of the timescales of the receiving and transmitting parts of an LFM ionosonde (not worse than 20–30 µs), can be reached by constructing a subsystem of exact time on the basis of GPS receivers and rubidium frequency standards. A frequency standard can also serve as a source of oscillation for the LFM signal synthesizer. The parameters of an LFM signal formed by the synthesizers in the receiving and transmitting parts of the complex should meet high requirements: a wide range of frequency change, high linearity of the frequency change law, spectral purity, and coherence. LFM signal synthesizers constructed by using the method of direct digital synthesis answer such requirements. The choice of the LFM signal frequency change speed is defined by the necessary frequency resolution and delay as well as by the necessity of compatibility with the LFM ionosondes of the world net. It should be noted that by changing the speed of the LFM signal frequency change, the width of the signal spectrum of difference frequency grows, too, which requires an increase of the receiver transmission band, that is, of the transmission in signal-to-noise ratio. The positive factor here is that, in this case, the influence on the communication systems working in decameter wave band (DKB) are decreased. A decrease of the sounding speed, on the contrary, improves signal-to-noise ratio, reduces the receiver transmission band, but increases the duration of the sounding session, and increases the influence on DKB of the communication system. Most transmitters of the world net of LFM ionosondes use the LFM signal change speed of 100 kHz/s, when the receiver transmission band is sufficiently wide within the limits of permissible duration of the sounding d which constitutes 280 s when sounding in the range 2–30 MHz.

The technical characteristics of LFM ionosondes for vertical, oblique, and vertical-oblique sounding of the ionosphere are represented in Table 8.5.

With the help of LFM ionosondes, extensive data on the condition of natural and artificially disturbed ionosphere were received, formation peculiarities of the field of short-wave signals on middle-latitude, sub-auroral, transequatorial routes of different length were investigated. The high effectiveness of using LFM sondes as part of adaptive short-wave radio communication system was experimentally shown.

In order to implement the possibilities of using LFM sounds when solving fundamental and applied problems in the field of ionosphere physics and radio wave propagation, it is necessary to widen the world net of LFM sondes and to place them, first of all, in high-latitude regions, where the effects of the sun–Earth connection in the system of the magnetosphere–ionosphere atmosphere are most strongly manifested. Additionally, the placement of LFM sondes in the Arctic and in the Antarctic regions will allow us to receive in real-time regime information about the conditions of radio wave propagation in regions strongly exposed to the influence of magneto-ionospheric disturbances. This data can be used to solve the practical problems of providing effective

Table 8.5 Technical Characteristics of LFM Ionosonde Operating with Continuous Signals

Main Characteristics	VS	OS	BOS
Radio Transmitter			
Frequency range of radiated LFM signal, MHz	2–16	2–40	2–40
Velocity of changes of frequency LFM signal, kHz/s	50	100–1000	25–100
Radiated power, Watt	2–10	2–100	10–1000
Level of side discrete constituents, dB	–50	–50	–5
Radio Receiver			
Frequency range of received LFM signal, MHz	2–16	2–0	2–40
Velocity of changes of frequency of LFM signal, kHz/s	50	100–1000	25–100
Frequency band of the receiver, Hz	500	500	1000
Dynamic band, dB	80	100	120
Aproval of the device on delay, μs	20	10	10–40
The range of the observed delays, μs	10	5	5–20

functioning of short-wave radio communication systems, above-the-horizon short-wave radio-location, radio-navigation, and radio direction-finding.

Further development of hardware–software devices of an LFM complex presupposes inclusion into the database of geophysical information received from Sputniks in an on-line regime and the exchange of sounding results over the Internet.

Thus, we can quite definitely assert that the development of an LFM ionosondes network, with software improvement and its inclusion into a single world network of LFM ionosondes will allow us to control and forecast on a new technological level the effects of space weather that play such an important role in life support on the Earth.

8.3 Monitoring of Ionospheric Communication Channels

At present, there is an increasing need for information transmission systems via ionosphere communication channels and for signal processing devices, with emphasis on their reliability and productivity [120,121]. It is impossible to ensure corresponding radio-channel parameters and normal conditions of information transmission under quickly changing conditions of ionosphere without regular monitoring of radio channels in real-time scale and without using operative and long-time forecast of the conditions of radio wave propagation. Included in operative forecast is the prediction of short-wave signals ranging from some minutes to one hour.

With regular monitoring of radio channels, the received information gives us the possibility for solving the following multiparameter tasks: defining the number of propagation modes; defining the optimal operating frequencies of single-mode channels; determining the possible speed of information transmission; and calculating the variation of each mode's amplitudes and phases, etc.

Modern methods of long-term forecast of the conditions of ionosphere radio-wave propagation err in ~20%. And in transition hours, when the parameters of the ionosphere change most sharply, the error increases up to 50% and even more. The reason of relatively big errors in long-term forecast is the changeableness of the flow of radiation from the sun from day to day and the complexity of the ionosphere reaction to geomagnetic perturbations [57]. In this connection, along with perfection of the methods of long-term forecast, considerable attention is paid to the methods of direct monitoring of the parameters of an ionospheric radio channel using different systems of oblique and vertical-oblique sounding, and to operative forecast of the conditions of radio-wave propagation. Introduction of the first systems of direct monitoring which worked in impulse regime, Pathfinder [122], CHEC [123], CURTS [124,125], increased considerably communication reliability. In particular, the operation of the CURTS system showed that communication reliability has grown up to 90%.

To monitor the ionospheric channels of short-wave communications, it is necessary to create a net of experimental–technological radio routes. Such a net can be created on the basis of operating radio routes, the equipment of which needs to be complemented with equipment for the diagnostics of radio communication channels and with corresponding software. The equipment for the diagnostics of a communication channel should measure the characteristics of the radio channel quality when transmitting test communications, record information about the changes of the ionospheric parameters, and suggest an optimal variant of uninterrupted work of the radio channel.

The use of small-power LFM ionosondes for oblique sounding of the ionosphere marks considerable progress in the development of the methods of the ionosphere channels of short-wave communication diagnostics. The problem of the ionosphere monitoring is solved with acceptable mass and dimension characteristics of the equipment, less energy consumption, and better electromagnetic compatibility. The availability of LFM ionosondes of oblique sounding on diagnostic radio routes allows to synthesize ionograms for operating radio routes, estimate the expected quality of radio communication, and effectively manage frequency resources. It reduces by half the chance of code error [126]. And the most important factor is that it becomes possible to use antennas working in the system of the operating radio routes to diagnose the channel. Diagnostic radio routes of different length (from 5 to 10,000 km and more), and orientation can be used to control the parameters of ionosphere radio channels.

At present, LFM ionosondes are widely used in adaptive systems of short-wave communication for dynamic control of operating frequencies. Because of its adaptivity, the system automatically supports the quality of short-wave radio communication during a communication session by changing the main parameters of transmission in accordance with the change of the current ionosphere condition.

Such an approach to constructing adaptive systems of short-wave communication with the use of LFM ionosondes allows to obtain data which is used for operative forecasting and extrapolation of information about ionospheric radio wave propagation to other regions.

The main task of the equipment monitoring the ionosphere radio channel consists in the choice and forecasting of optimal operative frequencies of communication by the results of the analysis of ionosphere radio wave propagation conditions and of the noise situation (conditions, environment) for the given radio channel. The radio channel monitoring system should consist of an LFM ionosonde, an analyzer of short-wave radio channel loading, and a packet of applied software on the choice of the main operating frequencies for the communication system. Different methods can be used for radio channel monitoring.

Here is an example of one method of short-wave radio channel monitoring, which includes the following stages:

1. Transmission of an informational communication by the transmitter of a communication system on the frequency chosen according to the data of long-term forecasting (in this case, the optimal operating frequency can be chosen close to the maximum of the used frequency (from the available resource) by the criterion of minimum noise level in radio channel with a set band)
2. Oblique LFM sounding of the ionosphere in the band ∼3–40 MHz
3. Analysis of the loading of the communication channel with noise and setting of optimal operating communication frequency from available resource (a strategy of the choice of the optimal operating frequency consists in defining the frequency range with maximum signal-to-noise ratio on condition that, in the region of multimode propagation, the amplitude of one mode exceeds the amplitude of another mode by not less than 10 dB)
4. Transmission of information via the operating radio channel on the optimal operating frequency chosen in the process of operative forecast

In all the cases it is necessary to define the percentage of error in the information communication. To compare the effectiveness of a connected radio line work on long-term forecast and on operative forecast we shall give here experiment data represented in the work [127].

Tests of the monitoring system based on LFM sound were done in October 1990 on the middle-latitude radio trace Alma-Ata–Moscow with a length of 3000 km. The same antennas were used for the transmission/reception of the LFM signal and for informational communication, which excluded the necessity of recalculating the energetics of connected and sounding radio traces.

The power of informational communication received two values: 5 and 100 W. In this case, the optimal operating frequency was chosen close to the maximum of the used frequency (from the

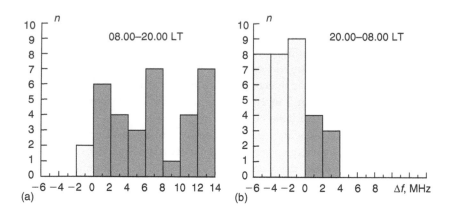

Figure 8.17 Histograms of distribution of deference frequency of the related signals, $\Delta f = f_1 - f_2$, selected by knowledge of the long-term (f_1) prediction and short-term (f_2) prediction for diurnal (a) and nocturnal (b) hours.

available resource) by the criterion of minimum noise level in the channel with a band of 3 kHz. The results of the tests are given in Figure 8.17a (daylight, evening hours—08.00–20.00 LT Moscow) and in Figure 8.17b (night, morning hours—20.00–08.00 LT Moscow) in the form of distribution histograms of the difference of communication operation frequencies $\Delta f = f_1 - f_2$, where f_1 was the frequency chosen for a long-term forecast, and f_2 the frequency chosen for an operative forecast.

As it is seen from Figure 8.17, in daylight hours, the difference of frequency $\Delta f > 0$, i.e., the frequency chosen by a long-term forecast, is higher than the one chosen by the operative forecast. In this case, for the long-term forecast, the work of connected short-wave radio line was carried out on frequencies close to maximum observed frequencies; for the operative forecast, it was done on frequencies with maximum signal-to-noise ratio, which in most cases fall on the double-shock mode of propagation. In night and morning hours, when the maximum observed frequency was considerably decreasing (by 10–15 MHz) and the range of the short wave signal passing was narrowing, the distribution of frequencies difference was biased to negative values, i.e., the operating frequency chosen by the operative forecast was higher than the one chosen by the one-term forecast. Such a choice of optimal operating frequencies is conditioned by the necessity of working at the given time on maximum high frequency where the level of station noise is lower.

Data concerning the effectiveness of informational communications reception transmitted by long-term forecast (black color) and by operative forecast (gray color) is given in Figure 8.18.

According to received data, correct reception of informational communications with the use of operative forecast occurred in 84% of the cases in daylight hours and in 90% of the cases in night hours. At the same time, the work of radio line based on long-term forecast data ensured correct reception of communications only in 54% of cases in daylight hours and in 46% of cases in night hours. Besides, in night and morning hours, when working by long-term forecast, considerable mistakes were registered in about 25% of cases ($\sim$19%–21%), connected with the fact that in transitional time of a 24 h period, a reconstruction of the ionosphere takes place, which is badly described by the algorithm of long-term forecast [57]. With an increase of the power of connection signal, the probability of correct reception when working by long-term forecast increased in proportion to the power logarithm. The dependence of communication reliability on power

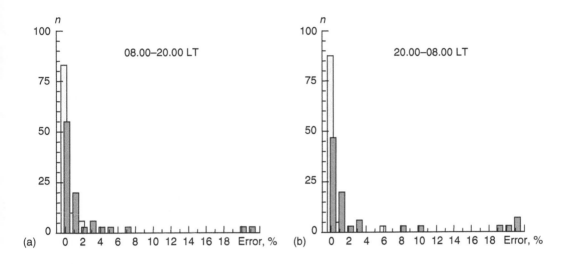

Figure 8.18 Histograms of distribution of the percentage of mistakes in the text information transmitted according to the long-term prediction (dark color) and according to short-term prediction (light color) during diurnal (a) and nocturnal (b) hours.

allowed to estimate the necessary power of a connection signal when the system worked by long-term forecast and on the basis of LFM monitoring on condition of equal reliability. It turned out that the work of a connection transmitter with the power of ∼600 W, when choosing optimal operating frequency of communication by long-term forecast, provided the same communication reliability as when choosing optimal operating frequency with the help of LFM monitoring in the case of using a connection signal with the power of 5 W.

In the United States and NATO countries, a great number of experiments were also conducted on the use of LFM sound in adaptive systems of short-wave radio communication on radio routes crossing high-latitudinal regions, northern lights zone, as well as on middle-latitudinal radio lines with transmitters power of 10 . . . 100 W [76,77,128,129]. Figure 8.19 shows the geometry of the radio traces in experiments conducted in the United States and in Western Europe. In the course of long experiments, data for more than 40 radio lines was received and analyzed. The investigations showed that communication reliability approaching 100% can be achieved on condition that several stations situated about great territory are accessible, and when there is a set of about ten operating frequencies. It is important to note that with a change of magnetic activity, the routing of communication organization can change.

The appearance of a sporadic layer E_S during strong disturbances in the F region of high-latitudinal ionosphere can be used to organize a communication channel via layer E_S. To get a full-scale estimation and a good quality forecast, it may be necessary to put together the information received from a great number of LFM sounds, as well as geophysical information.

In the cases when a communication line and a sounding route do not coincide, it is expedient to use the method of direct diagnostics of the communication channel characteristics, as suggested in Refs. [130,131]. The suggested method is based on the use of the adiabatic relation of the diagnostic signal characteristics and of the investigated radio channel with changes of ionosphere parameters. It has been established that when ionosphere parameters change within the limits of 20%, the following change little:

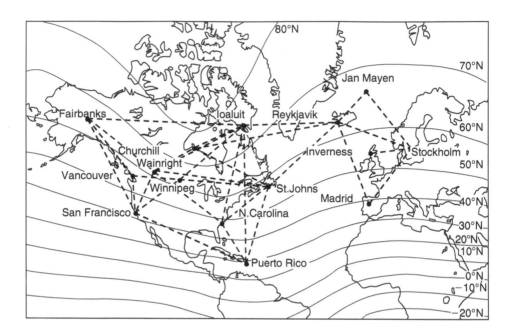

Figure 8.19 Geometry of radio paths for measurement of stability of short-wave radio communication using LFM ionosondes for diagnostics of the United States–Europe ionospheric channel.

■ The value η, which is equal to the ratio of the group path P_m, corresponding to the delay of the signal received on the maximum used frequency f_m, to the distance to the border of the lighted zone D_m

■ The value χ, equal to the ratio of maximum used frequencies of some modes for different radio routes

■ The ratio of the group path of the oblique sounding signal to the length of radio route on the relative frequency network (grid, lattice) $\beta = f \sim f_m$

When using the data of oblique sounding, the found adiabatic ratios allow extrapolation of the data on the route that does not coincide with the standard one by length or by azimuth. In this case, weak dependency of value χ is used, which equals the ratio of maximum used frequency of any modes for two routes. Value χ is calculated by the results of modeling of the ionograms of oblique sounding for the given routes. After this, in every instance, maximum usable frequency of the diagnosed route is defined as the product of the value χ and the value of maximum usable frequency of standard route.

The working capacity of the given algorithm was checked when comparing the results of the diagnostics of maximum usable frequencies on the radio traces Magadan–Irkutsk, Khabarovsk–Irkutsk [57], Khabarovsk–Balti and Dushanbe–Balti, which allowed considerably lowering of the error in calculation of maximum usable frequencies as compared with the long-term forecast.

The effectiveness of short-wave range resources management depends to a large extent on the exactness and operativeness of defining optimal operating frequencies of communication. Therefore, the development of new methods of operative forecast of optimal operating frequencies and of other characteristics of short-wave signals with the use of different kinds of ground-based sounding of the ionosphere and of automatization systems of data processing remains relevant to the present time [120,132].

When constructing graphs of diurnal variation of maximum usable frequencies by the experimental data of LFM-sounding of the ionosphere, short-periodic variations are observed, which are conditioned by both small-scale heterogeneity and different wave processes. The sign of such variations can change from session to session even after 10–15 min. For optimal solution of the problems of operative forecast it is advisable to point out the most significant variations of maximum usable frequencies and to smooth the variations connected with both small-scale heterogeneous structure of the ionosphere and experimental errors when defining maximum used frequencies but with preservation of most essential variations with the period constituting 1 h and more. To smooth short-period variations of experimental data, different modes can be used, for example, linear smoothing at three of five points and smoothing with polynomial in the third degree at seven points [57].

The use of liner smoothing of experimental data at three points is the most optimal for the solving of the problems of operative forecast. This method allows reduction of the number of errors, conditioned by inaccurate determination of maximum usable frequency, especially at night, because of F-scattering, and taking into account the general tendency of the change of maximum usable frequency in the course of time with preservation of the most significant variations. On the other hand, a forecast error is introduced a priori because small-scale heterogeneity is not taken into account.

Figure 8.20 shows examples of diurnal variation of experimental data of maximum usable frequencies and forecasted maximum usable frequencies (MUF) on the radio traces Ioshkar-Ola–Bălți, Dushanbe- Bălți and Khabarovsk- Bălți received on November 21 1991 [133]. To carry out an operative forecast, maximum usable frequencies of regular model of radio-signal propagation, received as a result of oblique sounding in different seasons of 1991, were used. The operative forecast was calculated by the formula of liner extrapolation using model calculation with a 10 min time internal.

It is seen that, on the whole, a long-term forecast describes in a qualitatively correct way, the diurnal variation of maximum usable frequencies. But in certain hours, the experimental values differed from the ones calculated by long-term forecast for more than 35%. An analysis of diurnal variation of relative errors showed that for magneto-quiet days, there are time intervals during sunrise and sunset hours when maximum forecast error can be observed (see Figure 8.20). Such error movement during evening time is probably connected with a generation of wave disturbances by the passing of the terminator though the route midpoint. This phenomenon leads to noticeable gradients of ionization and to variations of observed maximum usable frequencies, which are not taken into consideration in a long-term forecast model.

Methods of parameter extrapolation of the ionosphere short-wave channel by means of adapting the ionosphere model by the sounding results on the control radio trace were also approved on the routes of LFM-sounding. The application of such method is possible in the case of significant space–time correlation of the examined parameters. The investigations were conducted on two pairs of middle-latitude radio traces with different orientation of basic receiving stations spacing with regard to radiation station: Dushanbe–Bâlți, Dushanbe–Kiev and Khabarovsk–Bâlți, Khabarovsk–Kiev. Space correlation of maximum observed frequencies was examined. The investigations were conducted in 1988–1991. According to the experimental data, space correlation factor of maximum usable frequencies of the model $2F_2$, $3F_2$, and $4F_2$ for the radio lines oriented along basic receiving stations spacing was high and constituted ~0.92–0.96, with the spacing of receiving stations across the route, the space correlation factor constituted ~0.8. High values of space

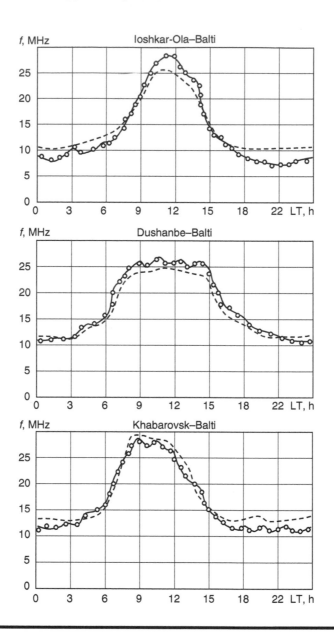

Figure 8.20 Diurnal variations of values of MUF for long-term and operative forecast obtained on November 21, 1991; solid line is the operative forecast, dashed curve is the long-term forecast, and circles are the experimental data.

correlation allowed to apply for the extrapolation of maximum usable frequencies the method of ionosphere model adaptation by the results of oblique sounding on control routes. In the experiments, the control radio traces Dushanbe–Bâlţi and Khabarovsk–Bâlţi were provided with diagnostic means, and the traces Dushanbe–Kiev and Khabarovsk–Kiev where considered operating. For the traces Dushanbe–Kiev and Khabarovsk–Kiev, the values of maximum usable frequencies were defined with the help of extrapolation, using the adaptation of an ionosphere model to the oblique-sounding data received on the control route.

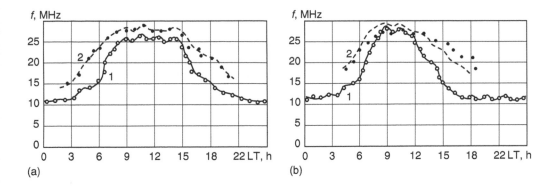

Figure 8.21 (a) Diurnal variations of maximum observed frequencies (MOF): ○—control trace Dushanbe–Beltsi (experiment), ●—work trace Dushanbe–Kiev (experiment); by 1 is denoted the corrected prediction of MOF using data of the oblique sounding for the trace Dushanbe–Beltsi, by 2 is denoted an extrapolation of MOF from the control route to the work trace Dushanbe–Kiev. (b) Diurnal variations of maximum observed frequencies (MOF): ○—control trace Habarovsk–Beltsi (experiment), ●—work trace Habarovsk–Kiev (experiment); by 1 is denoted the corrected prediction of MOF using data of the oblique sounding for the trace Habarovsk–Beltsi, by 2 is denoted an extrapolation of MOF from the control route to the working trace Habarovsk–Kiev.

The adaptation of the model consisted in the selection of an entrance of the model sun spots number W. Taking into account this parameter allowed to adjust the theoretical value of the maximum usable frequency on the control trace to the experimental value on the same trace with some error. A model adapted in such a manner was used in calculations to define maximum used frequencies on the work trace. The selection of a new value was done only when the deflection of maximum observed frequencies exceeded the prior set error. References model of the ionosphere SMI-88 (СМИ 88—Russian variant) was used as an empirical model. Figure 8.21a adduces data of the diurnal variation of maximum observed frequencies (experiment) for the control radio trace Dushanbe–Bâlţi (circles) and Dushanbe–Kiev (points). Full lines stand for forecasted maximum usable frequencies (1), and the results of extrapolation of maximum observed frequencies from the control to the work line (2). Analogical data are given in Figure 8.21b with the same symbols for the control radio trace Khabarovsk–Bâlţi (circles) and work radio trace Khabarovsk–Kiev (points). Data analysis showed that for the radio lines ~2000…3000 km long and quiet ionosphere conditions, the model correlation can be done on average in 6–7 h, and for the routes ~6000…7000 km long in 2–3 h. In this case, extrapolation errors of the values of maximum observed frequencies constitute ~4%–5%, which is considerably less than the errors of long-term forecast.

References

1. Mende, S.B., Eather, R.H., Rees, M.H., Vondrak, R.R., and Robinson, R.M., Optical mapping of ionospheric conductance, *J. Geophys. Res.*, Vol. 89, 1984, pp. 1755–1763.
2. Robinson, R.M., Mende, S.B., Vondrak, R.R., Kozyra, J.U., and Nagy, A.F., Radar and photometric measurements of an intense type a red Aurora, *J. Geophys. Res.*, Vol. 90, 1985, pp. 457–466.

3. Bird, J.C., Shepherd, G.G., and Tepley, C.A., Comparison of lower thermospheric winds measured by a polarizing Michelson interferometer and a Fabry–Perot spectrometer during the AIDA campaign, *J. Atmos. Terr. Phys.*, Vol. 55, 1993, pp. 313–324.

4. Feldman, P.D., Sahnow, D.J., Kruk, J.W., Murphy, E.M., and Moos, H.W., High resolution FUV spectroscopy of the terrestrial day airglow with the far ultraviolet spectroscopic explorer, *J. Geophys. Res.*, Vol. 106, 2001, pp. 8119–8130.

5. Wickwar, V.B., Meriwether, J.W., Hays, P.B., and Nagy, A.F., The meridional thermospheric neutral wind measured by radar and optical techniques in the auroral region, *J. Geophys. Res.*, Vol. 89, 1984, pp. 10,987–10,998.

6. Burnside, R.G. and Tepley, C.A., Optical observations of thermospheric neutral winds at Arecibo between 1980 and 1987, *J. Geophys. Res.*, Vol. 94, 1989, pp. 2,711–2,716.

7. Kerr, R.B., Cageao, R.P., Tepley, C.A., Atreya, S.K., and Donahue, T.M., High spectral resolution Fabry–Perot interferometer measurements of Comet Halley at H-alpha and 6300 Å, *Adv. Space Res.*, Vol. 5, 1986, pp. 283–287.

8. Tepley, C.A., Sargoytchev, S.I., and Hines, C.O., Initial Doppler Rayleigh lidar results from Arecibo, *Geophys. Res. Lett.*, Vol. 18, 1991, pp. 167–170.

9. Tepley, C.A., Neutral winds of the middle atmosphere observed at Arecibo using a Doppler Rayleigh lidar, *J. Geophys. Res.*, Vol. 99, 1994, pp. 25,781–25,790.

10. Beatty, T.J., Collins, R.L., and Gardner, C.S. et al., Simultaneous radar and lidar observations of sporadic *E* and Na layers at Arecibo, *Geophys. Res. Lett.*, Vol. 16, 1989, pp. 1,019–1,022.

11. Friedman, J.S., Tepley, C.A., Castleberg, P.A., and Roe, H., Middle-atmospheric Doppler lidar using an iodine-vapor edge filter, *Optics Letters*, Vol. 22, 1997, pp. 1,648–1,650.

12. Friedman, J.S., Tepley, C.A., and Raizada, S. et al., Potassium Doppler-resonance lidar for the study of the mesosphere and lower thermosphere at the Arecibo Observatory, *J. Atmos. Solar Terr. Phys.*, Vol. 65, 2003, pp. 1411–1424.

13. Grime, B.W., Kane, T.J., Collins, S.C., and Kelley, M.C., et al., Meteor trail advection and dispersion: Preliminary lidar observations, *Geophys. Res. Lett.*, Vol. 26, 1999, pp. 675–678.

14. Hecht, J.H., Kane, T.J., Walterscheid, R.L., Gardner, C.S., and Tepley, C.A., Simultaneous nightglow and Na lidar observations at Arecibo during the AIDA-89 campaign, *J. Atmos. Terr. Phys.*, Vol. 55, 1993, pp. 409–423.

15. Kane, T.J., Gardner, C.S., Zhou, Q., Mathews, J.D., and Tepley, C.A., Lidar, radar and airglow observations of a prominent sporadic Na/sporadic E layer event at Arecibo during AIDA-89, *J. Atmos. Terr. Phys.*, Vol. 55, 1993, pp. 499–511.

16. Raizada, S., Tepley, C.A., and Janches, D. et al., Lidar observations of Ca and K metallic layers from Arecibo and comparison with micrometeor sporadic activity, *J. Atmos. Solar Terr. Phys.*, Vol. 66, 2004, pp. 595–606.

17. Shiokawa, K., Katoh, Y., and Satoh, M. et al., Development of optical mesosphere thermosphere images (OMTI), *Earth Planets Space*, Vol. 51, 1996, pp. 887–896.

18. Zhou, Q.J. and Morton, Y.T., A case study of mesospheric gravity wave momentum flux and dynamical instability using the Arecibo dual beam incoherent scatter radar, *Geophys. Res. Lett.*, Vol. 33(10), doi:10.1029/2005GL025608, 2006.

19. Wen, C.H., Doherty, J.F., Mathews, J.D., and Janches, D., Meteor detection and non-periodic bursty interference removal for Arecibo data, *J. Atmos. Solar Terr. Phys.*, Vol. 67, 2005, pp. 275–281.

20. Luan, X., Liu, L., and Wan, W. et al., A study of the shape of topside electron density profile derived from incoherent scatter radar measurements over Arecibo and Millstone Hill, *Radio Sci.*, Vol. 41(4), doi:10.1029/2005RS003367, 2006.

21. Hysell, D.L., Incoherent scatter experiments at Jicamarca using alternating codes, *Radio Sci.*, Vol. 35, 2000, pp. 1425–1435.

22. Woodman, R.F. and Chau, J.L., Equatorial quasiperiodic echoes from field-aligned irregularities observed over Jicamarca, *Geophys. Res. Lett.*, 28, 2001, pp. 207–209.

23. Hysell, D.L. and Burcham, J.D., Long term studies of equatorial spread *F* using the JULIA radar at Jicamarca. *J. Atmos. Solar Terr. Phys.*, Vol. 64, 2002, pp. 1531–1543.

24. Caudal, G., de la Beaujardière, O., Alcaydé, D., Holt, J., and Lejeune, G., Simultaneous measurements of the electrodynamic parameters of the auroral ionosphere by the EISCAT, Chatanika, and Millstone Hill Radars, *Ann. Geophys.*, Vol. 2, 1984, pp. 369–376.

25. Lockwood, M., McCrea, I.W., Millward, G.H., Moffett, R.J., and Rishbeth, H., EISCAT observations of ion composition and temperature anisotropy in the high-latitude *F*-region, *J. Atmos. Terr. Phys.*, Vol. 55, 1993, pp. 895–906.

26. McCrae, I.W., Lockwood, M., and Moen, J. et al., ESR and EISCAT observations of the response of the cusp and cleft to IMF orientation changes, *Ann. Geophys.*, Vol. 18, 2000, pp. 1009–1026.

27. Zhang, S.-R., Holt, J.M., and Bilitza, D.K. et al., Multiple-site comparisons between models on incoherent scatter radar and IRI, *Adv. Space Res.*, Vol. 39, 2007, pp. 910–915.

28. Lathuillere, C., Wickwar, V.B., and Kofman, W., Incoherent scatter measurements of ion neutral collision frequencies and temperatures in the lower thermosphere of the auroral region, *J. Geophys. Res.*, Vol. 88, 1983, pp. 10,137–10,144.

29. Robinson, R.M., Rich, F., and Vondrak, R.R., Chatanika radar and S32 measurements of auroral zone electrodynamics in the didnight sector, *J. Geophys. Res.*, Vol. 90, 1985, pp. 8487–8499.

30. Rasmussen, C.E., Schunk, R.W., and Sojka, J.J. et al., Comparison of simultaneous Chatanika and Millstone Hill observations with ionospheric model predictions, *J. Geophys. Res.*, Vol. 91, 1986, pp. 6986–6998.

31. Hecht, J.H., Walterscheid, R.L., Sivjee, G.G., Christensen, A.B., and Pranke, J.B., Observations of wave driven fluctuations of OH nightglow emission from Sondre Stromfjord, Greenland, *J. Geophys. Res.*, Vol. 92, 1987, pp. 6091–6099.

32. Robinson, R.M., Clauer, C.R., and de la Beaujardière, O. et al., Sondrestrom and EISCAT radar observations of polewardmoving auroral forms, *J. Atmos. Terr. Phys.*, Vol. 52, 1990, pp. 411–420.

33. Johnson, R.M. and Virdi, T.S., High latitude lower thermospheric neutral winds at EISCAT and Sondrestrom during LTCS 1, *J. Geophys. Res.*, Vol. 96, 1991, pp. 1099–1116.

34. Kelley, J.D., Heinselman, C.J., Vickrey, J.F., and Vondrak, R.R., The sondrestrom radar and acoompanying ground-based instrumentation, *Space Sci. Rev.*, Vol. 71, 1995, pp. 797–813.

35. Bristow, W.A. and Watkins, B.J., Incoherent scatter observations of thin ionization layers at sondrestrom, *J. Atmos. Terr. Phys.*, Vol. 55, 1993, pp. 873–894.

36. Sofko, G.J., Koustov, A.V., McWilliams, K. et al., The super dual auroral radar network (SuperDARN): An international system for space weather determination, *Phys. Can.*, Vol. 54, 1998, pp. 6–23.

37. Oksavik, K., Greenwald, R.A., and Ruohoniemi, J.M. et al., First observations of the temporal/spatial variation of the sub-auroral polarization stream from the SuperDARN wallops HF radar, *Geophys. Res. Lett.*, Vol. 33, L12104, doi:10.1029/2006GL026256, 2006.

38. Ponomarenko, P.V., Menk, F.W., Waters, C.L., and Sciffer, M.D., Pc3–4 ULF waves observed by the SuperDARN TIGER radar, *Ann. Geophys.*, Vol. 23, 2005, pp. 1271–1280, *SRef-ID: 1432-0576/ag/ 2005-23-1271.*

39. Pinnock, M. and Rodger, A.S. On determining the noon polar cap boundary from the SuperDARN HF radar backscatter characteristics, *Ann. Geophys.*, Vol. 18, 2001, pp. 1523–1530.

40. Bristow, W.A. and Lummerzheim, D., Determination of field-aligned currents using the Super Dual Auroral Radar Network and the UVI ultraviolet imager, *J. Geophys. Res.*, Vol. 106, 2001, pp. 18,577–18,588.

41. Ponomarenko, P.V. and Waters, C.L., The role of Pc1-2 waves in spectral broadening of SuperDARN echoes from high latitudes, *Geophys. Res. Lett.*, Vol. 30, 2003, pp. 1122–1125.

42. Provan, G., Yeoman, T.K., Milan, S.E., Ruohoniemi, J.M., and Barnes, R., An assessment of the map-potential and beam-swinging techniques for measuring the ionospheric convection pattern using data from the SuperDARN radars, *Ann. Geophys*, Vol. 20, 2002, pp. 191–202.

43. Van Dierendonck, A.J., Signal specification for the future GPS civil signal at L5, *Proc. of ION Annual Meeting 2000,* San Diego, CA, June 26–28, 2000.

44. Shallberg, K., Shloss, P., Altashuler, E., and Tahmazyan, L., WAAS measurement prossesing, reducing the effects of multipath, *Proc. of ION Annual Meeting 2001,* Salt Lake City, UT, September 11–14, 2001.

45. Hajj, G.A. and Romans, L.J., Ionospheric electron density profiles obtained with the Global Positioning System: Results from the GPS/MET experiment, *Radio Sci.,* Vol. 33, 1998, pp. 175–190.

46. Barrile, V., Cacciola, M., Morabito, F.C., and Versaci, M., TEC measurements through GPS and artificial intelligence, *J. Electromag. Waves Appl.,* Vol. 20, 2006, pp. 1211–1220.

47. Valladares, C.E., Villalobos, J., Sheehan, R., and Hagan, M.P., Latitudinal extention of low-lattitude scintillations measured with a network of GPS receivers, *Ann. Gheophys.,* Vol. 22, 2004, pp. 3155–3175.

48. Afraimovich, E.L., Astafieva, E.I., and Voyeikov, S.V., Isolated ionospheric disturbances as deducted from global GPS network, *Ann. Geophys.,* Vol. 22, 2004, pp. 47–62.

49. Akhmedov, R.R. and Kunitsyn, V.E., Simulation of the ionospheric disturbances caused by earthquakes and explosions, *Geomagn. Aeronomy,* Vol. 44, 2004, pp. 95–101.

50. Kuo, Y.-H., Zou, X., and Huang, W., The impact of Global Positioning System data on the prediction of an extratropical cyclone: An observing system simulation experiment, *Dynamics of Atmospheres and Oceans,* Vol. 27, 1998, pp. 439–470.

51. Emardson, T.R., Elgered, G., and Johansson, J.M., Three months of continuous monitoring of atmospheric water vapor with a network of Global Positioning System receivers, *J. Geophys. Res.,* Vol. 103, 1998, pp. 1807–1820.

52. Kavak, A., Vogel, W.J., and Xu, G.H., Using GPS to measure ground complex permittivity, *Electron. Lett.,* Vol. 34, 1998, pp. 254–255.

53. Liu, Q., Time-dependent models of vertical crustal deformation from GPS-leveling data, *Surv. Land Inf. Syst.,* Vol. 58, 1998, pp. 5–12.

54. Kuk, Ch. and Bornfeld, M., *Radiolocation Signals. Theory and Applications,* Moscow: Soviet Radio, 1971, 567p.

55. Gnanaligam, S. and Weeks, K., *The Physics of the Ionosphere,* edited by J.A. Ratcliffe, The Physical Society, London, 1955, p.63.

56. Titheridge, J.E., The electron density in the lower ionosphere, *J. Atmos. Terr. Phys.,* Vol. 24, 1962, pp. 269–282.

57. Ivanov, V.A., Kurkin, V.I., Nosov, V.E., Uryadov, V.P., and Shumaev, V.V., Chirp ionosonde and its application in the ionospheric research, *Radiophysics Quantum Electron.,* Vol. 46, 2003, pp. 821–851.

58. Klauder, J.R., Price, A.C. Darlington, S., and Albersteim, W.J., The theory and design of chirp radars, *Bell Syst. Tech. J.,* Vol. 39, 1960, pp. 745–820.

59. Fenwick, R.B. and Barry, G.H., Step by step to a linear frequency sweep, *Electronics,* Vol. 38, 1965, pp. 66–67.

60. Barry, G.H. and Fenwick, R.B., Extraterrestrial and ionospheric sounding with synthesized frequency sweeps, *Hewlett-Packard J.,* Vol. 16, 1965, pp. 8–12.

61. Epstein, M.R., Polarization of ionospherically propagated HF radio waves with applications to radio communication, *Radio Sci.,* Vol. 4, 1969, pp. 53–61.

62. Fenwick, R.B. and Barry, G.H., Sweep-frequency oblique ionospheric sounding at medium frequencies, *IEEE Trans.,* Vol. BC-12, 1966, pp. 25–27.

63. Croft, T.A., Sky-wave backscatter: A means for observing our environment at great distance, *Rev. Geophys. Space Phys.,* Vol. 10, 1972, pp. 73–155.

64. Barnum, J.R., Skywave polarization rotation in swept-frequency sea backscatter, *Radio Sci.,* Vol. 8, 1973, pp. 411–419.

65. Washburn, T.W., Sweney, L.E., Barnum, J.R., and Zavoli, W.B., Development of HF skywave radar for remote sensing application. Special topics in HF propagation, *AGARD Conf. Proc.* No. 263, 28,05–1.06.1979, London, 32/1–32/17, New York, 1979.

66. The new VOS-1 vertical/oblique sounder, *Prospect Barry Research Palo Alto*, Calif., USA, 1970.

67. Chirpsounder receiver systems, *Prospect Barry Research Palo Alto*, Calif., USA, 1972.

68. Ionospheric chirpsounder transmitter TCS-4, *Prospect Barry Research Palo Alto*, Calif., USA, 1973.

69. HF chirpsounder receiver dodel RCS-5, *Prospect BR Communications*, USA, 1985.

70. HF chirpsounder transmitter model TCS-5, *Prospect BR Communications*, USA, 1990.

71. Poole, A.W.V., Advanced sounding (1): The FM-CW alternative, *Radio Sci.*, Vol. 20, 1985, pp. 1609–1616.

72. Poole, A.W.V. and Evans, G.P., Advanced sounding (2): first results from an advanced chirp ionosondes, *Radio Sci.*, Vol. 20, 1985, pp. 1617–1623.

73. Salous, S. and Shearman, E.D.R., Wideband measurements of coherence over an HF skywave link and implication for spread-spectrum communication, *Radio Sci.*, Vol. 21, 1986, pp. 463–472.

74. Salous, S., FMCW channel sounder with digital processing for measuring the coherence of wideband HF radio links, *IEE Proc.*, Vol. 133, Pt. F. 1986, pp. 456–462.

75. Basler, R.P., Price, G.H., Tsunoda, R.T., and Wong, T.L. Ionospheric distortion of HF signals, *Radio Sci.*, Vol. 23, 1988, pp. 569–579.

76. Goodman, J.M., Ballard, J.W., Sharp, E.D., and Trung, L., *Proc. of Session G5 at the XXVth GA URSI*, Published WDC-A, Boulder, 1998, pp. 64–70.

77. Bröms, M. and Lundborg, B., Results from Swedish oblique soundings campaigns, *Ann. Geofis.*, Vol. 37, 1994, pp. 145–152.

78. Lundborg, B., Broms, M., and Derblom, H., Oblique sounding of an auroral ionospheric HF channel, *J. Atmos. Terr. Phys.*, Vol. 57, 1995, pp. 51–63.

79. Arthur, P.C., Lissimore, M., Cannon, P.S., and Davies, N.C., Application of a high quality ionosonde to ionospheric research, *Proc. of Seventh International Conference on HF Radio Systems and Techniques*, IEE Conf. Publ., No. 441, 1997, pp. 135–139.

80. Arthur, P.C. and Cannon, P.S. ROSE—A high performance oblique ionosonde providing new opportunities for ionospheric research, *Annali di Geofisica*, Vol. 37, 1994, pp. 135–144.

81. Black, Q.R., Wood, J.F. Jr., Sonsteby, A.G., and Sherrill, W.M., A direction finding ionosondes for ionospheric propagation research, *Radio Sci.*, Vol. 28, 1993, pp. 795–809.

82. Millman, G.H. and Swanson, R.W., Comparison of HF oblique transmissions with ionospheric predictions, *Radio Sci.*, Vol. 20, 1985, pp. 315–318.

83. Anderson, S.J. and Lees, M.L., High-resolution synoptic scale measurement of ionospheric motions with the Jindalee sky wave radar, *Radio. Sci.*, Vol. 23, 1988, pp. 265–272.

84. Russell, C.J., Dyson, P.L., Houminer, Z., Bennett, J.A., and Li, L., The effect of large-scale ionospheric gradients on backscatter ionograms, *Radio Sci.*, Vol. 32, 1997, pp. 1881–1898.

85. Dandekar, B.S., Sales, G.S., Weijers, B., and Reynolds, D. Study of equatorial clutter using observed and simulated long-range backscatter ionograms, *Radio Sci.*, Vol. 33, 1998, pp. 1135–1158.

86. Earl, G.F. and Ward, B.D., The frequency management system of the Jindalee over-the-horizon backscatter radar, *Radio Sci.*, Vol. 22, 1987, pp. 275–291.

87. Baker, P.W., Clarke, R.H., Massie, A.D., and Taylor, D., Techniques for the measurement and decomposition of the time varying narrow bandwidth transfer function of an HF sky wave transmission, *Radio Sci.*, Vol. 32, 1997, pp. 1813–1820.

88. Lynn, K.J.W., Oblique sounding in Australia, *INAG-62*, 1998, pp. 14–18.

89. Belenov, A.F., Zinichev, V.A., and Ivanov, V.A. et al., Sounding of the ionosphere by quasi-continuous signals, *Proc. of XIII Conference of Radio Wave Propagation*, Gorky, Vol. 1, 1981, pp. 12–13.

90. Kochemasov, V.N., Belov, L.A., and Okoneshnikov, V.S., *Formation of Signals with Linear Frequency Modulation*, Moscow: Radio and Communication, 1983, 192p.

91. Erukhimov, L.M., Ivanov, V.A., and Mityakov, N.A. et al., *LFM-Method of Diagnostics of the Ionospheric Channel*, Moscow: VINITI, No. 9027–1386, 1986, 94p.

92. Ivanov, V.A., Frolov, V.A., and Shumaev, V.V., Sounding of the ionosphere by continuous LFM signals, *Izv. Vuzov, Radiophyzika*, Vol. 29, 1986, pp. 235–237.

93. Brynko, I.G., Galkin, I.A., and Grozov, V.P. et al., Automatically controlled data gathering and processing system using an FMCW ionosondes, *Adv. Space Res.*, Vol. 4, 1988, pp. 121–124.

94. Ivanov, V.A., Uryadov, V.P., Frolov, V.A., and Shumaev, V.V., Oblique sounding of the ionosphere by continuous LFM-signals, *Geomagn. Aeronomy*, Vol. 30, 1990, pp. 107–112.

95. Ivanov, V.A., Malishev, Yu.B., and Noga, Yu.V. et al., Automatic LFM-complex for ionospheric studies, *Radiotechnics*, No. 4, 1991, pp. 69–72.

96. Filipp, N.D., Blaunshtein, N.Sh., Erukhimov, L.M., Ivanov, V.A., and Uryadov, V.P., *Modern Methods of Investigation of Dynamic Processes in the Ionosphere*, Shtiintza, Kishinev, Moldova, 1991, 288p.

97. Zinichev, V.A., Ivanov, V.A., Frolov, V.A., and Shumaev, V.V., Use of LFM method of vertical sounding for study of modified upper layers of the ionosphere, *Izv. Vuzov. Radiophysika*, Vol. 29, 1986, pp. 629–631.

98. Erukhimov, L.M., Ivanov, V.A., Mityakov, N.A. et al., On frequency characteristics of effects of powerful radiation at the ionospheric F-layer, *Izv. Vuzov. Radiophysika*, Vol. 30, 1987, pp. 1055–1065.

99. Filip, N., Plohotniuc E., and Pascaru, M., Ionosphere prognosis using LFM-ionosound. Proc. of the *XVIII Congress of Româno-American Academy of Science and Arts*, Chisinău. Moldova, Vol. 2, 1993, pp. 135–137.

100. Ureadov, V., Ivanov, V., Plohotniuc, E., Eruhimov, L., Blaunshtein, N., and Filipp, N., *Dynamic Processes in Ionosphere and Methods of Investigation*, Iasi, Romania, Tehnopress, 2006, 284 p.

101. Boguta, N.M. and Ivanov, V.A. et al., Use of LFM ionosonde in adaptive system of Short Wave radio communication, *Radiotechniks*, No. 4, 1993, pp. 77–79.

102. Lynn, K.J.W., Harris, T.J., and Sjarifudin, M., Stratification of the F2 layer observed in Southeast Asia, *J. Geophys. Res.—Space Phys.*, Vol. 105, A12, 2000, pp. 27147–27156.

103. Eremenko, V.A., Erukhimov, L.M., and Ivanov, V.A. et al., Pedersen mode ducting in randomly-stratified ionosphere, *Wave. Random Media*, Vol. 7, 1997, pp. 531–544.

104. Vertogradov, G.G., Uryadov, V.P., and Vertogradov, V.G., Monitoring of wave disturbances by method of oblique sounding of the ionosphere, *Izv. Vuzov. Radiophysika*, Vol. 49, 2006, pp. 1015–1029.

105. Tyler, M.A., Round-the-world high frequency propagation: A synoptic study, *DSTO Research Report*, Commonwealth of Australia, 1995, 46p.

106. Fink, L.M., *Theory of Transmission of Discrete Information*, Moscow: M.: Soviet Radio, 1970, 727 p.

107. Ivanov, V.A., Peculiarities of propagation of short-wave LFM radio signals in the regular ionosphere, Moscow: VINITI, No. 3064–85. 1985, 41p.

108. Namazov, S.A., On dispersion distortions of signals with the limited spectrum during reflection from the ionosphere, *Radiotechniks Electron.*, Vol. 24, 1984, pp. 1280–1288.

109. Orlov, Yu.I., On geometrical theory of dispersion distortions of signals with the limited spectrum, *Izv. Vuzov. Radiophysika*, Vol. 25, 1982, pp. 772–783.

110. Terina, G.I., *Radiotechniks Electron.*, Vol. 12, 1967, pp. 124–132.

111. Ilyin, N.V., Khakhinov, V.V., and Kurkin, V.I. et al., *Proc. of the International Symposium on Antannas and Propagation*, Chiba, Japan, Vol. 3, 1996, pp. 689–692.

112. Budden, K.G., *Radio Waves in the Ionosphere*, Cambridge: Cambridge University Press, 1966, 542p.

113. Ginsburg, V.L., *Propagation of Waves in Plasma*, Moscow: Phys.-Math. Literature, 1967, 684p.

114. Davydenko, M.A., Ilyin, N.V., and Khakhinov, V.V., On the shape of measured spectra of the ionosphere sounding by an FMCW signal under dispersion case, *J. Atmos. Solar Terr. Phys.*, Vol. 64, 2002, pp. 1897–1902.

115. Ivanov, V.A., Ivanov, D.V., and Kolchev, A.A., Study of peculiarities of dispersive characteristics of radio channels with help of LFM ionosonde, *Izv. Vuzov. Radiophysika*, Vol. 44, 2001, pp. 241–253.

116. Barry, H.G., A low-power vertical-incidence ionosondes, *IEEE Trans. Geosci. Electron.*, Vol. GE-9, 1971, pp. 86–89.

117. Lynch, J.T., Fenwick, R.B., and Villard, O.G. Jr., Measurement of best time-delay resolution obtainable along east–west and north–south ionospheric paths, *Radio Sci.*, Vol. 7, 1972, pp. 925–929.

118. Sweeney L.E., Jr., Experimental measurements of amplitude and phase distributions across a 2.5 km HF array, *Proc. of URSI Spring Meeting*, Washington, D.C., April 1968.

119. Sweeney, L.E., Jr., A new swept-frequency technique for matching feedline lengths, *Proc. of the IEEE*, Vol. 59, Issue 8, 1971, pp. 1281–1282.
120. Golovin, O.V. and Prostov, S.P., *Systems and Devices of Short Wave Communication*, Goryachaya Liniya, Telecom, Moscow, 2006, 598p.
121. Cannon, P.S., Angling, M.J., and Lundborg, B., Characterization and modeling of the HF communications channel, *The Rewiew of Radio Science 1999–2002*, edited by W. Ross Stone, IEEE Press. 2002, pp. 597–623.
122. Baker, R.D., Synchronised oblique ionosphere sounding for real—time determination of HF optimum working frequencies, *Wescon Technical Papers*, No. 31/2, 1964, p. 21.
123. Stevens, E.E., *The CHEC Sounding System Ionospheric Radio Communication*, edited by K. Folkestad, Plenum Press, N.Y., 1968, p. 127.
124. Dayharsh, T.U., Application of CURTS concept to spectrum engineering, *Proc. of the National Electronics Conference*, Chicago, Vol. 24, 1968, p. 423.
125. Daly, R.F., The CURTS frequency selection and prediction system, *Proc. of the National Electronics Conference*, Chicago, Vol. 24, 1968, p. 410.
126. The Results of the Systems Trohy Dash 3 Tests, *Technical Reports*, USA, 1976, 80p.
127. Ivanov, V.A., Riabova, N.V., Uriadov, V.P., and Shumaev, V.V., Frequency provision equipment in adaptive short-wave radio communication system, *Electrocommunication*, No. 11, 1995, pp. 30–32.
128. Goodman, J., Ballard, J., and Sharp, E., A long-term investigation of the HF communication channel over middle- and high-latitude paths, *Radio Sci.*, Vol. 32, 1997, pp. 1705–1716.
129. Goodwin, R.J. and Harris, S.P., The design and simulation of an adaptive HF data network, HF radio systems and techniques, *Proc. of Fourth International Conference on HF Radio Systems and Techniques*, (Conf. Publ. No. 284), 1988, pp. 1–5.
130. Grozov, V.P., Kurkin, V.I., Nosov, V.E., and Ponomarchuk, S.N., An interpretation of data oblique-incidence sounding using the chirp-signal, *Proc. of ISAP'96*, Chiba, Japan, 1996, pp. 693–696.
131. Oinats, A.V., Kurkin, V.I., and Ponomarchuk, S.N., The technique for calculating of HF signals characteristics taking into consideration ionosphere waveguide propagation, *Proc. of MMET'02*, Kiev, Ukraine, Vol. 2, 2002, pp. 614–616.
132. Ivanov, V.A., Ryabova, N.V., Uryadov, V.P., and Shumaev, V.V. Forecasting and updating HF channel parameters on the basis of oblique chirp sounding, *Radio Sci.*, Vol. 32, 1997, pp. 983–988.
133. Plohotniuc, E.F. and Pascaru, M.D., Sounding of the ionosphere by LFM signals, *Scientific Anales of University A.Russo, Vol: Mathematics, Physics, Techniques*, Balti, State University of A. Russo, Vol. 20, 2004, pp. 52–61.

Chapter 9

Performance of Land–Sattelite Communication Links Passing through the Irregular Ionosphere

In recent decades demand has increased for mobile-satellite systems designed to provide global radio coverage and servicing using constellations of low-Earth-orbit (LEO) and medium-Earth-orbit (MEO) satellites, which have been intensively utilized in the operation of global positioning systems (GPS), in particular. Under these conditions, signals are typically received at high elevation angles by a moving or stationary vehicle placed at ground surface. Only ionospheric-layered structures with irregularities developed within these layers can affect the propagate radio signals with information through ionospheric radio channels. Unlike land communication channels, ionospheric channels are highly statistical in nature, because coverage across hundreds of kilometers has to be considered in terms of the large variations encountered in wide areas, caused by a broad spectrum of ionospheric geophysical features and processes.

The problem of wave propagation and scattering in the irregular inhomogeneous ionosphere has become increasingly important because these phenomena play a significant role in determining the service level and quality of service of the land–satellite and/or satellite–satellite communication links.

In Chapters 2, 5, and 6 we analyzed ambient geophysical phenomena which generate a broad spectrum of plasma instabilities and associated plasma irregularities directed mainly the reader to the dynamics of such irregularities and methods of investigation main physical parameters and characteristics of irregularities of various origins occurring in the ionospheric plasma of different altitudes and latitudes.

Below, we will discuss on spatial–temporal distribution of ionospheric irregularities and their effects on radio propagation through the ionosphere. It is well known that inhomogeneity and irregularity of the real ionospheric plasma are the main factors in determining conditions of VHF/X band wave propagation in land–satellite communication links [1–9].

Many experiments carried out using ground facilities (radars and ionosondes) [10–14] and methods of active modification of the ionosphere [15–18], as well as direct satellite measurements [19–23] show that in the real ionosphere there is a wide spectrum of irregular plasma structures, which cause main radiophysical effects, such as interference, scattering, diffraction, and refraction of radio waves passing through the disturbed layered ionosphere; variations of the incident angles of reflected waves, the multiray effect; and so on [24–34]. The small-angle scattering effects occur when waves propagate through an irregular medium, which is usually defined as *signal scintillations*, resulting in amplitude and phase fluctuations of radio signals (i.e., *fading*) near the ground surface. The following additional effects of fading can be also observed experimentally: changing in the duration and shape of radio signal, decreasing in signal-to-noise ratio (SNR), decreasing in capacity and spectral efficiency of the channel, and finally, increasing in bit-error-rate (BER) of information data inside the land-communication channel.

In Section 9.1, we briefly present information on refraction phenomena in the ionosphere, considered as a regular layered plasma with the refractive index, which continuously varies from layer to layer. This refraction affects rays passing through the ionosphere, finally leading to the loss of information as long as the satellite is in that position. Then, in Section 9.2, we discuss the effects of large-scale ionospheric plasma irregularities. The same is done in Section 9.3, but for small-scale ionospheric irregularities, which are much more accurate in predicting the key propagation parameters inside land–satellite communication links, as they result in significant signal amplitude and phase variations—that is, the effects of signal fading appear. Finally, in Section 9.4, we consider effects of magnetic storms on fading phenomena occurring in land–satellite communication links caused by effects of external cosmic sources (namely, solar activity, solar wind, "cosmic rain," high-energetic particles precipitation, and so on) on magnetosphere–ionosphere system coupling and electrodynamic interaction.

9.1 Refraction of Radio Waves in Quasi-Regular Ionospheric Plasma

The layered quasi-homogeneous ionospheric plasma changes the direction of radio waves transmitted from the Earth's surface due to changes of the wave effective refractive index. Depending on special conditions that are determined by factors such as the wave frequency, elevation angle of the terminal antennas at both the ends of land–satellite channel, plasma content, and so on, radio waves can leave Earth not along the straight line between the terminals (i.e., along radio path), but sufficiently far away from this direction (see Figure 9.1), even reflecting back to the Earth's surface. The refractive index n_r of an ordinary radio wave depends, according to Refs. [35,36], on the background plasma concentration N_0 and radiated wave frequency f

$$n_r^2 = 1 - \frac{f_0^2(h)}{f^2} \tag{9.1}$$

where $f_0(h)$ is the critical frequency of plasma at the given height h, defined as [35,36]

$$f_0(h) = 8.9788(N_0(h))^{1/2}[\text{Hz}] \tag{9.2}$$

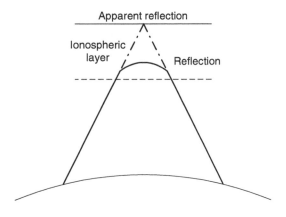

Figure 9.1 Schamatic presentation of reflection of radio wave from the inhomogeneous layered ionosphere.

Reflection from the ionosphere back to the Earth's surface occurs whenever the wave frequency is below this critical frequency f_0 at given altitude (see also Chapter 7). From this it follows that the *working frequencies* for satellite communications should be always above this critical frequency f_0.

9.2 Propagation Effects of Large- and Moderate-Scale Ionospheric Irregularities

In Chapter 6, we discussed methods of diagnostics of large- and moderate-scale artificially induced irregularities. We will now review the main effects of these irregularities on probing radio waves passing ionospheric communication links, without detailing how these irregularities were created, either naturally or artificially.

9.2.1 Effects of Large-Scale Irregularities on Radio Propagation

As mentioned in Chapter 7, radio waves propagate along the plane of a great circle, in the case of a spherically symmetric regular ionosphere [35,36]. However, the radio waves can change their direction from this plane in the presence of large-scale irregularities filling the ionosphere and having horizontal scale L, larger than the radius of the first Fresnel zone, $d_F = (\lambda R)^{1/2}$, that is, $L \gg d_F$. Here, as in Chapter 7, λ is the wavelength and R the distance from the ground station to the inhomogeneous area of the ionosphere containing large-scale plasma structures. In Refs. [35,36] it was found that the wave trajectory deviates from the plane of the great circle in the case of the real inhomogeneously layered ionosphere, consisting of irregularities with the horizontal scales of many hundreds of kilometers. Furthermore, additional detailed analysis presented in Refs. [35,36] showed that even small oscillations of the height of the level of equal plasma density led to the multimode reflection of radio signals, the growth of the angular spectrum width of waves, and their interference. Thus, even weakly ionized large-scale inhomogeneities can increase the thickness of the reflecting layer and the radio wave trajectory due to horizontal changes of the height of reflected layer. Finally, the reflected signal strength has a completely oscillatory character.

As follows from the previous section, if f_0 is the frequency of the ionospheric layer, radio wave reflection can occur only for frequencies that satisfy the constraint $f > f_0 \text{cosec} \theta_0$. At the same time, radio wave penetration can occur only for frequencies determined by the constraint $f < f_0 \text{cosec} \theta_0$. Here, f_0 is the maximum useful frequency (MUF) in the ionosphere related to the plasma frequency ω_{pe} at the corresponding ionospheric altitudes (see definitions in Chapter 7), and θ_0 is the angle of incidence of radio waves at the ionospheric layer filled by large-scale plasma irregularities. The first constraint presented above shows the possibility to perform successfully any ionospheric radio wave communication only for frequencies more than the MUF, f_0. According to the second constraint mentioned above, the thickness of the ionospheric layer becomes half-transparent for waves with $f > f_0 \text{cosec} \theta_0$.

Because for satellite–land or land–satellite communication it is important to investigate only those frequencies ω that exceed ω_{pe} in the ionosphere at altitudes of consideration, we will summarize some effects of large-scale irregularities for radio waves within the operation frequencies $\omega > \omega_{pe}$.

Following results obtained in Refs. [35,36] and summarized in Ref. [23], it can be shown that the large-scale plasma disturbances filling the E and lower F layer of altitudes $z < 200$ km may cause a *defocusing effect*, which leads to *wideband absorption* of radio waves. However, the areas below and above these layers, containing also large-scale plasma structures, conversely, may cause a *focusing* effect. Reflecting from different large-scale plasma structures located at different altitudes, the radio waves form an interference pattern, that is, a diffraction picture at the Earth's surface. The estimations presented in Ref. [23], however, show that irregularities in the ionospheric F region ($z = 200-400$ km) cause an interference pattern for radio waves with frequency $f < 40$ MHz and only for the relative changes of ionospheric plasma density satisfied the constraint $\delta N / \tilde{N}_0 > 0.2\%$, where $\tilde{N}_0 = \int_{Z_0}^{\infty} N_0 \, dZ$ is the plasma electrons content in the whole thickness of the ionosphere [23].

Hence, the propagation phenomena caused by the large-scale irregularities can be significant only for long-wave HF/VHF ionospheric propagation and for ionospheric sounding by various radio facilities (including LFM ionosondes described in Chapters 7 and 8), where the sporadic changes in the regular structure of ionograms are observed. However, radio waves of low frequencies are not so important for land–satellite communications. As for designers of land–ionosphere–land communication channels, here it has to be pointed out that when radio waves of low frequencies (HF or VHF) are reflected from the ionospheric layer filled by large-scale irregularities, a number of radiophysical features should be taken into account such as (a) deviation of the radio wave direction from a great circle plane, (b) increase of the vertical size of the ionospheric layer forming the reflected signal, and (c) the interference structure of reflected waves from the ionospheric layer and, as the result, additional irregular modulation of radio wave amplitude at the receiver.

9.2.2 Effects of Moderate-Scale Irregularities on Radio Propagation

We defined the moderate-scale irregularities, following Ref. [23], as those with a dimension $\ell_\perp$ of which transverse to $\mathbf{B}_0$ is smaller than the radius of the first Fresnel zone, i.e., $\ell_\perp \leq \sqrt{\lambda \bar{R}}$, where $\bar{R} = R_1 R_2 / (R_1 + R_2)$, R_1 and R_2 are distances from the transmitter and the receiver to the layer with irregularities. Usually, 0.1 km $< \ell_\perp < 10$ km (see also definitions in Chapter 7).

Difference between these irregularities compared to the large-scale ones, is that they not only change the structure of the signal reflected from the ionosphere during vertical and oblique sounding of the ionospheric layers filled by such kinds of irregularities, but also cause changes in phase and amplitude of signals of orbital and geostationary satellites. To understand these phenomena, we present below some analytical estimations following Refs. [10,11].

Fluctuations of signal amplitude A are caused by the so-called *ray-refraction effect* based on geometric-optics approximation, where the corresponding deviations of the normalized wave amplitude $\Delta A/A$ or intensity $\Delta I/I$ (because $\Delta I/I \approx 2\Delta A/A$, for $\Delta A/A \ll 1$ [10,11]) are determined by changes of the ray tube in the transverse to ray direction $\mathbf{r}(x, y)$ (i.e., to vector $\mathbf{k}_0 \| \mathbf{R}$). These changes can be defined by the value of $2(\partial\theta_R/\partial r_\perp)R$. Here θ_R is the angle of refraction, R the distance between the transmitter and receiver antennas introduced above, and $k_0 = 2\pi/\lambda$. At the same time, $\theta_R = k_0^{-1}(\partial\Delta\Phi/\partial r) \approx 2\Delta\Phi/k_0 l_\perp$ (for $R_1 \gg R_2$, i.e., for land–satellite communication where distance from satellite to the ionospheric layer is greater than that from the ground-based station). Here, $\Delta\Phi$ is the phase deviation of the wave caused by irregularities filling the ionospheric layer, and $l_\perp$ the characteristic scale of irregularities in the plane orthogonal to the trajectory of ray propagation. Then, the mean-square value of wave intensity fluctuations, $F_I = \langle(\Delta I)^2\rangle/\langle(I)^2\rangle$, approximately equals [10,11]:

$$F_I \approx q^2(\Delta\Phi_0)^2, \quad (\Delta\Phi_0)^2 = \langle(\Delta\Phi)^2\rangle \tag{9.1a}$$

The wave parameter q defines the ratio of the area of the first Fresnel zone, $\sim\lambda R$, and of the area filled by a irregularity, i.e., $q = \lambda R/S_I = 4R/k_0 l_\perp^2 \ll 1$. From Equation 9.1a it follows that for $q^2(\Delta\Phi_0)^2 \ll 1$ (i.e., for $l_\perp^2 > \lambda R$), fluctuations of the recorded signal amplitude are sufficiently small even the phase fluctuations, $(\Delta\Phi_0)^2$, are large significantly. Now it is clear that the large-scale irregularities with $l_\perp > \sqrt{\lambda R}$, affected by the influence of Fresnel filtering [23], cannot change conditions of radio propagation in the land–satellite communication links. Only moderate- and small-scale irregularities, for which $l_\perp \leq \sqrt{\lambda R}$, should be taken into account because they affect large fluctuations of amplitude and the phase of radio signals transmitted via the ionospheric channels [23,35,36].

The value $(\Delta\Phi_0)^2$ describing the radio signal phase changes can be easily estimated if we assume now that after passing only one single irregularity it equals $(\Delta\Phi_{00})^2 = k_0^2\langle(\delta\varepsilon)^2\rangle l_R$, where $\delta\varepsilon$ is fluctuation of plasma permittivity due to existence of moderate-scale irregularities, and l_R is the scale of irregularity along the ray. If now we also assume that the layer with thickness ΔR is filled by N irregularities, then we get:

$$(\Delta\Phi_0)^2 = (\Delta\Phi_{00})^2 N = k_0^2\langle(\delta\varepsilon)^2\rangle l_R\Delta R \tag{9.2a}$$

Taking into account that for $\omega \gg \omega_{pe}$ the parameter $(\delta\varepsilon) = \tilde{N}(\omega_{pe}^2/\omega^2)$ [35,36], we finally get:

$$(\Delta\Phi_0)^2 = \langle(\tilde{N})^2\rangle\frac{\omega_{pe}^2}{\omega^2}l_R\Delta R \tag{9.3}$$

The mean-square of plasma density fluctuations $\langle(\tilde{N})^2\rangle$ relates to the spectrum of these fluctuations $U_N(K)$ in the K domain as [35,36]

$$\langle(\tilde{N})^2\rangle = \int_{K_1}^{K_2} U_N(K')\,\mathrm{d}K' \tag{9.4}$$

where the wave numbers K_1 and K_2 correspond to maximum $l_{max} = 2\pi/K_1$ and minimum $l_{min} = 2\pi/K_2$ scales of plasma irregularities, respectively. If now, according to Ref. [23], we suggest the existence of a 3-D polynomial spectrum $U_N(K) \sim K^{-p}$, which yields that $\langle (\tilde{N})^2 \rangle \sim K^{-p+3}$ and $(\Delta\Phi_0)^2 \sim K^{-p+2}$, for which we can easily find the relative mean-square signal intensity deviations

$$F_I = \langle (\Delta I)^2 \rangle / \langle (I)^2 \rangle \sim K^{-p+6} \tag{9.5}$$

From these estimations it follows that the phase fluctuations of probing radio waves, passing a layer with moderate-scale irregularities, are sufficiently strong and the parameter of their degree of perturbations corresponds to the constraint $p > 2$. These fluctuations are increased with increase of the scale of irregularities along radio path, $l_R = 2\pi/K$. It should be noted that the contribution of large-scale irregularities with scales $l > \sqrt{\lambda R}$ in the wave intensity fluctuations F_I becomes essential only for the strong fluctuations of plasma density within each irregularity filled disturbed iono-spheric region (i.e., for $p \geq 6$). It has also been found [10,11] that for 2-D spectrum $U_N(K)$ of plasma density fluctuations, we can obtain $\langle (\tilde{N})^2 \rangle \sim K^{-p+2}$ and $(\Delta\Phi_0)^2 \propto K^{-p+1}$, which finally yields

$$F_I \sim K^{-p+5} \tag{9.6}$$

if the ray does not cross each irregularity along its biggest axis of anisotropy, in the direction of which $U_N(K)$ has Gaussian form. For smaller irregularities, $l \ll \sqrt{\lambda R}$ for weak fluctuations of signal phase, $(\Delta\Phi_0)^2 \ll 1$, and we get for fluctuations of signal intensity $F_I \approx 2(\Delta\Phi_0)^2$ and for fluctuation of signal amplitude $F_A \approx (1/2)(\Delta\Phi_0)^2$. Here, spectral characteristic of amplitude fluctuations is simply related to $U_N(K)$, for example [36]:

$$F_A(K_x) \sim \int U_N(K_x, K_y, K_z = 0)\, dK_y \tag{9.7}$$

Usually two cases should be investigated: (a) assuming "freezen" transport of irregularities which can be suggested only in the case of stationary source of radio waves (namely, in the case of geostationary satellite), and (b) in the case of fast movements of LEO or MEO satellites, where the velocity of propagation of ray at the altitude of the layer exceeds the chaotic velocities of irregularities. In case (b), which has much more practical application compared to case (a), we have:

$$F_A(\Omega = K_x V_x) \sim \int U_N(K_x, K_y, K_z = 0)\, dK_y \tag{9.8}$$

Here, $V_x \approx (R_2/R_1)V_h$, where V_h is the horizontal velocity of the moving satellite.

From Equation 9.8 it follows that if spectral characteristics of moderate-scale irregularities along x- and y-axis have different form, polynomial $U_N(K) \sim K_x^{-p} K_y^{-p'}$ or Gaussian $U_N(K) \sim K_x^{-p}$ $\exp\{-\alpha K_y^2 t\}$, from Equation 9.8 we obtain

$$F_A(\Omega = K_x V_x) \sim K_x^{-p} \tag{9.9}$$

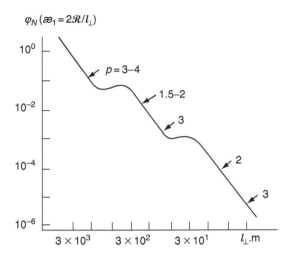

Figure 9.2 Dependence of the spectral characteristics of artificially induced turbulence (AIT) on its transverse scale relative to the ambient magnetic field.

That is, the amplitude fluctuations of the signal are characterized by the power of 1-D spectrum. The above mentioned allows for investigating the 3-D form of spectra of moderate-scale irregularities, $F_A(\Omega)$, depending on the angle Ω between the vector of the geomagnetic field $\mathbf{B}_0$ and the vector of satellite velocity, as well as on the angle between $\mathbf{B}_0$ and the visible ray at the satellite [37,38].

Thus, simultaneous study of spectra $F_A(\Omega)$ and $F_\Phi(\Omega)$ during sounding of the ionosphere by satellite allows for obtaining information about the spectrum of plasma density fluctuations $U_N(K)$ in the broad range of the wave numbers, as well as over whole altitudinal intervals where these plasma disturbances are localized in the perturbed ionosphere [35,36]. Another method of investigation of moderate-scale irregularities of various origins is the specular reflection and scattering described in detail in Chapter 7. Using this method together with satellite direct radio sounding of the ionosphere, the information on spectrum, $U_N(K_\perp)$ of artificially induced turbulences (AIT) can be found. Thus, dependence of this spectral characteristic on the transverse scale with respect to $\mathbf{B}_0, l_\perp = 2\pi/K_\perp$, is presented in Figure 9.2 according to Ref. [10]. It can be seen that for kilometer-scale irregularities the power of spectrum changes from $p=3$ to $p=4$, whereas for irregularities with hundreds to tens of meters, the power decreases slightly, i.e., $p=1.5$–2. For small-scale irregularities $p=2$–3. We will discuss below about effects of these kinds of ionospheric irregularities.

9.3 Fading Effects Caused by Small-Scale Ionospheric Irregularities

The influence of small-scale irregularities usually observed in the ionospheric E and F layer, results mainly in phenomena like diffraction and scattering. For these irregularities with characteristic scale $l \ll d_F$, (where, as above, $d_F = (\lambda\, R)^{1/2}$ and l represents the scale of inhomogeneity), and with $l < \lambda$, the geometrical optics approximation is not valid. In this case we should use the perturbation method described in detail in Refs. [23,35,36] by analyzing the following wave equation described as spatial distributions of the field strength E:

$$\Delta E + k_0^2(\tilde{\varepsilon}_0 + \tilde{\varepsilon}_1)E = 0 \tag{9.10}$$

In Equation 9.10, the total dielectric permeability ε of ionospheric plasma is presented as a sum of $\tilde{\varepsilon}_0 + \tilde{\varepsilon}_1$, where nonperturbed $\tilde{\varepsilon}_0$ and perturbed $\tilde{\varepsilon}_1$ parts can be defined as [35,36]

$$\tilde{\varepsilon}_0 = e^2 N_0(Z)/m_e\omega^2\varepsilon_0 \tag{9.11}$$

and

$$\tilde{\varepsilon}_1 = e^2 N_1(Z)/m_e\omega^2\varepsilon_0 \tag{9.12}$$

Here, $N_0(Z)$ and ε_0 are the density, as a function of the altitude Z, and the permittivity of undisturbed background plasma, respectively, e and m_e, are the charge and mass of the electron component of plasma. As was mentioned in Chapter 1, plasma usually is quasi-neutral and its electrons (or ions) fully present its spatial and temporal properties.

In Refs. [35,36], the problem was investigated, taking into consideration that the fluctuations of signal amplitude and phase are sufficiently small compared with those for undisturbed plasma. Following Refs. [35,36] and using the perturbation method, we finally can present the logarithm of signal amplitude, g, and signal phase, Φ, in the form of expansion into the perturbation sum:

$$g = g_0 + g_1 + \cdots, \quad \Phi = \Phi_0 + \Phi_1 + \cdots \tag{9.13}$$

In this case, Equation 9.10 has an exact solution which can be expressed by the Airy function $V(-\tau)$:

$$V(-\tau) = \frac{1}{\pi^{1/2}} \int_0^{+\infty} \cos(x^3/3 - x\tau)\, dx$$

where $\tau = k_0^{2/3} Z/D^{1/3}$. Then, the wave field strength can be finally presented in the following form:

$$E = E_0 \exp\{ik_0 g\} V(-k_0^{2/3}\Phi) \tag{9.14}$$

Using the perturbation approach to the first and second term of the expanded sum 9.13, we have according to Refs. [35,36]:

$$\Phi_0 = 2k_0 X_0 = 2k_0 \int_Z^{Z_0'} (\tilde{\varepsilon}_0)^{1/2}\, dZ \tag{9.15a}$$

$$\Phi_1 = 2k_0 X_1 = 2k_0 \int_Z^{Z_0'} \tilde{\varepsilon}_1/(\tilde{\varepsilon}_0)^{1/2}\, dZ \tag{9.15b}$$

Here, parameter Φ_0 defines the phase of a nonperturbed wave, and Φ_1 defines the disturbance of wave phase in the ionospheric layer filled by the moderate- and small-scale irregularities. The function g_1 in expansion 9.13 can be expressed as [35,36]

$$g_1 = (1/2k_0)\ln[|E(Z)|/E_0] = -(1/2k_0)\int\limits_Z^{Z_0'}[\Delta_\perp X_1/(\tilde{\varepsilon}_0)^{1/2}]\,dZ \qquad (9.16)$$

and describes the changes of the signal level in the process of scattering. In expression 9.16 the transverse component of the full Laplacian equals $\Delta_\perp = \frac{\partial^2}{\partial X^2} + \frac{\partial^2}{\partial Y^2}$.

Now using expressions 9.12 and 9.16, two actual cases of plasma permittivity perturbations were analyzed [23].

In the first case, the mean-square of phase perturbations Φ_1 was analyzed in terms of the assumption that the shape of the weakly ionized plasma density perturbations, $\delta N \equiv N_1 < N_0$, is distributed according to the Gaussian low inside inhomogeneous ionospheric layer with thickness D, from which it follows that

$$\langle\tilde{\varepsilon}_1^2\rangle = \frac{\omega_{pe}^4}{\omega^4}\left\langle\left|\frac{N_1}{N_0}\right|^2\right\rangle \qquad (9.17)$$

$$\langle\Phi_1^2\rangle = \pi^{1/2}k_0^2 lD\langle\tilde{\varepsilon}_1^2\rangle\ln\left(\frac{8D}{l}+\frac{C}{2}\right) \qquad (9.18)$$

where C is the Euler constant [35,36]. However, expression 9.18 is valid only in the sense that $\langle N_1^2\rangle$ is the same for all altitudes of the ionosphere. If relative fluctuations of plasma density $\langle|N_1/N_0|^2\rangle$ depend on the height, then

$$\langle\tilde{\varepsilon}_1^2\rangle = (1-\tilde{\varepsilon}_0)^2\left\langle\left|\frac{N_1}{N_0}\right|^2\right\rangle \qquad (9.19)$$

is the function of the altitude Z [35,36]. In this case:

$$\langle\Phi_1^2\rangle = \pi^{1/2}k_0^2 lD\left\langle\left|\frac{N_1}{N_0}\right|^2\right\rangle\ln\left(\frac{8D}{l}+C-\frac{3}{2}\right) \qquad (9.20)$$

In the *second case*, when perturbed plasma density is described by the polynomial function in the K-domain, the following form of spectrum of plasma density fluctuations $U_N(K)$ or of dielectric permeability fluctuations $U_\varepsilon(K)$, was obtained in Refs. [35,36]:

$$U_N(K) \sim U_\varepsilon(K) = M_\varepsilon\left[1+(K_X^2+K_Y^2+K_Z^2)L_0^2/4\pi^2\right]^{-p/2} \qquad (9.21)$$

where L_0 is an external (outer) scale of irregularities and M_ε determined by the condition $\langle\tilde{\varepsilon}_1^2\rangle = \int U_\varepsilon(K)\,dK_X\,dK_Y\,dK_Z$. Then, the correlation function of phase fluctuations can be presented as:

$$\Gamma_\Phi(\xi,\eta) = 2k_0^2 DM_\varepsilon \int\limits_{Z_0'-D}^{Z_0'} dK_X\, dK_Y\, dK_Z \int\limits_{Z_0'-D-\zeta/2}^{Z_0'-\zeta/2} dZ \exp\{i(\xi K_X + \eta K_Y + \zeta K_Z)\}$$

$$\times \left\{ \left[1 + (K_X^2 + K_Y^2 + K_Z^2)L_0^2/4\pi^2\right]^{-p/2}/Z - Z_0' - \zeta^2/4^{1/2} \right\} \tag{9.22}$$

In Ref. [38], it was found that formula 9.22 can not be integrated analytically. Moreover, for the case $L_0 \ll D$, analysis of the spectrum of phase fluctuations, Φ_1, showed that a spectrum $U_{\Phi 1}(K)$ does not reproduce the spectrum $U_N(K)$.

Using the *thin-screen* approximation [26–29], Booker with coauthors found the solution of Equation 9.22 only for weak fluctuations of the signal phase $(\langle\Phi_1^2\rangle \ll 1)$, i.e., for such inhomogeneities, the angle of scattering of which, $\Phi_s \sim (\lambda/l) < \Phi_1^2 >$, is sufficiently small. In this case, Φ_1 is distributed according to the low $\langle\exp(i\Phi_1)\rangle = \exp(-\langle\Phi_1^2\rangle/2)$, and we can easily obtain the average field $\langle|E_r|\rangle$ at the receiving point:

$$\langle|E_r|\rangle = \frac{E_0}{R}\exp\left(-\frac{\langle\Phi_1^2\rangle}{2}\right) \tag{9.23}$$

The average field attenuates since a part of the wave energy is transformed into the noncoherent component of the field, and the cross-correlation function of the received field can be presented as [35,36]:

$$\Gamma_E = \langle|(E_r - \langle E_r\rangle)(E_{r1} - \langle E_{r1}\rangle)|\rangle = \frac{E_0^2}{R^2}\left\{\frac{1}{\lambda^2\tilde{R}^2}\int\left\langle\exp\left[i\Phi_1(x',y') - i\Phi_1(x'',y'')\right]\right\rangle\right.$$

$$\times \exp\left[-\frac{i\pi}{\lambda\tilde{R}}\left(x'^2 - \left(x'' - \frac{R_2}{\tilde{R}}x_1'\right)^2 + y'^2 - y''^2\right)\right]dx'\,dy'\,dx''\,dy'' - \exp(-\langle\Phi_1^2\rangle)\right\} \tag{9.24}$$

Using Equation 9.24, the relationship between the correlation coefficient $\rho_{\Phi 1}$ and $\langle\Phi_1^2\rangle$ can be found as $\rho_{\Phi 1} = \Gamma_{\Phi 1}/\langle\Phi_1^2\rangle$. Then, the cross-correlation function of the wave field equals [25,36]:

$$\Gamma_E = \frac{E_0^2}{R^2}\left\{\exp\left[-\langle\Phi_1^2\rangle\left(1 - \rho_{\Phi 1}\left(\frac{R_1}{R}x_1'\right)\right)\right] - \exp(-\langle\Phi_1^2\rangle)\right\} \tag{9.25}$$

The parameter $R_1 x_1'/R = x_0' = |V_0|t$ is the coordinate of ray OP along the x'-axis at height Z_0 (see Figure 9.3), and V_0 is the speed of point O when the satellite moves.

Expression 9.25 enables the relationship between the cross-correlation function of phase fluctuations at the bottom boundary of the ionospheric layer and the correlation function of signal amplitude fluctuations, $\Gamma_E(t)$. The latter can be found for weak phase fluctuations $(\langle\Phi_1^2\rangle < 1)$ as:

$$\Gamma_E(t) = \frac{E_0^2}{R^2}\langle\Phi_1^2\rangle\rho_{\Phi 1}(x_0') \tag{9.26}$$

The time fluctuations of the field amplitude are found from the spatial phase fluctuations at the bottom boundary of the ionospheric layer with small scale inhomogeneities. Taking into account relationship 9.26 and following geometrical notations presented in Figure 9.3, we get:

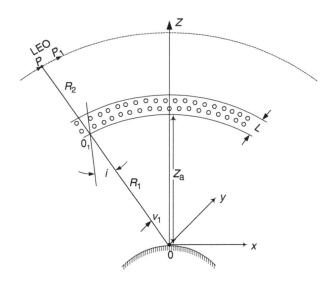

Figure 9.3 Geometry of satellite-land communication link.

$$\Gamma_E(\xi,\eta) = (E_0^2/R^2)k_0^2 L \sec i \int\limits_{-L\sec i}^{L\sec i} \Gamma_\varepsilon(x_0',\zeta') \, d\zeta' \qquad (9.27)$$

The Fourier transform 9.27 gives the spectrum of amplitude fast fading [35,36]:

$$U_g(\nu) \sim \int U_\varepsilon(K_{x'},K_{y'},K_{z'}) \, dK_{y'} \qquad (9.28)$$

We should note that expressions 9.25 and 9.27 were obtained for the small-scale irregularities ($l \ll d_F$). In real cases of satellite–land communications, a wide spectrum of scales was observed, from several centimeters up to few kilometers [1–4,40,41], that is, the small- and moderate-scale irregularities with various degree of plasma density perturbations. For this more general 2-D case, the spectrum of plasma density $U_N(\mathbf{K})$ of inhomogeneities was calculated by Shkarofsky [39] for perpendicular (to $\mathbf{B}_0$) wave numbers $|\mathbf{K}_\perp| \equiv K_\perp = 2\pi/l_\perp$:

$$U_N(\mathbf{K}_\perp) = \frac{\langle N_1^2 \rangle \ell_0^{(p+2)/2} K_0^{(p-2)/2} \left(\ell_0\sqrt{K_0^2+K_\perp^2}\right)^{-(p+1)/2}}{2\pi} \frac{K_{(p-1)/2}\left(\ell_0\sqrt{K_0^2+K_\perp^2}\right)}{K_{(p-2)/2}(K_0\ell_0)} \qquad (9.29)$$

Here, ℓ_0 and $L_0 = 2\pi/K_0$ are the inner and outer scales of plasma irregularities, $l_\perp$ is the scale of plasma structures across $\mathbf{B}_0$, K_x (•) is the function of Macdonalds of x-order, and p is the degree of perturbation of plasma density. At the same time, the spectrum of amplitude scintillations, measured during satellite experiments, for the spatial frequencies $L_0^{-1} \ll K_\perp \ll l_0^{-1}$ the spectrum of plasma density $U_N(K)$ can be presented in a simple manner similar to the 1-D case as [35,36]:

$$U_N(K_\perp) = \frac{\langle N_1^2 \rangle \Gamma\left(\frac{p-1}{2}\right)}{\pi \Gamma\left(\frac{p-2}{2}\right)} \frac{K_0^{p-2}}{(K_0^2+K_\perp^2)^{(p-2)/2}} \approx K_\perp^{1-p} \qquad (9.30)$$

where $\Gamma(w)$ is the gamma function.

It was also obtained from satellite observations that in the direction parallel to $\mathbf{B}_0$ for the longitudinal scales $l_\| > d_F$ the spectrum of inhomogeneities $U_N(K_\|)$ is Gaussian

$$U_N(K_\|) \sim \exp\left\{-K_\|^2 l_\|^2\right\} \tag{9.31}$$

The 1-D spectra of plasma density disturbances $U_N(K_\|) \sim K_\|^{-(p-2)} \equiv K_\|^{-p'}$ for various extinction parameters in exponent, $p' = p - 2$, can be found based on Booker's theory [25–29]. Combining the solution obtained by Shkarofsky by using a general expression (Equation 9.29), which covers different scales and degrees of perturbations of plasma irregularities with spectral characteristic of amplitude of fast fading (Equation 9.28), that is, analyzing effects of different parameters of the degree of plasma density perturbations, p, we have found the following. The combined solution gives adequate description of scattering phenomena occurring in the real ionosphere and caused by moderate- and small-scale plasma irregularities.

Going forward into the modification of the Booker's theory, described in Chapter 7, we should point out that a modified theory introduced in Ref. [23] replaces weak multiple scattering by strong single scattering in such a way that enables to understand the relation between *diffractive* and *refractive* scattering, considering the scintillation phenomena observed experimentally [25–29]. At the same time, results obtained by Booker [25–29] are valid only for weak signal scintillations, that is, for the case when the mean square fluctuations of phase $[(\Delta\Phi)^2] \equiv \langle\Phi_1^2\rangle$ is weak enough, but the root-mean-square (RMS) fluctuations of phase $[\langle\Phi_1^2\rangle]^{1/2}$ is larger than 1 rad. Moreover, in the case of 1-D ionospheric layer, as a thin screen, the power spectrum of phase fluctuations, denoted as $W(K)$, was proportional $K^{-p'}$, when $K \gg (1/L_0)$ and for $p' = p - 2$, defined by Booker as the *spectral index* [25–29]. For the practical applications, the parameter p was defined as the spectral index [35–41].

Observed values of the spectral index $p = \langle n \rangle + 1$ are ranged from 2 to 4, with values between 2.5 and 3.5 being much more common [25–29]. The average refractive index $\langle n \rangle$ is usually measured along a straight line.

We modified Booker's approach using Shkarofsky's power spectrum of density fluctuations (Equation 9.29), comparing it with the power spectrum of phase fluctuations $[(\Delta\Phi)^2]W(K)$, where again $[(\Delta\Phi)^2]$ is the mean square fluctuation of phase, and $W(K)$ is the phase spectra. In our computation, to illustrate results obtained by Booker [25–29], we take the outer scale to be equal to the scale height H of the region F, and also take the thickness of the region F, denoted by Booker as D, to be equal to H. In such notations, the mean square fluctuation of signal phase experienced on radio trace through the region F, can be presented in the same manner, as Equation 9.20, obtained from weak phase fluctuations when $\langle\Phi_1^2\rangle \leq 1$. The close results, except for a general case of $\langle\Phi_1^2\rangle > 1$, can be obtained by taking into account the relation between Γ_Φ from Equation 9.22 and $\rho_\Phi = (\langle\Phi_1^2\rangle)^{-1}$, as well as between Γ_E from Equation 9.27 and $U_N(K_\perp)$ from Equation 9.29. Introducing now the latter characteristics into Equation 9.28 and accounting for the above-mentioned relations, we finally get, according to the geometry of the problem shown in Figure 9.3, that

$$\langle(\Delta\Phi)^2\rangle \equiv \langle\Phi_1^2\rangle = 4r_e^2 N_0^2 \left\langle \left(\frac{N_1}{N_0}\right)^2 \right\rangle \lambda^2 H^2 \sec\chi \tag{9.32}$$

where all parameters are defined above; r_e is the radius of electron.

The corresponding autocorrelation function $\rho(x)$ of plasma perturbations is obtained by using the inverse Fourier transformation of $W(K)$. Direct calculations give us the following expressions for the case of $0 \leq \ell_0 \ll L_0$ (the case which is usually observed in the real ionosphere at the layer F) [17–19]:

(a) for weak plasma perturbations, i.e., for $p = 2$ ($p' = 0$),

$$W(K) = 4L_0/(1 + K^2 L_0^2) \quad \text{and} \quad \rho(x) = \exp\{-x/L_0\};$$ (9.33a)

(b) for moderate plasma perturbations, i.e., for $p = 3$ ($p' = 1$),

$$W(K) = 2\pi L_0/(1 + K^2 L_0^2)^{3/2} \quad \text{and} \quad \rho(x) = x K_1(x/L_0);$$ (9.33b)

(c) for strong plasma perturbations, i.e., for $p = 4$ ($p' = 2$),

$$W(K) = 8L_0/(1 + K^2 L_0^2)^2 \quad \text{and} \quad \rho(x) = (1 + x/L_0) \exp\{-x/L_0\}$$ (9.33c)

For the case when the inner scale ℓ_0 is much higher than zero, expressions for $\rho(x)$ and $W(K)$ are more complicated [23], but do not change significantly the main problem. Using the selected appropriate autocorrelation function $\rho(x)$, presented by Equation 9.33a through c for different inner and outer scales of the irregularity, we can find the average spectrum of signal intensity fluctuations in the layer filled by plasma irregularities:

$$\langle I(K) \rangle = 4 \int_0^\infty g(x,K) \cos(Kx) \, dx$$ (9.34)

where

$$g(x,K) = \exp\{-\lfloor (\Delta\Phi)^2 \rfloor [f(0,K) - f(x,K)]\} - \exp\{-\lfloor (\Delta\Phi)^2 \rfloor f(0,K)\}$$ (9.35)

and the additional function $f(x,k)$ is related to $\rho(x)$ through the following equation:

$$f(x, K) = 2\rho(x) - \rho(x - K d_F^2) - \rho(x + K d_F^2)$$ (9.36)

Using Equations 9.35 and 9.36, and the corresponding spectra $W(K)$ defined above, we can derive a general expression (Equation 9.34) for the spectrum of signal intensity fluctuations in the following form:

$$\langle I(K) \rangle = 4\langle (\Delta\Phi)^2 \rangle W(K) \sin^2 \left(0.5 \cdot K^2 d_F^2\right)$$ (9.37)

We note that the Fresnel oscillations associated with the $\sin^2(0.5K^2 d_F^2)$ term in Equation 9.37 describe the main lobe and the first side lobe of the random interference picture. For the remaining lobes, only the average value is considered, corresponding to replacement of $\sin^2(0.5K^2 d_F^2)$ by 0.5.

Now, substituting the spectral function $W(K)$ from above for different spectral indices p', following Ref. [23], yields:

for $p' = 0$ ($p = 2$):

$$I(K) = 16\langle (\Delta\Phi)^2 \rangle \frac{L_0}{1 + K^2 L_0^2} \sin^2 \left(\frac{1}{2} K^2 d_F^2\right)$$ (9.38a)

for $p' = 1$ $(p = 3)$:

$$I(K) = 8\pi \langle (\Delta\Phi)^2 \rangle \frac{L_0}{(1 + K^2 L_0^2)^{3/2}} \sin^2 \left(\frac{1}{2} K^2 d_F^2 \right) \tag{9.38b}$$

for $p' = 2$ $(p = 4)$:

$$I(k) = 32 \langle (\Delta\Phi)^2 \rangle \frac{L_0}{(1 + K^2 L_0^2)^2} \sin^2 \left(\frac{1}{2} K^2 d_F^2 \right) \tag{9.38c}$$

Using relations between square root of the scintillation index σ_I and the average intensity spectrum $\langle I(K) \rangle$

$$\sigma_I = \frac{1}{2\pi} \int\limits_0^\infty \langle I(K) \rangle \, dK \tag{9.39}$$

we can obtain σ_I by substituting expressions 9.38a through c into expression 9.39. Thus, for $p = 2$ $(p' = 0)$

$$\sigma_I^2 = \frac{2\sqrt{2}}{\sqrt{\pi} L_0} d_F \langle (\Delta\Phi)^2 \rangle \tag{9.40a}$$

for $p = 3$ $(p' = 1)$:

$$\sigma_I^2 = \frac{\pi}{2L_0^2} d_F^2 \langle (\Delta\Phi)^2 \rangle \tag{9.40b}$$

for $p = 4$ $(p' = 2)$:

$$\sigma_I^2 = \frac{8\sqrt{2}}{3\sqrt{\pi} L_0^3} d_F^3 \langle (\Delta\Phi)^2 \rangle \tag{9.40c}$$

Computation of the intensity spectrum and the scintillation index were performed in accordance with Equation 9.38a through c and Equation 9.40a through c, respectively, for weak $(p = 2)$, moderate $(p = 3)$, and strong $(p = 4)$ fluctuations for an outer scale $L_0 = (1.5 \div 50)d_F$ and inner scale $\ell_0 = 10^{-2} d_F$ (i.e., sufficiently small). Figure 9.4 represents the scintillation index calculated numerically for different scales of ionospheric irregularities and different phase fluctuations. It is seen from presented illustrations that for $p = 2$ $(p' = 0)$, the scintillation index with increase of phase fluctuations limits to the unit. For higher signal phase fluctuations, σ_I can exceed the unit, which explains the focusing properties of the ionospheric layer filled by various irregularities, causing strong variations of the signal phase after passing the inhomogeneous nonregular ionosphere.

We present now variations of phase and intensity of radio waves passing a layer of the inhomogeneous ionosphere containing small-scale irregularities of various degree of perturbation p. The computations were made for three different frequencies: 500 MHz, 1 GHz, and 2 GHz, found in actual land–satellite communication using standard wireless networks. Figure 9.5 presents

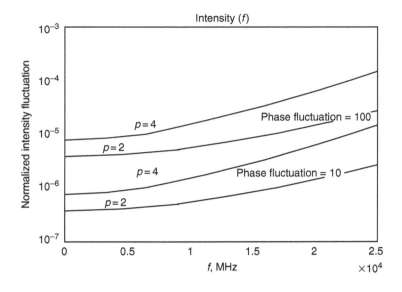

Figure 9.4 Normalized signal intensity fluctuation versus frequency time on 10^4 for different parameters of plasma density perturbations, $p = 2$ (weak perturbations) and 4 (strong perturbations), and for different RMS of signal phase fluctuations, $\langle \Phi_1^2 \rangle = 10, 100$.

signal phase variations $\langle \Phi_1^2 \rangle$, denoted simply by F_1, versus plasma density fluctuations. Here, for each frequency the corresponding graph was used. As it can be seen, with decrease of the radiated frequency, the signal phase fluctuations become much stronger; even the plasma density perturbations are changed insufficiently compared with the background plasma density.

Figure 9.6 describes variations in signal scintillation index due to variations of $\delta n/n$. The same effects, as was shown in Figure 9.5 for phase fluctuations, of increase of signal intensity fluctuations

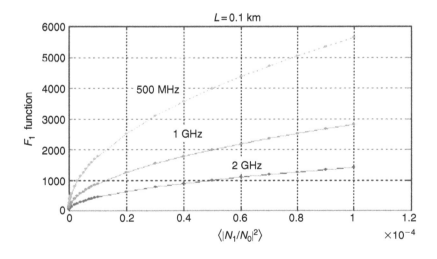

Figure 9.5 Variations in $\langle \Phi_1^2 \rangle$ versus $\langle |N_1/N_0|^2 \rangle$ time on 10^{-4}.

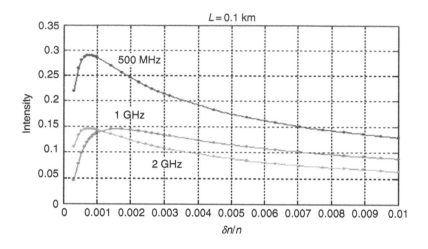

Figure 9.6 Variations in signal intensity versus $\delta n/n$.

with decrease of signal frequency are evident. Thus, for $f = 500$ MHz $\sigma_I^2 = 0.0575$, for $f = 1$ GHz $\sigma_I^2 = 0.0288$, and for $f = 2$ GHz $\sigma_I^2 = 0.0242$. The effects of the increase in the signal phase and intensity fluctuations with decrease of frequency (e.g., with increase of wavelength) can be explained by *diffractive scattering* (defined by Booker [25–29]) from small-scale plasma irregularities located in the ionospheric unstable region. This effect depends on local plasma perturbations within the ionospheric layer through which radio signals propagate. Thus, even for a dozen percentages of plasma density deviations, signal scintillations become sufficiently strong, with further increase of $\delta n/n$, obtained for all frequencies under consideration. Figure 9.7 shows variations in signal intensity against variations in $\langle \Phi_1^2 \rangle$ denoted simply by F_1. As follows from results of computations presented in Figure 9.7, signal intensity variations strongly depend on signal phase

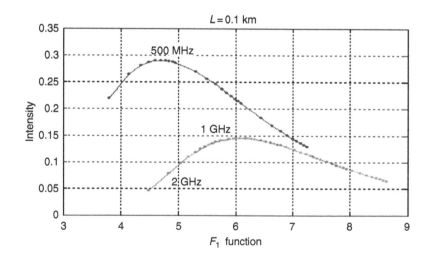

Figure 9.7 Variations in signal intensity versus $\langle \Phi_1^2 \rangle$.

variations, having a tendency to increase even for small phase deviations (of about 4–6 rad) with further tendency in decrease with increase of $\langle \Phi_1^2 \rangle$. Again, the occurrence of maximum signal intensity as function of F_1 depends strongly on radiated frequency.

Thus, for frequency $f = 500$ MHz, this maximum is two times larger than for $f = 1$ GHz and $f = 2$ GHz, where the maximum is the same and observed at stronger phase fluctuations. As for results presented in Figure 9.6, these effects can also be explained by the predominant role of small-scale inhomogeneities in *diffractive scattering* (comparing with *refractive scattering*, caused by moderate-scale irregularities) with an increase of wavelength with respect to inner inhomogeneity scale. Thus, for 500 MHz the wavelength is 60 cm and for 1 and 2 GHz are 30 and 15 cm, respectively, i.e., two–four times smaller, and for the latter frequencies the effect of diffractive scattering is predominant compared with those of refractive scattering. This tendency was obtained both theoretically and experimentally (see Refs. [23,35,36] and from the bibliography), which can be pointed as a evidence of the results presented in Figures 9.5 through 9.7. At the same time, obtained theoretical results describing signal phase and intensity fluctuations, with increase of plasma density perturbations and plasma total content deviations, are fully confirm by the results of satellite measurements carried out during periods of observations of the quiet [20,40,41] and the perturbed [42–46] ionosphere, notably, during magnetic storm events [47–51].

We now summarize some important results obtained above concerning effects of the small-scale inhomogeneities of the ionospheric plasma on radio wave propagation. Based on the original works [23–41], analysis of the field intensity and phase fluctuations by use of the perturbation method was made. Finally, we found an increase in the ionization density of ionospheric irregularities (i.e., their spectral index p) for a given frequency band, which causes an increase in signal phase fluctuations. Furthermore, for a given ionization density, using high frequencies for the satellite communication channel (more than UHF-band), a decrease in phase and amplitude fluctuations to values appropriate for weak scattering from ionospheric irregularities was observed. In addition, a distinction was made between weak scattering, $\sim \langle (\Delta\Phi)^2 \rangle = 10^{-1}$, 1, and strong scattering, $\sim \langle (\Delta\Phi)^2 \rangle = 100$. Below we present, as an example, effects of signal parameters variations during the magnetic storms occurring in the ionosphere due to the precipitation of high energetic cosmic and solar particles in the magnetosphere and coupling between magnetosphere and the high-latitude ionosphere, without going deeper into the physical description of the phenomena.

9.4 Effects of Magnetic Storm on Land–Satellite Communication

It is well known that during magnetic storm periods strong density irregularities occur in the ionosphere, causing phase and amplitude scintillations of radio signals and thus destroying satellite–satellite, satellite–land communications and navigations [42–51]. Usually this phenomenon occurs at polar and equatorial latitudes, respectively, where auroral and equatorial spread F phenomena take place. During magnetic storms the auroral zone expands equatorwards, overlapping the subauroral or middle-latitude ionosphere [42–45]. Thus, UHF radio signal scintillations (i.e., fading) due to equatorwards expanded phenomenon were observed during the magnetic storms of June 4–6, 1991 [42], September 22–23, 1999, and September 25–26, 2001, at Hanscom Observatory in the United States (42.5° N and 71.0° W), [43,44] and at Ithaca, USA (42.5° N and 48.0° W) [45], respectively.

It was found experimentally that a high-intensity polarized electric field generated during fluctuations of the ambient magnetic field in magnetic storm periods in the magnetosphere and due to coupling between magnetosphere and ionosphere, drives enhanced streams of westward convention currents with velocities $V_W \geq 500$ m/s in the subauroral zones. Such streams are called in the literature the subauroral polarization streams (SAPS). We do not go into complicated (and also interesting) physical explanations of all the processes occurring during these phenomena since this is beyond the main subject of this book and will introduce the reader to excellent works [46–51]. We will only mention, that such large velocities, as well as corresponding currents, can generate not only some specific instability in plasma (as discussed in previous subsections above analyzing high- and low-latitude irregularities), but a broad spectrum of instabilities, from two-stream instability to current-induced instabilities, such as two-wave and gradient-drift, and so on (see Chapter 2). Therefore, in our further explanations of the magnetic storm phenomenon, we will point out only characteristic parameters of irregular plasma structures associated with this phenomenon and the methods of their observations using satellite sounding of the ionosphere during magnetic storm periods.

Plasma within subauroral polarization streams (SAPS) has depleted densities and elevated temperatures. The density troughs follow accelerated ion–electron recombination stemming from enhanced chemical reaction rates in the non-equilibrium SAPS plasma [47].

Satellite and radar observations indicate the presence of wave-like plasma irregularities embedded within the SAPS [46–51]. It was found that during magnetic storms associated conventional fields driving ring-current ions penetrate deeply into the ionosphere. Sometimes exceptionally strong SAPS-wave structures were observed which collocate with precipitating ring current ions, elevated electron temperatures, and irregular plasma density troughs [47,49]. Thus, two additional special meteorological satellites (F13 at 22:33–22:40 UT and F14 at 00:56–00:59 UT) were launched near Hanscom on September 22, 1999. Figure 9.8 shows precipitating ring current ions and strongly oscillating the SAPS wave structures. Analysis of experimental data obtained at the satellite F14 observations reported similar SAPS wave structures, that is, strong fluctuations of plasma density $\delta n/n$ (see Figure 9.9) during the scintillation events near Hanscom at 00:57–00:59

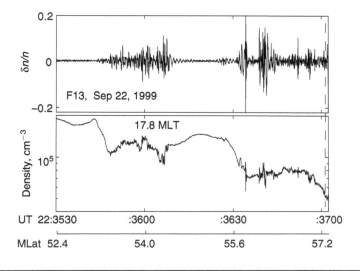

Figure 9.8 Time–latitude variations of plasma density perturbations $\delta n/n$ (top panel) and the total electronic content (TEC) of plasma (bottom panel) observed on September 22, 1999, during magnetic storm.

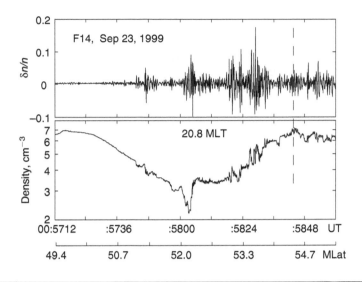

Figure 9.9 The same as in Figure 9.8, but observed on September 23, 1999, during magnetic storm.

UT, September 22, 1999. Thus, it appears reasonable to conclude that scintillation events at subauroral latitudes are related to SAPS wave structures, which in turn are strongly correlated with a highly irregular plasma density profile in the ionosphere during magnetic storms.

Thus, the analysis of variations in the plasma density, measured using Longmuir probe data, which are presented in Figure 9.10, were sampled at a rate of 24 Hz and show density variations within the SAPS wave structures near Hanscom from F13 and F14 near Ithaca, respectively. The corresponding power spectral densities (PSD) of plasma fluctuations are shown in Figure 9.11. As can be seen, strong deviations from PSD of usual density trough, which is proportional to $f^{-7/2}$ [46–48], are observed during a magnetic storm event. Finally, this leads to strong scintillations of the amplitude and phase of radio signals passing ionospheric channel with strong fading caused by magnetic storm. Figure 9.12 shows fast variations of $\delta n/n$ indicating the existence of unstable

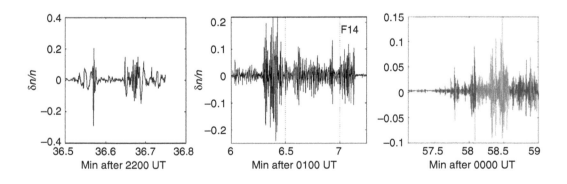

Figure 9.10 Temporal variations of plasma density perturbations $\delta n/n$ during two periods of magnetic storm observation. (Combined from Refs. [42–45].)

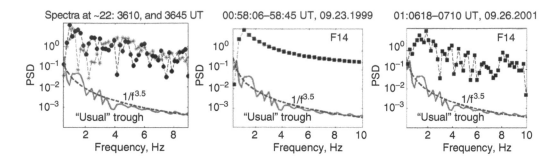

Figure 9.11 Signal power spectral density (PSD) versus Doppler shift frequency and its best fit approximation (dashed curves) during two periods of magnetic storm observations. (Combined from Refs. [43,44].)

irregular small-scale plasma structures at the altitudes of observation. Remember that a 24 Hz sampling rate yields the upper frequency-limit ~9.5 Hz. Figure 9.13 shows the power spectral densities of plasma fluctuation. The results correspond to later experiments with satellites F13 and F14 carried out in 2001. It can be seen that density oscillations with apparent frequencies 3–10 Hz are enhanced by a factor ~100. The enhanced fluctuations appeared at the poleward edges of large- and moderate-scale density irregularities, embedded within SAPS wave structures. Thus, when measuring the occurring Doppler-shift frequencies from satellite frames of reference, it was found that the signal oscillations correspond to irregularities with spatial scales of 0.7–2 km. It is most likely that these irregularities are responsible for radio-signal scintillations near 1 GHz. Therefore, a magnetic storm leads to the same radiophysical effects as observed experimentally in middle-latitude and equatorial zones with E spread and F spread layers consisting of large- and moderate-scale irregularities, the main ones being proved by:

> Strict observations of strong wave structures embedded within sub-aurora polarization streams (SAPS) found by F13, 14, and 15 satellites flying over the magnetic-storm-affected regions. Detailed analysis of highly irregular density disturbances associated with the strong SAPS-wave events using Longmuir probe data, sampled at a rate of 24 Hz.

We will now present measurements of relevant parameters made in space by different satellites. From these measurements we can estimate real-time sequences referring to the scintillation index σ_I^2, denoted in Refs. [44–48] by $\mathbf{S}_4$, and fluctuations of plasma density $\delta n/n$, which are important

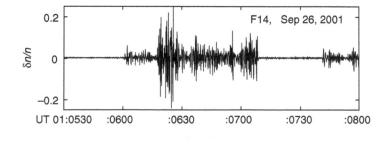

Figure 9.12 Temporal variations of plasma density perturbations $\delta n/n$ observed by satellite F14 during magnetic storm.

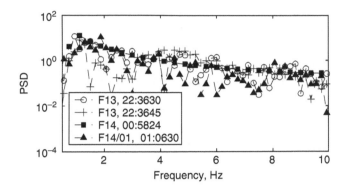

Figure 9.13 Cumulative distribution of PSD in the frequency domain observed by different satellites F13 and F14 during magnetic storm.

for further prediction of the key parameters of the land-communication links passing ionospheric regions disturbed by the magnetic storm event.

Results of observations of the scintillation index S_4 and the total electron content (TEC) measured at 250 MHz and 1.5 GHz are shown in Figures 9.14 and 9.15, respectively from different GPS satellites. Following on illustrations presented in these figures, during a magnetic storm deep depletions of the total plasma content (TEC) and the corresponding strong fluctuations of the spectral index of signal scintillations are observed.

Figure 9.16 shows more precise observations by satellite carried out on September 22, 1999, at 22:35-22:37 UT, concerning not only the TEC (the top graph), but also the plasma density

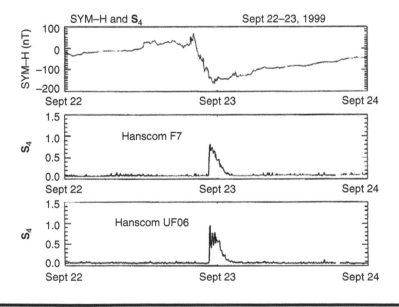

Figure 9.14 H-component of the ambient magnetic field variations and the corresponding index of signal scintillation S_4 observed during magnetic storm. (Combined from Refs. [44–46].)

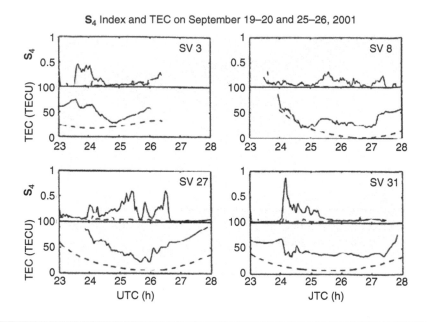

Figure 9.15 **S$_4$ index and TEC measurements for four different GPS satellites on September 19–20, 2001 (dashed lined) and September 25–26 (solid lines). (Combined from Refs. [44–46].)**

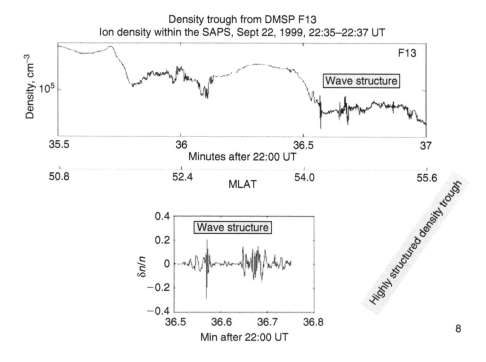

Figure 9.16 Plasma density fluctuations (TEC and $\delta n/n$) and wave structure registered during magnetic storm time. (Combined from Refs. [43,44].)

perturbations, denoted by $\delta n/n$ (the bottom graph). Not only great depletion of plasma density occurring during the magnetic storm occurred at 22:36–22:37 UT can be seen, but also strong fluctuations of plasma density $\delta n/n$.

Finally, this causes strong fast fading of the signal passing the disturbed ionosphere in the land–satellite channel.

Generally speaking, Figures 9.15 and 9.16 show significant variations in total plasma density content and in local plasma perturbations $\delta n/n$ occurring during the magnetic storm. The following features observed during satellite experiments regarding the scintillation index S_4 and the total electron content (TEC) measured at 250 MHz and 1.5 GHz, can be summarized as follows:

Small- and moderate-scale plasma density and electromagnetic oscillations were detected by F13, 14, and 15 satellites while flying near the regions of intense radio-signal scintillations.

Twenty-four hertz Longmuir probe data show that during the scintillation intervals the amplitudes of density oscillations in the frequency range of 3–10 Hz increased by a factor of 100.

Using Doppler shift measurements, it was found that the signal scintillations correspond to plasma irregularities of scales ~0.7–2 km, that is, to small-scale and moderate-scale plasma perturbations.

Most likely these irregularities are responsible for radio-signal scintillations near 1 GHz.

To summarize the efficiency of the theoretical framework presented in Section 9.3, let us compare theoretical results obtained there with observations of signal oscillations during the magnetic storm described in Section 9.4 based on results of measurements and theoretical frameworks presented in Refs. [42–51], where strict measurement of plasma density content, deviations of plasma density with respect to background, $\delta n/n$, and the index of signal intensity scintillations were precisely measured and analyzed. It is evident that the above-obtained theoretical prediction of signal phase and intensity fluctuations, with increase of plasma density and plasma total content deviations, fully confirm the results of satellite observations obtained in Refs. [42–46] and discussed in the previous section. Moreover, important results having practical applications for land–satellite (namely, GPS) links design during magnetic storm, as one of natural disaster, discussed in Section 9.4, can be fully predicted by theoretical background presented in the previous section. Thus, can be predicted the following phenomena observed experimentally in [42–46]:

Higher plasma content deviations and plasma density relative fluctuations (with respect to steady state background) occurring during the magnetic storm cause stronger phase and amplitude scintillations.

Smaller radiated frequencies of radio waves passing the disturbed ionospheric areas during the magnetic storm cause stronger signal phase and amplitude scintillations.

Even small plasma density deviations $\delta n/n \equiv \sqrt{\langle |\delta N/N_0|^2 \rangle}$ occurring during the magnetic storm (few dozen of the percentage) can generate significant amplitude and phase scintillations for frequencies less than 1 GHz and affect signal propagation much weakly for frequencies more than 2–5 GHz.

The above effect depends also on the ionospheric latitudes. Thus, in the high-latitude ionosphere where $\delta n/n$ can reach several percentages before the magnetic storm, the latter can significantly increase plasma density fluctuations, and consequently, can generate strong phase and amplitude scintillations, i.e., strong fast fading of the total signal registered at the Earth's surface.

9.5 Parameters of Data Signals in the Land–Satellite Communication Links with Fading

Based on the magnetic storm effects experimentally observed during sounding of the disturbed ionosphere using satellite in situ measurements, we will now analyze the effects of fading of probing radio signals on the main parameters of the ionospheric link and on the information data stream parameters passing through the multipath channels with fading. First, we will introduce the main parameters of the channel and the data stream. Then, we will show an example of how the magnetic storm perturbs the ionosphere, changing conditions of propagation of signals through it and affecting parameters of the information data stream within ionospheric channels with fading.

9.5.1 Main Parameters of the Information Data Stream

The capacity of a communication channel is defined in the classical communication theory as the traffic load of data in bits per second [23]. Two approaches are usually used to estimate the data capacity of the communication link using Equation 9.41.

According to the *first definition*, the capacity of an AWGN channel of bandwidth B_w is based on the Shannon–Hartley formula, in order to calculate the capacity of the channel, which finally is given by [52–55]:

$$C(r) = B_w \log_2 [1 + SNR] \tag{9.41}$$

where the signal to noise ratio, SNR equals (see also definitions in Chapter 3)

$$\text{SNR} = 10 \log \left(\frac{P_R}{N_R} \right) = P_{R[dB]} - N_{R[dB]} \tag{9.42}$$

and the spectral efficiency is given by:

$$\tilde{C}(r) = \frac{C(r)}{B_w} = \log_2 [1 + \text{SNR}] \tag{9.43}$$

In the *first approach* only the Gaussian noise (called *additive*) is taken into account with density N_0 inside the channel and with a broad bandwidth B_w of noise spectrum [52,53], i.e.,

$$C = B_w \log_2 \left[1 + \frac{S}{N_0 B_w} \right] \tag{9.44}$$

Here, again, C is the channel capacity in bits per second; B_w, the one-way transmission bandwidth of the channel in Hz; S is the signal power in W $=$ J/s; and N_0 the single-sided additive (white) noise power spectral density also in W/Hz, that is, $N_{add} = N_0 B_w$.

We should stress that the effects of interference can be accounted as another source of effective noise, which raises the noise level when calculating the capacity. In this case we must introduce in Equation 9.44 also the noise caused by interference N_{int}, together with N_{add} [56], i.e.,

$$C = B_w \log_2 \left[1 + \frac{S}{N_0 B_w + N_{int}} \right] \tag{9.45}$$

In Chapters 3 and 4 we defined a land–satellite communication channel as a multipath one with fast fading caused by significant Doppler effects occurring in the ionospheric channel due to the high speed of a moving satellite and/or movements of plasma irregularities with velocities exceeding hundreds of meters per second (see Chapter 6).

Therefore, we should introduce here the *second approach*, mentioned in [23], which can be used directly to describe the maximal data stream rate, that is, the capacity, inside dynamic satellite communication channels with fading, flat, or multiplicative, where an additional source of noise, called multiplicative noise, has to be introduced (see definitions in Chapters 3 and 4). Thus, taking into account the fading phenomena, we can estimate the multiplicative noise by introducing its spectral density, N_{mult}, and its frequency bandwidth, B_Ω, into Equation 9.45:

$$C = B_\omega \log_2 \left[1 + \frac{S}{N_0 B_\omega + N_{mult} B_\Omega} \right] \tag{9.46}$$

As was shown in Ref. [23], the formula 9.46 is valid when the LOS component of the total signal inside a channel exceeds the NLOS component; that is, for the K-factor, characterized the fading phenomena, exceeds the unit (see definitions in Chapters 3 and 4). We can now rewrite Equation 9.46 by introducing in it K-factor defined in Chapter 3 as $K = \frac{I_{co}}{I_{inc}}$, accounting for the following relation $\frac{S}{N_{mult}} = \frac{I_{co}}{I_{lm}}$ [23]:

$$C = B_w \log_2 \left(1 + \frac{S}{N_{add} + N_{mult}} \right) = B_w \log_2 \left(1 + \left(\frac{N_{add}}{S} + \frac{N_{mult}}{S} \right)^{-1} \right) \tag{9.47}$$

Then, using notations introduced above, we finally get the capacity as a function of the K-factor and the additive noise:

$$C = B_w \log_2 \left(1 + \left(\frac{N_{add}}{S} + K^{-1} \right)^{-1} \right) = B_w \log_2 \left(1 + \frac{K \cdot \frac{S}{N_{add}}}{K + \frac{S}{N_{add}}} \right) \tag{9.48}$$

Finally, from Equation 9.48 it is easy to obtain the spectral efficiency of the channel, $\tilde{C} = \frac{C}{B_w}$, where the bandwidth B_w can be changed according to the satellite system under investigation. We should point out here that the spectral efficiency is the most important parameter of the channel because it defines the maximum rate of information data stream inside the radio channel normalized on the constant bandwidth and, therefore, is measured in bits/second/Hz.

The next parameter of the information data passing through the satellite channel is the *bit error rate* (BER). The BER is the most important parameter of the data stream within the channel because it gives information about the rate of bit error inside the propagation link. For BER, in the percentage of bits that have errors relative to the total number of bits received at the receiver, we use the classical approach described in Refs. [52–54]:

$$\text{BER} = \frac{1}{2} \int_0^\infty p(x) \text{erfc} \left(\frac{\text{SNR}}{2\sqrt{2}} x \right) dx \tag{9.49}$$

where $p(x)$ is the probability density function, which in our case was taken as a Ricean one, and erfc($\bullet$) is the well-known error function (see definitions in Chapter 4). Using the BER definition

(9.49), where $p(x)$ is Ricean PDF, and where the SNR includes also the multiplicative noise, we finally get for a BER [23]:

$$
\mathrm{BER}\left(K, \frac{S}{N_{add}}, \sigma\right) = \frac{1}{2} \int_0^\infty \frac{x}{\sigma^2} \cdot e^{-\frac{x^2}{2\sigma^2}} \cdot e^{-K} \cdot I_0\left(\frac{x}{\sigma}\sqrt{2K}\right) \cdot \mathrm{erfc}\left(\frac{K \cdot \frac{S}{N_{add}}}{2\sqrt{2}\left(K + \frac{S}{N_{add}}\right)}x\right) dx \quad (9.50)
$$

This is an important formula, which gives the relation between the BER and the additive signal-to-noise ratio; the Ricean parameter K describing the multipath fading phenomena occurring within the multipath land–satellite communication link passing the ionosphere, and the probability of BER of the information data stream inside such a channel.

9.5.2 Data Stream Parameters for Channels with Fading Caused by Magnetic Storm

We now will analyze parameters of the information data stream in the land–satellite communication channel taking into account parameters of the ionosphere and the signal fading during magnetic storm described in Section 9.4, because radio observations of the disturbed ionosphere during this natural event were systematically carried out during the recent decade and described in detail in Refs. [42–51]. For these purposes, we introduce relations between the parameter of signal intensity scintillations and the K-factor describing the fast-fading phenomena within the channel following Ref. [23]. Thus, the main purpose of the theoretical analysis presented in Ref. [23] was to show that for a given frequency, and for the high perturbations of the ionosphere characterized by a spectral index p, an increase in the signal intensity fluctuations can be described by the Ricean K-factor as a ratio of coherent and incoherent components of the total signal intensity:

$$
K = \langle I_{co}\rangle / \langle I_{inc}\rangle \quad (9.51)
$$

Here $\langle I_{co}\rangle$ is the line-of-sight (LOS) component of the signal (called also the coherent component [23]), and $\langle I_{inc}\rangle$ is the incoherent component of the total signal intensity,

$$
\langle I_{inc}\rangle = \langle I_{co}\rangle \sqrt{\sigma_I^2} \quad (9.52)
$$

where σ_I^2 is the scintillation index described above in Section 9.2, which we will rewrite as

$$
\sigma_I^2 = \frac{\langle I^2\rangle - \langle I\rangle^2}{\langle I\rangle^2} = \frac{\langle I^2\rangle}{\langle I\rangle^2} - 1 \quad (9.53)
$$

where $\langle I\rangle = \langle I_{co}\rangle + \langle I_{inc}\rangle$. Then, combine Equation 9.53 with Equation 9.51, taking into account Equation 9.52, we finally have relations between K-factor and σ_I^2

$$
K = \left(\sigma_I^2\right)^{-1/2} \equiv (\mathbf{S}_4)^{-1/2} \quad (9.54)
$$

Taking into account Equation 9.54 and using formulas 9.48 through 9.50, we can analyze data stream parameters variations with changes in situation with signal scintillations and fading

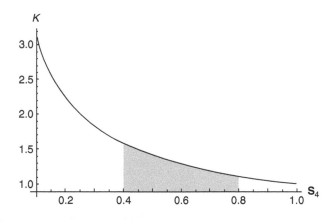

Figure 9.17 Dependence of *K*-parameter on S$_4$. The shadow area corresponds to data obtained experimentally by satellites during magnetic storm.

occurring within the ionospheric communication links during the magnetic storm. Results of numerical computations are shown in Figures 9.17 through 9.19 for various scintillation parameters $S_4 \equiv \sigma_I^2$ from Figures 9.14 and 9.15 and relation 9.54 between σ_I^2 and the *K*-factor. Thus, in Figure 9.17, the parameter *K* is presented versus $S_4 \equiv \sigma_I^2$, where the latter is taken from measurements obtained above from Figures 9.14 and 9.15. It can be seen that during the magnetic storm when the index of signal intensity scintillations varies from 0.4 to 0.9, the corresponding *K*-parameter varies, according to Equation 9.54, in the range of about 1.1 to 1.6, indicating the existence of direct visibility between the ground-based and the satellite antennas (i.e., the LOS component), accompanied by the additional effects of multipath phenomena (i.e., NLOS multipath component) caused by multiple scattering of radio signals at the small-scale plasma structures, $\delta n/n$, formed in the disturbed ionospheric regions during the magnetic storm, observed experimentally (see Figures 9.14 through 9.16). During the strong magnetic storms, where $S_4 = 0.7-0.9$, the *K*-factor, described as a multipath fading phenomenon within the propagation channel, is

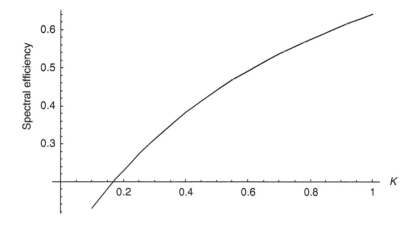

Figure 9.18 Spectral efficiency of the land–satellite channel versus *K*-parameter at the range corresponding to the shadow area depicted in Figure 9.17.

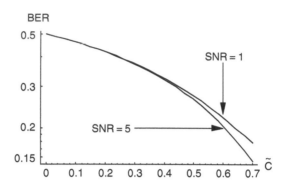

Figure 9.19 BER as a function of spectral efficiency of the channel for the range of parameters corresponding to the magnetic storm observed experimentally. Computations were carried for SNR = 5 dB (denoted SNR 5) and SNR = 1 dB (denoted SNR 1).

changed around the unit. With information about K-factor, we can predict deviations of the parameters of the data stream (i.e., the signal) in the multipath channels passing through the disturbed ionosphere. Thus, in Figure 9.18, the spectral efficiency of such a multipath channel is presented versus the parameter K. We made this parameter much wider, varying in the interval from 0.2 to 1.1, to cover the "worst" case, when $K \ll 1$, and the quasi-LOS case, when $K \approx 1$. As can be seen, during the strong magnetic storm, where $K \approx 1$, the spectral efficiency is around 0.6. As for BER, this parameter depends on ratio between the signal and the additive (white or Gaussian, see Chapter 4) noise, SNR $\equiv S/N_{add}$. We examined BER for different typical values of SNR, which is shown in Figure 9.19 for SNR $= 1$ dB and SNR $= 5$ dB. As it can be seen, for small $\tilde{C} = C/B_w$, BER is sufficiently high. At the range $\tilde{C} = 0.6$–0.7, interesting for us, the BER is twice smaller compared with the previous case, but it is high enough to lose information inside the channel with fading occurring during the magnetic storm. It should be noted that the increase of SNR inside the channel (from 1 to 5 dB) cannot decrease BER significantly even for high spectral efficiency. Generally speaking, with increase of the spectral efficiency from 0.1 to 0.7, the BER parameter decreases approximately three times.

Brief analysis carried out above concerning influence of the natural phenomena—such as the magnetic storm, usually occurring in the ionosphere, and the corresponding formulas obtained above that describe the key parameters of the channel and the signal data—allow us to predict the main K-parameter of fading for land–satellite radio communication links, using information about signal spectral index variations occurring in the radio channel passing through the natural ionosphere under disturbed ambient conditions, particularly, during the magnetic storm. At the same time, knowledge of the K-parameter for different situations observed in the real ionosphere allows us to obtain the key parameters of the channel and the signal data, such as the capacity, spectral efficiency, and bit error rate (BER), which describe real situations with fading occurring within the ionospheric radio channel.

Based on results presented above, the designers of the land–satellite radio communication links can, a priori, find relations between the real natural phenomena occurring in the ionosphere under perturbed natural or artificially made conditions, the parameters of the radio channel, and the parameters of the information data stream, allowing, finally, prediction of the types of fading (through the K-factor) that can be expected in the real disturbed situation in the ionosphere (e.g., during the magnetic storm), the effects of fading on the data stream parameters, such as the capacity as a maximum information rate, the spectral efficiency (for knowing, a priori, the bandwidth of the land–satellite system), and finally, the bit error rate (BER) as a measure of loss of digital information per second inside the channel, and so on.

References

1. Allnutt, J.A., *Satellite-to-Ground Radiowave Propagation*, IEEE Press, London, 1989.
2. Maral, G. and Bousquet, M., *Satellite Communication Systems, Technique and Technology*, Wiley and Son, Chichester, 1993.
3. Evans, B.G. (Ed.), *Satellite Communication Systems*, IEE Press, London, 1999.
4. Barts, R.M. and Stutzman W.I., Modeling and measurement of mobile satellite propagation, *IEEE Trans. Antennas Propagat.*, Vol. 40, no. 4, 1992, pp. 375–382.
5. International Communication Union ITU-R Recommendation, P.676-3, *Attenuation by Atmospheric Gases*, Geneva, 1997.
6. International Communication Union ITU-R Recommendation, P.618-5, *Propagation Data and Prediction Methods Required for Design of Earth-Space Telecommunication Systems*, Geneva, 1997.
7. Alpert, Ya.L., *Propagation of Electromagnetic Waves and Ionosphere*, Science, Moscow, 1972.
8. Brunt, P., IRIDIUM-overview and status, *Space Communic.*, Vol. 14, 1996, No. 1, pp. 61–68.
9. Pattan, B., *Satellite-Based Cellular Communications*, McGraw-Hill, NY, 1998.
10. Filipp, N.D., Blaunstein, N., Erukhimov, L.M., Ivanov, V.A., and Uryadov, V.P., *Modern Methods of Investigations of Dynamic Processes in the Ionosphere*, Shtiintsa, Kishinev, Moldova, 1991.
11. Ureadov, V., Ivanov, V., Plohotniuc, E., Eruhimov, L., Blaunstein, N., and Filip, N., *Dynamic Processes in Ionosphere and Methods of Investigations*, Tehnopress, Iasi, Romania, 2006.
12. Reinisch, B.W., *Modern Ionosondes*, in *Modern Ionospheric Science*, Ed. by Kohl, H., Ruster, R., and Schlegel, K., Copernicus GmbH, Katlenburg-Lindau, 1996, pp. 440–458.
13. Farley, D.T., *Incoherent Scatter Radar Probing*, in *Modern Ionospheric Science*, Ed. by Kohl, H., Ruster, R., and Schlegel, K., Copernicus GmbH, Katlenburg-Lindau, 1996, pp. 415–439.
14. Belikovich, V.V., Benediktov, E.A., Tolmacheva, A.V., and Bakhmet'eva, N.V., *Ionospheric Research by Means of Artificial Periodic Irregularities*, Copernicus GmbH, Katlenburg-Lindau, 2002.
15. Mishin, E.V., Ryzhin, Yu.Ya., and Telegin, V.A., *Interactions of Electronic Flows with Ionospheric Plasma*, Hydrometeoizdat, Leningrad, 1989.
16. Stubbe, P., *The Ionosphere as a Plasma Laboratory*, in *Modern Ionospheric Science*, Ed. by Kohl, H., Ruster, R., and Schlegel, K., Copernicus GmbH, Katlenburg-Lindau, 1996, pp. 274–321.
17. Philipp, N.D., Oraevsky, V.N., Blaunstein, N.Sh., and Ruzhin, Yu.Ya., *Evolution of Artificial Plasma Inhomogeneities in the Earth's Ionosphere*, Shtiintsa, Kishinev, Moldova, 1986.
18. Gurevich, A.V. and Shvartzburg, A.B., *Non-linear Theory of Radiowave Propagation in the Ionosphere*, Science, Moscow, 1973.
19. International Communication Union ITU-R Recommendation, P.531-4, *Ionospheric Propagation Data and Prediction Methods Required for the Design of Satellite Services and Systems*, Geneva, 1997.
20. Jakowski, N., *TEC Monitoring by Using Satellite Positioning Systems*, in *Modern Ionospheric Science*, Ed. by Kohl, H., Ruster, R., and Schlegel, K., Copernicus GmbH, Katlenburg-Lindau, 1996, pp. 346–370.
21. Briggs, B.H. and Parkin, I.A., On variation of radio star and satellite scintillations with zenith angle, *J. Atmos. Terr. Phys.*, Vol. 25, no. 3, 1963, pp. 339–365.
22. Sounders, S.R., *Antennas and Propagation for Wireless Communication Systems*, Wiley and Son, Chichester, 1999.
23. Blaunstein, N. and Christodoulou, Ch., *Radio Propagation and Adaptive Antennas for Wireless Communication Links: Terrestrial, Atmospheric and Ionospheric*, New Jersey: Wiley Intersience, 2007.

24. Booker, H.G. and Gordon, W.E., A theory of radio scattering in the ionosphere, *Proc. IRE*, Vol. 38, 1950, pp. 400–412.

25. Booker, H.G., Rateliffe, S.A., and Shinn, D.H., Diffraction from an irregular screen with applications to ionospheric problems, *Philos. Trans. Roy. Soc. London*, Ser. A., Vol. 242, 1950, pp. 579–607.

26. Booker, H.G., A theory of scattering by non-isotropic irregularities with application to radar reflection from the Aurora, *J. Atmos. Terr. Phys.*, Vol. 8, 1956, pp. 204–221.

27. Crain, C.M., Booker, H.G., and Fergusson, S.A., Use of refractive scattering to explain SHF scintillations, *Radio Sci.*, Vol. 14, 1974, pp. 125–134.

28. Booker, H.G. and Majidi Ahi, G., Theory of refractive scattering in scintillation phenomena, *J. Atmos. Terr. Phys.*, Vol. 43, 1981, pp. 1199–1214.

29. Booker, H.G., Application of refractive scintillation theory to radio transmission through the ionosphere and the solar wind and to reflection from a rough ocean, *J. Atmos. Terr. Phys.*, Vol. 43, 1981, pp. 1215–1233.

30. Backley, R., Diffraction by random phase screen with very large rms phase deviation. Two-dimensional screen, *Austral. J. Phys.*, Vol. 24, 1971, pp. 373–396.

31. Wernik, A.W. and Liu, C.H., Application of the scintillation theory to ionospheric irregularities studies, *J. Artificial Satellites*, Vol. 10, 1975, pp. 37–58.

32. Rino, C.L. and Fremouw, E.J., The angle dependence of single scattered wave fields, *J. Atmos. Terr. Phys.*, Vol. 39, 1977, pp. 859–868.

33. Blaunstein, N., Theoretical aspects of wave propagation in random media based on quanty and statistical field theory, *J. Electromagnetic Waves and Applications: Progress In Electromag. Research, PIER*, Vol. 47, 2004, pp. 135–191.

34. Titheridge, J.E., The diffraction of satellite signals by isolated ionospheric irregularities, *J. Atmos. Terr. Phys.*, Vol. 33, no. 1, 1971, pp. 47–69.

35. Gurevich, A.V. and Tsedilina, E.E., *Long-Range Propagation of Short Waves*, Moscow, Science, 1979.

36. Gel'berg, M.G., *Inhomogeneities of High-Latitude Ionosphere*, Novosibirsk: Science, USSR, 1986.

37. Denisov, N.G. and Erukhimov, L.M., Statistical properties of phase fluctuations during total reflection from a layer, *Geomagn. and Aeronomy*, Vol. 5, 1966, pp. 695–702.

38. Ga'lit, T.A., Gusev, V.D., Erukhimov, L.M., and Shpiro, P.I., About the spectrum of phase fluctuations during sounding of the ionosphere, *Radiophysics, Izv. Vuzov*, Vol. 26, 1983, pp. 795–801.

39. Shkarofsky, I.P., Generalized turbulence space-correlation and wave-number spectrum function pairs, *Can. J. Phys.*, Vol. 46, 1968, pp. 524–528.

40. Erukhimov, L.M. and Rizhkov, V.A., Stady of focusing ionospheric irregularities by methods of radio-astronomy at frequencies of 13–54 MHz, *Geomagn. and Aeronomy*, Vol. 5, 1971, pp. 693–697.

41. Erukhimov, L.M., Komrakov, G.P., and Frolov, V.L., About the spectrum of the artificial small-scale ionospheric turbulence, *Geomagn. and Aeronomy*, Vol. 20, 1980, pp. 1112–1114.

42. Burke, W., Rubin, A., and Mayanard, M. et al., Ionospheric disturbances observed by DMSP at mid to low latitudes during magnetic storm of June 4–6, 1991, *J. Geophys. Res.*, Vol. 105, 2000, pp. 18,391–19,402.

43. Basu, S., Basu, S., and Valladares, C.E. et al., Ionospheric aspects of major magnetic storms during the International Space Weather Period of September and October 1999: GPS observations, VHF/UHF scintillations, and in situ density structures at middle and equatorial latitudes, *J. Geophys. Res.*, Vol. 106, 2001, pp. 30,389–30,413.

44. Ledvina, B.M., Makela, J.J., and Kitner, P.M., First observations of intense GPS L1 amplitude scintillations at midlatitude, *Geophys. Res. Lett*, Vol. 29, 2002, pp. 1659–1662.

45. Mishin, E.V., Foster, J., and Potekhin, A. et al., Global ULF disturbances during a stormtime substorm on 25 December 1998, *J. Geophys. Res.*, Vol. 107, 2002, pp. 1486–1490.

46. Rich, F.J. and Hairston, M., Large-scale convection patterns observed by DMSP, *J. Geophys. Res.*, Vol. 79, 1994, pp. 3827–3835.

47. Mishin, E.V., Burke, W.J., and Viggiano, A., Stormtime subauroral density troughs: Ion-molecular kinetic effects, *J. Geophys. Res.*, Vol. 109, 2004, pp. 10,301–10,309.

48. Mishin, E.V. and Burke, W.J. Stormtime coupling of the ring current, plasmosphere and topside ionoaphere: Electromagnetic and plasma disturbances, *J. Geophys. Res.*, Vol. 110, 2005, pp. 7209–7216.

49. Foster, J. and Burke, W., A new categorization for subauroral electric fields, *EOS Trans. AGU*, Vol. 83, 2002, pp. 393–401.

50. Mishin, E., Burke, W., Huang, C., and Rich, F., Electromagnetic wave structures within subauroral polarization streams, *J. Geophys. Res.*, Vol. 108, 2003, pp. 1309–1315.

51. Maynard, N., Burke, W., and Basinska, E., et al., Dynamics of the inner magnetosphere near times of substorm onsets, *J. Geophys. Res.*, Vol. 101, 1996, pp. 7705–7712.

52. Papaulis, A., *Probability, Random Variables, and Stochastic Processes*, McGraw Hill, New York, 1991.

53. Wyner, A.D., Shannon theoretic approach to a Gaussian cellular multiple access channel, *IEEE Trans. on Information Theory*, Vol. 40, No. 6, 1994, pp. 1713–1727.

54. Biglieri, E., Proakis, J., and Shamai, S., Fading channels: Information theoretic and communication aspects, *IEEE Trans. on Information Theory*, Vol. 44, No. 6, 1998, pp. 2619–2692.

55. Krishnamurthy, S.V., Acampora, A.S., and Zorzi, M., On the radio capacity of TDMA and CDMA for broadband wireless packet communication, *IEEE Trans. on Vehicular Technol.*, Vol. 52, No. 1, 2003, pp. 60–70.

56. Rappaport, T.S., *Wireless Communications Principles and Practice*, Prentice Hall, New York, 1996.

Index

A

Absolute instability, 26
Absorption
  coefficient, 403
  index, 404
  intensity, 403
  magnitude, 405
  radiometric, 404
  standard deviation, 404
Acoustic ion wave
  speed, 52, 54, 62, 79, 98, 257
All-sky camera, 497
Altair radar, 332–336
Altitude-time domain, 329, 331, 339
Ambient electric field, 141
Ambient magnetic field, 141, 143, 150
Ambipolar
  field, 141
  diffusion, 142, 151, 164
  drift, 194–198
Amplitude-frequency characteristic, 479
Angle of arrival
  azimuth, 114
  elevation, 115
Angle PDF, 117
Anisotropic plasma irregularities
  large-scale, 258, 541
  moderate-scale, 261, 542
  small-scale, 262, 545
Appelton-Hartree (Appelton–Lassen)
  dispersion relation, 405
  magnetoionic theory, 404
Approaches
  linear, 55–57
  nonlinear, 64–66, 79–82, 95–98
  quasi-linear, 69–73, 75–78
Arecibo
  heating experiment, 258–260
  incoherent scatter radar, 494
  observatory, 494, 545

Artificial electromagnetic radiation, 256
Artificial periodic irregularity, 243, 245
Artificial instability
  parametric instability, 25, 36–37
  striction instability, 25, 29
A-scope, 306
Autocorrelation function, 119
Aulin's 3-D model, 115–116, 120

B

Back scattering, 439
Bandpass signal, 111
Baseband signal, 112
Bessel function, 132
Bit error rate (BER), 563
Bins, 122
Boltzman's collision integral, 53
Booker's theory, 432
Bragg's law, 362, 442
Bubbles, 347, 351–352
Buneman's hydrodynamic approach, 53

C

Capacity, 562
Capturing of radiowaves, 470–471
Carrier
  amplitude, 111, 115
  frequency, 115
  strength, 111
Channel
  dynamic, 137
  multipath, 127–128
  static, 136
Characteristic
  altitude-time, 433
  amplitude-frequency, 478, 480
  distance-frequency, 300

Chatanika incoherent radar, 360, 365
Conductivity
    Hall, 283
    Pedersen, 283
Coefficient
    ambipolar diffusion, 142, 144, 164
    ambipolar drift, 196–198
    of capture, 471
    of ion-molecular reactions, 19
    of recombination, 18, 20
    thermo–conductivity, 234
    thermo–diffusion, 233–234
    unipolar diffusion, 141, 144, 153
Coherence bandwidth, 124, 137
Coherent radar, 318
Coherence time, 125
Communication
    land-satellite, 114
    satellite-satellite, 354
Concentration
    barium ions, 271, 274, 279
    electron, 6, 141
    ion, 5, 141
    neutrals, 4
    plasma, 271, 274
Condor campaign, 308, 320
Consistent kinetic theory, 55–57
Convective instability, 48
Convergence instability, 57–58
Clarke's 2-D model, 114, 120
Clouds
    barium, 269
    ion, 272
Continuous wave, 114
Criterion of stretching, 229
Cumulative distribution function, 131
Current density
    electron, 141
    ion, 141
Current
    Hall, 93, 284
    Pedersen, 93, 284
    short-circuit, 150
Current-induced instabilities
    convergence instability, 57
    current-convective instability, 48
    dissipative instability, 58
    drift dissipative instability, 44, 58–59
    gradient-drift instability, 49–51
    hydrodynamic instability, 61–62
    Post-Rozenbluth instability, 44
    Rayleigh–Taylor instability, 60
    recombination-ionization instability, 42–43
    two-stream (Farley–Buneman) instability,
        52–54

D

Damping
    linear, 26, 55–57
    nonlinear, 64–65, 79–80, 95–96
Data stream parameters, 562–564
D-layer, 2
    middle-latitude, 472
Debye radius, 141
Decibels (dB), 129
Decrement of wave damping, 26
Degree of ionization, 164
Degree of magnetization, 164
Density profile, 146, 151
Depletion area, 149, 237–238
Depree's resonance broading theory, 69
Differential effective cross section, 439, 441
Diffraction effects, 545
Diffusion
    electron-controlled, 154, 178, 188
    ion-controlled, 154, 178, 188
Dimant–Sudan linear theory, 55–57
Directions
    along magnetic field, 16, 153
    across magnetic field, 16, 153
Dissociative recombination, 234
Distance-frequency characteristics, 479
Distributions
    Gaussian, 130
    log-normal, 130
    Lorenz, 135
    Nakagami-m, 134
    Rayleigh, 131
    Ricean, 132
    Suzuki, 135
Disturbed heated region, 239–240, 255, 262
Doppler Rayleigh lidar, 301, 303
Doppler
    effect, 116
    frequency, 116
    shift, 116, 302
        spectra, 120–121
        domain, 119
    spread, 137
Drift-dissipative instability, 58

E

Earth's magnetic field, 14, 142–143
Echo-signals, 306
Effective coefficient of recombination, 175
Effects
    defocusing, 542
    focusing, 542

E-layer
    equatorial, 303
    high-latitude, 359
    middle-latitude, 287
    sporadic, 371
Ebert-Fastie spectrometer, 495
EISCAT radar, 502
Electric potential, 156
Electron
    charge, 13, 18
    mass, 16
    temperature, 5, 7, 9
Electrojet, 304
Electric field strength, 115
E-field component, 115
Elementary theory, 141
Equatorial E-layer, 300–302
Equatorial F-layer, 322–325
Equatorial plasma bubbles, 349–350
Error function, 131
European incoherent scatter radar, 364
Extraordinary wave, 473–474

F

Fading
    fast fading, 131
        flat fast fading, 137
        frequency selective fast fading, 137
    multipath, 128
    slow fading
        flat slow fading, 137
        frequency selective slow fading, 137
Fabry–Perot interferometer, 494
F-layer, 2
    equatorial, 322
    high-latitude, 376–379
    middle-latitude, 477, 483
    sporadic, 335, 385
Faraday rotation, 321
Faraday angle, 322
Farley–Buneman (two-stream) instability, 52
Fading
    flat, 137
    frequency dispersive, 137
    frequency selective, 137
    large-scale, 127
    long-term (slow), 130
    short-term (fast), 131
    small-scale, 127
    time dispersive, 136
    time selective, 136
First Fresnel zone, 474
Field-aligned irregularities, 167, 178

Finger-like structures, 271–272
Filament-like structures, 280–281
Fluid theory, 13–16
Forecast, 525–527
    long-term, 527
    operative, 527
Fraction of energy loss
    during electron-ion interactions, 16
    during electron-neutral interactions, 16
    during ion-neutral interactions, 16
Free path length
    electrons, 12
    ions, 12
    neutral molecules (atoms), 12
Frequency
    blanking, 295
    limited, 297
Frequency of collisions
    electron-ion, 8
    electron-neutral, 8
    ion-neutral, 8
Friction force, 29
Functions
    Bessel, 134
    Gamma, 134–135
    Gaussian, 174

G

Gaussian
    form, 131
    profile, 130, 174
Global positioning system
    positioning, 507
    scintillations, 506
Gradient-drift instability, 48
Grazing angle, 469
Group propagation time, 453
Gyrofrequency
    electrons, 8, 239
    ions, 8
Guiding modes, 469

H

Hall current, 158, 284
Hall mobility, 284
Hall direction, 284
Hamza-St. Maurice nonlinear theory, 94–95
Heating
    adaptive, 242
    local, 237
    resonance, 241

Heating force, 30
Heating-induced instabilities, 255
Heating-induced irregularities, 256–257
Height of the neutral atmosphere, 2
H-field components, 121
HE–scattering, 288, 406
HE–irregularities, 289, 406
Homer radar, 360
Huancayo observatory, 340
Hydrodynamic instabilities
    electrostatic, 60
    electrodynamic, 61

I

IMAGE satellite campaign, 355
Increment of wave growth
    linear, 44, 47, 49, 52, 54, 60
    nonlinear, 70–71, 99
    quasi-linear, 76
Interference
    picture, 474
    random, 475
Ion
    beam, 279
    charge, 2
    cloud, 268, 284
    mass, 2
    temperature, 6–7
Ionosphere
    auroral, 364
    equatorial, 363
    high-latitude, 359
    middle-latitude, 287
    polar cup, 385
    sub-auroral, 378
    trans-equatorial, 325, 327
Ionospheric content
    electrons, 5–6
    ions, 5–6
    neutral molecules (atoms), 3
Ionospheric layers
    F1-layer, 2
    F2-layer, 2
    sporadic E-layer, 2, 296, 304, 422
    sporadic F-layer, 2, 332
Ionospheric waveguide, 469
Ionization
    balance of ionization, 19–20
    intensity of ionization, 18
    parameter of ionization, 12
Ionosonde
    continuous, 515
    digital, 514
    linear frequency modulated (LFM), 513

Ionozation-recombination balance, 17
Ion-molecular reactions, 18, 20
In-phase component, 117
Incoherent component, 564
Incoherent scatter radar, 300, 321, 499
Irregularities
    1st type, 305
    2nd type, 305
    3rd type, 366
    4th type, 366
    HE-irregularities, 410
    meteor-induced, 290–291
    strongly ionized, 290
    strongly anisotropic, 289
    weakly ionized, 290
    weakly anisotropic, 288
Interactions
    linear, 63
    nonlinear, 64, 66
    particle-wave, 68–69
    wave-wave, 63–64

J

Jicamarca
    coherent radar, 328, 331
    radar, 305
    radio observatory, 500
JULIA radar, 500
Joule heating, 236

K

K-factor, 133
Kadomtsev coupling theory, 83
Keskinen's nonlinear theory, 89
Kinetic instability, 44
Kinetic approach, 55, 57
Kinetic dissipative instability, 44
Kolmogorov's weak-turbulence theory, 64
Kraichnan's direct interaction approximation, 64, 66, 83
Kwajalein radar, 348

L

Landau damping (attenuation), 34
Larmor's radius
    electrons, 12
    ions, 12
Layered waveguide, 469
Life-time, 171–172, 190, 205

Line-of-sight (LOS), 127
Linear-frequency modulation
    ionosonde, 514, 517
    signals, 517, 519
Local heating, 235, 237
Long-distance radio propagation, 455, 457
Lower-earth-orbit satellite, 349, 355, 358, 385

M

Magnetic storm
    effects, 556–557
    observations, 558–560
Magneto-hydrodynamic approach, 13, 60
Magneto-hydrodynamic instabilities, 60
Magnitude of longitude, 2
Magnetosphere, 1
Magnetosphere-ionosphere coupling, 556
Maximum usable frequency (MUF), 529
Maximal data stream rate, 562
Maxwell's equations, 14
Mean value
    density fluctuations, 547
    excess delay, 123
    intensity fluctuations, 123, 564
    mean square root (rms), 132
    phase fluctuations, 547
Meteor trails
    highly-dense, 290, 412
    moderate-dense, 290, 411
    under-dense, 290, 411
Millstone Hill backscatter radar, 501
Mobility
    electron, 141
    Hall, 284
    ion, 141
    Pedersen, 284
Mode
    1F2, 481–483
    2F2, 481–483
    3F2, 481–483
    4F2, 483–485
    5F2, 483–485
Multibeam MU radar, 424

N

Neutral wind velocity, 14
Non-line-of-sight (NLOS), 127
Non-ordinary mode, 473
Nonlinearity
    striction, 29, 33
    thermal (heated), 30, 32

O

Oblique sounding, 407, 409–410, 454
One-hop propagation, 449
Ordinary wave, 473
Optical devices
    All-sky camera, 498
    CCD camera, 497
    Doppler Rayleigh lidar, 496
    Ebert-Fastie spectrometer, 494
    Fabry-Perot interferometer, 494
    resonance fluorescence lidar, 495
    tilting-filter photometer, 495

P

Parameter(s)
    of ionization, 10, 12
    of anisotropy, 183
Parametric instability, 24
Particle-wave interactions, 68–69, 71
Path loss, 129
Pedersen
    conductivity, 210, 283
    current, 284
    direction, 88
    mobility, 88
Permittivity, 454
Plasma
    frequency, 403, 444
    intrinsic field, 141, 210, 213
    quasi-neutral, 14
    strongly ionized, 11–12, 14
    strongly magnetized, 8, 14
    weakly ionized, 11–12
    weakly magnetized, 8
Plasma density
    depletions, 39, 237, 240, 242, 258, 260
    enhancements, 39, 239
Plasmoids
    electronic, 197, 284
    ionic, 197, 284
Plums, 332, 337
Plume-like structures, 339
Probability density function (PDF), 130–131, 133,
        135–136
Probing waves, 450, 454
Post-Rozenbluth instability, 44
Power delay spread, 124
Power spectral density, 119, 121
Puerto Rico
    observatory, 429
    radar experiment, 432
Pump wave, 257–258

Q

Quadrature component, 117
Quasi-neutral
    diffusion, 143
    drift, 213
    approximation, 457

R

Radiative recombination, 19
Radar echoes, 306
Radar (effective) cross section, 449
Radio
    frequency, 111
    path, 116
Range-time-intensity format, 423, 425–426
Ray trajectory, 459, 467–468
Ray tracing model, 457
Rayleigh-Taylor (two-stream) instability,
    52–53, 56
Recombination instability, 42
Refractive index, 540
Resonance fluorescence lidar, 494
Ricean parameter (factor), 133, 564
Rocket injection
    experiments, 269–270, 272
    Spolokh-1, 277
    Spolokh-2, 278

S

Saturation
    linear, 71
    nonlinear, 94
Scale
    inner, 549
    outer, 547, 549
Scattering
    diffractive, 550
    partial, 472
    refractive, 540, 543
Scintillation index, 564
Shakarofsky's power spectrum, 549
Signal
    amplitude, 111, 115
    burst-type, 291, 411
    envelope, 117
    linear frequency modulated, 516
    narrowband, 114
    power spectrum, 113
    quasi-continuous, 295, 412
    wideband, 122
Shadowing, 126
Short-circuit effect, 143, 196
Splitting mechanism, 283
Spreading
    electron-controlled, 178
    ion-controlled, 178
Steady-state (equilibrium) regime, 71, 76
Scintillation index (S_4), 559–560
STARE radar, 364
Strata, 281
Stratification, 282
Stretching criterion, 285
Striation mechanism, 286
Striated plasma structures(SPS), 231, 238,
    242–243, 267
Striction force, 29
Striction nonlinearity, 29
Sudan quasi-linear theory, 55–57
Supper dual radar network, 504
SuperDARN radar network, 503
Symbol period, 136

T

Temperature
    electron, 6–7, 14
    ion, 6–7, 14
    neutral, 3
    perturbations, 235
Tensors
    conductivity, 14–15
    diffusion, 14
    frequency interactions
        electron-ion, 14
        electron-neutral, 14
        ion-neutral, 14
    thermodiffusion, 14–15
    thermo-conductivity, 14–15
Time of splitting, 224–225
Time of striation, 227
Total electron content, 560
Temporal autocorrelation function, 119
Thermal parametric instability, 28
Thermal-induced instabilities, 40
Thermal velocity, 40
Thin-screen approximation, 548
Tilting-filter photometer, 494–495
Time of coherency, 137
Time delay spread, 136
Total electric field, 210

Traces
    Alma-Ata–Gorky, 477
    Beltsi–Dushanbe, 481
    Beltsi–Khabarovsk, 483
    Khabarovsk-Gorky, 479
Transport processes, 13
Triad-wave interaction, 64
Two-stream instability, 52–56

U

Unipolar diffusion, 148, 175–177
Unstable waves, 64–65

V

Velocity
    ambipolar, 195
    electron, 194, 213
    group, 195

    ion, 194, 213
    ion-sound wave, 52, 54, 60
    neutral wind, 14

W

Wavelength, 439, 449
Waveguide modes, 271
Wave number spread, 65
Wave-particle interactions, 68
Wave vector, 434, 437, 449
Wave-wave interactions, 63–64
Weak-coupling equations, 66
Wideband (pulse) signal, 122
Wind of neutrals, 14
Window spectrum, 82

Z

Z-mode (non-ordinary) mode, 473

Milton Keynes UK
Ingram Content Group UK Ltd.
UKHW051538141024
449569UK00028B/1518